The Automotive Chassis

Mechanical Engineering Series

Frederick F. Ling
Editor-in-Chief

The Mechanical Engineering Series features graduate texts and research monographs to address the need for information in contemporary mechanical engineering, including areas of concentration of applied mechanics, biomechanics, computational mechanics, dynamical systems and control, energetics, mechanics of materials, processing, production systems, thermal science, and tribology.

Advisory Board/Series Editors

Applied Mechanics	F.A. Leckie University of California, Santa Barbara
	D. Gross Technical University of Darmstadt
Biomechanics	V.C. Mow Columbia University
Computational Mechanics	H.T. Yang University of California, Santa Barbara
Dynamic Systems and Control/ Mechatronics	D. Bryant University of Texas at Austin
Energetics	J.R. Welty University of Oregon, Eugene
Mechanics of Materials	I. Finnie University of California, Berkeley
Processing	K.K. Wang Cornell University
Production Systems	G.-A. Klutke Texas A&M University
Thermal Science	A.E. Bergles Rensselaer Polytechnic Institute
Tribology	W.O. Winer Georgia Institute of Technology

For other titles published in this series, go to
http://www.springer.com/1161

Giancarlo Genta • Lorenzo Morello

The Automotive Chassis

Vol. 2: System Design

Springer

Prof. Dr. Giancarlo Genta
Politecnico Torino
Dipartimento di Meccanica
Corso Duca degli Abruzzi, 24
10129 Torino
Italy
giancarlo.genta@polito.it

Prof. Dr. Lorenzo Morello
Politecnico di Torino
Ingegneria dell'Autoveicolo
via Nizza, 230
10126 Torino
Italy
lorenzo.morello@polito.it.

ISBN: 978-94-024-0484-5   ISBN: 978-1-4020-8675-5 (eBook)
DOI 10.1007/978-1-4020-8675-5

Printed on acid-free paper

9 8 7 6 5 4 3 2 1

springer.com

CONTENTS

SYMBOLS LIST

a acceleration; generic distance; distance between center
of mass and front axle

b generic distance; distance between center of mass and rear axle

c viscous damping coefficient; specific heat

d generic distance, diameter

e base of natural logarithms

f rolling coefficient; friction coefficient

f_0 rolling coefficient at zero speed

$\mathbf{f}$ force vector

g gravitational acceleration

h wheel deflection

h_G center of mass height on the ground

k stiffness

l wheelbase; length

m mass

p pressure

r radius

s stopping distance, thickness

t temperature; time; track

$\mathbf{u}$ displacement vector

v slipping speed

z teeth number

A area

C cornering stiffness; damping coefficient

C_γ	camber stiffness
C_0	cohesiveness
E	energy; Young modulus
F	force
G	shear modulus
H	thermal convection coefficient
I	area moment of inertia
J	quadratic mass moment
K	rolling resistance coefficient; stiffness; thermal conductivity
$\mathbf{K}$	stiffness matrix
M	moment
M_f	braking moment
M_m	engine moment
M_z	self-aligning moment
P	power; tire vertical stiffness; force
P_d	power at the wheel
P_m	power at the engine
P_n	required power
Q	thermal flux
R	undeformed wheel radius; path radius
R_e	rolling radius
R_l	loaded radius
S	surface
T	temperature, force
V	speed; volume
W	weight
α	sideslip angle; road side inclination; angle
α_t	road transverse inclination angle
γ	camber angle
δ	steering angle
ϵ	toe-in, -out; brake efficiency; deformation
η	efficiency
θ	angle; pitch angle
μ	torque transmission ratio; adherence coefficient
μ_p	max friction coefficient
μ_x	longitudinal friction coefficient
μ_{x_p}	max longitudinal friction coefficient
μ_{x_s}	slip longitudinal friction coefficient
μ_y	transversal friction coefficient
μ_{y_p}	max transversal friction coefficient
μ_{y_s}	slip transversal friction coefficient
ν	speed transmission ratio; kinematic viscosity
ρ	density
σ	normal pressure; slip

τ transversal pressure; transmission ratio

ϕ angle; roll angle, friction angle

ω frequency; circular frequency

Φ diameter

Π tire torsional stiffness

χ torsional stiffness

Ω angular speed

Part III

FUNCTIONS
AND SPECIFICATIONS

INTRODUCTION TO PART III

A good knowledge of how a vehicle is made and manufactured, of how its components should be designed and integrated into a sound system, may not be enough to conceive and develop a successful car.

Such knowledge is only essential for obtaining a set of assigned targets for the product. We do not mean to belittle this knowledge, whose development is the main objective of these two books on the automotive chassis, but we must keep in mind that the success of a product is mostly dependent on how these targets are able to satisfy the customer's needs.

The combination of vehicle technical objectives and a brief description of its architecture make up what is everywhere known, with particular reference to the case of the car, as product *concept* or conceptual design.

The concept development is the starting point for development of a car. It can be expressed with a sketch or with a three dimensional simplified model of the car, suitable for explaining its appearance and its main functions.

This visual documentation must be accompanied with an exhaustive quantitative definition of technical and economical characteristics whose congruence and feasibility must be demonstrated.

These characteristics must be derived from a good understanding of customer needs and of how much the customer is inclined to spend for their satisfaction.

To define a concept means, therefore:

- to describe a product in terms of technical functions and specifications;

- to determine the product configuration and to choose its main components;

- to identify character, personality, feelings and other traits this new product will offer to the customer.

Each car manufacturer emphasizes different aspects of the product concept and determines, therefore, its characteristics and its potential success at the very beginning of the development process.

The central issue of the concept definition process is to obtain involvement throughout the company; the concept is partly driven by objective and measurable facts, whose determination will be the job of technicians, but also by insights that will be contributed not only by marketers, but by anyone who has gathered sufficient experience to contribute creatively.

People in charge of detailed product design, component specification, styling, production means development, sales and service should be involved, because their activity will condition customer satisfaction.

While ignoring the constrictions of the above operations during the concept development can cause large inconveniences, it is also true that an excessive involvement can cause premature conflicts and compromises, that can trivialize and flatten product characteristics.

A new vehicle cannot be simply the extrapolation of previous knowledge of customer needs, as assessed by the popularity of existing products; very often successful cars have been born as a response to needs that were unexpressed until the time of product launch.

We should remember, to underscore what we have said, the commercial success of the manufacturers that first launched sport utility vehicles, coupè-cabriolet, or minivans.

When defining a new concept it is important to proceed according to the following steps:

- to focus on customer needs;

- to identify latent or hidden needs, in addition to those demonstrated by existing products;

- to develop a common comprehension of customer needs.

There is an important difference between customer needs and product specification. Customer needs are independent of the product under development and are therefore not bound to the concept. Needs must be identified and focused on without necessarily knowing if and how they will be satisfied.

Product specifications depend, on the other hand, on the defined product architecture and the kind of components that have been defined; their linkage with customer needs can be understood only by knowing the success of similar products already in the market place.

If a good communication channel is established between customers and people in charge of product development, a sound knowledge of customer needs results.

Customers are identified through the following methods:

- market surveys;

- direct interviews;

- product use monitoring.

The purpose of the following chapters is not to develop a complete knowledge of these marketing techniques; rather, we would like to supply the future vehicle engineer with a view, complementary to the technical one, that enhances the interaction with marketing experts or customers.

We will examine the following topics, with particular reference to those aspects having a major impact on chassis and on chassis components design.

Transportation statistics

Transportation statistics represent the starting point for knowledge of vehicle use; these statistics make reference to Italy and to the European Union as far as passenger traffic transportation of goods are concerned.

In particular, we will analyze the volume breakdown between different means of transportation, in order to determine the average expected lifetime of vehicles.

Readers from different countries could take these data into consideration and adapt this method of analysis to their own interests.

Vehicle functions

In this chapter the vehicle system design which is the determination of the vehicle technical characteristics that condition the level of satisfaction of customer needs is tackled.

Since the field of our study is limited to the chassis, we will reduce system design to dynamic performance, in particular handling and ride comfort.

We will also outline the methods of correlation of objective technical characteristics with customer needs, which are subjective and only qualitatively expressed.

We will tackle vehicle mission definition as a description of the expected events during the vehicle's useful life; these data are necessary to forecast the age resistance of a vehicle and to define its scheduled maintenance.

Regulations

A not negligible influence on vehicle and, therefore, chassis characteristics is played by the legislative rules and regulations imposed by governments.

This matter is in continuous evolution and is now standardized in the European Union by general directives with which the technical regulations of the member states must comply; again, we will propose the European legislation as a reference, noting that other states have introduced legislation that is quite similar; existing rules address passive safety, active safety, pollution and energy consumption, among the topics relevant to chassis design.

Many other technical organizations, supported by consumer associations, and specialist magazines, according to their importance, influence product design with a power equal to legislation; an issue that must also be taken into account.

17
TRANSPORTATION STATISTICS

Data reported in this chapter were extracted from institutional documents of ANFIA, ACEA, ISTAT and Eurostat.

ANFIA (Associazione Nazionale Fra le Industrie Automobilistiche), the Italian national association of automotive manufacturers[1], was established in 1912 and is spokesman for its associates, on all issues (from technical, economic, fiscal and legislative to qualitative and statistical) regarding the mobility of people and goods.

Among several objectives, ANFIA has the task of gathering data and information, providing official statistical data for this segment of industry.

ANFIA publishes every year a report called *Autoincifre* (Figures of the Automobile), which is one of the fundamental references for statistical data on motoring in Italy and Europe. Much of the data collected in this report comes also from PRA (Pubblico Registro Automobilistico), the public vehicle register managed by ACI, the Association of Italian Motorists.

ISTAT (Istituto nazionale di STATistica) the Italian government institution for statistics [2]is well known. Established in Italy in 1926, ISTAT is the main producer of official statistics for citizens and public decision takers. It works in full autonomy while maintaining continuous interactions with the academic and scientific world.

This institution is fully involved in gathering European statistics (according to regulation R 322) and gathers data according to the fundamental rules of impartiality, reliability, efficiency, privacy and transparency.

---

[1]Web address: www.anfia.it.
[2]Web address: www.istat.it.

G. Genta, L. Morello, *The Automotive Chassis, Volume 2: System Design*,
Mechanical Engineering Series,
© Springer Science+Business Media B.V. 2009

The role of ACEA (Association des Constucteurs Européen d'Automobile[3]) in the European Union is similar to that of ANFIA in Italy; the 13 major vehicle manufacturers with headquarters in Europe are associated with ACEA.

This association represents European manufacturers in the European Union under a wide spectrum of activities, setting up research groups, supporting manufacturers with objective data and creating new legislative proposals in the fields of mobility, safety and environmental protection.

Eurostat [4] is the statistical office of the European Union. Its job is to supply the Union with statistics from corresponding national services. The European Statistic Service (ESS) adopts similar methods, allowing it to obtain comparable data. This service was established in 1953.

These data, accessible to the public, concern:

- key indicators of Union policies;

- general and national statistics;

- economy and finance;

- population and social conditions;

- industry, commerce and services;

- agriculture and fisheries;

- commerce with foreign nations;

- transportation;

- environment and energy;

- science and technology.

A further source of information within the European Union derives from the public documents of the different General Directions[5]; among these the Environment General Direction has set up a working group, including associations from the automotive and oil industries, that published the interesting report Auto-Oil II, on the impact of oil product combustion.

Since all data become obsolete quickly, we invite readers interested in updated details to consult the mentioned public sites, which allow access to the original archives.

In the interests of consistency, we will usually refer to the European Union as the original 15 countries, including Austria, Belgium, Denmark, Finland, France, Germany, Greece, Holland, Ireland, Italy, Luxemburg, Portugal, Spain, Sweden and United Kingdom.

---

[3] Web address: www.acea.be.

[4] Web address: epp.eurostat.cec.eu.int.

[5] General Directions are, for the European Union, the equivalent term for Department or Ministry.

17.1   TRAFFIC VOLUME

Traffic volumes are conventionally measured by the product of transported units times the distance covered by such transportation; therefore:

- passenger traffic is measured in passengers per kilometer [pass×km];

- the transportation of goods is measured in metric tons per kilometer [t×km].

It should be pointed out that the metric ton equivalent to 1,000 kg is a unit of mass; in any case what is relevant is the quantity of transported material, therefore mass and not weight. Nevertheless, the habit of considering the kilogram as a unit of weight and not of mass persists and therefore we sometimes see statements that traffic volume has the same dimensions as energy, which is only correct if the kilogram is assumed to be a unit of weight.

It is also true that if we assume a value for the acceleration of gravity and we know the vehicle coefficient of resistance, which will be explained later on, traffic volume is proportional to the energy spent to overcome motion resistance relative to the payload.

What was said could also apply to passenger traffic, if we substitute the number of passengers with the corresponding mass (conventionally 70 kg per passenger).

These considerations do not take into account the altitude difference between origin and destination or speed variations along the route, which are, instead, relevant for determining motion resistance and prime energy consumption.

17.1.1   Passenger transportation

Figure 17.1 reports passenger traffic volume in the European Union from 1970 to 2001, broken down according to the primary passenger transportation vehicles, such as cars, buses, urban railways with subways, trains and airplanes.

Cars definitely predominate over other means of transportation; car traffic represents in 2001 more than 78% of the total, and traffic on tires (cars and buses) is about 87%; this breakdown varies little during these years.

The total volume increased about 4% yearly during the first twenty years considered in this diagram; afterwards the growth slowed down to approximately zero in the last years for which data is available. Air transportation with its continuity of development is an exception.

A similar table is reported for Italy in Fig. 17.2.

The situation for Italy is not so different from that of Europe as a whole; in this case, car traffic represents about 82% of the total and traffic on tires (car and busses) is about 94%. This percentage has slightly increased during most recent years, mainly due to the reduction of railroad traffic.

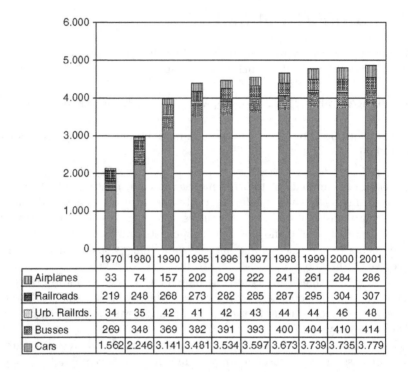

	1970	1980	1990	1995	1996	1997	1998	1999	2000	2001
Airplanes	33	74	157	202	209	222	241	261	284	286
Railroads	219	248	268	273	282	285	287	295	304	307
Urb. Railrds.	34	35	42	41	42	43	44	44	46	48
Busses	269	348	369	382	391	393	400	404	410	414
Cars	1.562	2.246	3.141	3.481	3.534	3.597	3.673	3.739	3.735	3.779

FIGURE 17.1. Passenger traffic volume in the European Union, from 1970 to 2001 (in billions of pass×km), broken down by main vehicle types: airplanes, railroads, urban railroads, including subways, buses and cars (Source: ANFIA).

The total traffic volume increased more than the average of the European Union, during the last years considered. Air transportation also increased during this period more than the average.

In Italy (source ISTAT) traffic volume is well correlated with the Gross Domestic Product. The total ground transportation system made use of a network of about 6,500 km of toll motorways, more than 46,000 km of national roads, about 120,000 km of country roads and about 20,000 km of railroads, interconnecting 8,100 communities, 146 harbors, 101 airports and many railroad stations.

On this network about 43 million vehicles were driven. Trains, ships and airplanes served about 57 million residents, whose total yearly distance travelled was about 15,000 km.

17.1.2   Transportation of goods

Figure 17.3 shows the volume of transportation of goods in the European Union from 1970 to 2001, broken down according to the main travel modes; in this case, road, rail, inland and sea navigation, and pipeline transportation are considered.

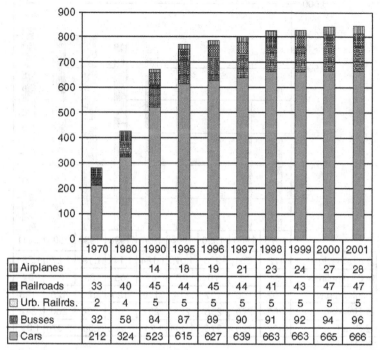

Passenger traffic volume [Gpass x km]

	1970	1980	1990	1995	1996	1997	1998	1999	2000	2001
Airplanes			14	18	19	21	23	24	27	28
Railroads	33	40	45	44	45	44	41	43	47	47
Urb. Railrds.	2	4	5	5	5	5	5	5	5	5
Busses	32	58	84	87	89	90	91	92	94	96
Cars	212	324	523	615	627	639	663	663	665	666

FIGURE 17.2. Passenger traffic volume in Italy, from 1970 to 2001 (in billions of pass×km), broken down by main vehicle types: airplanes, railroads, urban railroads, including subways, buses and cars (Source: ANFIA).

Here again road transportation is predominant: it accounted for 45% of the total in the last year of this period, starting from a percentage of 35% in 1970. The role of railroads has been reduced from 20%, at the beginning, to 8% in 2001. The contribution of sea navigation is relevant, considering the higher average distance travelled.

Figure 17.4 reports a similar table for Italy.

Road transportation plays a more important role in Italy than in the European Union: it carries 89% of the total in the last year considered, starting from 70% in 1970. In a similar way, railroad share has been reduced from an initial 16% to 6% in 2001. The contribution of sea navigation is not relevant, because the data include domestic transportation only.

In the most recent years, all developed countries have recorded continuous growth in transportation demand. Factors stimulating this growth have been many (economical and fiscal integration, market globalization, etc.) and seem likely to last in the medium term.

The most stimulating factor for Italy was the European economic integration process, implying free transfer of goods in the Union. Introduction of the

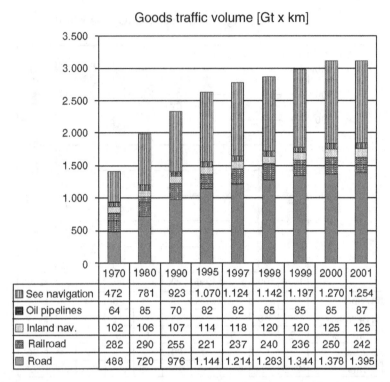

FIGURE 17.3. Transportation of goods volumes in the European Union from 1970 to 2001 (in billions of t×km, broken down according to the different kinds of carrier; road, railroad, inland navigation, oil pipes and sea navigation are considered (ANFIA).

Union currency and the prospect of a further enlargement of the European Union portend a continuation of this trend in the future.

17.1.3 Energy consumption

Energy consumption is usually measured in tons of equivalent petroleum [tep], corresponding conventionally to 41.87 GJ or 11.63 MWh; these values define the equivalent quantity of heat that is delivered by burning a ton of oil of average quality.

This unit is also used to measure energy from sources other than oil, evaluated at the energy cost for their production.

For instance, railroad transportation uses a combination of electric energy and oil refinery products; electric energy itself is produced partly in thermal power stations using oil products or natural gas and partly in hydroelectric power stations. Other contributions can come from geothermal energy or nuclear energy.

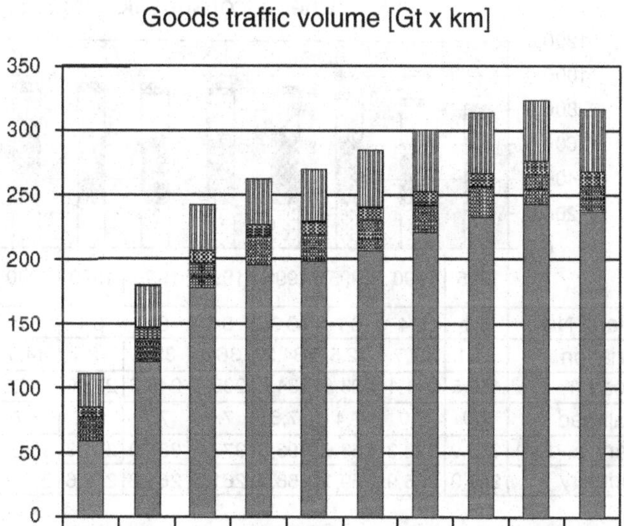

Goods traffic volume [Gt x km]

	1970	1980	1990	1995	1996	1997	1998	1999	2000	2001
See navigation	26	31	36	35	40	45	47	46	47	49
Oil pipelines	7	9	9	10	10	10	11	10	10	11
Inland nav.	0,4	0,2	0,1	0,1	0,1	0,2	0,1	0,2	0,2	0,2
Railroad	18	18	19	22	21	23	22	23	22	20
Road	59	120	178	195	198	207	220	233	243	237

FIGURE 17.4. Goods transportation volumes in Italy from 1970 to 2001 (in billions of t×km, broken down to the different carrier kinds; road, railroad, inland navigation, oil pipelines and sea navigation are considered (Source: ANFIA).

Every contribution is converted to an oil value, considering production losses and thermal equivalence.

Figure 17.5 displays a time series of energy consumption in Europe for most important means of transportation and other final applications.

The energy consumption of the transportation system is about 32% of the total; this share can be broken into:

- 2.4% for railroad transportation;

- 82.4% for road transportation;

- 13.6% for air transportation;

- 1.6% for inland navigation.

This last figure includes not only river, lake and channel navigation, but any maritime navigation in the European Union. The figure therefore includes sea navigation; this correction is particularly important for Italy because of its extensive coastline.

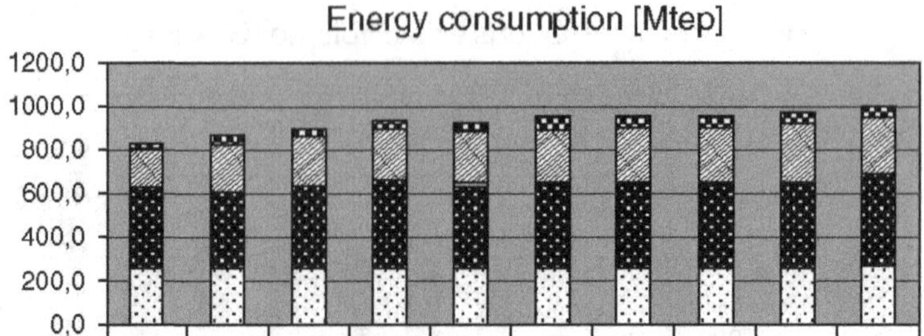

Energy consumption [Mtep]

	1985	1990	1995	1996	1997	1998	1999	2000	2001	
▦ Inland Nav.	5,1	6,4	6,7	6,9	6,5	6,5	6,1	5,3	4,9	
▨ Aviation	21,1	27,7	32,5	34,2	36,0	39,5	42,1	44,0	42,5	% in 2001
▨ Road tr.	169,5	211,1	228,6	234,6	238,4	246,3	252,0	253,6	257,0	
□ Railroad tr.	7,0	7,0	7,4	7,6	7,6	7,7	7,5	7,7	7,5	
▣ Serv. & dom.	356,0	342,3	363,6	393,9	377,6	384,0	381,3	369,7	388,9	
▫ Industry	267,9	265,9	259,1	258,7	261,8	261,2	262,6	270,8	269,5	

FIGURE 17.5. Energy consumption in the European Union for most important transportation systems and final applications; consumption is measured in millions of tep; percentages (for 2001 only) are multiplied by 10 to use the same scale (Source: Eurostat).

The energy used for sea navigation, the so-called bunkered quantity at the sailing harbour, is partially used for transportation to countries outside the European Union; it is conventionally treated as an oil export. In 2001 this quantity was estimated as 43.5 Mtep, about 14% of total transportation consumption.

The transportation system relies mainly on oil products; railroad transportation uses diesel fuel for 30% of its total energy consumption and a notable part of electric energy comes from oil combustion as well.

Road transportation uses primarily oil refinery products; Italy and Holland are an exception, consuming respectively 9% and 7% liquefied petroleum gas for traction (1999); the contribution of coal and natural gas is at this time negligible. Probably this situation will remain unchanged for the near future. Total consumption shows a leveling in recent years.

In Italy, road transportation seems to follow a different trend, as shown in Fig. 17.6, which concerns the consumption of oil products for ground transportation.

The following Fig. 17.7 shows the share between diesel fuel and gasoline.

The growth of diesel fuel over gasoline is evident; this trend is partly justified by the different retail prices of the two fuels and partly by the more efficient combustion of diesel engines. We should also remember that quantities are measured by mass units, but customers are paying by volume; at the same volume, diesel fuel contains 12% more energy than gasoline.

We have tried, using the available data, to compare the energy consumption of different means of transportation; we have defined as *energy efficiency* the

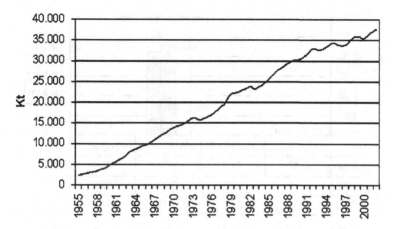

FIGURE 17.6. Total oil product consumption for ground transportation in Italy; the quantity in Ktep includes gasoline, diesel fuel and lubricants, these last accounting for about 1% of the total (Source: ANFIA).

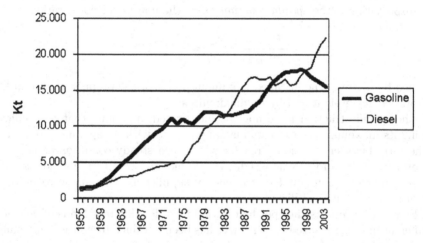

FIGURE 17.7. Gasoline and diesel fuel used in Italy by ground transportation, measured in Ktep (source ANFIA).

amount of energy necessary to perform a unit of traffic volume. We assume, as a common indicator, the goods traffic unit [t×km], which allows us to summarize with a single measurement goods and passengers transportation. We assumed an average mass of 70 kg for each passenger, including the transported baggage.

Accepting this questionable equivalence, we obtain the diagram of Fig. 17.8.

The values shown here display an increase over time of about 12% for road transportation and 16% for air transportation, covering a period of about ten years.

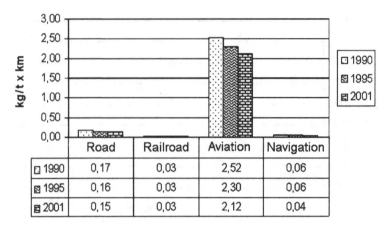

FIGURE 17.8. Comparison of energy efficiency of different transportation means; data are elaborated from time series of the European Union.

Another relevant parameter for evaluating the energy efficiency of a means of transportation is the *specific traction force*, the non-dimensional ratio:

$$\frac{P_{max}}{mgV_{max}} = \frac{F_t}{mg} = \frac{P_{max}t}{mgd},$$

obtained by dividing the installed maximum power of the propulsion system by the total vehicle weight and by its maximum speed.

The specific traction force may be interpreted, assuming that the vehicle is using its maximum power at its maximum speed with its maximum payload, as the ratio between traction force F_t, which in a steady condition equals the motion resistance, and vehicle weight; this is like an overall friction coefficient. Another interpretation could be the energy supplied by the engine to move a unit of mass for a unit of distance.

Figure 17.9 reports the specific traction force for different kinds of vehicles at different maximum speeds. Each curve has been obtained by considering many vehicles of the same family and charting them on the diagram. Curves on this diagram represent the lower envelope of the points represented.

This methodology may be questioned because only the top speed is taken into account and this may not reflect the most efficient condition for the traction engine; in addition, only the total weight is considered, instead of the payload alone.

All curves are superimposed on an ideal line that on logarithmic scales is straight, defined as the *limit for isolated vehicles*. This line can be interpreted as the optimum use condition for each vehicle, independent of its propulsion system.

The right side of Fig. 17.9 shows an enlargement of the part of this diagram regarding ground vehicles. It will be noticed that all vehicles including trailers are more efficient than isolated vehicles and are set below the limit line.

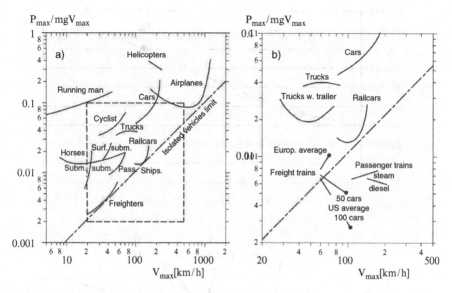

FIGURE 17.9. Specific traction force P_{max}/mgV_{max} as function of top speed. (a) General diagram; (b) enlarged portion concerning ground vehicles.

Although these are approximations, this parameter gives an immediate idea of the energy efficiency of different vehicles, and the distance to the limit lines suggests the amount of room for improvement.

17.2 OPERATING FLEET

17.2.1 Quantity

Vehicles owned by naturalized or legal residents of Europe totalled about 215 millions in 2002; they comprise the so-called vehicle *operating fleet*.

Figure 17.10 shows a time series for private vehicles, mainly cars; while Fig. 17.11 shows public service vehicles, including commercial vehicles, light, medium and heavy duty trucks and busses.

The year 2000 figures on total traffic volume are also available (source Eurostat):

- the railway fleet, included 40,000 engines and rail cars, about 76,000 cars for passenger transportation and about 500,000 freight cars;

- the navigation fleet, included about 15,000 vessels;

- the air fleet, included about 4,900 airplanes.

The private car fleet is predominant; about 469 cars for every 1,000 citizens were available in 2000. The fleet growth in thirty years was about 184%, with a yearly growth rate of about 3.5%; this growth has slowed but not stopped.

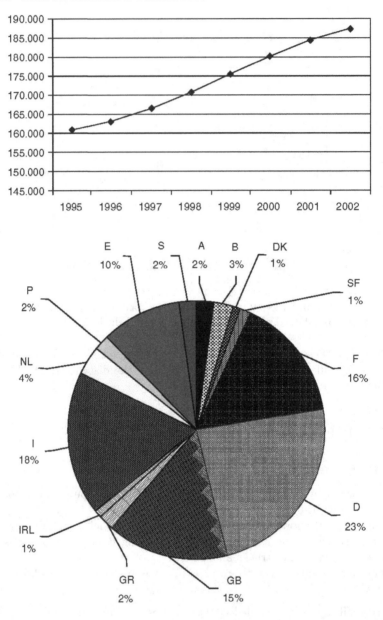

FIGURE 17.10. Time series of private cars in the European Union, in thousands; the lower pie chart shows the breakdown of the 2002 figures into the 15 considered countries, identified according to the international licence plate (Source: ACEA).

In the United States, car density has reached 750 cars per 1,000 citizens and is now steady; statistics show, in fact, that new car sales largely keep pace with written off units.

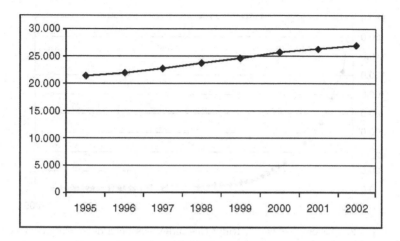

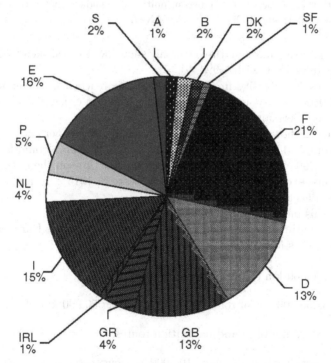

FIGURE 17.11. Time series of public service vehicles in the European Union, in thousands; the lower pie chart shows the breakdown of the 2002 figures into the 15 considered countries, identified according to the international licence plate (Source: ACEA).

Although this density is not inevitable for the European Union, the fleet there is still growing, and countries whose economies are growing fast are showing higher rates, such as Greece, with 9.2%, Portugal, with 7.3%, Spain, with 6.9%,

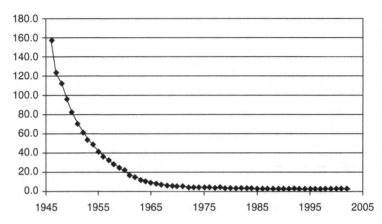

FIGURE 17.12. Citizens per car in Italy; this index has been decreasing continuously over time, with the exception of two discontinuities (not shown) at the time of the two world wars, in 1915-18 and 1939-45 (Source: ANFIA).

while countries with a more mature economy show lower rates, such as Denmark, with 1.8%, and Sweden, with 1.9%.

The highest car density in the year 2000 was reached in Luxemburg with 616 cars/1000 persons (corresponding to 1.62 cars per citizen), Italy with 563 cars/person and Germany with 522 cars/person.

Figure 17.12 shows a time series of the ratio of citizens per car in Italy; this diagram, if compiled from the beginning of the motoring era, would have shown a figure of 300,000 citizens per car in 1899; from that time on the index decrease was continuous, except during the two world wars in 1915 – 18 and 1939 – 45, when the total fleet decreased.

In 2003 this index has decreased to 1.5 citizen/vehicle.

At the same time the transportation infrastructures of the European Union can be described as follows:

- about 160,000 km of railroad network;

- about 3,250,000 km of road network, including 50,000 km of motorways;

- about 28,000 km of inland navigation routes;

- 204 airports with more than 100,000 passengers/year, with 30 of them treating 75% of the total air traffic.

17.2.2 Characteristics

If we want to better understand the contents of the car fleet, we can consider the histogram of Fig. 17.13, showing the breakdown of cars registered from 1995 to 2004 according to different market segments and body types.

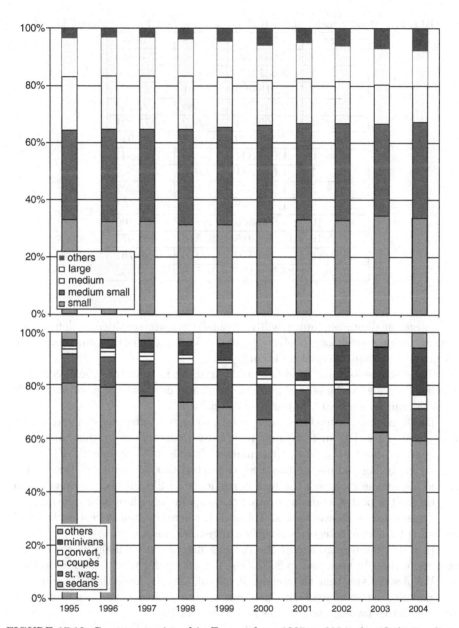

FIGURE 17.13. Car types registred in Europe from 1995 to 2004, classified according to market segments and body types; large, medium, medium small and small cars are defined, according to their length: larger than 4.5 m, 4 m, 3.5 m or equal to or smaller than 3.5 m (Source: ACEA).

TABLE 17.1. Road vehicles of the Italian operating fleet by age (Source: ISTAT).

Age	Cars	Busses	Trucks
0	2.033.296	3.819	222.443
1	2.541.933	6.056	260.116
2	2.518.499	5.381	243.297
3	2.391.709	5.485	197.700
4	2.399.014	4.569	173.021
5	2.381.400	3.936	136.940
6	1.667.344	3.409	142.841
7	1.619.341	2.610	136.149
8	1.533.972	1.898	108.402
9	1.497.088	1.866	102.005
10	1.993.566	2.852	134.100
>10	11.068.915	49.519	1.563.077
unknown	60.376	316	9.761
Total	33.706.153	91.716	3.429.852

Large, medium, medium small and small cars are considered, defined according to an overall length of more than 4.5 m, more than 4 m, more than 3.5 m and less than or equal to 3.5 m.

These segments show no substantial variation; in terms of type, a constant decline of sedans can be noticed, with a simultaneous growth of what were once considered niche segments, in particular minivans.

Diesel engines, introduced into mass production after the Second World War, have suffered from fiscal and regulatory intervention; at this time their fleet share is about 24%, while their market share is 44% (Source: ANFIA).

By analyzing and elaborating fleet characteristics, some information on expected life can be gathered; the task is particularly difficult because many data referring to the past are missing or are not comparable with present information.

As an example we can look at Table 17.1, where the Italian fleet is classified according to vehicle categories and their registration age.

In Italy, the average age of the running fleet is rather high: 32.8% of cars were more than 10 years old in 2002. The percentage for trucks reaches 46.1%.

We observe also that the weight of cars and trucks more than 10 years old has increased as compared with previous years, because they have benefitted from incentives favoring newer, less polluting vehicles.

Newer cars (less than one year of age) have moved from 7.2% in the year 2000 to 6.0% in 2002. A different trend appears in trucks, where newer vehicles went from 5.1% in 2000 to 6.5% in 2002.

ANFIA data, which reports on more age classes, estimate an average age of 8.85 years for cars in the year 2002. Analyzing these data, cars with gasoline engines appear to be the older category (9.35 years), but this must be weighed against the relatively recent success of diesel engines.

In the European Union the average age for cars is about 8 years; 70% of the fleet is younger than 10 years (source ACEA).

For industrial vehicles in Italy, the average age is about 10.5 years.

The estimate of vehicle expected life is hard to predict; but barring major changes, an expected life of 15 years for cars and 20 years for industrial vehicles would be a reasonable forecast.

The expected evolution of emission and passive safety regulations could promote a shorter life, increasing fleet obsolescence. Macroeconomics should not be forgotten.

The yearly distance covered by a vehicle can be estimated by dividing traffic volume by the number of operating vehicles and available places; in reality, not many vehicles operate at maximum capacity.

For example, in the European Union, cars deliver a traffic volume of 3.779 Gpass×km (see Fig. 17.1) with a working fleet of 187,400,000 units (see Fig. 17.10); by crediting each car with five places, we would obtain about 4,000 km/year.

The so-called *occupation factor* should be taken into account; it is defined as the ratio between occupied and available places; statistical surveys measured a mere 26.5% for this value, reducing total occupation to only 1.33 passengers per car.

The average yearly distance covered is therefore about 15,000 km/year (source ACEA).

A reasonable estimate for a car's life expectancy, therefore, should be close to 200,000 km.

Following the same process for other vehicle categories, we obtain:

- more than 400,000 km for busses;

- more than 800,000 km for long haul trucks.

17.3   SOCIAL IMPACT

As we have seen, transportation has a strong bearing on daily life. Every morning European Union services have to move more than 150 millions people to their working places and to return them to their homes in the evening, as well as serving longer routes; in addition, about 50,000,000 tons of freight are transported every day.

Considering passenger traffic only, each citizen travels approximately 12,700 km per year, using all available means of transportation; as a consequence, transportation is highly relevant to how people live.

We will consider in the following sections:

- accidents attributable to the use of transportation means;

- emissions of primary pollutant products;

- jobs offered by this economic sector;

- tax revenue generated by the transportation system.

As far as energy consumption is concerned, we refer to the previous section on this topic.

We will take into account mainly motor vehicles, our main field of interest, reporting also some reference data for other means of transportation.

17.3.1   Accidents

Like all human activities, road transportation involves risks and the number of accidents caused by the use of motor vehicles is remarkable in all countries of the world.

Their economic and human cost is high enough that the objective of increasing vehicle safety is generally considered a social and technical priority.

To evaluate the extent of these damages it may be useful to report a statistic on causes of death in the United States; these data could be similar to those in any other developed country.

These figures for 2002 are shown on Table 17.2. The number of fatalities connected to road transportation is higher than those caused by all remaining means of transportation and represents 44% of all fatalities from accidental causes[6].

In the European Union, transportation accidents caused 41,500 fatalities in the year 2000; 98% of these were due to road accidents. For people younger than 45, road accidents are the leading cause of death.

Figure 17.14 shows a summary of this worrying situation.

Total fatalities are decreasing, notwithstanding the traffic volume increase; this result is attributable to better driving education, infrastructure improvement and vehicle passive safety owing to the increasing severity of regulations.

TABLE 17.2. Death risk for different causes, referring to the USA population in 2002.

Cause	Total number of fatalities	Percentage
All causes	2.403.351	100
Heart troubles	936.923	39,0
Cancer	553.091	23,0
Total accidents	97.900	4,1
Motor vehicles	43.354	1,8
Generic accidents	17.437	0,73
Falls	13.322	0,55
Poisoning	12.757	0,53
Drowning	3.842	0,16
Burns	3.377	0,14

[6]Source: http:\\www.the-eggman.com/writings/death_stats.html.

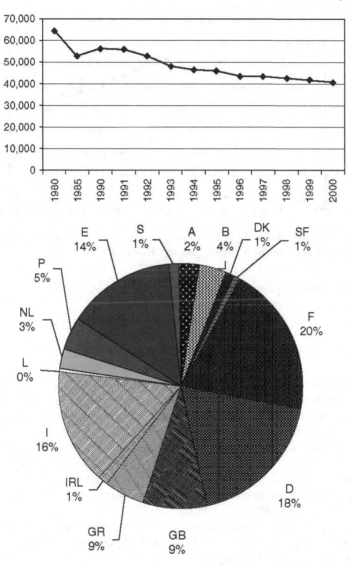

FIGURE 17.14. Time series of fatalities caused by road accidents in the European Union; the pie chart shows the contributions of the different States (Source: Eurostat).

The average mortality rate is about 109 deaths for each million residents $(1{,}09{\cdot}10^{-4})$; Italy and Ireland are close to the average; the lowest value is found in the United Kingdom $(0{,}60{\cdot}10^{-4})$, while the highest is in Greece $(1{,}98{\cdot}10^{-4})$.

55% of accident fatalities are represented by car occupants, 23% by bicycle occupants, 6% by bus occupants and the remainder by pedestrians.

Referring fatalities to different passenger traffic volumes, we obtain the following mortality rates:

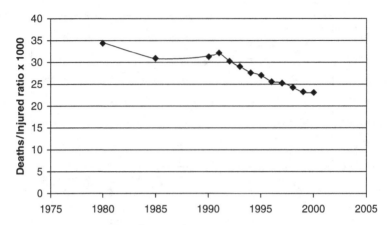

FIGURE 17.15. Time series of the ratio between deaths and non-fatal injuries due to road accidents in the European Union (Source: Eurostat).

- $10 \cdot 10^{-9}$ deaths/pass×km, for road transportation;

- $3 \cdot 10^{-9}$ deaths/pass×km, for railway transportation;

- $0.27 \cdot 10^{-9}$ deaths/pass×km, for air transportation.

Air transportation poses a challenge; one might consider accidents occurring inside the borders of the European Union, or accidents occurring to European airlines, whether inside the Union or abroad. Only European citizens might be considered or any person involved.

If we consider that most accidents occur near airports, the different counting policies give different conclusions. Moreover, these accidents, fortunately few, fluctuate over time and are difficult to average meaningfully.

The reported figure refers to all accidents occurring in 1999 within the borders of the European Union.

Returning to road transportation, accident severity has also decreased, as we can conclude by examining Fig. 17.15 showing the time series of the ratios between deaths and non-fatal injuries.

17.3.2 Emissions

The main pollutants emitted by the combustion of oil refinery products, in general, and by road traffic, in particular — all demonstrated to be harmful for public health — are the following:

- carbon monoxide (CO);

- nitrogen oxides (NO_x);

- non-methane organic compounds (NMOC);

- particulate matter (PM).

In recent times other gases have been added to the list; these are not directly harmful, but contribute to creating the so-called greenhouse effect. They are known, therefore, as greenhouse gases (GHG).

Carbon monoxide is a flavorless, colorless and poisonous gas; if exchanged with blood hemoglobin, in the lungs, it impairs the quantity of oxygen delivered to body organs and tissues.

A significant quantity of CO emission is produced by gasoline engine combustion and, therefore, by cars; all combustion processes of organic fuels that are incomplete for lack of oxygen contribute to the production of CO. Such contributions are many, including other gasoline engines (motorcycles, etc.), diesel engines, incinerators and homes.

Figure 17.16 shows a CO breakdown by source as estimated for the European Union in the year 2000.

These values are constantly decreasing because of the conversion to natural gas of many wood-burning furnaces, and because of the regulation on vehicle emissions that reduced the allowed limits, for example, for gasoline engine cars from 4.05 g/km, in 1992, to 1 g/km in 2005; the introduction of catalysts in 1992 had already reduced CO emission by ten times.

Nitrogen oxides (NO_x) are made by mixing NO and NO_2 and are the result of the combination of atmospheric nitrogen and oxygen due to combustion processes at high temperature and pressure; we can, therefore, say that the more efficient the combustion process, the higher the rate of nitrogen oxide formation.

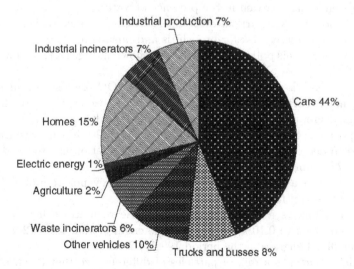

FIGURE 17.16. CO breakdown by source for the European Union for the year 2000 (source ACEA).

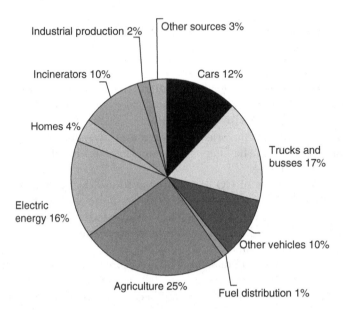

FIGURE 17.17. NO_x breakdown by source for the European Union in the year 2000 (Source: ACEA).

For this reason fuel consumption and CO_2 emission reductions conflict with reduced emissions of NO_x.

A second major source of NO_x is nitrate salts used in agriculture, which produce acids emitting nitrogen in the presence of water.

Nitrogen dioxide (NO_2) irritates the lungs and can reduce their resistance to infection, with increased risk for bronchitis and pneumonia.

Contributions to this pollutant are many, as shown in Fig. 17.17, again based on the year 2000.

We should remark that NO_x, together with NMOC are also precursors of complex chemical reaction leading to the formation of ozone (O_3) into the low altitude atmosphere, proven to be noxious to human health.

Anthropogenic sources of this pollutant are many; also in this case there is a clear trend to decrease. The evolution of vehicle regulation has reduced NO_x limits from 0.78 g/km in 1992, to 0.25 g/km in 2005.

Fig. 17.18 shows a similar diagram for NMOC; the evaporation of fuels and solvents is a major contributor.

NMOC also follows a decreasing trend; vehicle regulations have reduced levels from 0.66 g/km to 0.10 g/km for gasoline engines, and from 0.2 g/km to 0.05 g/km for diesel engines in the period 1992 to 2005.

Particulate matter is a mix of particles of different size that is harmful to health; it is also damaging to exposed materials and can reduce visibility. It is usually classified according to the average diameter of particles involved; these

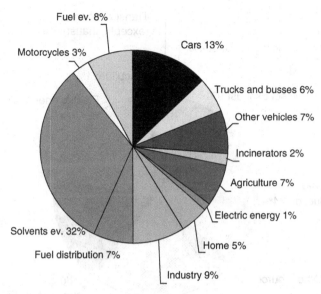

FIGURE 17.18. NMHC breakdown by source, in the European Union in the year 2000 (Source: ACEA).

are suspended in the atmosphere and precipitate very slowly. The most noxious particulates are those suspended in the atmosphere.

PM-10 indicates particles smaller than 10 μm, while PM-2,5 refers to sizes smaller than 2,5 μm.

The smaller the particle the greater the risk for human health. Extended exposure to these particles affects breathing, can worsen existing pulmonary diseases and increases cancer risk.

Beyond combustion products, particles also contain dust, ash, smoke and airborne droplets. If not washed away by rain or artificial means, powders on the ground can again become airborne due to natural wind or passing vehicles.

Fig. 17.19 shows a breakdown of the main sources of PM-10.

Gasses contributing to the greenhouse effect (GHG) include a basket of six chemical compounds that were identified in the final document of the Kyoto protocol; these are: carbon dioxide (CO_2), methane (CH_4), nitrogen dioxide (NO_2), chlorofluorocarbons (HFC), perfluorocarbons (PFC) and SF_6.

All these gasses, if diffused into the atmosphere, limit infrared radiation, contributing to an increase in the atmosphere's average temperature.

They are measured according to their heating potential, which is reported as CO_2 equivalent; their quantity is multiplied by weights p_i, which express the *carbon dioxide equivalent*.

The weights are the following: $p_{CO_2} = 1$, $p_{CH_4} = 21$, $p_{NO_2} = 310$, $p_{SF_6} = 23,900$. HFC and PFC include two large families of different gasses, each with its own weight.

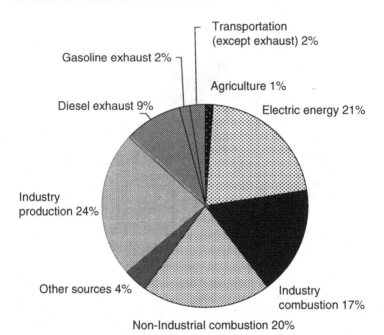

FIGURE 17.19. PM-10 particulate breakdown by source in the European Union in the year 2000 (Source: Auto-Oil II).

The pie chart showing the GHG breakdown by source is shown in Fig. 17.20.[7]

GHG emissions are strongly correlated to the population: France, Germany, Italy and the United Kingdom account for more than 50% of the total for Europe.

These emissions can be reduced only by reducing the burning of fossil fuels.

Air pollution is a phenomenon preceding development of the automobile, especially in urban environments and it has been declining for many years.

If public perception has increased, this is due more to the evolution of laws, than the problem per se.

Figure 17.21 shows an interesting diagram reporting SO_2 and smoke evolution recorded in London during the last four centuries; the period up to 1920 is an estimate based on coal consumption, while figures after that date are certain.

Urban pollution decreased considerably in the second half of the past century and is now lower than at any time in the last 400 years.

Similar diagrams are available for all major towns; on the other hand, towns of recent development show curves that are still increasing.

---

[7]These data are unfortunately not consistent with the others, because they refer to the European Union as extended to 25 countries.

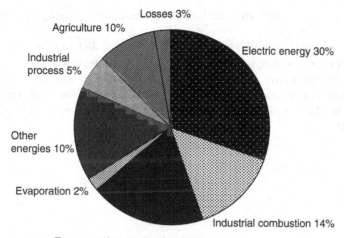

FIGURE 17.20. Greenhouse gasses broken down by source, measured as CO_2 equivalents (Source: Eurostat).

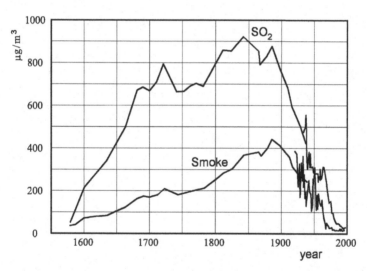

FIGURE 17.21. Diagram of SO_2 concentration and smoke, as functions of time, in London.

17.3.3 Economic figures

Limiting the figures to road transportation only, ACEA manufacturers produced in 2004 about 16,900,000 vehicles, 16,400,000 of them registered in the European Union with the remainder exported.

This production activity accounts for 1,200,000 jobs at vehicle manufacturers and 800,000 jobs at parts manufacturers; related activity, including distribution, service, parts and fuels distribution, car rentals, insurance, waste disposal, driving schools, dedicated press and infrastructure management, creates jobs for 12 millions people.

The yearly sales of vehicles alone have reached 452 billion euro in 2001; investments were about 8% of sales and research and development expenditures about 5%. The positive contribution to the trade balance was 33.4 billion euro.

Taxation on sales and property transfers, on vehicle ownership, on tolls and petroleum products was 346 billion euro in 2003, about 3.8% of the European Union GNP.

18
VEHICLE FUNCTIONS

18.1 SYSTEM DESIGN

The goal of a system approach to vehicle design is to define the technical specifications of each component, in such a way that the vehicle, as a whole, performs its functions according to assigned procedures and objectives.

By technical specifications, we mean a set of physical measurements that define each part, completely without the use of detailed drawings.

The system approach to design allows project, even the most elaborate one, to be carried out by assigning activities to teams working in parallel, each with comprehensible objectives that can be checked autonomously, and finalized to the overall performance of the vehicle. The system approach also allows a project to be developed, using standard components produced by suppliers, these components being developed for the purpose or chosen from a catalogue.

Finally, the system design is the initial phase of each project, when the feasibility of reaching the assigned targets is verified; this phase is usually called a *feasibility study*. The technical specifications of the main components are part of the concept documentation.

In the continuation of this section we will try to explain how to assign and measure overall vehicle performance and functions.

Unfortunately, performance and functions cannot be defined absolutely, because they are conditioned by the customer's expectations of the product he will buy, and these expectations depend on both objective and subjective parameters. Nor would we forget that these expectations are conditioned by product alternatives existing on the market when the product will be sold, alternatives which are usually unknown when the project is launched.

G. Genta, L. Morello, *The Automotive Chassis, Volume 2: System Design*,
Mechanical Engineering Series,
© Springer Science+Business Media B.V. 2009

Let us consider the climate system as an example.

Some functions of primary importance are easily identified.

We can assume that the objective of the climate system is to allow the passenger compartment to reach the assigned comfort temperature in a given time, notwithstanding the existing outside temperature. In reality, this time (cooldown or warm-up time) is only an approximate parameter; we would not want ideal comfort temperatures to be reached too quickly, causing passenger discomfort due to too strong a flow of air on the skin or high temperature gradients in the air.

Heat flux on passengers appears to be a matter of judgement, further complicating our understanding of the phenomenon.

In a luxury car, it might be mandatory to obtain comfort conditions automatically, requiring the passenger to adjust temperature control only. A few years ago it would have been a wild guess to design a multi-zone climate system for medium or small cars. Nevertheless this feature is beginning to appear in these cars and could became standard in the near future; when such system were designed, they may have seemed waste of time and money.

On next generation medium cars, automatic outside odor abatement could become indispensable. The abatement efficiency of various odorous reference substances should therefore be defined and measured.

In this connection, it should be remembered that too ambitious a specification could increase the product cost in ways that could prove to be unrecoverable in the final price.

Likewise, there are other functions that, although of secondary importance, cannot be neglected and, sometimes, impact other systems. They could be designed, for example, with goals like these:

- to reduce fuel consumption due to the climate system to a minimum;

- to reduce to a minimum the power required to the engine, especially during sudden acceleration;

- to minimize the noise of air flux into the passenger compartment;

- to control humidity, so as not to fog the windshield or side windows, etc.

Thus we see that in an apparently simple case, it can be very difficult to identify functions and specifications and that this identification should be done only after a careful study of customer expectations, which are in part a priori and in part depend on the competitor's products.

It is also difficult to define the boundaries of the climate system, or to identify which components influence the climate system by their behavior.

As a matter of fact, in a traditional approach, the climate system would be limited to:

- the heat exchange group, including the heater and the evaporator, the air channels and nozzles that direct the conditioned air to the passenger compartment, a part of the body system;

- the compressor, a part of the engine system;

- the electronic control box, a part of the electric system.

But, if we try to improve our control over climate system functions, this list should be expanded to include, for example, windows and windshield, responsible for an important part of the radiated heat; door panels, responsible for part of the transmitted heat; seats, as they influence human body heat exchange; and gearbox transmission ratios for their influence on engine warm-up.

The boundaries of the technical specification influencing certain function are wide and exceed the boundaries of the components dedicated to this function.

We suggest that a correct approach to system design should include, at least, the following steps.

1. To define the functions performed by the system.

2. To define the parameters that best measure those functions and the target values they should reach to obtain customer satisfaction.

3. To define which components are part of the system, because they influence the achievement of the target values.

4. To identify other system functions, if any, in competition with those of point 1.

5. For each component, to establish a set of technical specifications coherent with the system function target values.

Therefore system engineering implies the study of components that are normally classified under different automotive engineering disciplines; in the example of the climate system, we find components that are part of the body, the powertrain and of the electric system; these components are usually located in different sub-assemblies of the car.

The chassis system design we will study in this book is similar. We will study, for example, dynamic performance; a major factor here is the car's top speed. For a correct system approach, we will study not only the transmission as a component of the chassis, but also engine specifications (part of the powertrain), and aerodynamic resistance (part of the body system).

If engineering subjects are, by their nature, interdisciplinary, system engineering must likewise exceed the boundaries of its individual subjects.

A traditional, topological approach, on the other hand, classifies and studies vehicles according to three main subsystems.

- The chassis, as the cluster of components dedicated to vehicle path control, such as transmission, suspension, brakes, wheels and steering mechanism, with their dedicated supporting structures.

- The powertrain, as the cluster of components dedicated to traction power generation, such as engine, fuel supply, intake and exhaust plants.

- The body, as the structure supporting all other components including the passenger and payload compartment.

This book, dedicated to chassis system design, will consider all those functions that are primary to the chassis; nevertheless, while studying a function such as dynamic performance, we will also consider some aspects of the engine and the body or, in terms of automatic system controls, we will consider some issues interfacing with the electric and electronic system.

System design is necessarily rough, because it only studies the baseline specifications of the components included in the system; design details of the same components are left to specialists in each.

This approach will be useful not only to engineers, but to anyone involved in new vehicle development process.

18.1.1   Functions perceived by customers

Let us consider all functions performed by the vehicle, with particular reference to automobiles.

Vehicle functions can be defined as the categories by which the customer rates vehicle performance.

A complete list of functions, probably to be expanded in the future, can include the following:

- Appearance.

- Available space.

- Ergonomics.

- Climate comfort.

- Dynamic comfort.

- Dynamic performance.

- Handling.

- Safety.

- Resistance to age.

Each function can be explained through a certain set of requirements, which are qualitative and quantitative attributes that the vehicle must possess to perform each function correctly; we will describe these requirements shortly.

The first listed function (this list is not ranked by priority) is *appearance*, the ability to appeal to the customer; even if we these are beyond our scope, we will say the pertinent requirements involve the body, in terms of shape, volume, materials and details.

The contribution of the chassis to this function is marginal but not negligible and includes tire and wheel size, a part of car appearance; the engine also gives contributes through its proper lay-out at open hood, or through hood shapes and ventilation openings.

Roominess, or, from a designer point of view, the use of space, is important, because it embodies the primary objective of carrying people and goods.

Customers don't expect unlimited room, depending the class of the car, they are interested in; what is important is how much limited space can be rationally used. The room expended on car components is unavailable for this use; component lay-out should be designed to limit as much as possible any intrusion into the passenger compartment. This explains why we spent so much time on transmission and suspension bulk in the first volume.

Other important requirements also hinge on body design, such as roominess and availability of space to organize small objects; another important require-ment is adaptability (tilting or removable seats), in order to enable the customer to change the car interior to suit different transportation needs.

Car *ergonomics* can be defined as the ability to minimize the physical ac-tivity required by a given operation while using the car. Within this function, we usually include the pleasure of driving the car, including the many sensations the customer feels while driving.

The requirements of this function again involve the car body and include:

- the ease of entering and exiting the car for driver and passengers, of opening and closing doors, the glove compartment, hood, trunk, etc.;

- the ease of identifying and reaching the most important controls with min-imal reach;

- the comfort of the driver's posture;

- the ease of loading and unloading the transported goods.

Chassis design is primarily affected by the requirements established for car controls such as the steering wheel, gear shift stick, clutch and brake pedals; these have to do with their operating force, the placement of the controls and the feeling perceived by their operation. Any control, in fact, not only receives an input that should be minimally tiring, but returns a feed-back that should inform the driver about the correct accomplishment of the maneuvers.

We commented about *climate comfort* in our example. In this case, as well the related requirements affect body design and, partially, the engine.

The *dynamic comfort* function is evaluated by the ability to suppress all acoustic and vibration nuisances from outside (road pavement and other vehicles) and from inside (engine operation and component vibration).

The related requirements involve almost all vehicle components, as they participate as sources and potential transmitters of such disturbances.

Noise and vibration contain information useful for both driver and passen-gers. A totally silent vehicle, without vibrations, could prove to be dangerous, as has been demonstrated by experience with active noise suppression. In addition, some noise is peculiar to particular types of car, as, for example, sports cars.

The target is not total suppression but an acoustic environment compliant with customer expectations.

Filtering of components vibration is a task usually assigned to the body system, while filtering noise and vibrations from the road is usually assigned to the wheels and suspension.

Filtering powertrain noise (engine operation) involves powertrain suspension and the intake and exhaust system.

Unbalance specifications are assigned to any potential source of vibration.

The *dynamic performance* function includes requirements that are easy to measure, such as top speed, gradeability, acceleration and pick-up. Requirements that are more difficult to evaluate meaningfully are drivability and fuel economy.

Requirements involve all chassis components from the engine to the body and, in general, the entire vehicle.

Handling functions are usually defined as the vehicle's ability to follow driver inputs on the controls, when modifying car speed or trajectory; these controls include separate or combined operations on the steering wheel, brakes and accelerator.

Handling function requirements involve not only suspension, tires, steering mechanism and brakes, but also engine and transmission. Overall properties of inertia (mass and momentum) have vital importance for this function.

The *safety* function is usually classified in three ways:

1. *preventive safety*, such as the ability of the vehicle to keep the driver constantly updated on corrective maneuvers to be undertaken; a typical example of this category includes not only outside visibility, visibility of the main instruments (i.e. speedometer, outside thermometer, etc.) but also car trim variations;

2. *active safety*, such as the ability of the vehicle to react to driver inputs with a response that should be immediate, stable and proportional to the action, while avoiding obstacles or dangerous situations;

3. *passive safety*, such as the ability to limit, when a collision is unavoidable, the severity of injuries to car occupants, to pedestrians or to passengers of other cars, involved in the collision.

Safety cannot be, by definition, total, but requirements should be established for the most statistically relevant situations; homologation requirements are an important part of this approach, together with manufacturer's technical policies. In the passive safety category, repair cost limitations following low speed collisions have been added recently.

Safety involves all main vehicle components; the body is particularly involved in preventive (inside and outside visibility, lights) and passive safety (structures, passive and active restraint systems, component lay-out, surface materials and finishing).

The chassis must comply with all active safety requirements for suspension, brakes and tires, and with passive safety requirements, such as intrusion into the passenger compartment following a collisions.

The engine system is involved in passive safety as far as fuel spills after crashes and consequent fire hazards are concerned.

The *aging resistance* function is the ability of vehicle system and components to maintain their functions unchanged or to limit their degradation with aging within acceptable limits; reliability is a requirement of this function. Aging resistance involves all vehicle components and, obviously, all chassis components.

18.1.2   Technical specifications

Each vehicle function can be described through a coherent set of measurable requirements; compliance with them guarantees customer satisfaction with the vehicle.

These requirements determine the technical specifications of all components.

Each part or subassembly could be fully defined by an engineering drawing, containing all relevant geometric dimensions and materials. In reality, the detailed knowledge of complex components is irrelevant to the car manufacturer, who is more interested in performance than specifications. A very detailed cannot always guarantee complete fulfillment of the desired performance. In many cases, car manufacturers lack the technical competencies to understand complex details.

Technical specifications solve this problem by providing global and synthetic information only; it is therefore necessary to establish, for a certain component, what is relevant for system function. A technical specification should list:

- what physical properties describe the requirements requested for the component;

- in which conditions those properties must be measured;

- what values (with allowed tolerance) they must assume to obtain the desired system performance.

These technical specifications, together with simple outline drawings, represent the only technical documentation useful for managing the relationship between car and component manufacturer.

The component manufacturer's point of view is necessarily different, since he must create technical documents to produce the needed part consistently. After all it isn't rare that some second tier supplier is producing other parts that will be integrated into the final subassembly. The first tier supplier must therefore use his technical specifications to advantage.

Consider the example of a tire.

The vehicle system utilizes the tire to express forces on the wheel along the three directions (vertical, longitudinal and transverse); tire technical specifications will examine these three parameters first.

A reasonable approach to tire specification might be to determine maximum allowed values for these directions and magic formulae coefficients; their values can be calculated by mathematical models directly or interpreted by these models as applied to satisfactory results of experiments performed on the car.

Other specifications could describe the tire's age resistance, with acceptable values of tread wear using certain mission as reference.

The figures of vertical elasticity and damping close this specification list.

The common characteristics of these parameters are:

- they must be correlated with the function we wish to obtain on the vehicle;

- they must be overseen by the supplier, since he is the manufacturer, without implying a detailed knowledge of the application.

Other characteristics, such as, for example, the chemical composition of cord fibers applied to the tread, are not generally involved in vehicle operation and, if they are, the link between chemical composition and system behavior is part of the proprietary know-how of the supplier.

Technical specifications, therefore, define the performance that we want to obtain, but not the details that allow us to obtain this performance.

On the other hand, specifications should not be too superficial; for example, we should not make the mistake of providing a supplier specifications on road durability without referring to the driving conditions and the typical trip in question; a good specification should enable the supplier to evaluate for himself the results of his effort.

Continuing with the tire example, it is clear that the technical documents available to the supplier will be much more detailed than those used by the car manufacturer as technical specifications; the supplier will have available a complete set of drawings of the tread, including detailed dimensions, cord texture, materials, fixtures and production set up, etc. The design tools of the supplier will be able to correlate these parameters with tire performance on the vehicle system which are almost coincident with the technical specifications.

Some details, such as wires included in the body - whose performance is not only dependent on reference dimensions (diameter) but also on the manufacturing process in the steel mill - should be described by specifications including only diameter, yield and the physical properties of their surface.

Technical specifications represent a universal simplified language, allowing such different industrial organizations as final manufacturer and supplier to co-operate in reaching the same objective, the final customer satisfaction.

The same logic can be usefully applied within a company, particularly a vehicle manufacturer, to integrate the activities of different departments.

Although there is no conceptual obstacle to developing each component from scratch, it is always useful, before taking this decision, to clarify which function the component performs on the vehicle, how it can be quantified, and from which values the objective of satisfying the customer is obtained.

In this way it is possible to manage, with relative simplicity, complex activities involving a number of people, braking down each objective into sub-objectives that are measurable and understandable by the different parties involved.

18.1.3   Chassis system design

We have seen that the automotive chassis contributes to the following vehicle functions:

- dynamic performance;

- handling;

- ride and acoustic comfort;

- ergonomics;

- safety.

The engine and transmission relate to dynamic performance, in terms of available power; the body (aerodynamic resistance), tires (rolling resistance), transmission (mechanical efficiency) and mass properties of the vehicle involve dynamic performance in terms of absorbed power.

Handling and active safety are influenced by suspension and steering system geometry, by brake design and by the elastic properties of tires; the transmission determines the interaction between cornering and traction forces. Chassis control systems play a fundamental role.

Ride comfort is influenced by disturbances, essentially vibrations, coming from tire-ground contact and is affected by suspension geometry, by the elastic and damping properties of springs, bushings and shock absorbers, and by the vertical properties of tires.

Acoustic comfort, on the other hand, requires a notable development of our knowledge of body structure and trim. For this reason, this function is usually studied in body design.

As far as controls are concerned, ergonomics involves chassis design: the steering system, brake and transmission (clutch and shift stick); control systems contribute to this function through power assistance and automatic transmissions.

Passive safety involves chassis design and component intrusion into the passenger compartment and structure; since most cars have a body that includes in a single shell both chassis and body structures, we generally study this function as part of body design.

The objective of the design methods explained in this book dedicated to chassis design is therefore to design chassis components that satisfy the above functions at the vehicle system level.

The adequacy of these methods might appear at least partially unsatisfying, because we will explain how to verify which function an assigned vehicle is able to perform; we will also identify which components condition those functions but not how these components must be specified in order to perform the functions at the desired level. We are able to tackle this problem only *a posteriori*, while an *a priori* approach would be desirable.

This qualification could apply to all design courses, because if designing means to define a product that does not yet exist, what is really taught is to verify whether an already defined product is able to perform an assigned function.

The designer's job is, therefore, to assume an hypothesis and to verify the results that can be achieved; a deviation from the objective will guide to define a different hypothesis that will again be verified. The designer will be more efficient, if the first approximation hypothesis is close to correct, but, in any case, design will remain a trial and error process.

A technical specification definition is further complicated by the fact that the final judgement on the product will be issued by the customer and not by the designer, and customer judgments are sometimes difficult to express concretely, because they are influenced by unmeasurable parameters and alternative offers on the market that may be unknown at the beginning of the development process.

Technical specifications are developed and determined through different strategies, according to a process that can be divided into two parts, called *target setting* and *target deployment*. The target setting phase consists in setting objectives for each of the functions perceived by the customer; this job will be more fruitful if subjective judgements are avoided and only objective measurements are used. If this requirement appears easy to be meet for functions like top speed, acceleration, and gradeability, it will be difficult for functions like handling, where subjective feelings come into play.

We will see in the following paragraphs how subjective feelings can be transformed into objective measurements.

In the next phase of target deployment, as a first step, vehicle subsystems according to function are identified and their specifications tentatively set; the specifications adequacy to the targets will be verified, correcting any errors in the specification.

These verifications may be performed using mathematical models of the vehicle and in some cases also by building and testing simplified prototypes (*mule cars*) that will allow complex subsystems to be verified.

18.2   OBJECTIVE REQUIREMENTS

If we want to define vehicle functions and, particularly, measure the main requirements that determine those functions, we must refer to the test procedures used for this purpose; vehicle objectives are, in fact, set with reference to those procedures.

We commonly identify *objective* and *subjective* requirements. The first ones are directly measurable with the instruments of classic physics; the second are determined only by the satisfaction of the final vehicle user, but they can be converted into objective measurements through statistical investigations of customer groups.

A classic example of an objective target might be the time to accelerate the vehicle in top gear from one speed to a higher one. This is easily measurable, when reference conditions (road grade, wind speed, atmospheric pressure, etc.) and load conditions are set. This test can be performed by a professional driver who is able to achieve repeatable results; each test, even a simple one, is influenced by driving behavior.

If we want to define the customer satisfaction level, we should ask ourselves how it can be measured and if it depends on this requirement only; if that is the case, satisfaction will be influenced by the customer's expectations, depending on the class of the car, driving habits, etc.

The required objective follows from a statistical study of the reaction of a group of customers driving this car; the study of customer satisfaction on different questionnaires leads to significant data derived from subjective measurements. The customer sample must include only people likely to be final customers of the car under development.

We will refer in this paragraph only to objective measurements of vehicle performance involving the chassis, which are, as we have seen:

- dynamic performance;

- handling and active safety;

- ride comfort;

- ergonomics;

- passive safety.

For each of these we will comment on test procedures and measurable data; we will consider passive safety only when speaking about regulations.

18.2.1 Dynamic performance

For this kind of test it is necessary, for safety reasons, to use test tracks closed to public traffic.

Speed and acceleration tests should be performed on a flat straight road that is long enough to accomplish all tests reliably; a launch ramp should also be available that allows the vehicle to reach top speed before its measurement.

Sufficiently long constant slope roads, at different inclination angles, should be available for gradeability tests.

Loop tracks can be used to imitate of road trips that are particularly significant for vehicle use; according to the know how of each manufacturer, these tracks allow, while driving following certain rules, the measurement of average speeds and fuel consumption comparable to real values.

Because engine performance is influenced by air density and humidity, climate conditions during such tests are significant; a suitable condition is an outside temperature in the range between $10 \div 30°$C, with no wind and rain.

As an alternative to the test track, subject to variable climate conditions and, therefore, not always available, roller benches can be used, allowing electric brakes with electronic control to simulate vehicle driving resistance on the road; in this case, the car is driven according to an assigned speed time history. This practice is particularly useful for measuring fuel consumption.

A roller bench, when contained in pressure and temperature controlled chamber, allows dynamic performance at temperature and altitude conditions different from those available outside to be measured.

Test vehicles must be driven for a certain distance (about 5,000 km) after assembly to stabilize mechanical frictions and tire rolling resistance, since these parameters are subject to a certain settling depending on surface wear.

Since performance also depends on transported weight, it is necessary to control this value within a statistically meaningful tolerance; usually 2 passengers (including driver) and 20 kg of baggage are used for testing. For industrial and commercial vehicles the full load condition is considered.

The instruments used in these tests are quite simple, as far as speeds are concerned: they include optical devices to actuate stop watches that measure driving times, while space driven is determined by the position of these devices along the track.

Fuel consumption is measured by volumetric flow meters on the engine feed pipeline; in this case the recycled flow to the fuel reservoir must also be taken into account; sometimes an auxiliary tank is applied that is weighted before and after the test.

The best known dynamic performance is *top speed*, which is the maximum vehicle speed on a flat road, after a reasonably long launch ramp.

Acceleration is usually defined as the time necessary to reach a predetermined speed (usually 100 km/h or 60 mph), starting from a still condition, using the gearbox, at full throttle; sometimes it is also measured as the time necessary to cover a fixed distance (usually 1 km or $\frac{1}{4}$ mile), starting from a still position, using the gearbox, at full throttle. This kind of test must be repeated on a manual gearbox a number of times, to allow the driver to identify the best strategy to working the controls, because start-up and shift times influence the final result.

By contrast, *pick-up* time is instead the time needed to increase the vehicle speed, starting from an initial fixed value, without using the shift stick but at full throttle. The initial speed can be 50, 60, 70, or 80 km/h, while the final one is usually 100 km/h; the gear is usually the top one or, if different, the top speed gear. The distance driven could also be used to measure this performance.

Gradeability is the maximum road slope at which the vehicle is able to start up and be driven at constant speed, without slippage of the clutch; this value is approximated according to the available slopes on the test track. The grade is measured by the difference in elevation at the two end of the test track, divided by the horizontal projection of the track; this is the tangent of the longitudinal road inclination α.

Among the practices of manufacturers are reference loop drives on closed tracks or open roads which allow one to evaluate road performance under controlled conditions; in this case *average speed* or *driving time* are measured.

Increasingly congested traffic conditions have distracted customer attention from the performance obtained by intensive gearbox use, putting emphasis instead on pick-up time at low speed; the most recent statistical surveys correlating subjective judgements of performance, favor this measure on short test distances.

This trend increases the importance of low speed ($1,500 \div 2,500$ rpm) engine torque, with reference to maximum power. It is therefore not inaccurate to include *drivability* in the category of vehicle dynamic performance.

Drivability can be defined as the vehicle's ability to increase or decrease its traction force quickly, without fluctuation around the final desired value.

At the beginning of the test, the throttle pedal is depressed or released starting from a condition corresponding to the initial steady state reference speed.

Drivability is evaluated by examining the resulting car speed diagram as a function of time after the input time on the accelerator pedal, or by measuring the longitudinal vehicle acceleration. An objective evaluation parameter can be the number of peaks of this diagram before the asymptotic value.

Vehicle drivability is not only influenced by engine torque oscillations, induced by flow transients into the intake and exhaust ducts, but also by the elastic torsional stiffness of the driveline, from clutch to tires, and by the elasticity of powertrain and car suspensions mounts.

Fuel consumption at constant speed is usually measured between 50 km/h and top speed; the test is performed in top gear or, if different, on the top speed gear; this test is quite simple, but has a very low correlation with practical vehicle use, where speed variations and engine idling periods are very frequent.

For this reason, tests are always completed with a measurement of a driving cycle; this test is usually performed on a roller bench, able to simulate driving trips of different kinds; an important cycle will be described in the chapter on regulations.

It is a good practice to measure consumption at ambient temperatures different from the reference condition (usually 20°C) if the car is to be sold in countries where this condition is not significant; in this case, the effect of a cold start must also be investigated. On the road consumption measurements can also be performed, if there is sufficient control over ambient conditions.

18.2.2   Handling and active safety

Handling tests do not differ significantly from active safety tests and are therefore described together. This kind of test introduces a specific difficulty because on-road maneuvers can be many and their number is increased if different road pavement and conditions are to be considered.

Many manufacturers have adopted similar elementary maneuvers and most of these have been standardized by the ISO. Standardization applies to the execution of the maneuver only and sets no reference values for the output values to be measured.

The test track, usually a flat square that can be flooded under controlled conditions, provides marked courses that cars must follow; in this way the consequences of mistakes are not burdensome.

Cars are often equipped with roll over protection provided by additional wheels that contact the ground at high roll angles.

Vehicle instruments must be sophisticated because they have to measure dynamic values for the vehicle; the essential ones include:

- lateral acceleration;

- yaw velocity;

- vehicle side slip angle;

- roll angle;

- vehicle speed.

For the definition of each see the fourth part of this volume.

A fixed reference system is necessary to establish these values through instruments installed on the vehicle; an inertial platform is therefore used that measures the six components of rotation and displacement of the vehicle sprung mass with reference to the ground.

In many tests a particular steering wheel able to measure steering angle and torque is used.

Tests are classified as *open loop* and *closed loop* , with reference to the role of the driver during the maneuvers. In the first case, the driver manipulates vehicle controls (steering wheel, brake and accelerator pedals) according to a preset procedure, regardless of the result; in the second case the driver uses the controls as needed and tries to obtain specified objective, as, for example, driving along a course at the highest possible speed.

The simplest open loop maneuver is the *steering pad* (ISO 4138), where the vehicle is driven around a circle at constant speed.

This is an open loop maneuver because the controls are blocked during the test period, to guarantee a steady state motion.

Three different methods are considered depending on the skill of the driver, that are substantially equivalent in result; these are:

- constant curvature radius,

- constant steering wheel angle,

- constant speed.

Since these are the three independent variables that define motion, their test results can determine the remaining variables.

A typical value for the curvature radius could be 40 or 100 m; it is important that tests are performed in such a way as to obtain different values of lateral acceleration, starting from a very low one, which is useful to measure the Ackermann steering wheel angle.

This test allows the evaluation of the steering index of the vehicle and the determination of roll angle as a function of lateral acceleration; a maximum allowed lateral acceleration can be identified by a series of attempts.

We refer again to the fourth part of this volume for a definition of the parameters involved in this test.

To evaluate vehicle stability while entering a curve and the steering wheel re-alignment when exiting it, the *lateral transient test* (ISO 7401) is usually applied.

The car is stabilized on a straight road at 100 km/h or, if desired, at different speeds; in the step input version, the steering wheel is suddenly turned to a preset value; to simplify the maneuver, a steering wheel stop is set at the desired angle.

Without changing the accelerator pedal position, the steering wheel is kept turned for a specified time.

Important evaluation parameters are the gradient of lateral acceleration and yaw speed as a function of the steering wheel angle, the delay time between steering wheel angle peak, and yaw speed peak and the presence of overshoots on the yaw speed diagram (yaw speed peak higher than asymptotic value).

A variant of this maneuver is the application of a sinusoidal steering wheel input applying, as input:

- random function,

- triangular function,

- sinusoidal function,

at different frequencies.

The complexity of this test is evident despite the schematic simplicity of this transient between straight and curved steady state motions.

The *accelerator pedal release* maneuver (ISO 9816) is studies vehicle behavior when the accelerator pedal is released, while driving on a curve; this maneuver simulates what could happen if a driver attempts to drive at too high a speed.

It is possible to test vehicle stability and measure deviation from the original path. This test can be performed at the end of the steering pad test.

Two different methods are available.

- At a constant course by stabilizing the vehicle on the assigned curvature radius before releasing the accelerator pedal; the steady state speed can be increased as needed to investigate the influence of lateral acceleration.

- At constant speed, stabilizing at a certain speed on decreasing curvature radii.

The test output displays the interaction between the steering index, a function of lateral acceleration, and the varying cornering stiffness of tires due to the instantaneous change of traction caused by braking. The engine shows a braking effect increasing with initial rotation speed: the transient is affected by the selected gear ratio.

During this open loop test the steering wheel must be locked.

The evaluation parameters are the same as in the previous test, where longitudinal acceleration has to be added; because the steering wheel is blocked, there will be a deviation from the initial course after accelerator release; cars are usually designed so as to close the path after the transient slightly, without sensible discontinuity from the initial trajectory.

Still in the area of stability, *the braking in a curve* test (ISO 7975) has been designed, to add the application of brakes to the above procedure. Also in this case steering wheel is again locked.

To the parameters of the previous test, braking fluid pressure is added and the deviation from the initial course could be significant; the test is performed at increasing longitudinal accelerations until one of the wheels is blocked or the ABS system has started to work.

An important open loop maneuver is the *steering wheel release* (ISO 17288); the purpose of this test is to establish the vehicle's ability to return to a straight path after a curve.

The vehicle is stabilized on a steering pad at 100 km/h; path curvature is chosen so as to maintain a lateral acceleration of about 1 ms^{-2} and the test is repeated for growing acceleration values. At the beginning of the maneuver the steering wheel is left free to turn under the action of the existing forces on the contact points of the tires. The accelerator pedal is kept where it was at the beginning of the test.

The usual path parameters and the actual steering angle are acquired. Since steering wheel and lateral acceleration must show damped oscillations from an initial value to zero, damping factors measured on time histories are assumed as evaluating factors of these transients.

Side wind sensitivity tests for cars (ISO 12021) and for industrial vehicles (ISO 14793) are also available; for these vehicles specific tests can be used to examine the effect of trailers.

All the elementary maneuvers we have considered, although very complicated, do not correspond to real driver behavior; they are, nevertheless, very useful for understanding the natural vehicle response before any correction by the driver has been applied.

The opinion of the people that have designed these tests is that the first part of any real maneuver is always of the open loop type; in fact, drivers apply an action on controls (steering wheel , accelerator and pedal brake) whose amplitude is suggested by the desired response; the amplitude has been learned on previous similar maneuvers.

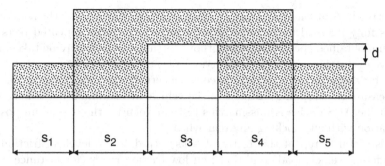

FIGURE 18.1. Course of a lane change maneuver, according to ISO 3888 standard.

After a short while corrections are applied, as soon as some deviation from expectation has been perceived. The simpler correction, the lower the deviation from the desired response.

A simple closed loop maneuver introduces difficulties in evaluation, as it depends on the different behavior or ability of different drivers.

The *lane change* maneuver (ISO 3888) studies vehicle stability by overtaking another vehicle or avoiding an obstacle; the relative course is shown in Fig. 18.1 and is driven from the left to the right.

The course is defined by rubber posts; the first stretch S_1 is the normal driving lane, the stretch S_3 represents the overtaking lane and S_5 again the normal driving lane.

The three stretches are joined by the zones S_2 and S_3; the right border of the overtaking lane is shifted by the deviation d to the left, with reference to the left border of the normal driving lane. The scheme of the course should be designed as follows (dimensions are in [m]).

Stretch S_1 length 12 m, width $l = 1, 1\ w + 0, 25$.
Stretch S_2 length 13,5 m.
Stretch S_3 length 11 m, width $l = w + 1$.
Stretch S_4 length 12,5 m.
Stretch S_5 length 12 m, width $l = 1, 3\ w + 0, 25$.
Deviation d 1 m.

Dimension w is the vehicle width, measured in [m] without taking into account side mirrors; the width of stretch S_3 cannot be less than 3 m.

This elementary maneuver is of the closed loop type, because driver course corrections are essential to avoid hitting any rubber post.

The car must be driven at 90 km/h in IVth gear and car speed should remain unchanged along the course; varying parameters may be used to compare different vehicles.

The maneuver is complicated and heavily influenced by driver skills; with professional drivers, interesting results can be gathered more by subjective impressions than objective measurements.

In order to assess vehicle stability while avoiding a sudden obstacle, the vehicle is tested many times on this course at increasing speeds; the vehicle enters stretch S_1 at the initial steady state speed, at which point the accelerator pedal is fully released to simulate obstacle detection. After repeated tests, the maximum obtained speed achieved without hitting any post may be taken as an objective evaluation of vehicle stability.

As far as the braking performance is concerned, the easiest repeatable test is to measure the stopping distance of the vehicle at maximum possible acceleration; if the ABS device is missing this test can identify the maximum possible deceleration without blocking any rear wheel.

It is useful to repeat this test on tracks with different levels of friction and with different vehicle load conditions; on low friction roads (for instance on icy roads) it is also a good practice to measure the vehicle path with different friction on the two sides of the vehicle (μ split). This maneuver, useful in evaluating ABS systems, simulates braking on a road with a wet or icy border.

The catalogue of maneuvers is not complete; many other tests are used to focus on specific problems.

18.2.3 Dynamic comfort

Comfort is correlated to passenger unease caused by vibrations between 1 and 100 Hz in frequency; higher frequency vibrations correlate solely with purely acoustic discomfort.

Vibrations in this range are caused by obstacles and hollows on the road surface, which are filtered by the elastic and damping properties of tires, suspensions and seats; these condition the influence of powertrain mass vibrations (caused by the road) and vibrations of the body, which enhance or reduce the effect of the road, according to their vibration modes.

Comfort tests are performed on closed tracks that reproduce the road surfaces the vehicle is most likely to encounter; the road must be maintained according to specific standards to allow repeatable results on these tests.

Ambient temperature must also be controlled and recorded because of its influence on the elastic and damping properties of elastomers, which are largely employed in the mechanical components connecting the ground profile with passengers; temperature also has a remarkable influence on oil viscosity of shock absorbers.

Measurements to be evaluated include the acceleration of the different parts in contact with the human body, such as floor, seat, steering wheel, etc.; other accelerometers could be set at different position in the mechanical chain, to monitor test accuracy and for diagnostic purposes.

It should be noticed that accelerations should be measured along the three main axes of the vehicle reference system, especially if the more important components are along the z and x axis. Vehicle speed measurement is also important, because it and the road profile validate the test.

Four profiles exist for elementary tests that replicate the most common road defects.

FIGURE 18.2. Typical defects of suburban and urban roads, relevant to vehicle comfort; at left a patched tarmac; at right a stone block pavement.

A motorway profile with perfect tarmac is characterized by a virtually flat surface, with peaks and hollows much further apart than the vehicle wheelbase; at speeds of 100÷120 km/h, this spacing may excite the natural vertical frequency of the powertrain and of the vehicle suspension. This kind of course is also used to identify and analyze vibrations coming from the shape of the tires and any defects in them.

A suburban road with a poor maintenance is characterized by hollows that are spaced closer than vehicle wheelbase, with different size pits, patches and ruptures of the wear layer; Fig 18.2, at the left supplies an example of these defects.

These kinds of defects involve a range of frequencies larger than the previous and excite the natural vibration modes of sprung and unsprung masses: comfort in this test is critical for customer satisfaction since these defects are widespread.

The stone block pavement (Fig. 18.2, at right) is still common in city centers because of its attractive appearance and relative immunity to ice damage; it is therefore a reference test for the urban environment and is associated with lower speeds than the previous tests.

Because of the nature of this surface the wave length spectrum is very wide, from a few centimeters to several meters. The consequent excitation encompasses the entire range of comfort frequencies, involving all suspension components and car structures.

The catalogue of comfort tests includes, usually, a single step obstacle, representing what happens when crossing a curb or a railway; this obstacle is represented by a steel bar of rectangular section across a flat tarmac road. This kind of obstacle generates a force pulse on the wheels and involves a wide range of frequencies, with vibrations along the z and x axis.

18.2.4   Ergonomics

Ergonomic functions involving the chassis are influenced by the position of controls and by the force required for their operation; the main controls include the steering wheel, brake and clutch pedals, gear shift stick and parking brake.

The current trend of widening the passenger compartment as much as possible makes the pedal board the starting point of preliminary studies of car habitability.

Installed into the passenger compartment, the pedal board is constrained by the following elements:

- front wheel well: its dimensions are determined by the front wheel steering envelope and by the suspension stroke; this volume should also take snow chains into account;

- the floor tunnel, for the transmission shaft on rear wheel driven cars and for the exhaust pipe in front wheel driven cars;

- the minimum clearance from the ground;

- the firewall, separating the passenger compartment from the engine, which is also used to attach the pedal board.

The most forward position possible for the pedal board is desired to increase the available space for driver and passengers. The limit to the front position is represented by the powertrain and by the steering box. The floor tunnel, from one side, and the wheel well, from the other side, limit lateral space for positioning the pedal board. These limits are especially critical on narrow cars.

The accelerator pedal is always in contact with the right foot of the driver, except when braking. It must be operated with minimal force and high precision: the foot requires a side rest to avoid interference from vertical vibrations.

The accelerator pedal stroke should be about $50 \div 60$ mm.

The reference point for positioning the accelerator pedal is the most rearward position of the driver's foot when resting on the floor, according to the projected comfort angles. It is called the *heel point*.

To avoid excess contact between shoe and pedal, the relative motions of these two parts should be minimized.

Because of pedal motion, the shoe sole changes direction. Since the hinge point of the pedal is fixed, this condition of no slip between pedal and shoe can be obtained only in one position; it is preferred that this position be the one most frequently used, usually at mid stroke. The slip can be reduced in the other positions by curving the pedal.

The brake pedal can be operated by relevant forces and stroke precision is not very important.

According to regulatory standards, the control force must not be higher than 500 N, but it is suggested that this control be designed to limit the maximum pedal force below $200 \div 250$ N, using the power assistance system.

To exert control forces easily, it is assumed that the driver's foot is angled on the floor to reduce the torque on the heel.

This kind of operation is allowed for emergency braking only, while for ordinary braking, the pedal is depressed in the same way as the accelerator.

The clutch pedal can also be operated in two ways, according to design choices and driver's habits:

- with the heel on the floor;

- with the foot at a higher position for the first part of the pedal stroke, and resting on the floor at the end of the stroke (clutch fully disengaged).

The first mode is used for precision modulation, as for starting up on a grade. In this phase the pedal stroke is limited.

Because the force on the pedal should stay below 100 N, the foot position could be advanced without negative consequences on heel torque.

On the most common pedal boards, the hinge axis of the accelerator and brake pedal are different, while they are the same for clutch and brake pedals. To allow different strokes for the last two pedals the clutch pedal in rest position can be placed higher.

To avoid interference with other pedals when depressing a pedal quickly, the distance between pedal centers should be as high as the sole width, not less than 100 mm.

Steering wheel positioning is more complex and must take into account:

- a minimum relative distance from the pedal board to allow correct operation of the pedals; this implies a minimum distance of about 650 mm between the highest pedal, in rest position, and the lower surface of the steering wheel;

- a comfortable inclination for the steering wheel of about $30° \div 35°$;

- a rotation axis placed at least 300 mm from the middle of the vehicle, to avoid interference with the front passenger during steering;

- interference with the driver's leg while entering and exiting the car.

All decisions on controls position should be taken at the same time the body is outlined. Because of this, such decisions are rarely made by chassis designers.

A relevant indicator of steering wheel ergonomics is the force needed to turn the steering wheel at low speed.

This evaluation should be made by executing steering cycles, at low car speed (about $5 \div 7$ km/h), from stop to stop; steering wheel rotation speed should be between 100 and $150°/s$.

The output of this test reveals the hysteresis cycle of the steering wheel, as explained in the first volume, in the section on power steering.

When electric by-wire transmissions have totally replaced mechanical controls, there will be much more freedom to position these controls than was possible earlier.

Major future developments include the possibilities of:

- using joy-sticks or other devices, instead of the traditional steering wheel;

- integrating other functions such as shift, brake and clutch control in the steering control;

- mounting the steering control on moving boards, to enhance vehicle accessibility and to allow driving from either side;

- personalizing controls depending on user needs, to allow disabled people, for example, to drive more easily.

Other information about controls is reported in the first volume.

18.3   SUBJECTIVE REQUIREMENTS

Vehicle testing by car manufacturers is not only the final verification of product competitiveness before the launch, but also a valuable instrument for establishing measurable system objectives.

Classic testing implies objective experiments that are defined by straightforward procedures that are not affected by the skill or the personality of the driver, and that lead to precise and repeatable results, allowing the immediate comparison of achieved with target values. This testing is by nature *objective*.

The limitation of this approach is that the correlation between these measurements and customer expectation is small.

Customer opinions are subjective, based upon an evaluation of the difference between what they actually obtain in a given car and what they think is a reasonable expectation; an acceptable result is conditioned by their experience with previous cars and what they learn from specialized magazines, discussions with others and advertisements.

Tests to evaluate these judgements are called *subjective* tests and do not require particular instruments or dedicated test facilities, because they simply reflect the customers' day by day experience.

Some function, to be interpreted, are simple such as dynamic performance; others, such as handling performance, are more complex.

The evaluation of dynamic performance consists of measuring variables (top speed, acceleration, pick-up, etc.), which, according to their type, satisfy the customer as their value is low or high. Nevertheless, it is very difficult to identify the correct value for each variable or to establish if a lower value in a particular variable (for instance, acceleration) can be tolerated, if it produces a better result in some other variable (for instance, fuel consumption).

The difficulty will be even greater if the optimum value for the understeering index of a new car has to be balanced against performance in the accelerator pedal release maneuver.

A simple way to overcome this difficulty is to perform *jury tests*, using potential costumers of the car under development; a jury test is a typically subjective test where subjective customer judgements on a homogenous cluster of cars are acquired and elaborated through the use of statistical methods. In this cluster of homogenous cars a prototype, representing the car under development, may also be included.

The proper execution of a jury test execution requires that cars to be evaluated are already built and have reached a satisfactory level of refinement. They would not be of use in the early stages of development as technical specifications are defined.

More often, jury tests and the analysis of their results are part of initiatives that are independent of the development of a specific product; they can be performed occasionally, to evaluate the development of competitors' products and consequent customer expectations. A test campaign like this can establish customer evaluation criteria and target values.

To develop technical specifications for a new car, a vehicle mathematical model will be applied that has been previously validated with experimental results.

Mathematical models will be used to assign to the components of the new vehicle technical specifications that match the results to be achieved in a virtual jury test, one that will be performed as soon as significant prototypes are available. Instead of potential customers, professional drivers from the company will be used, who will use the same evaluation methods as the customers in previous jury tests.

We will consider three examples, applied respectively to handling, to dynamic comfort and to fuel consumption; other requirements can be studied in a similar way.

18.3.1 Handling and active safety

In this section we will describe the approach that has been developed in many articles quoted in the references.

Manufacturers usually evaluate handling and active safety by using professional drivers, who are able to make comparisons, correct errors, and address chassis designers; the point is to correlate these judgements with subjective tests performed by potential costumers. These are eventually replicated with mathematical models to produce useful design tools.

In our references, dynamic behavior requirements are classified according to the scheme in Fig. 18.3.

The scheme suggests a classification that takes into account driving conditions involving lateral dynamics (driving in a curve), longitudinal dynamics (accelerating and braking) and the interaction between the two situations (braking and accelerating in a curve).

Driving conditions are classified as to driving ease or safety, as in emergency maneuvers; non professional drivers are able to evaluate maneuvers of the first kind, while only professional drivers are asked to judge the second kind.

		Driving easiness	Driving safety
Lateral Dynamics	Response	Steering activity Response speed Response progressiveness Roll Roll speed	Maximum lateral acceleration Stability Obstacle avoidance Roll-over
	Control	Selfalignment Center play Response graduality Reaction graduality	S. W. oscillation S. W. selfaligning speed
Longitudinal Dynamics	Response	Braking efficiency Response speed Traction Pitch Pitch speed	Maximum long. acceleration Stopping distance
	Control	Pedal graduality Force graduality Brake modulation	
Lat. / Long. Interaction	Response	Traction in curve	Braking in a curve Tip-in Tip-out

FIGURE 18.3. Fundamental requirements of longitudinal and lateral vehicle dynamics, which define handling and active safety.

For each driving condition it is important to be able to distinguish between vehicle response and control quality; for instance, power assistance is relevant to steering control quality, but not relevant for vehicle dynamic response.

In terms of lateral dynamics, steering activity is considered the quantity of work to be applied to this control (steering wheel) to obtain a certain result; response speed and response progressiveness[1] as proportional to the effects to the action on the command are also relevant. Roll angle and roll speed are also relevant for comfort and stability.

Viewing lateral dynamics in terms of control quality, again for lateral dynamics, self-alignement represents the ability of the vehicle to drive spontaneously on a straight line, while center play is relevant in evaluating the command insensitivity to small steering angles on a straight course.

Response and reaction graduality[2] are relevant to completing the judgement of steering wheel quality.

For the sake of brevity, we do not comment on requirements related to the interaction of longitudinal and lateral dynamics.

The sample of cars to be examined is finalized according to the result we want to obtain from this study; in the case we are summarizing, the sample is ample, in terms of car types (they were selected in different market segments), because this study is focused on the correlation between the subjective judgements of drivers and objective measures to be acquired during selected elementary standard maneuvers.

---

[1] We can define progressiveness of a control the derivative of its output (i.e. braking force), with reference to its input (i.e. brake pedal force).

[2] We can define graduality of a control the derivative of the force or torque applied with reference to its stroke.

As the correlation is demonstrated, the sample can be limited to existing cars that are more similar to the new product under development.

The test jury must include professional and non-professional drivers, in order to evaluate both emergency and normal driving conditions and also to find out if there is any systematic bias between the two categories; the size of the sample (at least 20 people) must be chosen to allow a sufficient confidence level.

All cars are evaluated in free driving conditions, on a test track offering the necessary safety and a course suitable for highlighting all requirements under evaluation.

A questionnaire must be set up, to gather test results; its questions address the requirements under consideration and must include numeric scores; an overall verdict is also required.

All scores are subjected to statistical techniques to:

- eliminate scoring bias; many jurors use only a part of the scoring scale, assigning scores that deviate constantly from the average;

- eliminate those scores that are too removed from the average of the jury.

After this treatment, the correlation between single values and overall judgment is investigated, using a multiple regression analysis.

Figure 18.4 shows the result of this analysis; all scores are normalized on a decimal scale. Histogram f refers to the overall judgement.

The same cars were evaluated on open loop elementary maneuvers that were felt to be more finalized to the requirements under evaluation.

The following elementary maneuvers were applied.

- Steering pad, according to the ISO 4138 standard, on a curvature 40 m in radius, starting from a low speed up to the maximum safe speed.

- Lateral transient test, according to the ISO 7401 standard, followed by the steering wheel release maneuver (ISO 17288). Cars were driven at 100 km/h on a straight path and received a steering input to the desired value at steering angular speed of at least $200\,°\mathrm{s}^{-1}$; this value was maintained for 3 s and the steering wheel released for other 3 s. The accelerator pedal was kept in place during the maneuvers. The test was repeated for incremental steering wheel values, until a value was identified at which the vehicle does not stabilize.

- Overtaking test according to ISO 3888 standards, at 90 km/h; although this maneuver is a closed loop and was not created for objective evaluation, an objective measurement has been obtained by dividing the square average of vehicle lateral acceleration by the steering work. Delays between acceleration and steering wheel peaks have also been measured.

There is no direct correlation between a single subjective judgement and a single elementary objective maneuver; nevertheless, the multiple regression

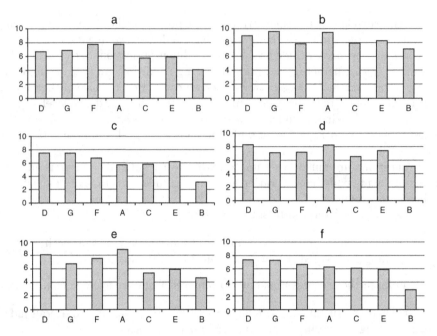

FIGURE 18.4. Average of subjective judgements on lateral dynamics of 7 cars; a: steering wheel activity; b: response speed; c: response progressiveness; d: roll speed; e: roll angle; f: overall judgement.

analysis relating each subjective judgement to all objective measurements under consideration returns an overall correlation factor of more than 0.85.

The correlation linear equations allow us to define partial *driving quality indices*, which correlate objective measurements, measured or calculated on elementary maneuvers, with subjective evaluations and, therefore, with customer satisfaction.

The *global driving quality index* in Fig. 18.4f is correlated to other indices with the same method and allows us to evaluate *a priori* the possible trade-off between these indices.

A similar process can be repeated for other groups of variables in Fig. 18.4, defining a global index that allows us to apply measurements to elementary maneuvers to forecast the customer's judgement on a car in terms of handling and active safety.

18.3.2 Dynamic comfort

This logical process has been also applied to the case of dynamic comfort, according to another article quoted in the references; although this work considers acoustic and vibration comfort as a whole, we will limit our analysis to vibrations only.

In this study drivers have driven their cars on open roads as they wish; the chosen road has been classified according to recorded acceleration on selected car positions.

The questionnaire was divided into three different parts.

The first one was addressed to vibrations perceived on different contact points with test cars:

- body floor;

- seat cushion;

- seat back rest;

- steering wheel.

The second part of this questionnaire solicited reactions to different perceived disturbances in the main movement of the vehicle; a follow-up variance analysis has suggested these results to not considered because of their excessive spread.

The third part was designed to gather an overall judgement; the scoring scale was again decimal, with scores of 10 assigned as the absolute optimum (no disturbances perceived).

Data on subjective measurements have not benefitted of standard elementary maneuvers, as in the case of handling; in their absence the following maneuvers have been used:

- random inputs on the motorway at 80, 100, 120, 140 km/h; on urban stone pavement, at 20, 40, 60 km/h; and on a low maintenance suburban road, at 20, 40, 60 km/h.

- shocks from a rectangular profile single obstacle across the road, at 30 and 50 km/h; on a rail level crossing at 20 and 40 km/h; and crossing a bump at 30 and 50 km/h; each obstacle has been reproduced on a track, by profiles elaborated statistically on open roads.

Acceleration measurements from random tests have been elaborated by calculating RMS in the domain of time and frequency. The same procedure has been applied to the shock test, including calculated magnitudes as, for instance, wasted energy, as well.

In this case, partial indices and a global index can be obtained that are well correlated to overall customer judgement and to the single measurements derived by elementary maneuvers.

18.3.3   Fuel consumption

Contrary to the previously described requirements, where customers have difficulty in formulating objective judgments about their satisfaction, fuel consumption is measured and recorded objectively by many customers. Even if this measurement is, sometimes, acquired without scientific methods, there is no doubt that these judgements about the car are more reliable than others.

The most difficult information to be obtained about fuel consumption is the effective driving conditions used by customers in evaluating it.

It is standard practice to include questions about fuel consumption on all questionnaires that car manufacturers send to their customer sample, to have feedback on their products after a short period of use.

According to a European Union law we will discuss in the next chapter, fuel consumption is measured on a roller bench on a simulated course reproducing urban, suburban and motorway traffic. Since this measure is the only allowed channel for customer information about fuel consumption, this procedure has been chosen instead of others to evaluate fuel consumption objectively.

This procedure is characteristically independent of vehicle performance and driving habits and imposes to all cars the same gear shifting speeds; if this first characteristic is justified by the high traffic density on our roads, the second and third have, as their justification, the need to make the test procedure objective and repeatable.

It is often the case that the results of this test, when compared with tests in actual traffic conditions, can suggest wrong judgements, when comparing different cars.

The interesting fact is that this result is not due to specific customer driving habits, but represents a phenomenon that can be detected with statistical procedures on a sample of homogeneous customers.

Similar criteria to those of previous examples have been applied to fuel consumption. We describe a recent research on medium size non-sporty cars.

Identifying potential customers is essential to defining the market for these cars. The customer's economic bracket influence the negative value assigned to high fuel consumption; relevant parameters may be car price (a high-income customer is less sensitive to fuel consumption than to comfort and consequent weight), yearly distance travelled and type of car. On a sporty car, for instance, driving habits are less mindful of consumption, while diesel cars are frequently driven by customers sensitive to this parameter.

This test campaign included 20 non-professional drivers using a homogeneous sample of recent cars. A driving mission was defined, including an urban, a suburban and a motorway section, representing real-world use of the car recorded over a significant period of time by each driver.

Departing from the standard driving cycle (correlation coefficient 0.77), a high correlation (correlation coefficient 0.97) was detected on this driving course with claimed fuel consumption, as previously determined by questionnaires yo potential costumers.

New driving schedules have been developed, to be performed on roller dynamometers, that are representative of each mission and driving habit.

The following approach has been adopted:

- for each car and driver, the statistical distributions of average speed, engine idling times, average positive and negative accelerations and average gear change speeds have been recorded and investigated;

- by comparing time histograms of the different gear speeds with car speed and acceleration, a map of gear shift speed has been obtained, as a function of longitudinal acceleration and speed;

- speed time histories on each mission have been analyzed from start to stop, identifying any cluster where similar sequences have been grouped.

A new urban and suburban cycle has been identified; the important characteristic of these new cycles is not relevant to car speeds which are slightly higher than standard speeds, but is remarkable that the speed shift criterion depends upon car pick-up time; the higher the available torque exceeding driving resistance, the lower the vehicle speed where the gear is upshifted.

It should be remembered that this first conclusion is not applicable to all cars, but is limited to the kind of car, customers and driving environment to be considered.

Figure 18.5 demonstrates our conclusion. On the two diagrams on the right, showing the upshift speeds, measured in [km/h], as a function of the longitudinal requested acceleration, measured in [g] a car with a peak torque at high engine speed is shown; on the left diagrams the same car with peak torque at lower

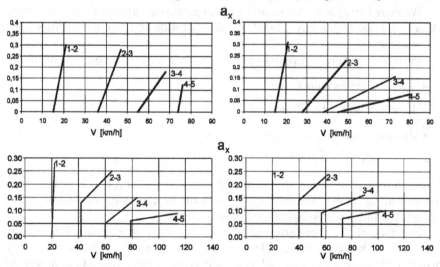

FIGURE 18.5. Comparison between upshift speeds for two cars of almost equal mass and displacement in urban traffic (above) and suburban traffic (below), as a function of the requested longitudinal acceleration a_x, measured in [g]. The engine of the car shown in the diagram on the right is designed to have its peak torque at higher engine revolutions.

engine speed is shown. The upper diagrams refer to urban driving, the lower to suburban driving.

The engine with higher elasticity induces drivers to lower upshift speeds for low accelerations; since accelerations between $0.12 \div 0.20$ g are frequent in urban traffic and between $0.02 \div 0.07$ g are frequent in suburban traffic, the more elastic engine is used, in the average, at lower rotational speeds, with higher gears.

Vehicle fuel consumption is strongly influenced by gear ratios; in general, we may say that in the prevalent conditions of use, fuel consumption is almost proportional to engine speed.

If in theory the availability of higher transmission ratios favours reduced fuel consumption in standard cycle, the conclusion of this work suggests that these reductions will not be achieved if the reduced pick-up capacity of the car stimulates drivers to use lower ratios more frequently.

Motorway consumption in this test campaign was sufficiently correlated with the traditional test procedure; under these conditions, the highest ratios were used.

Measured driving distances were equally divided over the three driving environments.

18.4   AGING RESISTANCE

A car's endurance, or resistance to age, is a function that can be evaluated objectively by driving it for a specified distance without damage.

We must clarify what we mean by damage; during the life of a car, customers ask not only for few failures, but little deterioration of those parameters found in a new car.

In the case of the automotive chassis, we mean also that requirements about:

- dynamic performance,

- handling and active safety,

- dynamic comfort,

- ergonomics,

can deteriorate only within an acceptable range of tolerance; in addition, nothing that can affect vehicle availability (to perform its function), can occur except through the fulfillment of scheduled maintenance.

Failure include, therefore, not only breakdowns of mechanical or electric parts, but also noise not detectable on a new vehicle, lubricant leakages, aesthetic corrosion, changes in dynamic behavior, fuel consumption, freedom of movement on controls, etc.

It is hard to forecast how the vehicle will be used, because use is conditioned by the life and driving style of the customer; in addition, applied loads can be determined by unforeseeable events.

Therefore, endurance specifications are assigned statistically often using the parameter B_{10}, which defines the endurance achieved without any failure by 10% of the population of vehicles produced.

As a reference for the value of this parameter, we can assume as adequate for today production, about:

- $B_{10} \geqslant 200,000$ km, for cars and commercial vehicles;

- $B_{10} \geqslant 400,000$ km, for buses;

- $B_{10} \geqslant 800,000$ km, for heavy duty trucks.

As technology and the market are evolving continuously, it is possible that these values will increase in the future.

These travelling distances make it almost impossible to perform, in the standard delay time of $3 \div 4$ years on the average devoted to a new vehicle development, the tests needed to assess endurance experimentally, after design and prototype manufacturing, with a sufficient confidence level; nor can this job be reasonably assigned to mathematical models.

In the case of cars, a life of about 200,000 km implies, on average, 4,000 h of driving time, assuming standard driving tasks; if we allocate six months for this task, as usually occurs, then to perform two complete sets of tests on two different prototypes generations (where the second carries corrections for the first), the test time must be shortened by at least $3 \div 4$ times.

Phenomena that can influence the endurance of a vehicle can be classified according to the following categories:

- fatigue;

- wear;

- corrosion;

- shocks and collisions.

External fatigue loads that can stress chassis components arise from two different sources: tires and engine.

Tires apply to the chassis longitudinal, lateral and vertical forces, changing over time; the first and second act with the frequency (on average low) of acceleration, braking and cornering events along the path of the vehicle; the last act with the higher frequencies given by the shape and spatial density of obstacles overcome.

The engine stresses the chassis at usually low frequencies, determined by the schedule of maneuvers (acceleration, releases, shifts) and at higher frequencies determined by its reciprocating parts.

Other periodic forces may be added when some of the above is working near the natural frequencies of the structures it is applied; this is particularly relevant for chassis structures and transmission.

Subject to fatigue are suspension and steering arms, wheels, bearings, springs, some braking parts (calipers, controls), transmission shafts, gears and chassis structures.

Most of these parts are made out of metals whose resistance can be described according to Wöhler's model; there is a threshold of load amplitude (fatigue limit) for these materials which can be applied indefinitely without any damage.

For this category of parts, test times can be reduced by applying techniques that remove the periods of load history below the fatigue limit. This can be done precisely by analyzing load time histories that will be applied to bench tests or by driving cars in more strenuous tests that apply loads that damage structures more severely.

Using this last approach the driving distance of a car's life can be condensed into about 50,000 km of heavy use.

Wear is determined by the friction of parts in relative motion; on the chassis, wear applies mainly to transmissions (bushings, rotary and sliding seals, synchro mesh and gears) and partly to suspensions.

Wear, the removal of material on sliding parts, depends on wasted friction, according to the hypothesis proposed by Theodor Reye about 140 year ago. Wear can therefore be accelerated by increasing loads, with attention to temperatures, that can affect the mechanical properties of materials.

A wear test for components can be reliably performed on test benches, where contact conditions are made more severe according to empirical procedures.

Corrosion is caused by the chemical action of many agents (humidity, salts, other chemical compounds and aerosols) on parts exposed to the atmosphere or splashed by the wheels; since this action is not constant throughout the life of the car, the test can be accelerated by exposing entire cars or components to corrosive humidostatic rooms for a certain period.

Another method, as effective as the first, is to drive through acid water pools during a certain portion of the fatigue course.

As we have seen, wear and corrosion test acceleration is totally empirical and is defined according to the manufacturer's experience.

Vehicle resistance to shocks and collisions must be examined through artificially reproduced events.

This applies to crash tests against barriers, requested by regulations, where chassis components must not interfere with occupants as a consequence of the collision.

There are also non-regulated shocks, where it is good practice to verify that there are no critical situations for occupants; one example of this category is the accidental collision of wheel against sidewalk, as a consequence of a mistaken maneuver; it is obvious that in these cases chassis structural integrity is not requested.

The designer must guarantee that there are no partial or hidden ruptures undetectable by the driver. Linkages and suspension arms must feature a rupture load at least 50% higher than the collapse load, where deformations become permanent; deformations, prior to rupture, must alter suspension geometry in such a way as to be easily noticeable by drivers, in order to suggest trip interruption.

Vehicle life is simulated by separate tests reproducing specific situations; fatigue tests are more difficult because they are determined not only by their duration but also by conditions of vehicle use.

Each manufacturer has decided to design vehicles for the most demanding conditions, accepting high safety margins for ordinary use; loop courses have been developed that are characterized by many bends, bumpy and uneven stretches (artificially damaged tarmac, stone pavement, dirty road, etc.) and rail crossings; a part of these courses is dedicated to water-crossing.

Such loops, if driven at high speed, can concentrate 200,000 km of real life into about 50,000 km, driven in about 1,000 h; this time corresponds to about two months, assuming three driving shifts on the same car, and including test interruptions to maintain and inspect the test prototype.

Load conditions to be considered in mathematical models or applied to test benches are also derived from this kind of loop.

A common test loop includes straight stretches long enough to allow the car to reach the highest acceleration and deceleration conditions; curves are driven at the slip limit.

Loop length is not relevant, because it will be repeated in both driving directions so as not to stress the vehicle in a selective way, until the total driving distance is reached; loop length is conditioned by the need to apply all tests suitable for simulating the most demanding driving situations. Usually these loops are between 20 and 30 km in length.

Pavement must offer a high friction coefficient to stress suspensions and chassis as much as possible. Sometimes, on long straight stretches, signals can be used to request additional maneuvers (decelerations, accelerations, slaloms).

Figure 18.6 shows a record of the main force components acting on a medium size car on a loop of this kind.

Instead of vertical forces suspension strokes have been measured; vertical forces may be calculated from suspension characteristics. Brake torque has also been added to separate transmission effects from those of the brakes in longitudinal forces.

These records originate from test bench load conditions, after mathematical elaboration; the same conditions can be applied to finite elements analyses, which are usually integrated into multibody models to simulate the entire vehicle.

Reorganization of the test cycles can be performed according to the rain flow method; the result is a set of load histograms, shown in Figs. 18.7, 18.8, 18.9, representing suspension stroke as well as longitudinal and lateral accelerations. Accelerations are derived by forces with reference to the sprung mass and can be applied to different cars as well, as a first approximation.

These histograms define the so called *load blocks* which correspond to driving the entire loop, about 30 km long, once clockwise and once counterclockwise; the load block is applied about 2,000 times to simulate the entire vehicle life.

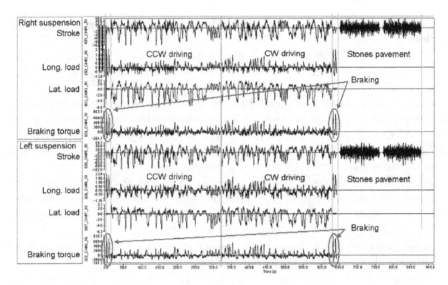

FIGURE 18.6. Records of suspension strokes, showing longitudinal and lateral loads on rear suspensions of a medium size car, driven on a fatigue loop clockwise (CW) and counterclockwise (CCW).

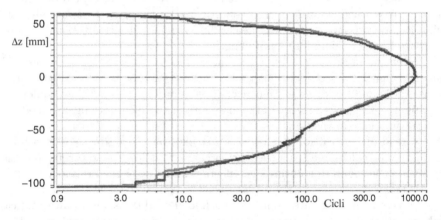

FIGURE 18.7. Histogram showing the cycle count of different suspension stroke classes Δz of right and left rear suspensions, on a fatigue loop.

It should be noted that accelerations refer to sprung mass; acceleration values apparently inconsistent with practical friction coefficients are not surprising, because vertical loads are increased by transfers due to lateral accelerations.

Figure 18.10 shows a bench for fatigue tests of a car body, complete in its main chassis components; actuators for applying loads and torques are also shown. The front wheel can be stressed similarly.

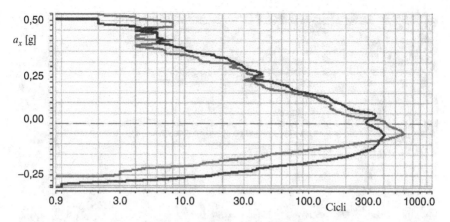

FIGURE 18.8. Histogram showing the cycle count of different longitudinal acceleration classes a_x of right and left rear suspensions, on a fatigue loop.

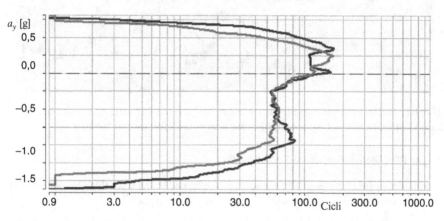

FIGURE 18.9. Histogram showing the cycle count of different suspension longitudinal acceleration a_y of right and left rear suspensions, on a fatigue loop.

Dynamic analysis of these forces assumes a particular importance if we want to determine fatigue load for particular components under specified driving conditions.

Usually the entire vehicle simulation applies multibody modeling techniques that are useful when displacement is relevant; these models allow not only displacement but also forces exchanged in any part of the system to be calculated.

In multibody modeling, system elements are considered as rigid bodies connected by elastic or viscoelastic couplings. In some cases, it is necessary, for a more precise determination of acting loads, to also take into account the flexibility of some part, such as the car body or the twist beam of a rear suspension.

Craig-Bampton's theory allows the flexibility of specific vehicle structures by their modal synthesis to be taken into account; this can be obtained by a finite element calculation of the modal deformations.

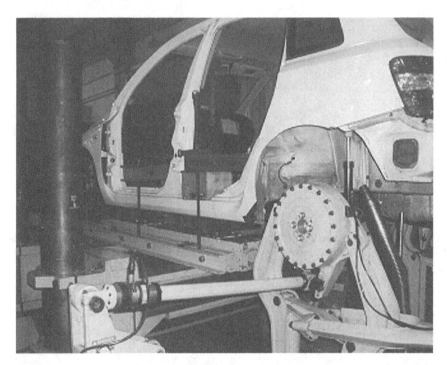

FIGURE 18.10. Fatigue test bench for a car body, complete with the most important chassis components; actuators supply rear suspension forces and torques.

In practice, for a specific structure:

- the displacements of the structure, compatible with its degrees of freedom, are evaluated as the result of a unitary force applied;

- the vibration modes are calculated, in a range up to about 100 Hz.

These data are applied to the multibody model, to determine the internal forces exchanged between the different components.

A multibody model of the vehicle system should include at least:

- the car body, as rigid body or as flexible body, according to the scenario under study;

- the powertrain mass and its suspension;

- the complete front and rear suspensions;

- tires, when loads are calculated from an open loop maneuver of interest.

The different parts are connected by joints, showing a suitable elastic and viscoelastic characteristic; mass and inertia properties are calculated for each of them.

If we refer to the previously described scenario, the same input for the test bench of Fig. 18.10 can be adopted as input of the multibody model.

The output of this analysis should be the forces that are exchanged at the articulation point of the suspension to the car body.

These forces may be applied to finite element models addressed to calculate the stress histories of the component under study, to determine the fatigue life of this element.

This calculation can be performed in two different ways, due to the fact that there is some overlap between force frequencies and natural frequencies of the component under investigation.

If there is no overlap, forces can be applied quasi-statically. A stress time history can be easily obtained by linear combination of the different effects.

If there is some overlap, dynamic modal techniques must be applied.

19
REGULATIONS

A vehicle cannot be sold and obtain the necessary registration for driving on public roads, unless it is built according to legal specifications. In Europe the agreement of these specifications with existing laws is demonstrated, by two official documents: the *certificate of homologation* and the *certificate of conformity*.

The first document proves that the vehicle is designed according to legal requirements. It is issued by a public authority in charge of this function; in Italy, for instance, that authority is the Department of Transportation. The homologation certificate is issued on the basis of a technical report of the manufacturer and the completion of given tests, performed on prototypes of that vehicle.

The second document proves that any produced vehicle is identical, in terms of homologation requirements, to the tested and approved prototype; this document is issued by an appointed representative of the manufacturer, for instance the general manager of the final assembly plant. Vehicle conformity can, at any time, be verified by the public authority in charge, by inspection of samples of produced vehicles and tests for requirements of the homologation certificate.

Homologation requirements are set by government laws, which impose minimum and maximum accepted values and the related test methods to be used for their verification; the manufacturer is free to identify the most suitable technologies to be employed for their fulfillment.

These requirements are relevant to part of the vehicle functions we have already introduced in previous chapters, particularly for:

- outside visibility;

- minimum dynamic performance necessary to grant a safe drive;

- occupant protection in case of collision;

- reduction of the environmental load caused by vehicle traffic, with particular reference to polluting gases, carbon dioxide, outside noise and waste produced by disposal of older vehicles.

The above laws are issued by each national government; in the past, some laws have also been developed by international institutions to enhance the free movement and sale of vehicles in countries other than that of their origin.

The European Community has already faced, in the 1960s, the problem of harmonization of national laws, to remove any impediment to the free circulation of goods within the Community and to grant the citizens of member states the availability to buy state of the art vehicles; this job has recently been carried over and completed by the European Union.

The European Union behaves like a supranational body, requiring all member states to develop laws complying with a common standard; these supranational laws are called *Directives* and will be cited with the letter D, followed by a figure showing the year of enactment and a following number. For instance the D70/156 directive was the hundred and fifty sixth law approved in 1970 regarding vehicle homologation; at the end of the last century and in the present one, the complete year figure is cited.

In parallel with Directives, *Regulations* have been also issued, summarizing in a single document all approved test procedures relevant to given homologation functions; these documents are quoted with the letter R, followed by a progressive number, unique for each title, independent of any addition or modification.

Since Directives must be established before national laws, they must be available in advance of their enforcement time; in their formulation, they provide an enforcement year for new homologations and one for new registrations.

No member state is allowed to prohibit sale, registration or circulation of any vehicle complying with the Directives in force.

In this chapter we will consider Directives and Regulations impacting chassis component design and vehicle system functions.

The following paragraphs will summarize laws regarding:

- vehicle system;

- wheels;

- steering system;

- braking system;

- chassis structures;

- gearbox.

The information discussed here is an updated summary at the time of writing this book and should be considered as a reference only; we suggest that anyone requiring sure guidelines look through the primary documents and check for new updates. This can be easily done by visiting the Internet sites of the European Union dedicated to this purpose; anyone can download excerpts of the Official Gazette.

The site:

$$\text{http://europa.eu.int/eur-lex/lex/it/index.htm}$$

contains a suitable research engine for looking into Directives using keywords or their identification numbers.

The site:

$$\text{http://www.unece.org/trans/main/wp29/wp29regs.htm}$$

contains all vehicle regulations and their updates.

An index of current documents is reported in Table 19.1; the content of these documents is summarized in the following sections.

The European Union situation should be considered a typical example of the government approach to vehicle homologation; different States could have slightly different legislation, which must be carefully considered in the case the vehicle is sold and registered in different countries.

19.1 VEHICLE SYSTEM

19.1.1 Homologation and general characteristics

The D 70/156 Directive defines the homologation procedure reported in summary in the previous section; according to this Directive, the manufacturer is obliged to submit an *information form*, reporting all vehicle characteristics that cannot be altered without a new homologation.

The information form reports the following information regarding the issues we are concerned with.

General data

These consist of the manufacturer's data, the position of the Vehicle Identification Number (VIN) identifying body, chassis and engine, and the vehicle category. The following vehicle categories are identified:

- M_1: vehicles for transporting people with fewer than eight seats, in addition to the driver's seat;

- M_2: vehicles for transporting people with more than eight seats, in addition to the driver's seat and with a maximum overall weight of 5 t;[1]

---

[1] Rules were issued before the compulsory introduction of the SI measuring system. Newer updates report masses, instead of weights.

TABLE 19.1. European Unions Directives and Regulations with impact on chassis design.

Theme	Reference	Contents
Vehicle system	D70/156	General information necessary for the homologation of vehicles and trailers.
	D83/403	Completion of the above with off road vehicles.
	D92/21	Modification of the above for mass and dimensions.
	D92/53	Modification and new text for D70/156
	D95/48	Modification of the above for conformity control.
	D98/14	Modification of the above for procedures.
	D70/220	Gaseous emissions measurement.
	D80/1268	Fuel consumption measurement.
	D1999/100	Modification of the above for new consumption cycles.
	R101	Fuel consumption measurement.
	D2000/53	Disposal of waste deriving from vehicle disposal.
Wheels	D78/549	Shape of fenders and wheel wells.
	D92/23	Homologations of tires.
	D94/78	Update of D92723 for procedures.
	D2001/43	Prescriptions on tire noise.
	R30	Tire homologation for M_1 vehicles.
	R54	Tire homologation for M_2 and M_3 vehicles.
	R64	T type spare wheel tire homologation.
Steering	D70/311	Admitted steering devices and operating forces.
	D74/297	Driver protection in case of collision.
	D91/662	Modification of the above.
	D1999/7	Update of the D70/311 for powersteering.
	R12	Steering system homologation for driver protection.
	R79	Further update of D70/311 for powersteering.
Brakes	D71/320	Braking systems for M, N, O vehicles.
	D91/422	Modification of the above for dates of enforcement.
	D98/12	Modification of the above for ABS introduction.
	R13	Braking system homologation for M, N, O vehicles.
	R90	Spare lining homologation.
Struct.	D96/79	Occupant protection for M_1 vehicles.
Gearbox	D75/443	Prescriptions for speedometer and reverse gear.
	D97/39	Update of the above.
	R39	Speedometer and reverse gear homologation.

- M_3: as above, but with an overall weight exceeding 5 t;

- N_1: vehicles for transportation of goods with maximum weight exceeding 1 t, but lower than 3.5 t;

- N_2: vehicles for transportation of goods with maximum weight exceeding 3.5 t, but lower than 12 t;

- N_3: vehicles for transportation of goods with a maximum weight exceeding 12 t;

- O_1: trailers with a maximum weight not exceeding 0.75 t;

- O_2: trailers with a maximum weight exceeding 0.75 t, but lower than 3.5 t;

- O_3: trailers with a maximum weight exceeding 3.5 t, but lower than 10 t;

- O_4: trailers with a maximum weight exceeding 10 t.

Vehicle characteristics

These consist of three pictures and a vehicle scheme showing the main dimensions; the number of axles and wheels is reported, showing permanent or part-time driving wheels. An outline scheme of the chassis frame, if any, should be included, showing the material used for side beams.

Weights and dimensions

Among the main dimensions, the wheelbase and interaxis (for vehicles of more than two axles) must be reported under full weight conditions; for trailers, the distance between the hook and first axle pivot must be declared; for road tractors, the saddle pivot longitudinal and elevation position must be referenced to the vehicle. All dimensions are defined by the ISO 586 standard. All tracks must also be declared.

The weight of the bare chassis frame (if any) must be declared, not including cabin, fluids, spare wheel, tools and driver; the weight breakdown on the axles must be also declared.

Also the weight of the vehicle completed with body or cabin (depending on the product sold by the manufacturer) and other items must be declared and its breakdown on the axles; if the vehicle is a semi-trailer, the weight on the hook must also be claimed.

Finally, maximum allowed weight has to be declared and its breakdown on axles and hook (if any).

Transmission

The transmission is described by a draft scheme, including data on its weight, architecture (single stage, double stage, etc.), type of control (manual or automatic), transmission ratios (gearbox and final drives) and the vehicle speed that can be obtained on the existing gears at an engine speed of 1,000 rpm.

Suspensions

Suspension schemes must be attached, including the damping and elastic characteristics of shock absorbers and springs; allowed tire sizes must be declared.

Steering system

A scheme of the steering mechanism and column must also be included; maximum design forces on the steering wheel and maximum steering angles at the wheel and the steering wheel must be declared. For these angles, the vehicle turning radii for right and left turns must be declared.

Brakes

As we will describe later, service, emergency and parking brakes must be fully described.

The engine, body and other vehicle systems not included in the chassis must also be described.

A *homologation form* certifies the released homologation and reports for each of the characteristics of the information form:

- the conformity of the presented prototypes to the described items;

- the conformity of those characteristics to legal requirements;

- the positive execution of tests;

- the existence of required drawings.

The D 70/156 Directive also reports all forms to be used for information and for the certificate of conformity.

The D 87/403 Directive completes the previous documents with the definition of off-road vehicles; these are vehicles of M_1 and N_1 categories, featuring these characteristics:

- at least one front and one rear driving axle, one of which can be disengaged by the driver;

- at least one self-locking or locking differential;

- the gradeability of at least 30%, with no trailer;

- at least one of the following requirements:

  1. angle of attack α_a of, at least, 25°;

  2. angle of exit α_u of, at least, 20°;

  3. ramp angle α_r of, at least, 20°;

  4. ground clearance h_2, under the front axle of, at least, 180 mm;

  5. ground clearance h_2, under the rear axle of, at least, 180 mm;

  6. ground clearance between the axles h_1 of, at least, 200 mm.

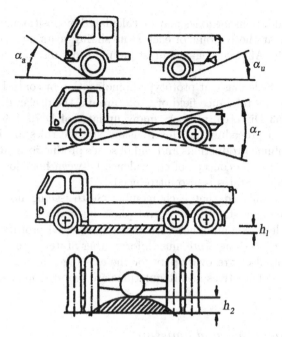

FIGURE 19.1. Front and rear attack angles α_a and α_p; α_r ramp angle; ground clearance h_1 and h_2.

Figure 19.1 defines with the help of sketches the reported dimensions.

Front and rear attack angles measure the capacity of a vehicle to face sudden slope changes in normal and reverse driving without any interference between chassis and ground; the ramp angle, on the other hand, refers to a sudden slope change in both directions.

The ground clearance between axles is the maximum height of an ideal parallelepiped that can be inserted between the axles and under the chassis; ground clearance under the axle refers to the lower point between the two contact points of the wheels on the same axle.

Other geometrical prescriptions are assigned to other kinds of vehicles.

The D 91/21 Directive updates the previous ones by specifying that an M_1 motor vehicle should feature a maximum speed of at least 25 km/h; in addition, it introduces mass as a measurement instead of weight.

The same Directive establishes maximum vehicle dimensions:

- 12,000 mm of overall length;

- 2,500 mm of overall width;

- 4,000 mm of overall height.

The maximum allowed vehicle mass must be, at least, the total of vehicle curb mass plus the product of offered passenger seats multiplied by 75 kg, which is assumed as the average weight of a passenger, including his hand baggage.

Mass breakdown on the axles may be calculated by positioning the passenger reference weight at the R point of each seat; sliding seats must be set at their rearmost position. Allowed baggage must be uniformly distributed on the trunk floor.

Measured vehicle mass, at prototype homologation or control of conformity, is admitted within a tolerance field of ± 5% around the declared values.

The D 92/53 Directive presents many updates of D 70/156 for the forms and homologation procedure. In this revision, specific rules are introduced regarding small volume productions and end of series productions, along with rules concerning waivers; the concept of equivalence between homologations granted by different member States is also introduced.

It is also established that each member State issuing homologation certificates must arrange statistical control plans on operating vehicles, suitable to detect possible non-compliance with the homologated prototypes; in case of non-conformity, the issuing State must inform other States of the event and must organize the compulsory recovery plan for the existing vehicles.

All applicable Directives and Regulations are reported in Attachment IV of this document.

19.1.2   Consumption and emissions

Directive D 70/220 and its following updates report the applicable rules of the European Union member states on the emission of polluting gases from motor vehicles; this Directive also subsumes the following ones about fuel consumption measurement.

The outstanding point of this Directive, in force for vehicles of the M_1 and N_1 types, consists in the definition of a transient driving cycle, which is defined to simulate vehicle usage in an urban environment. This cycle, also reported by the fuel consumption measurement Directive, consists of a speed-time history to be assigned to every vehicle to be homologated; it is related to high density urban traffic, where overtaking or slowing down is almost impossible.

The vehicle is tested on a dynamometer roller bench; a brake acting on the rollers is able to replicate vehicle driving resistance faithfully. The same rollers drive, in addition, a flywheel battery; each flywheel can be engaged or disengaged on the brake: a suitable flywheel combination can simulate vehicle inertia.

For this test, a *reference mass* is defined as the curb vehicle weight, with fuel supply, increased by 180 kg, corresponding to the average transported payload.

Table 19.2 reports the different reference mass classes and the corresponding rounded value for the equivalent inertia[2]; available flywheels must be able to replicate all reported equivalent inertia classes.

---

[2]Note the incongruity of measuring weight and inertia with the same units.

TABLE 19.2. Table for calculating the equivalent inertia of a vehicle, as a function of reference mass.

Reference mass P_r [kg]	Equivalent inertia [kg]
$P_r < 750$	680
$750 < P_r \leqslant 850$	800
$850 < P_r \leqslant 1{,}020$	910
$1{,}020 < P_r \leqslant 1{,}250$	1,130
$1{,}250 < P_r \leqslant 1{,}470$	1,360
$1{,}470 < P_r \leqslant 1{,}700$	1,590
$1{,}700 < P_r \leqslant 1{,}930$	1,810
$1{,}930 < P_r \leqslant 2{,}150$	2,040
$2{,}150 < P_r$	2,270

Brake torque absorption must be able to reproduce vehicle driving resistance at a constant speed of 50 km/h. For different speeds, only the parabolic relationship of torque with speed is requested.

To adjust the brake during a constant speed drive of the vehicle on a level road, in third gear or in D position for automatic transmissions, the intake manifold pressure is measured. Vehicles must be loaded with their reference weight and tires must be correctly inflated. To compensate for wind effect, the results of two measurements in opposite directions are averaged.

The same vehicle is set on the dynamometer and the brake is adjusted to reproduce the same manifold pressure.

The test bench is provided by a CRT monitor showing the actual vehicle speed on the bench, in combination with a band representing the driving cycle, with a tolerance of $\pm$ 2 km/h. The driver must follow this indicator, avoiding transient corrections that could affect the consumption measurement.

Figure 19.2 shows the speed-time diagram of the urban cycle for emission and consumption measurement; this cycle must be followed four more times.

All gases emitted by the exhaust pipe during the test are collected in bags, whose content is measured and analyzed after the test to determine HC, CO, NO_x and CO_2 levels; weighted gases are divided by the ideally travelled distance on the bench.

The D 80/1268 Directive prescribes the same cycle for fuel consumption measurement in an urban environment. This value was combined also with 90 and 120 Km/h constant speed fuel consumption, to supply the potential customer with more complete information.

This Directive was afterwards modified by introducing a second driving cycle of the suburban type, to be applied after the urban cycle. This cycle is reported in Fig. 19.3.

Table 19.3 shows a summary of the most important features of the two driving cycles.

Directive D 1999/100 imposes these cycles also for fuel consumption measurement; the consumption is calculated by standard formulas, depending on

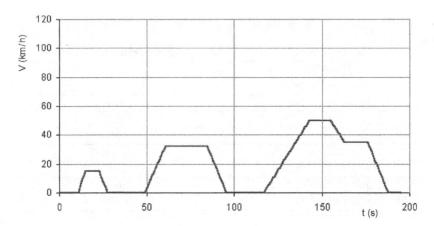

FIGURE 19.2. Speed-time diagram of the urban cycle for emissions and consumption measurement.

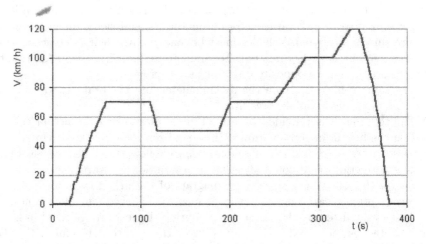

FIGURE 19.3. Speed-time diagram of the suburban driving cycle for emissions and fuel consumption measurement.

TABLE 19.3. Characteristics of the urban and suburban driving cycles.

Characteristic		Urban cycle	Suburban cycle
Travelled distance	[km]	4×1.013 = 4.052	6.955
Duration	[s]	4×195 = 780	400
Average speed	[km/h]	18.7	62.6
Maximum speed	[km/h]	50.0	120

the chemical composition of the fuel (gasoline, diesel, LPG, natural gas), starting with the emission value of CO_2 measured in g/km.

This value is the only allowed statement from manufacturers to customers on matters of fuel consumption; fuel consumption is also subject to conformity control from the issuer of the homologation certificate.

The same cycles must be used for electric energy consumption and range measurement in electric vehicles.

Regulation R 101 summarizes all topics on fuel consumption measurement.

19.1.3 Recyclability

Each year about 12 million vehicles are scrapped in the European Union; they correspond to about 0.5 % of the total production.

Directive D 2000/53 establishes rules to control waste products from these vehicles by component and materials recycling.

These rules also address improvements to the environmental operations of companies involved in this activity.

To prevent noxious waste formation, a set of laws has been introduced to limit the use of some substances for vehicle and component manufacturers, making recycling easier and avoiding dangerous waste treatment.

Member states must adopt laws suitable for reaching the following overall targets.

- The recovery percentage (materials not sent to a landfill) of scrapped vehicles must be at least 85% of the average vehicle weight, while at least 80% of the weight must be reemployed. For vehicles produced before 1980, lower targets can be set, but not lower than 75% for recovery and 70% for recycling.

- The recovery percentage must reach 95% by 2015 and the recycling percentage 85% of the average vehicle weight.

For this purpose, components and materials must be code labelled, to enhance identification and classification for selective recovery.

Scrapped vehicle treatment must include:

- batteries and LPG bottles removal;

- removal of explosive materials, such as air bags;

- removal and separated collection of fluids, like fuels, lubricants, cooling fluids, brake oil, air conditioning fluids and others, unless they are necessary to parts reemployment;

- removal of all components including mercury.

Other scheduled operations include:

- catalyst removal;

- selective removal of parts containing copper, aluminum and magnesium;

- tires and big plastic elements removal (bumper, dashboard, reservoirs);

- glass removal.

The economic accomplishment of these operations implies a number of additional design rules:

- banning certain materials, such as asbestos, lead, cadmium, esavalent chromium, etc.

- indelible material labelling of any component;

- designing components with a reduced number of materials;

- designing an easy disassembly;

- identification of components suitable for a second life.

In the near future, it is likely that manufacturers will be obliged to accept the burden of disassembly.

19.2 WHEELS

Directive D 78/549 is conceived as an implementation of D 70/156 for M_1 vehicles; it concerns fenders, sometimes part of the body, sometimes of the chassis (wheel wells).

The following rules refer to a running vehicle with wheel parallel to the longitudinal axis; Fig. 19.4 illustrates what we will report in this paragraph.

Within the sector defined by the radial planes through the wheel axis, which build up an angle of 30° before and 50° after the wheel with reference to the vehicle motion direction, the total width of the fender q must be sufficient to cover the width b of the tire, taking into account all possible combinations of tires and rims admitted by the manufacturer in the homologation information form.

In case of twin tires the total width of the two tires, as they are assembled on the wheel hub, must be taken into account.

The rear rim of the fender must be, at least, 150 mm higher than the wheel axis; in addition, the tangency point of the fender rim with a plane at 150 mm over the wheel axis (point A on Fig. 19.4) must be outside the equator plane of the tire, or, in case of twin tires, outside the equator plane of the outer wheel.

Profiles and location of fenders must be as close as possible to the tire; particularly with reference to the sector previously defined, the following prescriptions must be applied.

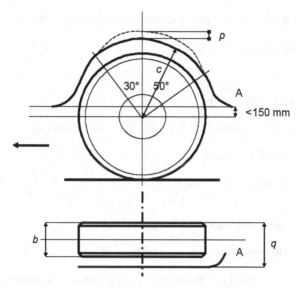

FIGURE 19.4. Dimensions for the fender and the wheel well, the object of homologation rules; the curved solid line represents the outer fender rim.

- The projection, on the tire equator plane, of distance p between the outer rim of the fender and its topmost point, must be at least 30 mm. This dimension can be progressively reduced to zero on the radial planes previously defined.

- The distance c between fender rims and wheel axis cannot be higher than twice the static radius R of the tire. In vehicles with adjustable trim, the above conditions must be satisfied when the vehicle is in the normal drive position requested by the manufacturer.

- Fenders can be built with more than one element; in this case gaps are not admitted between the different elements.

- Fender profile and tire position must be such that at least one type of snow chain can be mounted on the driving wheels, for one of the tire dimensions admitted by the manufacturer.

According to Directive D 92/23 and its updates on the D 94/78, member States are also required to homologate tires also as a component, independent of the vehicle where they should be installed.

This directive reports the standard identification system that was explained in the first volume. This system allows us to identify geometrical dimensions, inflation pressure and admitted load at reference conditions.

The homologation certificate of a tire must report the following information:

- application category;

- type of structure;

- maximum allowed speed;

- maximum admitted load with simple and twin assembly;

- the necessity of any tube;

- the type, as between the following: for cars; reinforced; for commercial and industrial vehicles; for temporary use on spare wheels (T type);

- description and dimensions of carcass structure;

- admitted rim dimensions for its application.

The Directive D 2001/43 limits the rolling noise produced by the tire, in order to contribute to the vehicle overall noise reduction; although we do not comment about Directives on outside noise, because this topic is part of the engine and its intake and exhaust systems design, we will describe briefly what pertains to the tire.

This Directive presents a reproducible test method to evaluate the noise produced by tires rolling on a paved road.

Tires are classified in the following categories:

- C1: tires for M_1 vehicles; class C1 is divided in subclasses, according to the dimension W;

- C2: tires for M_2 and M_3 vehicles with a load capacity index ≤ 121 and speed category $\geqslant$ N; class C2 is divided in subclasses, according to the tire application;

- C3: tires for M_2 and M_3 vehicles with a load capacity index $\leqslant 121$ and speed category $\leqslant$ M and with a load capacity index $\geqslant 122$ for twin assemblies; class C3 is divided in subclasses, according to the tire application.

The proposed test method consists of driving a vehicle with the test tire on a measurement course, as represented in Fig. 19.5; the vehicle must be launched and cross the course with idle gear and engine off.

This directive prescribes also the remaining test conditions, the specifications of the pavement and how to process signals from the microphones M.

The test speed is:

- 80 km/h for tires of C1 and C2 classes;

- 70 km/h for tires of C3 class.

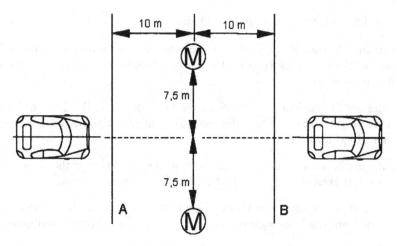

FIGURE 19.5. Test course for measuring tire rolling noise; on the course AB the vehicle is running at semi-constant speed, with idle gear and engine off. Points M represent the measuring microphones.

TABLE 19.4. Noise limits in [dB(A)] for the different tire classes; W is measured in [mm].

Class	Subclass	Limit [dB(A)]
C1	$W \leqslant 145$	72
C1	$145 < W \leqslant 165$	73
C1	$165 < W \leqslant 185$	74
C1	$185 < W \leqslant 215$	75
C1	$W > 215$	76
C2	normal	75
C2	snow	77
C2	special	78
C3	normal	76
C3	snow	78
C3	special	79

Noise limits are reported in Table 19.4. More severe limits are expected after 2007.

Regulations R30, R54, R64 summarize all matters respectively for M_1 vehicles, for M_2 and M_3 vehicles and for T type wheels.

A T type wheel is a spare wheel, complete with tire and disc, with differences only in design details, that limits its application to certain conditions only.

The development of these wheels is justified by the need for bulk containment in the trunk, when more than one tire size applies to a vehicle, or when the ordinary wheels are judged to be too expensive.

Four different categories are identified:

- category 1: consists of a wheel where the disc has a design different from that of the homologated tires; the wheel is inflated at a pressure prescribed for temporary use;

- category 2: consists of a wheel where both disc and tire have a different design from that of the homologated tires; the wheel is inflated at a pressure prescribed for temporary use;

- category 3: consists of a wheel of the same design as that of the homologated tires, but transported flat; the tire is inflated when necessary;

- category 4: consists of a wheel where both disc and tire have a design different from the homologated tires; the tire is inflated only when necessary.

These wheels must observe the rules of R30 regulation, implying that the tire is designed for the load of the heavily loaded wheel and for a maximum speed of at least 120 km/h; T wheels must show an indelible label prescribing a use limited to a short distance and 80 km/h maximum speed.

The same indications must be reported in the user manual, which also prohibits the simultaneous use of more than one T wheel on the same vehicle.

In addition, a full load braking test must be performed with the T wheel in all possible positions; the braking test is the same as prescribed by R13 regulation: with a pedal load lower than 500 N, the vehicle must stop from 80 km/h within the distance:

$$s \leqslant 0,1V + \frac{V^2}{150} , \qquad (19.1)$$

where V is the speed in [km/h].

The required performance must be obtained without wheel lock, deviation from the initial path or excessive corrections on the steering wheel.

19.3   STEERING SYSTEM

Directive D 70/311 is one of the oldest and requires that any motor vehicle, with more than 25 km/h as maximum speed and with at least four wheels, be equipped with a steering system; this rule applies also to trailers, but not to rail vehicles, agricultural and earth-moving machines.

For the steering system, the Directive means any device having the function of changing the driving path of a vehicle. It includes:

- the control (the steering wheel),

- the transmission from the control to the steering wheels,

- steering wheels,

- any device suitable for producing auxiliary energy for assistance; this auxiliary energy can be mechanic, hydraulic, pneumatic or any combination of the above, including energy storage.

Three categories of vehicles are identified with:

- manual steering, where steering energy is supplied by the driver's muscle power only;

- power steering, where this energy is partly supplied by the driver, partly by an assistance device;

- self-steering, where this energy is only supplied by sources different from the driver.

The control device must grant an easy and safe control of the vehicle; if necessary the steering system may be implemented by power assistance.

The control device must be manageable and ergonomic and must be conceived so as to allow an adjustable steering. The motion direction of the control device must be clearly correlated to the expected change of path of the vehicle.

The control force to be applied in order to obtain a turning radius of 12 m, starting from a straight direction, must in no case exceed 25 kg.

If a power assistance device is applied in order to observe this rule, the control force, when the assistance energy is missing, cannot exceed 60 kg. Vehicle steering must be guaranteed in the event of total or partial failure of the auxiliary system.

Self-steering devices are not admitted for vehicles as they are defined by this Directive.

It should be noticed that this condition applies only if all steering wheels are self-steering; for this purpose, hydrostatic steering (steering wheel moving a hydrostatic power distributor with wheels steered by hydrostatic actuators) is, for instance, not admitted, but an additional axis on a truck, steering only under the action of the cornering forces, is admitted.

Directive D 74/297 specifies the behavior of the steering wheel after a collision test against a barrier at 48.3 km/h; the test is done with no load or dummies. According to this directive the upper part of the steering column must not move back horizontally more than 12.7 cm with reference to any point of the body not involved in the collision.

In addition, the energy that the steering column must absorb during the secondary impact of the dummy against the steering wheel is specified; when the steering wheel is impacted by a test block of about 35 kg of mass, launched at a speed of 24.1 km/h, the reaction exerted by the test block cannot exceed 1,111 daN.

The steering wheel should be designed so as not to present any unevenness or sharp edges, which threaten to increase the danger or the severity of injuries to the driver after the impact. In addition, it should be designed, built and assembled so as not to present any accessory element able to snag driver's clothing during normal driving operation.

Directive D 1999/7 specifies minimum values for braking efficiency on those vehicles that use the same energy source for both power steering and power braking.

Regulation R 12 reports all matters regarding driver protection in case of collision and describes test conditions in detail.

Regulation R 79 summarizes all matters regarding homologation and introduces important updates, opening the possibility of applying *full power* systems, where the motion transmission between steering wheel and wheel is not mechanical.

In this document, the technical progress needed to make the following items possible is examined:

- the elimination of the steering column for safety,

- the easy transfer of controls to both sides of the vehicle,

- the automatic intervention of steering control, to avoid collisions or rollovers.

According to the above targets this document allows steering wheel and wheel to be connected by other than mechanical positive means.

Systems in which the driver has the primary vehicle control, but in which automatic control systems can also intervene, are defined as *Advanced Driver Assistance Steering Systems* (ADASS).

Steering systems assisted by means that are in part outside the vehicle are defined as *Lane Guidance, Lane Keeping or Heading Control* if they have the job of maintaining a preset trajectory; they are classified as *Automatically Commanded Steering Function.*

Systems like ADASS can also include devices able to monitor path deviation or correction in such a way as to improve the vehicle's dynamic behavior.

This Regulation allows the application of such systems if their presence does not degrade the operation of the conventional control system. They must be designed in such a way as to enable the driver to inhibit their operation deliberately; in case of emergency, a mechanical positive link between steering wheel and wheels must be reestablished.

If the same source of energy used for steering the car is used for different devices, steering must be guaranteed. If this source is shared with the braking system, steering must be given priority; in case of failure the braking efficiency must not decay below certain limits.

The system must also be designed so as not to allow speeds over 10 km/h in case of failure; if the energy source is not available or has failed, at least 24 double steering pad loops of 40 m must be driven at a limited speed of 10 km/h with the same performance as an undamaged system.

TABLE 19.5. Maximum forces S on the steering wheel, for a curve of radius R, starting from a straight path for a duration T, in case of damaged and undamaged system.

Category	Undamaged system			Damaged system		
	S [daN]	T [s]	R [m]	S [daN]	T [s]	R [m]
M_1	15	4	12	30	4	20
M_2	15	4	12	30	4	20
M_3	20	4	12	45	6	20
N_1	20	4	12	30	4	20
N_2	25	4	12	40	4	20
N_3	20	4	12	45	6	20

In case of failure of the control transmission, no sudden steering angle change is allowed.

The vehicle must be driven starting from a straight path to a constant radius curve, as prescribed in Table 19.5, at 10 km/h. The steering wheel force necessary to perform the maneuver in the prescribed time should be recorded for the undamaged system. The same test should be repeated with a damaged system by measuring same values.

19.4   BRAKING SYSTEM

The D 71/320 Directive, together with the updates introduced by D 91/422 and by D 98/12, specifies braking systems for vehicles of the M, N and O categories.

This directive considers as a braking device any mechanical system having the function of decreasing the speed or stopping a vehicle gradually, or preventing further motion when stopped.

This device is composed of a control, a brake and a transmission connecting the above elements together.

The brake is the device developing forces opposed to vehicle motion.

Brake types taken into consideration are:

- *friction brakes*;

- *electric brakes* (where braking forces are developed by any electro-magnetic action between parts with relative motion, but not in contact);

- *fluid brakes* (where braking forces are developed by a fluid interposed between parts with relative motion);

- *engine brake* (where braking forces are produced by an artificial increase of the engine braking effect);

- *inertia brakes* (where the braking forces on a trailer are produced by the reaction between the trailer and the tractor).

The Directive defines transmission as any device connecting control with brakes; this transmission can be mechanical, hydraulic, pneumatic, electric or mixed. When braking is performed or assisted by an energy source independent of the driver, but under his control, the energy storage applied to the system is also considered to be part of the transmission.

The following braking modes are defined:

- *adjustable braking*, when the driver can, at any time, increase or decrease the force on the control;

- *continuous braking*, when a train of vehicles is braked by a unique control that can be moderated by the driver, at his driver's seat, and when the braking energy comes from the same source (as well as the muscle power of the driver);

- *semi-continuous braking* similar to above, but with energy coming from more than one source (one source can be the muscle power of the driver);

- *automatic braking* is the braking of a trailer that occurs automatically when it is uncoupled from its tractor or when towing devices are broken; in this case, the braking efficiency of the rest of the train must remain unaffected;

- *retarder braking*, when a supplementary device is able to exert and maintain a braking force on the vehicle, for a long time, without reduction in efficiency; the term retarder includes the entire system and its control; at this time regenerative braking systems (used on electric and hybrid vehicles) are not considered to be part of the braking system.

A generic braking system, as described at the beginning of this paragraph, must perform one of the following functions.

- *Service brake*: must allow vehicle speed control, stopping it quickly and safely at any speed or road slope; it must be adjustable; it must be operated by the driver at the driver's seat, with both hands on the steering control.

- *Emergency brake*: must allow the vehicle to be stopped in a reasonable space, when the service brake is malfunctioning; it must be adjustable; it must be operated by the driver at the driver's seat, with one hand on the steering control.

- *Parking brake*: must allow the vehicle to remain unmoving on a climbing or descending slope, even when the driver is absent.

Devices providing service, emergency and parking braking can have common parts or devices, provided they fulfill the following specifications:

- two independent controls must be available; for all vehicle categories but M_2 and M_3, each control (except the retarder control) must return to rest position when released; this rule does not apply to the parking brake when locked in the braked position;

- service brake control must be independent of the emergency or parking brake control.

Braking system specifications are different for road pavement with high and low friction coefficients.

High friction coefficient

For service brakes, a stop test of 0 type, with cold brakes (when the brake temperature, measured on the disc or on the outside of the drum, is lower than 100°C) is specified to be performed with unloaded and fully loaded vehicle and engine disengaged.

In addition, a stop test of 1 type should be performed, including repeated braking, as for the following scheme, and of 2 and 2 A type, after long descents.

The type 1 test is performed on a loaded vehicle, after having warmed up the brakes, according to the rules shown in Table 19.6.

The typé 2 test provides that brakes are used when the vehicle is driving a course on a slope of 6%, 6 km long, at 30 km/h with the most suitable gear ratio and the retarder, when applicable.

Test type 2 A is similar to the previous one, but with a slope of 7%.

During this test, service, emergency and parking brakes cannot be used. The gear ratio must be chosen in order to run the engine at a speed not over the maximum value specified by the manufacturer. An integrated retarder (which can be operated by service brakes) can be used in such a way as not to operate the service brakes.

Test 2 A substitutes test 2 for tourism or long distance buses of category M_3, and for vehicles of category N_3 allowed to tow trailers of category O_4.

All above tests must be run on a high friction coefficient paved road.

The braking system efficiency is evaluated by measuring the stopping distance, the average deceleration and the response time.

The stopping distance is the distance travelled by the vehicle from the time the brake control is actuated to the time the vehicle is completely stopped; the initial speed of the test is defined as the speed at the time the driver starts to actuate the brake control.

TABLE 19.6. Rules for performing a type 1 test: V_1 and V_2 are the initial and final speed for the test, V_{max} is the vehicle maximum speed and Δt is the time between following brakings. The required number of repeated brakings is n.

Category	V_1 [km/h]	V_2 [km/h]	Δt [s]	n
M_1	$0.8 \cdot V_{max}$, $\leqslant 120$	$0.5 \cdot V_1$	45	15
M_2	$0.8 \cdot V_{max}$, $\leqslant 100$	$0.5 \cdot V_1$	55	15
M_3	$0.8 \cdot V_{max}$, $\leqslant 60$	$0.5 \cdot V_1$	60	20
N_1	$0.8 \cdot V_{max}$, $\leqslant 120$	$0.5 \cdot V_1$	55	15
N_2	$0.8 \cdot V_{max}$, $\leqslant 60$	$0.5 \cdot V_1$	60	20
N_3	$0.8 \cdot V_{max}$, $\leqslant 60$	$0.5 \cdot V_1$	60	20

TABLE 19.7. Minimum performance of the service brake system for 0 type tests, with disengaged engine.

Category	V [km/h]	s [m]	a_x [ms^{-2}]	F [N]
M$_1$	80	$\leqslant 0.1V + \frac{V^2}{150}$	$\geqslant 5.8$	500
M$_2$	60	$\leqslant 0.15V + \frac{V^2}{130}$	$\geqslant 5$	700
M$_3$	60	$\leqslant 0.15V + \frac{V^2}{130}$	$\geqslant 5$	700
N$_1$	80	$\leqslant 0.15V + \frac{V^2}{130}$	$\geqslant 5$	700
N$_2$	60	$\leqslant 0.15V + \frac{V^2}{130}$	$\geqslant 5$	700
N$_3$	60	$\leqslant 0.15V + \frac{V^2}{130}$	$\geqslant 5$	700

TABLE 19.8. Minimum performance of the service brake system for 0 type tests, with engaged engine.

Category	V [km/h]	s [m]	a_x [ms^{-2}]	F [N]
M$_1$	160	$\leqslant 0.1V + \frac{V^2}{130}$	$\geqslant 5$	500
M$_2$	100	$\leqslant 0.15V + \frac{V^2}{103,5}$	$\geqslant 4$	700
M$_3$	90	$\leqslant 0.15V + \frac{V^2}{103,5}$	$\geqslant 4$	700
N$_1$	120	$\leqslant 0.15V + \frac{V^2}{103,5}$	$\geqslant 4$	700
N$_2$	100	$\leqslant 0.15V + \frac{V^2}{103,5}$	$\geqslant 4$	700
N$_3$	90	$\leqslant 0.15V + \frac{V^2}{103,5}$	$\geqslant 4$	700

In the formulae below, suitable for measuring the braking efficiency, V is the initial speed, s the stopping distance, F the force on the brake pedal, a_x the average obtained deceleration.

The minimum result in Table 19.7 must be measured when the engine is disengaged, and in Table 19.8 when the engine is engaged. On the same tables, the maximum allowed value for F is reported.

Emergency brakes must be able to obtain:

$$s \leqslant 0.1V + \frac{2V^2}{150} , \tag{19.2}$$

for vehicles of category M$_1$,

$$s \leqslant 0.15V + \frac{2V^2}{130} , \tag{19.3}$$

for vehicles of categories M$_2$ and M$_3$ and:

$$s \leqslant 0.1V + \frac{2V^2}{115} , \tag{19.4}$$

for vehicles of categories N.

If the emergency brake is operated by a hand lever, the performance must be obtained with a force on the lever below 400 N for vehicles M$_1$, and below 600 N for the remaining categories.

TABLE 19.9. Minimum performance for service brakes, in case of transmission failure, in a 0 type test with engaged engine.

Category	V [km/h]	s [m], loaded vehicle	s [m], unloaded vehicle
M_1	80	$\leqslant 0.1V + \frac{100}{30}\frac{V^2}{150}$	$\leqslant 0.1V + \frac{100}{25}\frac{V^2}{150}$
M_2	60	$\leqslant 0.15V + \frac{100}{30}\frac{V^2}{130}$	$\leqslant 0.15V + \frac{100}{25}\frac{V^2}{130}$
M_3	60	$\leqslant 0.15V + \frac{100}{30}\frac{V^2}{130}$	$\leqslant 0.15V + \frac{100}{30}\frac{V^2}{130}$
N_1	70	$\leqslant 0.15V + \frac{100}{30}\frac{V^2}{115}$	$\leqslant 0.15V + \frac{100}{25}\frac{V^2}{115}$
N_2	50	$\leqslant 0.15V + \frac{100}{30}\frac{V^2}{115}$	$\leqslant 0.15V + \frac{100}{25}\frac{V^2}{115}$
N_3	40	$\leqslant 0.15V + \frac{100}{30}\frac{V^2}{115}$	$\leqslant 0.15V + \frac{100}{30}\frac{V^2}{115}$

In case of failure of any part of the transmission, residual system efficiency must allow the minimum values in Table 19.9, when a force not higher than 700 N is applied to the control, on a 0 type test with disengaged engine.

After a 1 type test, results on a stop test must be better than the limits shown in Table 19.7, reduced to 80%, or to the limits of 0 type test, reduced to 60%.

After a 2 type or 2 A type test, stopping distance must be better than:

$$s \leqslant 0.15V + \frac{1,33V^2}{130} , \qquad (19.5)$$

for category M_3,

$$s \leqslant 0.15V + \frac{1,33V^2}{115} , \qquad (19.6)$$

for category N_3.

Vehicles of categories M_3 and N_3 must pass the three tests; vehicles in the remaining categories must pass only type 0 and type 1 tests.

The parking system device must be able to hold a vehicle in place on a slope (both uphill and downhill slopes must be tested) of at least 18%; for vehicles allowed to tow trailers, the parking system device must hold vehicle and trailer in place on a slope of 12%.

If the parking brake is a hand brake, the force on the lever must not exceed 400 N for vehicles of M category and 600 N for the remaining categories. If it is a pedal brake, the limits are upgraded to 500 N and 700 N respectively. A device that must be actuated more than one time to achieve full performance is allowed.

Other limits are prescribed for vehicles of category O and for the response time of pneumatic brake systems.

The last Directive above also considers ABS or antilock systems; by *antilock system*, this Directive refers to all components regulating the slip of one or more wheels during braking.

Many prescriptions apply to this system; we will quote the main ones only.

Any failure of the electric system or to sensors, including the electric supply, wiring harness, control systems, or pressure modulator must be signaled to the driver by a warning light.

This light must be on when the ABS system is on and the vehicle is stopped; the light goes off, after a short period, to demonstrate that the system is working.

In case of any failure, the residual braking efficiency is defined by Table 19.9.

Only off-road vehicles of categories N_2 or N_3 can have devices able to switch off the ABS system or to modify its operational mode.

ABS systems must also maintain their efficiency when brakes are operated for a long period of time.

Low friction coefficient

For all categories of vehicles, when friction coefficient μ_x is between 0.2 and 0.8, the following relationship is applied:

$$z = \frac{a_x}{g} \geqslant 0.1 + 0,85(\mu_x - 0,2) \, , \qquad (19.7)$$

where a_x is the obtained longitudinal deceleration and g is the gravity acceleration; z is called the *braking level*.

If we call:
f_i the friction used by the axle i,
F_{xi} the braking force of the axle i,
F_{zi} the vertical force on axle i during braking,
P_i the static vertical force on axle i,
P the vehicle weight,
h_G the center of gravity height,
l the vehicle wheelbase,

we obtain:

$$f_1 = \frac{F_{x1}}{F_{z1}} = \frac{F_{x1}}{P_1 + z\frac{h_g}{l}Pg} \, , \qquad (19.8)$$

$$f_2 = \frac{F_{x2}}{F_{z2}} = \frac{F_{x2}}{P_2 - z\frac{h_G}{l}Pg} \, ,$$

for a two-axle vehicle.

These formulae consider the pitch rotation equilibrium of a symmetric vehicle during braking; a complete justification of these formulae is given in Part IV.

For every load condition, the friction applied to the front axle must be higher than applied to the rear one:

- for any braking level between 0.15 and 0.18, for vehicles of M_1 category; however in the range between 0.3 and 0.45 the opposite is admitted if the friction used by the rear axle does not exceed by 0.05 the value given by $\mu_x = z$ (equal friction relationship) on a diagram representing f_i as function of z;

- for any braking level between 0.15 and 0.18, for vehicles of N_1 category; this condition is considered fulfilled if, for braking levels between 0.15 and

0.30, the used friction curves of each axle lie between the two parallel lines given by the following equations:

$$\mu_x = z + 0.08 \quad \text{and} \quad \mu_x = z - 0.08, \qquad (19.9)$$

and if the curve of the friction used by the rear axle (which can cross the line $\mu_x = z - 0.08$), for braking levels between 0.3 and 0.5, respects the relationship:

$$z \gg \mu_x - 0.08, \qquad (19.10)$$

and, between 0.5 and 0.61, respects the relationship:

$$z \gg 0.5\mu_x + 0.21; \qquad (19.11)$$

- for any braking level between 0.15 and 0.30, for other vehicle categories; this condition is considered fulfilled if, for braking levels between 0.15 and 0.30, the used friction curves of each axle lie between the two parallel lines given by the following equations:

$$\mu_x = z + 0.08 \quad \text{and} \quad \mu_x = z - 0.08, \qquad (19.12)$$

and if the curve of friction used by the rear axle, for braking levels $z \geqslant 0.3$, follows the relationship:

$$z \gg 0.3 + 0.74(\mu_x - 0.38). \qquad (19.13)$$

This set of conditions is used to design the braking distributor, when the braking system is without ABS or has an electronic brake distributor (EBD).

For vehicles with ABS, the value of obtainable z must be at least 75% of that with an ideal brake distributor.

Brake lining

Directive D 98/12 also includes a test procedure that can be applied when a vehicle is modified by installing a new type of brake lining, when such vehicle is already homologated, according to the said Directive.

The new brake lining must be verified by comparison of its efficiency with that obtained with the original lining at the vehicle homologation test, and it must comply with the specifications on the information form.

In this case, a roller dynamometer is also allowed, where roller inertia meets vehicle values at the homologation test.

As for the entire vehicle, liners must be tested on test types 0, 1, 2 and 2 A; results are acceptable if the average deceleration, at the same force on the pedal, are included in a tolerance band as wide as the 15% value obtained by the vehicle homologation tests.

Lining friction tests are also specified to prove conformity of production; they are performed on a simplified bench, simulating a single brake.

The Regulation R13 summarizes all aspects of the above directives and introduces criteria about regenerative braking, by distinguishing between A type systems, which are not integrated with the braking system, and B type systems, which are integrated. The braking effect on A type systems is obtained by releasing the accelerator pedal only, on M_1 vehicles and with a separate control on N_1 vehicles.

To these systems all prescriptions must be applied; the 0 type test must be performed without regenerative braking.

Regulation R90 applies to braking linings used as spare parts.

19.5 STRUCTURES

The structural behavior of a vehicle is the subject of many Directives about what must and must not occur in a collision test, as regards front, rear and lateral impacts, for M_1 vehicles.

Chassis structures work together with the vehicle body in determining the behavior in front and rear impact, but the body plays a major role also if these structures are not limited to an underbody only, but there is a true separated chassis frame; the energy absorbed by the body deformation is fundamental. For side impact, the underbody plays a marginal role.

Nevertheless we will outline, for the sake of completeness, the D 96/79 Directive, which defines test procedures to guarantee occupant safety in case of front impact.

The vehicle impacts a deformable barrier, according to the scheme in Fig. 19.6; the vehicle under test must be at the reference weight condition.

On each of the front seats, a dummy Hybrid III type must be accommodated and wear the provided passive restraint systems. The accommodation geometry and seat adjustment is specified by this Directive.

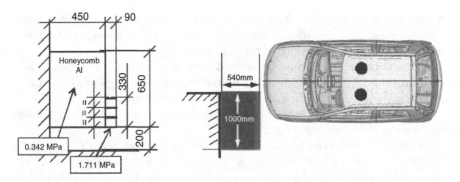

FIGURE 19.6. Scheme of the impact test against a deformable offset barrier, to demonstrate occupant protection capacity. At left is the side view of the barrier; at right, the upper view of the impact position.

The vehicle can be driven on its own or by an external device; at the impact time the steering system must be free and the propulsion system idle.

The impact speed must be 56 km/h, with a tolerance band of ±1 km/h. However, if the impact speed is higher and the test results comply with prescriptions, the test is considered as passed.

The vehicle path must be offset by 40% of the vehicle width (with a tolerance of ±20 mm), with reference to the vertical symmetry plane of the barrier. This offset was introduced, differently from the previous Directive, to take into account for the fact that most collisions between vehicles are not symmetric, but offset to their normal driving direction; cars homologated for countries with left-hand drive must have right-hand steering wheels and must be tested in symmetric position, as in the above scheme.

The barrier is positioned so as to have the first contact point with the vehicle on the driver's side.

The deformable barrier is designed to simulate a reference vehicle that is hit in the collision; for this purpose, it is made up of more elements:

- a main honeycomb aluminum structure;

- an element with the shape and position of a bumper, made of aluminum honeycomb;

- a reaction plate, bolted to a foundation;

- a cover plate for the bumper, made again of aluminum.

The front side of the barrier must be perpendicular to the vehicle impact direction with a tolerance of ±1°. The foundation mass must be at least 70,000 kg and its displacement must be limited by a concrete block. Barrier dimensions are shown on the left in Fig. 19.6.

During this test, many measurements are made to verify the performance criteria; different accelerometers are set on the dummy and on the car and the impact scene is shot by a high speed video camera.

The performance criteria involve the following limits:

- dummy head acceleration, processed according to a particular method taking into account the impact duration time;

- shear and traction forces on the dummy neck;

- thorax compression and compression speed;

- femur compression, as a function of time;

- tibia compression;

- rotula shear.

The values of these limits are derived by biomechanics studies on the human body and are continuously refined.

Beside the homologation test, rating tests by specialized independent laboratories and of specialized magazines become more and more relevant; the most famous of these is Euro NCAP (European New Car Assessment Programme).

Euro NCAP is a consortium between the main German Motoring Club (ADAC), the German, English and Dutch Transportation Ministries, the European Union and other partners, whose objective is to inform consumers about the passive safety of cars available on the market; information on Euro NCAP and available test results can be read on the internet site:

www.euroncap.com.

These tests results are scored and usually balanced in a final overall figure on vehicle safety, ranging from excellence to adequacy, measured by the number of stars.

There are tests similar to those reported by regulations but with additional severity in test procedures (for instance, by increasing impact speed) and in their evaluation scale, as well as new tests analyzing issues not already regulated (for instance, pedestrian impact was already rated before the issuance of the related law).

The peculiarity of rating tests is that they have reached a high level of reputation and have become important markers of product competitiveness. As a matter of fact, rating results have become an additional standard for car manufacturers.

Finally, the quest for higher rating evaluations contributes to raising the safety level of the operating fleet and promoting the improvement of the existing regulations.

19.6   GEARBOX

Regulations about the gearbox are few because this system has a low impact on functions covered by laws.

We also include in this section the prescription for the speedometer, because the pulse generator for the speed signal is usually within the gearbox. The final gear rotation speed is a simple indicator of wheel and car speed.

The Directive D 75/443 provides that all vehicles must have a reverse gear and a speedometer (with odometer).

A precision control is specified, according to the following procedure. The vehicle must have a set of homologated tires and the test must be repeated for any kind of speedometer included in the vehicle production.

The load on the axle having the speedometer installed (usually the driving axle) is defined by the D 70/156 Directive.

The vehicle must be tested at 40 km/h, 80 km/h and 120 km/h; the last value is substituted by 80% of the maximum speed, specified by the manufacturer, if this last is lower than 150 km/h.

The speed measured by the odometer cannot be lower than the actual speed.

At the specified speeds, the following relationship between indicated speed V_1 and actual speed V_2 must be verified:

$$0 \leqslant V_1 - V_2 \leqslant \frac{V_2}{10} + 4 \ . \tag{19.14}$$

The D 97/39 Directive updates the above for issues about the information form. The Regulation R39 is a summary of this topic.

Part IV

THE CHASSIS AS A PART OF THE VEHICLE SYSTEM

INTRODUCTION TO PART IV

After studying the main components of the vehicle chassis and defining the goals the designer of a motor vehicle must meet to comply with the standards and to satisfy the requirements of customers, the fourth part of this book is devoted to the study of the vehicle chassis as a system.

The behavior of the chassis as a whole has much more influence on the over-all characteristics of a motor vehicle in terms of performance, handling, safety and comfort than the characteristics of any single component. All complex machines must be studied and optimized at a system level, and this surely holds for motor vehicles as well.

After a short chapter devoted to the geometric and inertial properties of motor vehicles, properties that are highly dependent on those of the chassis, some basic ideas on vehicle aerodynamics and engines will be presented. Strictly speaking, both these subjects go beyond the study of the vehicular chassis, since they deal mainly with the other two systems constituting the motor vehicle (the body and the engine), but their influence on performance is too important to be neglected in the study of the chassis. They will be studied on in a synthetic way, examining only those aspects that are essential in the study of the chassis: the reader can find all the relevant details in specialized texts.

Two chapters dealing with the longitudinal dynamics of the motor vehicle, aimed at studying its performance both in acceleration and braking, will follow. In this study, many of the subsystems and components studied in Volume 1 are essential, namely the wheels and tires, the brakes and the driveline.

The following chapter is devoted to the study of lateral dynamics, an essential feature for assessing handling performance and active safety. In this chapter, only the lateral dynamics of the vehicle as a rigid body will be studied, and the effect of suspensions will not be dealt with, since it requires the use of more sophisticated mathematical models that will be introduced in Part V.

Suspensions will be introduced in the following chapter, which is devoted to the study of comfort. After a short section dealing with excitations due to the vehicle itself and to the road, and another one dealing with the response of the human body to vibration, the filtering characteristics of the chassis, due primarily to suspensions but also to tires, will be studied with the aim of obtaining satisfactory riding comfort.

The last chapter will be devoted to recently introduced automatic control devices. It is predicted that in the future they will become widespread with the aim of improving comfort and assisting the driver, especially when poor road conditions make his task more difficult.

20

GENERAL CHARACTERISTICS

20.1 SYMMETRY CONSIDERATIONS

Motor vehicles, like most machines, have a general bilateral symmetry. Only hypotheses can be advanced to explain why this occurs. Certainly to have a symmetry plane simplifies the study of the dynamic behavior of the system, for it can be modelled, within certain limits, using uncoupled equations. However, the reason is likely to be above all an aesthetic one: symmetry is considered an essential feature in most definitions of beauty.

All complex animals that evolved on our planet, including humans, have a symmetry plane defined by a vertical axis and an axis running in the longitudinal direction; symmetry is, however, not complete since some internal organs are positioned in an unsymmetrical way and some small deviations from symmetry are always present even in exterior appearance. When such lack of symmetry is too evident, it is felt to be incompatible with the aesthetic canons developed by all human civilizations.

A similar situation is encountered in all objects built by humans and, as in our interest here, in motor vehicles: a general outer symmetry and a certain lack of symmetry in the location of the internal components. Among the most common road vehicles, the only case where such a symmetry is not present is that of motorbikes with sidecar; these are, however, perceived to be made by a main unit, the motor bike, that has bilateral symmetry, plus a second unit, the sidecar, attached on a side, as its name suggests.

The sidecar often has its own symmetry plane, even if such characteristics are neither needed nor useful. This consideration may confirm the idea that symmetry has, in vehicles, purely an aesthetic justification.

A few other vehicles, built for very specialized use, like mobile cranes and building yard vehicles, have a non-symmetrical shape when strong functional reasons dictate it, but these are vehicles in which aesthetic considerations are utterly unimportant.

Many industrial vehicles may perhaps draw advantages from an asymmetrical architecture, for instance with the cab on one side and the loading surface on the other to use all the available length. Such a configuration seems, however, to be so unnatural as to be considered only if strictly needed. Imagination could at this point be set free to devise architectures that are not only without bilateral symmetry, but are even fully non-symmetrical, but this would likely be useless, since these configurations would seem to be unacceptable.

If the vehicle were completely symmetrical, the center of mass would lie in the symmetry plane. Actually, as already said, the mechanical systems and the load distribution are often not exactly symmetrical, so that the mass center can be displaced from it. In practice, the distance of the mass center from the symmetry plane is small.

20.2   REFERENCE FRAMES

The study of the motion of motor vehicles is usually performed with reference to some reference frames that are more or less standardized. They are (Fig. 20.1):

- *Earth-fixed axis system XYZ.* This is a right-hand reference frame fixed on the road. In the following sections, it will always be regarded as an inertial frame, even if strictly speaking it is not such as it moves along with the Earth: The inertial effects due to its motion (rotation about its axis, orbiting about the Sun, the galactic center, ...) are so small that they can be neglected in all phenomena studied in motor vehicle dynamics. Axes X and Y lie in a horizontal plane while axis Z is vertical, pointing upwards[1].

- *Vehicle axis system xyz.* This is a right-hand reference frame fixed to the vehicle's center of mass and moving with it. As already stated, if the vehicle has a symmetry plane, the center of mass is assumed to lie on it. The x-axis lies in the symmetry plane of the vehicle in an almost horizontal direction[2]. The z-axis lies in the symmetry plane, is perpendicular to the

---

[1] Recommendation SAE J670 and ISO/TC 22/SC9 standard state that the Z-axis is vertical and points downwards. Note that in the present text the direction of the Y and Z axes is opposite to that suggested in the mentioned standard.

[2] The mentioned standard states that the x-axis is contained in the plane of symmetry of the vehicle, is "substantially" horizontal and points forward.

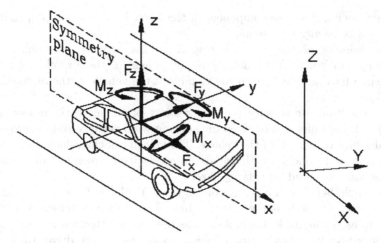

FIGURE 20.1. Reference frames, forces and moments used for the dynamic study of motor vehicles.

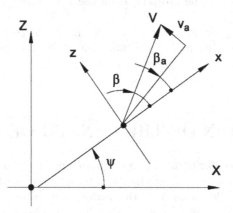

FIGURE 20.2. Projection of the axes of the vehiocle fixed frame in XY plane of the inertial frame.

x-axis, and points upwards. The y-axis is perpendicular to the other two[3]. The y-axis,then, points to the left of the driver. In the case of vehicles without a symmetry plane, the xz plane is identified by the direction of motion when the wheels are not steered and by a direction perpendicular to the road in the reference position of the vehicle.

The projections of the vehicle-fixed axes xyz and of the velocity on the plane XY of the inertial frame are shown in Fig. 20.2. The angle between the projection of the x-axis and the X-axis is the yaw angle ψ. The projection of the velocity of the center of mass on the XY-plane is here conflated with the

---

[3] Here again there is a deviation from the mentioned standard.

absolute velocity, since the component of the latter in a direction perpendicular to the road is usually very small.

The velocity of the air with respect to the ground, or ambient wind velocity v_a, is defined as the horizontal component of the air velocity relative to the earth-fixed axis system in the vicinity of the vehicle in the inertial frame XYZ.

The resultant air velocity (i.e. the velocity of the air with respect to the vehicle) V_r is the difference between the ambient wind velocity and the projection of the absolute velocity V of the vehicle. The velocity of the vehicle with respect to air is $-V_r$ and coincides with the absolute velocity V when the air is still, i.e. when the ambient wind velocity v_a is zero.

The angle between the projection on the XY plane of the x-axis and that of the velocity vector V is the sideslip angle β of the whole vehicle; sometimes referred to as the attitude angle. As it was the case for the sideslip angle of the tire α, it is positive when vector V points to the left of the driver (in forward motion). In a similar way it is possible to define an aerodynamic angle of sideslip β_a as the angle between the projections on the XY plane of the x-axis and the relative velocity $-V_r$.

Angles β and β_a usually refer to the velocity of the centre of mass, but can be referred to the velocity of any specific point of the vehicle.

20.3   POSITION OF THE CENTER OF MASS

The position of the center of mass is very important in determining the behavior of the vehicle and must be computed, or at least assessed, at the design stage and then experimentally determined. If the various components of the vehicle are designed using CAD techniques, it is generally possible to have their mass and the positions of their centers of mass among the outputs of the code. The position of the center of mass of the vehicle can then be assessed from the results obtained for the components, but this evaluation is generally approximate and at any rate is a long computation. Then what is of interest is not the position of the center of mass of the empty vehicle, but that of the vehicle in the various operating conditions (vehicle with all liquids, the driver, a variable number of passengers and possibly luggage).

The longitudinal position of the center of mass can be obtained by simply weighing the vehicle on a level road at the front and rear axles (Fig. 20.3a). If F_{z_1} and F_{z_2} are the vertical forces measured at the two axles, the equilibrium equations for translations in the z direction and for rotations about the y-axis can be written as:

$$\begin{cases} F_{z_1} + F_{z_2} = mg \\ lF_{z_1} = bmg \ , \end{cases} \qquad (20.1)$$

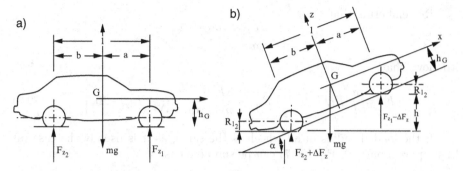

FIGURE 20.3. Sketch of the experimental determination of the longitudinal and vertical position of the center of mass.

and then:

$$\begin{cases} a = l\dfrac{F_{z_2}}{F_{z_1} + F_{z_2}} \\[3mm] b = l\dfrac{F_{z_1}}{F_{z_1} + F_{z_2}} \, . \end{cases} \qquad (20.2)$$

If the center of mass does not lie in the symmetry plane, it is possible to compute the transversal position of the center of mass by measuring the forces under the right and left wheels.

It is more difficult to determine the height of the center of mass. Once the longitudinal position is known (once a and b have been determined) it is possible to measure again the forces on the ground after the front (or the rear) axle has been raised from the ground (Fig. 20.3b).

Let the front axle be set on a platform with height h with respect to the platform on which the rear axle is located. If the height of the center of mass h_G is greater than the radius under load of the wheels, the force F'_{z_1} measured at the front axle will be smaller than that measured on level road. It is then possible to write:

$$F'_{z_1} = F_{z_1} - \Delta F_z. \qquad (20.3)$$

If the force at the rear axle is measured, it would be:

$$F'_{z_2} = F_{z_2} + \Delta F_z. \qquad (20.4)$$

The equilibrium equation for rotations about the center of the front axle is:

$$mg\left[a\cos\left(\alpha\right) + (h_G - R_{l_1})\sin\left(\alpha\right)\right] = \qquad (20.5)$$

$$= (F_{z_2} + \Delta F_z)\left[l\cos\left(\alpha\right) + (R_{l_2} - R_{l_1})\sin\left(\alpha\right)\right].$$

The height of the center of mass is then:

$$h_G = \frac{F_{z_2} + \Delta F_z}{mg}\left[\frac{l}{\tan\left(\alpha\right)} + R_{l_2} - R_{l_1}\right] - \frac{a}{\tan\left(\alpha\right)} + R_{l_1}. \qquad (20.6)$$

Remembering that:

$$\frac{F_{z_2}}{mg} = \frac{a}{l},$$

it follows that

$$h_G = \frac{a}{l} [R_{l_2} - R_{l_1}] + R_{l_1} + \frac{\Delta F_z}{mg} \left[\frac{l}{\tan(\alpha)} + R_{l_2} - R_{l_1} \right]. \qquad (20.7)$$

If the loaded radius of all wheels is the same, as it is usually the case (at least approximately), Eq. (20.7) can be simplified as:

$$h_G = R_l + \frac{\Delta F_z}{mg} \frac{l}{\tan(\alpha)}, \qquad (20.8)$$

that is:

$$h_G = R_l + \frac{\Delta F_z}{mg} \frac{l\sqrt{l^2 - h^2}}{h}. \qquad (20.9)$$

To measure accurately the height of the center of mass, the wheels must be completely free (brakes released and gear in neutral), and the vehicle must be restrained from rolling by chocks at one of the axles. Moreover, the suspensions must be locked at a height corresponding to the load distribution on level road, and the tires must be equally compressed.

This last condition can be obtained in an approximate way by increasing the inflation pressure. It is then possible to take several measurements of ΔF_z at different values of $\tan(\alpha)$ and to plot ΔF_z versus $\tan(\alpha)$. If the radii of the wheels are all equal, such a curve is a straight line, whose slope

$$\frac{\Delta F_z}{\tan(\alpha)} = \frac{mg}{l} (h_G - R_l) \qquad (20.10)$$

allows one to compute the height of the center of mass.

20.4   MASS DISTRIBUTION AMONG THE VARIOUS BODIES

In terms of dynamic behavior, vehicles are often modeled as a number of rigid bodies with different inertial properties. The simplest way to model a vehicle is to consider the body as a rigid body (sprung mass), to which a further rigid body is added to model each rigid axle, and two rigid bodies for each axle with independent suspension. This approach is approximate, since many vehicle elements (suspension linkages, springs, shock absorbers) do not belong to any of these bodies. Half of the mass of the elements located between two rigid bodies may be attributed to each one of them, but this is an approximation, although usually an acceptable one.

There is no alternative either to computing or experimentally determining the mass of the various components separately to evaluate the mass of the various subsystems. For a first approximation evaluation, the center of mass of a rigid axle can be assumed to lie on the line connecting the centers of the wheels in the symmetry planes. The center of mass of each independent suspension can be located at a distance from the symmetry plane equal to half of the track of the relevant axle. Better estimations can come from a detailed analysis of the drawings of the suspension.

Once the positions of the centers of mass of the whole vehicle and of the suspensions (unsprung masses) are known, it is straightforward to locate the center of mass of the sprung mass.

More detailed models can be obtained from computer codes based on multi-body modelling. Not only do they allow us to take into account the various parts constituting the vehicle in much greater detail (each element of the suspension can be introduced separately, with its mass, moments of inertia and exact kinematics), but their compliance can also be introduced using the finite element method (FEM).

20.5 MOMENTS OF INERTIA

The inertia tensor of a rigid body is:

$$\mathbf{J} = \begin{bmatrix} J_x & -J_{xy} & -J_{xz} \\ -J_{xy} & J_y & -J_{yz} \\ -J_{xz} & -J_{yz} & J_z \end{bmatrix} \qquad (20.11)$$

where the moments of inertia are:

$$J_x = \int_V \left(y^2 + z^2\right) dm \qquad J_y = \int_V \left(x^2 + z^2\right) dm \qquad J_z = \int_V \left(x^2 + y^2\right) dm$$

$$(20.12)$$

and the products of inertia[4] are:

$$J_{xy} = \int_V xy\,dm \qquad J_{xz} = \int_V xz\,dm \qquad J_{yz} = \int_V yz\,dm. \qquad (20.13)$$

If the vehicle has a plane of symmetry that coincides with the xz-plane, the products of inertia J_{xy} and J_{yz} vanish and the inertia tensor of the vehicle reduces to:

$$\mathbf{J} = \begin{bmatrix} J_x & 0 & -J_{xz} \\ 0 & J_y & 0 \\ -J_{xz} & 0 & J_z \end{bmatrix}. \qquad (20.14)$$

---

[4] Often products of inertia are defined as $J_{xy} = -\int_V xy\,dm$, etc. and signs $(-)$ are not included in Eq. (20.11).

The computation of the moments of inertia of the vehicle and of its parts is a complex operation, but if the design of the various components has been performed using CAD, the approximate values of the moments of inertia are among the outputs of the code.

Usually, the moments of inertia of interest are:

- the baricentric roll moment of inertia (about the x-axis) of the sprung mass J_{S_x},

- the baricentric pitch moment of inertia (about the y-axis) of the sprung mass J_{S_y} and

- the baricentric yaw moment of inertia (about the z-axis) of the whole vehicle J_z.

The first two are obviously referred to the center of mass of the sprung mass, while the latter is referred to the center of mass of the whole vehicle.

Some empirical formulae for their computation from the mass of the vehicle are often used. These usually yield the radius of gyration ρ instead of the corresponding moment of inertia:

$$J_i = m\rho_i^2, \qquad i = x,\ y,\ z. \qquad (20.15)$$

A first rough approximation for the yaw radius of gyration is:

$$\rho_z = \sqrt{ab}. \qquad (20.16)$$

First approximation values of the radii of gyration (in meters) that may be used for a medium size car are:[5]

Load condition	ρ_{S_x}	ρ_{S_y}	ρ_z
Empty	0,65	1,21	1,20
2 passengers	0,64	1,13	1,15
4 passengers	0,60	1,10	1,14
4 passengers + luggage	0,56	1,13	1,18

Note also that the values for the sprung mass are referred to the total mass of the vehicle. As a first approximation, inertia products may be neglected.

Remark 20.1 *The moments and products of inertia must be known with precision, since they are important in assessing the performance of the vehicle, both for its handling and its comfort. It is then important to measure them accurately once prototypes are built.*

The simplest way to measure the moments of inertia of any object is to suspend it in such a way that it can rotate as a pendulum about an axis that

---

[5] J. Reimpell, H. Stoll, *The Automotive Chassis*, SAE, Warrendale, 1996.

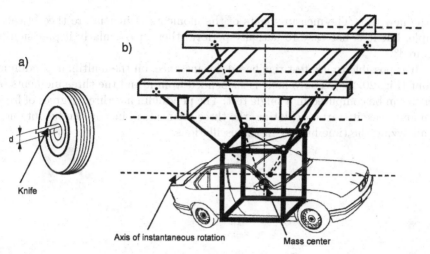

FIGURE 20.4. Measurement of the moments of inertia. (a): Oscillation on a knife located on an axis parallel to the baricentric axis. (b): Quadrifilar pendulum to measure the inertia tensor of a motor vehicle.

is parallel to the baricentric axis about which the moment of inertia must be measured (Fig. 20.4a). If d is the distance between the center of mass and the suspension axis, the moment of inertia about the latter is:

$$J = J_G + md^2 . \tag{20.17}$$

The period of oscillation of the pendulum made of the body suspended on the knife is

$$T = 2\pi \sqrt{\frac{J_G + md^2}{mgd}}. \tag{20.18}$$

By measuring the period of the small oscillations it is then possible to obtain the baricentric moment of inertia

$$J_G = m \left[\frac{gd}{4\pi^2} T^2 - d^2 \right]. \tag{20.19}$$

By repeating the measurement three times with the object suspended about three axes parallel to the x-, y- and z-axes, the three moments of inertia can be measured.

To measure the products of inertia, the measurements can be repeated measuring the moments of inertia about three axes different from the previous ones. Since the inertia tensor about the latter is

$$\mathbf{J'} = \mathbf{R}^T \mathbf{J} \mathbf{R}, \tag{20.20}$$

where $\mathbf{R}$ is a known rotation matrix, it is possible to compute the values of the elements of the unknown inertia tensor $\mathbf{J}$ from the three values of $\mathbf{J'}$ computed in the new test. The measurements of the moments of inertia are theoretically simple, but it is not easy to actually perform them, particularly if precision is required[6].

It is possible to use test rigs based, for instance, on the multifilar pendulum layout (Fig. 20.4b[7]). The vehicle is suspended from it and the three moments of inertia can be computed in a single test. The pendulum has three degrees of freedom and, once it starts oscillating, the three moments of inertia can be obtained by analyzing the time history of the oscillations.

[6]G. Genta, C. Delprete, *Some Considerations on the Experimental Determination of the Moments of Inertia*, Meccanica, 29, pp.125-141, 1994.

[7]G. Mastinu, M. Gobbi, C.M. Miano, *The Influence of the Body Inertia Tensor on the Active Safety and Ride Comfort of Road Vehicles*, SAE Paper 2002-01-2058, SAE, Warrendale, 2002.

21

AN OVERVIEW ON MOTOR VEHICLE AERODYNAMICS

The forces and moments the vehicle receives from the surrounding air depend more on the shape of the body than on the characteristics of the chassis. A detailed study of motor vehicle aerodynamics is thus beyond the scope of a book dealing with the automotive chassis.

However, aerodynamic forces and moments have a large influence on the longitudinal performance of the vehicle, its handling and even its comfort, so it is not possible to neglect them altogether.

Even if the goal of motor vehicle aerodynamics is often considered to be essentially the reduction of aerodynamic drag, the scope and the applications of aerodynamics in motor vehicle technology are much wider.

The following aspects are worth mentioning

- reduction of aerodynamic drag,

- reduction of the side force and the yaw moment, which have an important influence on stability and handling,

- reduction of aerodynamic noise, an important issue for acoustic comfort, and

- reduction of dirt deposited on the vehicle and above all on the windows and lights when driving on wet road, and in particular in mud or snow conditions. This aspect, important for safety, can be extended to the creation of spray wakes that can reduce visibility for other vehicles following or passing the vehicle under study.

G. Genta, L. Morello, *The Automotive Chassis, Volume 2: System Design*,
Mechanical Engineering Series,

The provisions taken to obtain these goals are often different and sometimes contradictory. A typical example is the trend toward more streamlined shapes that allow us to reduce aerodynamic drag, but at the same time have a negative effect on stability.

Another example is the mistaken assumption that a shape that reduces aerodynamic drag also has the effect of reducing aerodynamic noise. The former is mainly influenced by the shape of the rear part of the vehicle, while the latter is much influenced by the shape of the front and central part, primarily of the windshield strut (A-pillar). It is then possible that a change in shape aimed at reducing one of these effects may have no influence, or sometimes even a negative influence, on the other one.

At any rate, all aerodynamic effects increase sharply with speed, usually with the square of the speed, and are almost negligible in slow vehicles. Moreover, they are irrelevant in city driving.

Aerodynamic effects, on the contrary, become important at speeds higher than $60 \div 70$ km/h and dominate the scene above $120 \div 140$ km/h. Actually these figures must be considered only as indications, since the relative importance of aerodynamic effects and those linked with the mass of the vehicle depends on the ratio between the cross section area and the mass of the vehicle. At about $90 \div 100$ km/h, for instance, the aerodynamic forces acting on a large industrial vehicle are negligible when it travels at full load, while they become important if it is empty.

Modern motor vehicle aerodynamics is quite different from aeronautic aerodynamics, from which it derives, not only for its application fields but above all for its numerical and experimental instruments and methods. The shapes of the objects dealt with in aeronautics are dictated mostly by aerodynamics, and the aerodynamic fields contains few or no zones in which the flow separates from the body. On the contrary, the shape of motor vehicles is determined mostly by considerations like the possibility of locating the passengers and the luggage (or the payload in industrial vehicles), aesthetic considerations imposed by style, or the need of cooling the engine and other devices like brakes. The blunt shapes that result from these considerations cause large zones where the flow separates and a large wake and vortices result.

The presence of the ground and of rotating wheels has a large influence on the aerodynamic field and makes its study much more difficult than in the case of aeronautics, where the only interaction is that between the body and the surrounding air.

One of the few problems that are similar in aeronautical and motor vehicle aerodynamics is the study of devices like the wings of racing cars, but this is in any case a specialized field that has little to do with vehicle chassis design, and it will not be dealt with here in detail.

Traditionally, the study of aerodynamic actions on motor vehicles is primarily performed experimentally, and the wind tunnel is its main tool. The typical wind tunnel scenario is a sort of paradigm for interpreting aerodynamic phenomena, to the point that usually the body is thought to be stationary and the air moving around it, instead of assuming that the body moves through stationary air.

However, while in aeronautics the two wiewpoints are coincident, in motor vehicle aerodynamics they would be so only if, in the wind tunnel, the ground moved together with the air instead of being stationary with respect to the vehicle. Strong practical complications are encountered when attempting to allow the ground to move with respect to the vehicle, and allowing the wheels to rotate. Usually, in wind tunnel testing, the ground does not move, but its motion is simulated in an approximate way.

Along with wind tunnel tests, it is possible to perform tests in actual conditions, with vehicles suitably instrumented to take measurements of aerodynamic forces while travelling on the road. Measurements of the pressure and the velocity of the air at different points are usually taken.

Recently powerful computers able to simulate the aerodynamic field numerically have became available. Numerical aerodynamic simulation is extremely demanding in terms of computational power and time, but it allows us to predict, with increasing accuracy, the aerodynamic characteristics of a vehicle before building a prototype or a full scale model (note that reduced scale models, often used in aeronautics, are seldom used in vehicular technology).

There is, however, a large difference between aeronautical and vehicular aerodynamics from this viewpoint as well. Nowadays, numerical aerodynamics is able to predict very accurately the aerodynamic properties of streamlined bodies, even if wind tunnel tests are needed to obtain an experimental confirmation. The possibility of performing extensive virtual experimentation on mathematical models greatly reduces the number of experimental tests to be performed.

Around blunt bodies, on the other hand, it is very difficult to simulate the aerodynamic field accurately, given their large detached zones and wake. Above all, it is difficult to compute where the streamlines separate from the body. The impact of numerical aerodynamics is much smaller in motor vehicle design than has been in aeronautics.

As said, the aim of this chapter is not to delve into details on vehicular aerodynamics, but only to introduce those aspects that influence the design of the chassis. While the study of the mechanisms that generate aerodynamic forces and moments influencing the longitudinal and handling performance of the vehicle will be dealt with in detail, those causing aerodynamic noise or the deposition of dirt on windows and lights will be overlooked. In particular, those unstationary phenomena, like the generation of vortices that are very important in aerodynamic noise, will not be studied.

21.1   AERODYNAMIC FORCES AND MOMENTS

In aeronautics, the aerodynamic force acting on the aircraft is usually decomposed in the direction of the axes of a reference frame $Gx''y''z''$, usually referred to as the *wind axes system*, centered in the mass center G, with the x''-axis directed as the velocity of the vehicle with respect to air $-V_r$ and the z''-axis contained in the symmetry plane.

The components of the aerodynamic forces in the $Gx''y''z''$ frame are referred to as drag D, side force S and lift L. The aerodynamic moment is usually decomposed along the vehicle-fixed axes $Gxyz$.

In the case of motor vehicles, both the aerodynamic force and moment are usually decomposed with reference to the frame xyz: The components of the aerodynamic force are referred to as longitudinal F_{x_a}, lateral F_{y_a} and normal F_{z_a} forces while those of the moment are the rolling M_{x_a}, pitching M_{y_a} and yawing M_{z_a} moments.

In the present text, aerodynamic forces will always be referred to frame xyz, which is centred in the centre of mass of the vehicle. However, in wind tunnel testing the exact position of the centre of mass is usually unknown and the forces are referred to a frame which is immediately identified.

Moreover, the position of the centre of mass of the vehicle depends also on the loading, while aerodynamic forces are often assumed to be independent of it, although a change of the load of the vehicle can affect its attitude on the road and hence the value of aerodynamic forces and moments.

The frame often used to express forces and moments for wind tunnel tests is a frame centred in a point on the symmetry plane and on the ground, located at mid-wheelbase, with the x'-axis lying on the ground in the plane of symmetry of the vehicle and the y'-axis lying also on the ground (Fig. 21.1). Since the resultant air velocity V_r lies in a horizontal plane, angle α is the aerodynamic angle of attack. From the definition of the x axis, it is a small angle and is often assumed to be equal to zero.

Remark 21.1 *From the definitions here used for the reference frames it follows that α is positive when the x-axis points downwards.*

The forces and moments expressed in the xyz frame can be computed from those expressed in the $x'y'z'$ frame (indicated with the symbols F_x', F_y', F_z', M_x', M_y' and M_z') through the relationships

$$\begin{cases} F_x = F_x' \cos(\alpha) - F_z' \sin(\alpha) \\ F_y = F_y' \\ F_z = F_x' \sin(\alpha) + F_z' \cos(\alpha) \end{cases} \tag{21.1}$$

$$\begin{cases} M_x = M_x' + F_y h_G \\ M_y = M_y' - F_x h_G + F_z x_G' \\ M_z = M_z' - F_y x_G' \, . \end{cases} \tag{21.2}$$

Distance x_G' is the coordinate of the centre of mass with reference to the $x'y'z'$ frame and is positive if the centre of mass is forward of mid-wheelbase ($a < b$).

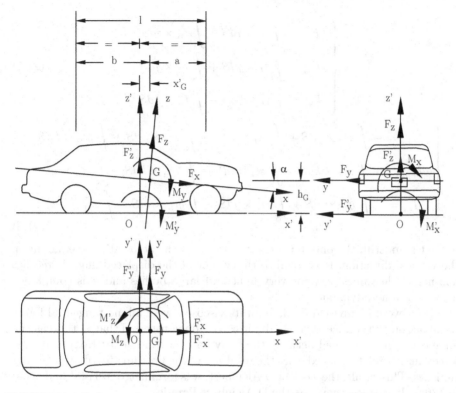

FIGURE 21.1. Reference frame often used to express aerodynamic forces in wind tunnel tests.

The air surrounding a road vehicle exerts on any point P of its surface a force per unit area

$$\vec{t} = \lim_{\Delta S \to 0} \frac{\Delta \vec{F}}{\Delta S} \, , \qquad (21.3)$$

where ΔS and $\Delta \vec{F}$ are respectively the area of a small surface surrounding point P and the force acting on it.

The force per unit area $\vec{t}$ can be decomposed into a pressure force acting in a direction perpendicular to the surface

$$\vec{t}_n = p\vec{n} \, , \qquad (21.4)$$

where $\vec{n}$ is a unit vector perpendicular to the surface and p is a scalar expressing the value of the pressure, and a tangential force $\vec{t}_t$ lying on the plane tangent to the surface. The latter is due to fluid viscosity.

These force distributions, once integrated on the entire surface, result in an aerodynamic force, which is usually applied to the centre of mass of the vehicle, and an aerodynamic moment. By decomposing the force and the moment in $Gxyz$ frame, it follows:

$$
\begin{cases}
F_{x_a} = \int_S \vec{t_t} \times \vec{i} dS + \int_S \vec{t_n} \times \vec{i} dS \\[2mm]
F_{y_a} = \int_S \vec{t_t} \times \vec{j} dS + \int_S \vec{t_n} \times \vec{j} dS \\[2mm]
F_{z_a} = \int_S \vec{t_t} \times \vec{k} dS + \int_S \vec{t_n} \times \vec{k} dS
\end{cases}
\tag{21.5}
$$

$$
\begin{cases}
M_{x_a} = -\int_S z\vec{t_t} \times \vec{j} dS + \int_S y\vec{t_t} \times \vec{k} dS - \int_S z\vec{t_n} \times \vec{j} dS + \int_S y\vec{t_n} \times \vec{k} dS \\[2mm]
M_{y_a} = -\int_S x\vec{t_t} \times \vec{k} dS + \int_S z\vec{t_t} \times \vec{i} dS - \int_S x\vec{t_n} \times \vec{k} dS + \int_S z\vec{t_n} \times \vec{i} dS \\[2mm]
M_{z_a} = -\int_S y\vec{t_t} \times \vec{i} dS + \int_S x\vec{t_t} \times \vec{j} dS - \int_S y\vec{t_n} \times \vec{i} dS + \int_S x\vec{t_n} \times \vec{j} dS \ .
\end{cases}
\tag{21.6}
$$

At standstill, the only force exerted by air is the aerostatic force, acting in the vertical direction. It is equal to the weight of the displaced fluid. It reaches non-negligible values only for very light and large bodies and it is completely neglected in aerodynamics.

If air were an inviscid fluid, i.e. if its viscosity were nil, no tangential forces could act on the surface of the body; it can be demonstrated that in this case no force could be exchanged between the body and the fluid, apart from aerostatic forces, at any relative speed since the resultant of the pressure distribution always vanishes. This result, the work of D'Alembert, was formulated in 1744[1] and again in 1768[2]. It is since known as the D'Alembert Paradox.

In the case of a fluid with no viscosity, the pressure p and the velocity V can be linked to each other by the Bernoulli equation

$$
p + \frac{1}{2}\rho V^2 = \text{constant} = p_0 + \frac{1}{2}\rho V_0^2 \ ,
\tag{21.7}
$$

where p_0 and V_0 are the values of the ambient pressure and of the velocity far enough upstream from the body[3]. The term

$$
p_d = \frac{1}{2}\rho V_0^2
\tag{21.8}
$$

is the so-called dynamic pressure. The sum

$$
p_{tot} = p_0 + p_d
\tag{21.9}
$$

is the total pressure.

---

[1]D'Alembert, *Traité de l'équilibre et du moment des fluides pour servir de suite un traité de dynamique*, 1774.

[2]D'Alembert, *Paradoxe proposé aux geometres sur la résistance des fluides*, 1768.

[3]Considering the actual case of the vehicle moving in still air, instead of the wind tunnel situation with air moving around a stationary object, V_0 is the velocity of the body relative to air $-V_r$.

TABLE 21.1. Pressure, temperature, density and kinematic viscosity of air at various altitudes, from the ICAO standard atmosphere. Only the part of the table related to altitudes of interest for road vehicles is reported.

z [m]	p [kPa]	T [K]	ρ [kg/m^3]	ν [m^2/s]
-500	107.486	291.25	1.2857	13.97×10^{-6}
0	101.325	288.16	1.2257	14.53×10^{-6}
500	95.458	284.75	1.1680	15.10×10^{-6}
1000	89.875	281.50	1.1123	15.71×10^{-6}
1500	84.546	278.25	1.0586	16.36×10^{-6}
2000	79.489	275.00	1.0070	17.05×10^{-6}
2500	74.656	271.75	0.9573	17.77×10^{-6}
3000	70.097	268.50	0.9095	18.53×10^{-6}

The values of the ambient pressure, together with those of the density, temperature, and kinematic viscosity at altitudes of interest in road vehicle technology, are reported in Table 21.1 from the ICAO standard atmosphere.

The density at temperatures and pressures different from p_a and T_a in standard conditions can be computed as

$$\rho = \rho_a \frac{p}{p_a} \frac{T_a}{T} , \qquad (21.10)$$

where temperatures are absolute.

The dynamic pressure is extremely low, when compared to the ambient pressure: consider, for instance, a vehicle moving air at the temperature and pressure equal to those indicated in Table 21.1 at sea level, at a speed of 30 m/s (108 km/h). The pressure is about 101 kPa, while the dynamic pressure is 0,55 kPa, corresponding to 0,5% of pressure.

The variations of pressure due to velocity variations are thus quite small with respect to atmospheric pressure; however, such small pressure changes, acting on surfaces of some square meters, yield non-negligible, and sometimes large, aerodynamic forces.

Note that the Bernoulli equation, which holds along any streamline, was written without the gravitational term, the one linked with aerostatic forces. It states simply that the total energy is conserved along any streamline.

An example of the D'Alembert Paradox is shown in Fig. 21.2, where the cross section of a cylinder of infinite length, whose axis is perpendicular to the direction of the velocity V_r of the fluid, is represented. The streamlines open around the body and the local velocity of the fluid increases on its sides, leading to a decrease of pressure as described by the Bernoulli Equation. On the front of the body there is a point (actually in the case of the cylinder it is a line) which divides the part of the flow which goes "above" the body from that going "below" it. At this point, known as the stagnation point, the velocity of the fluid reduces to zero and the pressure reaches its maximum, equal to the total pressure.

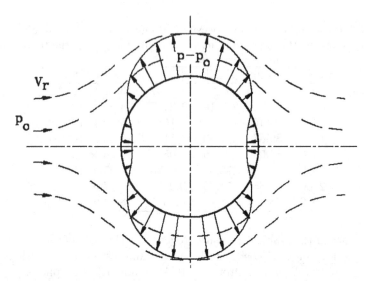

FIGURE 21.2. Streamlines and pressure distribution on a circular cylinder whose axis is perpendicular to the flow. This is a case of a fluid with no viscosity.

Since there is no viscosity, no energy is dissipated, and when the fluid slows down again, after reaching the maximum velocity at the point where the width of the body is maximum, the pressure is fully recovered: The pressure distribution is symmetrical and no net force is exchanged between the fluid and the body. This holds for any possible shape, provided that the viscosity is exactly nil.

No fluid actually has zero viscosity and the Paradox is not applicable to any real fluid. Viscosity has a twofold effect: It causes tangential forces creating so-called friction drag, and it modifies the pressure distribution, whose resultant is no longer equal to zero. The latter effect, which for fluids with low viscosity is generally more important than the former, generates the lift, the side force and the pressure drag. The direct effects of viscosity (i.e. the tangential forces) can usually be neglected, while the modifications of the aerodynamic field must be accounted for.

Owing to viscosity, the layer of fluid in immediate contact with the surface tends to adhere to it, i.e. its relative velocity vanishes, and the body is surrounded by a zone where there are strong velocity gradients. This zone is usually referred to as the "boundary layer" (Fig. 21.3) and all viscous effects are concentrated in it. The viscosity of the fluid outside the boundary layer is usually neglected and the Bernoulli equation can be used in this region.

Remark 21.2 *The thickness of the boundary layer increases as the fluid in it loses energy owing to viscosity and slows down. If the fluid outside the boundary layer increases its velocity, a negative pressure gradient along the separation line between the external flow and the boundary layer is created, and this decrease of pressure in a way boosts the flow within the boundary layer fighting its tendency*

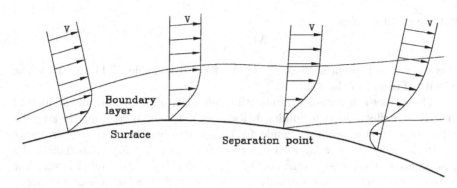

FIGURE 21.3. Boundary layer: Velocity distribution in direction perpendicular to the surface. The separation point is also represented.

to slow down. On the contrary, if the outer flow slows down, the pressure gradient is positive and the airflow in the boundary layer is hampered.

At any rate, at a certain point the flow in the boundary layer can stop and a zone of stagnant air can form in the vicinity of the body: The flow then separates from the surface, possibly starting the formation of a wake.

If the velocity distribution outside the boundary layer were known, the pressure distribution at the interface between the boundary layer and the external fluid could be computed. Provided that the boundary layer is very thin, and this is the case except where the flow is detached from the surface, the pressure on the surface of the body can be assumed to be equal to that occurring at the outer surface of the boundary layer, and then the aerodynamic forces and moments can be computed by integrating the pressure distribution. While this can be applied to computing the lift of streamlined objects, for blunt bodies, like the ones studied by road vehicle aerodynamics, and for drag, few results can be obtained along these lines.

To generalize the results obtained by experimental testing, performed mainly in wind tunnels, the aerodynamic force F and moment M are expressed as

$$F = \frac{1}{2}\rho V_r^2 S C_f , \qquad M = \frac{1}{2}\rho V_r^2 S l C_m , \qquad (21.11)$$

where forces and moments are assumed to be proportional to the dynamic pressure of the free current

$$\frac{1}{2}\rho V_r^2 ,$$

to a reference surface S (in the expression of the moment a reference length l is also present) and to nondimensional coefficients C_f and C_m to be experimentally determined.

Such coefficients depend on the geometry and position of the body, and on two non-dimensional parameters, the Reynolds number

$$\mathcal{R}_e = \frac{Vl}{\nu} ,$$

and the Mach number

$$\mathcal{M}_a = \frac{V}{V_s} ,$$

where ν is the kinematic viscosity of the fluid (see Table 21.1) and V_s is the velocity of sound in the fluid.

The former is a parameter indicating the relative importance of the inertial and viscous effects in determining aerodynamic forces. If its value is low, the latter are of great importance, while if it is high aerodynamic forces are primarily due to the inertia of the fluid. In this case (for vehicles, if $\mathcal{R}_e > 3{,}000{,}000$), the dependence of the aerodynamic coefficients on the Reynolds number is very low and can be neglected. This is usually the case for road vehicles, at least for speeds in excess of $30 \div 40$ km/h.

If, on the contrary, the Reynolds number is low, aerodynamic forces and moments are essentially due to viscosity. In this case, their dependence on the velocity V should be linear rather than quadratic or, to use equations (21.11), the aerodynamic coefficients should be considered as dependent on the speed, increasing with decreasing speed.

The Mach number is the ratio between the airspeed and the speed of sound[4]. When its value is low, the fluid can be considered as incompressible; aerodynamic coefficients are then independent of speed. Approaching the speed of sound ($\mathcal{M}_a \sim 1$), the compressibility of the fluid can no longer be neglected and aerodynamic drag increases sharply. It is commonly thought that the Mach number is irrelevant in automotive aerodynamics, since the speeds road vehicles may reach, with the exception of some vehicles built to set speed records, lead to Mach numbers low enough to have practically no influence on aerodynamic coefficients. Actually this is true for streamlined bodies, for which the influence of Mach number is negligible for values up to $0,5 \div 0,6$ (speeds up to $600 \div 700$ km/h), while for blunt bodies fluid compressibility starts to play a role at a lower speed, even for Mach numbers slightly larger than $0,2$ ($V = 70$ m/s $= 250$ km/h). As a consequence, the effects of the Mach number start to be felt at speeds that can be reached by racing cars. It is important to note that, owing to this effect of the Mach number, it is not possible to perform tests on reduced scale models by increasing the speed to increase the Reynolds number.

The reference surface S and length l are arbitrary, to the point that in some cases a surface not existing physically, like a power $2/3$ of the displacement for airships, is used. These references simply express the dependence of aerodynamic forces on the square of the dimensions of the body and that of the moments on their cube. It is, however, clear that the numerical values of the coefficients depend on the choice of S and l, which must be clearly defined. In the case of road vehicles, the surface is that of the cross section, with some uncertainty about whether the frontal projected area or that of the maximum cross section has been used (Fig. 21.4).

---

[4]For air at sea level in standard conditions $V_s = 330$ m/s $= 1.225$ km/h.

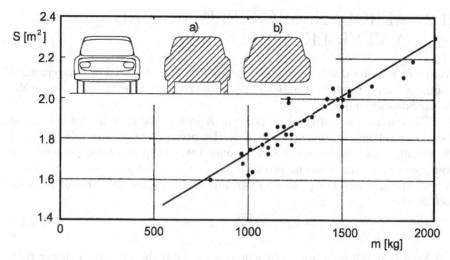

FIGURE 21.4. Area of the frontal projection of the vehicle as a function of its mass. a) Defnition of the frontal area; b) definition of the maximum cross-sectional area.

The mentioned SAE recommendation states that the frontal projected area, which should include the tires and the underbody parts, must be used. The reference surface is usually determined by using optical methods, by projecting a light beam on a screen, moving it to follow the outer shape of the vehicle. A simple but sometimes imprecise way of obtaining its value is

$$S = \psi b h \,, \tag{21.12}$$

where the value of coefficient ψ is about 0.81 and b and h are the width and the height of the vehicle. The area of the frontal area of various cars is reported as a function of their mass in Fig. 21.4. The points are well aligned on the straight line

$$S = 1.18 + 0.00056m \,, \tag{21.13}$$

where the surface is measured in m^2 and the mass in kg. The surface depends little on the mass, and its sensitivity can be measured by the derivative

$$\frac{dS}{dm} = 0.00056 \,. \tag{21.14}$$

The reference length l is usually the wheelbase, but in the expression of moment M_x the track t is often used.

The aerodynamic coefficients used in motor vehicle aerodynamics are those of the forces and moments decomposed along the vehicle axis system xyz: The longitudinal force coefficient C_x, the side force coefficient C_y, the normal force coefficient C_z, the rolling moment coefficient C_{M_x}, the pitching moment coefficient C_{M_y}, the yawing moment coefficient C_{M_z}.

21.2 AERODYNAMIC FIELD AROUND A VEHICLE

Consider a saloon car like the one sketched in Fig. 21.5. As usual in aerodynamics, assume a "wind tunnel" situation, i.e. consider the vehicle as stationary while the air flows around it.

The stream has a stagnation point at A, where the flow divides below and above the vehicle; in the vicinity of A the pressure takes the value p_{tot}. In the vicinity of B, the pressure takes values lower than the total pressure and even lower than the ambient pressure p_0, as the velocity increases, as shown in Fig. 21.6b, where the pressure distribution is reported in terms of pressure coefficient

$$c_p = \frac{p - p_0}{\frac{1}{2}\rho V_r^2} = 1 - \frac{V^2}{V_r^2} \ . \tag{21.15}$$

Note that the pressure coefficient is negative if the pressure is lower than the ambient pressure.

Remark 21.3 *As already stated, pressure variations $p - p_0$ are extremely small when compared to atmospheric pressure; however, their small value must not lead to the conclusion that aerodynamic forces are small: An overpressure equal to 0.5% of atmospheric pressure, like the one present at the stagnation point at 100 km/h acting on a surface of 1 m^2, yields a force of 500 N.*

After point C, located between B and the lower edge of the windshield, the flow detaches from the surface, to attach again at point D on the windshield.

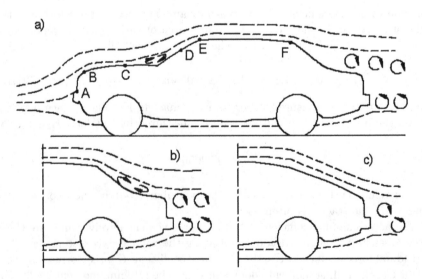

FIGURE 21.5. Streamlines about a passenger vehicle in the symmetry plane.

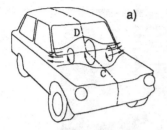
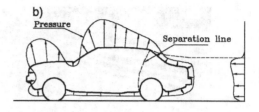

FIGURE 21.6. (a) Separation bubble on the windscreen of a car. (b) Pressure distribution on the symmetry plane of a saloon car and in the wake.

A separation bubble is formed between points C and D. The pressure in such a turbulent zone is fairly high, and it is reasonable to locate the intakes for ventilation of the passenger compartment there (Fig. 21.6). The separation bubble can be reduced by reducing the inclination of the windshield, which can be done only up to a limit since it may reduce visibility, or by increasing the transversal curvature of the windshield and of the hood. A curved windshield is effective in reducing drag but costs and also weighs more than a simple, flat one.

On the roof the pressure is again low, with a distribution that depends on its shape and curvature. At the end of the roof, the flow must slow down and the pressure should rise. In these conditions, the flow easily detaches and any surface irregularity can trigger the formation of the wake.

In Fig. 21.5a, the separation point has been located at the rear edge of the roof. There are cases in which the flow attaches again to the back of the trunk, giving way to a second separation bubble (Fig. 21.5b).

In the case of fastback cars with a sufficiently sloping back, the flow can remain attached up to the end of the body, giving way to a very small wake (Fig. 21.5c). The two situations are shown in the pictures of Fig. 21.7, obtained by visualizing the streamlines using smoke in a wind tunnel test.

The streamlines shown in Fig. 21.5 describe the situation occurring in the plane of symmetry. Outside this plane, the flow is no longer two-dimensional and tends to surround the vehicle at the sides as well.

This effect is generally beneficial and must be encouraged, as it tends to reduce all aerodynamic forces, giving a suitable curvature in the transverse direction to all surfaces. As already stated, point C can be moved further back by allowing the air to flow to the sides of the hood by lowering the fenders and giving them a curved shape; point D can be lowered by using a curved windshield. This results in a reduced separation bubble (Fig. 21.6).

The tridimensional flow on the back of the vehicle can cause vortices, as shown by tests on slanted blocks (Fig. 21.8). If angle α in the figure is lower than a critical value (about 62°), the flow separates abruptly, while for higher values the flow becomes strongly tri-dimensional and the streamlines which flow along the sides wind up in two large vortices while those flowing on the roof are deflected

FIGURE 21.7. Streamlines in the symmetry plane about two fastback cars. In (a) the flow detaches at the end of the roof while in (b) it remains attached up to the end of the trunk.

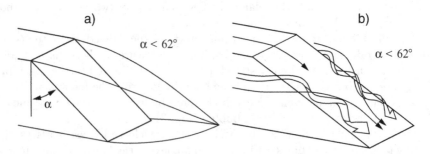

FIGURE 21.8. Flow on the back of slanted blocks.

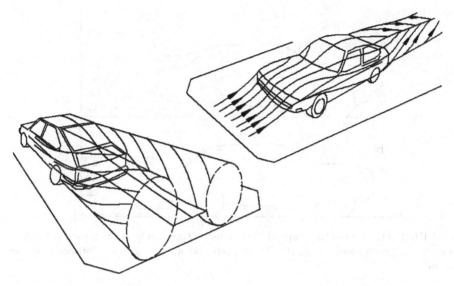

FIGURE 21.9. Qualitative pattern of the vortices behind a vehicle.

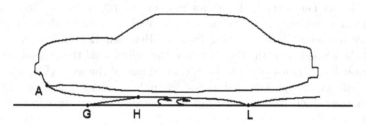

FIGURE 21.10. Flow below the vehicle. Boundary layer formation.

downwards and follow the tail of the vehicle. The flow in the symmetry plane, which is of the type shown in Fig. 21.9, is similar to that shown in Fig. 21.5c.

The wake is smaller, but this does not mean that the drag is lower: The pressure in the vortices is low, as is that on the centre of the tail since the flow is very fast in that zone: The overall pressure behind the vehicle can be even lower than that characterizing a large wake due to a small angle α.

The flow under the vehicle can be quite complicated and depends on many factors like the distance between vehicle and ground and the presence of a fairing under the body. Wind tunnel simulations can be misleading since in actual use the ground is stationary with respect to the air, at least if there is no wind, and not with respect to the vehicle, as occurs in wind tunnels.

In actual use, starting from the stagnation point A the boundary layer gradually thickens (Fig. 21.10). Outside the boundary layer, the velocity of the flow is different from that of the free stream, i.e., the flow is no longer at rest with respect to the ground, and from point G a second boundary layer appears on the ground as well.

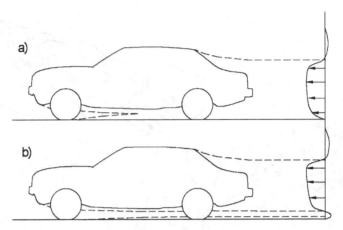

FIGURE 21.11. Effect of the shape of the bottom of the vehicle on the wake. (a) Bottom close to the ground and rough; (b) Streamlined bottom, at a greater distance from the ground.

Depending on the distance between the vehicle and the ground, the two boundary layers can meet in H or can remain separated. In the first case the flow is blocked and the air under the vehicle tends to move with it, giving way to another boundary layer starting from L. Between H and L a vortex may result. In the second case, the flow between the vehicle and the ground decreases aerodynamic lift, because of both the decreased size of the wake (Fig. 21.11) and the lower energy dissipation; if it is fast enough it causes a negative lift. The flow below the vehicle reduces the drag also, because the pressure in the wake is increased.

All improvements which facilitate the flow under the vehicle have these effects: Either the distance between vehicle and ground is increased or the bottom is given a curved shape, in the longitudinal or transverse direction, or the bottom is supplied with a smooth fairing covering the mechanical elements that are usually in the airflow. The last device may reduce the drag up to about $10 \div 15\%$, as shown in Fig. 21.12, but is seldom used in passenger cars as it is more difficult to reach the mechanical elements, making maintenance more costly.

These considerations cannot be generalized since any change of shape aimed at modifying the aerodynamic field at one point has an influence on the whole aerodynamic field, with effects that are difficult to predict.

Two effects can modify the airflow around the vehicle and make it more complicated: Wheel rotation and the presence of internal flows.

Consider a cylinder rotating and moving in directions consistent with those of a rolling wheel (Fig. 21.13a). It generates a drag and a lift (the Magnus effect)[5]

---

[5]A cylinder rotating with its axis perpendicular to the stream entrains, owing to viscosity, a certain quantity of air in rotation. On one side, the rotation velocity adds to the velocity of the stream; on the opposite side it subtracts. Where the velocity is higher the pressure is

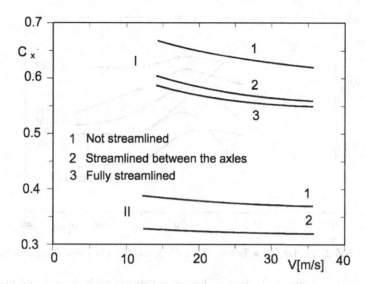

FIGURE 21.12. Effect of streamlining the bottom of the vehicle on the drag coefficient for two vehicles, I and II.

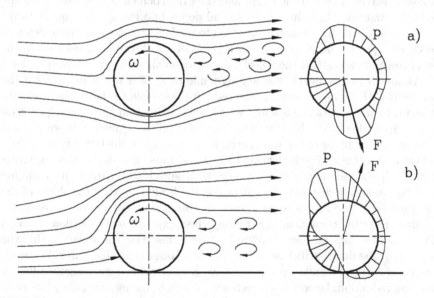

FIGURE 21.13. Streamlines, pressures and aerodynamic force acting on a wheel, modelled as a rotating cylinder, far from the ground (a) and in contact with it (b).

lower, with the effect on the other side. This pressure difference produces a force perpendicular to the axis of the cylinder and to the direction of motion. This effect is usually referred to as the Magnus effect.

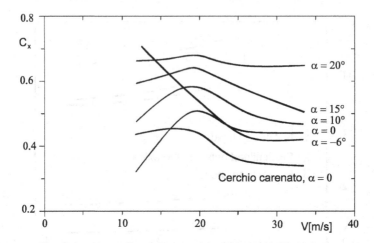

FIGURE 21.14. Drag coefficient of a rolling wheel. α is the sideslip angle.

which is directed downwards. If the wheel is in contact with the ground, however, the streamlines are completely changed by the presence of the latter and the lift becomes positive. The wake is larger and the drag coefficient increases; both the size of the first and the value of the second depend on the speed (Fig. 21.13b).

There is also an increase in drag owing to the larger wake, whose size depends also on the speed. The value of coefficient C_x of a rolling wheel, referred to the area of the cross section of the wheels, is plotted against the speed in Fig. 21.14.

As shown in the figure, the aerodynamic drag of a wheel increases if the wheel rolls with a sideslip angle measured with reference to the relative velocity of the air. In the case of the isolated wheel, this means that the drag depends on the sideslip angle of the wheel, while in normal conditions the flow is not parallel to the symmetry plane of the wheel even if the sideslip of the latter is zero, since the flow under the vehicle is deflected sideways. This effect is, in general, larger for front wheels and causes an increase in their aerodynamic drag. Streamlining the wheels in such a way to reduce drag has the limitation that the shape of the hubs must be studied so as to guarantee an appropriate cooling of the brakes.

Since the drag coefficient of a rolling wheel exposed to the airflow is about 0.45, it would seem that there is an advantage in inserting the wheels within the body only if the drag coefficient of the vehicle is lower than that value. However, all vehicles except formula racing cars have covered wheels for reasons different from drag reduction. Uncovered wheels are present in racing cars only when rules explicitly dictate. In Formula 1 racers, up to 45% of the aerodynamic drag can be ascribed to the wheels.

A sketch of the streamlines around a partially covered wheel is shown in Fig. 21.15, together with a plot of coefficients C_x and C_z versus the ratio h/D between the amount of wheel covered and its diameter. The curves are experimental and, particularly as related to C_z not very reliable owing to the method used to simulate the presence of the ground, but the results are at least qualitatively significant.

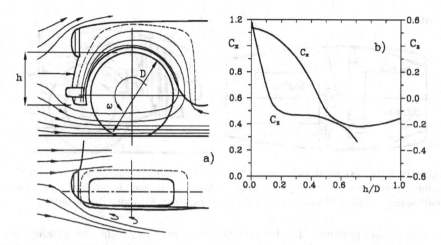

FIGURE 21.15. (a) Flow in the cavity around a covered wheel. (b) Aerodynamic coefficients of the wheel as functions of ratio h/D.

The advantage of covering the wheel, without exceeding a value $h/D = 0.5 \div 0.7$, is clear. The values of C_x are generally very high, particularly if compared with those of an isolated wheel, and the increase in drag when the wheel is largely covered can be explained by viscous effects within the fender.

Another reason for the deviation of the aerodynamic field from that shown in Fig. 21.5 is the presence of internal flows. There are usually two separate flows inside the vehicle: A cooling flow in the engine compartment, and a flow in the passenger compartment; other internal flows of lesser importance are those aimed at cooling mechanical devices such as brakes or the oil radiator, if it is located separately from the main radiator, etc.

The second flow is of lesser importance: If the intake is set at the base of the windshield and the outlet is in a zone in the wake, the result can be that of reducing the drag slightly, as this configuration reduces the pressure in the separation bubble and increases that in the wake.

A larger amount of air is needed for engine cooling. A good solution would be to use a radiator of the type common in water-cooled aircraft piston engines, in which a diffuser slows down the flow that is driven through the heat exchanger before being accelerated again in a converging duct (Fig. 21.16a). In motor vehicles, a fan allowing cooling with the vehicle stationary must also be provided. The diffuser should be long enough to allow the flow to be slowed down without separation (a slope of about 7° has been found to be a practical maximum) and the fan should operate only at speeds lower than those for which the system has been designed.

In practice, this solution cannot be used, at least on normal vehicles: A system of this type would be too long to be accommodated in the hood; instead there is a short diffuser whose opening is too large to allow a good attached flow, followed by a radiator. The flow then goes directly into the engine compartment

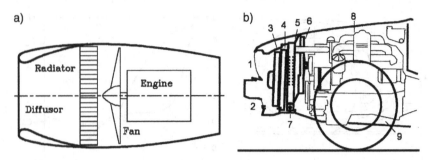

FIGURE 21.16. (a) Ideal radiator. (b) Actual layout of the cooling system in the engine compartment. 1) upper air intake; 2) lower air intake; 3) auxiliary fan; 4) air conditioning radiator; 5) radiator; 6) fan; 7) oil radiator; 8) engine; 9) air outlet.

without further guidance. The internal flow then mixes with the flow passing under the vehicle in a very disordered way. This situation is sketched in Fig. 21.16b. The complexities needed for obtaining a well guided flow, separated from the external flow, are considered not worth the added cost and weight and the difficulties they would add to maintenance operations in the engine compartment.

The presence of the internal flow in the engine compartment has a non-negligible influence on drag, lift, pitching moment and, although to a lesser extent, yawing moment. It can account for about 13% to 20% of the total drag; the increase in lift (generally positive, i.e. upwards) is even larger. The effect on moments is to move forward both lift (pitching moment) and side force (yawing moment). As will be seen later, both effects are detrimental to the overall behavior of the vehicle.

Aerodynamic testing should always be performed on models which reproduce the inside of the engine compartment as well or, better, on the actual vehicle, with open air intakes. Since the engine temperature affects the internal flow, aerodynamic testing should be done with the engine at running temperature.

21.3   AERODYNAMIC DRAG

As already stated, aerodynamic drag is the component of the aerodynamic force acting in the direction of the relative velocity, and thus the force that opposes the motion of the body in the fluid. If the relative velocity is confined to the symmetry plane (motion with no sideslip, and no lateral wind) the difference between drag and force F_x is quite small; this is due to the fact that the angle between the x-axis and the plane of the road is small, and that the aerodynamic efficiency, that is, the ratio between lift and drag, of motor vehicles is very low, if not equal to zero. In the case of road vehicles, the two are sometimes confused and force F_x is referred to as drag.

Remark 21.4 *In many cases, drag is considered positive when directed backwards, which is inconsistent with the general conventions on forces.*

Aerodynamic drag can be considered as the sum of three terms: Friction drag, shape drag and induced drag. Coefficient C_x can be similarly considered as the sum of the three corresponding terms

$$C_x = C_{x_a} + C_{x_f} + C_{x_i} \, . \tag{21.16}$$

While in aeronautics this subdivision is practically important, since the three terms can be computed separately in the various flight conditions, in the case of motor vehicles they cannot actually be separated. To consider them one by one is important only insofar it allows one to understand how the various components of the drag originate.

21.3.1 Friction drag

Friction drag is the resultant of the tangential forces acting on the surface

$$\int_S \vec{t}_t \times \vec{i} \, dS \, .$$

Since it is practically impossible to measure the friction drag on a body with complex geometry, reference is usually made to flat plates, where the only drag present is friction drag. Friction drag coefficient C_f, referring to the "wet" surface, i.e. to the surface exposed to the fluid, is plotted versus the Reynolds number, computed with reference to the length of the plate, as shown in Fig. 21.17.

The two straight lines (in the logarithmic plot) refer to a laminar and a turbulent flow in the boundary layer. They are approximated by the empirical relationships

$$C_f = \frac{1.328}{\sqrt{\mathcal{R}_e}} \, , \text{ or } C_f = \frac{0.074}{\sqrt[5]{\mathcal{R}_e}} \, , \tag{21.17}$$

respectively.

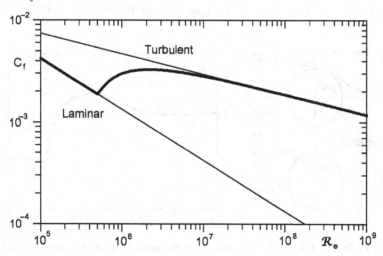

FIGURE 21.17. The friction coefficient referred to the wet surface versus the Reynolds number.

The flow is laminar if it is free from vorticity and there is no mixing between adjacent streamlines. The vortices which are present in a turbulent boundary layer are very small, but cause a mixing and a strong energy transfer within the layer. If the fluid is free from vorticity when it enters into contact with the plate, a laminar flow is maintained up to values of the Reynolds number of about 500,000, provided that surface irregularities do not trigger turbulence. If the Reynolds number is higher, at least a part of the plate experiences a turbulent flow; the transition is shown in Fig. 21.17, occurring where the local Reynolds number, computed with the distance from the leading edge, reaches a value of 500,000.

In the case of streamlined bodies, it is expedient to maintain a laminar boundary layer as long as possible to reduce friction drag. However, in the case of blunt bodies, it often happens that a laminar boundary layer results in higher drag than a turbulent one. This is due to the fact that in a laminar layer the fluid which is in immediate contact with the surface receives less energy from adjacent layers and tends to slow down more quickly. Particularly in cases where the flow outside the boundary layer slows down and the pressure subsequently increases, a thickening of the boundary layer which eventually results in the detachment of the flow and the formation of a wake takes place. This eventually occurs in the case of the turbulent layer as well, but the energy exchanges due to fluid mixing within the boundary layer help to maintain the flow attached to the surface for a longer distance.

The drag coefficient of a sphere is plotted as a function of the Reynolds number in Fig. 21.18, together with a sketch of the streamlines for the cases of laminar and turbulent flow.

The flow around motor vehicles is always turbulent, owing to the presence of vortices in the air near the ground due to other vehicles and, above all, if there is wind, to the ground and fixed obstacles. Vehicles actually move in what can be defined as the boundary layer of the Earth's surface. Even if it were

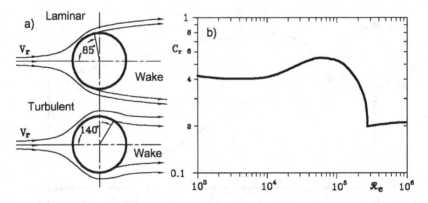

FIGURE 21.18. (a) Qualitative sketch of the streamlines around a sphere. (b) Drag coefficient of a sphere as a function of the Reynolds number.

expedient to keep the boundary layer laminar, it would be very difficult to do so. The percentage of the drag due to friction is usually low, on the order of 10% of the total aerodynamic drag.

21.3.2   Induced drag

Induced drag is that portion of aerodynamic drag that is linked with the generation of lift. In aeronautics, it plays the same role that rolling resistance plays in motor vehicle dynamics: It is responsible for the energy that is dissipated to support the vehicle during motion.

In the case of road vehicles, aerodynamic lift is not needed, and is actually a nuisance. The induced drag should be reduced to a minimum by reducing lift. An exception is the negative lift produced by aerodynamic devices aimed at increasing the normal force holding the vehicle to the ground: In this case, induced aerodynamic drag adds to increased rolling resistance.

To understand the origin of induced drag, reference can be made to the theory of high aspect ratio (the ratio between the span and the chord) wings attributed to Prandtl. This theory can be applied in many cases to the wings which produce negative lift in racing cars. The lift of a wing is directly linked with a difference of fluid velocity between the upper and the lower surface of the wing, which causes a difference of pressure and ultimately a lift force. The difference of velocity can be thought as a vortex superimposed on the uniform airflow (Figure 21.19a).

If the wing had an infinite span, all sections would experience a two-dimensional flow: No induced drag is produced. In the case of an actual finite-span wing, the vortex cannot vanish at the tips of the wing and its core is simply deflected backwards, creating a wake of vortices. To understand intuitively why this vorticity is generated, it must be considered that air under the wing, whose pressure has increased, tend to move toward the tip, where it goes around the end of the wing toward the upper surface, where pressure is lower. The vortex then winds around the tip edge of the wing, a motion that remains even after the wing has passed (or the stream flows beyond the wing, in the wing tunnel model, where the wing is stationary and the air flows around it), producing a trailing vortex.

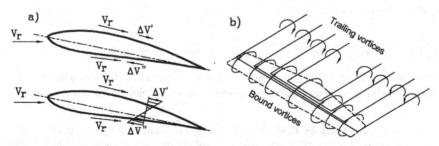

FIGURE 21.19. Vorticity in a lifting wing. (a) Bound vortex; (b) trailing vortices.

A bound vortex plus the two trailing vortices at its end constitute a horse-shoe vortex, like those shown in Fig. 21.19b. Since the vorticity is not constant along the wing, a set of such vortices is produced and the trailing vortices depart at different points along the wing. Actually, rather than of a *set* of vortices, we should speak of a *distribution* of vortices.

The energy dissipation needed for the creation of the trailing vortices explains the presence of the induced drag. Any device which reduces trailing vortices, such as tip plates or modified wing tips, is effective in reducing induced drag. Trailing vortices are sometimes easily visible at the tips of the wings of racing cars.

From the theory of high aspect ratio wings it can be deduced that induced drag is proportional to the square of the lift or, which is equivalent, that the induced drag coefficient is proportional to the square of the lift coefficient. However, in the case of low aspect ratio wings and, above all, blunt bodies, this proportionality no longer holds. The presence of the ground can also modify the pattern of vortices. It has been suggested in the case of road vehicles that it is not possible to define an induced drag and that the term vortex drag is preferable[6]. Whatever the case, the vortices which are created behind a vehicle (Fig. 21.9) are linked with the generation of lift, and a reduction of lift always causes a reduction of the overall aerodynamic drag.

21.3.3 Shape drag

Shape drag is what remains of the drag if the contributions due to friction and induced drag are removed and, in the case of road vehicles, it is mainly due to the wake. The pressure in the wake is low and fairly constant and hence shape drag can be approximately evaluated as the product of the wake pressure by the projection on yz plane of the area exposed to it: The shape of the part of the vehicle in the wake has little importance. This statement must not be misunderstood: The shape of the tail of the vehicle is important to assess where the wake starts, but once this issue is solved, only the extension of the wake matters.

Remark 21.5 *Any geometrical irregularity can precipitate the detachment of the flow and the wake formation, particularly if it is located in a zone in which the flow slows down.*

21.3.4 Aerodynamic drag reduction: passenger vehicles

Since the beginning of motor vehicle technology, several attempts aimed at reducing aerodynamic drag have been made. Shapes developed for aircraft and for

---

[6]R.T. Jones, Discussion on T. Morel, *The effect of base slant on the flow pattern and drag of three-dimensional bodies with blunt ends*, in Aerodynamic drag mechanism of bluff bodies and vehicles, Plenum Press, New York, 1978.

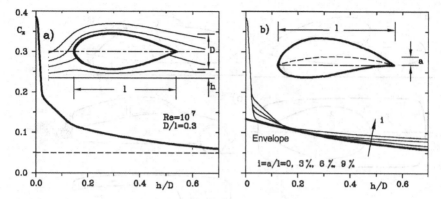

FIGURE 21.20. (a) Streamlines around a straight slender body and values of C_x versus the distance from the ground. (b) Values of C_x versus the distance from the ground for cambered slender bodies with different values of camber.

airships have been adopted, often in a naive way, as the attempt to streamline the body was often offset by mechanical parts completely exposed to the wind.

The shape which proved to produce the lowest ratio between aerodynamic drag and volume was the straight, circular, slender body with a diameter to length ratio equal to about 0.3. For a Reynolds number of about 10^7, its drag coefficient is of 0.05. However, it is difficult to produce a suitable car body from this specification and it is, moreover, optimal only if the vehicle's motion takes place far from the ground. The presence of the latter causes the flow to change substantially (Fig. 21.20a) and the value of C_x to be far higher, up to values of about 0.15 at distances from the ground typical of motor vehicles.

If the vehicle's distance from the ground were zero, the best shape would be half of a slender body, a consideration which seems to have inspired several designs of the past. However, the distance of the vehicle from the ground cannot be zero, and this solution leads to quite high values of the drag.

If the axis of the slender body is curved, a lower resistance to motion near the ground results (Fig. 21.20b), with an optimum value of the camber ratio a/l existing for each value of the distance from the ground. The optimum value of the camber ratio for a nondimensional ground clearance $h/D = 0.1$ is about 10%. However, the difficulties in adapting a slender body to a vehicle and housing the mechanical components and the wheels in it remain.

The results obtained by Lay in 1933 through a series of wind tunnel tests performed on modular models are still interesting. His basic shapes were a flat plate perpendicular to the current (he found a value of C_x larger than 1.2), a slender body (whose C_x was measured at 0.08), a rectangular box ($C_x = 0.86$), and a vehicle from his era ($C_x = 0.6$).

He then discovered that a slight rounding of the corners of the box resulted in a decrease of C_x to 0.46. By fitting different front and rear parts to the vehicle model, he saw not only that by shaping both parts suitably could aerodynamic drag be reduced, but also that the shaping of the rear part is more important

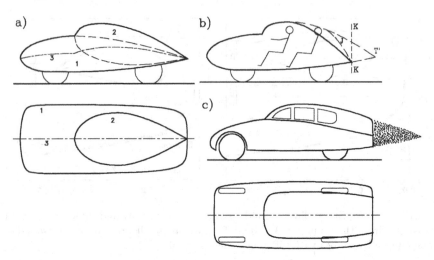

FIGURE 21.21. Streamlined car bodies following J and K shapes. (a) J-shape; (b) problems linked to the length of J-shapes and cutting of the tail; (c) K-shape: One of the drawings from the original patent by Koënig.

than that of the front part. Also, using the shapes tested by Lay, it is necessary to accept a very long vehicle to obtain low drag.

Ten years earlier, in 1923, Jaray obtained a patent in Germany for a car body made by a rectangular cambered stub wing with a slender body superimposed on it (Fig. 21.21a). This approach, named the J-shape, allowed the wheels and other mechanical components to be housed easily, but the problems related to the length of the vehicle, if a sufficient height for the passengers in the rear seats was required, were not solved. The centre line of the body was also quite curved, resulting in non-negligible lift and induced drag. However, the J-shape can be easily identified in many vehicles beginning in the 1950s, like the Lancia Stratos, Citroen DS and many coupé built by Porsche.

In 1937, a new approach was patented almost simultaneously by Kamm and by Koënig and Fachsenfeld Reinhard. From the observation that to obtain optimum height in the back of the J-shape without a very long vehicle, a shape which is prone to produce a large wake is obtained, and that the shape of the part of the vehicle in the wake has little significance, they suggested that the long streamlined tail of the J-shape could be truncated, following line KK of Fig. 21.21b. The result is shown in Fig. 21.21c, from the original patent[7].

Truncation does not affect shape drag, if the part cut off was already in the wake, and likely reduces lift and induced drag. This statement is a rough approximation, since any change in a part of the vehicle changes the entire aerodynamic field, but the use of the K-shape allowed designers to reduce the drag of many passenger vehicles. Many cars of the 1970s had essentially a K-shape, like

---

[7] Brevetto industriale n. 352583 - Carrozzeria per automobili.

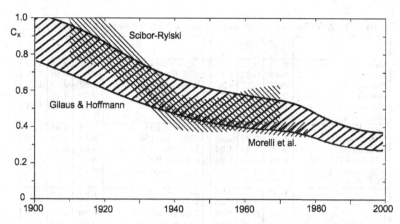

FIGURE 21.22. Experimental values of C_x for passenger vehicles versus the year of construction.

the Citroen GS and CX, Lancia Beta and Gamma, Alfa Romeo Alfasud, Rover 3 liters and many others.

The aerodynamic evolution of passenger vehicles, in terms of the C_x coefficient, is summarized in Fig. 21.22: The three hatched zones refer to plots obtained by different Authors: Gilhaus and Hoffmann[8] refer to cars of different types, Morelli et al.[9] report on the drag of cars with a sound aerodynamic shape, many of which were built following the J- or K-shape, while the data reported by Scibor Ryilski[10] are related to a larger sample containing cars of all types. There is no contradiction between them: While the best vehicles showed a constant progress towards low drag, the availability of more powerful engines and the low cost of energy caused a decrease in the movement towards better aerodynamics in the fifties, with an average increase in the drag of cars.

Remark 21.6 *Actually, what really matters is not the value of coefficient C_x, but the value of product SC_x: to reduce drag it is possible to search for a low value of the x-force coefficient or for a shape with low cross sectional area.*

Any device aimed at reducing the value of C_x that causes an increase of the frontal area, like a plate at the rear edge of the roof, is effective only if the decrease of the former is larger than the increase of the latter.

Some values of both coefficient C_x and of the product SC_x, allowing a more direct comparison, for more modern cars are reported in Table 21.2[11]. By comparing the values in the table with those reported in Fig. 21.22, the progress which occurred in the 1980s, mainly linked with the increase of the cost of energy

[8] A. Gilhaus and R. Hoffmannm, *Directional stability,* in W.H. Hucho (Ed.), *Aerodynamics of Road Vehicles*, SAE, Warrendale, 1998.

[9] A. Morelli, L. Fioravanti, A. Cogotti, *Sulla forma della carrozzeria di minima resistenza aerodinamica*, ATA, Nov. 1976.

[10] A.J. Scibor Ryilski, *Road Vehicle Aerodynamics*, Pentech Press, London, 1975.

[11] H.P. Leicht, *Kanal voll* , Auto-motor sport, 18/1986.

TABLE 21.2. Values of C_x. S and product SC_x for some European cars.

	$C_X S$ [m²]	C_X	S [m²]		$C_X S$ [m²]	C_X	S [m²]
Lancia Y10	0.57	0.33	1.76	Opel Corsa SR	0.61	0.35	1.73
Fiat Uno	0.62	0.34	1.83	VW Polo	0.65	0.38	1.70
Renault 5	0.67	0.37	1.80	Austin Metro	0.67	0.39	1.73
Peugeot 205	0.68	0.39	1.74	Fiat Panda	0.70	0.41	1.70
Citroen Visa	0.70	0.40	1.75	Ford Fiesta	0.73	0.41	1.76
Renault 4	0.90	0.49	1.83				
Opel Kadett GSi	0.60	0.32	1.88	Peugeot 309	0.64	0.34	1.86
VW Golf GL	0.65	0.34	1.89	Mercedes 190 E	0.65	0.34	1.89
Renault 21	0.66	0.34	1.94	Ford Sierra XR 4i	0.67	0.34	1.98
VW Golf GTI 16V	0.67	0.35	1.91	Citroen BX	0.68	0.36	1.91
VW Jetta CL	0.68	0.36	1.89	VW Passat GL	0.70	0.37	1.90
Fiat Ritmo	0.70	0.37	1.88				
Opel Omega	0.58	0.28	2.06	Mercedes 200	0.60	0.29	2.07
Audi 100	0.62	0.30	2.05	Renault 25	0.62	0.31	2.03
Ford Scorpio	0.70	0.35	2.02	Fiat Croma	0.70	0.34	2.04
Lancia Thema	0.73	0.36	2.06	Honda Prelude 16V	0.76	0.41	1.84
Alfa 90	0.77	0.40	1.92	Citroen CX	0.78	0.40	1.96
Mitsubishi Galant	0.79	0.40	1.98				
Ferrari Testarossa	0.61	0.33	1.85	Mercedes 190 E2.3	0.64	0.33	1.94
Porsche 944 turbo	0.65	0.35	1.89	VW Scirocco 16V	0.68	0.38	1.78
Porsche 911 Carrera	0.68	0.38	1.77	Mitsubishi Starion T	0.69	0.37	1.84
Alfa Romeo GTV	0.71	0.40	1.77	Jaguar XJ-S	0.73	0.40	1.83
Porsche 928 S	0.77	0.39	1.96	Audi Quattro	0.80	0.43	1.86
BMW M 635 CSi	0.80	0.40	2.00				

which took place a decade earlier, is clear. It must be noted, at any rate, that with a few exceptions, the values of C_x are rarely lower than 0.35, with many cars having a value between 0.35 and 0.45.

The search for a shape of minimum drag for any particular vehicle can be approached by identifying a number of critical details and optimizing them one by one. The principle of effects superimposition cannot be applied in aerodynamics; the drag of a body is not the sum of the drag of all its parts, and any change to one of them causes a change in the drag of all others. However, it is a common practice to obtain the drag of a body as the sum of the drag of its parts plus a further component, referred to as interference drag. This approach has been successfully used first in aeronautics and then in road vehicle technology.

A method, known as detail optimization is now widely used. It is based on subsequent detail modifications to achieve drag reductions which can be quite substantial. The drag coefficient of the base shape is measured and a number of specific details of the car body are chosen. One of them is modified and the wind tunnel test is repeated, with modifications continuing until a minimum of the drag is obtained. The work is then repeated for all the chosen details. This procedure can be thought as the search for the minimum of a function (drag) of many variables (the geometrical characteristics of the details), by modifying the value of each one of the variables one at a time, looking for a local minimum in a two-dimensional space and then proceeding to search for a local minimum in a multi-dimensional space. We are not saying that an absolute minimum can be reached in this way, nor even that a relative minimum can be obtained unless an iterative procedure is used, but it is certain that a decrease in drag is eventually obtained.

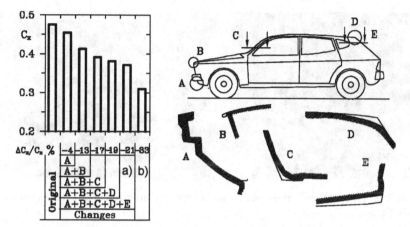

FIGURE 21.23. Detail optimization method. Definition of the five details used to optimize the shape and to reduce the C_x coefficient. (a) Optimized shape, (b) modified shape.

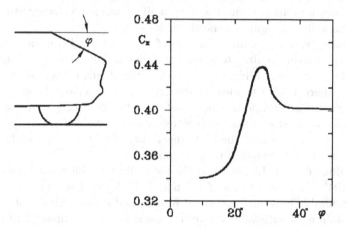

FIGURE 21.24. Coefficient C_x as a function of the angle between the rear window of a given car and the horizontal.

An example is shown in Fig. 21.23, where the base shape and five details are shown. The thin line gives the initial configuration while the thick one describes the optimized shape. By operating in the way above, the drag was reduced by about 21%, while a more substantial reduction of 33% could be achieved only by introducing modifications which changed to a larger extent the overall appearance of the car.

The results obtained by changing the inclination of the rear part of a hatchback car are shown in Fig. 21.24: If the angle between the rear window and the horizontal is larger than 35° the value of C_x is 0.4. With a low value of the angle it is possible to reduce the drag, with a coefficient of about 0.34, but there is a region, at about 28° where large vortices produce a substantial increase in drag,

FIGURE 21.25. Ideal and actual shape for a very low-drag research vehicle, shown in the wind tunnel.

up to $C_x = 0.44$. This result confirms the already mentioned results obtained from slanted rectangular blocks.

The advantage of this procedure is in leaving stylists free to design the vehicle without inhibiting their creativity, while obtaining a shape with low drag without drastically changing the aesthetic impact of the car.

Alternatively, it is possible to develop from theoretical considerations ideal shapes aimed at reducing drag to a minimum, modifying these ideal shapes when adapting them to motor vehicle use. An example of this procedure is shown in Fig. 21.25, where both the ideal shape and the car derived from it are represented[12]. The ideal shape has been obtained by specifying that the lift and pitching moment must be zero, the positive pressure gradients must be as low as possible, the cross section of the body must vary slowly in shape and area, and its contour must be rounded as much as possible.

The value of C_x of the ideal body in the vicinity of the ground proved to be as low as 0.049, the same as that of a slender body located at an infinite distance from the ground. The vehicle obtained from it had a value for C_x of only 0.23, while maintaining a satisfactory internal space for the occupants, the luggage and the mechanical components, not unlike a regular saloon car.

As a result of research performed in the eighties in many countries, it is possible to state that values of coefficient C_x as low as 0.25 can be achieved without excessive sacrifice to the internal space and general characteristics of the vehicle. The success of detail optimization procedures allowed designers to overcome fears, expressed several times in the 1970s, that aerodynamic studies would cause all cars to look alike, and that the image of individual manufacturers would be sacrificed to the need for good aerodynamic performance.

It is also clear, however, that the lower the drag coefficient is, the fewer are the advantages of further reductions in fuel consumption, particularly since

---

[12]CNR, Progetto Finalizzzato Energetica, Atti del primo seminario informativo, Torino, Aprile 1978 and A. Morelli, L. Fioravanti, A. Cogotti, *Sulle forme della carrozzeria di minima resistenza aerodinamica*, ATA, Nov. 1976.

the average use of road vehicles occurs at speeds lower than those at which aerodynamic drag is the most important form of resistance to motion.

As the actual average speed of driving depends mostly on issues unrelated to the design of vehicles themselves (road conditions, laws and their enforcement etc.), at present the search for very low drag is not pushed to extremes, at the expense of other characteristics, mainly aesthetic, in the design of passenger vehicles.

21.3.5   Aerodynamic drag reduction: industrial vehicles

Even in the recent past, industrial vehicles have usually been designed with little concern for their aerodynamic characteristics. The low speed and the high value of the ratio between the mass and the frontal area renders the power needed to overcome aerodynamic drag a small fraction of the total power needed for motion. However, the higher speeds industrial vehicles reach on highways and increased concern about energy saving have led to many studies aimed at reducing the drag in this field as well.

In the case of single-body vehicles, such as buses, vans and trucks without a trailer, the basic shape is essentially a square box. If the edges are blunt, the value of C_x is in the range 0.82–0.86, mostly owing to the fact that the flow is widely separated and the wake is very large (Fig. 21.26).

From this figure it is clear that any change of shape allowing flow separation to be reduced is very beneficial: Simply by rounding the edges slightly, the drag coefficient can be reduced almost by half, to about 0.45.

With an improved frontal profile, values in the range 0.4÷0.43 may be obtained for buses and vans. It is difficult to lower this value further. Fairings on the underside may make it possible to lower it by about 0.05, but many other devices aimed at increasing the pressure in the wake have been tried without success.

Tests on models have shown even greater reductions of drag: by simple rounding of the corners with a 150 mm radius, a value as low as $C_x = 0,36$ has been obtained.

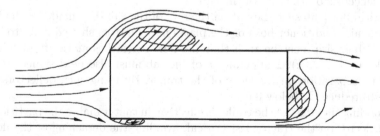

FIGURE 21.26. Streamlines around a square box with blunt edges, at a distance from the ground equal to 0.06 D, where D is the diameter of a circle having the same cross-sectional area of the body.

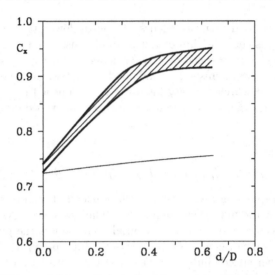

FIGURE 21.27. Values of the coefficient C_x versus ratio d/D for articulated vehicles with different cab shapes. d is the distance between the cabin and the semitrailer and D has the same meaning as in the previous figure.

Two flow stagnation points can be present in articulated vehicles, one on the cab and one on the trailer, with a flow occurring in the space between the two bodies. The drag depends largely on their distance d, as shown in Fig. 21.27, referring to articulated trucks. The dashed zone contains experimental points obtained with differently shaped cabs.

The value of C_x increases from about 0.72 to about 0.93 when the distance grows from $d = 0$ to $d = 0.6$. To reduce drag, the flow between cab and trailer must be blocked to create a single stagnation point; the simplest way is to put a vertical flat plate on the roof of the former (the thin line in Fig. 21.27).

The flow is similar to that characterizing the situation with $d = 0$; there is a single stagnation point on the tractor and a separation bubble between the two bodies. A second separation bubble is formed on the roof of the trailer, owing to the front edge of the latter. Further decrease of drag may be obtained using shaped deflectors on the roof of the cab.

Vertical flat plates can be put on the roof of the cab (e.g. made by the front side of the air conditioner box) or the plates may be true, shaped deflectors. At any rate, their size must be such that the flow reattaches at the front edge of the trailer (Fig. 21.28c). The planform of the cab must also be designed so that there is no separation on the sides of the trailer. By using such deflectors it is possible to reduce C_x below 0.6.

Rounding the edges of the trailer is effective, in that it makes the partitioning of the deflector on the tractor less critical, avoiding the detaching of the flow if the deflector is too low. At any rate it is more important designing correctly the tractor and rounding the edges of the trailer than streamlining the two parts independently.

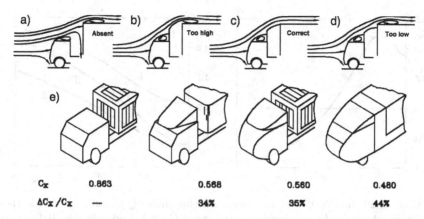

FIGURE 21.28. (a-d) Flow on the front part of an articulated truck with and without a flat plate (air deflector) on the cab. (e) Reduction of the drag of articulated vehicles; values obtained on half-size models.

Some results obtained on half scale models are shown in Fig. 21.28e; they show that with just few modifications, including a deflector, a large reduction is readily obtained, while further improvements are difficult to achieve and require a global streamlining of the vehicle. A value of C_x of about 0.5 can thus be obtained; lower values, down to 0.24 as obtained on a 1/10 scale model, could be achieved by using a complete fairing of the underside.

The above mentioned values of C_x all referred to vehicles driven with no aerodynamic sideslip angle. If β_a is not zero, the aerodynamic field and its associated aerodynamic forces and moments are quite different. This is true for all vehicles, but holds in particular for industrial vehicles, since the lower speed they usually travel increases the significance of side winds, even those of moderate strength. As an example, a wind perpendicular to the road blowing at $V_a = 10$ km/h causes an aerodynamic sideslip angle $\beta_a = 8°$ in the case of a vehicle travelling at $V = 70$ km/h.

A qualitative plot of C_x as a function of β_a for a generic industrial vehicle is shown in Fig. 21.29a. It must be stressed, however, that if sidewind is present, the sideslip angle β is not zero. The aerodynamic force F_x is then not exactly aligned with the velocity V, and the power needed for overcoming aerodynamic drag is $F_x V \cos(\beta)$ plus the component of the side force F_y in the direction of velocity V. Moreover, the rolling resistance of the tires increases due to the sideslip angle of the wheels.

The aerodynamic drag of articulated vehicles increases strongly with the aerodynamic sideslip angle β_a for angles between 0 and 20°. This increase is particularly noticeable for articulated trucks with aerodynamic devices aimed at preventing airflow between tractor and trailer. When lateral wind is present a flow of air in transverse direction between the cab and the trailer can be created and the advantages obtained by using deflectors end up disappearing for values of β_a of about 20°. Curved deflectors work better from this viewpoint than flat

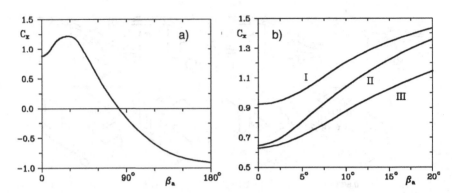

FIGURE 21.29. (a) Coefficient C_x versus sideslip angle β_a for an articulated truck. (b) Effects of drag reducing devices on the curve $C_x(\beta_a)$: I no deflector; II flat plate deflector; III deflector and round front edges of the trailer.

plates, which usually cause an increase of drag at even low values of β_a. Curves $C_x(\beta_a)$ for articulated trucks with different aerodynamic devices are shown in Fig. 21.29b. As shown, the use of deflectors must be accompanied by the rounding of the edges of the trailer.

21.4   LIFT AND PITCHING MOMENT

Apart from the drag increase due to induced drag, aerodynamic lift must be avoided since it reduces the load on the tires and consequently the forces the vehicle can exert on the ground; moreover, this reduction is strongly dependent on the speed. In the case of vehicles with high power/weight ratio, it is possible to use negative aerodynamic lift to enhance power transfer through the road-wheel contact. The same holds for increasing the cornering forces.

In addition, aerodynamic pitching moment M_y must be as small as possible, since it causes strong variations in the forces exerted by the wheels on the road, that depend on speed. With reference to Fig. 21.30, the pitching moment is positive when it acts to increase the load on the front wheels. As the aerodynamic drag is applied to the centre of mass, at a distance h_G from the ground, the longitudinal load transfer on a vehicle with two axles is

$$
\begin{cases}
\Delta Z_1 = \dfrac{1}{2}\rho V^2 S\left(C_{M_y} - \dfrac{h_G}{l}|C_x| - \dfrac{b}{l}C_z \right) \\[2mm]
\Delta Z_2 = \dfrac{1}{2}\rho V^2 S\left(-C_{M_y} + \dfrac{h_G}{l}|C_x| - \dfrac{a}{l}C_z \right).
\end{cases}
\tag{21.18}
$$

Instead of speaking of lift and pitching moment, the lift is often subdivided on the two axles and a front axle F_{z1} and rear axle F_{z2} lift are defined. Similarly, the lift coefficient C_z is split into two coefficients C_{z1} and C_{z2} with reference to the axles. The evolution in time of C_z in passenger cars, split into its components

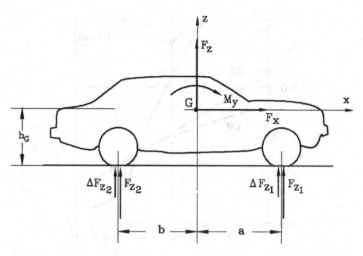

FIGURE 21.30. Longitudinal load transfer due to aerodynamic pitching moment and lift. Forces F_z and ΔF_{z_i} are the forces the vehicle receives from the ground; a positive ΔF_z indicates an increase of load.

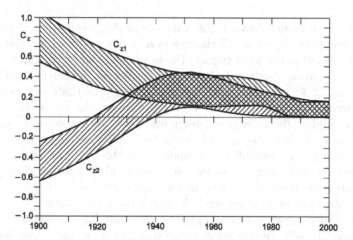

FIGURE 21.31. Evolution in time of coefficient C_z, split in its components on the front and rear axles.

on the axles, is summarized in Fig. 21.31[13]. Lift has remained small and positive (directed upwards), with a good reduction in recent years, while the average pitching moment was negative (tending to lower the load on the front axle) up to the 1940s, to decrease in later years.

Aerodynamic lift and pitching moment depend on the position of the vehicle on the ground, primarily on the aerodynamic angle of attack α, that in the

---

[13]A. Gilhaus and R. Hoffmannm, *Directional stability,* in W.H. Hucho (Ed.), *Aerodynamics of Road Vehicles,* SAE, Warrendale, 1998.

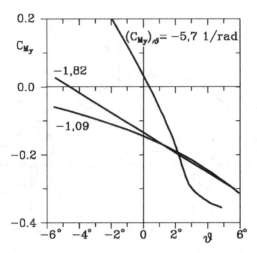

FIGURE 21.32. Pitching moment coefficient C_{M_y} as a function of angle θ for three different vehicles.

following pages is assumed to coincide with the pitch angle θ. Lift can be considered as varying linearly for small changes of α (or θ), and both C_z and $\partial C_z / \partial \theta$ must be measured in the wind tunnel. The same holds for the pitching moment. A plot of the moment coefficient C_{M_y} versus angle θ for 3 different vehicles is reported in Fig. 21.32; the values of the derivative $\partial C_{M_y} / \partial \theta$ (indicated as $(C_{M_y}),_\theta$) for small movements about the reference position are also reported. Note that the moment and its derivative are mostly negative; this is a general rule.

To reduce the lift, and in some cases to make it negative, many current passenger vehicles are provided with spoilers on the rear part of the body or on the front bumper. Apart from the obvious consideration that their position and size must be accurately studied in the wind tunnel since they are useless if located in the wake or in other zones in which the flow is detached, they must be placed in such a way as to avoid giving way to pitching moments.

A spoiler usually creates some shape drag, as it increases the size of the wake, but it can be effective in reducing the total aerodynamic force F_x owing to reduction of the induced drag (Fig. 21.33).

Since spoilers cause an increase of pressure on the tail of the vehicle, they usually create a positive pitching moment. This moment must be compensated by another surface positioned near the stagnation point, usually referred to as a bumper spoiler, air dam or apron. Its presence is also usually beneficial on drag.

Strong negative lift forces are obtained in racing cars both by the use of wings and by a suitable aerodynamic design of the whole body. Usually there is a rear wing, which may have a multiplane configuration, and a front wing, integrated with the nose of the vehicle. The airflow in the zone where these wings are located is strongly influenced by the rotation of the wheels and their

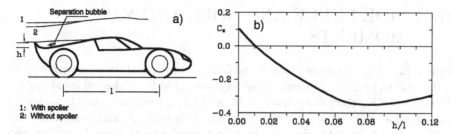

FIGURE 21.33. Lift reducing devices. Effect of a deck-lid spoiler on the streamlines (a) and on the values of the lift coefficient (b).

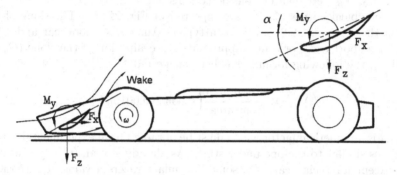

FIGURE 21.34. Forces and moments acting on the wings of a racing car. Interference between wings and rotating wheels.

actual angle of attack, i.e. the angle between the surface and the direction of the impinging current may be quite different from the geometrical angle of attack (Fig. 21.34). Each of the wings produces a negative lift, a drag, usually quite strong, and a pitching moment. These must combine in such a way that the total pitching moment acting on the car is as small as possible.

The entire body of a racing car can be designed to produce negative lift; in fact if the pressure under the vehicle is lower than atmospheric pressure, strong downwards forces may be exerted on the underside. Many racers were based on this concept, with the whole vehicle body designed as a sort of upside-down wing. Suitable side walls, almost reaching the ground, channelled air below the vehicle, producing an area of low pressure and then negative lift. Since aerodynamic devices have a strong impact on safety, racing regulations deal with them in detail. Since 1983, exploiting ground effect is no longer allowed and the bottom of the car must be flat. Since regulations change often, it is impossible to give general rules on the devices used on racers to produce negative lift.

21.5 SIDE FORCE AND ROLL AND YAW MOMENTS

If the vehicle has a symmetry plane and is in a symmetrical position with respect to the airflow, i.e. if the roll and the aerodynamic sideslip angles are equal to zero, the side force F_y, the rolling moment M_x and the yawing moment M_z vanish. In general, what matters is their rate of change with angle β_a and, sometimes, roll angle ϕ. In the case of racing cars with uncovered wheels, these forces can also be produced by offset steering wheels and it is important to study their variation with the steering angle δ.

For small variations of the mentioned parameters about zero, coefficients C_y, C_{M_x} and C_{M_z} can be approximated by linear functions and their derivatives $(C_y)_{,\beta_a}$, $(C_{M_x})_{,\beta_a}$, etc. can be considered constant.

Some typical curves $C_y(\beta_a)$ are reported in Fig. 21.35a. The slope of the curve in the origin is -2.2 rad^{-1} for a typical American saloon car and -2.85 rad^{-1} for a sport car. For a first approximation evaluation of the slope $(C_y)_{,\beta_a}$ (in rad^{-1}), the following formula has been suggested

$$(C_y)_{,\beta_a} = \frac{\text{lateral area}}{\text{front area}}\left(0.005 + 0.0019 n_f\right), \qquad (21.19)$$

where n_f is a numerical factor which must be obtained from experimental results on vehicles similar to the one under study. As already stated, $(C_y)_{,\delta_a}$ is usually small, except for racing cars. On some Formula 1 racers a value of 1.37 rad^{-1} has been recorded[14].

Even more important than the side force, the aerodynamic yawing moment M_z plays a key role in the dynamics of high speed driving. The evolution of coefficient C_{Mz} in passenger cars is summarized in Fig. 21.36[15].

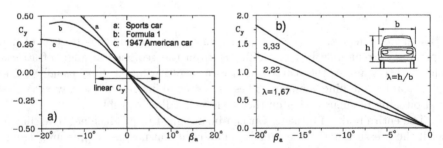

FIGURE 21.35. Coefficient C_y versus angle β_a. (a) Typical values for vehicles of different types; (b) dependence on the ratio width/height of the vehicle.

---

[14]A.J. Scibor Rylski, *Road Vehicle Aerodynamics*, Pentech Press, London, 1975.

[15]A. Gilhaus and R. Hoffmannm, *Directional stability*, in W.H. Hucho (Ed.), *Aerodynamics of Road Vehicles*, SAE, Warrendale, 1998.

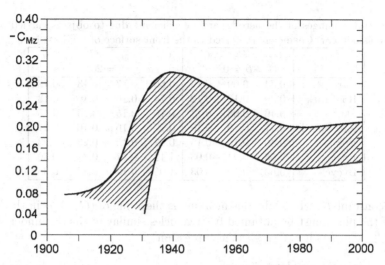

FIGURE 21.36. Evolution in time of coefficient C_{Mz}. Values for an aerodynamic sideslip angle $\beta_a = 20°$.

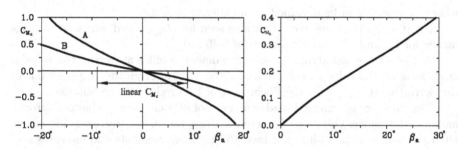

FIGURE 21.37. (a) Typical laws $C_{M_z}(\beta_a)$ for a well streamlined car (A) and a car with a less careful aerodynamic design (B). (b): Rolling moment coefficient C_{M_x} versus β_a for a typical saloon car.

Remark 21.7 *A reduction of aerodynamic drag, in particular if low drag is obtained even at high values of the sideslip angle, is usually detrimental for the lateral force and above all the yawing moment.*

Some typical laws $C_{M_z}(\beta_a)$ are reported in Fig. 21.37a: Usually the moment is negative for a positive sideslip angle, which amounts to saying that the side force is applied in a forward position with respect to the centre of mass. This arrangement, which occurs very often, is destabilizing at high speed; the value of C_{M_z} may be large enough to cause force F_y to be applied in front of the vehicle. A nonlinear expression for coefficient C_{M_z} is

$$C_{M_z} = -\left[K - \sin^2(\beta_a)\right]\tan(\beta_a),\qquad(21.20)$$

TABLE 21.3. Changes of the aerodynamic coefficients due to objects carried on the roof of a saloon car. Coefficients referred to the front surface of the base car.

	C_x	C_{z_1}	C_{z_2}	C_y	C_{M_z}	C_{M_y}	C_{M_x}
		$\beta = 0$			$\beta = 20°$		
Car	0.34	0.09	0.19	0.66	0.17	0.13	0.12
Roof rack	0.38	0.10	0.12	0.74	0.16	0.16	0.13
Ski	0.46	0.08	0.13	0.76	0.15	0.15	0.15
Surf board	0.47	0.10	0.13	0.77	0.16	0.16	0.17
Box	0.46	0.10	0.15	0.92	0.15	0.23	0.23
Boat	0.55	0.24	−0.03	1.12	0.17	0.37	0.30
Bycycle	0.55	0.19	0.03	1.00	0.12	0.32	0.38

where constant K, whose obvious meaning is the value of $(C_{M_z})_{,\beta_a}$ in the linear part of the plot, must be obtained from vehicles similar to the one under study. MIRA[16]   suggests for $(C_{M_z})_{,\beta_a}$ the expression

$$(C_{M_z})_{,\beta_a} = -\frac{1}{100}\left[\frac{\text{lateral area}}{\text{front area}}(0.208 + 0.0655 n_f - 0.00508 n_f^2)\right], \qquad (21.21)$$

where n_f must again be obtained from experimental data.

For $(C_{M_z})_{,\delta}$ the same considerations seen for $(C_y)_{,\delta}$ hold; an order of magnitude for Formula 1 racers is a value of 0.46 rad^{-1}.

A plot of the aerodynamic rolling moment coefficient versus the sideslip angle for a typical saloon car is shown in Fig. 21.37b. From the graph a value of the derivative $(C_{M_x})_{,\beta_a}$ for the linear part of 1.05 rad^{-1} can be obtained.

The presence of external loads on the roof of a car deeply changes aerodynamic performance. Obviously it increases drag, but it also decreases lift, above all at the rear axle; in addition, it increases pitching moments and increases sensitivity to side wind. The aerodynamic coefficients of a saloon car with different roof loads are reported as an example in Table 21.3.

21.6   EXPERIMENTAL STUDY
OF AERODYNAMIC FORCES

The traditional tool for the study of aerodynamic forces and moments is the wind tunnel. Modern wind tunnels used for road vehicle aerodynamics are specialized devices, quite different from those used in aeronautics. Wind tunnel tests of cars are almost always performed on full-size vehicles, since it is very difficult to simulate with the required precision internal flows on reduced scale models.

Wind tunnels for vehicular use are divided into aerodynamic and climatic tunnels.

---

[16]R.G.S. White, *A rating method for assessing vehicle aerodynamics side force and yawing moment coefficients*, MIRA Rep. n. 1, 1970.

The first of these are designed to simulate the aerodynamic field for the measurement of aerodynamic forces and moments, of pressure distribution, etc. and their main requirement is an accurate simulation of the aerodynamic field surrounding the moving vehicle.

Climatic tunnels are used to simulate motion in various climatic conditions, by controlling the temperature of the air and often its humidity. Provisions for producing rain, snow, various sun conditions, etc. are often included. Usually it is possible to run the engine and brake the driving wheels to study, in a realistic way, the temperature conditions in the engine compartment and the working of the cooling and air conditioning systems. The simulation of the aerodynamic field may be less accurate.

Sometimes the two functions are be combined in a single aerodynamic and climatic tunnel, but the convenience of building a single plant for the two tasks instead of two different tunnels must be assessed in each case, keeping in mind above all the compatibility of the balance that measures aerodynamic forces and moments with the rollers that brake the driving wheels.

Climatic tunnels allow the vehicle to be tested in extreme conditions without the need for moving instrumentation and personnel to distant, uncomfortable places, or waiting for extreme climatic conditions to occur in a given place.

The main tests that can be performed in aerodynamic tunnels are:

- Measurement of aerodynamic forces and moments. The three components of both force and moment are measured using a six-component balance.

- Visualization of the airflow around and inside the vehicle and measurement of the pressure at given points of the surface or in other points of the aerodynamic field. Other measurements aimed at understanding the aerodynamic field.

- Measurement of aerodynamic noise, possibly when other sources of noise (engine, transmission, etc.) are present.

Climatic tunnels allow tests to be performed on the behavior of the body (waterproofing, accumulation of dirt, ice formation, weathering, etc.) and of mechanical parts (cooling of engine, starting, air conditioning, operation of the electric devices, etc.) in various climates. Climatic tunnels must be large enough to allow full scale vehicles to be tested, and they must have rollers to brake the wheels and simulate inertias to properly simulate road loads on the driving wheels. The latter device makes it difficult to build tunnels that include a balance for the measurement of forces and moments.

Aerodynamic tunnels for motor vehicles also have a similar size, for as we have seen, the use of models in motor vehicle aerodynamics leads to much less accurate results.

Wind tunnels may be of the open or closed circuit type. While in the former the fan must supply all the kinetic energy of the stream, in closed circuit tunnels some of the kinetic energy of the air that flows around the vehicle is recovered, and the motor needs to supply to the flow only an energy equal to the losses.

The efficiency, defined as the ratio between the power of the airstream and the motor power

$$\eta = \frac{\frac{1}{2}\rho V^3 A}{P} \,,$$
(21.22)

of closed circuit tunnels is higher than that of open circuit ones, and is often higher than one.

Open circuit tunnels are simpler, smaller and cheaper but are energetically less efficient. A cost trade-off for the two types of plant must be performed in each case.

While in the aeronautical field most wind tunnels are of the closed circuit type, modern plants for vehicular use are mostly open, perhaps with air recirculation within the building to avoid the need of filtering large quantities of air and to reduce noise pollution to the environment. Recirculation can also increase energy efficiency.

To perform aerodynamic tests on vehicles, high airspeeds are not needed, since a speed of about 10 m/s is high enough to reach values of the Reynolds number of about 2 or 3 million. Tests are generally performed at a speed of 20 or 30 m/s, creating strong enough aerodynamic forces to allow easy and precise measurements.

Climatic tunnels must reach higher speeds, close to the maximum speed of the vehicle. Since the test chamber of wind tunnels for road vehicles has a sectional area of about $10 \div 30$ m^2, the motor must have a power of several hundred or thousand kW. In climatic closed circuit (or open circuit with recirculation of air) tunnels, the power of the refrigerators must be higher than that of the motors, at least if low temperature tests at top speed must be performed.

The type, shape and size of the test chamber is of primary importance in aerodynamic wind tunnels. The aerodynamic field around the test object must simulate as accurately as possible that occurring in free air, and this is easier if the test chamber is large, or rather if the value of the ratio between the area of the air jet and the area of the cross section of the test object is high.

On the other hand, it is clear that the smaller the size of the jet, the lower the power required, as well as the size and cost of the whole plant. The shape of the cross section of the test chamber must be accurately chosen to obtain the required precision with as small a jet as possible. This consideration is, however, mitigated by the fact that a test chamber with rounded walls is much more costly than a rectangular one, so that a rectangular jet is often used, even if by doing so a larger wind tunnel, with greater power requirements, is needed.

The designer of a wind tunnel must choose between an open or closed test chamber. Closed test chambers, like the ones used in most aeronautical wind tunnels, have lower losses but usually require larger jet areas to obtain the same precision in the simulation of the aerodynamic field. Taking both factors into account, closed test chambers require greater power, at least in automotive wind tunnels. An alternative is the use of closed test chambers with porous walls, that is walls allowing a certain quantity of air to be extracted from the flow, with

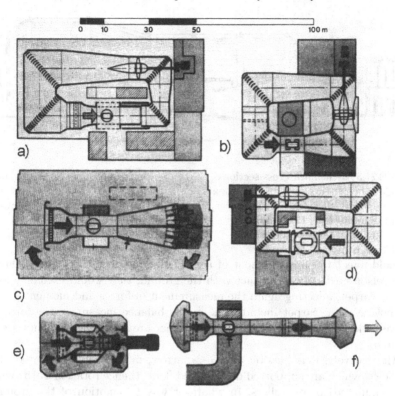

FIGURE 21.38. Plans of six wind tunnels for full-scale testing of road vehicles. (a)Volkwagen, Wolfsburg, Germany; (b) Ford, Detroit, USA; (c) F.K.F.S., Stuttgart, Germany; (d) Pininfarina, Torino, Italy; (e) M.I.R.A., Lindley Nuneaton, G.B., (f) Nissan, Oppama Yokosuka, Japan.

consequent reduction in the thickness of the boundary layer. Another possibility is the use of adaptive test chambers, in which a number of actuators allow the shape of the test chamber to be changed so that the streamlines at a certain distance from the vehicle are similar to those occurring in a free flow.

Sketches of six of the most important vehicular wind tunnels are plotted in Fig. 21.38. Three of them have a closed circuit and three are open, two with air recirculation. The cross section of the Pininfarina wind tunnel is shown in Fig. 21.39, as an example of a modern climatic aerodynamic tunnel with open circuit and air recirculation and open test chamber.

Peculiar difficulties encountered in wind tunnel testing of vehicles are linked with the presence of the ground. To simulate the motion of the ground with respect to the vehicle, a sort of carpet moving at the speed of the air should be used, with the wheels rolling on it. This approach has many drawbacks, particularly in terms of wheel rotation. If the vehicle is kept at a small distance from the ground and the wheels are moved by motors inside the vehicle or model, a flow of air between the wheels and the ground would result and the resulting negative

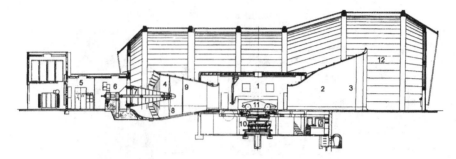

FIGURE 21.39. Schematic cross-section of the Pininfarina wind tunnel. 1) test chamber; 2) intake cone; 3) turbulence generators; 4) outlet cone; 5) power conditioning units; 6) motor; 7) propeller shaft; 8) propeller; 9) nets; 10) six-component force transducer; 11) vehicle under test; 12) building.

lift would affect the measurement of aerodynamic actions. If, on the contrary, the wheels are actually in contact with the ground, they would exchange forces with the carpet, affecting again the measurement of forces and moments, unless the whole system, carpet included, is on the balance measuring the forces. In that case, the forces between wheels and carpet are internal forces and are not measured.

Alternatively, it is possible to use a narrow, moving carpet located inside the wheels, which are supported on the fixed floor (hence rotation of the wheels is not simulated) or on rollers. In whatever way the motion of the ground is simulated, the problem of the boundary layer on the ground remains. As shown in Fig. 21.40a, if the ground does not move with respect to the vehicle, the latter is partially in the boundary layer of the ground.

The simplest way to simulate the presence of the ground is to use a second vehicle located in a mirror position with respect to the ground that physically does not exist (Fig. 21.40b). This approach is difficult to apply in full size testing, since the two vehicles must be suspended in a way that does not disturb the aerodynamic field, and the cross section of the jet is doubled. It is much easier to apply it in tests on models.

The approaches shown in Fig. 21.40c, d, g, h, i are based on aspiration, or blowing the boundary layer so that the velocity profile of the air close to the ground is similar to that occurring when the vehicle moves with respect to it. In Fig. 21.40e, f the vehicle is at a distance from the floor, on a surface (that does not move) simulating the ground, or on nothing. In the latter case, the wheels are in a different situation from the actual one. In Fig. 21.40l a small deflector deviates the air stream to the sides, to reduce the boundary layer of the ground.

In Fig. 21.40m, n, o, three ways to allow the rotation of the wheels are shown. One large roller, two small rollers or a belt supported by three rollers allow the wheels to rotate. They must be located on the balance, so that the forces they exchange with the vehicle are not measured along with the aerodynamic forces. An alternative is to substitute soft elements, similar to brushes, for the ground

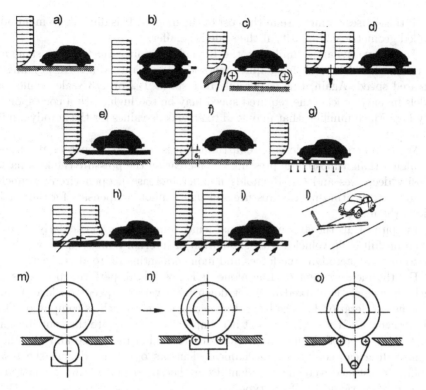

FIGURE 21.40. Simulation of the moving ground (a-l) and rotating wheels (m-o) in wind tunnel testing.

under the wheels, allowing them to rotate but blocking the air flow beneath them and producing no large forces that may affect measurements.

Another problem is that of air turbulence. Aeronautic wind tunnels are usually designed to have the lowest possible turbulence, while the air vehicles meet as they move, particularly on windy days, is more or less turbulent. Nets or other devices to increase air turbulence are therefore used in automotive wind tunnels.

Automotive aerodynamics is based on full scale tests, as we have seen; however, tests on reduced scale models are also used, especially for industrial vehicles or in the early development stages of cars. The main advantage of tests on models is their reduced cost, both in the construction and modifying of models and the use of small size tunnels. Nor is the cost reduction due to the reduced time needed to prepare and perform tests negligible. The disadvantages of models are linked to their limited geometric similitude and to the difficulty of obtaining a good dynamic similitude. Both these problems increase when model sizes are reduced.

In 1/2 or 1/3 scale models it is possible to reproduce tiny details (to study internal flows, internal details of the engine compartment must also be reproduced), even if this causes an increase in the cost of the models. It is difficult to maintain detailed geometrical similarity if the scale is smaller.

To achieve dynamic similarity, the Reynolds number must be the same as in the actual vehicle, and the simplest method is to operate at a correspondingly increased speed. Again, this is possible in tests at 1/2 or 1/3 scale, while for models in smaller scale the required speed may be too high, with a correspondingly high Mach number, that may lead to erroneous values for the aerodynamic coefficients.

While in aeronautic closed circuit tunnels it is possible to operate at a pressure higher than atmospheric pressure (by increasing the pressure the kinematic viscosity decreases and the Reynolds number increases), open circuit tunnels must operate at atmospheric pressure and thus cannot compensate for the small scale of the model.

In spite of all the above problems, wind tunnel tests, particularly if performed on full scale vehicles with simulation of internal flows, remain the best way to measure aerodynamic forces and moments acting on road vehicles.

For the measurement of drag alone, it is possible to perform road tests in which the vehicle is allowed to coast down from various speeds on level road: From the deceleration law the total drag acting on the vehicle is computed. The main drawback of these tests is that it is impossible to separate the contribution of drag due to aerodynamics from that due to rolling resistance and all other forms of drag caused by the mechanical elements operating during the test. If rolling resistance were independent from the speed and C_x were constant, function dV/dt would be of the type:

$$\frac{dV}{dt} = \left(-f_0 mg - \frac{1}{2}\rho S C_x V^2\right) \frac{1}{m_{at}} , \qquad (21.23)$$

and hence it would be enough to approximate the measured law dV/dt with a parabola and to compute C_x from the coefficient of the quadratic term.

The actual rolling resistance is of the type

$$F_r = -mg(f_0 + KV^2) ,$$

and both forms of drag are included in the quadratic term. Only using experimental data on rolling resistance in operating conditions is it possible to interpret coast-down tests correctly. It is possible to obtain realistic values of the aerodynamic drag, and hence of C_x, if tests of the type of those shown in Fig. 2.75 are performed on the whole vehicle.

Even if the aerodynamic field cannot achieve the correct configuration characterizing the steady state conditions at any speed during a cost down test, these tests are usually performed to compare the values of C_x obtained in the wind tunnel with those obtained in close to actual operating conditions, although the latter may be approximated,.

21.7 NUMERICAL AERODYNAMICS

In the last thirty years, numerical methods based on various discretization techniques applied to the Navier-Stokes equations were developed. The diffusion of such methods, usually referred to as numerical or computational aerodynamics, has been made possible by the availability of computers with increasing power and by the introduction of new computational techniques.

While in aeronautics computational aerodynamics has reached a point where it is possible to compute in a precise and reliable way the aerodynamic field and forces acting on aircraft (even if the validation of the results with experimental tests in the wind tunnel are still needed), in road vehicle aerodynamics the possibility of using computational aerodynamics to substitute at least partially for wind tunnel testing remains controversial, and we seem to be still far from achieving this goal.

This difference is linked with the difficulty of predicting numerically where the flow detaches from the body: computational aerodynamics yields very good results in case of attached flows, but when large wakes or separation bubbles are present the computation becomes more difficult and the results less reliable. Numerical methods may, on the contrary, be used without problems in automotive aerodynamics if the location where the flow detaches is known.

Numerical aerodynamics is based mainly on the discretization of the aerodynamic field, and on the use of a simplified form of the Navier-Stokes equations. The simplest approach is the use of the boundary elements method, which in aerodynamics is often referred to as the panels method. Only the surface of the body needs to be discretized (Fig. 21.41a). This is a linearized method, and can be used on an attached inviscid flow, although it may be applied also to zones where the separation zone is known and described geometrically with high precision. It is a technique widely used in aeronautics, but in the automotive case it requires the use of experimental techniques to define the zones where the flow separates.

The methods based on the discretization of the whole aerodynamic field up to a certain distance from the vehicle (finite differences method, finite elements method (FEM, Fig. 21.41b)) allow nonlinear forms of the Navier-Stokes equations to be discretized as well, and the zone where the flow detaches to be computed through iterative techniques. Combined approaches are also possible, to combine the advantages of the various methods.

An example of the computation of the pressure distribution, accurately obtained using the finite volumes methods, is shown in Fig. 21.42[17]. To obtain such a good result, the aerodynamic field was subdivided into more than 5 million cells.

---

[17]H. Wustenberg, B. Huppertz, reported in S.R. Ahmed, *Computational Fluid Dynamics*, in W.H. Hucho (Ed.), *Aerodynamics of Road Vehicles*, SAE, Warrendale, 1998.

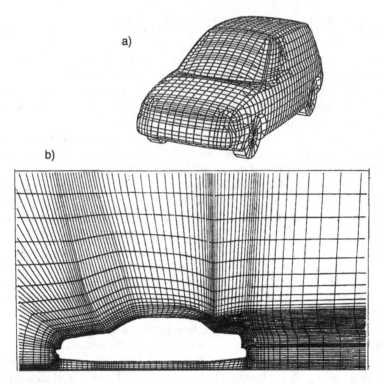

FIGURE 21.41. Discretization of the surface of a car to compute the aerodynamic field using the panels method (a) and of the whole aerodynamic field using the finite element approach (b).

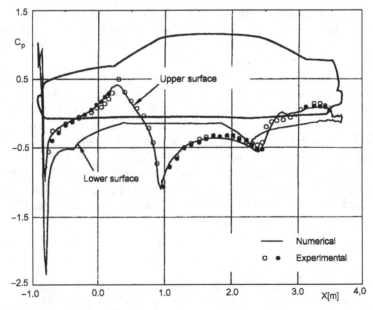

FIGURE 21.42. Pressure distribution in the symmetry plane computed using the finite volumes method.

The hopes that numerical aerodynamics raised must be at least partly down-sized, even if it is becoming (or better, may become in the future) a powerful tool. It allows the aerodynamic behavior of the vehicle to be studied with a reduced number of wind tunnel tests, simultaneously increasing the number of configurations studied and helping in the interpretation of the experimental results. However, it is not certain that it my become, in the foreseeable future, an alternative to experimental testing.

22

PRIME MOVERS FOR MOTOR VEHICLES

The motion of all vehicles requires the expenditure of a certain quantity of mechanical energy, and in motor vehicles the system that supplies such energy (in most cases an internal combustion engine) is on board. The lack of an adequate prime mover is the main reason that mechanical vehicles could be built only at the end of the industrial revolution, and enter mass production only in the Twentieth Century, in spite of attempts dating back to ancient times.

For a mechanical vehicle to be built, a prime mover able to move not only itself, but the vehicle structure and payload as well, was needed. Remembering that the power needed to move the mass m at the speed V on a level surface with coefficient of friction (sliding or rolling) f is equal to $P = mgfV$, it is easy to conclude that the minimum value of the power/mass ratio of a prime mover able to move itself is

$$\frac{P}{m} = \frac{gfV}{\eta\alpha} \ , \tag{22.1}$$

where α is the ratio between the mass of the engine and the total mass of the vehicle and η is the total efficiency of the mechanism which transfers the power and propels the vehicle.

Prime movers with an adequate power/mass ratio and transmission devices with a power rating and an efficiency high enough to allow the motion of the vehicle were not practical until the Nineteenth Century.

The engine must obtain the energy required for motion from an energy source that is usually on board the vehicle. Rail vehicles often receive such energy from outside, but the only road vehicles in which this occurs are trolleybusses.

In most cases, the energy is stored as the chemical energy of a fuel, but it can be stored in the form of electrochemical energy (electrical batteries) or,

TABLE 22.1. Onboard energy storage. Energy density e/m, power density P/m and general characteristics (data for electrochemical energy refer to lead-acid batteries).

Energy stored	Chemical	Electrochemical	Elastic	Kinetic
e/m [Wh/kg]	$10,000 - 12,000$	$10 - 40$	$2 - 10$	$6 - 20$
P/m [W/kg]	Engine dependent	$10 - 100$	High	Very high
Efficiency	$0.2 - 0.3$	$0.6 - 0.85$	$0.7 - 0.9$	$0.7 - 0.95$
Reversibility	None	Possible		
Pollution	In the site of utilization	In the site of generation		
Dependence on liquid hydrocarbons	Almost complete	The primary source can be different		

even if few attempts in this direction have been made, and even fewer vehicles of this type have a practical use, as kinetic energy (flywheels) or elastic energy (springs).

These forms of energy storage are compared in Table 22.1.

When two or more different types of energy are stored or supplied to a vehicle that can work either with energy supplied from the outside or with energy stored on board, and if the two modes of operation are used independently, the vehicle is said to be *bimodal*. A trolleybus with batteries that allow it to go on a part of its route where there is no power distribution is an example of a bimodal vehicle.

Vehicles with two or more methods of energy storage, in which one is used not only to supply energy but also to store energy coming from one of the other sources, are said to be *hybrid*. An example is a bus with an internal combustion engine and batteries, in which the electric energy is also used to transform the energy from the engine with greater efficiency and to recover braking energy.

It is also possible to have a bimodal hybrid vehicle if, in the previous example, the energy to charge the batteries is supplied not only by the thermal engine but also by the mains.

In vehicles there are huge quantities of energy that may be recovered. Theoretically, all energy not dissipated (by aerodynamic drag and rolling resistance, losses in the transmission and energy conversion) can be recovered.

If the kinetic energy or the gravitational potential energy of the vehicle is recovered when slowing down or travelling downhill, *regenerative braking* occurs.

When the only form of energy storage on board is chemical energy, regenerative braking is not possible, while it may be implemented in the other cases of Table 22.1. Energy recovery can, however, be only partial, not only due to the intrinsic losses of all energy transformations, but also because of the peculiar characteristics of braking.

The power involved in braking is hardly manageable by the device that has to convert the energy taken from the vehicle into usable energy, except in the case of slowing down with limited deceleration. Usually, to allow regenerative braking, there must be two braking systems, with the traction motors (in the case of electric vehicles) providing regenerative braking when slowing down or travelling downhill, while a conventional braking system performs, in a non-regenerative way, emergency or sudden decelerations

22.1   VEHICULAR ENGINES

The storage of energy in a liquid, less frequently gaseous, form of fuel has so many advantages that this form of energy storage has supplanted all others since the beginning of the Twentieth Century. The advantages of easy resupply (recharging) and above all the very high energy density are overwhelming.

The chemical energy of the fuel (gasoline, diesel fuel, but also liquefied petroleum gas (LPG), methane, alcohol, methylic or ethylic, etc.) is converted into mechanical energy by a thermal engine. In spite of the low conversion efficiency that characterizes all thermal engines, the actually available energy density is about $30 \div 50$ times greater than that of other energy storage devices. The power density is also very high.

The first self-propelled road vehicles were built at the end of the Eighteenth and above all at the beginning of the Nineteenth Century owing to the development of thermal engines, in this case reciprocating steam engines. However, while steam engines were adequate for ships and railway engines, their power/weight ratio was too low for road vehicles. This issue, together with other technical and non-technical factors, made steam coaches a commercial failure.

Only at the end of the Nineteenth Century did the development of reciprocating internal combustion engines allow the diffusion of motor vehicles.

As road vehicles began to spread, three competing types of engine were available: steam engines, that in the interim had undergone drastic improvements to become adapted to lightweight vehicles, the new internal combustion engines, and DC electric motors combined with recently developed lead acid accumulators. For a time it looked as though the electric motor would become the most common alternative, owing to its reliability, cleanliness, quietness and ease of control. The various types of engine were balanced in performance, as shown by the fact that the first car able to overcome the 100 km/h barrier in 1898 was an electric vehicle.

However, then as today, the main drawback of the electric vehicle, its unsatisfactory range, prevented its diffusion.

The reciprocating internal combustion engine become the main source of power for all road vehicles, and has remained so since the first decades of the Twentiethth Century.

In the 1960s, after the great success of turbojet and turboprop engines in aeronautics, which would quickly almost completely replace reciprocating engines in aircraft and helicopters, several attempts to introduce gas turbines in motor vehicles were made. They were unsuccessful, primarily because of the strong fuel consumption at idle.

At the same time, attempts to reintroduce the steam engine were also made, primarily for reducing pollution and for the scarcity, then more supposed than actual, of fuels suitable for reciprocating engines. Even if steam engines were much different from those of the previous century, the results were not satisfactory.

A further attempt to innovate, although less radical, was the introduction of rotary internal combustion engines. Some vehicles with this innovative engine were mass produced and had a limited commercial success, but this attempt was likewise another failure.

It is likely that the greatest advantage of the reciprocating automotive engine is a century of uninterrupted development, leading to performance, low cost and reliability that could not be imagined one century ago.

Practically, every attempt to substitute a different propulsion device to solve one of its many problems was answered with industry innovations that solved, in an equally (or more) satisfactory way, the same problems.

The issues that fuel today's drive to replace the internal combustion engine with a prime mover of a different kind remain its dependence on liquid hydrocarbons as fuel and the emission of pollutants and greenhouse gases.

The dependence on fuels derived from oil is characteristic of the whole economic system, particularly in Europe and even more in Italy. Even if electric vehicles became widespread or hydrogen took over as fuel, this problem would remain essentially unchanged if the primary energy used to produce electric energy or hydrogen came from the combustion of oil derivatives. More precisely, the problem would become worse, owing to lower overall energy efficiency (*from well to wheel*, as is usually said).

Only a massive use of nuclear energy, possibly with some contribution from renewable sources including hydrocarbons derived from biomasses, can radically solve this problem.

Environmental problems due to pollutants like carbon monoxide, nitrogen oxides, particulates, etc., all substances not necessarily produced by combustion, have already been tackled with success and modern internal combustion engines are much cleaner than older ones. This trend is bound to continue in the future.

Carbon dioxide, on the contrary, is the result of the type of fuel used and can be reduced only by using fuels with lower carbon content, like methane, and only completely eliminated by using hydrogen. However, the production of hydrogen must use a primary source that does not produce carbon dioxide, like nuclear energy.

Hydrogen can be used both in internal combustion engines and in fuel cells.

Fuel cells are electrochemical devices able to directly convert the energy of a fuel-oxidizer pair into electric energy, without a combustion process taking place. Since in this transformation there is no intermediate stage of thermal energy, the efficiency can be, theoretically, higher than that of any thermal engine, even if it is limited by losses of various kinds.

The reactions occurring in fuel cells are electrochemical reactions of the kind typical of batteries. The choice of fuel is severely limited, since the use of molecules that may be easily ionized is mandatory. Hydrogen is the most common choice, even if methane is an interesting alternative, while the oxidizer must be, in vehicular applications, atmospheric oxygen. The energy density of fuel cells using liquid fuels like methanol or formic acid is too low for vehicular applications.

The problems linked with the use of hydrogen as a fuel primarily relate to its low volume energy density (its mass energy density is, on the contrary, quite high) and to the subsequent need to use pressurized tanks, cryogenic storage at 20 K, or to resort to technologies like those based on metal hydrides. There are also problems involved in its supply network. The technological problems are being solved, since hydrogen is used in experimental vehicles as a fuel for internal combustion engines, and in many countries there are already a number of supply points. Safety does not seem to be a problem, since hydrogen is not much more dangerous than a highly flammable and volatile liquid such as gasoline.

Hydrogen may also be stored on board as methanol or methane, from which hydrogen is then obtained by chemical dissociation. This solution has the drawback of causing poisoning of the fuel cell catalyst if impurities due to this process remain in the hydrogen.

At present there are many types of fuel cells, based on different types of membranes and catalysts. They operate at different temperatures (from less than 100°C to more than 900°C, the latter being unsuitable for vehicular use), and each has its advantages and drawbacks. The technology developed in the aerospace field (fuel cells were developed in the 1960s for the *Apollo* programme and are now used on the *Space Shuttle*) cannot be used in road vehicles. Many problems are still to be solved, from cost to reliability, with added problems linked to their use under the conditions of much variable load and reduced maintenance that are typical of motor vehicles.

Until fuel cells suitable for vehicular use are available, the only way to use electric motors is by employing accumulators. Their worst drawback is the impossibility of obtaining high energy density and power density at the same time. This is particularly true for lead-acid accumulators, whose energy density decreases fast with increasing power density, that is, with increasing current.

Also, the duration and the efficiency of batteries decrease with increasing power density. The field of batteries for vehicular propulsion has seen much research activity, and the possibility of building electric vehicles with performance not much different from that of vehicles with internal combustion engines, especially in terms of range, may yet emerge.

The possibility of using different forms of energy accumulators in a single vehicle in a hybrid configuration is particularly interesting. There are many experimental vehicles of this kind and some of them have been mass produced.

22.2   INTERNAL COMBUSTION ENGINES

As stated in the previous section, most road vehicles are powered by reciprocating internal combustion engines. The performance of an internal combustion engine is usually summarized in a single map plotted in a plane whose axes are the rotational speed Ω_e and either the power P_e or the engine torque M_e (Fig. 22.2). Often the former is reported in rpm, the power in kW and the torque in Nm.

If a plot of the power as a function of speed is used, the plot is limited by the curve $P_e(\Omega_e)$ expressing the maximum power the engine can supply as a function of the speed. Such a curve is typical of any particular engine and must be obtained experimentally. However, when building a simple model of the vehicle, it is possible to approximate it with a polynomial, usually with terms up to the third power,

$$P_e = \sum_{i=0}^{3} P_i \Omega_e^i \,. \tag{22.2}$$

The values of coefficients P_i can easily be obtained from experimental testing. In the literature it is possible to find some values of the coefficients which can be used as a first rough approximation. M.D. Artamonov et al.[1] suggest the values

$$P_0 = 0 \,, \qquad\qquad P_3 = -\frac{P_{max}}{\Omega_{max}^3}$$

for all types of internal combustion engines and

$$P_1 = \frac{P_{max}}{\Omega_{max}} \,, \qquad\qquad P_2 = \frac{P_{max}}{\Omega_{max}^2} \,,$$

for spark ignition engines,

$$P_1 = 0.6 \, \frac{P_{max}}{\Omega_{max}} \,, \qquad\qquad P_2 = 1.4 \, \frac{P_{max}}{\Omega_{max}^2}$$

for indirect injection diesel engines and

$$P_1 = 0.87 \, \frac{P_{max}}{\Omega_{max}} \,, \qquad\qquad P_2 = 1.13 \, \frac{P_{max}}{\Omega_{max}^2}$$

for direct injection diesel engines.

In these formulae Ω_{max} is the speed at which the power reaches its maximum value P_{max}.

The driving torque of the engine is simply

$$M_e = \frac{P_e}{\Omega_e} \,, \tag{22.3}$$

or, if the cubic polynomial is used and coefficient P_0 vanishes,

$$M_e = \sum_{i=1}^{3} P_i \Omega_e^{i-1} \,. \tag{22.4}$$

At present, internal combustion engines for vehicular use are controlled by systems of increasing complexity and their performance is increasingly dependent on the control logic used. The power and torque maps are, then, not unique for a certain engine but may be changed simply by modifying the programming of

---

[1] M.D. Artamonov et al. *Motor vehicles, fundamentals and design*, Mir, Moscow, 1976.

the electronic control unit (ECU). If the above mentioned equations have always been just a rough approximation, today the situation is even more complex from this point of view, and in some cases the equations may supply results much different from those actually observed.

If experimental results on a similar engine are available, it is possible to obtain the maximum power curve from the power curve of that engine.

Remark 22.1 *The practice of correcting engine performance in a way proportional to the displacement is not correct, even if it is acceptable and often used for small changes of capacity. A scaling parameter that may be more correct is the area of the piston multiplied by the number of cylinders, that is, the ratio between capacity and stroke.*

The mean effective pressure p_{me}, i.e., the ratio between the work performed in a complete cycle and the capacity of the engine, is often used instead of the torque. In four-stroke engines it is defined as

$$p_{me} = \frac{4\pi M_e}{V} \, , \tag{22.5}$$

where V is the total capacity of the engine.

All points below the maximum power curve are possible working points for the engine, when it operates with the throttle partially open.

Remark 22.2 *Since the engine is seldom used at full throttle, usually only when maximum acceleration is required, the conditions of greatest statistical significance are those at much reduced throttle.*

A diagram of the specific fuel consumption of a direct injection diesel engine with a capacity of about 2 liters is shown in Fig. 22.1; on the same plot, the circles show the points at which the engine operates on the driving cycle used in Europe for computing fuel consumption for a car with a reference mass of 1600 kg.

The percentages shown close to the circles refer to the time the engine is used in the conditions related to their centers, with reference to the total time the engine is producing power (the time at idle is then not accounted for); the center of the circles represents the average of all utilization points in a rectangle with sides of 500 rpm on the speed axis and one bar on the p_{me} axis.

The curves below the one related to the maximum mean effective pressure in the plot of Fig. 22.1 are those characterized by various values of the specific fuel consumption q. The correct S.I. units for the specific fuel consumption, the ratio between the mass fuel consumption (i.e., the mass of fuel consumed in the unit time) and the power supplied, is kg/J, i.e. s²/m², while the common practical units are still g/HPh or g/kWh. If the thermal value of the fuel is equal to 4.4×10^7 J/kg, it follows that

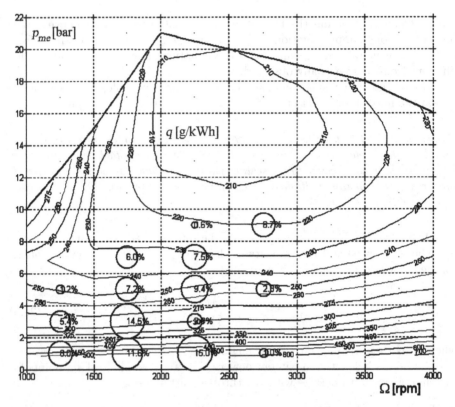

FIGURE 22.1. Map of a direct injection diesel internal combustion engine of about 2 liters capacity, with constant specific fuel consumption curves. The circles show the points where the engine operates on the driving cycle used in Europe for computing fuel consumption with a car with a reference mass of 1600 kg. The consumption of this engine at idle is about 0.62 l/h.

$$q = \frac{2.272 \times 10^{-8}}{\eta_e} \ \mathrm{kg/J} = \frac{60.16}{\eta_e} \ \mathrm{g/HPh} = \frac{81.79}{\eta_e} \ \mathrm{g/kWh} \ ,$$

where η_e is the efficiency of the engine.

This map allows the fuel consumption of the engine to be stated in various working conditions: at far left is the minimum speed at which the engine works regularly; at far right is the maximum speed. The speed axis shows conditions at idle, where the mean effective pressure (p_{me}) vanishes together with the efficiency and the specific fuel consumption is infinite.

The map can be represented in a different way, plotting power on the ordinates and using the efficiency η_e of total energy conversion, from chemical energy of the fuel to mechanical energy at the shaft, as a parameter.

A plot of this type is shown in Fig. 22.2.

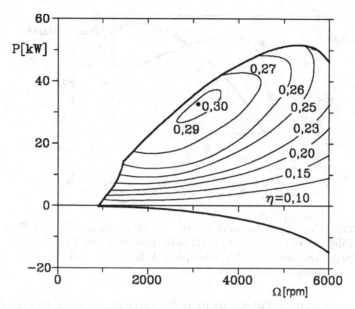

FIGURE 22.2. Map of a spark ignition internal combustion engine, with constant efficiency curves.

Remark 22.3 *The efficiency of a spark ignition engine reaches its maximum in conditions close to full throttle and at a speed close to the one where the torque is at its maximum. The efficiency decreases quickly as power is reduced at a fixed speed. This decrease is less severe in diesel engines.*

Efficiency and specific fuel consumption are linked by the relationship

$$q = \frac{1}{H\eta_e} \qquad (22.6)$$

where H is the thermal value of the fuel.

Example 22.1 *Compute the coefficients of a cubic polynomial approximating the power versus speed curve of the engine of the vehicle in Appendix E.1. Compare the curve so obtained with the experimental one and that obtained from the coefficients suggested by Artamonov. Plot on the same chart the engine torque and the specific fuel consumption. By taking from the plot points spaced by 250 rpm and using a standard least squares procedure, it follows that*

$$P = -10,628 + 0,1506\Omega - 9,5436 \times 10^{-5}\Omega^2 - 5,0521 \times 10^{-8}\Omega^3 \,,$$

where Ω is expressed in rad/s and P in kW. Using Artamonov's coefficients for a spark ignition engine, the equation becomes

$$P = 0,7024\Omega + 1,290 \times 10^{-4}\Omega^2 - 2,369 \times 10^{-7}\Omega^3 \,.$$

The two curves are plotted in Fig. 22.3. Both expressions approximate the experimental curve well, even if the coefficients are quite different.

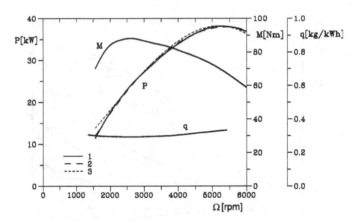

FIGURE 22.3. Engine power curve for the car of Appendix E.1. (1) Experimental curve, (2) third-power least square fit, (3) cubic polynomial with coefficients computed as suggested by Artamonov *et al.* The torque and the specific fuel consumption are also reported as functions of speed.

Two more examples of engine maps for two spark ignition engines of about 2 l capacity are reported in Figures 22.4 and 22.5. The first refers to an indirect injection engine (in the intake manifold), while the second one is for a direct injection (in the combustion chamber) engine. The latter is similar to the diesel engine shown earlier.

Remark 22.4 *When the fuel consumption is needed in points different from those shown in the plot, it is advisable not to interpolate in the map of specific fuel consumption, but on that of efficiency. The consumption changes in a strongly nonlinear way with both speed and mean effective pressure, and tends to infinity when the p_{me} tends to zero. The efficiency, on the contrary, tends to zero, when the p_{me} tends to zero.*

22.3   ELECTRIC VEHICLES

Batteries and electric motors are the most common alternative to internal combustion engines. As already stated, the performance obtainable is lower than that typical of vehicles with internal combustion engines, especially in terms of range, but also in terms of operating costs and vehicle availability. Studies on batteries for vehicular use are very active, and it is a common opinion that only through electric vehicles will some of the problems caused by the use of motor vehicles in urban areas be solved. The performance of some of the batteries suggested instead of the more common lead-acid batteries are reported in Table 22.2. Future progress seems to be linked more to the possibility of mass producing accumulators with sufficient performance at costs compatible with vehicular use than to an increase of performance.

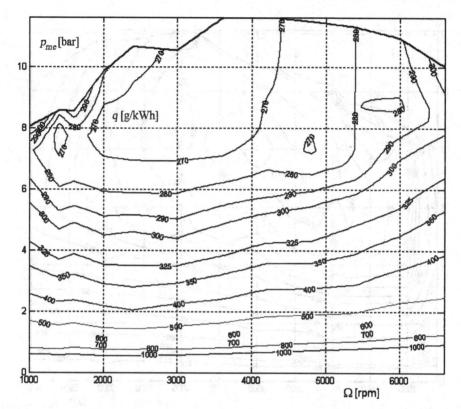

FIGURE 22.4. Map of the specific fuel consumption of an indirect injection spark ignition engine of about 2 liters capacity. The consumption of this engine at idle is about 0.92 l/h.

The advantages of electric vehicles are linked primarily to the possibility of moving the pollution from where the vehicle is used to where the power is generated, taking advantage of the better pollution control of power stations versus small engines. Another advantage is the possibility of regenerative braking. The performance of electric drives is, however, decreased by losses in both the engine and the batteries, and above all by the difficulties that batteries have in accepting the high power bursts occurring in braking. The disadvantages are also well known: The reduced range and duration of batteries and their high mass. However, even today, the performance of electric vehicles is sufficient for urban use.

From the point of view of energy the advantages of battery powered electric vehicles (BEV) are still in doubt: When the primary source is a fossil fuel, in spite of the greater efficiency of the primary conversion and regenerative braking, the overall consumption is comparable to that of internal combustion engines. The very fact that the thermo-mechanical conversion occurs far from the vehicle makes it impossible to use waste heat for heating, and this makes the energy balance worse.

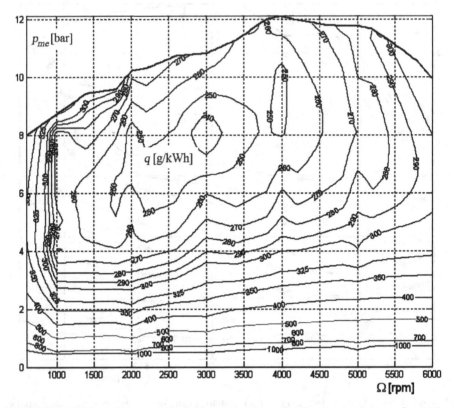

FIGURE 22.5. Map of the specific fuel consumption of a direct injection spark ignition engine of about 2 liters capacity. The consumption of this engine at idle is about 0.90 l/h.

TABLE 22.2. Main characteristics of some battery types for automotive use (M.J. Riezenman, *The great battery barrier*, IEEE Spectrum, Nov. 1992). a): Constant current 3 hours discharge. b): Cycles with 80% discharge depth. c): 100% discharge depth in urban cycle. d): 80% discharge..

Type	E/m [a] [Wh/kg]	P/m [b] [W/kg]	Efficiency	Life[c] [cycles]
Sodium-sulphur	81	152	91 %	592
Sodium-sulphur	79	90	88 %	795
Lithium-sulphides	66	64	81 %	163[d]
Zinc-bromine	79	40	75 %	334
Nickel-zinc	67	105	77 %	114
Nickel-metal hydrides	54	186	80 %	333
Nickel-metal hydrides	57	209	74 %	108
Nickel-metal hydrides	55	152	80 %	380
Nickel-iron	51	99	58 %	918

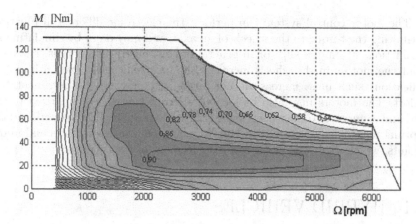

FIGURE 22.6. Map of the efficiency of an induction AC motor with a nominal power of 35 kW.

The traditional configuration is based on direct current (DC) or alternating current (AC) motors connected to the wheels through a transmission of more or less conventional type. Since the electric motor can start under load, there is no need for a clutch and usually no need for a gearbox with various transmission ratios; only a reduction gear and a differential are necessary. The motor is controlled with power electronic devices (*choppers*) whose efficiency is at present extremely high.

Instead of a DC motor (with brushes) it is possible to use an AC motor, controlled by an inverter.

The map of the efficiency of an induction AC motor with a nominal power of 35 kW is shown in Fig. 22.6.

Recently permanent magnet synchronous brushless motors with related control electronics have also been used in vehicular applications. The efficiency and control are generally better, and the cost is transferred from the motor to the power electronics.

As an alternative to the traditional architecture, with the motor operating the wheels through a mechanical transmission, it is possible to put two or more motors directly in the wheels. This is a configuration suggested and tried several times in the past with limited success except for special vehicles, and it is one that seems to be ready for large scale application today. Traditional CC or AC motors require a mechanical transmission in any case, since they supply an insufficient torque and operate at a speed that is higher than that of the wheels. At present, high torque motors (*torque motors*, both with internal and external rotor) are available; these can be connected directly to the wheels without interposing a reduction gear. Apart from the advantage, which may be important in some applications, of allowing an arbitrarily large steering angle, even up to 360°, putting the motor in the wheels without using a reduction gear leads to high efficiency, low noise and a large degree of freedom in placing the various subsystems of the vehicle.

The motor control system can perform the electronic differential function, distributing the torque to the wheels of an axle, and may do so by simulating all the functions of limited slip (or in general of controlled) differentials. However, to put the motors in the wheels increases the unsprung mass, even if in recent applications such mass increase is not large, and may not be detrimental to comfort. The motors may also be located close to the wheels but fixed to the body, and connected to the wheels using transmission shafts. The reduction of the unsprung mass is compensated by reintroducing transmission shafts and above all joints, which work with a relative displacement of the two parts.

22.4   HYBRID VEHICLES

While the only accumulators able to store all the energy required for motion are electrochemical, the quantity of energy to be accumulated in the secondary accumulators of hybrid systems is lower, and this may allow devices of other types to be used. The drawbacks of electrochemical batteries become also less severe.

Elastic energy can be stored in a solid or in a gas. In the first case, the energy density e/m of the device is

$$\frac{e}{m} = \alpha_1 K \frac{\sigma^2}{\rho E} \, , \tag{22.7}$$

where α_1 and K are coefficients linked to the ratio of the mass of the energy storage elements and that of the whole device, and to the shape of the storage element and the stress distribution, with σ the maximum stress in the energy storing element and E the Young's modulus of the material.

Material with very high strength (spring steel) or low stiffness (elastomers) must be used. The latter are particularly well suited, since some of them may be stretched up to 500% with a good fatigue life and limited energy losses.

The use of a compressed gas, while considered for fixed installations, has several disadvantages for vehicular uses, due to its lower efficiency, the high mass of the container of pressurized fluid, and burst danger. Hydraulic accumulators, in which the energy is stored in the walls of an elastomeric vessel full of fluid, have been suggested and tested in connection with hydraulic motors and pumps. The pressure of the oil, however, is controlled by the characteristics of the elastomeric material independent of driving or braking (in case of regenerative braking) torque. Reversible variable displacement motors, which are quite complex and costly, are then required.

Energy can be stored in the form of kinetic energy in a flywheel. The energy density of a kinetic energy accumulator can be expressed as

$$\frac{e}{m} = \alpha_1 \alpha_2 K \frac{\sigma}{\rho} \, , \tag{22.8}$$

where α_1, α_2 and K are coefficients linked with the ratio of the mass of the fly-wheel and that of the whole system, to the depth of discharge actually performed and to the shape and the stress distribution in the flywheel. σ is the maximum stress in the energy storing element and ρ is the density of the material.

Apart from some applications, like the city busses built by Oerlikon in the 1950s and actually used in public service, flywheels are now considered for use only in hybrid systems. Their potentially high power density makes them very suitable for supplying short bursts of power for acceleration or for storing braking energy.

Nor is the problem of designing an adequate transmission trivial: the veloc-ity of the vehicle must be variable at the will of the driver down to a full stop, while the angular velocity of the flywheel is proportional to the square root of the energy it contains. The flywheel reaches its maximum speed after recover-ing all the energy of the vehicle, when the latter stops. This demands complex continuous transmissions that may offset the advantages of this solution

Some possible schemes for hybrid vehicles are the following (Fig. 22.7):
a) internal combustion engine − electric accumulator,
b) internal combustion engine − elastic accumulator,
c) internal combustion engine − flywheel,
d) electric accumulator − flywheel,
e) internal combustion engine − electric accumulator − flywheel.

The first three systems are similar, at least in principle. The thermal engine supplies the average power, working in conditions that may be optimized in terms of efficiency or pollution. A trade-off between these requirements can be made.

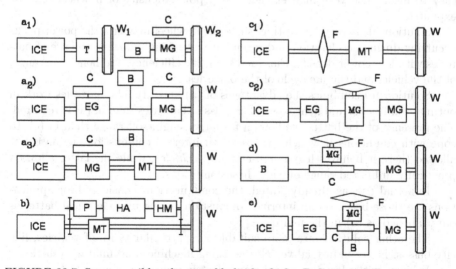

FIGURE 22.7. Some possible schemes of hybrid vehicles B, batteries; C, control unit; EG, electric generator; F, flywheel; HA, hydraulic accumulator; HM, hydraulic mo-tor; ICE, internal combustion engine; MG electric motor/generator; MT, mechanical transmission; P, pump; W wheels.

When the duty cycle includes frequent accelerations and braking, the advantages of disconnecting the instantaneous power requirements from the working conditions of the thermal engine, and of making regenerative braking possible, are large. The possibility of using a far smaller engine allows one to keep mass and cost within the limits of conventional systems or even to obtain mass and cost savings.

The solutions (a) above in Fig. 22.7 are based on an internal combustion engine and electric batteries. The Prius built by Toyota is an example of a hybrid vehicle of this type (see below).

Remark 22.5 *Hybrid vehicles with internal combustion engine and batteries appear the worst alternative in theory, since electric accumulators work exactly in the way which should be avoided, being called to supply high power for short periods; nevertheless this is the only system used in practice today.*

Solution (a_1) is the most interesting, since the presence of an axle controlled by a thermal engine and one controlled by electric motors allows side advantages, like fully controlled 4WD and an active differential on the electric axle, as effective as a VDC system, to be obtained.

Solution (b) may be used in hydrostatic transmissions; owing to the cost of the latter, it is mainly considered for large city buses.

Solution (c) allows the use of a mechanical transmission, although the requirement of an efficient CVT with a wide range of transmission ratios is not easy to meet. The very high efficiency and power density of flywheels can be exploited.

Solution (d) is very interesting, since the flywheel manages the power peaks occurring during acceleration and regenerative braking, allowing the use of batteries with low power density, thus increasing the efficiency, and hence the range, of the vehicle, and the life cycle of the batteries.

Solution (e) combines the advantages of (a) and (d): The batteries work in optimal conditions, and hence a smaller mass of batteries than in (a) is required. The presence of the batteries allows a far larger engine-off range than in (c), to cope with conditions in which the use of an internal combustion engine is not allowed (here it behaves like a *zero-emission vehicle*), while the latter allows a practically unlimited range outside these conditions.

In actual use, as already stated, the configurations considered for applications are those based on an internal combustion engine plus electrical batteries only, labelled as (a) in the figure.

The other solutions are more suitable for particular types of vehicles, like city busses, heavy industrial vehicles, working machines and military vehicles.

Vehicles with electric transmission (electric generator connected to the engine and electric motor driving the wheels) without any energy storage system are sometimes defined as hybrid. This configuration has been used for decades in diesel electric systems, and much used in rail transportation. The lack of a storage device makes it impossible to perform regenerative braking. Often a solution of this type is called a Fake Hybrid (FH).

True hybrid vehicles are subdivided into *parallel hybrids* (PH, Fig. 22.7a1) and *series hybrids* (SH, Fig. 22.7a2).

In series hybrids, at least a part of the energy generated by the thermal engine is transformed into electric energy and used to recharge the batteries, or stored in the designated way.

In parallel hybrids, the electric energy (or the accumulated energy) does not interact with the internal combustion engine, but comes only from recovered energy.

Remark 22.6 *The difference between the two types of hybrids does not depend so much on the configuration shown in Fig. 22.7, but mostly on the strategy of the controller.*

The advantage of parallel architecture is its simple layout and the possibility of being offered as an *option* on a conventional vehicle. A traditional rear wheel drive vehicle may be transformed into a parallel hybrid just by adding an electric motor operating the front wheels and a battery, along with the necessary control system.

Another possible advantage of parallel hybrid systems is the higher efficiency with which the power flowing through mechanical transmission is transferred to the wheels.

Another distinction is between *weak hybrids* (WH) and *strong hybrids* (SH).

In weak hybrids the vehicle usually works with the thermal engine, while the electric motor is used to increase performance, when needed, and above all to restart the engine, also working as a generator for regenerative braking. The internal combustion engine is thus switched off when the vehicle stops even for a short time, or supplies only a very small amount of power (*restart systems*).

The layout of Fig. 22.7a3 is that of a conventional vehicle with starter motor and generator integrated in a single unit and an oversized battery.

In the case of strong hybrids, the capacity of the battery is such as to allow both a non-negligible power increase and a certain engine off range. The instant needs of the vehicle can therefore be completely uncoupled from the power supplied by the engine, to the point that it is even possible to avoid a gearbox.

Finally, there are *Plug-capable Hybrid Electric Vehicles* (PHEV) that use a battery that can be recharged from an external source, so that the vehicle can operate like a true Battery Electric Vehicle (BEV) as well. It is possible to speak of a weak version, where the engine-off operation is limited to low speed and short range, and a strong version that may operate in engine-off mode at higher speed with a larger range.

The possibilities offered by the various hybrid layouts are summarized in Table 22.3

TABLE 22.3. From conventional to hybrid and electric vehicles.

Type	Regenerative braking	Battery operation	Rechargeable	Primary el. traction	Indep. of fossil fuels
Normal - FH	–	–	–	–	–
PH(W)	X	–	–	–	–
PH(S)	X	X	–	–	–
PHEV(W)	X	X	X	–	–
PHEV(S)	X	X	X	X	–
BEV	X	X	X	X	X

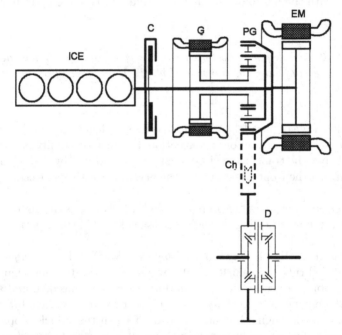

FIGURE 22.8. Layout of the hybrid power system of the Toyota Prius; ICE: thermal engine, C: automatic clutch, G: generator, EM: driving electric motor, PG: planetary gear, Ch: chain driving the final gear at the differential D.

One of the few hybrid vehicles that went beyond the research phase and entered the market at a reasonable price is the Toyota Prius. Its hybrid system is sketched in Fig. 22.8.

In the figure, ICE is the thermal engine, C is an automatic clutch, G a generator, EM a traction electric motor, PG a planetary gear, Ch a chain controlling the gear ratio of the final drive of the differential D. Note that there is no gearbox between the thermal engine and the transmission to the wheels[2].

---

[2]M. Duoba *et al.*, *In-situ mapping and Analysis of the Toyota Prius HEV Engine*, SAE Paper 2000-01-3096

If τ_o is the gear ratio on the planetary gear when the carrier is fixed (here the carrier is connected to the thermal engine), it follows that

$$\tau_o = -\frac{n_a}{n_s} ,$$
(22.9)

where n_a and n_s are the number of teeth of the crown and the sun ; the sign is minus since, when the carrier is fixed, the crown and the sun rotate in opposite directions.

The simple equation

$$-\frac{1}{\tau_o}\Omega_G + \Omega_{EM} = (1 - \frac{1}{\tau_o})\Omega_{ICE} ,$$
(22.10)

where the angular velocities of the various elements are Ω_G, Ω_{EM} and Ω_{ICE}, can be written.

In the same way, by indicating with M_G, M_{EM} and M_{ICE} the torques acting on the same elements, it is possible to write

$$M_T - M_{EM} = M_{ICE} - M_G ,$$
$$(M_T - M_{EM})\Omega_{EM} = M_{ICE}\Omega_{ICE} - M_G\Omega_G .$$
(22.11)

These equations have been obtained by stating the equilibrium for rotation of the gears and the conservation of the power that goes through it. M_T is the available torque on the gear wheel driving the chain Ch. By eliminating one of the three equations, it follows that

$$M_G = M_{ICE}\frac{1}{\tau_o - 1} ,$$
(22.12)

$$M_T = M_{EM} - \tau_o M_G .$$
(22.13)

This system works both as a parallel and a series hybrid.

The angular velocity of the thermal engine adapts to that of the vehicle by changing the speed of the generator, following Eq. (22.10), something that can occur only by subtracting a torque, through the generator, following Eq. (22.12). By doing this, some power from the thermal engine charges the battery, as in the series layout.

At low speed, a part of the power needed for motion is supplied by the electric motor, which takes it from the battery. Finally, at a very low speed only the electric motor operates, as in parallel layouts. This also occurs when the speed of the thermal engine can adapt itself to that of the vehicle, without the generator subtracting any power.

When the vehicle slows down, the available kinetic energy is recovered.

This method allows the working range of the thermal engine to be restricted to that where minimum fuel consumption is obtained, for a given power requirement. It is also possible to stop the engine when the vehicle stops and to restart it easily at a speed greater than those at which conventional starter motors operate, owing to the generator that is now used as a motor.

The batteries are never recharged from outside the vehicle.

The fuel consumption, obtained using a gasoline engine (Atkinson cycle), is similar to that of a diesel vehicle with similar performance; CO_2 emissions are lower, due to the lower quantity of carbon contained in the same volume of gasoline; other emissions are much lower, due to the reduced working of the engine in variable conditions owing to the more constant use of the thermal engine made possible by the hybrid layout.

23
DRIVING DYNAMIC PERFORMANCE

When computing the performance of a vehicle in longitudinal motion (maximum speed, gradeability, fuel consumption, braking, etc.), the vehicle is modelled as a rigid body, or in an even simpler way, as a point mass.

The presence of suspensions and the compliance of tires are then neglected and motion is described by a single equation, the equilibrium equation in the longitudinal direction. If the x-axis is assumed to be parallel to the ground, the longitudinal equilibrium equation reduces to

$$m\ddot{x} = \sum_{\forall i} F_{x_i}, \tag{23.1}$$

where F_{x_i} are the various forces acting on the vehicle in the longitudinal direction (aerodynamic drag, rolling resistance, traction, braking forces, etc.).

As will be seen later, Eq. (23.1) is quite a rough model for various reasons. For one thing, when the vehicle is accelerated, a number of rotating masses must be accelerated as well; this, however, can be accounted for easily. Other approximations come from the fact that the vehicle does not travel under symmetrical conditions, particularly when the trajectory is not straight and the direction of the x-axis does not coincide with the direction of the velocity or, in other words, the sideslip angle β is in general different from zero.

23.1 LOAD DISTRIBUTION ON THE GROUND

Longitudinal dynamics is influenced by the distribution of normal forces at the wheels-ground contact. A vehicle with more than three wheels is statically indeterminate, and the load distribution is determined by characteristics of the suspensions which, as seen in Part I, also have the task of distributing the load on the ground in proper way. However, if the system is symmetrical with respect to the xz plane, all loads are equally symmetrical, and the velocity is contained in the symmetry plane, then the two wheels of any axle are equally loaded. In this case, it is possible to think in terms of axles rather than wheels, and a two-axle vehicle may be considered as a beam on two supports which is, then, a statically determined system. In this case, the forces on the ground do not depend on the characteristics of the suspensions and the vehicle can be modelled as a rigid body.

23.1.1   Vehicles with two axles

Consider the vehicle as a rigid body and neglect the compliance of the suspensions and of the body. As previously stated, if the vehicle is symmetrical with respect to the xy-plane[1], it can be modelled as a beam on two supports, and normal forces F_{z_1} and F_{z_2} acting on the axles can be computed easily.

With the vehicle at a standstill on level road the normal forces are

$$\begin{cases} F_{z_1} = mg\epsilon_{0_1} \\ F_{z_2} = mg\epsilon_{0_2} \end{cases} \quad \text{where} \quad \begin{cases} \epsilon_{0_1} = b/l \\ \epsilon_{0_2} = a/l \ . \end{cases} \quad (23.2)$$

The forces acting on a two-axle vehicle moving on straight road with longitudinal grade angle α (positive when moving uphill) are sketched in Fig. 23.1. Note that the x-axis is assumed to be parallel to the road surface.

Taking into account the inertia force $-m\dot{V}$ acting in x direction on the centre of mass, the dynamic equilibrium equations for translations in the x and z direction and rotations about point O are

$$\begin{cases} F_{x_1} + F_{x_2} + F_{x_{aer}} - mg\sin(\alpha) = m\dot{V} \\ F_{z_1} + F_{z_2} + F_{z_{aer}} - mg\cos(\alpha) = 0 \\ F_{z_1}(a + \Delta x_1) - F_{z_2}(b - \Delta x_2) + mgh_G\sin(\alpha) - M_{aer} + |F_{x_{aer}}|h_G = -mh_G\dot{V} \ . \end{cases}$$
$$(23.3)$$

If the rolling resistance is ascribed completely to the forward displacement of the resultant F_{z_i} of contact pressures σ_z, distances Δx_i can be easily computed as

$$\Delta x_i = R_{l_i} f = R_{l_i}(f_0 + KV^2) \ . \quad (23.4)$$

Except in the case of vehicles with different wheels on the various axles, such as F-1 racers, the values of Δx_i are all equal.

---

[1]In the present section on longitudinal dynamics, a complete symmetry with respect to the xz plane is assumed: The loads on each wheel are respectively $F_{z_1}/2$ and $F_{z_2}/2$ for the front and the rear wheels. To simplify the equations, the x-axis is assumed to be parallel to the road surface.

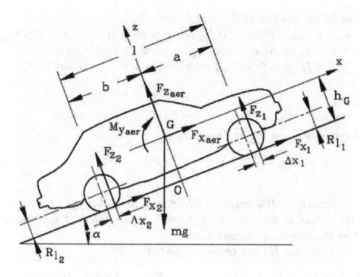

FIGURE 23.1. Forces acting on a vehicle moving on an inclined road.

The second and third equation (23.3) can be solved in the normal forces acting on the axles, yielding

$$
\begin{cases}
F_{z_1} = mg \dfrac{(b - \Delta x_2)\cos(\alpha) - h_G \sin(\alpha) - K_1 V^2 - \dfrac{h_G}{g}\dot{V}}{l + \Delta x_1 - \Delta x_2} \\[3mm]
F_{z_2} = mg \dfrac{(a + \Delta x_1)\cos(\alpha) + h_G \sin(\alpha) - K_2 V^2 + \dfrac{h_G}{g}\dot{V}}{l + \Delta x_1 - \Delta x_2}
\end{cases}
\tag{23.5}
$$

where

$$
\begin{cases}
K_1 = \dfrac{\rho S}{2mg}\left[C_x h_G - l C_{M_y} + (b - \Delta x_2) C_z \right] \\[3mm]
K_2 = \dfrac{\rho S}{2mg}\left[-C_x h_G + l C_{M_y} + (a + \Delta x_1) C_z \right].
\end{cases}
$$

The values of Δx_i are usually quite small (in particular, their difference is usually equal to zero) and can be neglected. If considered, they introduce a further weak dependence of the vertical loads on the square of the speed, owing to the term KV^2 in the rolling resistance.

Example 23.1 *Compute the force distribution on the ground of the small car of Appendix E.1 at sea level, with standard pressure and temperature, in the following conditions:*
a) at standstill on level road;
b) driving at 100 km/h on level road;

c) driving at 70 km/h on a 10% grade;

d) braking with a deceleration of 0.4 g on level road at a speed of 100 km/h.
    The air density in the mentioned conditions is 1.2258 kg/m^3.

a) Using Eq. (23.2), the static load distribution between the axles is

$$\epsilon_{0_1} = 0.597, \qquad \epsilon_{0_2} = 0.403.$$

The forces acting on the axles are then

$$F_{z_1} = 4863 \ N, \qquad F_{z_2} = 3280 \ N.$$

b) From Eq. (23.4), at 100 km/h = 27.78 m/s the value of Δx is 4.6 mm for all tires. This value is so small that it could be neglected; it will, however, be considered in the following computations.

Constants K_1 and K_2 are easily computed

$$K_1 = 8.505 \times 10^{-6} \ s^2/m, \qquad K_2 = -5.869 \times 10^{-5} \ s^2/m.$$

The forces acting on the axles are then

$$F_{z_1} = 4820 \ N, \qquad F_{z_2} = 3491 \ N.$$

c) A 10% grade corresponds to a grade angle $\alpha = 5.7°$. Operating in the same way, at 70 km/h = 19.44 m/s the value of Δx is 4.0 mm for all tires. The other results are

$$K_1 = 8.490 \times 10^{-6} \ s^2/m, \qquad K_2 = -5.867 \times 10^{-5} \ s^2/m,$$

$$F_{z_1} = 4643 \ N, \qquad F_{z_2} = 3542 \ N.$$

d) The acceleration is $\dot{V} = -3.924 \ m/s^2$. As the speed is the same as in case b), the same values for Δx, K_1 and K_2 hold. The forces are

$$F_{z_1} = 5498 \ N, \qquad F_{z_2} = 2813 \ N.$$

23.1.2 Vehicles with more than two axles

If more than two axles are present, even in symmetrical conditions the system remains statically indeterminate and it is necessary to take into account the compliance of the suspensions (Fig. 23.2a). The equilibrium equations (23.3) still hold, provided that the terms

$$F_{x_1} + F_{x_2} , \quad F_{z_1} + F_{z_2} , \quad F_{z_1}(a + \Delta x_1) - F_{z_2}(b - \Delta x_2)$$

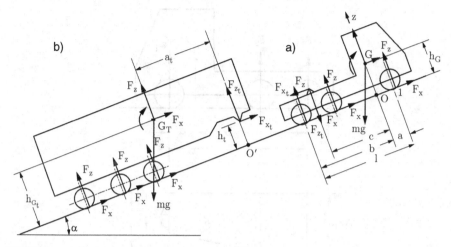

FIGURE 23.2. Forces acting on an articulated vehicle moving on an inclined road. (a) Tractor or vehicle with more than two axles; (b) trailer.

are substituted by

$$\sum_{\forall i} F_{x_i} \,,\quad \sum_{\forall i} F_{z_i} \,,\quad \sum_{\forall i} F_{z_1}(x_i + \Delta x_i)\,,$$

where distances x_i are positive for axles located forward of the centre of mass and negative otherwise.

For computation of normal loads on the ground a number $(n-2)$ of equations, where n is the total number of axles, must be added. Each one of them simply expresses the condition that the vertical displacement of the point where each intermediate suspension is attached to the body is compatible with the displacement of the first and the last.

To account for possible nonlinearities of the force-displacement curves of the suspension, it is advisable to compute a reference position in which each suspension exerts a force $(F_{z_i})_0$. The linearized stiffness of the ith suspension, possibly taking into account the compliance of the tires as well, is K_i. The vertical displacement of the point where the ith suspension is attached is

$$\Delta z_i = -\frac{1}{K_i}\left[F_{z_i} - (F_{z_i})_0\right]\,. \tag{23.6}$$

With reference to Fig. 23.3, the vertical displacement of the vehicle body in the point where the ith suspension is attached can be expressed as a function of the displacement of the first and nth suspension by the equation

$$\frac{1}{l}\left(\Delta z_n - \Delta z_1\right) = \frac{1}{a - x_i}\left(\Delta z_i - \Delta z_1\right)\,. \tag{23.7}$$

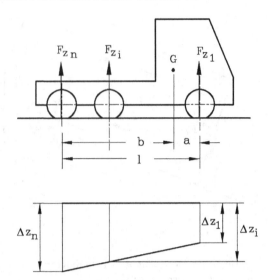

FIGURE 23.3. Compatibility condition for vertical displacements of the points where the suspensions are attached. In the figure, the ith axle is behind the center of mass and its coordinate x_i is negative.

By eliminating displacements Δz_i between equations (23.6) and (23.7), the required equation is obtained,

$$\frac{b + x_i}{K_1} [F_{z_1} - (F_{z_1})_0] + \frac{a - x_i}{K_n} [F_{z_n} - (F_{z_n})_0] - \frac{l}{K_i} [F_{z_i} - (F_{z_i})_0] = 0 , \quad (23.8)$$

$$\text{for } i = 2, \dots, n - 1 .$$

The mentioned reference condition can be referred to any value of the load or any position of the centre of gravity, provided that the values of the linearized stiffnesses are the same as those in the actual condition. Forces $(F_{z_i})_0$ can all be set to zero if the springs are linear and the suspensions are such that a position (i.e. a vertical and a pitch displacement) exists in which all wheels just touch the ground, exerting on it vanishing forces (neglecting the weight of the axles).

Equations (23.8), together with the second and third equation (23.3), form a set of n equations that can be solved to yield the n normal forces acting on the axles.

Remark 23.1 *Forces F_{z_i} can never become negative: If a negative value is obtained, it means that the relevant axle loses contact with the ground and the computation must be repeated after setting the force to zero due to the relevant axle. The procedure is repeated until no negative force is present.*

23.1.3   Articulated vehicles

In the case of articulated vehicles with a tractor with two axles and one or more trailers with no more than a single axle each (Fig. 23.2), the computation is straightforward. In this case, the equilibrium equations of the tractor are

$$
\begin{cases}
\displaystyle\sum_{i=1}^{n} F_{x_i} - F_{x_t} + F_{x_{aer}} - mg\sin(\alpha) = m\dot{V} \\[2mm]
\displaystyle\sum_{i=1}^{n} F_{z_i} - F_{z_t} + F_{z_{aer}} - mg\cos(\alpha) = 0 \\[2mm]
\displaystyle\sum_{i=1}^{n} F_{z_i}(x_i + \Delta x_i) + F_{z_t}c + F_{x_t}h_t + mgh_G\sin(\alpha) - M_{aer} + \\[2mm]
\qquad + |F_{x_{aer}}|h_G = -mh_G\dot{V} \,,
\end{cases}
\tag{23.9}
$$

where forces F_{x_t} and F_{z_t} are those the tractor exerts on the trailer, as in the figure, the number of axles of the tractor is assumed to be n (in the present case $n = 2$), the moments are computed with reference to point O, and the aerodynamic forces and moments are those exerted on the tractor only.

Similarly, the equilibrium equation of the trailer are

$$
\begin{cases}
\displaystyle\sum_{i=1}^{m} F_{x_i} + F_{x_t} + F_{x_{R_{aer}}} - m_R g\sin(\alpha) = m_R\dot{V} \\[2mm]
\displaystyle\sum_{i=1}^{m} F_{z_i} + F_{z_t} + F_{z_{R_{aer}}} - m_R g\cos(\alpha) = 0 \\[2mm]
\displaystyle\sum_{i=1}^{m} F_{z_i}(x_i + \Delta x_i) - F_{x_t}h_t + m_R g h_{G_R}\sin(\alpha) + m_R g a_R\cos(\alpha) - M_{R_{aer}} + \\[2mm]
\qquad + |F_{x_{R_{aer}}}|h_{G_R} = -m_R h_{G_R}\dot{V} \,,
\end{cases}
\tag{23.10}
$$

where the number of axles of the trailer is assumed to be m (in the present case $m = 1$), the moments are computed with reference to point O', the aerodynamic forces and moments are those exerted on the trailer only and x_i are the coordinates of the axle in the reference frame centred in O'. Note that all x_i are usually negative.

The last two equations (23.9), together with the last two equations (23.10) are sufficient only on level road at a standstill, when force F_{x_t} vanishes. If it is other than zero the first equation (23.9) must also be used. However, the forces F_{x_i} it contains are not known since they depend on the normal forces F_{z_i}. A simple iterative scheme can be used, to compute the normal forces with $F_{x_t} = 0$, repeating the computation until a stable solution is found. If the wheels of the trailer exert driving or braking forces, these forces must also be introduced into the computation.

If the tractor has more than two axles or the trailer has more than one, additional equations must be introduced. The additional $(n - 2)$ equations of the tractor (n is the number of axles of the tractor), are equations (23.8) while the additional $(m - 1)$ equations for the trailer, where m is the number of its axles, are

$$\frac{(a+c)(x_m - x_i)}{lK_{t_i}} [F_{z_1} - (F_{z_1})_0] + \frac{(b-c)(x_m - x_i)}{lK_n} [F_{z_n} - (F_{z_n})_0] +$$

$$+ \frac{x_m}{K_{R_m}} [F_{zR_m} - (F_{zR_m})_0] - \frac{x_m}{K_{R_i}} [F_{zR_i} - (F_{zR_i})_0] = 0 , \qquad (23.11)$$

$$\text{for } i = 1, \ldots, m-1 ,$$

where K_{t_i} and F_{zt_i} are the linearized stiffness of the ith suspension of the trailer and the force acting on it and $(F_{zt_i})_0$ is the normal force in the same axle in any reference condition.

The first two terms of Eq. (23.11) are linked to the vertical displacement of the hitch, and the equation expresses the displacements of the hitch, the last axle and the relevant axle.

The number of unknowns and equations is then equal to the total number of axles plus one, since the normal force the two parts of the vehicle exchange is also unknown. When force F_{x_t} does not vanish, it must be computed iteratively, as seen above.

Example 23.2 *Compute the force distribution on the ground of the five-axle articulated truck of Appendix E.9 at sea level, with standard pressure and temperature, in the following conditions:*
a) at standstill on level road;
b) at standstill on a 10% grade;
c) driving at 70 km/h on a 10% grade;
The air density in the mentioned conditions is 1.2258 kg/m³.
a) The static load distribution on level road can be computed directly, as the horizontal force exchanged between the two parts of the vehicle vanishes. The unknowns are six, the loads of the five axles and the vertical force exchanged between tractor and trailer. These can be computed from the set of linear equations

$$\begin{bmatrix} 1.000 & 1.000 & 0 & 0 & 0 & -1.000 \\ 1.175 & -2.310 & 0 & 0 & 0 & 1.860 \\ 0 & 0 & 1.000 & 1.000 & 1.000 & 1.000 \\ 0 & 0 & -6135 & -7.395 & -8.715 & 0 \\ -1,070 & -0,1109 & 4,054 & 0 & -4,446 & 0 \\ -0,5474 & -0,06087 & 0 & 4,054 & -5,359 & 0 \end{bmatrix} \times 10^{-3}$$

$$\times \begin{Bmatrix} F_{z_1} \\ F_{z_2} \\ F_{zR_1} \\ F_{zR_2} \\ F_{zR_3} \\ F_{zt} \end{Bmatrix} = \begin{Bmatrix} 70.100 \\ 0 \\ 313.900 \\ -1.597.900 \\ 0 \\ 0 \end{Bmatrix} .$$

The forces acting on the axles are then 58.660 kN, 105.700 kN, 80.060 kN, 83.600 kN and 56.050 kN. The force at the tractor-trailer connection is 94.210 kN.

b) A 10% grade corresponds to a grade angle $\alpha = 5.7°$. In this case the load distribution can also be computed directly, since the horizontal force exchanged between the two parts of the vehicle does not depend on the normal forces. Operating as in the previous case, the forces acting on the axles are 45.050 kN, 115.720 kN, 78.900 kN, 84.490 kN and 58.000 kN. The forces at the tractor-trailer connection are 90.980 kN in the vertical direction and 31.236 kN in the horizontal direction.

c) At 70 km/h = 19.44 m/s the value of Δx is 3.7 mm for all tires. In this case, owing to rolling resistance, an iterative solution must be obtained. However, the convergence is very fast, as only five iterations are needed to reach a difference between the results at the i -th and at the $(i-1)$-th iteration smaller than 10^{-6} in relative terms. The other results are not dissimilar to those obtained in the previous case: The forces on the axles are 43.980 kN, 116.970 kN, 78.710 kN, 84.440 kN and 58.060 kN; those at the tractor-trailer connection are 91.150 kN in the vertical direction and 33.440 kN in the horizontal direction.

Note that the matrix of the coefficients of the relevant set of equations is the same in all cases.

23.2 TOTAL RESISTANCE TO MOTION

Consider a vehicle moving at constant speed on a straight and level road. The forces that must be overcome to maintain a constant speed are aerodynamic drag and rolling resistance.

By using the simplified formula seen in Part I to express the dependence of rolling resistance on speed, the modulus of the first is

$$R_r = \sum_{\forall i} F_{z_i} \left(f_0 + KV^2 \right) , \tag{23.12}$$

where F_{z_i} is the force acting in a direction perpendicular to the ground on the ith wheel.

Assuming that the rolling coefficient f is the same for all wheels[2], the sum of all normal forces can be brought out from the sum and, taking into account aerodynamic lift as well, it follows that

$$R_r = \left(f_0 + KV^2 \right) \sum_{\forall i} F_{z_i} = \left[mg\cos(\alpha) - \frac{1}{2}\rho V_r^2 SC_z \right] \left(f_0 + KV^2 \right) . \tag{23.13}$$

Aerodynamic drag (or, better, the aerodynamic force in the x direction, Eq.(21.11)) has a value (always as an absolute value) of

$$R_a = \frac{1}{2}\rho V_r^2 SC_x . \tag{23.14}$$

---

[2]This assumption holds only as a first approximation, since it does not take into account the dependence of f on the driving or braking conditions or other variables.

With increasing speed, the importance of the former grows; at a given value of the speed aerodynamic drag becomes more important than rolling resistance. This speed is lower for small cars while for larger vehicles, particularly for trucks at full load, rolling resistance is the primary form of drag. Another factor is that usually the mass of the vehicle grows with its size more rapidly than the area of its cross section.

If the road is not level, the component of weight acting in a direction parallel to the velocity V, i.e. the grade force

$$R_p = mg \sin(\alpha) \qquad (23.15)$$

must be added to the resistance to motion.

The grade force becomes far more important than all other forms of drag even for moderate values of grade (Fig. 23.1). Since the force acting in a direction perpendicular to the ground on a sloping road is only the component of weight perpendicular to the road, the total resistance to motion, or road load, as it is commonly referred to, can be written in the form

$$R = \left[mg \cos(\alpha) - \frac{1}{2} \rho V^2 S C_z \right] (f_0 + K V^2) + \frac{1}{2} \rho V^2 S C_x + mg \sin(\alpha) , \quad (23.16)$$

where, assuming that the air is still, the velocity with respect to air V_r becomes conflated with velocity V.

To highlight its dependence on speed, the road load can be written as

$$R = A + B V^2 + C V^4 , \qquad (23.17)$$

where

$$A = mg \left[f_0 \cos(\alpha) + \sin(\alpha) \right] ,$$

$$B = mgK \cos(\alpha) + \frac{1}{2} \rho S [C_x - C_z f_0] ,$$

$$C = -\frac{1}{2} \rho S K C_z .$$

The last term in Eq. (23.17) becomes important only at very high speed in the case of vehicles with strong negative lift: It is usually neglected except in racing cars.

Since the grade angle of roads open to vehicular traffic is usually not very large, it is possible to assume that

$$\cos(\alpha) \approx 1 , \qquad \sin(\alpha) \approx \tan(\alpha) \approx i ,$$

where i is the grade of the road. In this case coefficient B is independent of the grade of the road and

$$A \approx mg(f_0 + i)$$

depends linearly on it. C never depends on grade.

23.3   POWER NEEDED FOR MOTION

The power needed to move at constant speed V is obtained simply by multiplying the road load given by Eq. (23.17) by the value of the velocity

$$P_n = VR = AV + BV^3 + CV^5 . \qquad (23.18)$$

Example 23.3 *Plot the curves of the road load of the car of Appendix E.1 on level road and on a 10% grade. Plot the curve of the power needed for constant speed driving on level road.*

   The results obtained through Eq. (23.16) are shown in Fig. 23.4.

Example 23.4 *Plot the curves of the road load of the articulated truck of Appendix E.9 on level road and on a 10% grade.*

   The results obtained through Eq. (23.16) are shown in Fig. 23.5. Note that in this case aerodynamic drag amounts to a relatively small part of the road load and that on a 10% grade the grade force is very high.

   Motion at constant speed is possible only if the power available at the wheels at least equals the required power given by Eq. (23.18). This means that the

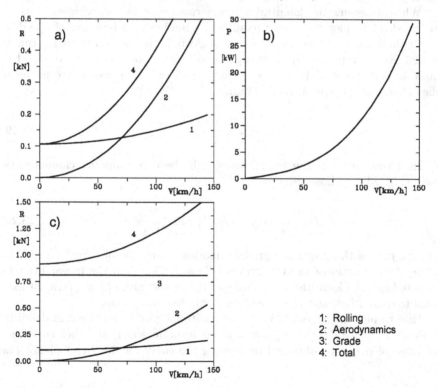

FIGURE 23.4. Resistance (a), and power (b) needed for motion at constant speed for the small car of Appendix E.1. Road load on the same car driving on a 10% slope (c).

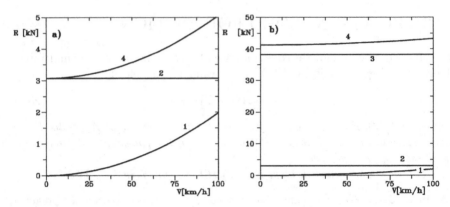

FIGURE 23.5. Aerodynamic drag (curve 1), rolling resistance (2), grade force (3) and total road load (4) for the articulated truck of Appendix E.9 on level road (a) and on a 10% grade (b).

engine must supply sufficient power, taking into account losses in the transmission as well, and that the road-wheel contact must be able to transmit this power.

When assessing the longitudinal performance of motor vehicles, it is often expedient to plot the power required for motion as a function of speed on a logarithmic plot. If the term CV^5 is neglected, the two remaining terms of Eq. (23.18) are represented by straight lines with slopes respectively equal to 1 and 3. The two straight lines cross in a point whose x-coordinate is the so called characteristic speed, easily obtained from Eq. (23.18)

$$V_{car} = \sqrt{\frac{A}{B}} \,. \tag{23.19}$$

Its y-coordinate is the logarithm of half the corresponding characteristic power (Fig. 23.6), whose value is:

$$P_{car} = AV_{car} + BV_{car}^3 = 2A\sqrt{\frac{A}{B}} \,. \tag{23.20}$$

The plot of the power required for motion cannot be obtained directly by adding the ordinates of the two straight lines of Fig. 23.6: the power must be computed from its logarithm, the values of the powers given by the straight lines added to each other, and the logarithm computed once again.

How to plot the curve $P_n(V)$ on a logarithmic plot is described in detail by M. Bencini[3]. Once the curve for $\alpha = 0$ has been obtained, all other curves for any value of α can be obtained by moving the curve on the $P(V)$ plane. The

---

[3]M. Bencini, *Dinamica del veicolo considerato come punto*, Tamburini, Milano, 1956.

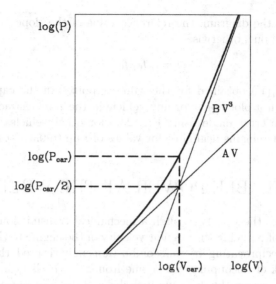

FIGURE 23.6. Power needed for motion. Logarithmic plot.

curve must be moved by

$$\Delta_1 = \frac{1}{2} \log \left[\frac{f_0 + \tan(\alpha)}{f_0} \right] + \frac{1}{2} \log \left[\cos(\alpha) \frac{2gK + \lambda}{2gK \cos(\alpha) + \lambda} \right] , \qquad (23.21)$$

where

$$\lambda = \rho S C_x / m ,$$

in the horizontal direction, and by

$$\Delta_2 = \frac{3}{2} \log \left[\frac{f_0 + \tan(\alpha)}{f_0} \right] + \frac{1}{2} \log \left[\cos^3(\alpha) \frac{2gK + \lambda}{2gK \cos(\alpha) + \lambda} \right], \qquad (23.22)$$

in the vertical direction. For values of the road slope small enough to accept the approximation $\cos(\alpha) \approx 1$, equations (23.21) and (23.22) simplify as

$$\Delta_1 = \frac{1}{2} \log \left(\frac{f_0 + i}{f_0} \right) , \qquad \Delta_2 = \frac{3}{2} \log \left(\frac{f_0 + i}{f_0} \right) , \qquad (23.23)$$

where $i = \tan(\alpha)$. The curves move along a straight line with slope 3 and the displacement depends only on the value of f_0, and obviously on the slope i of the road.

It is then possible to obtain a single logarithmic plot containing a set of curves $P_n(V)$ for different values of α that can be used for any vehicle in the range where $\cos(\alpha) \approx 1$. Such a plot must be adapted to any particular vehicle by computing the values of characteristic speed and power (V_{car}, P_{car}) on level road. If the value of the rolling coefficient f_0 coincides with the reference value

f_{0_r} used to plot the diagram, the reference value of the slope i_r shown on the plot is the actual one, otherwise

$$i = i_r f_0 / f_{0_r} \ . \tag{23.24}$$

The plot $P_n(V)$ obtained for the values reported in the caption is shown in Fig. 23.7. Such a plot holds for any vehicle in the zone characterized by low values of α. The error made using Fig. 23.7 for other vehicles increases with increasing α, but remains negligible for values of i up to $0.3 \div 0.4$.

23.4   AVAILABLE POWER AT THE WHEELS

The engine drives the wheels through a mechanical transmission whose task is essentially that of reducing the angular velocity of the engine to that required at the wheels. If a reciprocating internal combustion engine is used, the transmission also has the task of uncoupling the engine from the wheels at a stop or at low speed, for which reason the driveline includes a clutch or a torque converter as well.

It is possible, at least as a first approximation, to state a value of the efficiency η_t for any type of driveline. The power available at the wheels is then

$$P_a = P_e \eta_t \ . \tag{23.25}$$

Depending on the type of transmission, the efficiency can be considered as a constant (obviously only as a first approximation), or may be computed, but only after assessing the working conditions of the driveline and above all of the torque converter, if present.

To compute the efficiency of the driveline, or the power it dissipates, see the relevant section of Part II.

The equation linking the speed of the engine to that of the wheels is simply

$$V = \Omega_e R_e \tau_t \ , \tag{23.26}$$

where τ_t is the overall gear ratio, defined as the ratio between the speed of the wheels and that of the engine shaft, and is usually smaller than 1. Once the gear ratios of all parts of the transmission are known, the power available at the wheels can be plotted as a function of the speed of the vehicle on the same plot where the power needed for motion at constant speed is reported.

If the curves of the required and available power are plotted on a logarithmic plot, any change of the gear ratio causes a translation of the curve related to the available power along the V-axis, while a change in the efficiency of the driveline causes a translation of the same curve along the P-axis (Fig. 23.8). If a continuous transmission (CVT) is present, the position of the curve is a function of the gear ratio, and then it is possible to define a zone on the VP-plane where all possible working conditions are included.

See Part II for situations where a torque converter is present.

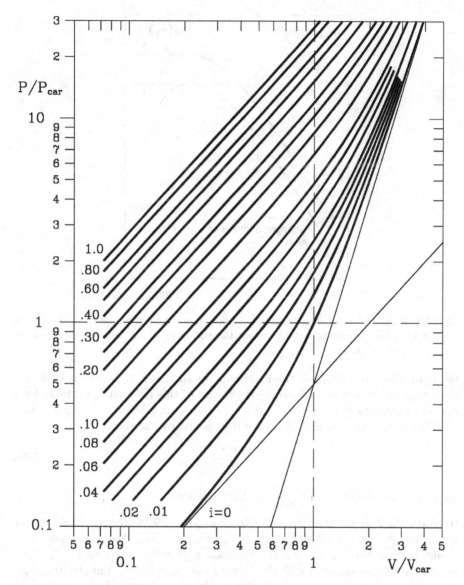

FIGURE 23.7. Logarithmic plot of the power needed for motion. The characteristic speed and power refer to a level road with $f_0 = 0.013$. For high values of the slope the plot holds only if $m = 1000$ kg; $S = 1.7\,\text{m}^2$; $C_x = 0.42$; $K = 6.5 \times 10^{-6}\,\text{s}^2/\text{m}^2$; $g = 9.81$ m/s^2; $\rho = 1.22$ kg/m^3.

23.5   MAXIMUM POWER THAT CAN BE TRANSFERRED TO THE ROAD

The power needed to overcome the road load must be transferred through the road-wheel contact. As it increases with both increasing speed and the grade of

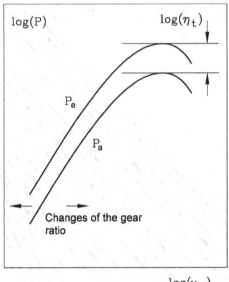

$$\log(\omega_m)$$

FIGURE 23.8. Curves of maximum engine power and power available at the wheels plotted with logarithmic scales. Changing the efficiency of the transmission and the overall transmission ratio.

the road, there is a limit on the maximum speed that can be reached and the maximum grade that can be managed because of this limit on the driving force the vehicle can exert, even if no limit to the power supplied by the engine exists.

The maximum power that can be transferred by the vehicle is

$$P_{max} = V \sum_{\forall i} F_{z_i} \mu_{i_p} \,, \tag{23.27}$$

where the sum is extended to all the driving wheels.

Remark 23.2 *If the maximum longitudinal force coefficient μ_{i_p} and the load acting on the driving wheels were independent of speed, the maximum power would increase linearly with V. The optimum engine characteristic $P_e(\Omega_e)$ for a vehicle with a transmission with fixed ratio would be a linear characteristic. This is not the case, however, as the situation is far more complicated.*

To begin with, consider the case of a vehicle with two axles, all of which are driving and assume that all wheels work with the same longitudinal slip, i.e. that the values of μ_i are equal. This situation will be referred here to as "ideal driving force".

The maximum power that can be transferred to the road is then

$$P_{max} = V \mu_{i_p} \left(mg \cos(\alpha) - \frac{1}{2} \rho V^2 S C_z \right) . \tag{23.28}$$

23.5.1   Vehicles with all wheels driving

When all wheels are driving wheels, it is possible to assume that the rolling resistance of the wheels is due only to the forward displacement of force F_z at the road-wheel contact, and hence is overcome directly by the driving torque exerted by the engine.

To compute the maximum grade that can be managed at low speed, it is possible to assume that the only road load that must be overcome at the wheel-road contact at such speeds is the grade load. By equating the power required for motion (Eq. 23.18) to the maximum power that can be transferred (Eq. 23.28), the maximum grade is readily obtained,

$$\tan(\alpha_{max}) = i_{max} = \mu_{x_p} . \tag{23.29}$$

To compute the maximum speed that can be reached, the decrease of traction at the road-wheel contact occurring with increasing speed must be modelled. A very simple way is to assume a linear law

$$\mu_{i_p} = c_1 - c_2 V . \tag{23.30}$$

By equating the power required for motion at constant speed (except that related to rolling resistance) to Eq. (23.28) and using Eq. (23.30) for expressing the decrease of the available driving force with the speed, the maximum speed can be obtained from the cubic equation

$$C_z c_2 V^3 + \left(C_z c_1 + C_x \right) V^2 - \frac{2mg}{\rho S} \left[(c_1 - c_2 V) \cos(\alpha) - \sin(\alpha) \right] = 0 . \tag{23.31}$$

The values of the maximum grade and of the maximum speed can only be achieved in ideal conditions, since the longitudinal slip of all wheels has been assumed to be the same. The forces acting on the driving axles can be computed by using the procedure already seen: They generally depend not only on the static load distribution but also on the speed and the acceleration.

In the case of a vehicle with two axles, both driving, the ratio

$$K_T = \frac{F_{x_1}}{F_{x_2}}$$

between the driving force at the front wheels and that at the rear wheels is usually a constant. If the wheels all have the same diameter, it coincides with the ratio between the driving torques supplied to the two axles. Assume that the two axles have tires of the same type and operate on patches of road with the same characteristics. If

$$K_T \frac{F_{z_2}}{F_{z_1}} > 1 ,$$

the limit conditions occur at the front wheels. At the onset of slipping, the power that can be transferred to the road is then

$$P_{max} = V \mu_{x_p} F_{z_1} \frac{1 + K_T}{K_T} . \tag{23.32}$$

By plotting the maximum power obtained by Eq. (23.32) versus the speed together with the power required given by Eq. (23.17) multiplied by V, the maximum speed at which the vehicle is able to transfer enough power to maintain its speed is readily obtained. It must be remembered that rolling resistance must be neglected in the computation of the required power.

If, on the contrary,

$$K_T \frac{F_{z_2}}{F_{z_1}} < 1 ,$$

the limit conditions occur at the rear wheels and the maximum power that can be transferred to the road is

$$P_{max} = V \mu_{x_p} F_{z_2} (1 + K_T) . \tag{23.33}$$

The maximum grade that can be managed is also easily obtained. Since in this case the speed can be set to zero, it follows that, for an all-wheel drive vehicle,

$$\tan(\alpha_{max}) = i_{max} = \frac{b c_1 \mu_{x_p} \left(1 + \frac{1}{K_T}\right)}{l + z_G \mu_{x_p} \left(1 + \frac{1}{K_T}\right)} ,$$

$$\tan(\alpha_{max}) = i_{max} = \frac{a \mu_{x_p} (1 + K_T)}{l + z_G \mu_{x_p} (1 + K_T)} , \tag{23.34}$$

respectively if the rear wheels slip first, i.e. if

$$K_T < \frac{b - h_G \tan(\alpha_{max})}{a + h_G \tan(\alpha_{max})} ,$$

or if this condition does not hold. It is, however, unlikely that the rear wheels are in a critical condition on a very steep grade, since this would require an abnormally low value of K_T. The value of μ_{x_p} is that for a vanishingly small speed.

23.5.2 Vehicles with a single driving axle

If not all axles are driving, the power that can be transferred to the ground is smaller. Aerodynamic drag increases the load on the rear wheels, as does a positive grade of the road: The power that can be transferred by a rear-wheels drive vehicle thus increases with speed, due to drag, and with the slope. Aerodynamic moment and lift have different effects depending on the sign of the moments and the position of the centre of mass. The maximum power is then

$$P_{max} = V \mu_{x_p} F_{z_1} , \qquad P_{max} = V \mu_{x_p} F_{z_2} , \tag{23.35}$$

respectively for the cases of front and rear wheel drive. Only the rolling resistance of the free wheels must be accounted for in the computation of the power needed for constant speed driving ; this is easily done by introducing F_{z_2} or F_{z_1}, respectively for front- and rear-wheels drive, in the expression of the road load instead of the total load on the ground.

Equations (23.34) could still be used for the computation of the maximum grade that can be managed. The first equation holds for front-wheel drive $(1/K_T = 0)$ and the second for rear-wheel drive, $(K_T = 0)$ but they do not include the rolling resistance of the free wheels.

If this effect is accounted for, the equations are modified as

$$\tan(\alpha_{max}) = i_{max} = \frac{b\mu_{x_p} - af_0}{l + z_G(\mu_{x_p} + f_0)} ,$$

$$\tan(\alpha_{max}) = i_{max} = \frac{a\mu_{x_p} - bf_0}{l - z_G(\mu_{x_p} + f_0)} . \tag{23.36}$$

Example 23.5 *Plot the curves of maximum transmissible and required power for the vehicle of Appendix E.2 on dry and wet road. Compute the maximum speed and the maximum grade that can be managed.*

Repeat the computations assuming that the same vehicle has rear-wheel drive, without changing the static load distribution on the ground.

Assume that $c_1 = 1.1$ and $c_2 = 6 \times 10^{-3}$ s/m on dry road and $c_1 = 0.8$ and $c_2 = 8 \times 10^{-3}$ s/m on wet road.

The curves of the maximum transmissible power are shown in Fig. 23.9, together with those of the required power. The vehicle of the example can thus reach a maximum speed of 225 km/h (wet road) or 308 km/h (dry road) for reasons linked only to the wheel driving force.

The computations were repeated assuming that the driving wheels are the rear ones. In this case, the maximum power that can be transferred to the ground

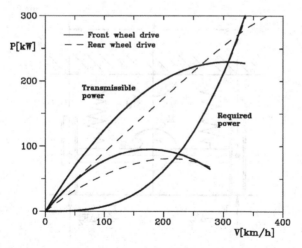

FIGURE 23.9. Maximum transmissible power and required power on level road in the case of Example 23.5.

at low speed is lower than in the previous case, since the static load distribution was specified in order to obtain good performance with front-wheel drive.

The load on the rear wheels increases with increasing speed and eventually gets larger than that on the front wheels. On dry road, the maximum speed is then higher for the vehicle with rear-wheel drive, despite the fact that at standstill the front wheels are loaded by about 60% of the weight.

The values of the maximum speed for the rear-wheel drive vehicle are of 218 km/h on wet road or 328 km/h on dry road.

Note that the required power curve includes only the rolling resistance of the non-driving wheels, and is slightly different in the two cases.

Also note that the curves do not take into consideration the load shift due to acceleration, so for speeds lower than the maximum speed, where the vehicle would accelerate if the maximum power is applied, they are not realistic.

The maximum grade angle that can be managed when only the wheel driving force is considered may be computed using Eq. (23.36), obtaining 28.0° on dry road and 22.1° on wet road, corresponding to grades of 53.1% and 40.6% respectively. If the driving wheels were the rear ones the values would have been 29.4° (56.3%) and 20.6° (37.6%).

In the case of rigid axles in which the final gear is directly mounted on the axle and the propeller shaft is in the longitudinal direction, the drive torque M_d applied to the axle causes a transversal load shift between the driving wheels of the same axle.

With reference to Fig. 23.10 the load shift ΔF_z could be determined easily as

$$\Delta F_z = \frac{M_d}{t_i} \qquad \text{where} \qquad M_d = F_x R_l \tau_f , \qquad (23.37)$$

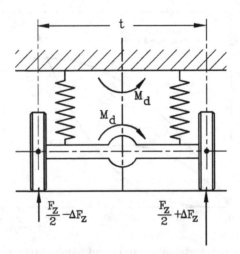

FIGURE 23.10. Transversal load shift due to the driving torque T_d.

F_x is the longitudinal force exerted by the axle on the ground and τ_f is the gear ratio of the final drive, defined as the ratio between the speed of the wheels and that of the propeller shaft (it is usually smaller than unity).

Equation (23.37) is not, however, usually correct as under the action of the driving torque the vehicle body is subject to a roll rotation, which in turn produces an added torque on the axle through the suspension system. If the roll stiffness of the ith suspension is K_{t_i}, the roll angle is

$$\phi = -\frac{M_d}{\sum_{\forall i} K_{t_i}} .$$

The torque exerted on the axle is then equal to

$$\phi K_t = -\frac{M_d K_t}{\sum_{\forall i} K_{t_i}} ,$$

where K_t is the roll stiffness of the relevant suspension.

The load shift is thus

$$\Delta F_z = \frac{F_x R_l \tau_f}{t_i} \left(1 - \frac{K_t}{\sum_{\forall i} K_{t_i}} \right) . \qquad (23.38)$$

If the vehicle has a standard differential gear, the maximum driving force which can be exerted by the driving axle is equal to twice that which can be exerted by the less loaded wheel, i.e.

$$F_{x_{max}} = \mu_p (F_z - 2\Delta F_z) . \qquad (23.39)$$

If, on the contrary, a locking differential is used, within the limits of the assumption that the force coefficient μ_p is independent of the load, the transversal load shift does not affect the maximum driving force.

Example 23.6 *Consider the articulated truck of Appendix E9. Compute*
a) the maximum driving force at a constant speed of 70 km/h on level road;
b) the same as in a), but on a 10% grade;
c) the maximum grade that can be managed at 10 km/h.
All the above computations must be performed taking into account the transversal load shift and repeated for the case of a locking differential. Assume that the maximum longitudinal force coefficient is $\mu_p = 1$.
a) At 70 km/h = 19.44 m/s the load on the driving axle is 106.940 kN while the required driving force is 3.187 kN. Taking into account the gear ratio of the final drive, the driving torque on the axle is 344 Nm, yielding a roll angle of 2.67°. The transversal load shift is $\Delta F_z = 96.4$N, and the maximum longitudinal force is 106.75 kN. This value compares with the 106.94 kN that could be exerted if a locking differential were used, showing that the latter would improve the ability to exert longitudinal forces only marginally in this case.
b) At 70 km/h on a 10% grade the load on the driving axle is 116.97 kN and the required driving force is 91.15 kN, corresponding to a driving torque on the

axle of 4453 Nm. A very large roll angle, namely 34.6°, results from the values of the stiffness of the axles, but this is an unrealistic result as for large torques the nonlinear nature of the suspensions would limit rotations. Assuming that the stiffness distribution between the suspensions in the nonlinear range is the same as in the linear range, the transversal load shift is $\Delta F_z = 1249N$, yielding a maximum longitudinal force of 114.47 kN; if a locking differential were used, a force of 116.97 kN would have been exerted.

c) By computing the force required for motion and the maximum force that can be exerted by the driving wheels at 10 km/h = 2.78 m/s for different values of the grade, it is possible to find the value of the latter at which the two are equal. This procedure allows one to find the maximum value of the grade as 34.9%, i.e. a grade angle of 19.2°.

Note that the driving torque is very large on that grade and the suspensions operate clearly outside their linear range: The load shift can thus be far smaller than that computed. If no load shift was accounted for, a value of the grade of 37.8%, i.e. a grade angle of 20.7°, would have been found.

23.6 MAXIMUM SPEED

The maximum speed that can be reached on level road with a given transmission ratio can be found by intersecting the curve of the available power at the wheels with that of the required power on level road. The transmission ratio causing this intersection to occur at the maximum available power allows the highest speed that can be attained by a given vehicle-engine combination (curve 1 in Fig. 23.11) to be reached.

The computation of the maximum speed and of the overall gear ratio τ_t necessary to reach it is straightforward. By intersecting the required power curve with the horizontal straight line

$$P = P_{a_{max}} = P_{e_{max}}\eta_t \, ,$$

a fifth degree equation is obtained

$$AV + BV^3 + CV^5 = P_{e_{max}}\eta_t \, , \qquad (23.40)$$

whose solution directly yields the maximum value of the speed.

If aerodynamic lift is neglected (actually it is sufficient to neglect the contribution to rolling resistance proportional to the square of the speed due to lift), C vanishes and the equation is cubic. Its solution can be obtained in closed form

$$V_{max} = A^* \left(\sqrt[3]{B^* + 1} - \sqrt[3]{B^* - 1} \right) \, , \qquad (23.41)$$

where

$$A^* = \sqrt[3]{\frac{P_{e_{max}}\eta_t}{2mgK + \rho SC_X}} = \sqrt[3]{\frac{P_{e_{max}}\eta_t}{2B}} \, ,$$

$$B^* = \sqrt{1 + \frac{8m^3 g^3 f_0^3}{27 P_{m_{max}}^2 \eta_t^2 (2mgK + \rho S C_X)}} = \sqrt{1 + \frac{4A^3}{27 P_{m_{max}}^2 \eta_t^2 B}} \, .$$

Once the maximum speed has been obtained, the gear ratio allowing the vehicle to reach it is

$$\tau_t = \frac{V_{max}}{R_e(\Omega_e)_{P_{max}}} \, , \tag{23.42}$$

where $(\Omega_e)_{P_{max}}$ is the engine speed at which the peak power is obtained.

If the transmission is of the mechanical type, the overall gear ratio is the product of the gear ratio at the gearbox (in the relevant gear) and that of the final drive

$$\tau_t = \tau_g \tau_f \, .$$

The transmission ratio of the gearbox, which in top gear is usually close to 1, can be stated and consequently the gear ratio τ_f at the final drive can be computed.

Note that this procedure is based on the assumption that the intersection in Fig. 23.11 occurs at the peak of the engine power curve. This can, however, occur in only one given condition, since not only the load, but also the rolling resistance coefficient and even the air density, affect the road load curve. Air density also affects the engine power curve. If the intersection occurs in the descending branch of the curve (situation 2 in Fig. 23.11) the vehicle is said to be "undergeared", i.e., the overall transmission ratio is "too short". Conversely, if the intersection occurs in the ascending branch of the curve (situation 3 in Fig. 23.11), the vehicle is "overgeared" and the overall transmission ratio is "too long".

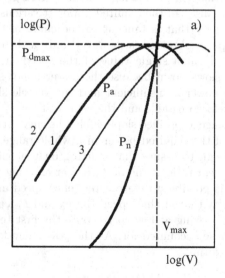

FIGURE 23.11. Maximum speed for a vehicle with internal combustion engine.

The first situation can be purposely obtained to improve the acceleration and grade performance of the vehicle, while the second allows fuel consumption to be reduced. The degree of undergearing λ_u can be defined as

$$\lambda_u = \frac{(\Omega)_{V_{max}}}{(\Omega)_{P_{max}}} . \tag{23.43}$$

It is greater than unity if the vehicle is undergeared and smaller than 1 in case of overgearing.

There are thus two ways of choosing the top gear ratio: One has already been stated, namely a "fast" gear ratio, with a degree of undergearing equal to about unity, i.e., chosen in order to reach the maximum speed. A different approach is to use a longer overgeared ratio, with the goal of reducing fuel consumption (see below). In practical terms, this trade-off is typical of five or six speed transmissions: Either the maximum speed is reached in fifth (sixth) gear or in fourth (fifth) gear, the longest one being an overdrive gear.

Remark 23.3 *This strategy works only in the case of vehicles with high power/ weight ratio: In low powered vehicles, this "economy" gear would be very difficult to use since any increase of the required power, e.g., due to a slight grade, headwind, etc., would compel a shift to a shorter gear. Undergearing may be a necessity in this case.*

23.7   GRADEABILITY AND INITIAL CHOICE OF THE TRANSMISSION RATIOS

The maximum grade that can be managed with a given gear ratio may be obtained by plotting the curves of the required power at various values of the slope and looking for the curve that is tangent to the curve of the available power (Fig. 23.12). The slope so obtained is, however, only a theoretical result, since it can be managed only at a single value of the speed: If the vehicle travels at a higher speed, it slows down because the power is not sufficient, but if its speed is reduced the power is insufficient and the vehicle slows down further: The condition is therefore unstable and the vehicle stops.

To be able to manage a specified slope safely, the curve of the available power must be above that of the required power in a whole range of speeds, starting from a value low enough to assure that starting on that slope is possible. To choose a value of the gear ratio of the bottom gear allowing the vehicle to start on a given grade, it is possible to state a reference speed and to compute the gear ratio in such a way that at that speed the P_a and P_n curves intersect.

As the vehicle is moving at low speed, only the first term of the required power curve needs to be accounted for. As the power developed by the engine can be written in the form

$$P_e = M_e \Omega_e = M_e \frac{V}{R_e \tau_g \tau_f} , \tag{23.44}$$

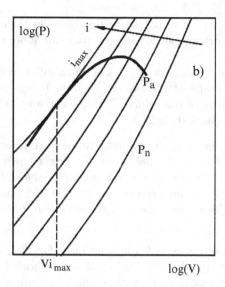

FIGURE 23.12. Maximum slope for a vehicle with internal combustion engine.

where M_e is the engine torque, the equilibrium condition allows the overall gear ratio to be computed as

$$\tau_t = \frac{M_e \eta_t}{R_e mg [f_0 \cos(\alpha) + \sin(\alpha)]} . \qquad (23.45)$$

The value of the engine torque to be introduced into Eq. (23.45) can be the maximum torque available at the minimum engine speed, possibly multiplied by a number smaller than 1 for safety. The mass of the vehicle must be that at full load, including the maximum trailer mass the vehicle is allowed to tow. For the grade, values of 25% or even 33% for road vehicles can be considered, but it must be kept in mind that in some cases, as in ferry ramps or private garage ramps, very steep grades may be encountered. For off-road vehicles values up to 100% can be considered.

Another consideration in the choice of the gear ratio for the bottom gear is to assure a regular working of the engine at a speed chosen so as to avoid a prolonged use of the clutch in very low speed driving. A reference value may be 6 or 8 km/h. Both criteria must be satisfied.

Once the ratios of the bottom and top gears have been chosen, the intermediate ones can be stated using different criteria. The simplest is to set them in geometric sequence, i.e., stating that the ratios between two subsequent gear ratios are all equal. Operating in this way, the available power curves on the $P(V)$ logarithmic plot are all equispaced.

There may be some advantages in having the curves a bit closer to each other in the high speed range, so that the third gear (in a four speed gearbox) is closer to the fourth. If this is required, it is possible to set in a geometric sequence not the gear ratios τ_i but the ratios between them τ_i/τ_{i+1}. This can give a feeling of sport driving, since the gear ratios are more crowded in the zone of most common use.

The choice of the transmission ratios is much influenced by considerations that are beyond the scope of the present section, being mostly linked to the acceleration performance of the vehicle. This aspect was introduced in Part III and will be dealt with in Section 23.10.

Remark 23.4 *Since the values of the gear ratios have a large influence on the performance of the vehicle and above all on the driver's perception of them, the trade-off dominating their choice is also a matter of subjective impressions and the traditions of various manufacturers. The market sector a manufacturer aims at may have more influence in deciding the matter than technical considerations alone*

Example 23.7 *Choose the overall top gear ratio for the car of Appendix E.2 to reach the maximum speed in the load condition indicated. Choose the bottom gear ratio to start on a 33% grade with a safety margin of 1.1 with respect to the maximum engine torque. Compare the ratio obtained with those listed in the Appendix.*

Equation (23.40), solved numerically, yields a maximum speed of 42.6 m/s = 153.4 km/h.

The overall transmission ratio $\tau_g\tau_f$ allowing the intersection between the two curves on the $P(V)$ plane to occur at the peak power is 0.3044. If a value of $22/21 = 1.048$ is accepted for the top gear ratio, the transmission ratio of the final drive is 0.2906, which can be approximated as $18/62$ with an error of about 0.08%.

The actual ratio of the final drive is 0.284. By computing the maximum speed with this value of the transmission ratio, a value of 41.2 m/s = 148.36 km/h is found. The top speed is reached at 5147 rpm, yielding a degree of undergearing $\lambda_u = 0.99$.

The overall transmission ratio of the bottom gear can be found using Eq. (23.45). By dividing the maximum engine torque by a factor of 1.1, a value of 0.1056 is obtained, corresponding to a value for the gearbox ratio of 0.3639. This value is far longer than the actual one (0.2154), since the computation was performed with the vehicle unloaded.

23.8   FUEL CONSUMPTION AT CONSTANT SPEED

The energy needed to travel at constant speed can be immediately computed by multiplying the power required for constant speed driving by the time

$$e = P_n t = \frac{P_n d}{V} ,$$

(23.46)

where d is the distance travelled. Note that Eq. (23.46) gives the energy required at the wheels: To obtain the energy actually required, it must be divided by the various efficiencies (transmission, engine, etc.).

If the efficiency of the engine η_e and the thermal value H of the fuel are known, the fuel consumption can be computed. Introducing the expression derived from Eq. (23.16) for the total road load into the expression for the power, the fuel consumption per unit distance Q is

$$Q = \frac{A + BV^2 + CV^4}{\eta_t \eta_e H \rho_f}, \qquad (23.47)$$

where ρ_f is the density of the fuel, introduced to obtain the consumption in terms of volume of fuel per unit of distance. In S.I. units it is measured in m³/m, while liters per 100 km is a more practical, although not consistent, unit. Often the reciprocal of Q, expressed in km per liter or miles per gallon, is used.

From Eq. (23.47), if the aerodynamic lift is neglected, the fuel consumption would be a quadratic function of the speed if the efficiency of the engine could be considered as a constant. The plot $Q(V)$ for a car with a mass of 1,000 kg, with $H = 4.4 \times 10^7$ J/kg, $\rho = 730$ kg/m³ and $\eta_e = 0.25$ is shown in Fig. 23.13.

This is not the case, however, as the efficiency of the engine is strongly influenced by its rotational speed and above all by the power the engine is required to supply.

To compute the consumption Q, the simplest procedure is to obtain the power required at the wheels as a function of the speed and hence to compute the power the engine must supply to travel at constant speed

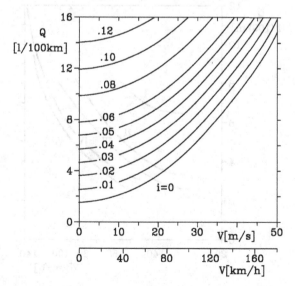

FIGURE 23.13. Fuel consumption at constant speed as a function of the speed, assuming that the efficiency of the engine is constant.

$$P_e = \frac{P_n}{\eta_t} \ .$$

Once the transmission ratio has been stated, the rotational speed of the engine is known and thus the working point on the map of the engine is located. From it the efficiency η_e or, which is the same, the specific fuel consumption

$$q = \frac{H}{\eta_e}$$

is obtained and the fuel consumption can be computed as

$$Q = \frac{qP_n}{\eta_t V \rho_f} \ . \tag{23.48}$$

The curves $Q(V)$ are of the type shown in Fig. 23.14. They usually have a minimum at low speed, obtained in conditions in which the engine works at low power with low efficiency.

Since the conditions in which the engine works depend on the overall transmission ratio, the fuel consumption is also largely influenced by the value of the gear ratio. Usually the longer the ratio, the lower the consumption, as a "long" ratio allows the engine to be used at low speed in conditions which are close to the maximum power, where the specific fuel consumption is low.

As already stated, a transmission ratio longer than that needed to reach the maximum speed can be used. It is possible to choose it in such a way that the

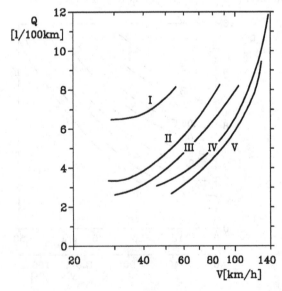

FIGURE 23.14. Fuel consumption with different gear ratios at constant speed on level road. Passenger vehicle with five-speed gearbox.

curve of the required power crosses that of the maximum efficiency at a given cruise speed, e.g. at a speed equal to 3/4 of the top speed. The fuel consumption at that speed is consequently the minimum possible value, with the added advantages of a reduction in noise and engine wear due to the reduced engine speed. Obviously, the performance in terms of maximum speed, acceleration and gradeability is reduced with respect to that available with a shorter gear ratio.

If a CVT is used, it is possible to control it in such a way that the engine works at conditions of maximum efficiency at all speeds, i.e. at all speeds the working point on the map lies on the curve of the maximum efficiency. This is really expedient, however, only if the increase in efficiency so obtained is greater than the loss of efficiency, with respect to that of a simpler transmission, due to the use of the CVT. Moreover, the control law for the transmission ratio of the CVT is a trade-off among different requirements, which also take into account acceleration and gradeability.

Example 23.8 *Plot the fuel consumption curve in top gear for the car of Appendix E.2.*

The map of the engine is shown in Fig. 23.15a. The curves of the power required at the engine, i.e. of the power required at the wheels divided by the transmission efficiency, are plotted for the different gear ratio in the same figure.

The curves identify the working conditions of the engine.

The points at which the curve of the power required in top gear intersects the curves at constant specific fuel consumption are reported in the first two columns of the following table

Ω [rpm]	P [kW]	q [g/HPh]	V [km/h]	Q [l/100km]	1/Q [km/l]
2083	3.819	400	60.05	4.65	21.52
3157	9.711	300	91.02	5.85	17.10
4135	19.610	250	119.20	7.51	13.31
4664	27.152	240	134.46	8.85	11.30
5320	37.876	250	153.36	11.28	8.87

The other columns list the specific fuel consumption, the speed and the fuel consumption (in l/100 km) and its reciprocal (in km/l). A value of 730 kg/m^3 has been used for the density of the fuel. The fuel consumption is also reported in Fig. 23.15b.

The experimental data do not allow the fuel consumption to be computed directly in the other gears, since the required power curves do not cross the curves of the map. Although there is no difficulty in repeating the tests and plotting the specific fuel consumption in the relevant zone of the map, not having other experimental data available it is still possible to interpolate linearly the values of the efficiency between the lowest curve and the Ω-axis where the efficiency is zero.

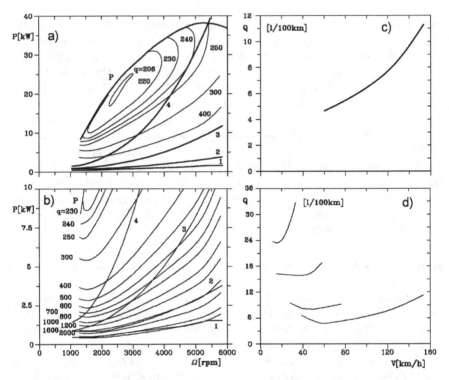

FIGURE 23.15. Fuel consumption for the car of Appendix E.2. (a) Map of the engine with superimposed curves of power required at the engine in various gears (1: bottom gear; 2, 3: intermediate gears; 4: top gear). The specific fuel consumption is reported in g/CVh. (b) Fuel consumption in l/100 km as a function of the speed (top gear). (c) Zone of the engine map for low-power operation, with curves of power required at the engine in various gears. (d) Fuel consumption in l/100 km as a function of the speed in the various gears.

To interpolate the efficiency means to interpolate the reciprocal of the specific fuel consumption[4]; consequently, the curve midway between the curve at 400 gCV/h and the Ω-axis is that related to a doubling of the fuel consumption, 800 gCV/h, and so on. The lower part of the plot of Fig. 23.15a, obtained in this way, is shown in Fig. 23.15c.

The fuel consumption curves (Fig. 23.15d) were then obtained in the same way seen above for the top gear. The results are only a rough approximation, but at any rate their pattern is realistic.

---

[4] To interpolate directly on the specific fuel consumption has little meaning, since the latter tends to infinity on the Ω-axis.

23.9 VEHICLE TAKE-OFF FROM REST

Since internal combustion engines cannot operate below a minimum speed Ω_{min}, the vehicle cannot slow down below the speed

$$V_{min} = \Omega_{min} R_e \tau_f \tau_g$$

with the engine connected to the driving wheels. Either a torque converter or a friction clutch must be used, both for starting and stopping the vehicle and to facilitate the shifting of gears.

The starting manoeuvre may be easily simulated in an approximate way by accepting the following assumptions:

1. The manoeuvre is started with the engine running at a speed Ω_{e_0} and the clutch control is released gradually from time $t = 0$ to time $t = t_i$ in such a way that the torque M_c it transmits increases linearly in time from 0 to the maximum value it can handle in slipping conditions M_c^*, and then remains constant until time t_s when no more slipping occurs;

2. the engine torque is maintained constant at the value M_e;

3. if the vehicle starts on a sloping road, it is kept stationary by some external means until the clutch torque is sufficient to produce motion;

4. the longitudinal slip of the wheels is small;

5. the terms in V^2 and V^4 of the road load are neglected owing to the low speed at which the manoeuvre is performed.

The vehicle can be modelled in terms of two moments of inertia, one to model the engine J_e and one to model the vehicle J_v (Fig. 23.16a). The first includes

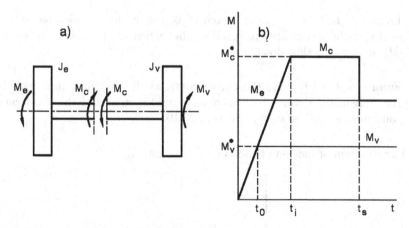

FIGURE 23.16. (a) Model of the vehicle for the starting manoeuvre. (b) Time history of the torques acting on the vehicle.

the moment of inertia of the engine, up to the flywheel, while the moments of inertia of the clutch disks, of the shaft entering the gearbox, of all the rotating parts (reduced to the engine shaft), and the mass of the vehicle as "seen" from the engine are included in the second. For the computational details, see Section 23.10.

Torque M_e, which has been assumed to be constant, acts on the moment of inertia J_e. On J_v a drag torque M_v is acting, whose value is simply

$$M_v^* = mg\left[f_0\cos(\alpha) + \sin(\alpha)\right]R_e\frac{\tau_f\tau_g}{\eta_t} , \tag{23.49}$$

when the vehicle is moving. When the vehicle is stationary, at the beginning of the starting manoeuvre, the drag torque is simply equal to the torque the clutch is supplying, if it is smaller than M_v^*,

$$M_v = \min(M_v^*, \ M_c) . \tag{23.50}$$

The maximum torque the clutch can transfer to the vehicle M_c^* is usually slightly larger, by 10% to 20%, than the maximum engine torque.

The torques acting on the system are plotted versus time in Fig. 23.16b. The manoeuvre can thus be subdivided into three phases:

1. From $t = 0$ to

$$t = t_0 = t_i\frac{M_v^*}{M_c^*} ,$$

in which the vehicle is at a standstill, since the torque transferred by the clutch is not yet sufficient to overcome the drag. The engine speeds up.

2. From $t = t_0$ to $t = t_i$, the clutch slips, the vehicle accelerates and the engine initially continues to speed up, but when M_c^* becomes greater than M_e, it starts to slow down.

3. From $t = t_i$ to $t = t_s$, the clutch continues to slip until time t_s, when the transmission starts to behave as a rigid system and the acceleration continues, as will be seen in Section 23.10.

The equation of motion of the system is simply

$$\begin{cases} \dot{\Omega}_e = \dfrac{M_e - M_c}{J_e} \\[3mm] \dot{\Omega}_v = \dfrac{M_c - M_v}{J_v} \end{cases} . \tag{23.51}$$

23.9.1   First phase

In the first phase the moments are

$$\begin{cases} M_e = M_e^* \\ M_c = M_v = M_c^* \dfrac{t}{t_i} \, , \end{cases} \tag{23.52}$$

and then the equations of motion are

$$\begin{cases} \dot\Omega_e = \dfrac{M_e^*}{J_e} - \dfrac{M_c^*}{J_e}\dfrac{t}{t_i} \\ \dot\Omega_v = 0 \, , \end{cases} \tag{23.53}$$

with the initial conditions

$$\begin{cases} \Omega_e = \Omega_{e_0} \\ \Omega_v = 0 \end{cases} \quad \text{for } t = 0. \tag{23.54}$$

23.9.2   Second phase

In the second phase the moments are

$$\begin{cases} M_e = M_e^* \\ M_v = M_v^* \\ M_c = M_c^* \dfrac{t}{t_i} \, . \end{cases} \tag{23.55}$$

The equations of motion are then

$$\begin{cases} \dot\Omega_e = \dfrac{M_e^*}{J_e} - \dfrac{M_i^*}{J_e}\dfrac{t}{t_i} \\ \dot\Omega_v = \dfrac{M_c^*}{J_v}\dfrac{t}{t_i} - \dfrac{M_v^*}{J_v} \, , \end{cases} \tag{23.56}$$

with the initial conditions that can be obtained at the end of the first phase.

23.9.3   Third phase

In the third phase the moments are all constant and their values are M_e^*, M_c^* and M_v^*. The equations of motion are

$$\begin{cases} \dot\Omega_e = \dfrac{M_e^* - M_c^*}{J_e} \\ \dot\Omega_v = \dfrac{M_c^* - M_v^*}{J_v} \, , \end{cases} \tag{23.57}$$

while the initial conditions can be obtained from those at the end of the second phase.

The manoeuvre ends when the condition $\Omega_e = \Omega_v$ holds, i.e., when the clutch does not slip any more.

By integrating Eq. (23.51) separately for the three phases, the following time histories for the engine and for the vehicle are obtained:

$$
\begin{cases}
\Omega_e = \Omega_{e_0} + \dfrac{1}{J_e}\left(M_e t - \dfrac{M_c^*}{2t_i}t^2\right) & \text{for } 0 < t < t_i \\[3mm]
\Omega_e = \Omega_{e_0} + \dfrac{1}{J_e}\left[t(M_e - M_c^*) - \dfrac{M_c^*}{2}t_i\right] & \text{for } t_i < t < t_s ,
\end{cases}
\tag{23.58}
$$

$$
\begin{cases}
V = 0 & \text{for } 0 < t < t_0 \\[3mm]
V = \dfrac{R_e}{J_v}\left(\dfrac{M_c^*}{2t_i}t^2 - M_v^* t + \dfrac{M_v^{*2}t_i}{2M_c^*}\right) & \text{for } t_0 < t < t_i \\[3mm]
V = \dfrac{R_e}{J_v}\left[t(M_c^* - M_v^*) + \dfrac{M_v^{*2}t_i}{2M_c^*}\left(M_v^{*2} - M_c^{*2}\right)\right] & \text{for } t_i < t < t_s .
\end{cases}
\tag{23.59}
$$

The starting time t_s can be defined as the time at which the clutch stops slipping: $\Omega_v = \Omega_e$. By equating the two angular velocities it follows that

$$
t_s = \frac{2J_e J_v M_c^* \Omega_{e_0} + M_c^{*^2} t_i(J_v - J_e) - t_i M_v^{*^2} J_e}{2M_c^*\left[J_e(M_c^* - M_v^*) + J_v(M_c^* - M_e)\right]} .
\tag{23.60}
$$

To make the subsequent acceleration of the vehicle possible, the angular velocity of the engine at time t_s must be in excess of the minimum velocity at which it can work regularly; otherwise, it stops. This can occur if the values of Ω_{e_0} or of M_e are too low or if the clutch engages too quickly (t_i too low).

If $t_s < t_i$ the vehicle completes the starting manoeuvre before the clutch is fully engaged: This poses no problem, but Eq. (23.60) fails to yield a correct value of t_s.

During the manoeuvre, the engine delivers an energy equal to the difference between its kinetic energy at times 0 and t_s added to the energy it produces in the time interval

$$
e_e = \int_0^{t_s} M_e \Omega_e dt + \frac{1}{2}J_e\left(\Omega_{e_0}^2 - \Omega_{e_s}^2\right) .
\tag{23.61}
$$

Similarly, the vehicle receives the energy

$$
e_v = \int_0^{t_s} M_v \Omega_v dt + \frac{1}{2}J_v \Omega_{v_s}^2 .
\tag{23.62}
$$

The difference

$$
e_c = e_e - e_v
$$

yields the energy which is dissipated by the clutch during the starting manoeuvre. It is strictly linked to the quantity of friction material removed from the disc of the clutch, i.e. with the wear of that element.

The overall efficiency of the clutch is

$$\eta_c = \frac{e_v}{e_e} \, . \tag{23.63}$$

The space travelled during the take-off manoeuvre may be computed by integrating the speed in time. As the vehicle speed follows a pattern that is roughly quadratic, it may be approximated as

$$\frac{V_s t_s}{3} \, .$$

Example 23.9 *Simulate a starting manoeuvre for the car of Appendix E.2. As-sume that the manoeuvre is started at 2000 rpm with the engine supplying a torque equal to 60% of the maximum torque while the clutch can transfer a torque equal to 120% of the maximum torque. Assume that $t_i = 0.5$ s, but repeat the computations for $t_i = 0.2$ s and $t_i = 0.8$ s.*

With simple computations it follows that the moment of inertia simulating the vehicle is $J_v = 0.2113$ kg m^2 and that $M_v^* = 1.829$ Nm, $M_e = 52.2$ Nm, $M_c^* = 104.4$ Nm and $\Omega_{e_0} = 209.4$ rad/s. The results are shown in Fig. 23.17.

The angular velocity of the flywheel simulating the vehicle at the end of the manoeuvre is 160.4 rad/s, corresponding to a vehicle speed $V = 2.561$ m/s = 9.22 km/h.

The engine speed, 1532 rpm, is low but sufficient for accelerating the vehicle.

The results obtained for the three cases are

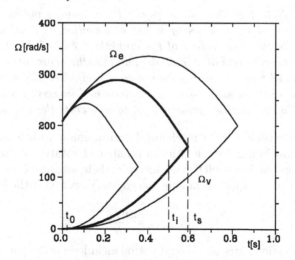

FIGURE 23.17. Angular velocities of the engine and flywheel, simulating the vehicle during a starting manoeuvre. Results for $t_i = 0.5$ s, 0.2 s and 0.8 s.

t_i	t_s	Ω_{e_s}	Ω_{v_s}	V_s	V_s	e_e	e_v	e_c	η	s_s
[s]	[s]	[rpm]	[rad/s]	[m/s]	[km/h]	[kJ]	[kJ]	[kJ]		[m]
0.2	0.36	1147	120.1	1.918	6.90	5.08	1.63	3.45	36%	0.257
0.5	0.59	1532	160.4	2.561	9.22	8.52	2.78	5.74	33%	0.494
0.8	0.83	1910	200.0	3.193	11.5	12.8	4.45	8.37	35%	0.863

Remark 23.5 *The efficiency of the clutch is lower than the value 0.5 which is often assumed. Actually, it would be 0.5 if the engine rotates at constant speed with no drag acting on the inertia that has to be accelerated.*

The assumptions made are quite rough, particularly those on the laws $M_e(t)$ and $M_c(t)$. However, the results do allow one to obtain reference values that are independent of the actual behavior of the driver.

In cases where the transmission has a torque converter instead of a clutch, the torque entering the gearbox may be computed using the equations discussed in Part II. It is then possible to integrate the equations of motion numerically and to obtain the time history of the speed.

In the case of a servo-controlled clutch, a procedure similar to the one shown can be followed by introducing the relevant control laws.

23.10   ACCELERATION

If the curve of the required power lies, at a certain speed, below that of the power available at the wheels, the difference $P_a - P_n$ between the two is the power available to accelerate the vehicle.

Remark 23.6 *Note that the engine power P_e is usually measured in steady-state running, in which case using it for acceleration is arbitrary; however, the time scales of the acceleration of the crankshaft and of the thermodynamic cycle are different by orders of magnitude, and thus the error introduced by using the values obtained from the steady-state map is negligible. The driving torque is then almost the same in steady-state conditions and in acceleration, but in the latter case part of the engine torque is used to accelerate the engine itself.*

Consider a vehicle with a mechanical transmission with a number of different gear ratios. During acceleration a number of rotating elements (wheels, transmission, the engine itself) must increase their angular velocity, and it is expedient to write an equation linking the engine power with the kinetic energy $\mathcal{T}$ of the vehicle

$$\eta_t P_e - P_n = \frac{d\mathcal{T}}{dt} . \tag{23.64}$$

The transmission efficiency should not be included for the part or the engine power needed to accelerate the engine, but the error created as a result is usually negligible.

Once the transmission ratio has been chosen, Eq. (23.26) gives the relationship between the speed of the vehicle and the rotational speed of the engine. Similar relationships may be used for the other rotating elements that must be accelerated when the vehicle speeds up.

The kinetic energy of the vehicle can then be expressed as

$$T = \frac{1}{2}mV^2 + \frac{1}{2}\sum_{\forall i} J_i\Omega_i^2 = \frac{1}{2}m_e V^2 \,, \tag{23.65}$$

where the sum extends to all rotating elements which must be accelerated when the vehicle speeds up. The term m_e is the equivalent or apparent mass of the vehicle, i.e., the mass of an object that, when moving at the same speed as the vehicle, has the same total kinetic energy. Usually it is written in the form

$$m_e = m + \frac{J_w}{R_e^2} + \frac{J_t}{R_e^2\tau_f^2} + \frac{J_e}{R_e^2\tau_f^2\tau_g^2} \,, \tag{23.66}$$

where J_w is the total moment of inertia of the wheels, which are assumed to have the same radius and hence to rotate at the same speed, and of all elements rotating at their speed, J_t is the moment of inertia of the propeller shaft and of all elements of the transmission, and J_e is the moment of inertia of the engine, the clutch and all the elements rotating at speed Ω_e.

To account for the fact that the engine is accelerated directly, at least in an approximate way, the last term is sometimes multiplied by η_t. The modifications to Eq. (23.66) to take the presence of different wheels on different axles into account are obvious.

Of the three last terms the first is usually small, the second negligible, while the third may become very important, particularly in low gear. As only the last term depends on the transmission ratio at the gearbox, the equivalent mass can be written in the form

$$m_e = F + \frac{G}{\tau_g^2} \,, \tag{23.67}$$

where

$$F = m + \frac{J_w}{R_e^2} + \frac{J_t}{R_e^2\tau_f^2} \,, \qquad G = \frac{J_e}{R_e^2\tau_f^2}$$

or, possibly

$$G = \frac{J_e\eta_t}{R_e^2\tau_f^2} \,.$$

As the equivalent mass is a constant, once the gear ratio has been chosen, Eq. (23.64) yields

$$\eta_t P_e - P_n = m_e V \frac{dV}{dt} \,. \tag{23.68}$$

Equation (23.68) holds only in the case of constant equivalent mass. If a CVT or a torque converter is used, the overall transmission ratio, and hence the equivalent mass, changes in time and the equation should be modified as

$$\eta_t P_e - P_n = m_e V \frac{dV}{dt} + \frac{1}{2} V^2 \frac{dm_e}{dt} , \tag{23.69}$$

and then

$$\eta_t P_e - P_n = \left(m_e + \frac{1}{2} V \frac{dm_e}{dV} \right) V \frac{dV}{dt} . \tag{23.70}$$

The correction present in Eq. (23.69) is, however, usually very small, since the equivalent mass does not change very quickly.

From Eq. (23.68), the maximum acceleration the vehicle is capable of at various speeds is immediately obtained

$$\left(\frac{dV}{dt} \right)_{max} = \frac{\eta_t P_e - P_n}{m_e V} , \tag{23.71}$$

where the engine power P_e is the maximum power the engine can deliver at the speed Ω_e, corresponding to speed V.

The plot of maximum acceleration versus speed for a passenger vehicle with a four speed gearbox is shown in Fig. 23.18.

The minimum time needed to accelerate from speed V_1 to speed V_2 can be computed by separating the variables in Eq. (23.71)

$$dt = \frac{m_e V dV}{\eta_t P_e - P_n} \tag{23.72}$$

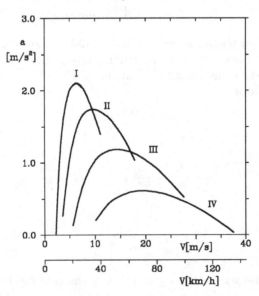

FIGURE 23.18. Maximum acceleration as a function of speed. Vehicle with a 4-speed gearbox.

and integrating

$$T_{V_1 \rightarrow V_2} = \int_{V_1}^{V_2} \frac{m_e}{\eta_t P_e - P_n} V \, dV \, . \qquad (23.73)$$

The integral must be performed separately for each velocity range in which the equivalent mass is constant, i.e. the gearbox works with a fixed transmission ratio. Although it is possible to integrate Eq. (23.73) analytically if the maximum power curve is a polynomial, numerical integration is usually performed.

A graphical interpretation of the integration is shown in Fig. 23.19: The area under the curve

$$\frac{V m_e}{\eta_t P_e - P_n} = \frac{1}{a}$$

versus V is the time required for the acceleration.

The speeds at which gear shifting must occur to minimize acceleration time are readily identified on the plot $1/a(V)$. Since the area under the curve is the acceleration time or the time to speed, the area must be minimized and gears must be shifted at the intersection of the various curves. If they do not intersect, the shorter gear must be used up to the maximum engine speed.

A criterion for choosing the gear ratios can also be evolved. The lower envelope of the curves (dashed line in the figure) does not depend on the transmission ratios and may be thought of as the curve that can be followed using a CVT

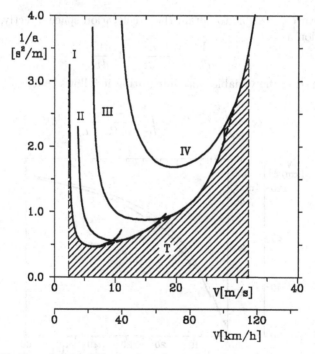

FIGURE 23.19. Function $1/a(V)$ showing the optimum speeds for gear shifting. The hatched area is the time to speed.

having the same efficiency as the gearbox and optimized to obtain the maximum acceleration. The area under the dashed curve is the minimum time to speed under ideal conditions.

The areas between the dashed and the continuous lines account for the time which must be added due to the presence of a finite number of speeds: The transmission ratios can be chosen in such a way as to minimize this area.

By increasing the number of speeds the acceleration time is reduced, since the actual curve gets closer to the ideal dashed line. However, at each gear shifting there is a time in which the clutch is disengaged and, consequently, the vehicle does not accelerate: Increasing the number of speeds leads to an increase in the number of gear shifts and thus of the time wasted without acceleration. This restricts the use of a high number of gear ratios.

The speed-time curve at maximum power can be easily obtained by integrating Eq. (23.73). An example is shown in Fig. 23.20. The actual curve, obtained by adding the time needed for gear shifting, is also reported. The speed is assumed to be constant during gear shift.

By further integration it is possible to obtain the distance needed to accelerate to any value of the speed

$$s_{V_1 \to V_2} = \int_{t_1}^{t_2} V \, dt . \tag{23.74}$$

It is, however, possible to obtain the acceleration space directly, by writing the acceleration as

$$a = \frac{dV}{dt} = \frac{dV}{dx}\frac{dx}{dt} = V\frac{dV}{dx} . \tag{23.75}$$

By separating the variables and integrating it follows that

$$s_{V_1 \to V_2} = \int_{V_1}^{V_2} \frac{V}{a} dV = \int_{V_1}^{V_2} \frac{m_e}{\eta_t P_e - P_n} V^2 dV . \tag{23.76}$$

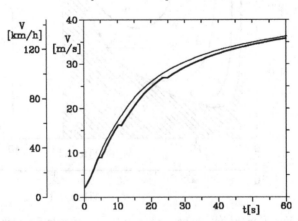

FIGURE 23.20. Speed versus time curve for the vehicle studied in the previous figures.

Instead of modelling the vehicle as an equivalent mass accelerated along the road, it is sometimes modelled as an equivalent moment of inertia attached to the flywheel of the engine, as seen in the previous section. Its value is

$$J_e = F'\tau_g^2 + G' , \qquad (23.77)$$

where

$$F' = FR_e^2\tau_f^2 , \qquad G' = J_e .$$

The acceleration curves can thus be obtained in terms of acceleration of the engine instead of acceleration of the vehicle.

It is possible to choose the gear ratio of the bottom gear to optimize the acceleration at low speed. When the transmission ratio is shortened, the torque available at the wheels increases; however, the equivalent mass also increases and it is not convenient, from the viewpoint of acceleration, to use transmission ratios that are too short.

Assuming that the engine torque M_e is constant and discarding the terms in V^3 and V^5 in the required power since at low speed their contribution is negligible, Eq. (23.71), written for the case of level road, yields

$$\left(\frac{dV}{dt}\right)_{max} = \frac{\eta_t M_e \Omega_e - AV}{m_e V} = \frac{\eta_t M_e - AR_e \tau_f \tau_g}{R_e \tau_f \tau_g \left[F + \frac{G}{\tau_g^2}\right]} . \qquad (23.78)$$

By differentiating Eq. (23.78) with respect to τ_g and equating the derivative to zero, a quadratic equation in τ_g, yielding the value of the gear ratio which maximizes the acceleration, is obtained. If the road load is neglected, which is reasonable on level road when dealing with strong accelerations, the value of the optimum gear ratio is

$$(\tau_g)_{opt} = \sqrt{\frac{G}{F}} \approx \sqrt{\frac{J_e}{mR_e^2}} . \qquad (23.79)$$

The last value has been obtained by neglecting the terms representing the inertia of the wheels and transmission in the expression of the equivalent mass. Note that the value so obtained leads to equal contributions for the mass of the vehicle and the inertia of the engine in the equivalent mass.

The value of the transmission ratio obtained with this criterion is, however, too short: It usually yields driving torques exceeding the maximum torque that may be transmitted by the driving wheels without slipping.

Example 23.10 *Plot the acceleration curve for the vehicle in Appendix E.2 and compute the time needed to reach 100 km/h. Compute also the time needed to travel for 1 km from standstill.*

Assume that the time needed for gear shifting is 0.5 s and that the takeoff manoeuvre follows the results obtained in the previous example.

Constants F and G are $F = 855.2$ kg and $G = 15.96$ kg, leading to the following values of the equivalent mass and moment of inertia:

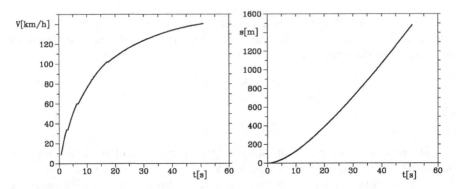

FIGURE 23.21. Speed and distance travelled as functions of time during a full power acceleration. The initial take-off manoeuvre has also been considered.

$$
\begin{array}{lll}
m_e = 1199\ kg = 1.45\ m & J_e = 0.296\ kg\ m^2 & \text{in first gear,} \\
m_e = 975\ kg = 1.18\ m & J_e = 0.692\ kg\ m^2 & \text{in second gear,} \\
m_e = 897\ kg = 1.08\ m & J_e = 1.823\ kg\ m^2 & \text{in third gear,} \\
m_e = 870\ kg = 1.05\ m & J_e = 5.085\ kg\ m^2 & \text{in fourth gear.}
\end{array}
$$

The results of the numerical integration yielding the speed and the distance travelled as functions of time during an acceleration are shown in Fig. 23.21. They were computed based on the results obtained in the previous example with a time $t_i = 0.5$ s, namely a time of 0.59 s, a speed of 9.22 km/h and a distance of 0.494 m.

The engine power was introduced in the computation through the best-fit third degree polynomial found in Example 4.5, and the speeds at which gear shifting occurs were determined as the minimum value between that corresponding to the maximum speed of the engine (6000 rpm) and the speed at which the acceleration obtainable in the following gear equals that obtainable with the gear under consideration. They are

$$
\begin{array}{ll}
5784\ rpm\ (V=34.3\ km/h) & \text{for the first gear,} \\
6000\ rpm\ (V=60.6\ km/h) & \text{for the second gear,} \\
6000\ rpm\ (V=102.3\ km/h) & \text{for the third gear.}
\end{array}
$$

The time to reach a speed of 100 km/h is 16.3 s and that needed to reach the 1 km mark is 38.1 s.

23.11   FUEL CONSUMPTION IN ACTUAL DRIVING CONDITIONS

Fuel consumption at variable speed gives the customer a more reliable estimate of the actual fuel consumption, but its determination is much more difficult. For this reason, several simplifications are usually accepted.

The first describes the actual use of the vehicle through a cycle, i.e. a time history of the speed of the vehicle, that takes into account neither the behavior of an actual driver nor the actual road and traffic conditions. This time history is used for all vehicles.

This simplification is implicitly accepted by European standards, which establish an urban and a suburban cycle to evaluate fuel consumption. These cycles were described in Part I. Fuel consumption measured in these cycles is the only value that may be supplied to the customer, and must be expressly stated on the data sheet of the vehicle.

Different cycles, obtained directly by manufacturers on similar vehicles in actual operating conditions may give more realistic values, but they are useful only to designers; this subject has also been covered in Part I.

A second simplification is that of computing the fuel consumption in a cycle as the sum of the partial consumption obtained in the various parts that approximate the chosen cycle assuming quasi-steady-state operation.

It is clear that the error so introduced decreases with decreased duration of the parts of the cycle that are assumed to be steady state; the error, however, is also due to the fact that fuel consumption in variable conditions is different from the fuel consumption obtained by approximating them with a sequence of steady state operations. This is due to the following reasons:

- In non-stationary operation, the thermal conditions of the engine are variable, so that the thermal energy losses are different from those occurring when the temperature has reached its steady-state value;

- in non-stationary conditions, part of the fuel burns with a lower efficiency due to condensation of the vapor on the intake manifold in indirect injection engines, or to a different evaporation rate of the fuel droplets in direct injection engines.

The difference is never very large, particularly if the instant power required by the cycle is much lower than the maximum engine power; this occurs often in statistically relevant cycles, since traffic conditions are always such as to prevent the engine from obtaining maximum performance. The comparison between measured and computed data always shows differences between 2 and 5%, with the computed consumption always lower than the actual one, due to the mentioned causes.

Following the above mentioned simplifications, the reference cycle is subdivided into a series of short time intervals; experience shows that a duration of about 1 s is acceptable for the intervals.

If V_i is the speed at instant t_i of the cycle, the fuel consumption in the time interval from t_i to t_{i+1} will be

$$\Delta e_i = \frac{1}{\eta_t} \left(A + BV_{mi} + CV_{mi}^4 + m_e V_{mi} \frac{V_{i+1} - V_i}{t_{i+1} - t_i} \right) (t_{i+1} - t_i) \frac{1}{\eta_e H \rho_f} ,$$

$$(23.80)$$

where:

$$V_{mi} = \frac{V_{i+1} + V_i}{2} \, . \tag{23.81}$$

This equation holds if the value of the first term in brackets is positive or vanishes, that is if the vehicle accelerates or decelerates with a rate low enough to compensate for the road load with inertia forces. If its value is negative, the contribution must be set to zero, since controllers on all modern engines cut off the fuel supply as the vehicle slows.

The contribution to fuel consumption when $V_i = 0$ is

$$\Delta e_i = Q_i(t_{i+1} - t_i) \, , \tag{23.82}$$

where Q_i is the fuel consumption at idle in liters/s.

The total fuel consumption in the cycle is

$$Q = \sum_{i=1}^{n} \Delta e_i \, . \tag{23.83}$$

An idea of the relative importance of the various forms of resistance to motion on fuel consumption in actual conditions is given by Fig. 23.22, where two different driving conditions are considered. Although the figure was obtained for a particular car (a medium sized saloon car), the results are typical. While in motorway driving aerodynamic drag is important, most of the energy in urban driving is expended to accelerate the vehicle.

The average was computed by using statistical data on average European conditions. From the average it is clear that reducing the mass of the vehicle, which affects both rolling resistance and the power needed to accelerate, is more important than reducing aerodynamic drag, and that the possibility of recovering

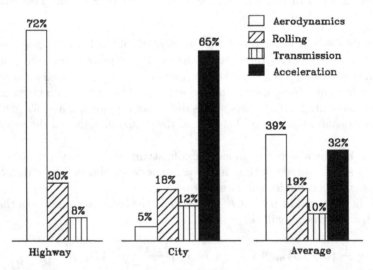

FIGURE 23.22. Energy required for motion in two different driving conditions.

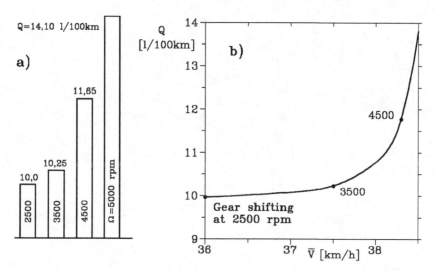

FIGURE 23.23. Effect of the engine speed at which gear shifting occurs (a) and of the average speed (b) on the fuel consumption in city driving.

braking energy, which allows a part of the energy used to accelerate the vehicle to be recovered, allows important energy savings to be obtained.

Numerical simulations can be used to study the effects of driving style on fuel consumption. In city driving, it is expedient to use the engine at the lowest speed consistent with its regular operation and particularly to maintain it near the speed of maximum torque and maximum efficiency. Prolonged use of low gears increases consumption without increasing the average speed appreciably (Fig. 23.23).

24

BRAKING DYNAMIC PERFORMANCE

The study of braking on straight road is performed using mathematical models similar to those seen in Chapter 23 for longitudinal dynamics. But in this case, the presence of suspensions and the compliance of tires are neglected and the motion is described by the longitudinal equilibrium equation (23.1) alone

$$m\ddot{x} = \sum_{\forall i} F_{x_i}.$$

Apart from cases in which the vehicle is slowed by the braking effect of the engine, which can dissipate a non-negligible power (lower part of the graph of Fig. 22.2), and by regenerative braking in electric and hybrid vehicles, braking is performed in all modern vehicles on all wheels. Subscript i thus extends to all wheels or, when thinking in terms of axles, as is usual for motion in symmetrical conditions, on all axles.

24.1 BRAKING IN IDEAL CONDITIONS

Ideal braking can be defined as the condition in which all wheels brake with the same longitudinal force coefficient μ_x.

The study of braking forces the vehicle can exert will follow the same scheme seen in Section 23.5, the only obvious difference being that braking forces, like the corresponding longitudinal force coefficients and the longitudinal slip, are negative. Normal forces between road and tires can be computed using the equations seen in Chapter 23.1, remembering here as well that the acceleration is negative.

The total braking force F_x is thus

$$F_x = \sum_{\forall i} \mu_{x_i} F_{z_i} \, , \tag{24.1}$$

where the sum extends to all the wheels. The longitudinal equation of motion of the vehicle is then

$$\frac{dV}{dt} = \frac{\sum_{\forall i} \mu_{x_i} F_{z_i} - \frac{1}{2}\rho V^2 S C_X - f \sum_{\forall i} F_{z_i} - mg\sin(\alpha)}{m} \, , \tag{24.2}$$

where m is the actual mass of the vehicle and not the equivalent mass, and α is positive for uphill grades. The rotating parts of the vehicle are slowed directly by the brakes, and hence do not enter into the evaluation of the forces exchanged between vehicle and road. These parts must be accounted for when assessing the required braking power of the brakes and the energy that must be dissipated.

Aerodynamic drag and rolling resistance can be neglected in a simplified study of braking, since they are usually far smaller than braking forces. Also, rolling resistance can be considered as causing a braking moment on the wheel more than a direct braking force on the ground.

Since in ideal braking all force coefficients μ_{x_i} are assumed to be equal, the acceleration is

$$\frac{dV}{dt} = \mu_x \left[g\cos(\alpha) - \frac{1}{2m}\rho V^2 S C_Z \right] - g\sin(\alpha) \, . \tag{24.3}$$

On level road, for a vehicle with no aerodynamic lift, Eq. (24.3) reduces to

$$\frac{dV}{dt} = \mu_x g \, . \tag{24.4}$$

The maximum deceleration in ideal conditions can be obtained by introducing the maximum negative value of μ_x into Eq. (24.3) or (24.4).

The assumption of ideal braking implies that the braking torques applied on the various wheels are proportional to the forces F_z, if the radii of the wheels are all equal.

As will be seen later, this may occur in only one condition, unless some sophisticated control device is implemented to allow braking in ideal conditions. If μ_x can be assumed to remain constant during braking, the deceleration of the vehicle is constant, and the usual formulae hold for computing the time and space needed to slow from speed V_1 to speed V_2:

$$t_{V_1 \to V_2} = \frac{V_1 - V_2}{|\mu_x|g} \, , \qquad s_{V_1 \to V_2} = \frac{V_1^2 - V_2^2}{2|\mu_x|g} \, . \tag{24.5}$$

The time and the space to stop the vehicle from speed V are then

$$t_{arr} = \frac{V}{|\mu_x|g} \, , \qquad s_{arr} = \frac{V^2}{2|\mu_x|g} \, . \tag{24.6}$$

The time needed to stop the vehicle increases linearly with the speed while the space increases quadratically.

To compute the forces F_x the wheels must exert to perform an ideal braking manoeuvre, forces F_z on the wheels must be computed first. This can be done using the formulae in Section 23.1. However, for vehicles with low aerodynamic vertical loading, such as all commercial and passenger vehicles with the exception of racers and some sports cars, aerodynamic loads can be neglected. Drag forces can also be neglected and, in the case of a two-axle vehicle, the equations reduce to

$$F_{z_1} = \frac{m}{l}\left[gb\cos(\alpha) - gh_G\sin(\alpha) - h_G\frac{dV}{dt}\right] , \qquad (24.7)$$

$$F_{z_2} = \frac{m}{l}\left[ga\cos(\alpha) + gh_G\sin(\alpha) + h_G\frac{dV}{dt}\right] . \qquad (24.8)$$

Since the values of μ_x are all equal in ideal braking, the values of longitudinal forces F_x can be immediately computed by introducing Eq. (24.3)

$$\frac{dV}{dt} = \mu_x g\cos(\alpha) - g\sin(\alpha)$$

into equations (24.7) and (24.8)

$$F_{x_1} = \mu_x F_{z_1} = \mu_x \frac{mg}{l}\cos(\alpha)\left(b - h_G\mu_x\right) , \qquad (24.9)$$

$$F_{x_2} = \mu_x F_{z_2} = \mu_x \frac{mg}{l}\cos(\alpha)\left(a + h_G\mu_x\right) . \qquad (24.10)$$

By adding Eq. (24.9) to Eq. (24.10), it follows that:

$$F_{x_1} + F_{x_2} = \mu_x mg\cos(\alpha) , \qquad (24.11)$$

and then:

$$\mu_x = \frac{F_{x_1} + F_{x_2}}{mg\cos(\alpha)} . \qquad (24.12)$$

By introducing the value of μ_x into equations (24.9) and (24.10) and subtracting the second equation from the first, it follows that

$$F_{x_1} - F_{x_2} = \frac{b-a}{l}(F_{x_1} + F_{x_2}) - \frac{2h_G}{lmg\cos(\alpha)}(F_{x_1} + F_{x_2})^2 . \qquad (24.13)$$

A relationship between F_{x_1} and F_{x_2} is readily obtained. It is an equation expressing the relationship between the forces at the front and rear axles that must hold to make ideal braking possible,

$$(F_{x_1} + F_{x_2})^2 + mg\cos(\alpha)\left(F_{x_1}\frac{a}{h_G} - F_{x_2}\frac{b}{h_G}\right) = 0 . \qquad (24.14)$$

The plot of Eq. (24.14) in the F_{x_1},F_{x_2} plane is a parabola whose axis is parallel to the bisector of the second and fourth quadrants if $a = b$ (Fig. 24.1).

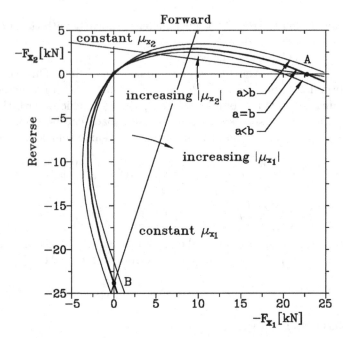

FIGURE 24.1. Braking in ideal conditions. Relationship between F_{x_1} and F_{x_2} for vehicles with the centre of mass at mid-wheelbase $(a = b)$, forward $(a < b)$ and backward $(a > b)$ of that point. Plots obtained with $m = 1000$ kg; $l = 2.4$ m, $h_G = 0.5$ m, level road.

The parabola is thus the locus of all pairs of values of F_{x_1} and F_{x_2} leading to ideal braking.

Only a part of this plot is actually of interest: That with negative values of the forces (braking in forward motion) and with braking forces actually achievable, i.e. with reasonable values of μ_x (Fig. 24.2).

On the same plot it is possible to draw the lines with constant μ_{x_1}, μ_{x_2} and acceleration. On level road, the first two are straight lines passing, respectively, through points B and A, while the lines with constant acceleration are straight lines parallel to the bisector of the second quadrant.

Remark 24.1 *All forces here relate to the axles and not to the wheels: In the case of axles with two wheels their values are then twice the values referred to the wheel.*

The moment to be applied to each wheel is approximately equal to the braking force multiplied by the loaded radius of the wheel: If the wheels have equal radii, the same plot holds for the braking torques as well. If this condition does not apply, the scales are simply multiplied by two different factors and the plot, though distorted, remains essentially unchanged.

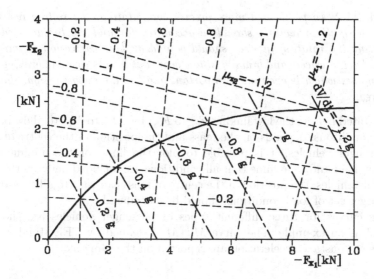

FIGURE 24.2. Enlargement of the useful zone of the plot of Fig. 24.1. The lines with constant μ_{x_1}, μ_{x_2} and acceleration are also reported.

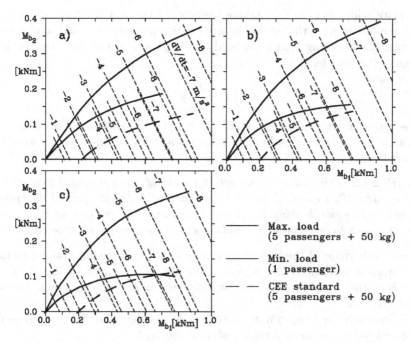

FIGURE 24.3. Plots $M_{b_2}(M_{b_1})$ for ideal braking. (a) typical plot for a rear drive car with low ratio h_G/l; (b) typical plot for a front drive saloon car with higher ratio h_G/l; (c) plot for a small front drive car, sensitive to the load conditions and with high value of ratio h_G/l.

Remark 24.2 *To perform a more precise computation, the rolling resistance, which is a small correction, should be accounted for and the torque needed for decelerating the rotating inertias should be added. This correction is important only for driving wheels and braking in low gear, but in this case the braking effect of the engine, which is even more important and has the opposite sign, should be considered.*

As stated before, the law linking F_{x_1} to F_{x_2}, i.e. M_{b_1} to M_{b_2} to allow braking in ideal conditions, depends on the mass and the position of the centre of mass. For passenger vehicles, it is possible to plot the lines for the minimum and maximum load and to assume that all conditions are included between them; for industrial vehicles, the position of the centre of mass can vary to a larger extent, and a larger set of load conditions should be considered.

The curves for three different types of passenger vehicles are shown in Fig. 24.3 as an example. The curve $M_{b_2}(M_{b_1})$ defined by CEE standards and the lines at constant acceleration are reported on the same plot.

24.2 BRAKING IN ACTUAL CONDITIONS

The relationship between the braking moments at the rear and front wheels is in practice different from that stated in order to comply with the conditions needed to obtain ideal braking, and is imposed by the parameters of the actual braking system of the vehicle.

A ratio

$$K_b = \frac{M_{b_1}}{M_{b_2}}$$

between the braking moments at the front and rear wheels can be defined. If all wheels have the same radius, its value coincides with the ratio between braking forces.

Remark 24.3 *This statement neglects the braking moment needed to decelerate rotating parts. This can be adjusted by considering M_b as the part of the braking moment that causes braking forces on the ground; the fraction of the braking moment needed to decelerate the wheels and the transmission must be added to it.*

For each value of the deceleration a value of K_b allowing braking to take place in ideal conditions can be easily found from the plot of Fig. 24.2. K_b depends on the actual layout of the braking system, and in some simple cases is almost constant.

In hydraulic braking systems, the braking torque is linked to the pressure in the hydraulic system by a relationship of the type

$$M_b = \epsilon_b(Ap - Q_s) \,, \tag{24.15}$$

where ϵ_b, sometimes referred to as the efficiency of the brake, is the ratio between the braking torque and the force exerted on the braking elements and hence has the dimensions of a length. A is the area of the pistons, p is the pressure and Q_s is the restoring force due to the springs, when they are present.

The value of K_b is thus

$$K_b = \frac{\epsilon_{b_1}(A_1 p_1 - Q_{s_1})}{\epsilon_{b_2}(A_2 p_2 - Q_{s_2})}, \tag{24.16}$$

or, if no spring is present as in the case of disc brakes,

$$K_b = \frac{\epsilon_{b_1} A_1 p_1}{\epsilon_{b_2} A_2 p_2}. \tag{24.17}$$

In disc brakes, ϵ_b is almost constant and is, as a first approximation, the product of the average radius of the brake, the friction coefficient and the number of braking elements acting on the axle, since braking torques again refer to the whole axle. If the pressure acting on the front and rear wheels is the same, the value of K_b is constant and depends only on geometrical parameters.

The behavior of drum brakes is more complicated, as restoring springs are present and the dependence of ϵ_b on the friction coefficient is more complex. As stated in Part I, shoes can be of the *leading* or of the *trailing* type. If leading, the braking torque increases more than linearly with the friction coefficient and there is even a value of the friction coefficient for which the brake sticks and the wheel locks altogether.

The opposite occurs with trailing shoes and ϵ_b increases less than linearly with the friction coefficient.

The efficiency of the brakes is a complex function of both temperature and velocity and, during braking, it can change due to the combined effect of these factors. When the brake heats up there is usually a decrease of the braking torque, at least initially. Later an increase due to the reduction of speed can restore the initial values. This "sagging" in the intermediate part of the deceleration is more pronounced in drum than in disc brakes. With repeated braking, the overall increase of temperature can lead to a general "fading" of the braking effect.

If K_b is constant, the characteristic line on the plane M_{b_1}, M_{b_2} is a straight line through the origin (Fig. 24.4).

The intersection of the characteristics of the braking system with the curve yielding ideal braking defines the conditions in which the system performs in ideal conditions. On the left of point A, i.e. for low values of deceleration, the rear wheels brake less than required and the value of μ_{x_2} is smaller than that of μ_{x_1}. If the limit conditions occur in this zone, i.e. for roads with poor traction, the front wheels lock first.

On the contrary, all working conditions beyond point A are characterized by

$$\mu_{x_2} > \mu_{x_1}$$

and the rear wheels brake more than required, i.e., the braking capacity of the front wheels is underexploited. In this case, when the limit conditions are reached, the rear wheels lock first, as in the case of Fig. 24.4.

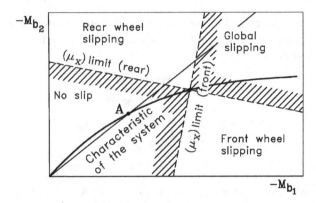

FIGURE 24.4. Conditions for ideal braking, characteristic line for a system with constant K_b and zones in which the front or the rear wheels lock. In the case shown the value of μ_p is high enough to cause sliding beyond point A.

From the viewpoint of handling, it is advisable that

$$\mu_{x_2} < \mu_{x_1} \; ,$$

since this increases the stability of the vehicle; the characteristics of the braking system should lie completely below the line for ideal braking. Locking of the rear wheels is a condition that must be avoided since it triggers directional instability.

In A the ideal conditions are obtained: If the limit value of the longitudinal force coefficient occurs at that point, simultaneous locking of all wheels occurs.

The values of ratio K_b for which the ideal conditions occur at a given value of the longitudinal force coefficient μ_x^* are immediately computed,

$$K_b^* = \frac{b + h_G|\mu_x^*|}{a - h_G|\mu_x^*|} \; . \tag{24.18}$$

It is possible to define an efficiency of braking as the ratio between the acceleration obtained in actual conditions and that occurring in ideal conditions, obviously at equal value of the coefficient μ_x of the wheels whose longitudinal force coefficient is higher,

$$\eta_b = \frac{(dV/dt)_{actual}}{(dV/dt)_{ideal}} = \frac{(dV/dt)_{actual}}{\mu_x g} \; , \tag{24.19}$$

where the last expression holds only on level road for a vehicle with negligible aerodynamic loading.

The total braking force acting on the vehicle when the rear wheels lock is

$$F_{x_1} + F_{x_2} = F_{x_2} \left(1 + K_b\right) \; , \tag{24.20}$$

and thus the deceleration on level road is

$$\frac{dV}{dt} = \frac{F_{x_2} \left(1 + K_b\right)}{m} . \tag{24.21}$$

Eq. (24.8) yields

$$F_{x_2} = \frac{\mu_{x_2} g}{l} \left[am + h_G F_{x_2} \left(1 + K_b \right) \right] ,$$ (24.22)

and then

$$F_{x_2} = \frac{\mu_{x_2} g a m}{l - \mu_{x_2} h_G \left(1 + K_b \right)} ,$$ (24.23)

$$\frac{dV}{dt} = g \frac{\mu_{x_2} a \left(1 + K_b \right)}{l - \mu_{x_2} h_G \left(1 + K_b \right)} .$$ (24.24)

If on the contrary the front wheels lock, the total braking force acting on the vehicle is

$$F_{x_1} + F_{x_2} = F_{x1} \left(1 + \frac{1}{K_b} \right) .$$ (24.25)

Operating as already seen with rear wheels lock, the value of the acceleration can be found,

$$\frac{dV}{dt} = g \frac{\mu_{x_1} b \left(1 + K_b \right)}{l K_b - \mu_{x_1} h_G \left(1 + K_b \right)} .$$ (24.26)

The braking efficiency is then

$$\eta_b = \min \left\{ \frac{a(K_b + 1)}{l - \mu_p h_G (K_b + 1)} , \frac{b(K_b + 1)}{l K_b + \mu_p h_G (K_b + 1)} \right\} .$$ (24.27)

The first value holds when the rear wheels lock first (above point A in Fig. 24.4), the second when the limit conditions are reached at the front wheels first.

A typical plot of the braking efficiency versus the peak braking force coefficient is plotted in Fig. 24.5.

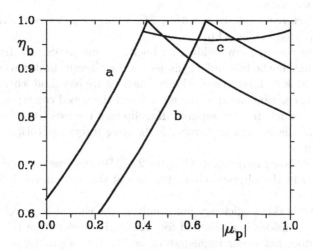

FIGURE 24.5. Braking efficiency η_b as a function of the limit value of μ_x for a vehicle without (a) and (b) and with (c) pressure proportioning valve.

The value of the maximum longitudinal force coefficient μ_p at which the condition $\eta_b = 1$ must hold can be stated and the value of ratio K_b can be easily computed. For values of $|\mu_p|$ lower than the chosen one, the rear wheels lock first while for higher values locking occur at the front wheels.

Once K_b is known, the braking system can easily be designed. The curve $\eta_b(\mu_x)$ can be plotted by assigning increasing values to the pressure in the hydraulic system, computing K_b and then the values of μ_x and η_b referred to the front and rear wheels. The result is of the type shown in Fig. 24.5, curve (a) or (b).

Operating in this way, the rear wheels lock when the road is in good condition. To postpone the locking of the rear wheels, curves of the type of line (b) can be used, but this reduces efficiency when the road conditions are poor.

To avoid locking of the rear wheels without lowering efficiency at low values of μ_x, a pressure proportioning valve, i.e. a device that reduces the pressure in the rear brake cylinders when the overall pressure in the system increases above a given value, may be used. A linear reduction of the pressure on the rear brakes with increasing pressure in the front ones above a certain pressure p_i,

$$\begin{cases} p_2 = p_1 & \text{for } p_1 \leq p_i, \\ p_2 = p_1 + \rho_c\,(p_1 - p_i) & \text{for } p_1 > p_i, \end{cases} \tag{24.28}$$

where ρ_c is a characteristic constant of the valve, can be assumed.

Pressure p_i and constant ρ_c must be chosen in such a way that the device starts acting when the efficiency η_b gets close to unity. The reduction of the rear pressure must be such that it does not cause locking of the rear wheels; nor should it be so high as to substantially lower the efficiency (see Fig. 24.5, curve (c)).

To comply with these conditions in all load conditions of the vehicle, p_i and, possibly, ρ_c must vary following the load. A possible way to achieve this is to monitor the load on the rear axle, e.g. by monitoring the vertical displacement of the rear suspension.

The characteristic line in the M_{b_1}, M_{b_2} plane of a device operating along this line is reported in Fig. 24.6.

To prevent wheels from locking, antilock systems (ABS) act directly to reduce the pressure in the hydraulic cylinders of the relevant brakes when the need to reduce the braking force arises. Modern devices are based on wheel speed sensors allowing the actual speed of the wheels and the speed corresponding to the velocity of the vehicle to be compared. If a slip that exceeds the allowable limits is detected, the device acts to reduce the braking torque, restoring appropriate working conditions.

As will be shown in detail in Chapter 27, ABS systems may work in different ways, both in the physical characteristics of the system and in the control algorithms.

The above braking efficiency holds only in the case of rigid vehicles. If the presence of suspensions is accounted for, the load transfer from the rear to the front wheels does not occur immediately, and at the beginning of the braking manoeuvre the vertical loads on the wheels are the same as those at constant

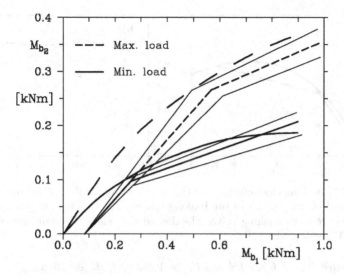

FIGURE 24.6. Characteristic of a braking system in which a pressure proportioning valve operating following Eq. (24.28) is present. To take into account the variability of the parameters of the system, specifically the friction coefficient, a band of characteristics has been considered instead of a single line. The ideal braking lines at the two different load conditions have also been plotted.

speed. The body of the vehicle then starts to dive and the load on front wheels increases, until steady state conditions are reached and the loads take the values given by Equations (24.7) and (24.8). This effect actually depends largely on the characteristics of the suspensions: the rotation of the body can be very small and load shift is almost immediate when antidive arrangements are used.

The load on the rear wheels is higher and the locking of the rear wheels is more difficult at the beginning of the manoeuvre: This consideration explains the practice of giving short brake pulses, effective when modern braking systems designed to avoid rear wheel locking were not available.

Example 24.1 *Plot the braking efficiency of the car of Appendix E.2, assuming that the braking system is designed to reach the ideal conditions for a longitudinal force coefficient* $\mu_x = -0.4$. *Use a pressure proportioning valve in such a way that the front wheels lock before the rear ones up to a value of* μ_x *equal to unity. Neglect aerodynamic forces and rolling resistance.*

The curve characterizing the conditions for ideal braking in plane F_{x_1}, F_{x_2} *is plotted (Fig. 24.7a). In order to obtain the ideal conditions at a value of the longitudinal force coefficient* $\mu_x^* = -0.4$, *ratio* K_b *is immediately computed from Eq. (24.18):* $K_b = 2.283$. *The braking forces corresponding to the ideal conditions are* $F_{x_1} = 2.265$ *kN and* $F_{x_2} = 0.992$ *kN.*

The pressure proportioning valve is assumed to start acting when values of the forces, equal to 90% of those for ideal conditions, are reached: $F_{x_1} = 2.038$ *kN and* $F_{x_2} = 0.893$ *kN. As the point at which the ideal conditions with* $\mu_x = 1$ *are reached is*

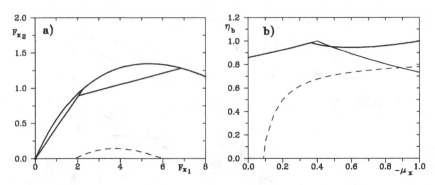

FIGURE 24.7. Braking characteristics of the vehicle of Appendix E.2. (a) Ideal braking conditions and characteristics of the braking system. (b) Braking efficiency with and without pressure proportioning valve. The dashed lines show the minimum conditions stated by CEE standards.

easily computed ($F_{x_1} = 6.861$ kN and $F_{x_2} = 1.282$ kN), the equation expressing force F_{x_2} as a function of F_{x_1} when the valve is operating is immediately found. From its slope, the value of constant $\rho_c = 0.184$ is obtained. The characteristic of the braking system is plotted in Fig. 24.7a.

At each point a pair of values F_{x_1} and F_{x_2} are obtained. From them the deceleration and the maximum value of the longitudinal force coefficient may be computed, ultimately obtaining the braking efficiency. The results are plotted in Fig. 24.7b.

In the same figures the curves related to the CEE standards are also plotted (dashed lines). Note that the position of the centre of mass results in the very low position of the dashed line in Fig. 24.7a.

24.3   BRAKING POWER

The instantaneous power the brakes must dissipate is

$$|P| = |F_x|V = V\left|\frac{dV}{dt}m_e + mg\sin(\alpha)\right| , \qquad (24.29)$$

where all forms of drag have been neglected.

The brakes cannot dissipate this power directly; they usually work as a heat sink, storing some of the energy in the form of thermal energy and dissipating it in due time. Care must obviously be exerted to design the brakes in such a way that they can store the required energy without reaching excessively high temperatures and so that adequate ventilation for cooling is ensured. The average value of the braking power must, at any rate, be lower than the thermal power the brakes can dissipate.

Two reference conditions are usually considered: Driving in continuous acceleration-braking cycles, and downhill running in which the speed is kept constant with the use of brakes.

In the first case, neglecting all resistance to motion, the energy to be dissipated during braking from speed V to zero is equal to the kinetic energy of the vehicle. The worst case is a number of accelerations from standstill to speed V, performed in the lowest possible time, followed by braking to standstill. The average power on an acceleration-deceleration cycle is

$$|P| = \frac{m_e V^2}{2(t_a + t_b)} .$$ (24.30)

The acceleration time t_a increases with V and can be computed with the method used in the previous chapter. Braking time t_b is, at least,

$$t_b = \frac{V}{g\eta_b |\mu_{x_{max}}|} .$$

The average braking power first increases with the speed V, and then decreases again since the acceleration time increases far more than the braking energy. When the vehicle is approaching its maximum speed t_a tends to infinity and the average power tends to zero.

In the case of downhill driving, the speed is assumed to be held constant by the use of brakes. The average power is then coincident with the power to be dissipated in each instant, since it is not possible that in the long run large quantities of heat are stored in the brakes. It then follows:

$$|P| = |V mg \sin(\alpha)| .$$ (24.31)

The power that must be dissipated increases linearly with V. The speed must then be limited, and the braking effect of the engine must be exploited on long downhill slopes.

Industrial vehicles are sometimes supplied with devices to maintain constant the speed when driving downhill to prevent the brakes from over-heating. By limiting the speed as a function of α, that is by stating a function $V = V(\alpha)$, the average power can be expressed as a function of the speed of the type shown in Fig. 24.8.

Acceleration-deceleration cycles are usually the critical condition for passenger vehicles and, above all, for sports cars, while for industrial vehicles the worst condition is downhill driving. Plots of the type seen in Fig. 24.8 give an indication of the maximum value of the average power the brakes must dissipate, making them useful for designing their cooling system.

If the road conditions or the driving style require significant use of the brakes, they may be required to store much heat and become very hot, with consequent thermo-mechanical problems. To give an idea of the magnitude of the temperatures reached by some components of the braking system, some experimental temperature readings obtained on mountain and hill roads are reported in Fig. 24.9.

In vehicles with regenerative braking capabilities, the average power computed above gives an idea of how much energy can be stored, and thus determines

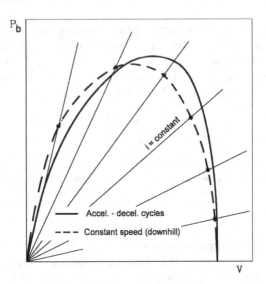

FIGURE 24.8. Power to be dissipated by brakes.

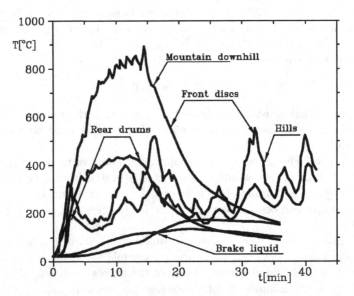

FIGURE 24.9. Time history of the temperature of the brakes and of the braking fluid during testing of a car on different roads.

the capacity of the accumulator. An accumulator able to store braking energy is large enough to provide true hybrid capabilities, i.e. to uncouple the requirements of the vehicle from the instantaneous power of the engine.

At any rate, vehicles with regenerative braking must have a conventional braking system as well. Regenerative braking is usually performed on only one axle, usually the driving axle, with the exception of schemes such as that shown in

Fig. 22.7a1. Braking power is limited by both the transmission and the ability of the accumulator to accept high power levels. The conventional braking system works in less demanding conditions, since it provides emergency braking only rather than frequent slowing in normal use.

25

HANDLING PERFORMANCE

25.1 LOW SPEED OR KINEMATIC STEERING

25.1.1 Two-axle vehicles without trailer

Low speed or kinematic steering is, as already stated, defined as the motion of a wheeled vehicle determined by pure rolling[1] of the wheels. The velocities of the centres of all the wheels lie in their midplane, that is the sideslip angles α_i are vanishingly small. In these conditions, the wheels cannot exert any cornering force to balance the centrifugal force due to the curvature of the path. Kinematic steering is possible only if the velocity is vanishingly small.

Kinematic steering of two-axle vehicles without trailer was dealt with in detail in Chapter 4 (Section 4.2). Here only the value of the *path curvature gain* needs be recalled

$$\frac{1}{R\delta} = \frac{1}{l} . \tag{25.1}$$

Remark 25.1 *The path curvature gain is a linearized value, holding only if the radius of curvature of the path R is much larger than the wheelbase. It is independent of the steering angle and of the curvature of the path.*

---

[1]The term 'pure rolling' is often used to indicate rolling without slip. 'Free rolling', as opposed to 'tractive rolling', is used to indicate rolling without exerting tangential forces (K.L. Johnson, *Contact Mechanics*, Cambridge University Press, Cambridge, 1985). Here the two terms are considered as equivalent, because a tire must operate in slip (longitudinal or side slip) conditions to produce a tangential force.

G. Genta, L. Morello, *The Automotive Chassis, Volume 2: System Design*,
Mechanical Engineering Series,
© Springer Science+Business Media B.V. 2009

Another important transfer function of the vehicle is ratio β/δ, usually referred to as *sideslip angle gain*. The sideslip angle of the vehicle, referred to the centre of mass, may be expressed as a function of the radius of the path R as

$$\beta = \arctan\left(\frac{b}{\sqrt{R^2 + b^2}}\right) . \qquad (25.2)$$

By linearizing Eq. (25.2) and introducing the expression (25.1) linking R to δ, it follows:

$$\frac{\beta}{\delta} = \frac{b}{l} . \qquad (25.3)$$

As seen in Chapter 6, the optimal condition for kinematic steering of a 4 wheel steering vehicle (4WS) is equal and opposite steering angles of the two axles: the radius of the path is thus halved with respect to the same vehicle with a single steering axle.

Particularly in the case of long vehicles, the off-tracking distance, i.e. the difference of the radii of the trajectories of the front and the rear wheels, is an important parameter. If R_a is the radius of the path of the front wheels, the off-tracking distance is

$$R_a - R_1 = R_a\left\{1 - \cos\left[\arctan\left(\frac{l}{R_1}\right)\right]\right\} . \qquad (25.4)$$

If the radius of the path is large when compared to the wheelbase, Eq. (25.4) reduces to

$$R_a - R_1 \approx R\left[1 - \cos\left(\frac{l}{R}\right)\right] \approx \frac{l^2}{2R} . \qquad (25.5)$$

In the same way, it is possible to define a minimum steering radius between walls, that is the diameter of the largest circle described by any point of the vehicle at maximum steering. If the point following the curve with the largest radius is point A in Fig. 25.1 (note that the figure refers to a vehicle with 3 axles), the minimum steering radius is

$$D_v = 2\sqrt{(R_1 + y_A)^2 + x_A^2}. \qquad (25.6)$$

25.1.2 Vehicles with more than two axles without trailer

True kinematic steering of vehicles with more than two axles is possible only if the wheels of several axles (all except one) can steer, and if the steering angles comply with conditions similar to those seen in Chapter 6 for the steering axle of a two-axle vehicle. In order to avoid serious wear to the tires, it is possible to lift one axle from the ground in certain conditions: In some countries it is legal to design the suspensions in such a way that not all axles are on the ground when the vehicle is unloaded, while in others this is not allowed. Some axles can be lifted for low-speed manoeuvring while being in contact with the ground in normal driving.

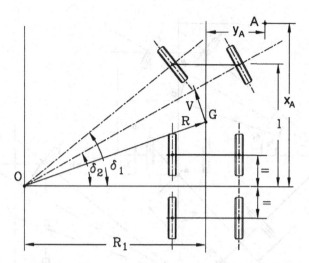

FIGURE 25.1. Low speed steering of industrial vehicles; approximate kinematic condition for a truck with three axles.

Some axles may also be self-steering, i.e. the wheels are allowed to orient themselves to minimize sideslip. An axle of this type clearly cannot exert side forces and reduce the overall cornering ability of the vehicle. Different laws hold in different countries, sometimes allowing the use of self-steering axles in normal driving and sometimes specifying that self-steering axles be blocked except in low speed manoeuvres. In the case of a three-axle vehicle with non-steering axles close to each other, an approximation such as the one shown in Fig. 25.1 can be used to study low speed steering.

25.1.3   Vehicles with trailer

If the vehicle has a trailer with one or two axles, with the front axle on a dolly attached to the draw bar, kinematic steering is always possible if the tractor allows it.

Generally speaking, if the wheels of the trailer are fixed, the trailer follows a path which is internal to that of the tractor. In the case of the vehicle of Fig. 25.2a radius R_R is

$$R_R = \sqrt{R_1^2 + l_A^2 - l_R^2} \ . \tag{25.7}$$

If these equations can be linearized, the value of ratio θ/δ, i.e. the *trailer angle gain*, is

$$\frac{\theta}{\delta} = \frac{l_A + l_R}{l} \ , \tag{25.8}$$

where l_A is positive if point A is outside the wheelbase. Distance $l_A + l_R$ is the distance between the axle of the trailer and the rear axle of the tractor.

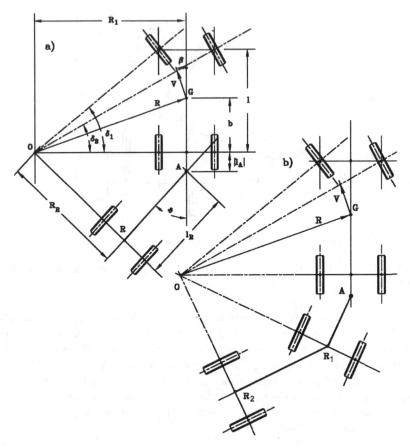

FIGURE 25.2. Low speed steering of vehicles with trailer. (a) steering of a vehicle with a trailer with one axle or an articulated vehicle; (b) steering of a vehicle with a trailer with two axles.

In the case of Fig. 25.2b, the radius of the path of the trailer can be obtained by considering the latter as two subsequent trailers of the type already considered.

The radii of the trajectories of the centers of the axles of the two trailers are

$$
\begin{aligned}
R_{R_1} &= \sqrt{R_1^2 + l_A^2 - l_{R_1}^2}, \\
R_{R_2} &= \sqrt{R_{R_1}^2 - l_{R_2}^2} = \sqrt{R_1^2 + l_A^2 - l_{R_1}^2 - l_{R_2}^2}.
\end{aligned}
\tag{25.9}
$$

The only way to prevent the trailer from following a path internal to that of the tractor is to provide its wheels with a steering mechanism (Fig. 25.3). The steering angle of the last axle must be opposite to the one of the tractor.

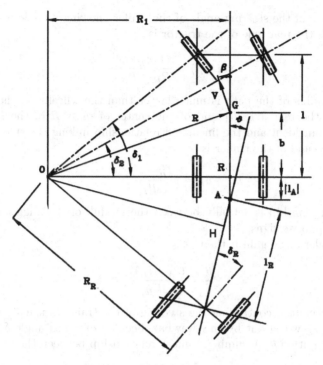

FIGURE 25.3. Kinematic steering of a vehicle with a trailer with a steering angle.

If the average steering angle of the wheels of the trailer is δ_R, the relationship linking the radii of the trajectories of points A and R is

$$R_A = \sqrt{R_R^2 + l_R^2 - 2l_R R_R \sin(\delta_R)} \,. \tag{25.10}$$

The radius of the path of the trailer is then

$$R_R = \sqrt{R_1^2 + l_A^2 - l_R^2 + 2l_R R_R \sin(\delta_R)} \,. \tag{25.11}$$

The difference between the radii of the trajectories of the trailer and the tractor can thus be reduced, allowing the space needed by the vehicle in a bend to be reduced. However, this method is not free from drawbacks, since the driver cannot visually control the rear part of the trailer that, at the beginning of the bend, seems to moves outwards.

This last problem is sometimes solved by placing a second driver in the rear of the trailer to control the relevant steering mechanism, or better, by using an actuator controlled by a suitable control law from the steering control, to steer the trailer. The dynamic problems linked with the steering of trailers will be dealt with later.

The trailer angle gain is

$$\frac{\theta}{\delta} = \frac{l_A + l_R}{l} - \frac{\delta_R}{\delta} \,. \tag{25.12}$$

The value of the steering angle of the trailer allowing its axle to follow the same path as the rear axle of the tractor is

$$\sin{(\delta_R)} = \frac{1}{R_R} \frac{l_R^2 - l_A^2}{2l_R} . \tag{25.13}$$

If the radius of the path is much larger than the wheelbase, the radius of the path of the rear axle R_1 and of the center of mass R of the tractor are practically coincident and the linearized relationship linking the steering angles of the tractor and of the trailer is

$$\delta_R = \delta \frac{l_R^2 - l_A^2}{2ll_R} . \tag{25.14}$$

This relationship is actually between the moduli of the angles, since they must have opposite signs.

The trailer angle gain is then

$$\frac{\theta}{\delta} = \frac{(l_A + l_R)^2}{2ll_R} . \tag{25.15}$$

The mechanism controlling the steering of the trailer is usually not driven by the steering wheel but by the drawbar, because of which angle δ_R does not depend on δ but on θ. Assuming a linear relationship between the two angles

$$\delta_R = K_R \theta , \tag{25.16}$$

the trajectories of the trailer and of the tractor are the same if

$$K_R = \frac{l_R^2 - l_A^2}{(l_A + l_R)^2} . \tag{25.17}$$

Remark 25.2 *The path of the trailer is circular only after a certain time: When the tractor starts to follow a circular path there is an initial transient in which the path of the trailer starts to bend, followed by the period of time needed to reach the steady state conditions.*

The path of the trailer, or better of point R in Fig. 25.2a, can be computed as follows. In Fig. 25.4a the vehicle is sketched in its initial configuration with the trailer and tractor aligned; the generic configuration at time t is shown in Fig. 25.4b. In the second figure, the tractor is rotated by an angle α and the trailer is rotated by an angle β. Note that angle ϕ is positive if A lies between B and C.

The positions of the centre of rotation of the tractor O and of the trailer O_1 at time t and $t+dt$ are shown in Fig. 25.5. Distances $\overline{RR'}$, $\overline{AA'}$ and $\overline{RR''}$ are very small if compared with $\overline{AR}$ and $\overline{A'R'}$. Neglecting vanishingly small quantities, it follows that

$$\begin{aligned} \overline{AA'} &= R_A d\alpha \\ \overline{A'A''} &= l_R d\beta = \overline{AA'} \sin(\alpha + \phi - \beta) . \end{aligned} \tag{25.18}$$

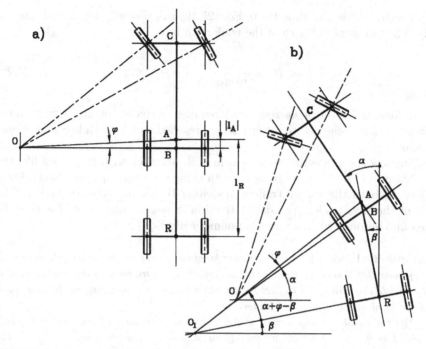

FIGURE 25.4. Vehicle with two axles pulling a trailer with one axle. (a) Situation at time $t = 0$ with the vehicle in straight position; (b) Situation at time t.

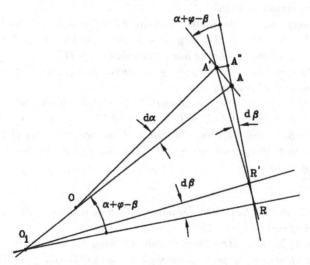

FIGURE 25.5. Position of the vehicle of Fig. 25.4 at time t and $t + dt$.

Equations (25.18) yield

$$\frac{d\beta}{d\alpha} = \frac{R_A}{l_R} \sin(\alpha + \phi - \beta) .$$

(25.19)

Since $\alpha = \beta = 0$ at time $t = 0$, Eq. (25.19) can be easily integrated numerically. The radius of the path of the trailer R_R is

$$R_R = \frac{l_R}{\tan(\alpha + \phi - \beta)} \,. \tag{25.20}$$

A long trailer on a narrow bend requires a change of direction of more than 90° before steady state conditions are reached and its path becomes almost circular.

The low-speed steering of a vehicle with a trailer with two axles like the one shown in Fig. 25.1b can be dealt with using the same equations seen above, applied to both the simple trailers modelling the actual two-axle trailer. The path of the first trailer (the dolly) is initially not circular, and this must be taken into account while integrating numerically Eq. (25.19).

Example 25.1 *Study the conditions for kinematic steering of the articulated vehicle of Appendix E.9. Assume a value of the radius of the centre mass of the tractor of 10 m and compute the path of the trailer. Assume that the trailer has a single axle, coinciding with the third axle of the actual trailer.*

The radius of the trajectories of the front and rear axles of the tractor is easily computed as 9.730 and 10.335 m; the off-tracking of the tractor is thus 605 mm. The approximated expression (25.5) for the off-tracking yields 607 mm, very close to the correct value even if the radius of the path is not actually very large compared to the wheelbase (10 m versus 3.485 m).

The steering angles of the front wheels are 17.99° and 21.77°, with an average value of 19.71°. This value is also very close to the correct value of 19.88°, obtained without any linearization, and to the linearized value of 19.77°.

The steady state radius of the path of the trailer is 5.446 m, yielding a value of 4.889 m for the total off-tracking distance.

The path of the trailer has been computed by numerically integrating Eq. (25.19) for α included between 0 and 450°, with a step of 0.5°. The values of ϕ and R_A are, respectively, of 2.648° and 9.740 m. The path and the locus of points O' are shown in Fig. 25.6. Note that after a rotation of 90° the radius of the path is still larger than that in steady state conditions.

Example 25.2 *Repeat the previous example, assuming that the trailer axle is steering with a mechanism realizing law (25.17).*

The value of K is 1,118. The equation allowing the path of the trailer to be computed is the same as in the previous example, the only difference being that reference is made to point H in Fig. 25.3 instead of point R in Fig. 25.4.

The radius of the steady-state path of the trailer is 9.942 m , very close to that of the trailer. The steering angle of the trailer is $\delta_R = 20.02°$ and the angle between the trailer and the tractor is $\theta = 19.76°$. The path of the trailer was computed by numerically integrating the relevant equation for values of α from 0 and 450°, with increments of

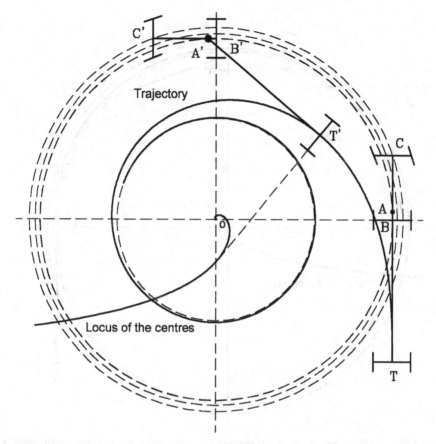

FIGURE 25.6. Path and locus of the centres of curvature of the path of the trailer for an articulated vehicle. The positions of the vehicle before starting on the curved path and after a rotation of the tractor of 90° are reported.

0.5°, as in the previous example. The path and the locus of points O′ are plotted in Fig. 25.7. Note that steady-state conditions are quickly reached and that at the beginning the trailer moves outwards.

25.2   IDEAL STEERING

If the speed is not vanishingly small, the wheels must move with suitable sideslip angles to generate cornering forces. A simple evaluation of the steady state steering of a vehicle in high-speed or dynamic[2] steering conditions may be performed

---

[2]The term *dynamic steering* is used here to denote a condition in which the path is determined by the balance of forces acting on the vehicle, as opposed to *kinematic steering* in which

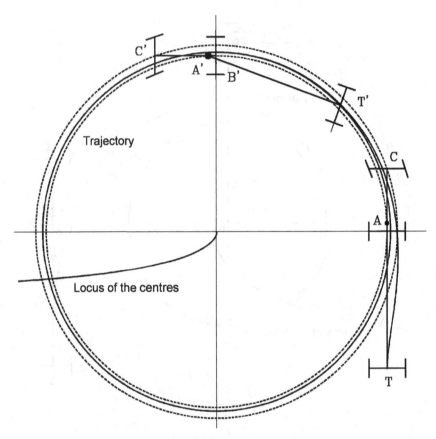

FIGURE 25.7. Path and locus of the centers O' of the path of the trailer with steering axle. The positions at the beginning of the maneouvre and after a 90° rotation are also reported.

as follows. Consider a rigid vehicle moving on level road with transversal slope angle α_t and neglect the aerodynamic side force. Define a η-axis parallel to the road surface, passing through the centre of mass of the vehicle and intersecting the vertical for the centre of the path, which in steady-state condition is circular (Fig. 25.8). Axis η does not coincide with the y axis, except at one particular speed.

25.2.1   Level road

Assume that the road is flat and neglect aerodynamic forces. The equilibrium equation in η direction can be written by equating the centrifugal force mV^2/R to forces P_η due to the tires

---

the path is determined by the direction of the midplane of the wheels. Note that dynamic steering applies to both steady state and unstationary turning.

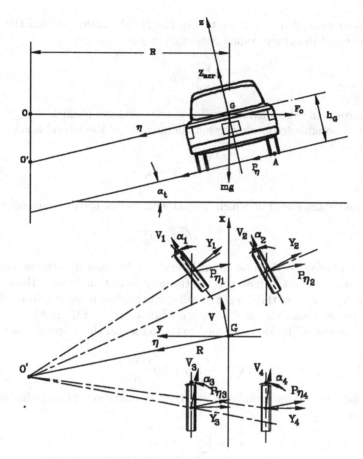

FIGURE 25.8. Simplified model for dynamic steering.

$$\frac{mV^2}{R} = \sum_{\forall i} P_{\eta_i} \; . \tag{25.21}$$

For a first approximation study, forces P_η may be conflated with the cornering forces F_y of the tires and all wheels may be assumed to work with the same side force coefficient μ_y. As the last assumption is similar to that seen for braking in ideal conditions, this approach will be referred to as *ideal steering*. These two assumptions lead to substituting the expression $\sum_{\forall i} P_{\eta_i}$ with $\mu_y F_z$.

Force

$$F_z = \sum F_{z_i}$$

exerted by the vehicle on the road is

$$F_z = mg \; . \tag{25.22}$$

By introducing Eq. (25.22) into Eq. (25.21) the ratio between the lateral acceleration and the gravitational acceleration g is

$$\frac{V^2}{Rg} = \mu_y \tag{25.23}$$

By introducing the maximum value of the side force coefficient μ_{y_p} into Eq. (25.23), it is possible to obtain the maximum value of the lateral acceleration

$$\left(\frac{V^2}{R}\right)_{max} = g\mu_{y_p} . \tag{25.24}$$

The maximum speed at which a bend with radius R can be negotiated is

$$V_{max} = \sqrt{Rg}\sqrt{\mu_{y_p}} , \tag{25.25}$$

The limitation to the maximum lateral acceleration due to the cornering force the tires can exert is, however, not the only limitation, at least theoretically. Another can come from the danger of rollover occurring if the resultant of forces in the yz plane crosses the road surface outside point A (Fig. 25.8).

The moment of the forces applied to the vehicle in the ηz-plane about point A is

$$M_A = -\frac{t}{2}mg + h_G \frac{mV^2}{R} . \tag{25.26}$$

The limit condition for rollover can then be computed by equating moment M_A to zero

$$\left(\frac{V^2}{R}\right)_{max} = g\frac{t}{2h_G} . \tag{25.27}$$

The rollover condition is identical to the sliding conditions, once ratio

$$\frac{t}{2h_G}$$

has been substituted for μ_{y_p}.

The maximum lateral acceleration is then

$$\left(\frac{V^2}{R}\right)_{max} = g \min \left\{\mu_{y_p}, \frac{t}{2h_G}\right\} . \tag{25.28}$$

Whether the limit condition first reached is that related to sliding, with subsequent spin out of the vehicle, or related to rolling over depends on the relative magnitude of μ_{y_p} and $\frac{t}{2h_G}$. If the former is smaller than the latter, as often occurs, the vehicle spins out. This condition can be written in the form

$$\mu_{y_p} < \frac{t}{2h_G} .$$

25.2.2   Effect of aerodynamic lift

If aerodynamic lift is accounted for, Eq. (25.22) becomes:

$$F_z = mg - \frac{1}{2}\rho V^2 S C_Z \ . \tag{25.29}$$

By introducing ratio

$$M = \frac{\rho S C_z}{2mg} \ ,$$

expressing the ratio between aerodynamic lift at unit speed and weight, it follows that

$$F_z = mg\left(1 - MV^2\right) \ . \tag{25.30}$$

Note that M is negative if the lift is directed downwards. To take aerodynamic lift into account it is sufficient to multiply the expressions seen in the previous section by $1 - MV^2$.

The maximum lateral acceleration is now

$$\left(\frac{V^2}{R}\right)_{max} = g\left(1 - MV^2\right)\min\left\{\mu_{y_p}, \frac{t}{2h_G}\right\} \ . \tag{25.31}$$

Term MV^2 is usually very small and often negligible, with the exception of racing cars. For instance, let $\rho = 1.22$ kg/m^3 (value at sea level in standard atmosphere), $S = 1.7$ m^2, $C_z = -0.5$ (an already high value) and $m = 1000$ kg. It follows that $M = -5.3 \times 10^{-5}$ s^2/m^2 and thus, at 100 km/h, the value of the additional term is 0.05. To change things radically high speeds must be reached: at 300 km/h the additional term becomes $-MV^2 = 0.37$, i.e. the maximum lateral acceleration increases by 37%.

The negative value of C_z is very high in racing cars, and at high speed strong lateral accelerations are possible.

25.2.3   Transversal slope of the road

The equilibrium equation in η direction may be written by equating the components of weight mg and of the centrifugal force mV^2/R acting in that direction with forces P_η due to the tires

$$\frac{mV^2}{R}\cos(\alpha_t) - mg\sin(\alpha_t) = \sum_{\forall i} P_{\eta_i} \ . \tag{25.32}$$

By introducing the previously discussed assumptions characterizing ideal steering, substituting expression $\sum_{\forall i} P_{\eta_i}$ with $\mu_y F_z$, force $F_z = \sum F_{z_i}$ exerted by the vehicle on the road becomes

$$F_z = mg\cos(\alpha_t) + \frac{mV^2}{R}\sin(\alpha_t) - \frac{1}{2}\rho V^2 S C_Z \ . \tag{25.33}$$

By introducing Eq. (25.33) into Eq. (25.32), the latter yields the following value for the ratio between the lateral acceleration and the gravitational acceleration g

$$\frac{V^2}{Rg} = \frac{\tan(\alpha_t) + \mu_y(1 - MV^2)}{1 - \mu_y \tan(\alpha_t)} . \tag{25.34}$$

Ratio M can be redefined as

$$M = \frac{\rho S C_z}{2mg \cos(\alpha_t)}$$

so that MV^2 is the ratio between the aerodynamic lift and the component of weight in a direction perpendicular to the road.

By introducing the maximum value of the side force coefficient μ_{y_p} into Eq.(25.34), the maximum value of the lateral acceleration is obtained

$$\left(\frac{V^2}{R}\right)_{max} = g f_s , \tag{25.35}$$

where the so-called *sliding factor* f_s can be defined as[3]

$$f_s = \frac{\tan(\alpha_t) + \mu_{y_p}(1 - MV^2)}{1 - \mu_{y_p} \tan(\alpha_t)} ; \tag{25.36}$$

and is in general a function of the speed, if the aerodynamic lift is accounted for.

Note that on level road and with no aerodynamic lift the sliding factor reduces to μ_{y_p}.

The sliding factor is reported as a function of μ_{y_p} for different values of the transversal slope of the road in Fig. 25.9a and for different values of ratio MV^2 in Fig. 25.9b. Note that if the road is flat and the aerodynamic lift is neglected it reduces to the maximum value of the side force coefficient μ_{y_p}.

The maximum speed at which a bend with radius R can be negotiated is

$$V_{max} = \sqrt{Rg}\sqrt{\frac{\tan(\alpha_t) + \mu_{y_p}}{1 - \mu_{y_p}[\tan(\alpha_t) - RgM]}} , \tag{25.37}$$

i.e.

$$V_{max} = \sqrt{Rg}\sqrt{f_s} . \tag{25.38}$$

The rollover condition can also be modified to take into account the transversal slope of the road and aerodynamic lift. The moment of all forces applied to the vehicle in ηz plane about point A (Fig. 25.8) is

---

[3]The sliding factor is more commonly defined as the square root of the same quantity considered here. The present definition, which refers directly to the lateral acceleration instead of the speed at which a given turn may be negotiated, is here preferred as in particular conditions it reduces to the side force coefficient.

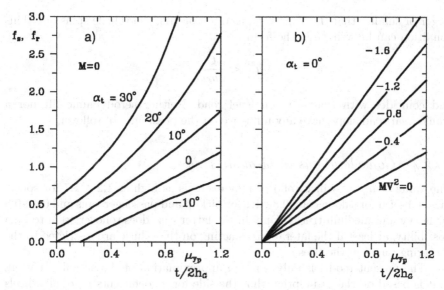

FIGURE 25.9. Sliding and rollover factors as functions of μ_{y_p} and of $t/2h_G$ respectively for roads with different transversal slope (a) and for vehicles with different values of ratio MV^2 (b).

$$M_A = -\frac{t}{2}\left[mg\cos(\alpha_t) + \frac{mV^2}{R}\sin(\alpha_t) - \frac{1}{2}\rho V^2 S C_Z\right] +$$
$$+h_G\left[\frac{mV^2}{R}\cos(\alpha_t) - mg\sin(\alpha_t)\right] . \qquad (25.39)$$

The limit condition for rollover can be obtained by equating moment M_A to zero, obtaining

$$\left(\frac{V^2}{R}\right)_{max} = gf_r , \qquad (25.40)$$

where the *rollover factor* can be defined as

$$f_r = \frac{\tan(\alpha_t) + \frac{t}{2h_G}(1 - MV^2)}{1 - \frac{t}{2h_G}\tan(\alpha_t)} . \qquad (25.41)$$

The expression of the rollover factor is identical to that of the sliding factor, once ratio $t/2h_G$ has been substituted for μ_{y_p} (Fig. 25.9). It depends on speed because of the effects of aerodynamic lift.

The maximum lateral acceleration is then

$$\left(\frac{V^2}{R}\right)_{max} = g\min\{f_s, f_r\} . \qquad (25.42)$$

Whether the limit condition first reached is that related to sliding, with subsequent spin out of the vehicle, or rolling over depends on whether f_s is larger or smaller than f_r. If $f_s < f_r$, as often occurs, the vehicle spins out. This condition can be written in the form

$$\mu_{y_p} < \frac{t}{2h_G} \, ,$$

and coincides with that seen on level road. Neither aerodynamic lift nor a transversal road slope have any influence on the possibility of rollover.

25.2.4 Considerations in ideal steering

The value of μ_{y_p} at which rollover may occur is as high as $1.2 \div 1.7$ for sports cars, $1.1 \div 1.6$ for saloon cars, $0.8 \div 1.1$ for pickup and passenger vans and $0.4 \div 0.8$ for heavy and medium trucks. Only in the latter case does rollover seem to be a possibility, at least if the lateral forces acting on the vehicle are restricted to the cornering forces of the tires.

The present model is only a rough approximation of the actual situation, as it is based on the assumption that the side force coefficients μ_y of all wheels are equal, implying that all wheels work with the same sideslip angle α. It also ignores the effect of the different directions of the cornering forces of the various wheels, which should be considered as perpendicular to the midplanes of the wheels and not directed along the η axis. The load transfer between the wheels of the same axle and the presence of the suspensions have also been neglected, two other assumptions contributing to the lack of precision of this model.

If the maximum speed at which a circular path can be negotiated is measured in a steering pad test and the value of the lateral force coefficient is computed through Eq. (25.25), a value of μ_{y_p}, well below that obtained from tests on the tires, is obtained.

Remark 25.3 *The cornering force coefficient obtained in this way is that of the vehicle as a whole, and the difference between its value and that related to the tires gives a measure of how well the vehicle is able to exploit the cornering characteristics of its wheels.*

The side force coefficient measured on the whole vehicle also depends on the radius of the path, with a notable decrease on narrow bends. The majority of industrial and passenger vehicles are able to use only a fraction, from 50% to 80%, of the potential cornering force of the tires, with higher values found only in sports cars. This reduction of the lateral forces makes the danger of rollover more remote.

Actually rolling over in a quasi-static condition is impossible for most vehicles, notwithstanding the fact that rollover actually occurs in many road accidents. Rollover can usually be ascribed to dynamic phenomena in nonstationary conditions or to lateral forces caused by side contacts, e.g. of the wheels with the curb of the road, that rule out the possibility of side slipping while causing far stronger lateral forces to be exerted on the wheels. The presence of

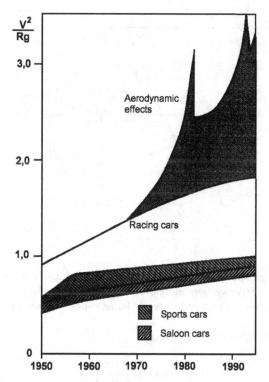

FIGURE 25.10. Evolution in time of the maximum lateral acceleration for saloon cars, sports cars and racers. Note that for the latter the change of racing rules caused sharp changes in the maximum lateral acceleration

the suspensions also contributes to this picture, making rollover a likely outcome of many accidents.

From the equations it is also clear that only the use of aerodynamic devices able to exert a strong negative lift allows high values of lateral acceleration, well above 1 g in the case of racers, to be reached (Fig. 25.10).

25.2.5   Vehicles with two wheels

The cornering dynamics of a vehicle with two wheels are radically different from those of four wheeled vehicles (Fig. 25.11). If the gyroscopic moments of the wheels are neglected, the equation expressing rolling equilibrium can be used to compute the roll angle the vehicle must maintain in order not to capsize, since a two-wheeled vehicle is a system underconstrained in roll.

The limitation on lateral acceleration and speed on a curved path is solely the result of lateral sliding, with a further geometric limitation on the maximum roll angle that can be reached before the vehicle or the driver touches the road on one side. Equation (25.24) yielding the maximum lateral acceleration still holds, the difference being that the global side force coefficient is usually higher.

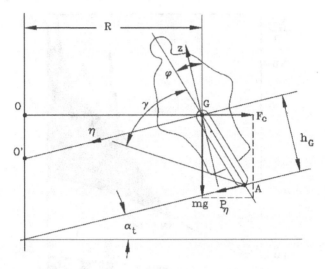

FIGURE 25.11. High speed steering of a two wheeled vehicle. Point G is the centre of mass of the vehicle-driver system and can be displaced from the plane of symmetry of the former if the latter is displaced to one side, as usually occurs in bends.

The roll angle is easily computed

$$\phi = \arctan\left(\frac{V^2}{Rg}\right) , \tag{25.43}$$

and the geometrical limitation

$$\phi \leq \pi/2 + \alpha_t - \gamma$$

(Fig. 25.11) usually does not induce further limitations.

Remark 25.4 *Since motor cycles roll into the curve, the lateral forces due to camber add to those due to sideslip, instead of subtracting as in the case of motor vehicles that roll towards the outside of the curve.*

Further terms must be introduced into the relevant equations if gyroscopic moments of the wheels are considered . When the vehicle runs on a circular path with radius R, the gyroscopic moment, due to the ith wheel with radius R_i and moment of inertia J_{p_i} about its spin axis, is equal to

$$\frac{J_{p_i} V^2 \cos(\phi)}{RR_i} .$$

The equation expressing the equilibrium for rolling motions is then

$$mgh_G \sin(\phi) - \frac{V^2}{R}\cos(\phi)\left[mh_G + \sum_{\forall i}\left(\frac{J_{p_i}}{R_i}\right)\right] = 0 . \tag{25.44}$$

The roll angle is

$$\phi = \arctan\left\{\frac{V^2}{Rg}\left[1 + \frac{1}{mh_G}\sum_{\forall i}\left(\frac{J_{p_i}}{R_i}\right)\right]\right\} . \qquad (25.45)$$

The added term in Eq. 25.45 is positive and thus the roll angle needed to manage a certain bend at a certain speed is increased by gyroscopic moments.

Remark 25.5 *Generally speaking, the effect of the gyroscopic moment of the wheels on the dynamic behavior of the whole vehicle is small even in the case of vehicles with two wheels. Gyroscopic moments are usually important only in the dynamics of the steering device.*

25.3   HIGH SPEED CORNERING: SIMPLIFIED APPROACH

To go beyond the extremely simplified model of ideal steering, the distribution of cornering forces between the axles, the sideslip angle of the vehicle on the path and the sideslip angles of the wheels must be taken into account.

Assume that the vehicle is moving at constant speed on a circular path and that the road is level. Moreover, assume that the radius of the path R is much larger than the wheelbase l and, as a consequence, all sideslip angles are small. The small size of all angles allows the "monotrack" or "bicycle" model to be used.

Neglecting aerodynamic forces and aligning torques, the forces acting in the xy plane at the tire-road interface in a monotrack vehicle are shown in Fig. 25.12.

The equilibrium equation in the direction of the y axis is similar to Eq.(25.21), except for the presence of the sideslip and steering angles

$$\frac{mV^2}{R}\cos(\beta) = \sum_{\forall i}F_{x_i}\sin(\delta_i) + \sum_{\forall i}F_{y_i}\cos(\delta_i) . \qquad (25.46)$$

The equilibrium to rotations about point G can be expressed as

$$\sum_{\forall i}F_{x_i}\sin(\delta_i)x_i + \sum_{\forall i}F_{y_i}\cos(\delta_i)x_i = 0 . \qquad (25.47)$$

Since angles β and δ_i are assumed to be small, the terms containing the longitudinal forces of the tires can be neglected and the equilibrium equations reduce to

$$\begin{cases} \sum_{\forall i}F_{y_i} = \frac{mV^2}{R} \\ \\ \sum_{\forall i}F_{y_i}x_i = 0 . \end{cases} \qquad (25.48)$$

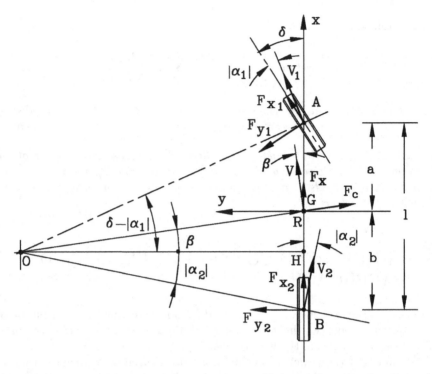

FIGURE 25.12. Simplified model (monotrack vehicle) for studying the handling of a two axle vehicle.

For a two axle vehicle, they can be immediately solved, yielding

$$F_{y_1} = \frac{mV^2}{R}\frac{b}{l} , \quad F_{y_2} = \frac{mV^2}{R}\frac{a}{l} . \tag{25.49}$$

Assuming that the cornering forces of the axles are proportional to the sideslip angles through their cornering stiffness, it follows that

$$\alpha_1 = -\frac{mV^2}{R}\frac{b}{lC_1} , \quad \alpha_2 = -\frac{mV^2}{R}\frac{a}{lC_2} , \tag{25.50}$$

where C_i is the cornering stiffness of the ith axle, and is equal to the cornering stiffness of the wheels multiplied by the number of wheels of the axle.

A relationship between the sideslip and steering angles can be found with simple geometrical considerations from Fig. 25.12

$$\delta - \alpha_1 + \alpha_2 = \frac{l}{R} . \tag{25.51}$$

Introducing the expressions of the sideslip angles into Eq. (25.51), it follows that

$$\delta = \frac{l}{R} + \frac{mV^2}{Rl}\left(\frac{b}{C_1} - \frac{a}{C_2}\right) , \tag{25.52}$$

or, in terms of path curvature gain,

$$\frac{1}{R\delta} = \frac{1}{l} \frac{1}{1 + K_{us}\frac{V^2}{gl}} , \qquad (25.53)$$

where

$$K_{us} = \frac{mg}{l^2} \left(\frac{b}{C_1} - \frac{a}{C_2} \right) \qquad (25.54)$$

is the so-called *understeer coefficient* or *understeer gradient* of the vehicle. The understeer coefficient is a non-dimensional quantity, and is often expressed in radians.

As already stated, in kinematic conditions

$$\left(\frac{1}{R\delta} \right)_{kin} = \frac{1}{l} . \qquad (25.55)$$

The expression $1 + K_{us}V^2/gl$ can be considered as a correction factor giving the response of the vehicle in dynamic conditions as opposed to kinematic conditions.

From Eq.(25.52) it follows that

$$\delta - \delta_{kin} = \frac{V^2}{Rg}K_{us} , \qquad (25.56)$$

i.e.

$$K_{us} = \frac{g}{a_y} (\delta - \delta_{kin}) . \qquad (25.57)$$

The understeer coefficient can thus be interpreted as the difference between the steering angles in kinematic and dynamic conditions divided by the centrifugal acceleration expressed as a multiple of the gravitational acceleration.

Sometimes, instead of the understeer coefficient, a *stability factor*

$$K = \frac{m}{l^2} \left(\frac{b}{C_1} - \frac{a}{C_2} \right) . \qquad (25.58)$$

is defined.

As a first approximation, K and K^* may be considered as constant for a given vehicle and load condition. As will be seen below, however, in many cases their dependence on speed cannot be neglected for more precise assessments.

It is possible to define a *lateral acceleration* gain as the ratio between the lateral acceleration and the steering input:

$$\frac{V^2}{R\delta} = \frac{V^2}{l} \frac{1}{1 + K_{us}\frac{V^2}{gl}} . \qquad (25.59)$$

The sideslip angle can be obtained through simple geometrical considerations, yielding

$$\beta = \frac{b}{R} - \alpha_2. \qquad (25.60)$$

A *sideslip angle gain*, expressing the ratio between the sideslip angle and the steering angle can be defined as well. Its value is

$$\frac{\beta}{\delta} = \frac{b}{l}\left(1 - \frac{maV^2}{blC_2}\right)\frac{1}{1 + K_{us}\frac{V^2}{gl}} ,$$ (25.61)

25.4 DEFINITION OF UNDERSTEER AND OVERSTEER

If $K_{us} = 0$ the value of $1/R\delta$ is constant and equal to the value characterizing kinematic steering; i.e. the response of the vehicle to a steering input is, at any speed, equal to that in kinematic conditions. This does not mean, however, that the vehicle is in kinematic conditions, since the value of the sideslip angle β is not equal to its kinematic value and the values of the sideslip angles of the wheels are not equal to zero.

A vehicle behaving in this way is said to be *neutral-steer* (Fig. 25.13a).

If $K_{us} > 0$ the value of $1/R\delta$ decreases with increasing speed. The response of the vehicle is then smaller than in kinematic conditions and, to maintain a constant radius of the path, the steering angle must be increased as speed increases.

A vehicle behaving in this way is said to be *understeer*.

A quantitative measure of the understeering of a vehicle is given by the *characteristic speed*, defined as the speed at which the steering angle needed to

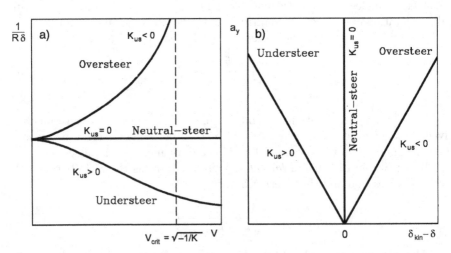

FIGURE 25.13. Steady state response to a steering input. Plot of the path curvature gain as a function of speed (a) and handling diagram (b) for an oversteer, an understeer and a neutral steer vehicle. The understeer factor is assumed to be independent of speed.

negotiate a turn is equal to twice the Ackerman angle, i.e. the path curvature gain is equal to $1/2l$.

Using the simplified approach outlined above, the characteristic speed is

$$V_{car} = \sqrt{\frac{gl}{K_{us}}} = \sqrt{\frac{1}{K}}. \tag{25.62}$$

If $K_{us} < 0$ the value of $1/R\delta$ increases with increasing speed until, for a speed

$$V_{crit} = \sqrt{-\frac{gl}{K_{us}}} = \sqrt{-\frac{1}{K}} \tag{25.63}$$

the response tends to infinity, i.e., the system develops an unstable behavior.

A vehicle behaving in this way is said to be *oversteer*, and the speed given by Eq. (25.63) is called the *critical speed*. The critical speed of any oversteer vehicle must be well above the maximum speed it can reach, at least in normal road conditions.

Instead of plotting the path curvature gain as a function of the speed, it is possible to plot the *handling diagram*, i.e. the plot of the lateral acceleration a_y as a function of $\delta_{kin} - \delta$ (Fig. 25.13b). If the vehicle is neutral steer, the plot is a vertical straight line, if it is oversteer it is a straight line sloping to the right, while in case of an understeer vehicle it slopes to the left.

The value of β, or better, of β/δ, decreases with the speed from the kinematic value up to the speed

$$(V)_{\beta=0} = \sqrt{\frac{blC_2}{am}}. \tag{25.64}$$

at which it vanishes. At higher speed it becomes negative, tending to infinity when approaching the critical speed for oversteer vehicles and tending to

$$\frac{aC_1}{aC_1 - bC_2}$$

when the speed tends to infinity in the case of understeering vehicles.

The sideslip angles of the front and rear wheels are equal in neutral-steer vehicles. In oversteer vehicles, the rear wheels have a larger sideslip angle (in absolute value, since the sideslip angles are negative when the radius of the path is positive), while the opposite holds in understeer vehicles. It follows that oversteer vehicles can be expected to reach limit conditions at the rear wheels and understeer vehicles at the front wheels, even if the present model cannot be applied when the sideslip angles increase, approaching limit conditions.

A graphical interpretation of this result, for a vehicle with a single steering axle, is shown in Fig. 25.14. The vehicle is modelled as a steering front axle and a fixed rear axle. Kinematic steering applies if the speed tends to zero: the sideslip

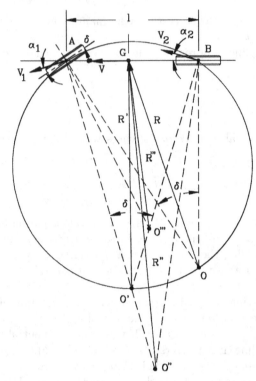

FIGURE 25.14. Geometrical definition of the behavior of a vehicle with a single steering axle.

angles vanish and the center of path is point O. It follows immediately that

$$\frac{l}{R} = \tan(\delta) \approx \delta .$$

With increasing speed the wheels work with increasing sideslip angles α_1 and α_2. If $\alpha_1 = \alpha_2$ angle BO'A is still equal to δ (its value is $|\alpha_2| + \delta - |\alpha_1|$) and thus O' lies on a circle through points A, B and O.

Since $l \ll R'$, O' is in a position almost opposite to A and B and then $R' \approx R$. The radius of the path is still equal to that characterizing kinematic steering, and the vehicle is neutral steer.

If $|\alpha_1| > |\alpha_2|$ the center of the path moves to O" and radius R'' is larger thus R. The vehicle is then understeer. If, on the other hand, $|\alpha_1| < |\alpha_2|$, the center of the path is O''', radius R''' is smaller than R' and the vehicle is oversteer.

25.5   HIGH SPEED CORNERING

25.5.1   Equations of motion

The study of the handling of the vehicle seen in the previous sections was based on the assumption of steady-state operation. Moreover, only the cornering forces acting on the tires were considered.

A simple mathematical model for the handling of a rigid vehicle that overcomes the above limitations can, however, be built.

To keep the model as simple as possible, the following assumptions may be made

1. The sideslip angle of the vehicle β and of the wheels α are small. The yaw angular velocity $\dot{\psi}$ can also be considered a small quantity.

2. The vehicle can be assumed to be a rigid body moving on a flat surface, i.e. roll and pitch angles are neglected as well as the vertical displacements due to suspensions.

If a motor vehicle is considered as a rigid body moving on a surface, a model with three degrees of freedom is needed for the study of its motion. If the road is considered as a flat surface, the motion is planar. By using the inertial reference frame[4] XY shown in Fig. 25.15, it is possible to use the coordinates X and Y of the centre of mass G of the vehicle and the yaw angle ψ between the x and X axes as generalized coordinates.

The equations of motion of the vehicle are

$$\begin{cases} m\ddot{X} = F_X \\ m\ddot{Y} = F_Y \\ J_z\ddot{\psi} = M_z , \end{cases} \tag{25.65}$$

where F_X, F_Y and M_z are the total forces acting in the X and Y directions and the total yawing moment. For the latter, subscript z has been used instead of Z since the directions of the two axes coincide.

Equations (25.65) are very simple but include the forces acting on the vehicle in the direction of the axes of the inertial frame. They are clearly linked with the forces acting in the directions of axes x and y of the vehicle by the obvious relationship

$$\left\{ \begin{array}{c} F_X \\ F_Y \end{array} \right\} = \left[\begin{array}{cc} \cos(\psi) & -\sin(\psi) \\ \sin(\psi) & \cos(\psi) \end{array} \right] \left\{ \begin{array}{c} F_x \\ F_y \end{array} \right\} . \tag{25.66}$$

If the model is used to perform a numerical integration in time, they can be used directly without any difficulty.

---

[4] As already stated, such a reference frame is not, strictly speaking, inertial, since it is fixed to the road surface and hence follows the motion of Earth. It is, however, inertial "enough" for the problems here studied, and this issue will not be dealt with any further.

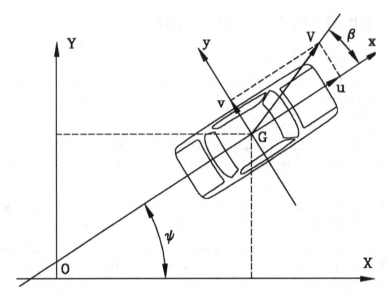

FIGURE 25.15. Reference frame for the study of the motion of a rigid vehicle. The vehicle has three degrees of freedom, and the coordinates X and Y of the centre of mass G and the yaw angle ψ can be used as generalized coordinates.

However, if the model has to be used to obtain linearized equations in order to gain a general insight into the behavior of the vehicle, it is better to write the equations of motion with reference to the non-inertial xy frame, to avoid dealing with the trigonometric functions of angle ψ, which in general is not a small angle, and would make linearizations impossible.

To write the equations of motion with reference to the body-fixed frame xyz, it is expedient to use the components u and v of the speed in the directions of the x and y axes and the yaw angular velocity

$$r = \dot{\psi} \ .$$

There are many ways to obtain the mathematical model, but perhaps the simplest is to remember that the derivative with respect to time of a generic vector $\vec{A}$, expressed in the body-fixed frame, but performed in the inertial frame

$$\left. \frac{d\vec{A}}{dt} \right|_i$$

can be expressed starting from the derivative performed in the body fixed frame

$$\left. \frac{d\vec{A}}{dt} \right|_m ,$$

as

$$\left. \frac{d\vec{A}}{dt} \right|_i = \left. \frac{d\vec{A}}{dt} \right|_m + \vec{\Omega} \wedge \vec{A} \ , \tag{25.67}$$

where $\vec{\Omega}$ is the absolute angular velocity of the body fixed frame.

In the present case, the velocity and the angular velocity vectors, in the body-fixed frame, are

$$\vec{V} = \left\{ \begin{array}{c} u \\ v \\ 0 \end{array} \right\} , \qquad \vec{\Omega} = \left\{ \begin{array}{c} 0 \\ 0 \\ r \end{array} \right\} . \qquad (25.68)$$

The derivative of the velocity of the vehicle is then

$$\left. \frac{d\vec{V}}{dt} \right|_i = \left. \frac{d\vec{V}}{dt} \right|_m + \vec{\Omega} \wedge \vec{V} = \left\{ \begin{array}{c} \dot{u} - rv \\ \dot{v} + ru \\ 0 \end{array} \right\} . \qquad (25.69)$$

The equations of motion of the vehicle, expressed with reference to the xyz frame, are

$$\left\{ \begin{array}{l} m(\dot{u} - rv) = F_x \\ m(\dot{v} + ru) = F_y \\ J_z \dot{r} = M_z . \end{array} \right. \qquad (25.70)$$

As an alternative, a procedure based on Lagrange equations can be followed. Although apparently more complicated, it will be shown here, since it is consistent with what will be done later for more complex models. In the present case other approaches are more straightforward.

The kinetic energy of the vehicle is

$$T = \frac{1}{2}m\left(u^2 + v^2\right) + \frac{1}{2}J_z r^2 . \qquad (25.71)$$

The rotational kinetic energy of the wheels has been neglected: No gyroscopic effect of the wheels will be obtained in this way.

Velocities u, v and r are linked to the derivatives of the generalized coordinates $\dot{X}$, $\dot{Y}$ and $\dot{\psi}$ by the relationship:

$$\left\{ \begin{array}{c} u \\ v \\ r \end{array} \right\} = \left[\begin{array}{ccc} \cos(\psi) & \sin(\psi) & 0 \\ -\sin(\psi) & \cos(\psi) & 0 \\ 0 & 0 & 1 \end{array} \right] \left\{ \begin{array}{c} \dot{X} \\ \dot{Y} \\ \dot{\psi} \end{array} \right\} , \qquad (25.72)$$

i.e.

$$\mathbf{w} = \mathbf{A}^T \dot{\mathbf{q}} , \qquad (25.73)$$

where

$$\mathbf{w} = \left[\begin{array}{ccc} u & v & r \end{array} \right]^T$$

is the vector containing the generalized velocities, and

$$\dot{\mathbf{q}} = \left[\begin{array}{ccc} \dot{X} & \dot{Y} & \dot{\psi} \end{array} \right]^T$$

is the vector containing the derivatives of the generalized coordinates.

Since matrix $\mathbf{A}$ is a rotation matrix,

$$\mathbf{A}^T = \mathbf{A}^{-1} \tag{25.74}$$

and the inverse transformation is

$$\dot{\mathbf{q}} = \mathbf{A}\mathbf{w} . \tag{25.75}$$

The equations of motion are

$$\frac{d}{dt}\left(\frac{\partial T}{\partial \dot{q}_i}\right) - \frac{\partial T}{\partial q_i} = Q_i , \tag{25.76}$$

where coordinates q_i are X, Y and ψ and forces Q_i are the corresponding generalized forces F_X, F_Y and M_z.

The derivatives needed to write the equations of motion are

$$\begin{cases} \dfrac{\partial T}{\partial \dot{X}} = \dfrac{\partial T}{\partial u}\dfrac{\partial u}{\partial \dot{X}} + \dfrac{\partial T}{\partial v}\dfrac{\partial v}{\partial \dot{X}} + \dfrac{\partial T}{\partial r}\dfrac{\partial r}{\partial \dot{X}} \\[2ex] \dfrac{\partial T}{\partial \dot{Y}} = \dfrac{\partial T}{\partial u}\dfrac{\partial u}{\partial \dot{Y}} + \dfrac{\partial T}{\partial v}\dfrac{\partial v}{\partial \dot{Y}} + \dfrac{\partial T}{\partial r}\dfrac{\partial r}{\partial \dot{Y}} \\[2ex] \dfrac{\partial T}{\partial \dot{\psi}} = \dfrac{\partial T}{\partial u}\dfrac{\partial u}{\partial \dot{\psi}} + \dfrac{\partial T}{\partial v}\dfrac{\partial v}{\partial \dot{\psi}} + \dfrac{\partial T}{\partial r}\dfrac{\partial r}{\partial \dot{\psi}} . \end{cases} \tag{25.77}$$

i.e.

$$\left\{\frac{\partial T}{\partial \dot{q}}\right\} = \mathbf{A}\left\{\frac{\partial T}{\partial w}\right\} , \tag{25.78}$$

where

$$\left\{\frac{\partial T}{\partial \dot{q}}\right\} = \left[\begin{array}{ccc} \dfrac{\partial T}{\partial \dot{X}} & \dfrac{\partial T}{\partial \dot{Y}} & \dfrac{\partial T}{\partial \dot{\psi}} \end{array}\right]^T$$

is the vector containing the derivatives with respect to the derivatives of the generalized coordinates, while

$$\left\{\frac{\partial T}{\partial w}\right\} = \left[\begin{array}{ccc} \dfrac{\partial T}{\partial u} & \dfrac{\partial T}{\partial v} & \dfrac{\partial T}{\partial w} \end{array}\right]^T$$

is the vector containing the derivatives with respect to the generalized velocities. By differentiating with respect to time, it follows that

$$\frac{\partial}{\partial t}\left(\left\{\frac{\partial T}{\partial \dot{q}}\right\}\right) = \mathbf{A}\frac{\partial}{\partial t}\left(\left\{\frac{\partial T}{\partial w}\right\}\right) + \dot{\mathbf{A}}\left\{\frac{\partial T}{\partial w}\right\} , \tag{25.79}$$

where

$$\dot{\mathbf{A}} = \dot{\psi}\left[\begin{array}{ccc} -\sin(\psi) & -\cos(\psi) & 0 \\ \cos(\psi) & -\sin(\psi) & 0 \\ 0 & 0 & 0 \end{array}\right] . \tag{25.80}$$

The computation of the derivatives with respect to the generalized coordinates $\left\{\frac{\partial T}{\partial q}\right\}$[5] is more complex. The generic derivative $\frac{\partial T}{\partial q_k}$ is

$$\frac{\partial T^*}{\partial q_k} = \frac{\partial T}{\partial q_k} + \sum_{i=1}^{n} \frac{\partial T}{\partial w_i} \frac{\partial w_i}{\partial q_k} = \frac{\partial T}{\partial q_k} + \sum_{i=1}^{n} \frac{\partial T}{\partial w_i} \sum_{j=1}^{n} \frac{\partial A_{ij}}{\partial q_k} \dot{q}_j \,, \qquad (25.81)$$

where T^* is the kinetic energy expressed as a function of the generalized coordinates and their derivatives (the expression to be introduced into Lagrange equations in their original form), while T is expressed as a function of the generalized coordinates and the velocities in the body fixed frame. It is possible to show that

$$\frac{\partial T^*}{\partial q_k} = \frac{\partial T}{\partial q_k} + \mathbf{w}^T \mathbf{A}^T \frac{\partial \mathbf{A}}{\partial q_k} \left\{ \frac{\partial T}{\partial w} \right\} . \qquad (25.82)$$

Note that product

$$\mathbf{w}^T \mathbf{A}^T \frac{\partial \mathbf{A}}{\partial q_k}$$

is a row matrix of order n (3 in the present case) that, multiplied by the column $\left\{ \frac{\partial T}{\partial w} \right\}$, yields the required number.

To use a more synthetic notation, those row matrices can be superimposed, yielding a square matrix

$$\left[\mathbf{w}^T \mathbf{A}^T \frac{\partial \mathbf{A}}{\partial q_k} \right] .$$

and thus

$$\left\{ \frac{\partial T^*}{\partial q_k} \right\} = \left\{ \frac{\partial T}{\partial q_k} \right\} + \left[\mathbf{w}^T \mathbf{A}^T \frac{\partial \mathbf{A}}{\partial q_k} \right] \left\{ \frac{\partial T}{\partial w} \right\} . \qquad (25.83)$$

The equation of motion is thus

$$\mathbf{A} \frac{\partial}{\partial t} \left(\left\{ \frac{\partial T}{\partial w} \right\} \right) + \dot{\mathbf{A}} \left\{ \frac{\partial T}{\partial w} \right\} - \left\{ \frac{\partial T}{\partial q_k} \right\} - \left[\mathbf{w}^T \mathbf{A}^T \frac{\partial \mathbf{A}}{\partial q_k} \right] \left\{ \frac{\partial T}{\partial w} \right\} = \left\{ \begin{array}{c} F_X \\ F_Y \\ M_z \end{array} \right\} . \qquad (25.84)$$

By premultiplying all terms of matrix $\mathbf{A}^T = \mathbf{A}^{-1}$, it follows that

$$\frac{\partial}{\partial t} \left(\left\{ \frac{\partial T}{\partial w} \right\} \right) + \mathbf{A}^T \left(\dot{\mathbf{A}} - \left[\mathbf{w}^T \mathbf{A}^T \frac{\partial \mathbf{A}}{\partial q_k} \right] \right) \left\{ \frac{\partial T}{\partial w} \right\} + \qquad (25.85)$$

$$- \mathbf{A}^T \left\{ \frac{\partial T}{\partial q_k} \right\} = \mathbf{A}^T \left\{ \begin{array}{c} F_X \\ F_Y \\ M_z \end{array} \right\} .$$

---

[5]For details on this part of the analysis, see L. Meirovitch, *Methods of Analytical Dynamics*, Mc Graw-Hill, New York, 1970.

By performing the derivatives of the kinetic energy and all products, the equation becomes

$$
\left\{ \begin{array}{c} m\dot{u} \\ m\dot{v} \\ J_z r \end{array} \right\} + \left[\begin{array}{ccc} 0 & -r & 0 \\ r & 0 & 0 \\ 0 & 0 & 0 \end{array} \right] \left\{ \begin{array}{c} mu \\ mv \\ J_z\dot{\psi} \end{array} \right\} = \left\{ \begin{array}{c} F_x \\ F_y \\ M_z \end{array} \right\} , \tag{25.86}
$$

where forces F_x and F_y refer to the body fixed frame. The final expression of the equations of motion is then

$$
\left\{ \begin{array}{l} m(\dot{u} - rv) = F_x \\ m(\dot{v} + ru) = F_y \\ J_z \dot{r} = M_z , \end{array} \right. \tag{25.87}
$$

which obviously coincides with that obtained previously.

Velocities u and v are not derivatives of true coordinates, but nevertheless they can be used to write the equations of motion. They are actually derivatives of pseudo-coordinates, and the procedure here followed can also be used in cases where the kinematic equation (25.72) is more complicated, and where, in particular, the equation contains a matrix $\mathbf{A}^T$ that does not satisfy the relationship

$$
\mathbf{A}^T = \mathbf{A}^{-1} .
$$

Equations (25.87) are nonlinear in the velocities u, v and r but, since the sideslip angle β is small and its trigonometric functions can be linearized, the linearization of the equations is possible. The components of velocity V can be written as

$$
\left\{ \begin{array}{l} u = V \cos(\beta) \approx V \\ v = V \sin(\beta) \approx V\beta . \end{array} \right. \tag{25.88}
$$

Product $\dot{\psi}v$ can be considered the product of two small quantities and it is thus of the same order as the first term ignored in the series for the cosine. It is therefore cancelled.

The speed V can be considered a known function of time, which amounts to studying the motion with a given law $V(t)$ (in many cases at constant speed) and assuming as an unknown the driving or the braking force needed to follow such a law. The unknown for the degree of freedom related to translation along the x-axis in this case is the force F_{x_d} exerted by the driving wheels. When braking, force F_{x_d} is the total braking force exerted by all wheels.

Equation (25.87) reduces to the linear form in F_x, v and r:

$$
\left\{ \begin{array}{l} m\dot{V} = F_x \\ m\,(\dot{v} + rV) = F_y \\ J_z \dot{r} = M_z . \end{array} \right. \tag{25.89}
$$

If the interaction between longitudinal and transversal forces due to the tires is neglected or accounted for in an approximate way, the first equation of motion, which has already been studied in the section dealing with the longitudinal performance of the vehicle, uncouples from the other two.

This amounts to saying that the lateral behavior is uncoupled from the longitudinal behavior and can be studied using just two variables, either velocities v and r:

$$\begin{cases} m\left(\dot{v}+rV\right)=F_y \\ J_z\dot{r}=M_z\ , \end{cases} \tag{25.90}$$

or β and r if the equations are written in the equivalent form

$$\begin{cases} mV\left(\dot{\beta}+r\right)+m\beta\dot{V}=F_y \\ J_z\dot{r}=M_z\ . \end{cases} \tag{25.91}$$

25.5.2 Sideslip angles of the wheels

The sideslip angles of the wheels may be expressed easily in terms of the generalized velocities. With reference to Fig. 25.16, the velocity of the centre P_i of the contact area of the ith wheel, located in a point whose coordinates are x_i and y_i in the reference frame of the vehicle, is

$$\vec{V}_{P_i}=\vec{V}_G+\dot{\psi}\Lambda\overline{(P_i-G)}=\left\{\begin{array}{c} u-\dot{\psi}y_i \\ v+\dot{\psi}x_i \end{array}\right\}\ . \tag{25.92}$$

Angle β_i between the direction of the velocity of point P_i and x-axis is

$$\beta_i=\arctan\left(\frac{v_i}{u_i}\right)=\arctan\left(\frac{v+\dot{\psi}x_i}{u-\dot{\psi}y_i}\right)\ . \tag{25.93}$$

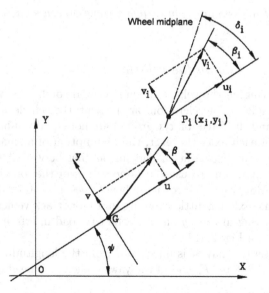

FIGURE 25.16. Position and velocity of the centre P_i of the contact area of the i-th wheel.

If the ith wheel has a steering angle δ_i, its sideslip angle is

$$\alpha_i = \beta_i - \delta_i = \arctan\left(\frac{v + \dot{\psi}x_i}{u - \dot{\psi}y_i}\right) - \delta_i \ . \tag{25.94}$$

Equation (25.94) can be easily linearized. By noting that $y_i\dot{\psi}$ is far smaller than the speed V, it follows that

$$\alpha_i = \beta_i - \delta_i \approx \frac{v + rx_i}{V} - \delta_i = \beta + \frac{x_i}{V}r - \delta_i \ . \tag{25.95}$$

Coordinate y_i of the centre of the contact area of the wheel does not appear in the expression for the sideslip angle α_i. If the differences between the steering angles δ_i of the wheels of the same axle are neglected, the values of their sideslip angles are then equal. This allows one to work in terms of axles instead of single wheels and to substitute a model of the type of Fig. 4.1b to that of Fig. 4.1a. This approach is very common and is often referred to as the monotrack vehicle or bicycle model.

The explicit expressions of the sideslip angles of the front and rear axles of a vehicle with two axles are then

$$\begin{cases} \alpha_1 = \beta + \dfrac{a}{V}r - \delta_1 \\[2mm] \alpha_2 = \beta - \dfrac{b}{V}r - \delta_2 \ . \end{cases} \tag{25.96}$$

In the majority of cases only the front axle can steer and $\delta_2 = 0$.

Remark 25.6 *The assumption of a rigid vehicle prevents one from considering roll steering.*

25.5.3 Forces acting on the vehicle

Normal forces acting on the vehicle in symmetrical conditions were obtained in Chapter 23. When lateral accelerations are present, the vehicle is not in symmetrical conditions and the forces on the ground are not equally subdivided between the two wheels of each axle. However, the assumption of a small sideslip angle β and the subsequent linearization and uncoupling between lateral and longitudinal behavior allow one to use the same values of the forces on the ground previously seen. Moreover, to investigate how forces are subdivided between the wheels of the same axle has little meaning in a monotrack vehicle.

The forces acting in the xy plane at the tire-road interface in a monotrack vehicle are shown in Fig. 25.12

Since the lateral behavior is uncoupled from the longitudinal one, only the resultants of the side force F_y and of the yaw moment M_z need to be computed:

$$F_y = \sum_{\forall i} F_{x_i}\sin(\delta_i) + \sum_{\forall i} F_{y_i}\cos(\delta_i) + \frac{1}{2}\rho V_r^2 SC_y + F_{y_e} \ , \tag{25.97}$$

where the external force F_{y_e} may be $mg\sin(\alpha_t)$ in the case of a road with transversal slope α_t, and

$$M_z = \sum_{\forall i} F_{x_i}\sin(\delta_i)x_i + \sum_{\forall i} F_{y_i}\cos(\delta_i)x_i + \sum_{\forall i} M_{z_i} + \frac{1}{2}\rho V_r^2 SC_{M_z} + M_{z_e} , \quad (25.98)$$

where x_i and y_i are the coordinates of the center of the contact zone, M_{z_i} represents the aligning moments of the wheels and M_{z_e} is a yawing moment applied to the vehicle. Subscript i indicates the axle, and thus if the vehicle has two axles $i = 1, 2$. If the rear axle does not steer, $\delta_2 = 0$.

Cornering forces

Owing to linearization, equation (25.97) reduces to

$$F_y = \sum_{\forall i} F_{x_i}\delta_i + \sum_{\forall i} F_{y_i} + \frac{1}{2}\rho V_r^2 SC_y + F_{y_e} , \quad (25.99)$$

where products $F_{x_{i_t}}\delta_i$ can usually be neglected, since they are far smaller than the other forces included in the equation.

Since the model has been linearized, cornering forces can be expressed as the product of the cornering stiffness by the sideslip angle

$$F_{y_i} = -C_i\alpha_i = -C_i\left(\beta + \frac{x_i}{V}r - \delta_i\right) . \quad (25.100)$$

Equation (25.100) is written in terms of axles. The cornering stiffness is then that of the axle and not of the single wheel. In this way no allowance is taken for the camber force as, owing to the assumption of a rigid vehicle, no roll is considered and the wheels of a given axle have opposite camber. The camber forces then cancel each other.

Nor is allowance made for toe in and transversal load transfer. If the dependence of the cornering stiffness were linear with the load F_z, this would be correct since the increase of cornering stiffness of the more loaded wheel would exactly compensate for the decrease of the other wheel. As this is not exactly the case, the load transfer causes a decrease of the cornering stiffness of each axle, but this effect is usually considered negligible, at least for lateral accelerations lower than 0.5 g[6]. Toe in causes an increase of the cornering stiffness of the axle if it is positive, a decrease if it is negative.

By linearizing also the value of the aerodynamic coefficient C_y

$$C_y = (C_y)_{,\beta}\beta$$

and assuming that the steering angles of the various axles can be expressed as

$$\delta_i = K_i'\delta, \quad (25.101)$$

---

[6]L. Segel, *Theoretical Prediction and Experimental Substantiation of the Response of the Automobile to Steering Control*, Cornell Aer. Lab., Buffalo, N.Y.

the expression of the total lateral force (25.99) can be reduced to the linear equation

$$F_y = Y_\beta \beta + Y_r r + Y_\delta \delta + F_{y_e},$$ (25.102)

where

$$
\begin{cases}
Y_\beta = -\sum_{\forall i} C_i + \dfrac{1}{2}\rho V_r^2 S(C_y)_{,\beta} \\[2mm]
Y_r = -\dfrac{1}{V}\sum_{\forall i} x_i C_i \\[2mm]
Y_\delta = \sum_{\forall i} K_i' (C_i + F_{x_i}) \ .
\end{cases}
$$ (25.103)

Equation (25.102) can be considered as a Taylor series for the force $F_y\,(\beta,\,r,\,\delta)$ about the condition $\beta = r = \delta = 0$, truncated after the linear terms. Coefficients Y_β, Y_r and Y_δ are the derivatives of the force with respect to the three variables β, r and δ and may be obtained in any way, even experimentally, if possible.

In the case of vehicles with only one steering axle, all K_i' vanish except $K_1' = 1$, while in other cases they can be functions of many parameters. If the variables of motion β or r enter such equations the model is no longer linear.

The first Eq. (25.98) has been obtained conflating the sideslip angle of the vehicle β with the aerodynamic sideslip angle β_a, as occurs when no side wind is present, and in the third equation the terms in $F_{x_{i_t}}$ are usually neglected.

Yawing moments

Equation (25.98) can be linearized yielding

$$M_z = \sum_{\forall i} F_{x_i}\delta_i x_i + \sum_{\forall i} F_{y_i} x_i + \sum_{\forall i} M_{z_i} + \frac{1}{2}\rho V_r^2 SC_{M_z} + M_{z_e} \ .$$ (25.104)

The aligning torque can be expressed as a linear function of the sideslip angle,

$$M_z = (M_z)_{,\alpha}\,\alpha \ ,$$ (25.105)

holding only in a range of α smaller than that for which the side force can be linearized.

The same considerations seen for the cornering force hold here; moreover, the aligning torque is far less important and the errors in its evaluation affect the global behavior of the vehicle far less than errors in the cornering force. In the following equations the values of $(M_z)_{,\alpha}$ are referred to the whole axle.

Acting similarly to what seen for the cornering forces, the linearized expression for the yawing moments is

$$M_z = N_\beta \beta + N_r r + N_\delta \delta + M_{z_e} \ ,$$ (25.106)

where

$$\begin{cases} N_\beta = \sum_{\forall i} [-x_i C_i + (M_{z_i})_{,\alpha}] + \frac{1}{2}\rho V_r^2 S(C_{M_z})_{,\beta} \\ N_r = \frac{1}{V} \sum_{\forall i} [-x_i^2 C_i + (M_{z_i})_{,\alpha} x_i] \\ N_\delta = \sum_{\forall i} K_i' [C_i x_i - (M_{z_i})_{,\alpha} + F_{x_i} x_i] \ . \end{cases} \qquad (25.107)$$

In this case the terms in $F_{x_{i_t}}$ are usually neglected.

25.5.4 Derivatives of stability

As already stated, the terms Y_β, Y_r, Y_δ, N_β, N_r and N_δ are nothing but the derivatives $\partial F_y/\partial\beta$, $\partial F_y/\partial r$, etc. They are usually referred to as derivatives of stability. N_r is sometimes referred to as yaw damping, as it is a factor that, multiplied by an angular velocity, yields a moment, like a damping coefficient.

In a simplified study of the handling of road vehicles, aerodynamic forces are usually neglected, as is the interaction between the longitudinal and transversal forces of the tires. In these conditions, Y_β, Y_δ, N_β and N_δ are constant while Y_r and N_r are proportional to $1/V$. Note that they are strongly influenced by the load and road conditions through the cornering stiffness of the tires.

If aerodynamic forces are considered, the airspeed V_r is often substituted by the groundspeed V. These forces introduce a strong dependence with V^2 in Y_β and N_β and with V in N_r.

Example 25.3 *Compute the derivatives of stability at 100 km/h of the vehicle of Appendix E.2, using the simplified and the complete formulations. Plot the derivatives of stability as functions of the speed for the same vehicle. In the whole computation neglect the longitudinal forces on the tires.*

The normal forces on the ground are first computed. At 100 km/h, at constant velocity on level road, they are 4.804 and 3.536 kN for the front and rear axles respectively.

From these values the cornering and aligning stiffness can be computed as $C_1 = 67,369$ N/rad, $C_2 = 63,411$ N/rad, $(M_{z_1})_{,\alpha} = 2,010$ Nm/rad and $(M_{z_2})_{,\alpha} = 1,366$ Nm/rad.

These values refer to the axles; the normal load on each wheel must be first computed and introduced into the "magic formula"; the results are then multiplied by the number of wheels on the axles.

By taking into account only the cornering forces of the tires, the following values of the derivatives of stability at 100 km/h are obtained:

Y_β N/rad	Y_r Ns/rad	Y_δ N/rad	N_β Nm/rad	Y_r Nms/rad	Y_δ Nm/rad
$-130,570$	824.62	67,374	22,906	$-5,622$	58,615

If the complete expressions, including aligning torques, aerodynamic forces and load shift between the wheels of the same axle are used, the values of the derivatives of stability at 100 km/h are:

Y_β N/rad	Y_r Ns/rad	Y_δ N/rad	N_β Nm/rad	Y_r Nms/rad	Y_δ Nm/rad
$-132,340$	824.62	67,374	26,488	$-5,630$	55,962

The derivatives of stability are plotted as functions of the speed in Fig. 25.17. The values obtained from the complete expressions are reported as full lines while the dashed lines are the constant values (proportional to $1/V$ for Y_r and N_r) obtained when considering the cornering forces only, computed at 100 km/h.

Note that N_β is the only derivative of stability strongly affected by load shift, aligning torques and the other effects. Here an apparently strange result is obtained: From the formula a decrease in N_β seems to occur with increasing speed, as the aerodynamic term is negative, while the plot shows an increase.

The latter is due to the longitudinal load shift which, while causing an increase of the load on the rear axle, produces an increase of N_β that is larger than the decrease due to the aerodynamic moment M_z.

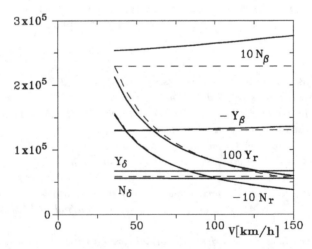

FIGURE 25.17. Derivatives of stability as functions of the speed. Full lines: Values obtained from the complete expressions; dashed lines: Constant values (proportional to $1/V$ for Y_r and N_r) obtained considering the cornering forces only, computed at 100 km/h.

25.5.5   Final expression of the equations of motion

The final expression of the linearized equations of motion for the handling model is thus

$$\begin{cases} mV\left(\dot{\beta}+r\right)+m\dot{V}\beta = Y_\beta\beta + Y_r r + Y_\delta\delta + F_{y_e} \\ J_z\dot{r} = N_\beta\beta + N_r r + N_\delta\delta + M_{z_e} \; . \end{cases} \tag{25.108}$$

These are two first order differential equations for the two unknown β and r.

These equations are apparently first order equations: the variables β and r are actually an angular velocity (r) or a quantity linked with a velocity (β was introduced instead of velocity v); their derivatives are thus accelerations. The missing term is therefore not the second derivative (acceleration), but the displacement.

Alternatively, a set of two first order differential equations in v and r could be written.

The steering angle δ can be considered as an input to the system, together with the external force and moment F_{y_e} and M_{z_e}. This approach is usually referred to as the "locked controls" behavior.

Alternatively, it is possible to study the "free controls" behavior, in which the steering angle δ is one of the variables of the motion and a further equation expressing the dynamics of the steering system is added.

In the first case, β and r can be considered as state variables and Eq. (25.108) can be written directly as a state equation

$$\dot{\mathbf{z}} = \mathbf{A}\mathbf{z} + \mathbf{B}_c\mathbf{u}_c + \mathbf{B}_e\mathbf{u}_e \; , \tag{25.109}$$

where the state and input vectors $\mathbf{z}$, $\mathbf{u}_c$ and $\mathbf{u}_e$ are

$$\mathbf{z} = \left\{ \begin{array}{c} \beta \\ r \end{array} \right\} , \qquad \mathbf{u}_c = \delta \; , \qquad \mathbf{u}_e = \left\{ \begin{array}{c} F_{y_e} \\ M_{z_e} \end{array} \right\} ,$$

the dynamic matrix is

$$\mathbf{A} = \begin{bmatrix} \dfrac{Y_\beta}{mV} - \dfrac{\dot{V}}{V} & \dfrac{Y_r}{mV} - 1 \\[3mm] \dfrac{N_\beta}{J_z} & \dfrac{N_r}{J_z} \end{bmatrix}$$

and the input gain matrices are

$$\mathbf{B}_c = \begin{bmatrix} \dfrac{Y_\delta}{mV} \\[3mm] \dfrac{N_\delta}{J_z} \end{bmatrix} , \quad \mathbf{B}_e = \begin{bmatrix} \dfrac{1}{mV} & 0 \\[3mm] 0 & \dfrac{1}{J_z} \end{bmatrix} .$$

The block diagram corresponding to the state equation is shown in Fig. 25.18.

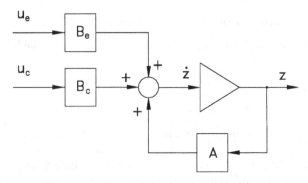

FIGURE 25.18. Block diagram for the rigid vehicle handling model.

The study of the system is straightforward: The eigenvalues of the dynamic matrix allow one to see immediately whether the behavior is stable or not, and the study of the solution to given constant inputs yields the steady state response to a steering input or to external forces and moments.

There is, however, an interesting analogy. If the speed is kept constant in such a way that the derivatives of stability are constant in time, there is no difficulty in obtaining r from the first Eq. (25.108) and substituting it into the second, which becomes a second order differential equation in β. Similarly, solving the second in β and substituting it in the first one, an equation in r is obtained. The result is

$$P\ddot{\beta} + Q\dot{\beta} + U\beta = S'\delta + T'\dot{\delta} - N_r F_{y_e} + J_z \dot{F}_{y_e} - (mV - Y_r)M_{z_e} \qquad (25.110)$$

or

$$P\ddot{r} + Q\dot{r} + Ur = S''\delta + T''\dot{\delta} + N_\beta F_{y_e} - Y_\beta M_{z_e} + mV\dot{M}_{z_e} , \qquad (25.111)$$

where

$$\begin{cases} P = J_z mV \\ Q = -J_z Y_b - mV N_r \\ U = N_\beta (mV - Y_r) + N_r Y_\beta \end{cases} \qquad \begin{cases} S' = -N_\delta (mV - Y_r) - N_r Y_\delta \\ S'' = Y_\delta N_\beta - N_\delta Y_\beta \\ T' = J_z Y_\delta \\ T'' = mV N_\delta . \end{cases}$$

If the simplified expressions of the derivatives of stability are used, the expressions for P, Q, etc., for a vehicle with two axles reduce to

$$\begin{cases} P = J_z mV \\ Q = J_z(C_1 + C_2) + m(a^2 C_1 + b^2 C_2) \\ U = mV(-aC_1 + bC_2) + C_1 C_2 \dfrac{l^2}{V} \end{cases} \qquad \begin{cases} S' = C_1\left(-amV + C_2\dfrac{bl}{V}\right) \\ S'' = lC_1 C_2 \\ T' = J_z C_1 \\ T'' = mVaC_1 . \end{cases}$$

Each of equations (25.110) and (25.111) is sufficient for the study of the dynamic behavior of the vehicle.

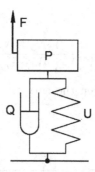

FIGURE 25.19. Formal analogy of the motor vehicle with a mass-spring-damper system (mass P, stiffness U, damper Q). Force F includes the different forcing functions.

The equations are formally identical to the equation of motion of a spring-mass-damper system (Fig. 25.19).

The linearized behavior of a rigid motor vehicle at constant speed is thus identical to that of a mass P suspended from a spring with stiffness U and a damper with damping coefficient Q, excited by the different forcing functions stated above (the command δ and the external disturbances).

Remark 25.7 *The analogy here suggested is only a formal one: as already stated, the state variables β and r are dimensionally an angular velocity (r) or are related to velocities (β has been introduced to express the lateral velocity v) and not displacements, and thus P, Q and U are dimensionally far from being a mass, a damping coefficient and a stiffness.*

25.6   STEADY-STATE LATERAL BEHAVIOR

In steady state driving the radius of the path is constant, i.e. the path is circular. The relationship linking the angular velocity r to the radius R of the path is thus

$$r = \frac{V}{R} . \tag{25.112}$$

Computing the steady state response to a steering angle δ is the same as computing the equilibrium position of the equivalent mass-spring-damper system under the effect of a constant force $S'\delta$ or $S''\delta$ since in steady state motion $\dot{\delta} = 0$

$$\begin{cases} \beta = \dfrac{S'}{U}\delta = \dfrac{-N_\delta\left(mV - Y_r\right) - N_r Y_\delta}{N_\beta\left(mV - Y_r\right) + N_r Y_\beta}\delta \\[3mm] r = \dfrac{S''}{U}\delta = \dfrac{Y_\delta N_\beta - N_\delta Y_\beta}{N_\beta\left(mV - Y_r\right) + N_r Y_\beta}\delta \, . \end{cases} \tag{25.113}$$

The transfer functions of the vehicle are thus the

- *path curvature gain*

$$\frac{1}{R\delta} = \frac{Y_\delta N_\beta - N_\delta Y_\beta}{V\left[N_\beta\left(mV - Y_r\right) + N_r Y_\beta\right]} , \qquad (25.114)$$

expressing the ratio between the curvature of the path and the steering input; the

- *lateral acceleration gain*

$$\frac{V^2}{R\delta} = \frac{V\left[Y_\delta N_\beta - N_\delta Y_\beta\right]}{N_\beta\left(mV - Y_r\right) + N_r Y_\beta} , \qquad (25.115)$$

expressing the ratio between the lateral acceleration and the steering input: the

- *sideslip angle gain*

$$\frac{\beta}{\delta} = \frac{-N_\delta\left(mV - Y_r\right) - N_r Y_\delta}{N_\beta\left(mV - Y_r\right) + N_r Y_\beta} , \qquad (25.116)$$

expressing the ratio between the sideslip angle and the steering angle; and the

- *yaw velocity gain*

$$\frac{r}{\delta} = \frac{Y_\delta N_\beta - N_\delta Y_\beta}{N_\beta\left(mV - Y_r\right) + N_r Y_\beta} , \qquad (25.117)$$

expressing the ratio between the yaw velocity and the steering angle.

If a simplified expression of the derivatives of stability, including only the cornering forces of the tires, is introduced in the above expressions, the same values of the gains reported in equations from (25.53) to (25.61) are obtained.

When the dependence of the derivatives of stability on the speed is accounted for, the law $1/R\delta$ as a function of V is no more monotonic as those shown in Fig. 25.13a and the behavior may change from understeer to oversteer (or viceversa)

The aerodynamic yawing moment produces a strong effect. If $\partial C_{M_z}/\partial\beta$ is negative (the side force F_y acts forward of the centre of mass), the effect is increasing oversteer or decreasing understeer, at increasing speed. If a critical speed exists, such an aerodynamic effect lowers it and has an overall destabilizing effect, increasing with the absolute value of $(C_{M_z})_{,\beta}$. The opposite occurs if $(C_{M_z})_{,\beta}$ is positive.

The longitudinal load shift produces another important effect. If the load on the rear axle increases more, or decreases less, than that on the front axle, the understeer increases with increasing speed.

The case of a vehicle that is oversteer at low speed and understeer at high speed, as it can be caused by a positive value of $(C_{M_z})_{,\beta}$, is shown in Fig. 25.20.

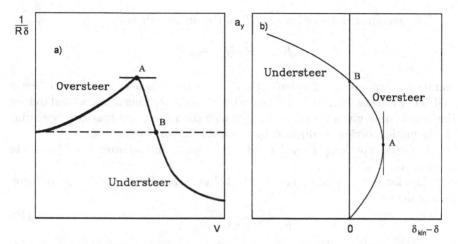

FIGURE 25.20. Steady state response to a steering input. Plot of the path curvature gain as a function of speed (a) and handling diagram (b) for a vehicle that at low speed is oversteer and then becomes understeer at high speed.

Following the definition seen above, the speed at which neutral-steer is obtained is identified by point B.

If the simplified expressions for the derivatives of stability are not used, a new definition of a neutral-steer, and hence under- and oversteer, vehicle may be introduced. Instead of referring to the condition

$$\frac{1}{R\delta} = \frac{1}{l} \, ,$$

neutral-steering can be defined by the relationship

$$\frac{d}{dV}\left(\frac{1}{R\delta}\right) = 0 \, . \tag{25.118}$$

On the plot of Fig. 25.20 the speed at which neutral-steering is obtained is point A, where the curve reaches its maximum.

Remark 25.8 *In case the derivatives of stability are constant (Y_r and N_r are proportional to $1/V$) the first definition, which can be said to be absolute and the second, which can be said to be incremental, coincide.*

Remark 25.9 *The incremental definition corresponds to the sensations of the driver, who perceives the vehicle as oversteering if an increase of speed is accompanied by a decrease in radius of the path and vice versa. The driver clearly has no way of sensing the kinematic value of the radius of the path and hence the absolute definition has little meaning for him. From the viewpoint of the equations of motion, on the other hand, the absolute definition is more significant.*

The generalized definition (25.57) of the understeer fator

$$K_{us} = \frac{g}{a_y} \left(\delta - \delta_{\text{kin}} \right) ,$$

and the corresponding definition of the stability factor holds in the present case as well. They are essentially the difference between the steering angle needed to keep the vehicle on a given trajectory in dynamic conditions and that corresponding to kinematic steering, multiplied by a suitable factor proportional to $1/V^2$.

Generally speaking, they depend on the speed and on other conditions, like acceleration.

Also for the understeer factor it is, however, possible to introduce an incremental definition

$$\frac{1}{K_{us}} = \frac{1}{g} \frac{da_y}{d\left(\delta - \delta_{\text{kin}} \right)} . \tag{25.119}$$

In this case the point in which the understeer factor vanishes and the vehicle is neutral steer is point A in Fig. 25.20b instead of being point B

25.7   NEUTRAL POINT AND STATIC MARGIN

The neutral-steer point of the vehicle is usually defined as the point on the plane of symmetry on which is applied the resultant of the cornering forces due to the tires as a consequence of a sideslip angle β, obviously with $\delta = 0$ and $r = 0$. The cornering forces under these conditions, computed through the linearized model, are simply $-C_1\beta$ and $-C_2\beta$ and the x coordinate of the neutral point is

$$x_N = \frac{aC_1 - bC_2}{C_1 + C_2} . \tag{25.120}$$

A better definition of neutral-steer point may, however, be introduced. If all forces and moments due to a sideslip angle β, with $\delta = 0$ and $r = 0$ are considered, the resultant force and moment are simply $Y_\beta\beta$ and $N_\beta\beta$ respectively[7]. The x coordinate of the neutral-steer point, defined as the point of application of the resultant of all lateral forces is thus

$$x_N = \frac{N_\beta}{Y_\beta} . \tag{25.121}$$

The static margin $\mathcal{M}_s$ is the ratio between the x coordinate of the neutral point and the wheelbase

$$\mathcal{M}_s = \frac{x_N}{l} . \tag{25.122}$$

An external force applied to the neutral-steer point does not cause any steady-state yaw velocity, as will be seen when dealing with the response to

---

[7]Y_β may be considered as a sort of cornering stiffness of the vehicle.

TABLE 25.1. Directional behavior of the vehicle.

Behavior	K	K_{us}	$\mathcal{M}_s$	x_N	$|\alpha_1| - |\alpha_2|$	N_β
Understeer	> 0	> 0	< 0	< 0	> 0	> 0
Neutral steer	0	0	0	0	0	0
Oversteer	< 0	< 0	> 0	> 0	< 0	< 0

external forces and moments. Owing to the mathematical model used in the present chapter, the height of the neutral-steer point cannot be defined.

Note that to obtain a neutral-steer response, the neutral-steer point must coincide with the centre of mass, i.e.

$$x_N = 0 \ , \ \mathcal{M}_s = 0 \ , \ N_\beta = 0 \ .$$

If they are positive the vehicle is oversteer[8] (centre of gravity behind the neutral point); the opposite applies to understeer vehicles.

The signs of parameters K, K_{us}, $\mathcal{M}_s$, x_N, $|\alpha_1| - |\alpha_2|$ and N_β corresponding to oversteer, understeer or neutral-steer behavior are reported in Table 25.1.

Since $N_\beta = 0$ in case of neutral-steer, the second equation of motion (25.108) uncouples from the first and simplifies as

$$J_z \dot{r} = N_r r + N_\delta \delta + M_{z_e} \ . \tag{25.123}$$

The behavior of a neutral-steer motor vehicle is thus that of a first order system rather than a second order system.

Example 25.4 *Study the directional behavior of the vehicle of Appendix E.2, using the simplified and the complete formulations.*

The value of N_β is positive and hence the vehicle is understeer. Using the values of the derivatives of stability computed from the cornering stiffness at 100 km/h, the values of the coordinate of the neutral-steer point and of the static margin are $x_N = -175$ mm, $\mathcal{M}_s = -0.081$, while the values obtained, always at 100 km/h, using a complete expression of the derivatives of stability are $x_N = -200$ mm, $\mathcal{M}_s = -0.093$.

The path curvature gain, the lateral acceleration gain, the sideslip angle gain and the yaw velocity gain are plotted as functions of the speed in Fig. 25.21. The values obtained from the complete expressions of the derivatives of stability are shown as full lines, while the dashed lines refer to the simplified expressions for the derivatives of stability (constant or proportional to $1/V$ for Y_r and N_r) obtained by considering only the cornering forces computed at 100 km/h. The dotted lines refer to a neutral-steer vehicle.

The vehicle has a strong understeer behavior, even more so if the complete expression of the derivatives of stability is considered. However, the simplified approach allows one to obtain a fair approximation of the directional behavior of the vehicle.

---

[8]Sometimes the position of the neutral-steer point and the static margin are defined with different sign conventions: Instead of referring to the position of the neutral- steer point with respect to the centre of mass, the position of the latter with respect to the former is given. In this case the signs of x_N and $\mathcal{M}_s$ are changed and an understeer vehicle has a positive static margin.

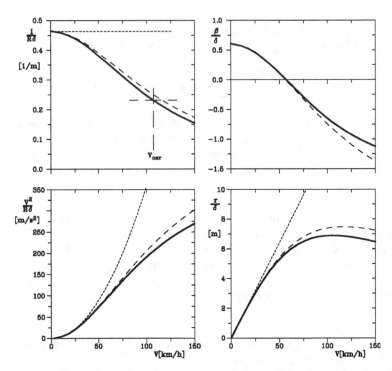

FIGURE 25.21. Example 25.4: path curvature gain, lateral acceleration gain, sideslip angle gain and yaw velocity gain as functions of the speed. Full lines: Values obtained from the complete expressions of the derivatives of stability; dashed lines: Simplified approach (constant derivatives of stability, Y_r and N_r proportional to $1/V$, obtained considering only the cornering forces computed at 100 km/h); dotted lines: Neutral-steer vehicle.

25.8 RESPONSE TO EXTERNAL FORCES AND MOMENTS

From the equivalent mass-spring-damper model the steady state response to an external force F_{y_e} or an external moment M_{z_e} is immediately obtained. The relevant gains are

$$
\begin{cases}
\dfrac{1}{RF_{y_e}} = \dfrac{N_\beta}{VU} \\[2mm]
\dfrac{V^2}{RF_{y_e}} = \dfrac{VN_\beta}{U} \\[2mm]
\dfrac{\beta}{F_{y_e}} = \dfrac{-N_r}{U}
\end{cases}
\qquad
\begin{cases}
\dfrac{1}{RM_{z_e}} = \dfrac{-Y_\beta}{VU} \\[2mm]
\dfrac{V^2}{RM_{z_e}} = \dfrac{-VY_\beta}{U} \\[2mm]
\dfrac{\beta}{M_{z_e}} = \dfrac{-mV + Y_r}{U}
\end{cases}
\qquad (25.124)
$$

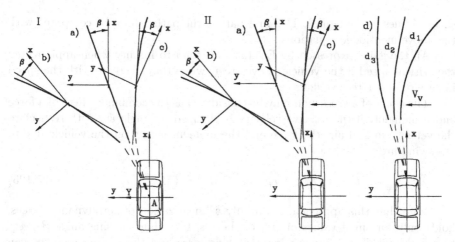

FIGURE 25.22. I. Response to a force F_{y_e} applied to the centre of mass; (a) neutral-steer, (b) understeer and (c) oversteer vehicle. II. Response to a lateral wind; point of application of the side force in the neutral-steer point (a), forward (b) and after the neutral-steer point (c) and (d).

If the vehicle is neutral-steer, $N_\beta = 0$ and consequently

$$\frac{1}{RF_{y_e}} = 0 \ .$$

In neutral steer vehicles, then, the path remains straight under the effect of an external force (Fig. 25.22Ia). This may be easily understood considering that the neutral-steer point lies in the centre of mass, i.e. in the point of application of the external force.

Actually, this condition can be used to define the neutral-steer point as the point in which the application of an external force does not cause a yaw rotation of the vehicle. If the presence of the suspension is accounted for, it is possible to define, instead of a neutral-steer point, a neutral-steer line as the locus of the points in the xz plane in which an external force applied in the y direction does not cause any yaw rotation.

The path is, however, changed from the one preceding the application of force F_{y_e}: The deviation is equal to angle β, i.e. to $-F_{y_e}/Y_\beta$. The lateral velocity of the vehicle is simply

$$v = V\beta = -V\frac{F_{y_e}}{Y_\beta} \ .$$

Remark 25.10 *It is very important that Y_β be as large as possible in order to avoid large lateral velocities, particularly in the case of fast vehicles.*

If the vehicle is understeer, the neutral-steer point is behind the centre of mass and the path bends as in Fig. 25.22Ib. The opposite effect can be found in the case of oversteer vehicles. Note that the trajectories so computed are steady-state trajectories, and when the force is applied an unstationary motion occurs (dashed lines in the figure). This first part of the path cannot be computed with the above mentioned equations.

All the gains expressed by Eq. (25.124) tend to infinity when approaching the critical speed if the vehicle is oversteer, while they decrease with the speed in case of understeer vehicles.

The effect of a crosswind may be considered as the combined effect of a force and a moment. If the relative velocity is changed by angle ψ_w with respect to the velocity in still air, the force and the moment acting on the vehicle due to crosswind are

$$F_{y_w} = (F_{y_{aer}})_{,\beta}\psi_w , \qquad M_{z_w} = (M_{z_{aer}})_{,\beta}\psi_w . \tag{25.125}$$

Note that this approach, essentially a linearization of aerodynamic forces, holds only for small values of ψ_w, or, better, for values causing angle $\beta + \psi_w$ to remain within the range where the side force and the yawing moment can be linearized. This occurs either for feeble crosswinds or for head- or tailwinds. If the wind velocity is not small, the aerodynamic terms of the derivatives of stability must be computed using V_r instead of V.

The response in terms of curvature of the path, computed as the sum of the response to a force and to a moment, is

$$\frac{1}{R} = \frac{F_{y_w} N_\beta - M_{z_w} Y_\beta}{V \left[N_\beta \left(mV - Y_r \right) N_r Y_\beta \right]} = \frac{F_{y_w} Y_\beta}{VU} \left(\frac{N_\beta}{Y_\beta} - \frac{M_{z_w}}{F_{y_w}} \right) . \tag{25.126}$$

Ratio M_{z_w}/F_{y_w} is nothing but the distance of the point of application of the aerodynamic side force from the centre of mass. If it is equal to N_β/Y_β, the aerodynamic force is applied to the neutral steer point and a straight path occurs. The deviation angle is

$$\beta = \frac{M_{z_w}}{Y_\beta} = -\frac{F_{y_w} x_N}{N_\beta} . \tag{25.127}$$

In general, the value of β is

$$\beta = \frac{F_{y_w}}{U} \left[-\frac{M_{z_w}}{F_{y_w}} (mV - Y_r) - N_r \right] . \tag{25.128}$$

The trajectories are shown in Fig. 25.22II.

Usually the point of application of the aerodynamic force is in front of both the centre of mass and the neutral-steer point. In this case the path bends downwind (curve b).

The path bends upwind (curves c and d) on the other hand, if the point of application of aerodynamic forces is behind the neutral-steer point . If this effect is not too strong (curve d_3), it is beneficial since very little correction is needed, but if the result resembles that of curve d_1 a large correction may be required in a direction opposite to the instinctive reaction of the driver.

Remark 25.11 *It must be noted again that the present steady-state model has limited application in the case of wind gusts, which involve primarily unsteady phenomena.*

The application of a side force to the centre of mass is easy: It is sufficient to use a road with a transversal slope fashioned in a proper way. Wind gusts may be simulated using jet engines and suitable ducts to distribute the gust with the required profile.

25.9 SLIP STEERING

As stated in Chapter 4, the trajectory of a vehicle on pneumatic tires may be controlled by applying differential longitudinal forces to the tires on the right and left side instead of steering some of the wheels. This method of driving a vehicle is usually referred to as slip steering: While it is the usual strategy for controlling tracked vehicles, it is used as a primary strategy for wheeled vehicles only on some light construction machines. In the automotive field, however, it is increasingly used as an additional control in connection with VDC (Vehicle Dynamics Control) systems (see Chapter 27).

Consider the mathematical model of the vehicle expressed by equations (25.108), and add a control yawing torque M_{z_c} to the second equation

$$
\begin{cases}
mV\left(\dot{\beta}+r\right) + m\dot{V}\beta = Y_\beta\beta + Y_r r + Y_\delta\delta + F_{y_e} \\
J_z\dot{r} = N_\beta\beta + N_r r + N_\delta\delta + M_{z_e} + M_{z_c} .
\end{cases}
\tag{25.129}
$$

If the two wheels of the ith axle, whose track is t_i, produce a longitudinal force

$$
F_{x_{iL,R,}} = \frac{F_{x_i}}{2} \pm \Delta F_{x_i} ,
\tag{25.130}
$$

where subscripts L and R designate the left and right wheel, the control torque is

$$
M_{z_c} = \sum_{\forall i} \Delta F_{x_i} t_i .
\tag{25.131}
$$

If the longitudinal slip σ of the tires is small enough, the longitudinal force is proportional to the slip through the slip stiffness C_σ (see Section 2.6). Assuming that the differential longitudinal slip $\Delta\sigma$ is the same on all axles, the yawing moment can thus be expressed as

$$
M_{z_c} = N_\sigma\Delta\sigma ,
\tag{25.132}
$$

where

$$
N_\sigma = \sum_{\forall i} C_{\sigma_i} t_i .
\tag{25.133}
$$

The equation of motion is still Eq. (25.109)

$$\dot{\mathbf{z}} = \mathbf{A}\mathbf{z} + \mathbf{B}_c\mathbf{u}_c + \mathbf{B}_e\mathbf{u}_e ,$$

but now

$$\mathbf{u}_c = \left\{ \begin{array}{c} \delta \\ \Delta\sigma \end{array} \right\} , \quad \mathbf{B}_c = \left[\begin{array}{cc} \dfrac{Y_\delta}{mV} & 0 \\[2ex] \dfrac{N_\delta}{J_z} & \dfrac{N_\sigma}{J_z} \end{array} \right] .$$

In steady-state conditions, it is possible to define a path curvature gain for slip steering

$$\frac{1}{R\Delta\sigma} = \frac{-N_\sigma Y_\beta}{V\left[N_\beta\left(mV - Y_r\right) + N_r Y_\beta\right]} , \tag{25.134}$$

expressing the ratio between the curvature of the path and the differential longitudinal slip. If the simplified expressions for the derivatives of stability are accepted, it follows that

$$\frac{1}{R\Delta\sigma} = \frac{C_1 + C_2}{C_1 C_2 l^2} \frac{\displaystyle\sum_{\forall i} C_{\sigma_i} t_i}{1 + K_{us}\frac{V^2}{gl}} , \tag{25.135}$$

Remark 25.12 *This approach to slip steering assumes that the differential longitudinal slip is imposed. Different equations would be obtained for cases in which the differential velocity of the wheels is imposed.*

Remark 25.13 *The formulae above are based on the assumption that the radius of the trajectory is much larger than the wheelbase: They do not hold when slip steering is used for very sharp turns, or even for turning on the spot.*

Remark 25.14 *Even when the speed tends to zero no kinematic conditions exist: By definition slip steering implies that the wheels operate with both longitudinal and side slip.*

25.10   INFLUENCE OF LONGITUDINAL FORCES ON HANDLING

A vehicle's directional behavior is strongly influenced by the presence of longitudinal forces between tires and road. Any longitudinal force causes a reduction of cornering stiffness: If applied to the front axle, it reduces the value of C_1 and consequently makes the vehicle more understeer or less oversteer. The opposite effect is caused by a longitudinal force applied to the rear axle.

In the linearized model this can be easily accounted for by using the elliptical approximation which, if a complete linearization of the behavior of the tires is assumed, can be applied directly to each axle

$$C_i = C_{0_i} \sqrt{1 - \left(\frac{F_{x_i}}{\mu_p F_{z_i}} \right)^2}. \tag{25.136}$$

Note that the forces and the cornering stiffness refer to the whole axle.

The driving force needed to maintain a constant speed increases with the latter and, as a consequence, the cornering stiffness of the tires of the driving axle decreases. The effect is felt particularly if road conditions are poor, since in Eq. (25.136) the ratio between the actual and the maximum value of the driving force is present.

The variation of static margin for a front-wheel and a rear-wheel drive saloon car with the speed due to the effect of the driving forces is shown in Fig. 25.23. It is clear that the effect is minor in the whole practical speed range of the car if the road conditions are good while, if μ_p is low, the change in handling of the car due to traction is quite strong.

In the case of rear-wheel drive vehicles driving forces increase oversteer or decrease understeer. The critical speed, if it exists, decreases or a critical speed may appear. In bad road conditions, a rear-wheel drive vehicle may have a very low critical speed and the driver may be required to limit the speed for stability reasons, to avoid spinout. Starting and accelerating the vehicle may be difficult and the driver has to exert a great care in operating the accelerator control; antispin or TCS devices are very useful in these conditions.

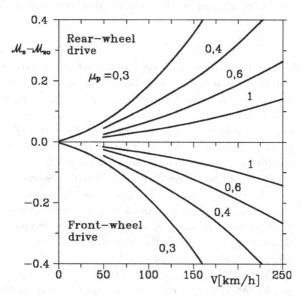

FIGURE 25.23. Variation of the stability margin due to the longitudinal forces on the tires in the cases of front- and rear-wheel drive saloon cars. Various values of μ_p; a completely linearized model has been used.

Front-wheel drive vehicles, on the other hand, have a tendency toward understeering and become more stable with increasing speed or decreasing μ_p and an increasingly large steering angle is needed to maintain the vehicle on a given path. The limit condition is that of an infinitely stable vehicle, i.e. a vehicle that can only move on a straight line.

In vehicles with more than one driving axle, and when braking, handling depends upon how the longitudinal forces are distributed between the axles. If the front axle is working with a larger longitudinal force coefficient μ_x than the rear axle, which does not necessarily imply that force F_x is larger but that the ratio F_x/F_z of the front wheels is larger than that of the rear wheels, the vehicle becomes more understeering and is, in a sense, more stable. When the limit conditions are reached and the front wheels slip (lock in braking or spin in traction) the vehicle cannot be steered and follows a straight path.

A larger ratio F_x/F_z at the rear wheels makes the vehicle more oversteer and readily introduces a critical speed. When reaching limit conditions a spinout occurs, unless the driver promptly reduces the longitudinal forces and countersteers, a manoeuvre that can be expected only from very proficient drivers. To avoid this situation the braking system must be such that the working point on the F_{x_1}, F_{x_2} plane is not found above the curve for ideal braking. Antispin and antilock devices are very important from this viewpoint.

When all values of μ_x are equal, the behavior should theoretically not be affected by the longitudinal forces; however, when limit conditions occur, the vehicle can spin out or go straight depending on small changes in many parameters, such as the conditions of the individual wheels and brakes, the load transfer, etc.

Example 25.5 *Study the directional behavior of the vehicle of Appendix E.2, taking into account the reduction of the cornering stiffness of the driving wheels caused by the longitudinal forces needed to move at constant speed. Repeat the computation for two values of μ_p, namely 1 and 0.2.*

The study is performed by computing, at each speed, the values of the longitudinal and normal component of the tire forces, using the "magic formula" for the cornering stiffness and then reducing it through the elliptic expression (25.136).

The results, in terms of path curvature gain, lateral acceleration gain, sideslip angle gain and yaw velocity gain, are plotted as functions of the speed in Fig. 25.24 for both values of the maximum longitudinal force coefficient. The dashed lines refer to the simplified expressions for the derivatives of stability (constant or proportional to $1/V$ for Y_r and N_r) obtained considering only the cornering forces computed at 100 km/h; the dotted lines refer to a neutral-steer vehicle.

By comparing Fig. 25.24 with Fig. 25.21, it is clear that the effect of the driving force is almost negligible throughout the entire speed range if the road conditions are good ($\mu_p = 1$): The lines of the two figures are almost completely superimposed. However, if μ_p is lowered to 0.2, the understeer behavior becomes much more marked, particularly at high speed.

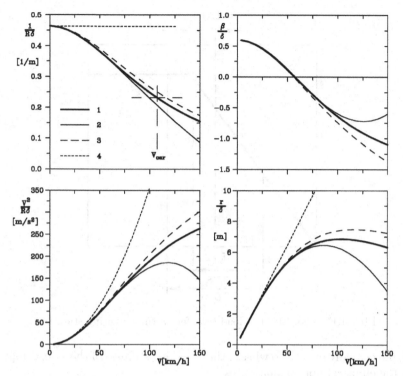

FIGURE 25.24. Example 25.5: path curvature gain, lateral acceleration gain, sideslip angle gain and yaw velocity gain as functions of the speed. Values obtained from the complete expressions of the derivatives of stability, with the effect of the driving forces accounted for; (1): $\mu_t = 1$; (2): $\mu_t = 0.2$; (3): Simplified approach (constant derivatives of stability, Y_r and N_r proportional to $1/V$, obtained considering only the cornering forces computed at 100 km/h, assuming no longitudinal force effects); (4): Neutral-steer vehicle.

25.11   TRANSVERSAL LOAD SHIFT

No allowance has yet been taken for the transversal load shift. If the dependence on the load of the cornering stiffness of a single wheel is of the type shown in Fig. 25.25, this does not introduce errors if the load transfer ΔF_z is small, lower than $(\Delta F_z)_{lim}$ in the figure (condition a).

But if the load shift is larger, as in the case of ΔF_{zb}, the increase in stiffness of the more loaded wheel cannot compensate for the decrease in the other wheel and the cornering stiffness of the axle is reduced. This effect introduces a nonlinearity in the behavior of the vehicle.

The simultaneous presence of longitudinal forces and load transfer makes things more complicated. Even if the cornering stiffness is still in the linear part of the plot of Fig. 25.25, i.e. the load transfer is smaller than $(\Delta F_z)_{lim}$, the combined effect yields a nonlinear behavior. Assuming that the longitudinal

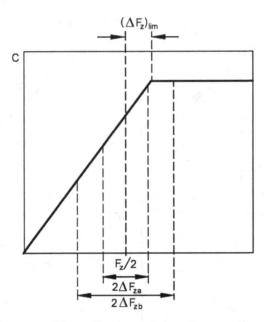

FIGURE 25.25. Effect of load transfer on the cornering stiffness.

force splits equally on its two wheels, the cornering stiffness of the axle, computed using the elliptical approximation, is

$$
C = \frac{1}{2}\left(C_0 + \Delta F_z \frac{\partial C}{\partial F_z}\right)\sqrt{1 - \left[\frac{F_x}{\mu_p(F_z + 2\Delta F_z)}\right]^2} +
$$

$$
+\frac{1}{2}\left(C_0 - \Delta F_z \frac{\partial C}{\partial F_z}\right)\sqrt{1 - \left[\frac{F_x}{\mu_p(F_z - 2\Delta F_z)}\right]^2} ,
$$

(25.137)

where forces F_x and F_z refer to the whole axle.

Owing to the presence of the square root, the decrease in cornering stiffness of the less loaded wheel is greater, particularly if μ_x is low, than the increase at the other wheel.

Load transfer on the driving axle thus increases the effect of longitudinal forces; this combined action can be reduced by introducing an anti-roll bar on the other axle. Operating in this way, the increased load transfer on the non-driving axle also reduces its cornering stiffness, reducing the overall effect of longitudinal forces on handling.

Anti-roll bars affect the distribution of transversal load shift between the axles, increasing the load shift on the relevant one while decreasing that on the other axles. They can be used to correct the behavior of the vehicle, particularly in conditions approaching the limit lateral acceleration, as their effect on the cornering stiffness increases when the latter increases.

Remark 25.15 *A large rear-wheel drive saloon car can benefit from the application of an anti-roll bar at the front axle to correct the strong oversteering tendency when the rear wheels approach their traction limit, while a small front wheel car can use an anti-roll bar at the rear axle to reduce its understeering behavior.*

It is impossible to state the effect of anti-roll bars on the gains defined in the previous sections since they introduce a strong nonlinearity into the mathematical model of the vehicle and the very definition of the gains is based on a complete linearization. It is only possible to study a number of specific cases where the lateral acceleration is defined, and to compute the response of the vehicle in such conditions.

25.12   TOE IN

Consider an axle (e.g., the front axle), in which the midplanes of the wheels are not exactly parallel and assume that the x-axes of the reference frames of the wheels converge in a point lying forward with respect to the axle [9].

Let α_c be the angle each wheel makes with the symmetry plane of the vehicle, positive when the toe-in is positive. With reference to Fig. 4.1, the steering angle of the wheel on the right side of the vehicle is increased by an angle equal to α_c, while the steering angle of the wheel on the left side is decreased by the same quantity.

If the usual linearization assumptions are accepted, the sideslip angles of the two wheels of the axle are then

$$\begin{cases} \alpha_{i_r} = \beta + \dfrac{x_i}{V}r - \delta_i - \alpha_c = \alpha_i - \alpha_c \\[3mm] \alpha_{i_l} = \beta + \dfrac{x_i}{V}r - \delta_i + \alpha_c = \alpha_i + \alpha_c\,, \end{cases} \tag{25.138}$$

where subscripts r and l refer to the right and left wheels respectively and i refers to the ith axle.

Consider a vehicle negotiating a bend to the left; the sideslip angle α_i is negative while the side force is positive. The transversal load shift causes an increase of the load on the wheels on the right, the sideslip angle α_i is negative and the side force is positive.

If C is the total stiffness of the axle, the cornering force the axle exerts is

$$F_y = -\frac{1}{2}\left[(\alpha_i - \alpha_c)\left(C + \Delta F_z \frac{\partial C}{\partial F_z}\right) + (\alpha_i + \alpha_c)\left(C - \Delta F_z \frac{\partial C}{\partial F_z}\right)\right], \tag{25.139}$$

---

[9]Toe in is usually defined as the difference between the distance of the front part and the rear part of the wheels of an axle, measured at the height of the hub, when the steering is in its central position. It is positive when the midplanes converge forward.

i.e.

$$F_y = C|\alpha_i| + \alpha_c \Delta F_z \frac{\partial C}{\partial F_z} \,. \qquad (25.140)$$

If transversal load shift is not taken into account, and the two wheels have the same cornering stiffness, toe in has no effect within the validity of the linearized model. The situation is different if load shift is included into the model: then toe in causes an increase of the cornering force due to the axle. This has the effect of increasing the cornering stiffness of the axle, depending on the load shift. Toe-in at the front wheels or toe-out of the rear ones thus has an oversteer effect.

The effect of toe in is complicated since α_c depends on the steering angle due to steering error, on suspension geometry and on the relative roll stiffness of the suspensions that affect the total shift of the various axles.

25.13   EFFECT OF THE ELASTO-KINEMATIC BEHAVIOR OF SUSPENSIONS AND OF THE COMPLIANCE OF THE CHASSIS

In the present chapter the vehicle is modelled as a rigid body moving on a plane. Suspensions, apart from causing inertial effects that cannot be studied using the present model, also change all working angles of the tires and thus affect the forces acting on the vehicle. The effects introduced by the elasto-kinematic characteristics of suspensions may be of two different types: some of these effects may be studied using linearized models, at least for small motion about a nominal configuration, while others must be studied by considering their nonlinear effects, even for small displacements

An example of the first type is roll steer. The characteristics $\delta(\phi)$ can be linearized, and a steering angle

$$(\delta)_{,\phi} \, \phi$$

can easily be added to the steering angle of the various wheels, or the various axles in monotrack models.

When, on the contrary, the compliance of the suspensions is accounted for, the characteristic angles of the wheels depend in a nonlinear way on the variables of motion and the resulting effects are nonlinear. No general results can thus be obtained and numerical simulation must be used.

Even if it is possible to remain within linearity limits, the mathematical models seen in this chapter are too simplified to depict how the elasto-kinematic characteristics of the suspensions affect the behavior of the vehicle. Some more complex models taking suspensions into account will be seen in Part V.

Similar considerations also hold for the compliance of the chassis or the body. In this case, the displacements due to compliance are usually considered small and the models describing their flexibility are linearized. However, although linear, these models are complex owing to the large number of deformation degrees of freedom involved, together with the rigid body degrees of freedom typical of the rigid-body models. Some models of this kind will be studied in Part V.

In general, we can say that the compliance of the chassis in its plane has little influence on the handling of the vehicle. On the other handy, its torsional deformations can strongly affect handling and lateral behavior.

25.14   STABILITY OF THE VEHICLE

It is customary to define a static and a dynamic stability. A system is statically stable in a given equilibrium condition if, when its state is perturbed, it tends to return to the previous situation. If the motion following this tendency towards the previous state of equilibrium succeeds, at least asymptotically, at restoring it, then the system is dynamically stable. This motion can tend to the equilibrium condition monotonically or through a damped oscillation. If, on the contrary, the equilibrium conditions are not reached, usually because a divergent oscillation takes place, the system is dynamically unstable. If an undamped oscillation occurs, as in the case of an undamped spring-mass system, the dynamic stability is neutral.

Remark 25.16 *If the system is linear, such definitions hold in the entire range in which the state variables are defined. If, on the contrary, the system is nonlinear, this definition holds "in the small", i.e. for small variations of the state variables about the values corresponding to an equilibrium point in the state space. The linearized model here studied is then a linearization suitable for the stability "in the small".*

The definition of stability above refers to the state of the system; in the case of the handling model with two degrees of freedom the state variables are β and r (or v and r). A motor vehicle is thus stable if, when in motion with given values β_0 and r_0 of β and r, after a small external perturbation, it follows that

$$\beta(t) \to \beta_0 , \qquad r(t) \to r_0 .$$

No reference is made to the path: After a perturbation the vehicle cannot return to the previous path, and a correction by the driver or by an automatic control system is required in order to maintain the vehicle on the road.

25.14.1   Locked controls

If the steering wheel is kept in a position that allows the vehicle to maintain the required path, the stability can be studied simply by using the homogeneous equation of motion

$$\dot{\mathbf{z}} = \mathbf{A}\mathbf{z} .$$

The eigenvalues of the dynamic matrix $\mathbf{A}$ are readily found and the stability is assessed from the sign of their real part, which must be negative. If the imaginary part is nonzero the behavior is oscillatory, which does not necessarily imply that the path is oscillatory but only that the time histories $\beta(t)$ and $r(t)$ are.

The analogy with the spring-mass-damper system allows a simpler approach to the study of the stability at constant speed.

Assuming a solution of the type

$$\beta(t) = \beta_0 e^{st} , \qquad r(t) = r_0 e^{st} ,$$

the characteristic equation yielding the poles of the system is

$$Ps^2 + Qs + U = 0 . \tag{25.141}$$

Since P, Q and U depend in general on the speed V, it is possible to compute the roots locus at various speed. By using the simplified expression for the derivatives of stability, the characteristic equation reduces to

$$J_z mV s^2 + \left[J_z(C_1 + C_2) + m(a^2 C_1 + b^2 C_2) \right] s+$$

$$+mV(-aC_1 + bC_2) + C_1 C_2 \frac{l^2}{V} = 0 . \tag{25.142}$$

At any rate, the analogy allows to state that

• to ensure static stability the stiffness U must be positive,

• to ensure dynamic stability the damping coefficient Q must be positive,

• if Q is lower than the critical damping $2\sqrt{PU}$ the system has an oscillatory behavior.

Using the simplified expression of the derivatives of stability, the following expression of the "stiffness" U can be readily obtained

$$U = \frac{C_1 C_2 l^2}{V}(1 + KV^2) , \tag{25.143}$$

where K is the stability factor defined by Eq. (25.58).

U is thus always positive for understeer and neutral-steer vehicles, and in the latter case it tends to zero when the speed tends to infinity. In the case of oversteer vehicles, it is positive up to the critical speed, where it vanishes to become negative at higher speed. The critical speed is thus the threshold of instability for oversteer vehicles. Similar results are obtained if the complete expressions for the derivatives of stability are used.

It is also easy to verify that Q is always positive: If the vehicle is statically stable it is also dynamically stable. If the simplified expression for the derivatives of stability is accepted, the value of Q is independent of the speed

$$Q = J_z(C_1 + C_2) + m(a^2 C_1 + b^2 C_2) . \tag{25.144}$$

The *critical damping* of the equivalent system Q_{crit} is, under the same simplifying assumptions

$$Q_{crit} = 2\sqrt{PU} = 2\sqrt{C_1 C_2 J_z m l^2 (1 + KV^2)} \,. \qquad (25.145)$$

It is a constant in the case of neutral-steer vehicles, increases with speed for understeer vehicles, and decreases, vanishing at the critical speed, in case of oversteer ones.

By comparing the actual with the critical damping, it follows that understeer vehicles tend to develop an oscillatory behavior with a frequency which increases with the speed (similar to a spring-mass-damper system with constant damping and increasing stiffness). Oversteer vehicles, on the other hand, tend to return to the original state without oscillations, but in a way that slows with increasing speed, similar to a spring-mass-damper system with constant damping and decreasing stiffness.

In a neutral-steer vehicle, under the same assumptions seen above, when $K = 0$ and

$$C_1 a = C_2 b \,,$$

the values of Q_{crit} and Q are

$$Q_{crit} = 2\frac{C_1 l J_z}{b}\sqrt{\frac{mab}{J_z}} \,,$$

$$Q = \frac{C_1 l J_z}{b}\left(1 + \frac{mab}{J_z}\right) \,. \qquad (25.146)$$

In many cases ratio

$$\frac{mab}{J_z}$$

is not far from unity. By writing

$$\frac{mab}{J_z} = 1 + \epsilon,$$

and expanding the above expressions in a power series in ϵ it follows

$$Q = \frac{C_1 l J_z}{b}(2 + \epsilon) \,. \qquad (25.147)$$

$$Q_{crit} = 2\frac{C_1 l J_z}{b}\sqrt{1 + \epsilon} = \frac{C_1 l J_z}{b}\left(2 + \epsilon - \frac{\epsilon^2}{4} + ...\right) \,.$$

Thus it is clear that the damping coefficient Q has its critical value with an error as small as a term in ϵ^2. A neutral-steer vehicle is then critically damped, at least in an approximate way, while understeer and oversteer vehicles are, respectively, underdamped and overdamped: The free behavior of the former can then be expected to be oscillatory. It must be noted, however, that the

TABLE 25.2. Example 25.6. Values of P, Q, U, Q_{crit} and of the real and imaginary parts of the roots at 100 km/h (27.78 m/s). Column 1: Simplified expression of the derivatives of stability; 2: Complete expressions, with no allowance for the effect of driving forces; 3: With driving forces with $\mu_p = 1$; 4: With driving forces with $\mu_p = 0.2$.

	1	2	3	4
P [kg^{2}m^3/s]	2.790×10^7	2.790×10^7	2.790×10^7	2.790×10^7
Q [kg^{2}m^3/s^2]	2.876×10^8	2.899×10^8	2.897×10^8	2.829×10^8
U [kg^{2}m^3/s^3]	1.243×10^9	1.334×10^9	1.335×10^9	1.369×10^9
Q_{crit} [kg^{2}m^3/s^2]	3.725×10^8	3.858×10^8	3.860×10^8	3.908×10^8
$\Re(s)$ [1/s]	-5.155	-5.196	-5.192	-5.070
$\Im(s)$ [1/s]	±4.242	±4.562	±4.573	±4.834

issue of whether a given vehicle has an oscillatory behavior or not cannot be satisfactorily resolved using the present rigid body model since the presence of rolling motions, which are neglected here and are almost always underdamped and thus oscillatory, can also induce an oscillatory behavior for β and r. This is particularly true for vehicles whose suspensions exhibit roll steer.

Example 25.6 *Study the stability with locked controls of the vehicle of Appendix E.2, taking into account the reduction of the cornering stiffness of the driving wheels caused by the longitudinal forces needed to move at constant speed.*

The parameters of the equivalent spring-mass-damper system are evaluated first and then the poles of the system are computed. The values obtained at 100 km/h (27.78 m/s) are reported in Table 25.2.

It is clear that the effect of driving forces on stability at 100 km/h is not great, even if the available traction is quite low, and that the simplified formulae already yield satisfactory results.

The values of P, Q and U are reported, together with that of Q_{crit}, as functions of the speed in Fig. 25.26a. In the same figure, the real and imaginary parts of s and the roots locus are also shown. The figure has been obtained using the complete expressions of the derivatives of stability, but neglecting the effect of driving forces.

Note that the stiffness U reduces with speed without tending to zero as in the case of neutral vehicles, and that the vehicle is almost always underdamped, except for very low speed, when $Q > Q_{crit}$.

25.14.2   Free controls

If the steering wheel is not controlled, motion of the vehicle with free controls occurs. The steering angle δ then becomes not an input to the system but one of its state variables, and a new equation stating the equilibrium of the steering system has to be included.

The same approach could be followed in the study of motion with locked controls, since what is locked is actually not the steering angle δ but the position of the steering wheel and, if the compliance of the steering system is accounted for, steering angle and position of the wheel do not coincide.

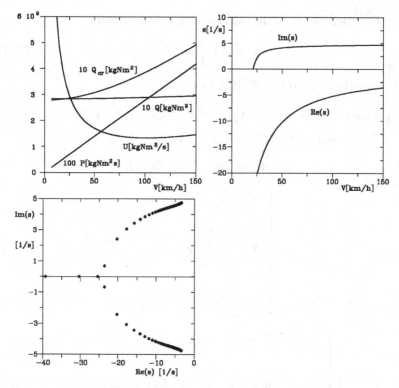

FIGURE 25.26. Example 5.5: Study of the stability. (a) Parameters of the equivalent spring-mass-damper system as functions of the speed. (b) Real and imaginary parts of the eigenvalues as functions of the speed. (c) Roots locus at varying speed. Complete expressions of the derivatives of stability, with the effect of driving forces neglected.

However, if the compliance of the steering system is considered, oscillatory motions with high frequency can usually be found, and it is unrealistic to consider the driver as a device that inputs a position signal δ to the vehicle. It is more correct to consider the driver as a device supplying a driving torque on the steering wheel. The motion thus occurs in conditions closer to a free than a locked control situation.

The actual situation is mixed: at low frequencies, such as those typical of the motion of the vehicle as a whole, the locked control model is adequate, while for high frequency modes the free control model is more suitable.

At any rate, since the motion of the vehicle includes high frequency components, the dynamic behavior of the tires cannot be neglected. The simplest way to include it into a linearized model is to use relationships of the type

$$F_y = -C \left(\alpha - B\dot{\alpha} \right),$$
$$M_z = (M_z)_{,\alpha} \left(\alpha - B'\dot{\alpha} \right), \tag{25.148}$$

for the cornering force and the aligning torque.

The time derivatives of the sideslip angles are obviously

$$\dot{\alpha}_i = \dot{\beta} + \frac{x_i}{V}\dot{r} - \dot{\delta}_i \ . \tag{25.149}$$

The equations of motion (25.109) modify as

$$\begin{cases} mV\left(\dot{\beta}+r\right) + m\dot{V}\beta = Y_\beta\beta + Y_r r + Y_\delta\delta + Y_{\dot{\beta}}\dot{\beta} + Y_{\dot{r}}\dot{r} + Y_{\dot{\delta}}\dot{\delta} + F_{y_e} \\ J_z\dot{r} = N_\beta\beta + N_r r + N_\delta\delta + N_{\dot{\beta}}\dot{\beta} + N_{\dot{r}}\dot{r} + N_{\dot{\delta}}\dot{\delta} + M_{z_e} \ , \end{cases} \tag{25.150}$$

where the expressions of the derivatives of stability already seen still hold while those of the others are

$$\begin{cases} Y_{\dot{\beta}} = \displaystyle\sum_{\forall i} C_i B_i \\[2ex] Y_{\dot{r}} = \dfrac{1}{V}\displaystyle\sum_{\forall i} x_i C_i B_i \\[2ex] Y_{\dot{\delta}} = -\displaystyle\sum_{\forall i} K_i' C_i B_i \ , \end{cases} \tag{25.151}$$

$$\begin{cases} N_{\dot{\beta}} = \displaystyle\sum_{\forall i}\left[x_i C_i B_i - (M_{z_i})_{,\alpha} B_i' \right] \\[2ex] N_{\dot{r}} = \dfrac{1}{V}\displaystyle\sum_{\forall i}\left[x_i^2 C_i B_i - (M_{z_i})_{,\alpha} x_i B_i' \right] \\[2ex] N_{\dot{\delta}} = \displaystyle\sum_{\forall i}\left[-K_i' C_i x_i B_i + K_i'(M_{z_i})_{,\alpha} B_i' \right] \ . \end{cases}$$

The equation that must be added to equations (25.150) states the equilibrium to rotation of the steering system, assumed to be a rigid system. The geometry of the steering system is sketched in Fig. 25.27. The wheel rotates about an axis, the kingpin axis, which is neither perpendicular to the ground nor passing through the centre of the contact area: The caster angle ν, the lateral inclination angle λ and the longitudinal and lateral offset at the ground d_l and d_t are reported in the figure. In the figure, the kingpin axis intersects with the rotation axis of the wheel, a very common situation. The case in which the two axes are skewed will not be dealt with here.

If the kingpin axis were perpendicular to the ground and no offset were present, the torque acting on the wheel as a consequence of the road-tire interaction forces would be the aligning torque alone. The actual situation is different, however, and the torque about the kingpin axis contains all forces and moments acting on the wheel.

With geometrical reasoning, assuming that all angles are small, the total moment M_k about the kingpin axis of both wheels of a steering axle may be

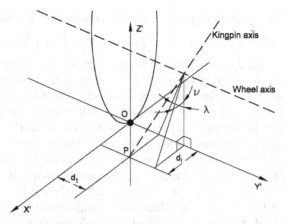

FIGURE 25.27. Simplified geometry of the steering system and definition of the caster angle ν, the lateral inclination angle λ and the offset at the ground d_l and d_t. The right wheel is sketched and ν, λ and d_t are positive. The kingpin axis is assumed to intersect the rotation axis of the wheel.

approximated as[10]

$$
M_{sr} = -(F_{z_l} + F_{z_r})d_t \sin(\lambda)\sin(\delta) + (F_{z_l} - F_{z_r})d_t \sin(\nu)\cos(\delta)+
$$
$$
+(F_{y_l} + F_{y_r})r_s \tan(\nu) + (F_{x_l} - F_{x_r})d_t + (M_{z_l} + M_{z_r})\cos\left(\sqrt{\lambda^2 + \nu^2}\right) ,
$$
$$(25.152)$$

where r and l indicate the right and left wheels respectively.

In symmetrical conditions, the forces on the ground at the two wheels are equal. By assuming that the steering angle is small, Eq. (25.152) reduces to

$$
M_{sr} = -F_z d_t \sin(\lambda)\delta + F_y r_s \tan(\nu) + M_z \cos\left(\sqrt{\lambda^2 + \nu^2}\right) , \qquad (25.153)
$$

where forces and moments refer to the whole axle.

By introducing expressions (25.148) into Eq. (25.153) the following linearized expression of the moment about the kingpin is obtained

$$
M_{sr} = M_{\dot\beta}\dot\beta + M_{\dot r}\dot r + M_{\dot\delta}\dot\delta + M_\beta\beta + M_r r + M_\delta\delta , \qquad (25.154)
$$

where

$$
M_{\dot\beta} = CBr_s \tan(\nu) - (M_z)_{,\alpha}B'\cos\left(\sqrt{\lambda^2 + \nu^2}\right) ,
$$
$$
M_{\dot r} = M_{\dot\beta}\frac{a}{V} , \qquad M_{\dot\delta} = -M_{\dot\beta} ,
$$
$$
M_\beta = -Cr_s \tan(\nu) + (M_z)_{,\alpha}\cos\left(\sqrt{\lambda^2 + \nu^2}\right) , \qquad (25.155)
$$
$$
M_r = M_\beta \frac{a}{V} , \qquad M_\delta = -M_\beta - F_z d_t \sin(\lambda) .
$$

---

[10]   T. D. Gillespie, *Fundamentals of Vehicle Dynamics*, SAE, Warrendale, 1992.

The linearized equation of motion of the steering system is then

$$J_s\ddot{\delta} + c_s\dot{\delta} = M_{\dot{\beta}}\dot{\beta} + M_{\dot{r}}\dot{r} + M_{\dot{\delta}}\dot{\delta} + M_\beta\beta + M_r r + M_\delta\delta + M_s\tau_s, \qquad (25.156)$$

where M_s, τ_s, c_s and J_s are, respectively, the torque exerted by the driver on the steering wheel, the steering ratio (the ratio between the rotation angle of the wheel and that of the kingpin), the damping coefficient of the steering damper and the moment of inertia of the whole system, the latter two reduced to the kingpin. Note that the steering ratio is often not constant and that the compliance of the mechanism, here neglected, may have a large effect on it.

No gyroscopic effect of the wheels has been accounted for, which is consistent with the assumption of a rigid vehicle, even if a weak gyroscopic effect should be present if the kingpin axis is not perpendicular to the road.

Equation (25.156) holds also when more complicated geometries are accounted for, provided that a linearization about a reference position is performed. In this case, the expressions of the derivatives of stability M_β, M_r etc. also contain the longitudinal offset at the ground.

Since the second derivative of the state variable δ enters the equations of motion, a further state variable

$$v_\delta = \dot{\delta}$$

must be introduced and a further equation stating the mentioned identity must be added. The state equation is still Eq. (25.109)

$$\dot{\mathbf{z}} = \mathbf{A}\mathbf{z} + \mathbf{B}_c\mathbf{u}_c + \mathbf{B}_e\mathbf{u}_e ,$$

where the state and input vectors $\mathbf{z}$, $\mathbf{u}_c$ and $\mathbf{u}_e$ are

$$\mathbf{z} = \left\{ \begin{array}{c} \beta \\ r \\ v_\delta \\ \delta \end{array} \right\}, \qquad \mathbf{u}_c = M_s , \qquad \mathbf{u}_e = \left\{ \begin{array}{c} F_{y_e} \\ M_{z_e} \end{array} \right\},$$

the dynamic matrix is

$$\mathbf{A} = \begin{bmatrix} mV - Y_{\dot{\beta}} & -Y_{\dot{r}} & -Y_{\dot{\delta}} & 0 \\ -N_{\dot{\beta}} & J_z - N_{\dot{r}} & -N_{\dot{\delta}} & 0 \\ -M_{\dot{\beta}} & -M_{\dot{r}} & J_s & 0 \\ 0 & 0 & 0 & 1 \end{bmatrix}^{-1}$$

$$\times \begin{bmatrix} Y_\beta & -mV + Y_r & 0 & Y_\delta \\ N_\beta & N_r & 0 & N_\delta \\ M_\beta & M_r & (M_{\dot{\delta}} - c_s) & M_\delta \\ 0 & 0 & 1 & 0 \end{bmatrix}$$

and the input gain matrices are

$$\mathbf{B}_c = \begin{bmatrix} mV - Y_{\dot{\beta}} & -Y_{\dot{r}} & -Y_{\dot{\delta}} & 0 \\ -N_{\dot{\beta}} & J_z - N_{\dot{r}} & -N_{\dot{\delta}} & 0 \\ -M_{\dot{\beta}} & -M_{\dot{r}} & J_s & 0 \\ 0 & 0 & 0 & 1 \end{bmatrix}^{-1} \begin{bmatrix} 0 \\ 0 \\ \tau_s \\ 0 \end{bmatrix},$$

$$\mathbf{B}_e = \begin{bmatrix} mV - Y_{\dot\beta} & -Y_{\dot r} & -Y_{\dot\delta} & 0 \\ -N_{\dot\beta} & J_z - N_{\dot r} & -N_{\dot\delta} & 0 \\ -M_{\dot\beta} & -M_{\dot r} & J_s & 0 \\ 0 & 0 & 0 & 1 \end{bmatrix}^{-1} \begin{bmatrix} 1 & 0 \\ 0 & 1 \\ 0 & 0 \\ 0 & 0 \end{bmatrix}.$$

The state equation can be used to study the stability of the vehicle and the response to any given law $M_s(t)$. In a similar way, it is possible to study the steady-state performance simply by assuming that all derivatives are vanishingly small (the last state equation may then be dropped, since it reduces to the identity $0 = 0$)

$$\begin{bmatrix} -Y_\beta & mV - Y_r & -Y_\delta \\ -N_\beta & -N_r & -N_\delta \\ -M_\beta & -M_r & -M_\delta \end{bmatrix} \begin{Bmatrix} \beta \\ r \\ \delta \end{Bmatrix} = \begin{Bmatrix} F_{y_e} \\ M_{z_e} \\ M_s \tau_s \end{Bmatrix}. \tag{25.157}$$

The *steering wheel torque gain* M_s/δ with reference to the steering angle and that referring to the curvature of the path $M_s R$, may be easily computed.

The eigenproblem

$$\det(\mathbf{A} - s\mathbf{I}) = 0 \tag{25.158}$$

allows one to study stability. Since the size of the dynamic matrix $\mathbf{A}$ is only four, it is possible to write the characteristic equation and to solve it using the formula for 4-th degree algebraic equations. However, no closed form solution from which to draw general conclusions is available. The eigenvalues are either a pair of complex conjugate solutions – yielding damped oscillations (if both real parts are negative), one usually at low frequency and the other at high frequency – or two nonoscillatory solutions and one high frequency oscillation. The high frequency solution is usually linked with the dynamics of the steering device while the others are linked primarily to the behavior of the vehicle.

The vibrations of the steering system were of concern in the past, particularly in the 1930s, when they were referred to as *steering shimmy*. Such vibrations were also present in the tailwheel of aircraft undercarriages. The use of tires with lower pneumatic trail and, above all, the introduction of damping in the steering mechanism has completely rectified the problem. Both viscous damping and dry friction have been used with success, but the latter decreases the reversibility of the steering system and thus decreases its precision and its centering characteristics.

The, now common, use of servosystems in the steering control implies the presence of non-negligible damping with viscous characteristics in the steering device.

The present model is, however, too imprecise for a detailed study of this phenomenon, since the compliance of the steering system and the lateral compliance of the suspension are important causal factors in this type of vibration that may become self-excited.

If only the low-frequency overall behavior of the vehicle is studied, it is possible to neglect the dependence of the tire forces on the time derivative of the sideslip angle. In this case, the expressions of the dynamic matrix and of the input gain matrix simplify as follows

$$
\mathbf{A} =
\begin{bmatrix}
\dfrac{Y_\beta}{mV} & \dfrac{Y_r}{mV} - 1 & 0 & \dfrac{Y_\delta}{mV} \\[2ex]
\dfrac{N_\beta}{J_z} & \dfrac{N_r}{J_z} & 0 & \dfrac{N_\delta}{J_z} \\[2ex]
\dfrac{M_\beta}{J_s} & \dfrac{M_r}{J_s} & \dfrac{-c_s}{J_s} & \dfrac{-M_\delta}{J_s} \\[2ex]
0 & 0 & 1 & 0
\end{bmatrix},
$$

$$
\mathbf{B}_c =
\begin{bmatrix}
0 \\
0 \\
\dfrac{\tau_s}{J_s} \\
0
\end{bmatrix},
\qquad
\mathbf{B}_e =
\begin{bmatrix}
\dfrac{1}{mV} & 0 \\[2ex]
0 & \dfrac{1}{J_z} \\[2ex]
0 & 0 \\
0 & 0
\end{bmatrix}.
$$

If the inertia and the damping of the steering system are likewise neglected, Eq. (25.156) can be solved in δ. By introducing this value into the equations of motion, an approximate model for the behavior of the vehicle with free controls is obtained.

By assuming that the speed V is constant, the homogeneous state equation for a vehicle with front axle steering only is then

$$
\left\{ \begin{array}{c} \dot{\beta} \\ \dot{r} \end{array} \right\} =
\begin{bmatrix}
\dfrac{Y_\beta + Y_\delta}{mV} & \dfrac{Y_r + Y_\delta \frac{a}{V}}{mV} - 1 \\[3ex]
\dfrac{N_\beta + N_\delta}{J_z} & \dfrac{N_r + N_\delta \frac{a}{V}}{J_z}
\end{bmatrix}
\left\{ \begin{array}{c} \beta \\ r \end{array} \right\}.
\tag{25.159}
$$

The equation is formally identical to the homogeneous Eq. (25.108) and in this case as well, it is possible to resort to a spring-mass-damper analogy and to study the constant speed stability in a simple way. It can be shown that both the stiffness and the damping coefficient are always positive, denoting both static and dynamic stability.

By introducing only the cornering forces due to the tires, the vehicle is overdamped at low speed, up to

$$
V = \frac{1}{2} \left(b^2 + \frac{J_z}{m} \right) \sqrt{\frac{C_2}{J_z b}}.
$$

Above that speed the behavior becomes more and more underdamped, with an increasingly oscillatory behavior.

Note, however, that the last simplification is usually too rough: In most cases, the high value of the steering ratio τ_s makes the inertia of the steering wheel when reduced to the kingpin axis non-negligible and the use of equation (25.159) can lead to non-negligible errors. Other errors may be introduced by neglecting steering damping since a certain amount of damping is present in the system, the neglect of which may cause dynamic instability.

Example 25.7 *Compute the torque that must be exerted on the steering wheel necessary to maintain the vehicle of Appendix E.2 on a circular path with a radius of 100 m and to counteract a transversal slope of 1° at constant speed.*

The additional data for the steering system are: $\lambda = 11°$, $\nu = 3°$, $d = 5$ mm and $\tau_s = 16$.

The steering wheel torque gain $M_s R$ can be computed from Eq. (25.157). By stating $F_{y_e} = 0$, $M_{z_e} = 0$ and $M_s = 1$, it is possible to obtain the yaw velocity r that follows the application of a unit torque to the steering wheel.

Since $R = V/r$, the gain $M_s R$ may be immediately computed and thus the value of the torque needed to maintain any given circular path. The results for $R = 100$ m are reported in Fig. 25.28a.

To obtain the steering torque needed to counteract a transversal road slope, Eq. (25.156) needs to be rearranged. The slope α_t is felt by the vehicle as a side force

$$F_{y_e} = mg \sin(\alpha_t).$$

If the path is straight, $r = 0$ and also M_{z_e} is equal to zero, as no external moment acts on the vehicle. The unknowns are β, δ and M_s.

The equation is rearranged as

$$\begin{bmatrix} -Y_\beta & -Y_\delta & 0 \\ -N_\beta & -N_\delta & 0 \\ -M_\beta & -M_\delta & \tau_s \end{bmatrix} \begin{Bmatrix} \beta \\ \delta \\ M_s \end{Bmatrix} = \begin{Bmatrix} mg \sin(\alpha_t) \\ 0 \\ 0 \end{Bmatrix}.$$

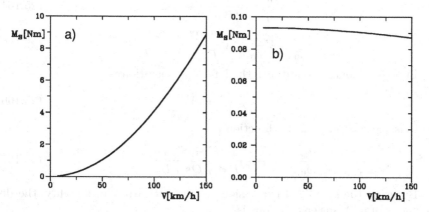

FIGURE 25.28. Example 25.7: Steering wheel torque needed to maintain the vehicle on a circular path with a radius of 100 m (a) and to counteract a transversal slope of 1° at constant speed (b).

25.15 UNSTATIONARY MOTION

The response to a steering input in unstationary conditions may be computed using the constant-speed linearized model expressed by equations (25.110) or (25.111), reported here without the terms due to external forces and moments

$$P\ddot{\beta} + Q\dot{\beta} + U\beta = S'\delta + T'\dot{\delta},$$
$$P\ddot{r} + Q\dot{r} + Ur = S''\delta + T''\dot{\delta}.$$

(25.160)

If the variable for motion in the y direction is the lateral velocity v instead of the sideslip angle β, the first equation becomes

$$P\ddot{v} + Q\dot{v} + Uv = VS'\delta + VT'\dot{\delta}.$$

(25.161)

If an input of the type

$$\delta = \delta_0 e^{st}$$

is assumed, the solution takes the form

$$\beta = \beta_0 e^{st}, \quad r = r_0 e^{st}, \quad v = v_0 e^{st}.$$

The algebraic equations into which the differential equations transform are

$$\begin{pmatrix} Ps^2 + Qs + U \end{pmatrix} \beta_0 = (T's + S')\,\delta_0,$$
$$\begin{pmatrix} Ps^2 + Qs + U \end{pmatrix} r_0 = (T''s + S'')\,\delta_0,$$
$$\begin{pmatrix} Ps^2 + Qs + U \end{pmatrix} v_0 = V\,(T's + S')\,\delta_0.$$

(25.162)

The transfer functions are then

$$\frac{\beta_0}{\delta_0} = \frac{T's + S'}{Ps^2 + Qs + U},$$

(25.163)

$$\frac{r_0}{\delta_0} = \frac{T''s + S''}{Ps^2 + Qs + U},$$

(25.164)

$$\frac{v_0}{\delta_0} = V\frac{\beta_0}{\delta_0} = V\frac{T's + S'}{Ps^2 + Qs + U}.$$

(25.165)

In non-stationary conditions, the lateral acceleration is

$$a_y = \dot{v} + rV$$

(25.166)

and thus the relevant transfer function is

$$\frac{a_{y0}}{\delta_0} = V\frac{T's^2 + (T'' + S')\,s + S''}{Ps^2 + Qs + U}.$$

(25.167)

By using the simplified expressions of the derivatives of stability, the denominator of all transfer functions is

$$\Delta = J_z m V s^2 + \left[J_z(C_1 + C_2) + m(a^2C_1 + b^2C_2) \right] s +$$
$$+ mV(-aC_1 + bC_2) + C_1 C_2 \frac{l^2}{V}.$$

(25.168)

The equation $\Delta = 0$ allows the poles of the system to be computed, as seen in section 6.13.1.

Assuming only front axle steering, the transfer functions are

$$\frac{r_0}{\delta_0} = \frac{mVaC_1s + lC_1C_2}{\Delta}, \tag{25.169}$$

$$\frac{a_{y0}}{\delta_0} = \frac{J_zVC_1s^2 + C_1C_2bls + lVC_1C_2}{\Delta}. \tag{25.170}$$

By equating the numerator of the transfer functions (25.169) and (25.170) to zero it is possible to find their zeros. For functions (25.169) the result is straightforward, and the only zero is real and negative

$$s = -\frac{lC_2}{mVa}. \tag{25.171}$$

The computation for function (25.170) is not as simple. The zeros are

$$s = \frac{-blC_2 \pm \sqrt{b^2l^2C_2^2 - 4V^2lJ_zC_2}}{2J_zV}. \tag{25.172}$$

At low speed, i.e. if

$$V \le \sqrt{\frac{b^2lC_2}{4J_z}}, \tag{25.173}$$

the two solutions are both real and negative. They are distinct if Eq. (25.173) holds with ($<$), coincident if it holds with ($=$).

At higher speeds, the two solutions are complex conjugate

$$s = \frac{-blC_2}{2J_zV} \pm \sqrt{\frac{4V^2lJ_zC_2 - b^2l^2C_2^2}{4J_z^2V^2}}, \tag{25.174}$$

with a negative real part: the zeros lie in the left part of the Argand plane.

The situation may be different for the sideslip angle: S' may be either positive or negative depending on the values of the parameters. By using the simplified expressions of the derivatives of stability, the value of the relevant transfer function is

$$\frac{\beta_0}{\delta_0} = \frac{J_zVC_1s + C_1C_2bl - maV^2C_1}{V\Delta}. \tag{25.175}$$

The expression of the zero is obtained by equating to zero the numerator

$$s = \frac{maV^2C_1 - C_1C_2bl}{J_zVC_1}. \tag{25.176}$$

At low speed the zero is negative and real, but if

$$V > \sqrt{\frac{blC_2}{am}} \tag{25.177}$$

it moves to the positive part of the Argand plane and then the system is a non-minimum phase system.

From Eq. (25.110) and following it is clear that the response to steering is a linear combination of the laws $\delta(t)$ and $\dot{\delta}(t)$. If the numerator of the transfer function is linear in s, and if the zero of the transfer function (which is always real since the numerator is linear) is negative, the coefficients of the linear combination have the same sign and the sign of the response does not change in time.

Example 25.8 *Plot the roots locus of the transfer function related to the lateral acceleration at varying speed for the vehicle in Appendix E.2, taking into account both the simplified and the complete expressions of the derivatives of stability used in Example 25.5. Compute the speed at which the transfer function β_0/δ_0 becomes a non-minimum phase function.*

Then compute the response to a step steering input at a speed of 100 km/h.

The transfer function a_{y_0}/δ_0 has two real zeros up to a speed of 24.67 km/h; it then has two complex conjugate poles. The locus of the zeros is reported in Fig. 25.29a.

The two formulations yield practically the same results. Function β_0/δ_0 has a negative real zero up to a speed of 56.22 km/h; than has a positive real zero.

If function $\delta(t)$ is a unit step function

$$\begin{cases} \delta = 0 & per \ t < 0 \\ \delta = 1 & per \ t \geq 0 \ , \end{cases}$$

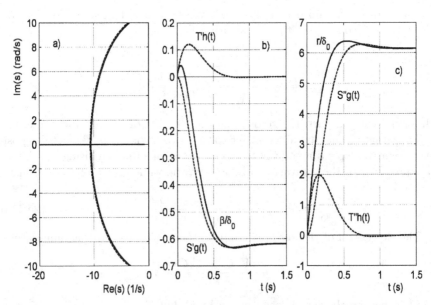

FIGURE 25.29. a): Locus of the zeros of the transfer function a_{y_0}/δ_0. Full line: complete expression of the derivatives of stability; dashed line: simplified expression. b) and c): response to a step steering input computed in closed form.

its derivative $\dot{\delta}$ is an impulse function(Dirac's δ):

$$\begin{cases} \dot{\delta} = 0 & per\ t < 0 \\ \dot{\delta} = \infty & per\ t = 0 \\ \dot{\delta} = 0 & per\ t > 0 \end{cases}$$

$$\int_{-\infty}^{\infty} \dot{\delta}dt = 1 \ .$$

Since the vehicle is understeer, the step and impulse responses $g(t)$ and $h(t)$ are both oscillatory and are

$$h(t) = \frac{1}{m\omega_n\sqrt{1-\zeta^2}}e^{-\zeta\omega_n t}\sin\left(\sqrt{1-\zeta^2}\omega_n t\right)$$

$$g(t) = \frac{1}{k} - \frac{e^{-\zeta\omega_n t}}{k}\left[\cos\left(\sqrt{1-\zeta^2}\omega_n t\right) + \frac{\zeta}{\sqrt{1-\zeta^2}}\sin\left(\sqrt{1-\zeta^2}\omega_n t\right)\right] \ ,$$

where

$$m = P \ , \quad \omega_n = \sqrt{\frac{U}{P}} \ , \quad \zeta = \frac{Q}{2\sqrt{PU}} \ .$$

The total response is a linear combination of the step and impulse responses

$$\beta(t) = S'g(t) + T'h(t)$$
$$r(t) = S"g(t) + T"h(t) \ .$$

At 100 km/h the mass-spring-damper system is underdamped, since the damping ratio has a value $\zeta = 0.77$. The natural frequency of the undamped system is $\omega_n = 6.67$ rad/s $= 1.06$ Hz, while the frequency of the free damped oscillations is $\omega_p = 4.24$ rad/s $= 0.68$ Hz.

The results are reported in Fig. 25.29b) and c).

The step and impulse responses have the same sign for the yaw velocity, and they simply add in modulus. In the response for the sideslip angle they have opposite sign and initially the second one prevails. When, after some time, the first one begins to prevail, the sign of the response changes.

This is typical for non-minimum phase systems: the system initially reacts in a direction opposite to that of the steady state response, then goes to zero and changes its sign.

Once law $r(t)$ has been obtained, it is possible to integrate it to yield the yaw angle

$$\psi(t) = \int_0^t r(u)du \ . \tag{25.178}$$

The path can then be obtained directly in the inertial frame X, Y. The velocities $\dot{X}$ and $\dot{Y}$ can be expressed in terms of angles β and ψ

$$\left\{\begin{array}{c} \dot{X} \\ \dot{Y} \end{array}\right\} = V\left[\begin{array}{cc} \cos(\psi) & -\sin(\psi) \\ \sin(\psi) & \cos(\psi) \end{array}\right]\left\{\begin{array}{c} \cos(\beta) \\ \sin(\beta) \end{array}\right\} \ . \tag{25.179}$$

By integrating equations (25.179) the path is readily obtained

$$
\begin{cases}
X = \displaystyle\int_0^t V \left[\cos(\beta) \cos(\psi) - \sin(\beta) \sin(\psi) \right] du \\
Y = \displaystyle\int_0^t V \left[\cos(\beta) \sin(\psi) + \sin(\beta) \cos(\psi) \right] du \ .
\end{cases}
\tag{25.180}
$$

The integration to obtain the path must actually be performed numerically even in the simplest cases where laws $\beta(t)$ and $r(t)$ may be computed in closed form owing to the fact that angle ψ is usually too large to allow linearizing its trigonometric functions even when using the linearized model. In general, it is more convenient to integrate the equations of motion numerically, since there is no difficulty in doing so for Eq. (25.109) once laws $\delta(t)$, $F_{y_e}(t)$, $M_{z_e}(t)$ and $V(t)$ have been stated. Nowadays numerical integration is so straightforward that closed form solutions that are too complicated to allow a quick qualitative understanding of the phenomena to be obtained are considered of little use.

Example 25.9 *Study the motion with locked controls of the vehicle of Appendix E.2 following a step steering input.*

Assume that the value of the steering angle is that needed to obtain a circular path with a radius of 200 m at a speed of 100 km/h.

At 100 km/h the path curvature gain $1/R\delta$ is equal to 0.2472 $1/m$. To perform a curve with a radius of 200 m a steering angle $\delta = 0.0202$ rad $= 1.159°$ is needed.

In kinematic conditions, the radius of the path corresponding to the same value of δ is 106.8 m. The fact that it is almost half the above was easily predictable, since 100 km/h is only slightly less than the characteristic speed.

The steady state values of r and β are respectively 0.1389 rad/s and -0.0131 rad $= -0.749°$.

The equation of motion of the vehicle was integrated numerically for a duration of 30 s. The results are plotted in Fig. 25.30. The time histories of the yaw velocity and sideslip angle are shown along with the path.

The steady-state conditions are reached after a few seconds, with a slightly underdamped behavior.

Example 25.10 *Study the motion with locked controls of the vehicle of Appendix E.2 following a wind gust. Assume a step lateral gust, like the one encountered when exiting a tunnel. Assume an ambient wind velocity $v_a = 10$ m/s and a vehicle speed of 100 km/h.*

The driver does not react to the gust and the steering angle is kept equal to zero.

The presence of a cross-wind is accounted for by adding a side force F_{y_e} and a yawing moment M_{z_e} equal to

$$
\begin{cases}
F_{y_e} = \dfrac{1}{2}\rho V^2 S(C_y)_{,\beta}\psi_w \\[2mm]
M_{y_e} = \dfrac{1}{2}\rho V^2 Sl(C_{M_z})_{,\beta}\psi_w \ ,
\end{cases}
$$

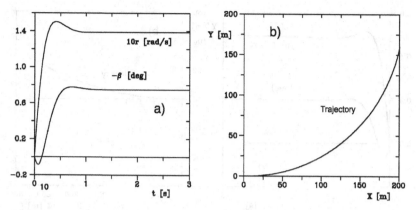

FIGURE 25.30. Example 25.9: Response to a step steering input. (a) Time histories of the yaw velocity and sideslip angle and (b) path.

where ψ_w is the angle between the direction of the relative velocity and the tangent to the path. This is clearly an approximation since it relies on the linearity of the aerodynamic forces and moments with the aerodynamic sideslip angle and holds only if angle $\beta + \psi_w$ remains small.

As the path of the vehicle curves after the manoeuvre, the components of the relative velocity along the path and in a direction perpendicular to it are

$$\begin{cases} V_\parallel = V - v_a \sin(\psi + \beta) \\ V_\perp = -v_a \cos(\psi + \beta) \, , \end{cases}$$

yielding

$$\psi_w = \arctan\left(\frac{-v_a \cos(\psi + \beta)}{V - v_a \cos(\psi + \beta)} \right) .$$

The above relationships may be approximated by neglecting angle β.

Another approximation is neglecting the contribution of the wind velocity to the airspeed, which is always considered at 100 km/h.

The equation of motion of the vehicle has been integrated numerically for a duration of 10 s. The results are plotted in Fig. 25.31. The time histories of the yaw velocity and sideslip angle are shown along with the path.

Quasi steady-state conditions are again reached after a few seconds, with a slightly underdamped behavior. The conditions are not actually steady-state since the direction of the wind is fixed, while the direction of the vehicle axes change. However, this effect is minimal for the duration of the manoeuvre, and a good approximation could have been obtained by assuming a constant value for angle ψ_w (ψ_w increases from 19.8° to 20.9° for $t = 0$ to $t = 10$ s).

At the end of the manoeuvre, the values of r and β are, respectively, 0.0505 rad/s and -0.0036 rad $= -0.2073°$. The errors linked to neglecting β in the above expression are thus negligible. The response in terms of β in this case is that typical of a non-minimum phase system.

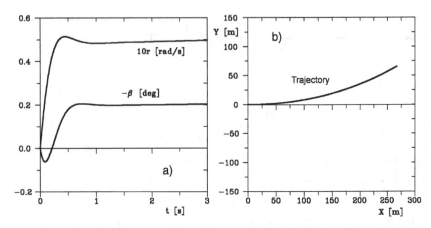

FIGURE 25.31. Example 25.10: Response to a cross-wind gust. (a) Time histories of the yaw velocity and sideslip angle and (b) path.

Example 25.11 *The following manoeuvre is often performed by test drivers to assess the handling and stability of a car: A step steering input is supplied and the steering wheel is kept in position for a short time. The driver then releases the wheel and the vehicle returns to a straight path. The whole manoeuvre is performed at constant speed.*

Study the motion of the vehicle of Appendix E.2 following a manoeuvre of this kind with a 45° steering wheel input held for 1.5 s at 100 km/h.

The data for the steering system are $J_s = 15$ kg m², $c_s = 150$ Nms/rad, $\lambda = 11°$, $\nu = 3°$, $d = 5$ mm and $\tau_s = 16$.

The first part of the manoeuvre is the same as in Example 25.9, only with a greater value of δ: 2.81°.

The integration in time is performed in two parts: A locked controls model is used for the first 1.5 s; a free control model is used after the driver releases the wheel.

This second part of the simulation is performed using two alternative models: One in which the dependence of tire forces on the derivative $\dot{\alpha}$ is neglected, and a second in which the inertia and damping of the steering system are also not considered.

The time histories of the yaw velocity, sideslip angle and steering angle are reported together with the path in Fig. 25.32.

The inertia of the steering system plays an important role in the response, since it slows the recovery of the vehicle, thus affecting the path. It also increases the oscillatory behavior of the vehicle, and if no damping is considered, an unstable behavior emerges.

The effect of neglecting the inertia of the steering system can be verified by comparing the poles of the system: If neither inertia nor damping is accounted for, the two eigenvalues are $-3.011 \pm 7.709i$, while the more complete model yields four eigenvalues $-9.129 \pm 8.2921i$ and $-1.065 \pm 5.563i$. The first is quite damped and is not important in the motion, but the second is clearly different from that obtained from the simpler model. The high value of the steering ratio, whose square enters the computation of the equivalent inertia of the steering wheel, is responsible for this effect.

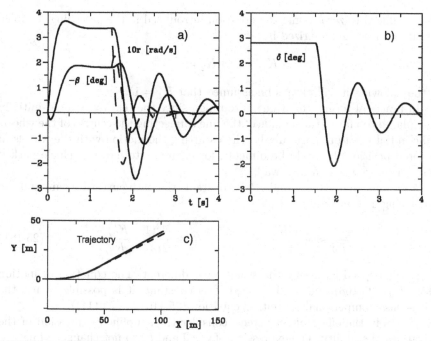

FIGURE 25.32. Example 25.11: Response to a step steering input and a subsequent recovery of the straight path with free controls. (a) Time histories of the yaw velocity and sideslip angle and (b) of the steering angle; (c) path. The inertia and damping of the steering system are considered (full lines) and then neglected (dashed lines).

25.16   VEHICLES WITH TWO STEERING AXLES (4WS)

In the majority of vehicles with two axles, only the front wheels are provided with a steering system. However, beginning in the 1980s, an increasing number of cars with steering on all four wheels (4WS) appeared on the market, in the beginning most of them Japanese. The primary goal was to improve manoeuvrability and handling characteristics both in low- and high-speed steering. 4WS system were dealt with in Part I, Chapter 6.

Simple four-wheel steering may be implemented by equipping the rear axle with a compliance purposely designed to provide the required steering action under the effect of road loads without adding an actual steering device. This approach is defined as passive steering. Active steering occurs when the rear axle is provided with a second steering device, operated by the driver along with that of the front axle through adequate actuators.

To reduce the radius of the path in low-speed (kinematic) conditions, the rear axle must steer in a direction opposite to the front; if the absolute values of the steering angles are equal, the radius is halved and the off-tracking of the rear axle is reduced to zero. Using the notation introduced in the preceding sections, this situation is characterized by

$$K_1' = 1 \ , \ \ K_2' = -1$$

(in the following it will always be assumed that $K_1' = 1$).

In practical terms, this value is too high since the rear axle would initially be displaced too far to the outside of the line connecting the centres of the wheels in the initial position, particularly when starting the motion with the wheels in a steered position. It would be difficult, for example, to move a vehicle parked near a curb or, worse, near a wall.

Assuming that $K_1' = 1$ and K_2' is constant, the path curvature gain and the off-tracking distance are

$$\frac{1}{R\delta} \approx \frac{1 + K_2'}{l} \ , \qquad R_a - R_1 \approx \frac{l^2(1 - K_2')}{2R(1 + K_2')} \ . \tag{25.181}$$

In high-speed cornering the situation is different. The equation of motion is still Eq. (25.108) and, if the speed V is constant, it is possible to use the spring-mass-damper analogy (either equation (25.110) or (25.111)).

To study the effect of rear steer, consider the simplified expression of the derivatives of stability. The expression of P, Q and U do not change, while

$$\begin{cases} S' = mV\left(-aC_1K_1' + bC_2K_2'\right) + C_1C_2\dfrac{l}{V}\left(K_1'b + K_2'a\right) \\ S'' = lC_1C_2\left(K_1' - K_2'\right) \\ T' = J_z\left(K_1'C_1 + K_2'C_2\right) \\ T'' = mV\left(aK_1'C_1 - bK_2'C_2\right) \ . \end{cases}$$

The expressions of the gains in steady state conditions become:

- path curvature gain

$$\frac{1}{R\delta} = \frac{1}{l}\frac{(K_1' - K_2')}{1 + KV^2} \ , \tag{25.182}$$

- lateral acceleration gain

$$\frac{V^2}{R\delta} = \frac{V^2}{l}\frac{(K_1' - K_2')}{1 + KV^2} \ , \tag{25.183}$$

- sideslip angle gain

$$\frac{\beta}{\delta} = \frac{b}{l}\left[K_1' + K_2'\frac{a}{b} - \frac{mV^2}{l}\left(\frac{aK_1'}{bC_2} - \frac{K_2'}{C_1}\right)\right]\frac{1}{1 + KV^2} \ , \tag{25.184}$$

- yaw velocity gain

$$\frac{r}{\delta} = \frac{V}{l}\frac{(K_1' - K_2')}{1 + KV^2} \ . \tag{25.185}$$

From the equations above it is clear that opposite steering (K'_1 and K'_2 with opposite signs) produces an increase of the gains related to the curvature of the path, while steering with the same sign allows larger cornering forces to be produced for the same steering angle

However, the most important advantages of 4WS are felt in non-steady state conditions, making it important to assess the transfer functions in these conditions. Equations from (25.162) to (25.167) still hold. If the simplified expressions of the derivatives of stability are used, it follows that

$$\frac{r_0}{\delta_0} = \frac{mV\left(aK'_1C_1 - bK'_2C_2\right)s + lC_1C_2\left(K'_1 - K'_2\right)}{\Delta}, \tag{25.186}$$

$$\frac{a_{y_0}}{\delta_0} = \frac{J_zV\left(K'_1C_1 + K'_2C_2\right)s^2 + C_1C_2l\left(aK'_1 + bK'_2\right)s + lVC_1C_2\left(K'_1 - K'_2\right)}{\Delta}, \tag{25.187}$$

where Δ is still expressed by Eq. (25.168).

Opposite steering also makes the vehicle more responsive about the yaw axis in non-steady state conditions. The second transfer function shows how steering in the same direction increases the response at the highest frequencies, in particular for lateral acceleration due to motion in the y direction, while opposite steering increases the contribution due to centrifugal acceleration, especially at low frequency.

Strong rear axle steering may cause some of the zeros of the transfer functions to lie into the positive half-plane of the complex plane, making the system a non-minimum phase system. This will be studied in greater detail in Chapter 27.

The limiting case of same sign steering is, for a vehicle with the center of mass at mid-wheelbase, that with equal steering angles

$$K'_1 = K'_2 = 1.$$

Remark 25.17 *This is, however, too theoretical, since the vehicle would be quick in a lane change, moving sideways, but would never be able to move on a curved path. Instead of turning, it would move sideways.*

Thus it is clear that the steering mechanism must adapt the value of K'_2 to the external conditions and to the requests of the driver. As seen in Part I, the simplest strategy is to use a device, possibly mechanical, to link the two steering boxes with a variable gear ratio: When angle δ is small, as typically occurs in high speed driving, K'_2 is positive and the steering angles have the same direction while when δ is large, as occurs when manoeuvring at low speed, K'_2 is negative. Obviously, K'_2 must be much smaller than K'_1.

However, more complicated control laws for the steering of the rear axle must be implemented to fully exploit the potential advantages of 4WS. The parameters entering such laws are numerous, e.g. the speed V, the lateral acceleration, the sideslip angles α_i, etc. Such devices must be based on electronic controllers and actuators of different types, and their implementation enters into the important field of autronics (Chapter 27).

From the viewpoint of mathematical modelling, the situation is, at least in principle, simple. There is no difficulty in introducing a suitable function $K_2'(V, \delta, \dots)$ into the equations (actually it would appear only in the derivatives of stability Y_δ and N_δ) and in modifying the equations of the rigid-body model seen above accordingly. The more advanced models of the following sections can be modified along the same lines. If function K_2' includes some of the state variables, the modifications can be larger but no conceptual difficulty arises.

Except in the latter case, locked control stability is not affected by the introduction of 4WS, while stability with free controls can be affected by it.

Generally speaking, the advantages of 4WS are linked with an increase in the quickness of the response of the vehicle to a steering input, but this cannot be true for all types of manoeuvres: Steering all axles in the same direction may make the vehicle quick in lane change manoeuvres but slower in acquiring a given yaw velocity. The sensations of the driver may be strange and, at least at the beginning, unpleasant. A solution may be a device that initially steers the rear wheels in the opposite direction for a short time, to initiate a yaw rotation, and then steers them in the same direction as the front wheels, to generate cornering forces. This requires a more complicated control logic, possibly based on microprocessors.

As a final consideration, most applications are based on vehicles already designed for conventional steering to which 4WS is then added, normally as an option. In this case, the steering of the rear wheel is limited to $1° \div 2°$ or even less since the rear wheel wells lack the space required for larger movement. Even if the car is designed from the beginning for 4WS, a trade-off between its advantages and the loss of available space in the trunk due to 4WS will take place.

25.17   MODEL WITH 4 DEGREES OF FREEDOM FOR ARTICULATED VEHICLES

25.17.1   Equations of motion

An articulated vehicle modelled as two rigid bodies hinged to each other has, in its motion on the road surface, four degrees of freedom (Fig. 25.33). The assumption of rigid bodies implies that the hinge is cylindrical and that its axis is perpendicular to the road: In practice different setups are used, but if rolling is neglected the present one is the only possible layout.

There is no difficulty in writing the six equations of motion of the two rigid bodies (each has three degrees of freedom in the planar motion on the road) and then in introducing the two equations for the constraints due to the hinge to eliminate two of the six. The forces exchanged between the two bodies are explicitly introduced.

Here a different approach is followed and the equations of motion are obtained through Lagrange equations. To this end, a set of four generalized

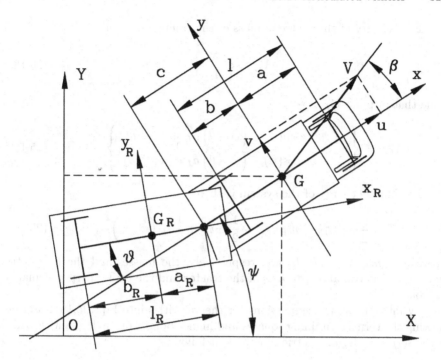

FIGURE 25.33. Articulated vehicle. Reference frames and generalized coordinates.

coordinated is first stated: X and Y are the inertial coordinates of the centre of mass of the tractor and ψ is its yaw angle. They are the same coordinates used in the study of the insulated vehicle. The added coordinate is angle θ between the longitudinal axes x of the tractor and x_R of the trailer. Positive angles are shown in Fig. 25.33.

Instead of angle θ, it is possible to use the yaw angle of the trailer ψ_R, i.e. the angle between the inertial X-axis and the body-fixed axis x_R.

The model can be simplified and linearized, as seen for the model of the isolated vehicle, by assuming that the motion occurs in a condition not much different from the symmetrical, which implies that the trailer angle θ and the sideslip angles are small. Moreover, the vehicle will be assumed to be a monotrack vehicle, i.e. the sideslip angles of the wheels of each axle will be assumed to be equal. The model will be built in terms of axles rather than wheels.

As a damper with damping coefficient Γ may be attached to the hinge between tractor and trailer, a Raleigh dissipation function must be written along with the kinetic energy. No conservative forces act in the plane of the road, assuming the hinge has no elastic restoring force, and hence no potential energy need be computed.

The position of the centre of mass of the trailer is

$$(\overline{G_R - O}) = \left\{ \begin{array}{l} X - c\cos(\psi) - a_R\cos(\psi - \theta) \\ Y - c\sin(\psi) - a_R\sin(\psi - \theta) \end{array} \right\} . \tag{25.188}$$

The velocity of the centre of mass of the tractor is simply

$$V_G = \left\{ \begin{array}{c} \dot{X} \\ \dot{Y} \end{array} \right\} , \qquad (25.189)$$

while that of point G_R is

$$V_{G_R} = \left\{ \begin{array}{c} \dot{X} + \dot{\psi} c \sin(\psi) + \left(\dot{\psi} - \dot{\theta} \right) a_R \sin(\psi - \theta) \\ \dot{Y} - \dot{\psi} c \cos(\psi) - \left(\dot{\psi} - \dot{\theta} \right) a_R \cos(\psi - \theta) \end{array} \right\} . \qquad (25.190)$$

The kinetic energy of the system is then:

$$T = \frac{1}{2} m_T V_G^2 + \frac{1}{2} m_R V_{G_R}^2 + \frac{1}{2} J_T \dot{\psi}^2 + \frac{1}{2} J_R \left(\dot{\psi} - \dot{\theta} \right)^2 , \qquad (25.191)$$

where m_T, m_R, J_T and J_R are, respectively, the masses and the baricentric moments of inertia about an axis of the tractor and the trailer perpendicular to the road.

By introducing the expressions for the velocities into Eq. (25.191) and neglecting the terms containing squares and higher powers of small quantities, also in the series for trigonometric functions, it follows

$$\begin{aligned} T = \quad & \frac{1}{2} m \left(\dot{X}^2 + \dot{Y}^2 \right) + \frac{1}{2} J_1 \dot{\psi}^2 + \frac{1}{2} J_3 \dot{\theta}^2 - J_2 \dot{\psi} \dot{\theta} + \\ & + m_R \left[c \dot{\psi} + a_R \left(\dot{\psi} - \dot{\theta} \right) \right] \left[\dot{X} \sin(\psi) - \dot{Y} \cos(\psi) \right] + \\ & - m_R a_R \theta \left(\dot{\psi} - \dot{\theta} \right) \left[\dot{X} \cos(\psi) - \dot{Y} \sin(\psi) \right] , \end{aligned} \qquad (25.192)$$

where

$$\left\{ \begin{array}{l} m = m_T + m_R , \\ J_1 = J_T + J_R + m_R \left[a_R^2 + c^2 + 2 a_R c \right] , \\ J_2 = J_R + m_R \left[a_R^2 + a_R c \right] , \\ J_3 = J_R + m_R a_R^2 . \end{array} \right.$$

The components of the velocity in the tractor reference frame may be used

$$\left\{ \begin{array}{c} u \\ v \\ r \\ v_\theta \end{array} \right\} = \left[\begin{array}{cccc} \cos(\psi) & \sin(\psi) & 0 & 0 \\ -\sin(\psi) & \cos(\psi) & 0 & 0 \\ 0 & 0 & 1 & 0 \\ 0 & 0 & 0 & 1 \end{array} \right] \left\{ \begin{array}{c} \dot{X} \\ \dot{Y} \\ \dot{\psi} \\ \dot{\theta} \end{array} \right\} , \qquad (25.193)$$

where r is the yaw angular velocity of the tractor and v_θ is the relative yaw angular velocity of the trailer with respect to the tractor. The relationship between angular velocities and derivatives of the generalized coordinates is

$$\mathbf{w} = \mathbf{A}^T \dot{\mathbf{q}} , \qquad (25.194)$$

where the structure of $\mathbf{A}$ is that of a rotation matrix, and then

$$\mathbf{A}^T = \mathbf{A}^{-1} \tag{25.195}$$

The final expression of the kinetic energy is then

$$
\begin{aligned}
\mathcal{T} = \ & \tfrac{1}{2}m\left(u^2 + v^2\right) + \tfrac{1}{2}J_1 r^2 + \tfrac{1}{2}J_3 v_\theta{}^2 - J_2 r v_\theta + \\
& - m_R v \left[cr + a_R\left(r - v_\theta\right) \right] - m_R a_R \theta u \left(r - v_\theta\right) \ .
\end{aligned}
\tag{25.196}
$$

The rotation kinetic energy of the wheels has been neglected: No gyroscopic effect of the wheels will be obtained in this way.

The Raleigh dissipation function due to the above mentioned viscous damper is simply

$$\mathcal{F} = \frac{1}{2}\Gamma\dot{\theta}^2 \ . \tag{25.197}$$

The equations of motion obtained in the form of Lagrange equations are

$$\frac{d}{dt}\left(\frac{\partial \mathcal{T}}{\partial \dot{q}_i}\right) - \frac{\partial \mathcal{T}}{\partial q_i} + \frac{\partial \mathcal{F}}{\partial \dot{q}_i} = Q_i \ , \tag{25.198}$$

where the coordinates q_i are X, Y, ψ and θ and Q_i are the corresponding generalized forces F_X, F_Y and the moments related to rotations ψ and θ.

The velocities in the reference frame fixed to the tractor can be considered as derivatives of pseudo-coordinates. Operating in the same way as for the isolated vehicle, and remembering that the kinetic energy does not depend on coordinates X and Y:

$$\left(\frac{\partial \mathcal{T}}{\partial X} = \frac{\partial \mathcal{T}}{\partial Y} = 0\right) \ ,$$

that the dissipation function does not depend on the linear velocities

$$\left(\frac{\partial \mathcal{F}}{\partial \dot{X}} = \frac{\partial \mathcal{F}}{\partial \dot{Y}} = 0\right)$$

and that angular velocities r and v_θ coincide with $\dot\psi$ and $\dot\theta$, the equation of motion can be written in the form (A.126), with the derivatives of the dissipation function added[11]

$$\frac{\partial}{\partial t}\left(\left\{\frac{\partial T}{\partial w}\right\}\right) + \mathbf{A}^T\left(\dot{\mathbf{A}} - \left[\mathbf{w}^T\mathbf{A}^T\frac{\partial \mathbf{A}}{\partial q_k}\right]\right)\left\{\frac{\partial T}{\partial w}\right\} +$$

$$-\mathbf{A}^T\left\{\frac{\partial T}{\partial q_k}\right\} + \left\{\frac{\partial \mathcal{F}}{\partial w}\right\} = \mathbf{A}^T\left\{\begin{array}{c} F_X \\ F_Y \\ Q_\psi \\ Q_\theta \end{array}\right\}. \tag{25.199}$$

The terms included in the equation of motion are

$$\left\{\frac{\partial T}{\partial w}\right\} = \left\{\begin{array}{c} mu - m_R a_R \theta\,(r - v_\theta) \\ mv - m_R\left[(c + a_R)r - a_R v_\theta\right] \\ J_1 r - J_2 v_\theta - m_R a_R \theta u - m_R v\,(c + a_R) \\ J_3 v_\theta - J_2 r + m_R v a_R + m_R a_R \theta u \end{array}\right\}, \tag{25.200}$$

$$\frac{d}{dt}\left(\left\{\frac{\partial T}{\partial w}\right\}\right) = \left\{\begin{array}{c} m\dot u - m_R a_R v_\theta\,(r - v_\theta) - m_R a_R \theta\,(\dot r - \dot v_\theta) \\ m\dot v - m_R\left[(c + a_R)\dot r - a_R \dot v_\theta\right] \\ J_1\dot r - J_2\dot v_\theta - m_R a_R \theta\dot u - m_R a_R u v_\theta - m_R\dot v\,(c + a_R) \\ J_3\dot v_\theta - J_2\dot r + m_R \dot v a_R + m_R a_R \theta\dot u + m_R a_R u v_\theta \end{array}\right\}, \tag{25.201}$$

$$\mathbf{A}^T\left(\dot{\mathbf{A}} - \left[\mathbf{w}^T\mathbf{A}^T\frac{\partial \mathbf{A}}{\partial q_k}\right]\right)\left\{\frac{\partial T}{\partial w}\right\} =$$

$$= \left\{\begin{array}{c} -r\left\{mv - m_R\left[(c + a_R)r - a_R v_\theta\right]\right\} \\ r\left[mu - m_R a_R \theta\,(r - v_\theta)\right] \\ -v\left[mu - m_R a_R \theta\,(r - v_\theta)\right] + u\left\{mv - m_R\left[(c + a_R)r - a_R v_\theta\right]\right\} \\ 0 \end{array}\right\}, \tag{25.202}$$

$$\mathbf{A}^T\left\{\frac{\partial T}{\partial q_k}\right\} = \left\{\begin{array}{c} 0 \\ 0 \\ 0 \\ -m_R a_R u\,(r - v_\theta) \end{array}\right\}, \tag{25.203}$$

$$\left\{\frac{\partial \mathcal{F}}{\partial w}\right\} = \left\{\begin{array}{c} 0 \\ 0 \\ 0 \\ \Gamma\dot\theta \end{array}\right\}, \quad \mathbf{A}^T\left\{\begin{array}{c} F_X \\ F_Y \\ Q_\psi \\ Q_\theta \end{array}\right\} = \left\{\begin{array}{c} Q_x \\ Q_y \\ Q_\psi \\ Q_\theta \end{array}\right\}. \tag{25.204}$$

---

[11] In this case, the equation of motion is not written in its general form, but only for the case with $\mathbf{A}^T = \mathbf{A}^{-1}$.

The first two equations are then

$$
\begin{cases}
m\left(\dot{u}-vr\right)-m_{R}a_{R}\theta\left(\dot{r}-\dot{v}_{\theta}\right)-2m_{R}a_{R}rv_{\theta}+m_{R}a_{R}v_{\theta}{}^{2}+ \\
\qquad +m_{R}\left(c+a_{R}\right)r^{2}=Q_{x} \\
m\left(\dot{v}+ur\right)-m_{R}\left[c+a_{R}\right]\dot{r}+m_{R}a_{R}r\dot{v}_{\theta}-m_{R}a_{R}\theta r\left(r-v_{\theta}\right)=Q_{y}\,.
\end{cases}
$$
$$(25.205)$$

Remembering that, owing to the assumption of small angles, $V \approx u$ and also that v is small, equations (25.205) may be linearized as

$$
\begin{cases}
m\dot{V}=Q_{x} \\
m\left(\dot{v}+Vr\right)-m_{R}\left(c+a_{R}\right)\dot{r}+m_{R}a_{R}r\ddot{\theta}=Q_{y}\,.
\end{cases}
$$
$$(25.206)$$

The third and forth equations, those for generalized coordinates ψ and θ, once linearized, are

$$
\begin{cases}
J_{1}\dot{r}-J_{2}\dot{v}_{\theta}-m_{R}\left(c+a_{R}\right)\left(\dot{v}+Vr\right)-m_{R}a_{R}\dot{V}\theta=Q_{\psi} \\
J_{3}\dot{v}_{\theta}-J_{2}\dot{r}+m_{R}a_{R}\left(\dot{v}+Vr\right)+m_{R}a_{R}\theta\dot{V}=Q_{\theta}\,.
\end{cases}
$$
$$(25.207)$$

where the damping term $\Gamma\dot{\theta}$ is included in term Q_{θ}.

25.17.2 Sideslip angles of the wheels

The sideslip angles of the wheels of the tractor are the same as for the insulated vehicle. In a similar way, it is possible to write the sideslip angles of the wheels of the trailer.

With reference to Fig.25.34, the coordinates of point P_i, the centre of the contact zone of the ith wheel of the trailer, are

$$
\begin{cases}
X_{P_{i}}=X-c\cos(\psi)-l_{i}\cos\left(\psi-\theta\right)-y_{R_{i}}\sin\left(\psi-\theta\right) \\
Y_{P_{i}}=Y-c\sin(\psi)-l_{i}\sin\left(\psi-\theta\right)+y_{R_{i}}\cos\left(\psi-\theta\right)\,.
\end{cases}
$$
$$(25.208)$$

The velocity of the same point may be obtained by differentiating the expressions of the coordinates. For the computation of the sideslip angle the velocity of point P_i must be expressed in the reference frame $G_{R}x_{R}y_{R}$ of the trailer

$$
\left\{\begin{array}{c}\dot{X}_{P_{i}}\\\dot{Y}_{P_{i}}\end{array}\right\}_{R}=\left[\begin{array}{cc}\cos\left(\psi-\theta\right)&\sin\left(\psi-\theta\right)\\-\sin\left(\psi-\theta\right)&\cos\left(\psi-\theta\right)\end{array}\right]\left\{\begin{array}{c}\dot{X}_{P_{i}}\\\dot{Y}_{P_{i}}\end{array}\right\}\,.
$$
$$(25.209)$$

The velocity of the centre of the contact area can thus be expressed in the reference frame of the trailer as

$$
\begin{cases}
\dot{V}_{x_{R}}(P_{i})=u\cos(\theta)-v\sin(\theta)+c\dot{\psi}\sin(\theta)-y_{R_{i}}\left(\dot{\psi}-\dot{\theta}\right) \\
\dot{V}_{y_{R}}(P_{i})=u\sin(\theta)+v\cos(\theta)-c\dot{\psi}\cos(\theta)-l_{i}\left(\dot{\psi}-\dot{\theta}\right)\,,
\end{cases}
$$
$$(25.210)$$

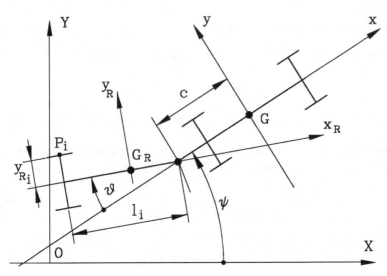

FIGURE 25.34. Position of the centre P_i of the contact area of the i-th wheel of the trailer.

or, remembering that some of the quantities are small,

$$\begin{cases} \dot{V}_{x_R}(P_i) = V - y_{R_i}\left(\dot{\psi} - \dot{\theta}\right) \\ \dot{V}_{y_R}(P_i) = V\theta + v - c\dot{\psi} - l_i\left(\dot{\psi} - \dot{\theta}\right) . \end{cases} \qquad (25.211)$$

Since the sideslip angle of a steering wheel can be obtained as the arctangent of the ratio of the y and x components of the velocity minus the steering angle δ, it follows that

$$\alpha_i = \arctan\left[\frac{V\theta + v - c\dot{\psi} - l_i\left(\dot{\psi} - \dot{\theta}\right)}{V - y_{R_i}\left(\dot{\psi} - \dot{\theta}\right)}\right] - \delta_i . \qquad (25.212)$$

Using the monotrack vehicle model ($y_{R_i} = 0$) and remembering that the sideslip angle is small, it follows that

$$\alpha_i = \theta + \beta - \frac{r}{V}(c + l_i) + \frac{\dot{\theta}}{V}l_i - \delta_i . \qquad (25.213)$$

The term in y_{R_i} does not enter the expression of the sideslip angle: The wheels of the same axle have the same sideslip angle, and it is also possible to work in terms of axles instead of single wheels for the trailer.

The steering angle δ_i is either 0 or, if the axle can steer, is usually not directly controlled by the driver but is linked with the variables of the motion, e.g. with angle θ. If law $\delta_i(\theta)$ is simply

$$\delta_i = -K_i'\theta ,$$

the expression for the sideslip angle is

$$\alpha_i = \theta(1 + K_i') + \beta - \frac{r}{V}(c + l_i) + \frac{\dot{\theta}}{V}l_i .$$

(25.214)

If some of the wheels of the trailer are free to pivot about their kingpin, an equilibrium equation for the relevant parts of the steering system of those axles must be written, similar to the procedure for the study of motion with free controls.

25.17.3 Generalized forces

The contributions to the generalized forces Q_x, Q_y and Q_ψ due to the tractor are the same as those of the insulated vehicle. The tractor does not give any contribution to force Q_θ. To compute the contributions due to the ith wheel of the trailer and the aerodynamic forces of the latter, the easiest method is to write their virtual work $\delta\mathcal{L}$ due to a virtual displacement

$$\{\delta s\} = [\delta x, \delta y, \delta\psi, \delta\theta]^T .$$

Using the assumption of small angles, it follows that

$$\begin{cases} \delta x_R(P_i) = \delta x - \theta\delta y + c\theta\delta\psi - y_{R_i}(\delta\psi - \delta\theta) \\ \delta y_R(P_i) = \theta\delta x + \delta y - c\delta\psi - l_i(\delta\psi - \delta\theta) . \end{cases}$$

(25.215)

If the ith wheel has a steering angle δ_i, the forces it exerts in the reference frame $G_R x_R y_R z_R$, the same in which the virtual displacement has been written, are simply

$$F_{x_{iR}} = F_{x_{ip}}\cos(\delta_i) - F_{y_{ip}}\sin(\delta_i) \approx F_{x_{ip}} - F_{y_{ip}}\delta_i,$$
$$F_{y_{iR}} = F_{x_{ip}}\sin(\delta_i) + F_{y_{ip}}\cos(\delta_i) \approx F_{x_{ip}}\delta_i + F_{y_{ip}},$$

(25.216)

where $F_{x_{ip}}$ and $F_{y_{ip}}$ are the forces in the reference frame of the tire.

The virtual work can be computed by multiplying the forces and moments (the aligning torque M_{z_i}) by the corresponding virtual displacement (for the moment, rotation $\delta\psi - \delta\theta$)

$$\delta\mathcal{L} = \left[F_{x_{ip}} + F_{y_{ip}}(\theta - \delta_i)\right]\delta x + \left[-F_{x_{ip}}(\theta - \delta_i) + F_{y_{ip}}\right]\delta y +$$
$$+ \left\{F_{x_{ip}}[c(\theta - \delta_i) - y_{R_i} - l_i\delta_i] + F_{y_{ip}}(-c + y_{R_i}\delta_i - l_i) + M_{z_i}\right\}\delta\psi +$$
$$+ \left\{F_{x_{ip}}(y_{R_i} + l_i\delta_i) + F_{y_{ip}}(-y_{R_i}\delta_i + l_i) - M_{z_i}\right\}\delta\theta.$$

(25.217)

The generalized forces due to the ith wheel of the trailer can be obtained by differentiating the virtual work $\delta \mathcal{L}$ with respect to the virtual displacements δx, δy, $\delta \psi$ and $\delta \theta$:

$$
\begin{cases}
Q_{x_i} = \dfrac{\partial \delta \mathcal{L}}{\partial \delta x} = F_{x_{i_p}} + F_{y_{i_p}}(\theta - \delta_i) \\[2mm]
Q_{y_i} = \dfrac{\partial \delta \mathcal{L}}{\partial \delta y} = -F_{x_{i_p}}(\theta - \delta_i) + F_{y_{i_p}} \\[2mm]
Q_{\psi_i} = \dfrac{\partial \delta \mathcal{L}}{\partial \delta \psi} = F_{x_{i_p}}\left[c(\theta - \delta_i) - y_{R_i} - l_i \delta_i \right] + F_{y_{i_p}}\left(-c + y_{R_i}\delta_i - l_i \right) + M_{z_i} \\[2mm]
Q_{\theta_i} = \dfrac{\partial \delta \mathcal{L}}{\partial \delta \theta} = F_{x_{i_p}}\left(y_{R_i} + l_i \delta_i \right) + F_{y_{i_p}}\left(-y_{R_i}\delta_i + l_i \right) - M_{z_i} \, .
\end{cases}
\tag{25.218}
$$

In a similar way, the generalized forces resulting from the aerodynamic forces and moments acting on the trailer can be accounted for. It is usually difficult to distinguish between the forces acting on the tractor and those acting on the trailer, as what is measured in the wind tunnel are the forces acting on the whole vehicle. In the following equations, it will be assumed that the forces acting on the tractor are measured separately from those acting on the trailer, and that they are applied at the centre of mass of the relevant rigid body and decomposed along the axes fixed to it. The forces acting on the trailer are so decomposed along axes $x_R y_R z_R$.

The generalized forces due to aerodynamic forces acting on the tractor contribute to Q_x, Q_y and Q_ψ just as they do for the insulated vehicle, while the expression of the generalized aerodynamic forces applied on the trailer can be obtained from equations (25.218), by substituting $F_{x_{R_{aer}}}$, $F_{y_{R_{aer}}}$, $M_{z_{R_{aer}}}$ and a_R to $F_{x_{i_p}}$, $F_{y_{i_p}}$, M_{z_i} and l_i and by setting both y_{R_i} and δ_i to zero.

The external force $F_{y_{e_R}}$ acting on the centre of mass or the trailer and the component of the weight $m_R g \sin(\alpha)$ due to a longitudinal grade α of the road will be assumed to act in the directions of axes x and y of the tractor; consequently the relevant equations must be modified accordingly.

25.17.4 Linearized expressions of the forces

The linearized expressions of the generalized forces Q_x, Q_y, Q_ψ and Q_θ can be obtained with the methods used for the isolated vehicle. Linearization can be performed by introducing the cornering and aligning stiffnesses C_i and $(M_{z_i})_{,\alpha}$ of the axles (subscript i refers now to the ith axle and not to the ith wheel). In the same way, the derivatives of the aerodynamic coefficients $(C_y)_{,\beta}$, etc. can also be introduced.

A simple expression for Q_x is thus obtained:

$$Q_x = X_m - \left(f_0 + KV^2\right)\left[mg\cos(\alpha) - \tfrac{1}{2}\rho V^2\left(SC_z + S_RC_{z_R}\right)\right] +$$

$$-\tfrac{1}{2}\rho V^2\left(SC_x + S_RC_{x_R}\right) - mg\sin(\alpha), \tag{25.219}$$

where, as usual, X_m is the driving force of the driving axle, but may also be the total braking force.

By substituting the sideslip angle β of the vehicle for ratio v/V, the expressions of the forces appearing in the handling equations are

$$\begin{cases} Q_y = (Q_y)_{,\beta}\beta + (Q_y)_{,r}r + (Q_y)_{,\dot\theta}\dot\theta + (Q_y)_{,\theta}\theta + (Q_y)_{,\delta}\delta + F_{y_e} + F_{y_{e_R}} \\ Q_\psi = (Q_\psi)_{,\beta}\beta + (Q_\psi)_{,r}r + (Q_\psi)_{,\dot\theta}\dot\theta + (Q_\psi)_{,\theta}\theta + (Q_\psi)_{,\delta}\delta + M_{z_e} + \\ \quad + M_{z_{e_R}} - (c + a_R)F_{y_{e_R}} \\ Q_\theta = (Q_\theta)_{,\beta}\beta + (Q_\theta)_{,r}r + (Q_\theta)_{,\dot\theta}\dot\theta + (Q_\theta)_{,\theta}\theta - M_{z_{e_R}} + a_RF_{y_{e_R}} \ . \end{cases} \tag{25.220}$$

The derivatives of stability entering the expression for Q_y are

$$\begin{cases} (Q_y)_{,\beta} = Y_\beta - \sum_{\forall i_R} C_i + \dfrac{1}{2}\rho V_r^2 S_R(C_{Y_R})_{,\beta} \\[2mm] (Q_y)_{,r} = Y_r + \dfrac{1}{V}\left[\sum_{\forall i_R}(c + l_i)C_i + \dfrac{1}{2}\rho V_r^2 S_R(c + a_R)(C_{Y_R})_{,\beta}\right] \\[2mm] (Q_y)_{,\dot\theta} = -\dfrac{1}{V}\left[\sum_{\forall i_R} l_iC_i - \dfrac{1}{2}\rho V_r^2 S_R a_R(C_{Y_R})_{,\beta}\right] \\[2mm] (Q_y)_{,\theta} = -\sum_{\forall i_R} C_i + \dfrac{1}{2}\rho V_r^2 S_R(C_{Y_R})_{,\beta} \\[2mm] (Q_y)_{,\delta} = Y_\delta \ , \end{cases} \tag{25.221}$$

where Y_β, Y_r and Y_δ are the derivatives of stability of the tractor expressed by equations (25.103).

The derivatives of stability entering the expression for Q_ψ and Q_θ are respectively

$$\begin{cases} (Q_\psi)_{,\beta} = N_\beta + \sum_{\forall i_R} C_1 + (c + l_i)C_i + (M_{z_i})_{,\alpha} \\[2mm] (Q_\psi)_{,r} = N_r - \dfrac{1}{V}\left[\sum_{\forall i_R}(c + l_i)^2 C_i + (c + l_i)(M_{z_i})_{,\alpha} + (c + a_R)C_{a1}\right] \\[2mm] (Q_\psi)_{,\dot\theta} = \dfrac{1}{V}\left[\sum_{\forall i_R} l_i(c + l_i)C_i + l_i(M_{z_i})_{,\alpha} + a_RC_{a1}\right] \\[2mm] (Q_\psi)_{,\theta} = \sum_{\forall i_R}(c + l_i)C_i + (M_{z_i})_{,\alpha} + C_{a1} \\[2mm] (Q_\psi)_{,\delta} = N_\delta \ , \end{cases} \tag{25.222}$$

$$
\begin{cases}
(Q_\theta)_{,\beta} = \sum_{\forall i_R} C_2 - l_i C_i - (M_{z_i})_{,\alpha} \\[2mm]
(Q_\theta)_{,r} = \dfrac{1}{V}\left[\sum_{\forall i_R}(c+l_i)l_i C_i + (c+l_i)(M_{z_i})_{,\alpha} + (c+a_R)C_{a2}\right] \\[2mm]
(Q_\theta)_{,\dot\theta} = -\dfrac{1}{V}\left[\sum_{\forall i_R} l_i^2 C_i + l_i(M_{z_i})_{,\alpha} + a_R C_{a2}\right] - \Gamma \\[2mm]
(Q_\theta)_{,\theta} = (Q_\theta)_{,\beta} \\[1mm]
(Q_\theta)_{,\delta} = 0\,,
\end{cases}
\tag{25.223}
$$

where aerodynamic terms C_{a1} and C_{a2} are:

$$
C_{a1} = \frac{1}{2}\rho V_r^2 S_R \left[l_R(C_{N_R})_{,\beta} - (c+a_R)(C_{Y_R})_{,\beta}\right]\,,
$$

$$
C_{a2} = \frac{1}{2}\rho V_r^2 S_R \left[l_R(C_{N_R})_{,\beta} - a_R(C_{Y_R})_{,\beta}\right]\,.
$$

N_β, N_r and N_δ are the derivatives of stability of the tractor expressed by equations (25.107). All axles of the trailer have been assumed as non-steering

If the axles of the trailer can steer and their steering angles δ_i are linked with angle θ by the law

$$
\delta_i = -K_i'\theta\,,
$$

the expressions of the derivatives of stability reported above still hold, except for $(Q_y)_{,\theta}$, $(Q_\psi)_{,\theta}$ and $(Q_\theta)_{,\theta}$ in which all terms in C_i and $(M_{z_i})_{,\alpha}$ must be multiplied by $(1 + K_i')$.

25.17.5   Final expression of the equations of motion

As with the equations of the insulated vehicle, the linearization of the equations allows the longitudinal behavior (first equation of motion) to be uncoupled from the lateral, or handling behavior, which can be studied using only the three remaining equations. This occurs if the law $u(t)$, which can be confused with $V(t)$, is considered as a stated law, while the unknowns are the driving or braking forces F_x for the longitudinal behavior and β, r and θ for handling.

The linearized equation for the longitudinal behavior

$$
m\dot V = Q_x
\tag{25.224}
$$

can thus be studied separately.

The linearized equations for the lateral behavior of the articulated vehicle can be expressed in the space of the configurations as

$$
\mathbf{M}\ddot{\mathbf{x}} + \mathbf{C}\dot{\mathbf{x}} + \mathbf{K}\mathbf{x} = \mathbf{F}\,,
\tag{25.225}
$$

where the vectors of the generalized coordinates and of the forces are

$$
\mathbf{x} = \left\{\begin{array}{c} y \\ \psi \\ \theta \end{array}\right\}\,,\qquad
\mathbf{F} = \left\{\begin{array}{c} (Q_y)_{,\delta}\delta + F_{y_e} + F_{y_{e_R}} \\ (Q_\psi)_{,\delta}\delta + M_{z_e} + M_{z_{e_R}} - (c+a_R)F_{y_{e_R}} \\ -M_{z_{e_R}} + a_R F_{y_{e_R}} \end{array}\right\}
\tag{25.226}
$$

and the matrices are

$$\mathbf{M} = \begin{bmatrix} m & -m_R(c+a_R) & m_R a_R \\ -m_R(c+a_R) & J_1 & -J_2 \\ m_R a_R & -J_2 & J_3 \end{bmatrix},$$

$$\mathbf{C} = \begin{bmatrix} -\dfrac{(Q_y)_{,\beta}}{V} & mV - (Q_y)_{,r} & -(Q_y)_{,\dot{\theta}} \\ -\dfrac{(Q_\psi)_{,\beta}}{V} & -m_R V(c+a_R) - (Q_\psi)_{,r} & -(Q_\psi)_{,\dot{\theta}} \\ -\dfrac{(Q_\theta)_{,\beta}}{V} & m_R V a_R - (Q_\theta)_{,r} & -(Q_\theta)_{,\dot{\theta}} \end{bmatrix},$$

(25.227)

$$\mathbf{K} = \begin{bmatrix} 0 & 0 & -(Q_y)_{,\theta} \\ 0 & 0 & -(Q_\psi)_{,\theta} \\ 0 & 0 & -(Q_\theta)_{,\theta} \end{bmatrix}.$$

The set of differential equations (25.225) is actually of the fourth order and not of the sixth, since variables y^{12} and ψ appear in the equation only as first and second derivatives (the first two columns of matrix $\mathbf{K}$ vanish). The equation can thus be written in the state space in the form of a set of four first order differential equations by introducing a fourth state variable $v_\theta = \dot{\theta}$

$$\dot{\mathbf{z}} = \mathbf{A}\mathbf{z} + \mathbf{B}_c \mathbf{u}_c + \mathbf{B}_e \mathbf{u}_e \,..$$

The state vector $\mathbf{z}$ is simply

$$\mathbf{z} = \begin{bmatrix} \beta & r & v_\theta & \theta \end{bmatrix}^T,$$

the dynamic matrix is

$$\mathbf{A} = \begin{bmatrix} -\mathbf{M}^{-1}\mathbf{C} & \mathbf{M}^{-1} \left\{ \begin{matrix} (Q_y)_{,\theta} \\ (Q_\psi)_{,\theta} \\ (Q_\theta)_{,\theta} \\ 0 \end{matrix} \right\} \\ \begin{bmatrix} 0 & 0 & 1 \end{bmatrix} \end{bmatrix},$$

the input gain matrices are

$$\mathbf{B}_c = \begin{bmatrix} \mathbf{M}^{-1} \begin{bmatrix} (Q_y)_{,\delta} \\ (Q_\psi)_{,\delta} \\ 0 \end{bmatrix} \\ 0 \end{bmatrix},$$

$$\mathbf{B}_e = \begin{bmatrix} \mathbf{M}^{-1} \begin{bmatrix} 1 & 1 & 0 & 0 \\ 0 & -(c+a_R) & 1 & 1 \\ 0 & a_R & 0 & -1 \end{bmatrix} \\ \begin{bmatrix} 0 & 0 & 0 & 0 \end{bmatrix} \end{bmatrix},$$

---

[12] Actually, as already stated, v is the derivative of a pseudo-coordinate and thus y has no physical meaning. It has been introduced into the equations only for completeness and, since it is always multiplied by 0, its presence can be accepted.

and the input vector is

$$\mathbf{u}_c = \delta. \qquad \mathbf{u}_e = \begin{bmatrix} F_{y_e} & F_{y_{eR}} & M_{z_e} & M_{z_{eR}} \end{bmatrix}^T.$$

25.17.6   Steady-state motion

To study the steady-state behavior of the vehicle, Eq. (25.225) can be used, along with the assumption that $\dot{v} = \dot{r} = \dot{\theta} = \ddot{\theta} = 0$. The following equation is thus obtained

$$\begin{bmatrix} -(Q_y)_{,\beta} & mV - (Q_y)_{,r} & -(Q_y)_{,\theta} \\ -(Q_\psi)_{,\beta} & -m_R V(c + a_R) - (Q_\psi)_{,r} & -(Q_\psi)_{,\theta} \\ -(Q_\theta)_{,\beta} & m_R V a_R - (Q_\theta)_{,r} & -(Q_\theta)_{,\theta} \end{bmatrix} \begin{Bmatrix} \beta \\ r \\ \theta \end{Bmatrix} = $$

$$= \begin{Bmatrix} (Q_y)_{,\delta}\delta + F_{y_e} + F_{y_{eR}} \\ (Q_\psi)_{,\delta}\delta + M_{z_e} + M_{z_{eR}} - (c + a_R)F_{y_{eR}} \\ -M_{z_{eR}} + a_R F_{y_{eR}} \end{Bmatrix}. \tag{25.228}$$

There is no difficulty in solving such a set of equations. For instance, after stating that $\delta = 1$ and setting all other inputs to zero, the gains $1/R\delta$, β/δ etc. can be computed.

A particularly simple solution is obtained for a two-axle vehicle with a one-axle trailer if only the cornering forces of the wheels are accounted for

$$\begin{cases} \dfrac{1}{R\delta} = \dfrac{1}{l} \dfrac{1}{1 + KV^2} \\[3mm] \dfrac{\theta}{\delta} = \dfrac{a + c + K'V^2}{l(1 + KV^2)}, \end{cases} \tag{25.229}$$

where the stability factor K and K' are

$$\begin{cases} K = \dfrac{1}{l^2}\left[\left(m_T + m_R \dfrac{l_R - a_R}{l_R} \right) \left(\dfrac{b}{C_1} - \dfrac{a}{C_2} \right) + \right. \\[3mm] \left. \qquad - m_R \dfrac{c(l_R - a_R)}{l_R} \left(\dfrac{1}{C_1} + \dfrac{1}{C_2} \right) \right] \\[4mm] K' = \dfrac{1}{l}\left\{ m\dfrac{a}{C_2} + \dfrac{m_R}{l_R}\left[\dfrac{(a + c)(l_R - a_R)}{C_2} - \dfrac{l_R a_R}{C_1} \right] \right\}. \end{cases} \tag{25.230}$$

The same definitions used for the insulated vehicle also hold in this case and, if the derivatives of stability are constant or proportional to $1/V$, the sign of the stability factor allows one to state immediately whether the vehicle is oversteer, neutral-steer or understeer.

The simplified expression of the stability factor (25.230) is composed of two terms: The first usually has the same sign of $bC_1 - aC_2$, i.e. of the factor that decides the behavior of the tractor alone. The second term is negative, unless the product $c(l_R - a_R)$ is negative, i.e. the centre of mass of the trailer is behind its axle.

If

$$c(l_R - a_R) > 0 \ ,$$

the trailer increases the understeering character of the vehicle, more so if the hinge is far from the centre of mass of the tractor and the centre of mass of the trailer is close to the hinge. In the case of trailers with a single axle, like caravans, this effect can be reduced by reducing the distance between its centre of mass and the axle.

If the centre of mass is exactly on the axle, that is, if

$$l_R - a_R = 0 \ ,$$

the trailer has no effect on the steady state behavior of the tractor; it does, however, affect its dynamic behavior and stability.

If the centre of mass of the trailer is behind its axle,

$$l_R - a_R < 0 \ ,$$

the trailer increases the oversteer behavior of the tractor. If the vehicle is oversteer, the presence of a critical speed can be expected.

Remark 25.18 *This way of comparing the behavior of the tractor alone with that of the complete vehicle is not correct however: The presence of the trailer can change the loads on the wheels of the former, thus affecting their cornering stiffness.*

Example 25.12 *Study the steady state directional behavior of the articulated truck of Appendix E.9. Compare the results obtained using the complete expressions of the derivatives of stability with those computed considering only the cornering forces of the tires.*

The computation is straightforward. At each value of the speed the normal forces on the ground must be computed, although they change little with the speed. The cornering stiffness and the aligning stiffness of the axles are readily obtained from the normal forces.

At 100 km/h, for instance, the normal forces on the axles are 57.25, 107.28, 79.83, 83.56 and 56.14 kN, yielding the following values for the cornering stiffness and the aligning stiffness: 422.05, 806.64, 641.34, 665.89, 416.42 kN/rad and 22.724, 41.472, 26.102, 28.175, 22.116 kNm/rad.

The path curvature gain, the sideslip angle gain and the trailer angle gain θ/δ are plotted as functions of the speed in Fig. 25.35. The values obtained from the complete expressions of the derivatives of stability are shown as full lines while the dashed lines refer to the simplified expressions for the derivatives of stability obtained by considering only the cornering forces.

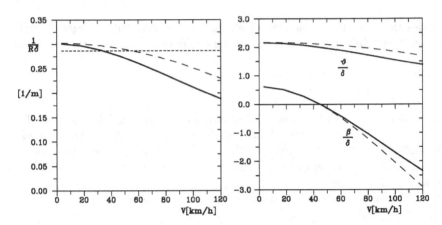

FIGURE 25.35. Example 25.12: Path curvature gain, sideslip angle gain and trailer angle gain as functions of the speed. Full lines: Values obtained from the complete expressions of the derivatives of stability; dashed lines: Simplified approach obtained considering only the cornering forces.

When the speed tends to zero, the path curvature gain does not tend to the kinematic value $1/l$ of the tractor: The trailer has a number of axles greater than one and correct kinematic steering is impossible. The vehicle is understeer, even if weakly.

The simplified approach allows one to obtain a fair approximation of the directional behavior of the vehicle, the differences between the two results being due mostly to the aligning torques of the tires and only marginally to aerodynamic forces and moments.

25.17.7 Stability and nonstationary motion

The study of the stability in the small, i.e., for small changes of the state of the system around the equilibrium conditions, may be performed by computing the eigenvalues of the dynamic matrix. The plot of the eigenvalues (their real and imaginary parts) as functions of the speed and the plot of the roots locus give a picture of the stability of the system that can be easily interpreted.

The eigenvalues of the system are four; two of these are usually complex conjugate showing an oscillatory behavior; the corresponding eigenvector shows that the motion of the trailer is primarily involved. These oscillations are usually lightly damped, and can become, mainly at high speed, self excited leading to a global instability of the vehicle.

Remark 25.19 *The presence of an eigenvalue with positive real part, and hence of an instability in the mathematical sense, is felt by the driver as a source of discomfort rather than an actual instability. If the values of both the imaginary and the real parts of the eigenvalue are small enough, i.e. if the frequency is low and the amplitude grows slowly, the driver is forced to introduce continuous steering corrections without actually recognizing the instability of the vehicle.*

The introduction of a damper at the trailer-tractor connection can solve this problem, while the use of steering axles on the trailer makes things worse. A steering axle, controlled so that the wheels steer in the direction opposite to those of the tractor with a magnitude proportional to angle θ, provides a restoring force to keep the trailer aligned with the tractor. The effect is similar to that of increasing the stiffness of a system: If the damping is not increased the underdamped character is magnified, while the natural frequency is also increased.

For the study of motion in nonstationary conditions, the considerations already seen for the insulated vehicle still hold. The more complicated nature of the equations of motion, however, compels us to resort to numerical integration in a larger number of cases.

Example 25.13 *Study the stability with locked controls of the articulated truck of Appendix E.9.*

The plot of the real and imaginary parts of s and the roots locus are shown in Fig. 25.36.

The figure has been obtained using the complete expressions of the derivatives of stability, but neglecting the effect of driving forces. At 100 km/h the eigenvalues are

$$-2.3364 \pm 1.5896i \; , \; -2.2698 \pm 3.4037i;$$

the corresponding eigenvectors are

$$
\left\{
\begin{array}{c}
-0.8723 \pm 0.4849i \\
0.0305 \pm 0.0424i \\
-0.0037 \mp 0.0346i \\
0.0058 \pm 0.0109i
\end{array}
\right\}
\; , \;
\left\{
\begin{array}{c}
-0.6448 \mp 0.6533i \\
-0.0521 \pm 0.0862i \\
-0.1322 \pm 0.3429i \\
0.0518 \mp 0.0734i
\end{array}
\right\}
\; .
$$

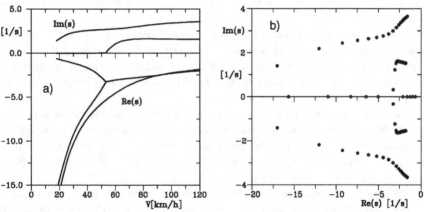

FIGURE 25.36. Example 25.13: Study of the stability. (a) Real and imaginary parts of s as functions of the speed. (b) Roots locus at varying speed. Complete expressions of the derivatives of stability, with the effect of driving forces neglected.

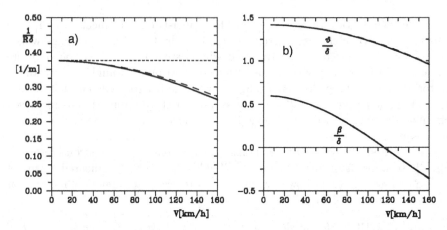

FIGURE 25.37. Example 25.14: Path curvature gain, sideslip angle gain and trailer angle gain as functions of speed. Full lines: Values obtained from the complete expressions of the derivatives of stability; dashed lines: Simplified approach obtained considering only the cornering forces.

The vehicle has a strong oscillatory behavior, even if both modes are well damped and no dynamic instability occurs; both modes involve the tractor as well as the trailer.

Example 25.14 *Study the directional response and the stability with locked controls of the car of Appendix E.5 with a caravan with a single axle. Assume the following data for the caravan: Mass $m_R = 600$ kg, moment of inertia $J_R = 800$ kg m^2, $c = 2.87$ m, $a_R = l_3 = 2.5$ m, $h_R = 1$ m, $S_R = 2.5$ m^2; $(C_{Y_R})_{,\beta} = -1.5$, $(C_{N_R})_{,\beta} = -0.6$. Assume that the trailer has the same tires used on the tractor.*

The path curvature gain, sideslip angle gain and trailer angle gain are plotted against the speed in Fig. 25.37. Both the complete and simplified expressions of the derivatives of stability have been used, while the effect of driving forces has been neglected.

Note that the curve obtained from the simplified expressions of the derivatives of stability is completely superimposed on that describing the behavior of the insulated vehicle, as was predictable since $a_R = l_3$. Note also that the path curvature gain tends to the kinematic value for a speed tending to zero, since the trailer has a single axle and correct kinematic steering is possible.

The plot of the real and imaginary parts of s and the roots locus are reported in Figs. 25.38a and b. Here only the complete expressions of the derivatives of stability have been used. The vehicle is stable, but the absolute value of the real part of the Laplace variable s is quite low at high speed, denoting a strong and little damped oscillatory motion, which occurs at low frequency.

To compare the behavior of the vehicle with and without trailer the computation has been repeated without the latter and the results are shown in Figures 25.38c and d.

The comparison shows that the modes affecting primarily the vehicle are fairly uncoupled from those primarily affecting the trailer, although a correct analysis of such coupling demands a through analysis of the eigenvectors.

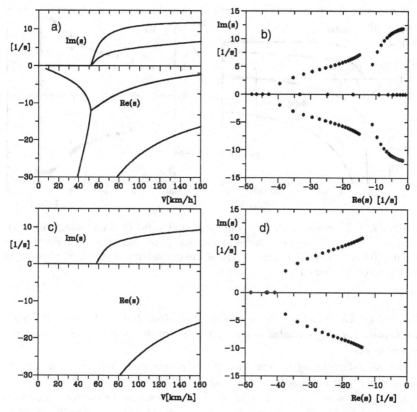

FIGURE 25.38. Example 25.14: Study of the stability. (a) Real and imaginary parts of s as functions of the speed. (b) Roots locus at varying speed. (c), (d): Same as (a), (b) but for the vehicle without trailer. Complete expressions of the derivatives of stability, with the effect of driving forces neglected.

The trailer mode with low frequency and low dynamic stability is superimposed on the more stable tractor mode, which is not strongly affected by the presence of the trailer. The motion of the tractor in the trailer mode can also be quite large, as this mode affects the whole system.

Example 25.15 *Study the stability with locked controls of the car of Appendix E.2 with the caravan of Example 25.14. Assume that the tires of the caravan are the same as those used on the tractor. Then study the motion with locked controls of the same vehicle following a step steering input at 80 and 140 km/h. Assume that the value of the steering angle is that needed to obtain a circular path with a radius of 200 m, computed neglecting the presence of the trailer.*

The plot of the real and imaginary parts of s and the roots locus computed using the complete expressions of the derivatives of stability are shown in Figures. 25.39a

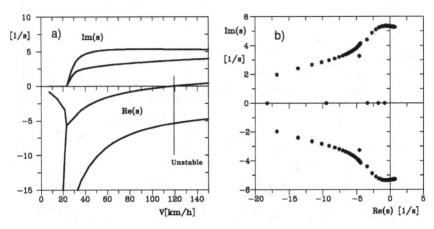

FIGURE 25.39. Example 25.15: Study of the stability. (a) Real and imaginary parts of s as functions of speed. (b) Roots locus at varying speed. Note the instability threshold at about 120 km/h. Complete expressions of the derivatives of stability, with the effect of driving forces neglected.

and b. The vehicle is stable only up to a speed of about 120 km/h, where the real part of the Laplace variable s related to one of the two modes vanishes, to become positive at higher speed.

The absolute value of the real part of s is always quite low, denoting a marginal dynamic stability at low speed and a marginal instability at higher speed.

This type of behavior is quite evident in the response to a step steering input. The steering angle needed to obtain a radius of the path of 200 m is $0.9659°$ at 80 km/h and $1.7271°$ at 140 km/h. The integration of the equation of motion was performed numerically. At 80 km/h the response is stable but the step input excites a strong, slowly damped, oscillatory behavior (Fig. 25.39a).

The strong oscillatory behavior is primarily due to the trailer, and the time history showing more pronounced oscillations is that of the trailer angle θ. After 6 s the values of $r/V\delta$, $\beta\delta$ and $\theta\delta$ are almost stabilized at the values of 0.3018, -0.4056 and 0.3098 that characterize the steady state behavior (the former two are almost the same as those obtained for the vehicle without trailer, except for a small difference due to the difference in aerodynamic drag, which influences the loads on the road and hence the cornering stiffness). The path is, however, not oscillatory.

At 140 km/h the vehicle is unstable and the oscillations of r, β and θ quickly diverge. The path reported in Fig. 25.40b, however, is not strongly oscillatory.

This example is a limiting case since the trailer is not correctly matched to the vehicle, nor are the tires correct for the trailer; it has been shown as an example of unstable behavior occurring in an incorrectly designed vehicle with trailer.

Note that a step input is prone to excite strongly an unstable behavior and is the worst thing to do with a marginally stable vehicle. The oscillations have a low frequency, and it is possible that the driver may be able to stabilize the vehicle even at speeds at which the real part of s is positive: A test driver would probably find the handling and

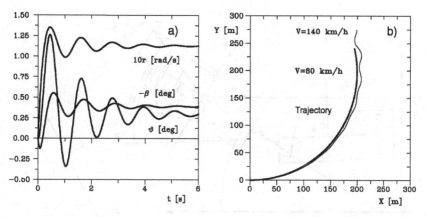

FIGURE 25.40. Example 25.15: Response to a step steering input. (a) Time histories of the yaw velocity, sideslip angle β and trailer angle θ at 80 km/h and (b) path at 80 and 140 km/h.

comfort of the vehicle poor rather than seeing the vehicle as unstable, owing to the need for continuous steering corrections.

On the other hand, it is possible that a vehicle with a low negative real part of s becomes unstable because of the action of the driver. Ultimately, the stability of the vehicle-driver system is what counts, but intrinsic stability of the vehicle is necessary, so that the driver is not forced to stabilize a system that is itself unstable.

25.18   MULTIBODY ARTICULATED VEHICLES

25.18.1   Equations of motion

Consider a vehicle with a trailer with two axles, one connected to its body, the other connected to the draw bar (Fig. 25.41a). Its dynamic behavior may be studied using the same kind of model seen in the previous section, where the trailer is modelled as two simple trailers connected in sequence. The model has five degrees of freedom, and the five generalized coordinates may be X Y, ψ, θ_1 and θ_2. The first two coordinates can be substituted by displacements x and y referred to the frame of the tractor and the first equation for longitudinal motion may be decoupled from the others, if the equations of motion are linearized. The transversal behavior can be studied using a set of four differential equations that can be linearized under the usual conditions, yielding a set of linear differential equations whose order is six.

This procedure can be generalized to a generic multibody vehicle made of a tractor and a set of n trailers (Fig. 25.41b). Note that while in Europe no vehicle with multiple trailers is legal for road use, in America and Australia such vehicles are legal but subject to restrictions (Fig. 25.42). The model here described, leading to a set of $n + 3$ differential equations ($n + 2$ for the lateral behavior if the first equation is uncoupled), allows one to study the behavior of any vehicle of this type.

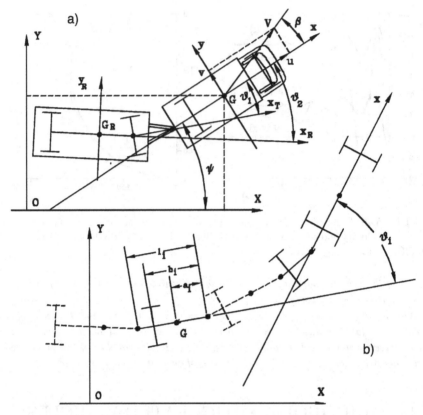

FIGURE 25.41. (a) Vehicle with a trailer with two axles. (b) model of a multibody articulated vehicle; parameters for the i-th trailer.

With reference to Fig. 25.41b, the position of the centre of mass of the ith trailer is

$$
(\overline{G_i - O}) = \left\{
\begin{array}{l}
X - c\cos(\psi) - \displaystyle\sum_{k=1}^{i-1} l_k \cos(\psi - \theta_k) - a_i \cos(\psi - \theta_i) \\[2mm]
Y - c\sin(\psi) - \displaystyle\sum_{k=1}^{i-1} l_k \sin(\psi - \theta_k) - a_i \sin(\psi - \theta_i)
\end{array}
\right\}.
$$

$$(25.231)$$

The velocity of point G_i is

$$
V_{G_i} = \left\{
\begin{array}{l}
\dot{X} + \dot{\psi}c\sin(\psi) + \displaystyle\sum_{k=1}^{i-1}\left(\dot{\psi}-\dot{\theta}_k\right) l_k \sin(\psi - \theta_k) + \left(\dot{\psi}-\dot{\theta}_i\right) a_i \sin(\psi - \theta_i) \\[2mm]
\dot{Y} - \dot{\psi}c\cos(\psi) - \displaystyle\sum_{k=1}^{i-1}\left(\dot{\psi}-\dot{\theta}_k\right) l_k \cos(\psi - \theta_k) - \left(\dot{\psi}-\dot{\theta}_i\right) a_i \cos(\psi - \theta_i)
\end{array}
\right\}.
$$

$$(25.232)$$

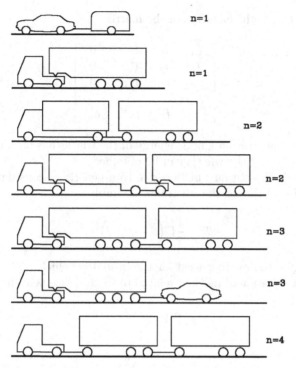

FIGURE 25.42. Examples of multibody vehicles; note that only the first three are road legal in Europe.

The contribution to the kinetic energy due to the ith trailer with mass m_i and moment of inertia J_i about a baricentric axis parallel to z-axis is then

$$T_i = \frac{1}{2}m_i V_i^2 + \frac{1}{2}J_i \left(\dot{\psi} - \dot{\theta}_i\right)^2 , \qquad (25.233)$$

i.e.

$$T_i = \frac{1}{2}m \left[\dot{x}'^2 + \dot{y}'^2 - 2\left(\dot{X}\alpha_i + \dot{Y}\beta_i\right)\cos(\psi) + \\ +2\left(\dot{X}\beta_i - \dot{Y}\alpha_i\right)\sin(\psi)\right] + \frac{1}{2}J_i \left(\dot{\psi} - \dot{\theta}_i\right)^2 , \qquad (25.234)$$

where

$$\alpha_i = \sum_{j=1}^{i} l_{ij}\left(\dot{\psi} - \dot{\theta}_j\right)\sin(\theta_j) ,$$

$$\beta_i = c\dot{\psi} + \sum_{j=1}^{i} l_{ij}\left(\dot{\psi} - \dot{\theta}_j\right)\cos(\theta_j),$$

and constants l_{ij} are the elements of the matrix

$$
1 = \begin{bmatrix}
a_1 & 0 & 0 & 0 \\
l_1 & a_2 & 0 & 0 \\
l_1 & l_2 & a_3 & 0 \\
\cdots & \cdots & \cdots & \cdots \\
l_1 & l_2 & l_3 & a_n
\end{bmatrix} .
$$

Here again the rotation kinetic energy of the wheels has been neglected and no gyroscopic effect of the wheels can be obtained.

The Raleigh dissipation function due to a generic viscous damper located between the $(i-1)$-th and the ith trailer is simply

$$
\mathcal{F} = \frac{1}{2}\Gamma \left(\dot\theta_i - \dot\theta_{i-1}\right)^2 . \tag{25.235}
$$

Operating in the manner used for the insulated vehicle and linearizing the result, the first equation of motion, related to the displacement in x direction, is

$$
m\dot V = Q_x , \tag{25.236}
$$

where

$$
m = m_T + \sum_{i=1}^{n} m_i
$$

is the total mass of the vehicle.

The second equation of motion, related to the displacement in the y direction, is

$$
m\left(\dot v + V\dot\psi\right) + \sum_{i=1}^{n}\left[-\ddot\psi\left(c + \sum_{j=1}^{i} l_{ij}\right) + \sum_{j=1}^{i} l_{ij}\ddot\theta_j\right] = Q_y . \tag{25.237}
$$

The third equation refers to the degree of freedom ψ

$$
\left\{J_T + \sum_{i=1}^{n}\left[m_i\left(S_i^2 + C_i^2\right) + J_i\right]\right\}\ddot\psi + \sum_{i=1}^{n} m_i\Bigg\{\left(-\dot u + v\dot\psi\right) S_i +
$$

$$
-\left(\dot v + u\dot\psi\right) C_i - S_i \sum_{j=1}^{i} l_{ij}\left[\ddot\theta_j \sin(\theta_j) - \dot\theta_j\left(\dot\psi - \dot\theta_j\right)^2 \cos(\theta_j)\right] +
$$

$$
-C_i \sum_{j=1}^{i} l_{ij}\left[\ddot\theta_j \cos(\theta_j) + \dot\theta_j\left(\dot\psi - \dot\theta_j\right)^2 \sin(\theta_j)\right] +
$$

$$
+\dot\psi S_i \sum_{j=1}^{i} l_{ij}\dot\theta_j \cos(\theta_j) - \left(\dot\psi C_i + v\right)\sum_{j=1}^{i} l_{ij}\dot\theta_j \sin(\theta_j)\Bigg\} = Q_\psi , \tag{25.238}
$$

where

$$S_i = \sum_{j=1}^{i} l_{ij} \sin(\theta_j) ,$$

$$C_i = c + \sum_{j=1}^{i} l_{ij} \cos(\theta_j) .$$

The third equation can also be linearized, yielding

$$\left\{ J_T + \sum_{i=1}^{n} \left[m_i C_i^2 + J_i \right] \right\} \ddot{\psi} + \tag{25.239}$$

$$\sum_{i=1}^{n} m_i \left[\dot{V} S_i - \left(\dot{v} + V\dot{\psi} \right) C_i - C_i \sum_{j=1}^{i} l_{ij} \ddot{\theta}_j \right] = Q_\psi ,$$

where

$$S_i = \sum_{j=1}^{i} l_{ij} \theta_j , \quad C_i = c + \sum_{j=1}^{i} l_{ij} .$$

The following n equations refer to the rotational generalized coordinates θ_j (for $j = 1, 2, \ldots, n$). The generic equation for θ_k, i.e. the $(3+k)$-th equation, is

$$\sum_{i=k}^{n} m_i l_{ik} \left\{ \sin(\theta_k) \left[\dot{u} - v\dot{\psi} + \dot{\psi}^2 C_i - \ddot{\psi} S_i - \sum_{j=1}^{i} l_{ij} \ddot{\theta}_j \sin(\theta_j) + \right. \right.$$

$$\left. -2\dot{\psi} \sum_{j=1}^{i} l_{ij} \dot{\theta}_j \cos(\theta_j) + \sum_{j=1}^{i} l_{ij} \dot{\theta}_j^2 \cos(\theta_j) \right] +$$

$$+ \cos(\theta_k) \left[\dot{v} + u\dot{\psi} - \ddot{\psi} C_i + -\dot{\psi}^2 S_i + \sum_{j=1}^{i} l_{ij} \ddot{\theta}_j \cos(\theta_j) + 2\dot{\psi} \sum_{j=1}^{i} l_{ij} \dot{\theta}_j \sin(\theta_j) + \right.$$

$$\left. \left. - \sum_{j=1}^{i} l_{ij} \dot{\theta}_j^2 \sin(\theta_j) \right] \right] \right\} + J_k \left(\ddot{\theta}_k - \ddot{\psi} \right) = Q_{\theta_k} .$$

$$\tag{25.240}$$

By linearizing also these equations, it follows that

$$J_k \left(\ddot{\theta}_k - \ddot{\psi} \right) + \sum_{i=k}^{n} m_i l_{ik} \left(\theta_k \dot{V} + \dot{v} + V\dot{\psi} - \ddot{\psi} C_i + \sum_{j=1}^{i} l_{ij} \ddot{\theta}_j \right) = Q_{\theta_k} . \tag{25.241}$$

The derivatives of the Raleigh dissipation function have not been included in the equations: The generalized forces due to the dampers, if they exist at all, will be included in the forces Q_{θ_k}.

25.18.2   Sideslip angles of the wheels and generalized forces

The sideslip angles of the wheels of the trailer can be computed as they were for the articulated vehicle. If the rth wheel of the ith trailer has a steering angle δ_{i_r}, using the monotrack vehicle model, the sideslip angle is

$$\alpha_{i_r} = \theta_i + \beta - \frac{\dot{\psi}}{V}\left(c + \sum_{j=1}^{i} l_{ij}^*\right) + \sum_{j=1}^{i} l_{ij}^* \frac{\dot{\theta}_i}{V} - \delta_{i_r} . \tag{25.242}$$

where l_{ij}^* are equal to l_{ij}, but defined using distance b_{i_r} of the axle instead of a_i.

The contributions to the generalized forces Q_x, Q_y and Q_ψ due to the tractor are the same as for the insulated vehicle. As usual, the tractor does not give any contribution to the forces Q_{θ_k}. To compute the contributions due to the rth wheel of the ith trailer and to the aerodynamic forces of the latter, it is possible to proceed as for the previous models, by writing their virtual work and then differentiating with respect to the virtual displacements.

The results obtained for the wheels are

$$\begin{cases} Q_{x_{i_r}} = F_{x_{i_{rt}}} + F_{y_{i_{rt}}}(\theta_i - \delta_{i_r}) \\ Q_{y_{i_r}} = -F_{x_{i_{rt}}}(\theta_i - \delta_{i_r}) + F_{y_{i_{rt}}} \\ Q_{\psi_{i_r}} = F_{x_{i_{rt}}}\left[c(\theta_i - \delta_{r_i}) + \sum_{j=1}^{i} l_{ij}^*(\theta_i - \theta_j - \delta_{r_i}) - y_{i_r}\right] + \\ \qquad\quad + F_{y_{i_{rt}}}\left[-c - \sum_{j=1}^{i} l_{ij}^* + y_{i_r}(\delta_{r_i})\right] + M_{z_{r_i}} \\ Q_{\theta_{k_{i_r}}} = F_{x_{i_{rt}}} l_{ik}^*(\theta_i - \theta_k - \delta_{r_i}) + F_{y_{i_{rt}}} l_{ik}^* \qquad\qquad \text{if } k < i \\ Q_{\theta_{k_{i_r}}} = F_{x_{i_{rt}}}[y_{i_r} + l_{ik}^*\delta_{r_i}] + F_{y_{i_{rt}}}[-y_{i_r}\delta_{r_i} + l_{ik}^*] - M_{z_{r_i}} \qquad \text{if } k = i \\ Q_{\theta_{k_{i_r}}} = 0 . \qquad\qquad\qquad\qquad\qquad\qquad\qquad\qquad\qquad \text{if } k > i \end{cases} \tag{25.243}$$

The generalized forces due to the aerodynamic forces and moments acting on the trailers can be accounted for in a similar way. Assuming that it is possible to distinguish between the forces acting on the various rigid bodies, the generalized forces can be computed immediately from equations (25.243), using l_{ij} instead of l_{ij}^*, setting δ_{i_r} to zero and using the aerodynamic forces and moments instead of the forces acting between road and wheels.

The generalized forces due to dampers located between the various bodies are

$$\begin{cases} Q_x = Q_y = 0 \\ Q_\psi = -\Gamma_1\dot{\psi} + \Gamma_1\dot{\theta}_1 \\ Q_{\theta_1} = \Gamma_1\dot{\psi} - (\Gamma_1 + \Gamma_2)\dot{\theta}_1 + \Gamma_2\dot{\theta}_2 \\ Q_{\theta_k} = \Gamma_k\dot{\theta}_{k-1} - (\Gamma_k + \Gamma_{k+1})\dot{\theta}_k + \Gamma_{k+1}\dot{\theta}_{k+1} \qquad k = 2, ..., n . \end{cases} \tag{25.244}$$

The external forces $F_{y_{e_i}}$ acting in the centres of mass of the trailers and the components of the weight $m_i g \sin(\alpha)$ are assumed to act in the directions of axes x and y of the tractor; the expressions of the generalized forces must therefore be modified accordingly.

The equations of motion are $n + 3$; together with the equations yielding the sideslip angles of the wheels, those expressing the forces and moments of the tires as functions of the sideslip angles, the load, and the other relevant parameters, they allow one to study the handling of the vehicle.

As was the case for all the previous models, the linearization of the equations allows one to uncouple the longitudinal behavior (first equation of motion) from the lateral behavior, which can be studied using the remaining $n + 2$ equations:

$$
\begin{cases}
m\left(\dot{v} + u\dot{\psi}\right) - \ddot{\psi}\sum_{i=1}^{n} m_i d_i + \sum_{i=1}^{n} \ddot{\theta}_i \left(\sum_{j=i}^{i} m_j l_{ji}\right) = \\
\qquad = (Q_y)_\beta \beta + (Q_y)_\psi \psi + \sum_{i=1}^{n}\left[(Q_y)_{\theta_i}\theta_i\right] + (Q_y)_{\dot{\theta}_i}\dot{\theta}_i + (Q_y)_\delta \delta + F_{y_e} + \sum_{i=1}^{n} F_{y_{e_i}} \\[2mm]
J'\ddot{\psi} - \sum_{i=1}^{n} J_i'\ddot{\theta}_i + \sum_{i=1}^{n} m_i \left\{ -\dot{V}\sum_{j=1}^{i} l_{ij}\theta_j - \left(\dot{v} + V\dot{\psi}\right) d_i \right\} = \\
\qquad = (Q_\psi)_\beta \beta + (Q_\psi)_\psi \psi + \sum_{i=1}^{n}\left[(Q_\psi)_{\theta_i}\theta_i + (Q_\psi)_{\dot{\theta}_i}\dot{\theta}_i\right] + (Q_\psi)_\delta \delta + \\
\qquad + M_{z_e} + \sum_{i=1}^{n} M_{z_{e_i}} - \sum_{i=1}^{n} F_{y_{e_i}} d_j \\[2mm]
\sum_{i=k}^{n} m_i l_{ik}\left(\theta_k \dot{V} + \dot{v} + V\dot{\psi} - d_i\ddot{\psi} + \sum_{j=1}^{i} l_{ij}\ddot{\theta}_j\right) = (Q_{\theta_k})_\beta \beta + (Q_{\theta_k})_\psi \psi + \\
\qquad + \sum_{i=1}^{n}\left[(Q_{\theta_k})_{\theta_i}\theta_i + (Q_{\theta_k})_{\dot{\theta}_i}\dot{\theta}_i\right] + (Q_{\theta_k})_\delta \delta - M_{z_{e_i}} + \sum_{i=1}^{k} F_{y_{e_i}} l_{ik} ,
\end{cases}
$$
$$\tag{25.245}$$

where

$$
d_i = \sum_{j=1}^{i} l_{ij}, \quad J' = J_T + \sum_{i=1}^{n}(J_i + m_i d_i^2)
$$

and

$$
J_i' = \sum_{j=1}^{i}(J_i + m_i d_j l_{ji}) .
$$

By linearizing the generalized forces Q_x, Q_y, Q_ψ and Q_{θ_k} as for the previous models, the derivatives of stability entering Eq. (25.245) are readily computed.

The set of $(n + 2)$ differential equations (25.245) is of the $(2n + 2)$-th order, since variables y and ψ appear in the equation only as first and second derivatives. The equation can thus be written in the state space in the form of a set of $2n + 2$ first order differential equations by introducing the state variables $v_{\theta_i} = \dot{\theta}_i$.

25.19 LIMITS OF LINEARIZED MODELS

Linearized models have some features that make them particularly useful. These are namely

- They allow us to simplify the equations of motion to obtain closed form solutions which, when simple enough, provide a general insight into the dynamic behavior of the vehicle, particularly in terms of the effect of changes to its parameters.

- The possibility of studying the stability with the usual methods of linear dynamics.

The disadvantages are also clear: Linearized models can be applied only within a limited range of sideslip angles and lateral acceleration, and used for trajectories whose radius is large with respect to the size of the vehicle. They can thus be applied with confidence to the conditions corresponding to normal vehicle use, while they fail for sport driving and above all for the motions involved in road accidents.

Another consideration for the models seen in the present chapter is that they are based on rigid body dynamics, with the presence of the suspensions neglected. This assumption is well suited to describe the behavior of a vehicle driven in a relaxed way: Although dependent on the stiffness of the suspensions, the roll and pitch angles under these conditions are very small and may be assumed to have little effect on the dynamic behavior.

It must, however, be stated that a linearization carried too far will lead to results contradicting experimental evidence.

If the cornering stiffness is assumed to be proportional to the load F_z acting on the wheel not only for the small load variations acting on each wheel but also for the differences of load between front and rear axle, in the case of a vehicle with two axles with equivalent tires it follows that

$$\frac{C_1}{C_2} = \frac{F_{z_1}}{F_{z_2}} = \frac{b}{a} \, . \tag{25.246}$$

If only the cornering forces of the tires are included in the formula for the neutral-steer point, it follows that this point always coincides with the centre of mass, leading to the conclusion, clearly incorrect, that all vehicles with four equivalent wheels are neutral-steer.

26

COMFORT PERFORMANCE

The definition of comfort in a motor vehicle is at once complex and subjective, changing not only with time (cars considered comfortable just twenty years ago are nowadays considered unsatisfactory) but also from user to user. The same user may change his appraisal depending on circumstances and his psycho-physical state. But comfort remains an increasingly important parameter in customer choice and strongly competitive factor among manufacturers.

This chapter will deal primarily with vibrational comfort, although it is difficult to separate it from acoustic comfort without entering into details linked more with the driveability and handling of the vehicle. Not just driving comfort, but vibrational and acoustic comfort as well (the latter deeply affects the conditions in which the driver operates), all have a strong impact on vehicle safety.

It is possible to distinguish between *vibrational and acoustic comfort* − linked with the vibration and noise produced inside vehicles by mechanical devices or on its surface by the air − and *ride comfort*, which is linked primarily with the ability of the tires and the suspensions to filter out vibration caused by motion on a road that is not perfectly smooth.

With this distinction in mind, SAE defines:

- *ride*, low frequency (up to 5 Hz) vibration of the vehicle body

- *shake*, vibration at intermediate frequency (between 5 and 25 Hz), at which some natural frequencies of subsystems of the vehicle occur

- *harshness*, high frequency vibration (between 25 and 100 Hz) of the structure and its components, felt primarily as noise

- *noise*, acoustic phenomena occurring between 100 Hz and 22 kHz, i.e. up to the threshold of human hearing.

26.1 INTERNAL EXCITATION

The sources of vibration on board a vehicle are essentially three: The wheels, the driveline and the engine. All contain rotating parts and, as a consequence, a first cause of dynamic excitation is imbalance. A rotor is perfectly balanced when its rotation axis coincides with one of its principal axes of inertia; however, this condition can only be met approximately and balancing tolerances must be stated for any rotating object[1]. As a consequence of the residual imbalance a rotating object exerts on its supports, a force whose frequency is equal to the rotational speed Ω and its amplitude is proportional to its square Ω^2. Because the engine, the driveline and the wheels rotate at different speeds, the excitations they cause are characterized by different frequencies.

Apart from the excitation due to imbalance, there are other effects that are peculiar to each element. Wheels may show geometrical and structural irregularities. The outer shape of the tires cannot be exactly circular and is characterized by a runout (eccentricity) having the same effect as mass imbalance, exciting vibrations with a frequency equal to the rotational speed, plus other harmonic components which excite higher harmonics. An ovalization of the shape excites a vibration with frequency equal to 2Ω, a triangular shape with frequency 3Ω, etc. The very presence of the tread excites higher frequencies, which are usually found in the acoustic range; to avoid a strong excitation with a period equal to the time of passage of the single tread element, the pattern of the tread is usually made irregular, with randomly spaced elements.

The same effect occurs for variations of stiffness; these induce dynamic forces with frequencies equal to the rotational speed and its multiples. As various harmonics are present in differing degrees in different tires, the spectrum of the dynamic force exerted by the tire on the unsprung mass depends upon each tire. As is common in the dynamics of machinery, such a typical spectrum is referred to as the mechanical signature of the tire.

When the wheel is called upon to exert longitudinal and transversal forces, the irregularities, both geometrical and structural, also introduce dynamic components in these directions. The tire-wheel assembly, however, is a complex mechanical element with given elastic and damping properties that can filter out some of the frequencies produced at the road-tire interface. High frequencies are

---

[1]G. Genta, *Vibration of structures and machines*, Springer, New York, 1995, G. Genta, *Dynamics of rotating systems*, Springer, New York, 2005.

primarily filtered out by the tire itself, before being further filtered by the suspension. These frequencies are felt onboard primarily as airborne vibration, i.e. noise.

The excitation due to driveline imbalance is usually transferred to the vehicle body through its soft mountings. The transmission is, however, made of flexible elements and, particularly at high frequency, these may have resonances. A long drive shaft has its own critical speeds, and in the case of a two-span shaft with a central joint (common in front-engine, rear-drive layouts), a critical speed, corresponding to a mode in which the two spans behave as rigid bodies on a compliant central support, is usually located within the working range. If the balancing of the central joint is poor, strong vibration occurs when crossing this critical speed.

When Hooke's joints are present, torque pulsations occurring when the input and output shaft are at an angle can be a major problem. In modern front wheel drive cars, the joints near the wheels are of the constant-speed type to avoid vibration, but care must be taken to design the driveline layout to avoid excitations from these joints.

The engine is a major source of vibration and noise caused by imbalance of rotating parts, inertial forces from reciprocating elements and time variations of the driving torque. The excitation due to imbalance of rotating parts, mostly the crankshaft, has the frequency of the engine speed Ω. To reduce it, the crankshaft must be balanced accurately. The reciprocating masses produce forcing functions with frequencies that are equal to Ω and its multiples, in particular 2Ω and 4Ω.

The components with frequency Ω interact with those due to imbalance and can be reduced by using counter-rotating shafts with eccentric masses. Their compensation depends on the architecture of the engine, and above all on the number of cylinders; they are particularly strong in single cylinder engines, such as those used on many motor cycles. The simplest way to partially compensate for them is to use a counterbalance slightly larger than that used to compensate for the imbalance of rotating masses (this technique is usually referred to as *overbalancing*). To reduce components with frequency 2Ω, it is possible to use shafts counter-rotating at a speed twice the speed of the crankshaft, a practice fairly common on the engines of luxury cars.

Torsional vibration of the engine is another important source of vibration. Torsional vibrations of the engine were traditionally regarded as having little effect on comfort, important only for the structural survival of the mechanical components of the engine, in particular the crankshaft. This is, however, increasingly unrealistic, and the excitation caused by torsional vibration is increasingly seen as important for vehicle comfort.

The reason for this is the increasing number and mass of the ancillary devices, such as larger generators for coping with the increasing electrical needs of the vehicle, air conditioning compressors, power steering pumps, etc., that are located on brackets and driven by belts. Torsional vibration from the engine can set the system made by accessories, their brackets, belts covers, etc. into vibration These vibrations are then transferred to both engine and structure, producing noise both inside the vehicle and outside, because these accessories are usually located close to the cooling air intakes.

The use of diesel engines makes things worse, because the more abrupt changes of pressure in the combustion chamber lead to strong high order harmonics in torsional vibration. All vehicular diesel engines, and nowadays also spark ignition engines, have torsional vibration dampers, of the viscous (on industrial vehicles) or elastomeric type (passenger vehicles), but they may be not enough. More complex dampers have been introduced, both for reducing vibration of the crankshaft and for insulating the accessories. Moreover, the geometry of the engine is such that it is impossible to distinguish, at least as a first approximation, between torsional, axial and flexural vibration of the crankshaft.

Vibrations linked to the thermodynamic cycle have a fundamental frequency which, in four-stroke engines, is equal to half the rotational speed but a large number of harmonics are usually present. Because a reciprocating engine usually has a number of torsional critical speeds, its dynamics is quite complicated. It has been the object of many studies and the subject of many books[2].

The harmonics whose order is equal to the number of cylinders and its multiples are usually referred to as major harmonics; these often are the most dangerous. In the case of a four-in-line engine the frequency of the lowest major harmonics is 2Ω, coinciding with one of the forcing functions due to reciprocating masses. A partial compensation is often performed by setting the shaft counterrotating at a speed 2Ω in unsymmetrical position.

The design of the engine suspension system is a complex issue. The elimination of the sources of vibration, e.g. using dampers on the crankshaft or counterbalance shafts spinning at twice the rotational speed, properly insulating the engine from the vehicle structure by using adequate soft mountings and dampers, and insulating the passenger compartment for noise, are all useful provisions for increasing ride comfort. The engine suspension should be soft, to insulate the vehicle from vibration due to the engine, but must be stiff enough to avoid large relative motion between engine and vehicle.

The engine suspension is subject to a constant load, the weight of the engine, and to variable loads, such as inertia forces due to reciprocating parts and the motion of the vehicle, and a torque equal and opposite to the engine torque. The latter changes rapidly from zero to its maximum value, but can also change its sign when the engine is used to brake the vehicle.

Engine and gearbox are often in one piece, and in this case the torque acting on the engine suspension is the torque at the output of the gearbox (3 − 5 times the engine torque). If the differential is also inside the gearbox, the torque is that at the output of the final reduction, which may be as large as 10 − 14 times the engine torque.

---

[2]See, for instance, W. Thompson, *Fundamental of automobile engines balancing*, Mech. Eng. Publ. Ltd., 1978.

The solution once universally accepted, based on three elastomeric supports, is nowadays often replaced by a solution based on two elastomeric supports plus a connecting rod, hinged at its ends by two elastomeric supports, that reacts to the driving torque

The engine suspension must be designed with the aim of reducing, as much as possible, the transmission of engine vibration to the vehicle, but also allowing for the fact that the stresses in the engine components are influenced by how the engine is attached to the vehicle. The transmission the commands to the engine and, in particular, to the gearbox, is important in reducing the transmission of vibration to the vehicle. Instead of using rigid rods to transfer commands to the gearbox, it is convenient to use flexible cables or even to avoid mechanical transmission of commands altogether (servo-controlled gearbox).

Together with the conventional solutions based on elastomeric supports, more advanced and even active solutions in which it is possible to change the relevant parameters are now used.

The engine suspension can be used as a kind of dynamic vibration absorber. The engine mass, the compliance and the damping of its support constitute a damped vibration absorber that can be tuned on the main wheel hop resonance, about 12 – 15 Hz, to control vertical shake vibration due to wheel excitation.

The contribution to overall noise due to aerodynamics can be large and often, as discussed in Chapter 21, has specific causes that may be different from those causing aerodynamic drag. Aerodynamic noise is primarily caused by vortices and detached flow on the front part of the vehicle, generally in the zone close to the windshield and the first strut (pillar A). The wake and aerodynamic field at the rear of the vehicle, important in causing aerodynamic drag, usually make a limited contribution to overall noise.

Active noise cancellation is a promising way to increase acoustic comfort. Already applied in aeronautics, the first automotive applications in the form of active engine mufflers and passenger compartment noise control are due to appear soon. With the introduction of active noise control, more advanced goals than pure noise suppression can be achieved. As an example, experience in the field of rail transportation has shown that complete noise suppression is not considered satisfactory by most passengers, as it decreases privacy by allowing others to listen to what people are saying. A completely noiseless machine may seem unnatural (in the field of domestic appliances there have been cases of dishwashers considered too quiet by their users), and may even be dangerous in some automotive applications. The ultimate goal may not be to suppress noise, but to achieve a noise that users find pleasant.

Similarly, absolute vibration suppression may be undesirable because vibration conveys useful information to the driver and can give warning symptoms of anomalies. Here again the goal seems more to supply a vibrational environment the user finds satisfactory than to completely suppress all vibrational input.

26.2 ROAD EXCITATION

Knowledge of the excitation due to motion on uneven road is important for the study of riding comfort. Road excitation reduces the ability of the tires to exert forces in the x and y direction, because it causes a variable normal load F_z, and increases the stressing of the structural elements. Because such excitation cannot be studied with a deterministic approach, the methods used for random vibrations must be applied.

A number of studies have been devoted to characterizing road profiles experimentally and interpreting the results statistically. From experimental measurements of the road profile (Fig. 26.1), a law $h(x)$ can be defined and its power spectral density obtained through harmonic analysis. Note that the profile is a function of space and not of time, and the frequency referred to space $\overline{\omega}$ is expressed in rad/m or cycles/m and not in rad/s or in Hz. The power spectral density $\overline{S}$ of law $h(x)$ is thus expressed in $\text{m}^2/(\text{rad/m})$ or in $\text{m}^2/(\text{cycles/m})$.

The law $\overline{S}(\overline{\omega})$ can be expressed by a straight line on a logarithmic plot, i.e. by the law

$$\overline{S} = c\overline{\omega}^{-n} , \tag{26.1}$$

where n is a nondimensional constant while the dimensions of c depend on n (if $n = 2$, c is expressed in $\text{m}^2(\text{cycles/m})$, for instance).

An old I.S.O. proposal[3] suggested $n = 2$ for road undulations, i.e. for disturbances with a wavelength greater than 6 m, and $n = 1.37$ for irregularities, with a wavelength smaller than the mentioned value. The proposal stated various values of c depending on the type of road.

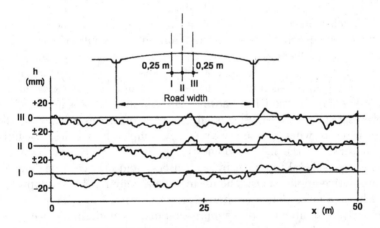

FIGURE 26.1. Examples of road profiles

---

[3]B.S.I. Proposal for Generalized Road Inputs to Vehicles, ISO/TC 108/WG9 Document 5, 1972.

A more recent approach is to abandon the distinction between undulations and irregularities. Often used values are

$$
\begin{array}{llll}
c = 4.7 \times 10^{-6} \ \text{m}^3 & n = 2.1 & \text{highway,} \\
c = 8.1 \times 10^{-7} \ \text{m}^3 & n = 2.1 & \text{road in poor conditions}
\end{array}
\tag{26.2}
$$

A[4] recent ISO proposal subdivides road profiles into 8 classes, indicated by letters from A to H, stating an exponent always equal to 2 for Eq. (26.1). The values of constant c for the various profiles are shown in Table 26.1. Classes from A to D are for hard-surfaced roads, with A for very smooth roads. Classes E and F are for natural surface roads or roads in bad conditions, such as a badly maintained pavé. G and H are for highly irregular surfaces. The power spectral density is defined in a frequency range from 0,01 to 10 cycles/m (wavelength from 100 m to 100 mm).

Some examples of power spectral density $\overline{S}$ for tarmac, concrete and pavé roads[5] are shown in Fig. 26.2 as functions of $\overline{\omega}$ together with the old ISO recommendation and ISO 8606:1995 standard.

If the vehicle travels with velocity V, it is possible to transform the law $h(x)$ into a law $h(t)$ and compute a frequency ω and a power spectral density S (measured in $\text{m}^2/(\text{rad/s})$ or m^2/Hz) referred to time from $\overline{\omega}$ and $\overline{S}$ defined with respect to space

$$
\begin{cases}
\omega = V\overline{\omega} \ , \\[2mm]
S = \dfrac{\overline{S}}{V} \ .
\end{cases}
\tag{26.3}
$$

The dependence of S from ω is thus

$$
S = cV^{n-1}\omega^{-n} \ .
\tag{26.4}
$$

TABLE 26.1. Minimum, average and maximum values of constant c for the various classes of road following ISO 8606:1995 standard.

Class	c_{min} (m^2cycles/m)	$c_{average}$ (m^2cycles/m)	c_{max} (m^2cycles/m)
A	–	1.6×10^{-7}	3.2×10^{-7}
B	3.2×10^{-7}	6.4×10^{-7}	1.28×10^{-6}
C	1.28×10^{-6}	2.56×10^{-6}	5.12×10^{-6}
D	5.12×10^{-6}	1.024×10^{-5}	2.048×10^{-5}
E	2.048×10^{-5}	4.096×10^{-5}	8.192×10^{-5}
F	8.192×10^{-5}	1.6384×10^{-4}	3.2768×10^{-4}
G	3.2768×10^{-4}	6.5536×10^{-4}	1.31072×10^{-3}
H	1.31072×10^{-3}	2.62144×10^{-3}	–

---

[4]ISO 8606:1995, Mechanical vibration - Road surface profiles - Reporting of measured data, 1/9/1995.

[5]G.H. Tidbury, *Advances in Automobile Engineering*, part III, Pergamon Press, Londra, 1965.

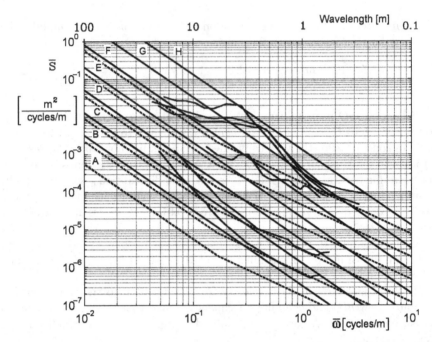

FIGURE 26.2. Power spectral density of some road profiles, ISO/TC 108/RS9 Document 5, 1972 recommendation (dashed lines) and ISO 8606:1995 standard (full lines)

Remark 26.1 *If $n = 2$, as is suggested by the most recent ISO standards, the power spectral density of the displacement is proportional to ω^{-2} and thus the power spectral density of the vertical velocity is constant: Road excitation is then equivalent to white noise in terms of vertical velocity of the contact point.*

The law $S(\omega)$ at various speeds for a road at the limit between the B and C classes (a fair but not very good road) following ISO standards is plotted in Fig. 26.3.

Once the power spectral density $S(\omega)$ of the excitation (namely of function $h(t)$) and the frequency response $H(\omega)$ of the vehicle are known, the power spectral density of the response $S_r(\omega)$ is easily computed as

$$S_r(\omega) = H^2(\omega)S(\omega) . \tag{26.5}$$

The root mean square (r.m.s.) value of the response is the square root of the power spectral density integrated in the relevant frequency range. If, for instance, the frequency response $H(\omega)$ is the ratio between the amplitude of the acceleration of the sprung mass and that of the displacement of the contact point, the response in terms of r.m.s. acceleration in the frequency range between ω_1 and ω_2 is

$$a_{rms} = \sqrt{\int_{\omega_1}^{\omega_2} S_r(\omega)\,d\omega} . \tag{26.6}$$

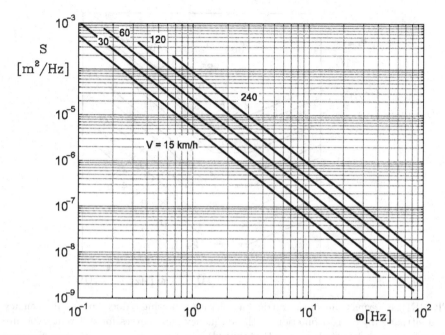

FIGURE 26.3. Power spectral density of the displacement $h(t)$ as a function of the frequency ω at various speeds for road at the border between the B and C classes following ISO standards

To summarize the quality of a road profile in a single figure an International Roughness Index (IRI), referring not to the road itself but to the way a given standard *quarter car model* (see below) reacts to it, was introduced. Because it refers to a particular model, the so-called quarter car with two degrees of freedom, it will be dealt with when we discuss suspension models.

26.3 EFFECTS OF VIBRATION ON THE HUMAN BODY

The ability of the human body to withstand vibration and related discomfort has been the object of countless studies and several standards on the subject have been stated. ISO 2631 standard (Fig. 26.4)[6], distinguishes between vibrations with a frequency in the range between 0,5 Hz and 80 Hz that may cause a reduction of comfort, fatigue, and health problems, and vibrations with a frequency in the range between 0,1 Hz and 0.5 Hz that may cause motion sickness.

---

[6]ISO Standards 2631, 1997, Mechanical vibration and shock - Evaluation of human exposure to whole-body vibration. The standards are older, but were revised in 1997.

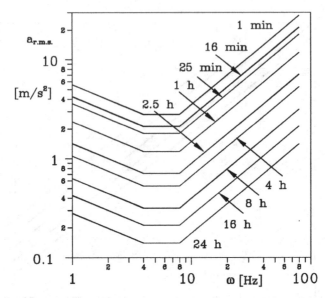

FIGURE 26.4. rms value of the vertical acceleration causing reduced physical efficiency to a sitting subject as a function of the frequency. The curves for different exposure times have been reported (ISO 2631 standard)

Standards refer to the acceleration due to vibration and suggest weighting functions of the frequency to compute the root mean square values of the acceleration. Such functions depend both on the point of the body where the acceleration is applied and the direction along which it acts.

Figure 26.4, shows the r.m.s. value of the acceleration causing, in a given time, a reduction of physical efficiency. The exposure limits can be obtained by multiplying the values reported in the figure by 2, while the "reduced comfort boundary" is obtained by dividing the same values by 3.15 (i.e., by decreasing the r.m.s. value by 10 dB). From the plot it is clear that the frequency range in which humans are more affected by vibration lies between 4 and 8 Hz.

As already stated, frequencies lower than 0.5 − 1 Hz produce sensations that may be associated with motion sickness. They depend on many parameters other than acceleration and vary among individuals. Between 1 and 4 Hz, the ability of humans to tolerate acceleration decreases with the frequency, reaching a minimum between 4 and 8 Hz. Between 8 and 80 Hz this tolerance increases again in a practically linear law with frequency. In practice, what creates discomfort in that range is not so much acceleration, but the ratio between acceleration and frequency.

Above 80 Hz the effect of vibration depends upon the part of the body involved, as local vibrations become the governing factor, making impossible to give general guidelines. There are also resonance fields at which some parts of the body vibrate with particularly large amplitudes. As an example, the thorax-abdomen system has a resonant frequency at about 3−6 Hz, although all resonant

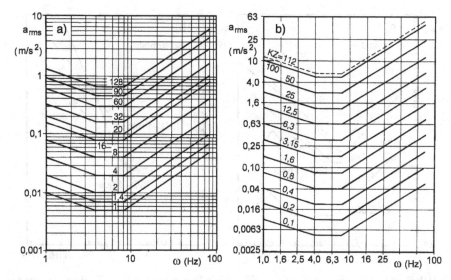

FIGURE 26.5. Curves of constant discomfort, following BSI 6472 (a) and VDI 2057 (b)

frequency values depend strongly upon individual characteristics. The head-neck-shoulder system has a resonant frequency at about $20 - 30$ Hz, and many other organs have more or less pronounced resonances at other frequencies (e.g., the eyeball at $60 - 90$ Hz, the lower jaw-skull system at $100 - 220$ Hz, etc.).

Essentially similar results are plotted in Fig. 26.5. The plots are related to equal discomfort lines following BSI 6472 (a) and VDI 2057 (b).

Other curves related to vertical and horizontal vibration from various sources and reported by M.W. Sayers, S.M. Karamihas, *The Little Book of Profiling*, The University of Michigan, 1998, are plotted in Fig. 26.6.

Remark 26.2 *The lower natural frequencies, those linked with the motion of the sprung mass, must be high enough to avoid motion sickness, but low enough to be well below 4 Hz. A common choice is to locate them in the range between 1,2 and 1,6 Hz. Higher frequencies, those due to the motion of unsprung masses, should be well above $8 - 10$ Hz. A good choice could be to locate them around $15 - 20$ Hz.*

26.4 QUARTER-CAR MODELS

The simplest model for studying the suspension motions of a vehicle is the so-called *quarter car model*, including a single wheel with related suspension and the part of the body whose weight is imposed on it. Often, in four wheeled vehicles, the quarter vehicle including the suspension and the wheels is called a *corner* of the vehicle. The quarter car model may be more or less complex, including not

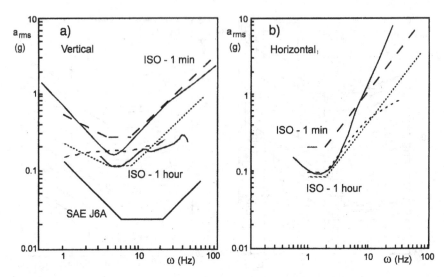

FIGURE 26.6. Comparison between discomfort limits from ISO and other sources for vertical (a) and horizontal (b) vibration

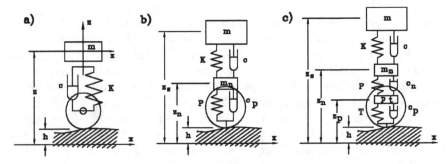

FIGURE 26.7. Quarter-car models with one (a), two (b) and three (c) degrees of freedom

only the compliance of the suspension, but the compliance of the tire and also that of the rubber mounts connecting the frame carrying the suspension to the body. It may even include the inertia of the tire.

Three models based on the quarter car approach are shown in Fig. 26.7.

The first model has a single degree of freedom. The tires are considered rigid bodies and the only mass considered is the sprung mass. This model holds well for motions taking place at low frequency, in the range of the natural frequency of the sprung mass (in most cases, up to $3 - 5$ Hz, in the range defined as *ride* by SAE).

The second model has two degrees of freedom. The tire is considered as a massless spring, and both the unsprung and the sprung masses are considered.

This model holds well for frequencies up to the natural frequency of the unsprung mass and slightly over (in most cases, up to $30 - 50$ Hz, including the ranges *ride*, *shake* and partially *harshness*).

The third model has three degrees of freedom. The tire is modelled as a spring-mass-damper system, representing its dynamic characteristics in the lowest mode. This model allows us to study motions taking place at frequencies in excess of the first natural frequency of the tires (up to $120 - 150$ Hz, including then *harshness*).

If higher frequencies must be accounted for, it is possible to introduce a higher number of tire modes by inserting other masses. These models, essentially based on the modal analysis of the suspension-tire system, are clearly approximated because a tire can be considered a damped system and one that is usually nonlinear.

26.4.1 Quarter-car with a single degree of freedom

Consider the simplest quarter-car model shown in Fig. 26.7a. It is a simple mass-spring-damper system that, among other things, has been used in the past to demonstrate that the shock absorber must be a linear, symmetrical viscous damper[7].

The equation of motion of the system is simple. Using the symbols shown in the figure it is

$$m\ddot{z} + c\dot{z} + Kz = c\dot{h} + Kh , \qquad (26.7)$$

where $z(t)$ is the displacement from the static equilibrium position, referred to an inertial frame, and $h(t)$ is the vertical displacement of the supporting point due to road irregularities[8].

The frequency response of the quarter car can be obtained simply by stating a harmonic input of the type

$$h = h_0 e^{i\omega t} .$$

The output is itself harmonic and can be expressed as

$$z = z_0 e^{i\omega t} ,$$

where both amplitudes h_0 and z_0 are complex numbers to account for the different phasing of response and excitation due to damping.

By introducing the time histories of the forcing function and the excitation into the equation of motion, an algebraic equation is obtained:

$$\left(-\omega^2 m + i\omega c + K\right) z_0 = (i\omega c + K) h_0 . \qquad (26.8)$$

---

[7]Bourcier De Carbon C.: *Théorie mathématiques et réalisation pratique de la suspension amortie des vehicules terrestres*, Proceedings SIA Conference, Paris, 1950.

[8]The z coordinate must be considered as the displacement from the static equilibrium position. By doing this, the static problem of finding the equilibrium position is separated from the dynamic problem here dealt with. This can only be done because of the linearity of the system.

It links the amplitude of the response to that of the excitation, and yields

$$z_0 = h_0 \frac{iwc + K}{-\omega^2 m + iwc + K} .$$

(26.9)

If h_0 is real (that is, if the equation is written in phase with the excitation), the real part of the response (the in phase component of the response) can be separated easily from its imaginary part (in quadrature component)

$$\begin{cases} \dfrac{\text{Re}(z_0)}{h_0} = \dfrac{K\left(K - m\omega^2\right) + c^2\omega^2}{\left(K - m\omega^2\right)^2 + c^2\omega^2} \\ \dfrac{\text{Im}(z_0)}{h_0} = \dfrac{-cm\omega^3}{\left(K - m\omega^2\right)^2 + c^2\omega^2} . \end{cases}$$

(26.10)

The amplification factor, i.e. the ratio between the absolute values of the amplitudes of the response and the excitation and the phase of the first with respect to the second, can be easily shown to be (Figures 26.8a and 26.8c)

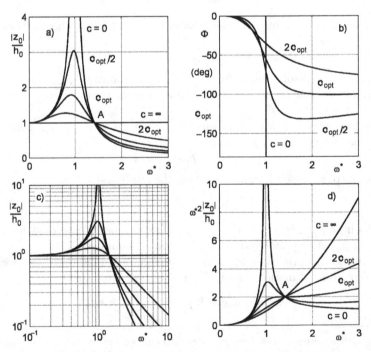

FIGURE 26.8. Quarter car with a single degree of freedom, response to harmonic excitation. Ratios between the amplitudes of the displacement (linear (a) and logarithmic (c) scales) and (d) the acceleration of the sprung mass, and the amplitude of the ground displacement and (b) phase, for different values of shock absorber damping. The responses are plotted as functions of the non-dimensional frequency $\omega^* = \omega\sqrt{m/K}$

$$\begin{cases} \dfrac{|z_0|}{|h_0|} = \sqrt{\dfrac{K^2 + c^2\omega^2}{(K - m\omega^2)^2 + c^2\omega^2}} \\ \Phi = \arctan\left(\dfrac{-cm\omega^3}{K(K - m\omega^2) + c^2\omega^2}\right) . \end{cases} \quad (26.11)$$

More than the frequency response expressing the ratio between the amplitudes of response and excitation, what matters in motor vehicle suspensions is the ratio between the amplitudes of the acceleration of the sprung mass and that of the displacement of the supporting point. Because in harmonic motion the amplitude of the acceleration is equal to the amplitude of the displacement multiplied by the square of the frequency, it follows that:

$$\frac{|(\ddot{z})_0|}{|h_0|} = \omega^2 \frac{|z_0|}{|h_0|} .$$

Both frequency responses are plotted in Fig. 26.8 for different values of the damping of the shock absorber, together with the phase Φ. The responses are plotted as functions of the nondimensional frequency

$$\omega^* = \omega\sqrt{\frac{m}{K}}$$

and damping is expressed as a function of the optimum damping defined below.

All curves pass through point A, located at a frequency equal to $\sqrt{2K/m}$. Because the acceleration of the sprung mass must be kept to a minimum to produce a comfortable ride, a reasonable way to optimize the suspension is to choose a value of shock absorber damping that leads to a relative maximum, or at least a stationary point, at point A on the curve related to acceleration. By differentiating the expression of

$$\omega^2 \frac{|z_0|}{|h_0|}$$

with respect to ω and equating the derivative to zero at point A, the following value of the optimum damping is obtained

$$c_{opt} = \sqrt{\frac{Km}{2}} = \frac{c_{cr}}{2\sqrt{2}} , \quad (26.12)$$

where

$$c_{cr} = 2\sqrt{Km}$$

is the critical damping of the suspension.

Although this method for optimizing the suspension can be readily criticized, because the comfort of a suspension is far more complex than simple reduction of the vertical acceleration (the so-called "jerk", i.e. the derivative of the acceleration with respect to time d^3z/dt^3 also plays an important role), it nonetheless gives important indications.

The dynamic component of the force the tire exerts on the ground is

$$F_z = c\left(\dot{z} - \dot{h}\right) + K(z - h) = -m\ddot{z} \ . \tag{26.13}$$

Remark 26.3 *Minimizing the vertical acceleration leads to minimizing the dynamic component of the vertical load on the tire, which has a negative influence on the ability of the tire to exert cornering forces. The condition leading to optimum comfort seems, then, to coincide with that leading to optimum handling performance.*

Equation (26.12) allows one to choose the value of the damping coefficient c. For the value of the stiffness K there is no such optimization: To minimize both the acceleration and the dynamic component of the force, K should be kept as low as possible. The only limit to the softness of the springs is the space available.

Remark 26.4 *This reasoning has, however, the following limitation: The softer the springs, the larger the oscillations of the sprung mass. Large displacements must be avoided because they may cause large errors in the working angles of the tire, causing the tires to work in conditions that may be far from optimal.*

An empirical rule states that soft suspensions improve comfort while hard suspensions improve handling. This is even more true if aerodynamic devices are used to produce negative lift: suspensions allowing a large degree of travel cause major changes of vehicle position with respect to the airflow, producing changes of the aerodynamic force that are detrimental.

Moreover, at a fixed value of the sprung mass, the lower the stiffness of the spring, the lower the natural frequency. Very soft suspensions easily lead to natural frequencies of about 1 Hz or even less, which may cause motion sickness and a reduction in comfort that varies from person to person. Cars with soft suspensions with large travel, typical of some manufacturers, are popular with some customers but considered uncomfortable by others.

The optimum value of the damping expressed by Eq. (26.12) is lower than the critical damping. The quarter car is then underdamped and may undergo free oscillations, even if these generally damp out quickly because the damping ratio $\zeta = c/c_{cr}$ is not very low:

$$\frac{c_{opt}}{c_{cr}} = \frac{1}{2\sqrt{2}} \approx 0,354 \ . \tag{26.14}$$

Example 26.1 *Consider a quarter car model with the following characteristics: sprung mass $m = 250$ kg; stiffness $K = 25$ kN/m; damping coefficient $c = 2,150$ Ns/m.*

Compute the natural frequency of the suspension and its frequency response. Assume that the vehicle travels at 30 m/s on a road that may be classified, following ISO standards, at the limit between class B and class C, and compute the power spectral density of the acceleration of the sprung mass and its root mean square value. Assess the performance of the quarter car in these conditions.

Frequency response. *The natural frequency is*

$$\omega = \sqrt{\frac{K}{m}} = 10 \ \ rad/s \ = 1.59 \ Hz$$

The value of the optimum damping is

$$c_{opt} = \sqrt{\frac{mK}{2}} = 1,770 \ \ Ns/m$$

and thus the suspension has a damping higher than the optimum value.

The dynamic compliance, that is the ratio between the displacement of the sprung mass and that of the supporting point,

$$|H(\omega)| = \frac{|z_0|}{|h_0|} = \sqrt{\frac{K^2 + c^2\omega^2}{(K - m\omega^2)^2 + c^2\omega^2}} \tag{26.15}$$

is plotted in Fig. 26.9, together with the inertance $\omega^2 H$, that is the ratio between the acceleration of the sprung mass and the displacement of the supporting point.

Response to road excitation. *The power spectral density is plotted in Fig. 26.2 in $m^2/(cycles/m)$ as a function of the frequency ω' in cycles/m. The equation of the line dividing zone B from zone C is*

$$S' = c\omega'^n \ ,$$

where the frequency with respect to space is expressed in cycles/m and the power spectral density in $m^2/(cycles/m)$; constants c and n are

$$c = 1.28 \times 10^{-6} \ m \ , \qquad n = -2$$

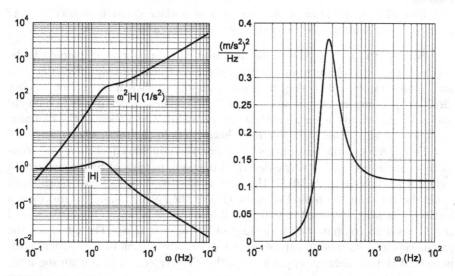

FIGURE 26.9. Dynamic compliance and inertance of the quarter car model with a single degree of freedom. Power spectral density of the acceleration due to motion on a road at the boundary between ISO classes B and C at a speed of 30 m/s

Computing the quantities referred to time from those referred to space at a speed of 30 m/s it follows that

$$S = c'\omega^n$$

where, expressing ω in Hz and S in m^2/Hz,

$$c' = \frac{c}{V^{n+1}} = 3.84 \times 10^{-5} \ m^3 \ s^{-2}$$

The power spectral density of the acceleration of the sprung mass can thus be immediately computed by multiplying the square of the inertance by the power spectral density of the road profile, obtaining the result reported in Fig. 26.2. The r.m.s. value of the acceleration can be computed by integrating the power spectral density. The limits of integration referred to the space frequency are 0,01 and 10 cycles/m. By referring them to time, the frequency range extends from 0,3 to 300 Hz, obtaining

$$a_{rms} = 5.84 \ m/s^2 \ = \ 0,60 \ g$$

This is a high value that causes reduced physical efficiency in less than 1 s at a frequency between 1 and 2 Hz, where the resonance of the sprung mass is located. This result should not surprise us, for the quarter car model has only one degree of freedom and no tire. From the power spectral density it is clear that the largest contribution to the integral is due to the range between 10 and 300 Hz, because the response is still quite high even with increasing frequency. The computation, performed by neglecting the ability of the tire to filter out the excitation at medium-high frequency, has little meaning.

What the example shows is that the suspension alone is unable to filter out road excitation, and that the presence of the tire is compulsory.

Because the quarter car model is linear, the damper was assumed to be acting both in the jounce and in the rebound stroke (double effect damper) and to be symmetrical (having the same damping coefficient for motion in both directions). Dampers used in early automotive suspensions acted only in rebound and are today double effect, but they are not symmetrical because the damping coefficient in the jounce stroke is much lower than in the rebound stroke.

To understand the advantages of symmetrical double effect dampers[9], consider a quarter car with a single degree of freedom, passing at high speed over a bump or pothole (Fig. 26.10). If the time needed to cross the road irregularity is far shorter than the period of oscillation of the sprung mass (for instance, a bump 0.3 m long is crossed in 0.01 s at a speed of 30 m/s), an impulsive model can be used to compute the trajectory of the sprung mass. The effects of the perturbation to its motion can be considered as a variation of the vertical component of the momentum applied instantly. The trajectory of the sprung mass

---

[9]Bourcier De Carbon C.: *Théorie mathématiques et r éalisation pratique de la suspension amortie des vehicules terrestres*, Proceedings SIA Conference, Paris, 1950.

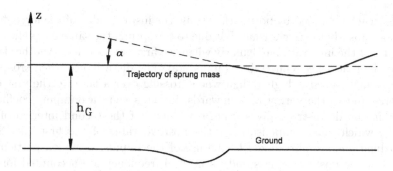

FIGURE 26.10. Quarter car with a rigid wheel crossing a ditch

deviates by an angle α:

$$\tan(\alpha) = \frac{w}{u} \,,$$

where w is the vertical component of the velocity. It can be computed using the momentum theorem

$$mw = \int_{t_1}^{t_2} F dt \,. \qquad (26.16)$$

The integral in Eq. (26.16) is the impulse of the forces due to the spring and to the damper from time t_1, when the wheel enters the ditch, to time t_2 when it gets out

$$w = \frac{1}{m} \left(\int_{t_1}^{t_2} F_m dt + \int_{t_1}^{t_2} F_a dt \right) \,. \qquad (26.17)$$

The force due to the spring is the part exceeding the static value compensating for the weight. If the integrals on the right hand side of Eq. (26.17) vanish, the suspension completely absorbs the irregularity, without any perturbation being transmitted to the sprung mass. The first of the two integrals is assumed to be far smaller than the second, because in a small amplitude, high frequency disturbance the force due to the spring, which is proportional to the displacement, is negligible compared to the force due to the shock absorber, which is proportional to the velocity. By neglecting the first integral and assuming that the damper is symmetrical, the expression for the vertical velocity becomes

$$w = \frac{1}{m} \int_{t_1}^{t_2} -c\dot{h} dt = -\frac{c}{m}(h_2 - h_1) \,, \qquad (26.18)$$

where $h(x)$ is the law expressing the road profile.

From Eq. (26.18) it is clear that if $h_2 = h_1$ the suspension is able to insulate the sprung mass perfectly from the road irregularity, a result that is due to the fact that the damping coefficient in rebound is equal to that in jounce.

This result, however, is compromised by the oversimplification of the model. It is well known that, while the shock absorber must act in both the up- and the down-stroke, the damping coefficients must be unequal for best performance.

This is easily explained by noting that while the instant value of the force due to the shock absorber is larger than that due to the spring, the same inequality does not hold for the integrals, particularly when the first one vanishes. Another factor is that the road-wheel constraint is unilateral. The disturbance when crossing a bump at high speed is higher than when crossing over a hole. In the first case, the force due to the spring acts upwards; in the second a damping coefficient higher in the downstroke gives a negative value of the second integral of Eq. (26.17), which may compensate for the positive value of the first one. Some approximations are also linked to the use of the impulsive model, particularly because if the unsprung mass and its natural frequency are accounted for, the time needed to cross the obstacle is no longer much smaller than the period of the free oscillations of the system.

Example 26.2 *Consider the quarter car with a single degree of freedom studied in Example 26.1, crossing at a speed V=30 m/s over an obstacle with harmonic profile similar to the usual obstacles (Fig.26.11c). Let the profile be*

$$h = h_0 \sin\left[\pi \frac{x - x_1}{x_2 - x_1}\right] . \tag{26.19}$$

with $h_0 = 100$ mm and a length $(x_2 - x_1)$ of 300 mm. Because

$$x = Vt , \tag{26.20}$$

the vertical velocity is

$$\dot{h} = \frac{\pi V h_0}{x_2 - x_1} \cos\left[\pi \frac{x - x_1}{x_2 - x_1}\right] . \tag{26.21}$$

Eq. (26.7) was numerically integrated, with the results shown in Fig.26.11a in terms of displacements and in Fig. 26.11b in terms of velocity. The quarter car with

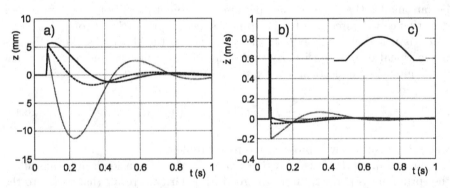

FIGURE 26.11. Response in terms of displacement (a) and velocity (b) of a single degree of freedom quarter car model crossing at 30 m/s a bump 300 mm wide and 100 mm high (c). Full line: symmetric damper; dashed line: the contribution of the spring to the impulse has been neglected; dotted line: asymmetric damper

symmetric damper (full line curve) succeeds well at filtering out the obstacle, with a maximum displacement of 6 mm. (6% of the displacement of the supporting point). The plot of the velocity shows that the negative impulse in the second part of the obstacle practically balances the positive impulse in the first one.

The dashed line was computed by neglecting the contribution of the spring to the impulse.

The dotted line was computed assuming that the damper has a coefficient in the jounce stroke equal to 80% of that in the rebound stroke. It is clear that the best performance are obtained with a symmetric damper. This consideration is, however, dependent on the extreme simplification of the model. Among other simplifications, the mono-lateral nature of the road-wheel contact has been neglected.

26.4.2 Quarter-car with two degrees of freedom

The following model is that shown in Fig. 26.7b. It is well suited for the study of the behavior of vehicle suspensions in a frequency range beyond the natural frequency of the unsprung mass.

With reference to Fig. 26.7b, the equation of motion of the model is

$$
\begin{bmatrix} m_s & 0 \\ 0 & m_u \end{bmatrix} \begin{Bmatrix} \ddot{z}_s \\ \ddot{z}_u \end{Bmatrix} + \begin{bmatrix} c & -c \\ -c & c + c_p \end{bmatrix} \begin{Bmatrix} \dot{z}_s \\ \dot{z}_u \end{Bmatrix} +
$$
$$
+ \begin{bmatrix} K & -K \\ -K & K + P \end{bmatrix} \begin{Bmatrix} z_s \\ z_u \end{Bmatrix} = \begin{Bmatrix} 0 \\ c_p \dot{h} + Ph \end{Bmatrix} ,
$$

(26.22)

where z_s and z_u are the displacements from the static equilibrium position and are referred to an inertial frame.

The response to a harmonic excitation $h(t)$ is readily obtained in the same way used for the previous model and yields a harmonic oscillation, not in phase with the excitation. The relationship linking the complex amplitudes of the response and the excitation is

$$
\left\{ -\omega^2 \begin{bmatrix} m_s & 0 \\ 0 & m_u \end{bmatrix} + i\omega \begin{bmatrix} c & -c \\ -c & c + c_p \end{bmatrix} + \right.
$$
$$
\left. + \begin{bmatrix} K & -K \\ -K & K + P \end{bmatrix} \right\} \begin{Bmatrix} z_{s0} \\ z_{u0} \end{Bmatrix} = h_0 \begin{Bmatrix} 0 \\ i\omega c_p + P \end{Bmatrix} ,
$$

(26.23)

By neglecting the damping of the tire c_p, which is usually small, the amplification factors of the sprung and unsprung masses are

$$
\begin{cases} \dfrac{|z_{s0}|}{|h_0|} = P\sqrt{\dfrac{k^2 + c^2\omega^2}{f^2(\omega) + c^2\omega^2 g^2(\omega)}} \\[3ex] \dfrac{|z_{u0}|}{|h_0|} = P\sqrt{\dfrac{(k - m\omega^2)^2 + c^2\omega^2}{f^2(\omega) + c^2\omega^2 g^2(\omega)}} , \end{cases}
$$

(26.24)

where

$$
\begin{cases}
f\left(\omega\right) = m_s m_u \omega^4 - \left[Pm_s + K(m_s + m_u)\right]\omega^2 + KP \\
\\
g\left(\omega\right) = (m_s + m_u)\omega^2 - P\,.
\end{cases}
$$

The dynamic component of the force exerted by the tire on the ground in the z direction may be easily computed in a similar way. The force in the z direction is

$$
F_z = -P\left(z_u - h\right) \tag{26.25}
$$

and thus

$$
|F_{z_0}| = P\left(|z_u - h|\right)\,. \tag{26.26}
$$

The modulus of $z_u - h$ is not coincident with the difference between the modulus of z_u and that of h because the two time histories are out of phase with each other. By performing the relevant computations, it follows that

$$
\frac{|F_{z_0}|}{|h_0|} = P\omega^2 \sqrt{\frac{\left[K(m_s + m_u) - m_s m_u \omega^2\right]^2 + c^2(m_s + m_u)\omega^2}{f^2\left(\omega\right) + c^2\omega^2 g^2\left(\omega\right)}}\,. \tag{26.27}
$$

The frequency responses related to both the sprung and the unsprung masses are plotted in Fig. 26.12a and b for a system with $P = 4K$ and $m_s = 10m_u$. The plots, shown using the non-dimensional frequency

$$
\omega^* = \omega\sqrt{\frac{m}{K}}\,, \tag{26.28}
$$

include curves obtained with different values of damping c; all curves lie in the non-shaded region of the graph.

If $c = 0$ the natural frequencies are two and the peaks are infinitely high. Also for $c \to \infty$ the peak, corresponding to the natural frequency of the whole system, which is now rigid, over the spring simulating the tire, goes to infinity.

The frequency responses of Fig. 26.12a and b multiplied by ω^{*^2} are shown in Fig. 26.12c and d; they give the non-dimensional ratio between the accelerations of the two masses and the displacement of the supporting point (suitably made non-dimensional). All curves pass through points O, A, B and C. Between O and A and between B and C the maximum acceleration of the sprung mass increases with decreasing damping, while between A and B and from C upwards it increases with damping

An optimum value of damping can be found by trying to keep the acceleration as low as possible in a large range extending up to the natural frequency of the unsprung mass, i.e. by looking for a curve having a relative maximum or a stationary point in A. Operating as seen with the previous model, the following value is obtained

$$
c_{opt} = \sqrt{\frac{Km}{2}}\sqrt{\frac{P + 2K}{P}}\,. \tag{26.29}
$$

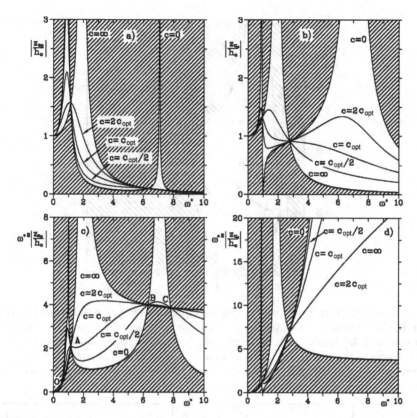

FIGURE 26.12. Quarter car with two degrees of freedom, response to harmonic excitation. Ratios between the amplitudes of the displacements of the sprung and the unsprung masses (a, b) and of the accelerations (c, d) to the amplitude of the displacement of the ground, for different values of the damping of the shock absorber. The responses are plotted as functions of the nondimensional frequency $\omega^* = \omega\sqrt{m/K}$

Because P is much larger than K, the value of $\sqrt{(P+2K)/P}$ is close to unity (in the case of Fig. 26.12 $\sqrt{(P+2K)/P} = 1.22$) and the optimum damping is only slightly larger than that computed for the model with a single degree of freedom (Eq. (26.12)). From Fig. 26.12c it is clear that this value of damping is effective in keeping the acceleration low in a wide frequency range.

The amplitude of the dynamic component of force F_z (Eq. (26.29)) is plotted in non-dimensional form (divided by $P|h_0|$) as a function of the nondimensional frequency in Fig. 26.13. The value of the optimum damping expressed by Eq. (26.29) is also effective in keeping the maximum value of the dynamic component of force F_z as low as possible, at least at low frequencies. At higher frequencies, a slightly higher value of damping could be effective, even if it would result in a larger acceleration of the sprung mass.

The maximum value of the non-dimensional amplitude of force F_z has been plotted as a function of ratio c/c_{opt} in Fig. 26.14a. When the damping goes

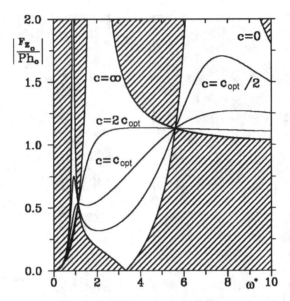

FIGURE 26.13. Quarter car with two degrees of freedom, response to harmonic excitation. Ratio between the amplitude of the dynamic component of force F_z between tire and road and the displacement of the ground, made non-dimensional by dividing it by the stiffness of the tire P, for different values of the damping of the shock absorber. The response is plotted as a function of the non-dimensional frequency $\omega^* = \omega\sqrt{m/K}$

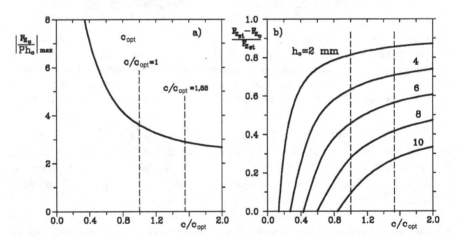

FIGURE 26.14. (a) Maximum value of the amplitude of the dynamic component of force F_z in a frequency range between 0 and 30 $\sqrt{K/m}$ as a function of ratio c/c_{opt}. Same characteristics as the system studied in the previous figures. (b) Minimum value of the ground force (static force minus amplitude of the dynamic component) as a function of ratio c/c_{opt} for a quarter car model with parameters typical for a small car: $m_s = 238$ kg; $m_u = 38$kg; $K = 15.7$ kN/m; $P = 135$ kN/m; actual value of c/c_{opt} equal to 1.53

beyond the optimum value computed above, a certain decrease of the maximum amplitude of the force at high frequency is clearly obtained.

The minimum value of the force on the ground (computed as the static component minus the amplitude of the dynamic component) has been plotted as a function of ratio c/c_{opt} in Fig. 26.14b, using data similar to those related to the front suspension of a small car. The curves refer to different amplitudes of the excitation h_0. If the damping is small enough the wheel can bounce on the road. Clearly, when this occurs the present linear model is no longer applicable.

Example 26.3 *Repeat the computations of example 26.1 using a quarter car model with two degrees of freedom. To the data already considered (sprung mass $m_s = 250$ kg; stiffness of the spring $K = 25$ kN/m; damping coefficient of the damper $c = 2,150$ Ns/m.) add the following: unsprung mass $m_u = 25$ kg; stiffness of the tire $k_t = 100$ kN/m.*

Compute the r.m.s. value of the acceleration as a function of the speed.

Natural frequencies. The optimum damping is

$$c_{opt} = 2,170 \ Ns/m,$$

a value very close to the actual one.

The characteristic equation allowing the natural frequencies of the undamped system to be computed is

$$\omega^4 - 5.100\omega^2 + 400.000 = 0.$$

The values of the natural frequencies are then

$$\begin{cases} \omega_1 = 8.93 \ rad/s = 1.421 \ Hz \\ \omega_2 = 70.85 \ rad/s = 11.28 \ Hz. \end{cases}$$

Frequency response. The dynamic compliance $H(\omega)$ and the inertance are plotted in Fig. 26.15.

Response to road excitation. The power spectral density can be computed in a way similar to what seen in the previous example. The result is plotted in Fig. 26.15.

The r.m.s. value of the acceleration is

$$a_{rms} = 1.34 \ m/s^2 = 0.136 \ g.$$

By comparing the results of the two examples it is clear that the presence of the tire is effective in filtering out high frequency disturbances, while the transmissibility at the natural frequency is slightly higher. At any rate, the r.m.s. value of the acceleration is now acceptable, even if it is not optimal. From the plot of Fig. 26.4, it is clear that this value causes reduced physical efficiency in 3 hours; however, the driver and the passengers are also insulated from vibration from the road by the seats.

Considering the integral between a wavelength of 0,1 and 100 m is practically equivalent to considering the whole spectrum from 0 to infinity, because the power spectral density vanishes outside this range.

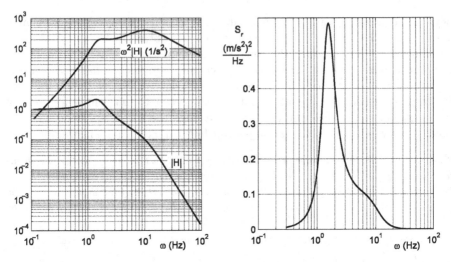

FIGURE 26.15. Dynamic compliance and inertance of the quarter car model with two degrees of freedom. Power spectral density of the acceleration due to motion on a road at a speed of 30 m/s

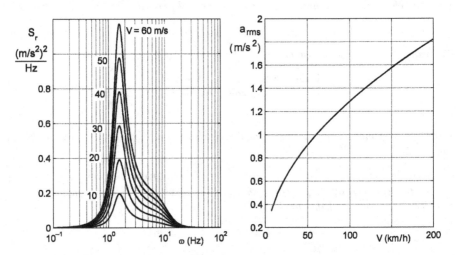

FIGURE 26.16. Power spectral density of the acceleration of the sprung mass (a) and its r.m.s. value (b) at different values of the speed of the vehicle

The computation has been repeated for various values of the speed and the results are shown in Fig. 26.16. The r.m.s. value of the acceleration grows with increasing speed.

From the considerations above it is possible to draw the conclusion that the value of the damping coefficient expressed in Eq. (26.29) is optimal both from the viewpoint of comfort and handling, because it leads to low variations in the forces on the ground. A slightly higher damping may, however, somewhat improve handling because it slightly reduces the variable component of the force on the ground.

This conclusion, obtained from a highly simplified model, is not in good accordance with experimental evidence stating that the value of damping for optimizing riding comfort is lower than that for optimizing handling.

To optimize the value of the damping coefficient of the suspension it is possible to consider motion on a road profile of the type defined by ISO 8606:1995 standard and expressed by Eq. (26.1) with $n = 2$. An excitation of this type may be considered, as already stated, as a white noise in terms of velocity, defined in the frequency range between 0,01 and 10 cycles/m.

The power spectral density of the vertical displacement of the contact point with the ground can be expressed as

$$S = cV\omega^{-2} \ . \tag{26.30}$$

In S.I. units (rad/s), the value of coefficient c is that reported in Table 26.1 multiplied by 2π. The frequency range in which the spectrum is defined is between frequencies ω_1 and ω_2, where:

$$\omega_1 = 0,01 * 2\pi V \ , \qquad \omega_2 = 10 * 2\pi V \ . \tag{26.31}$$

The r.m.s. value of the vertical acceleration of the sprung mass is

$$a_{rms} = \sqrt{\int_{\omega_1}^{\omega_2} \omega^4 H^2 S \ d\omega} = \sqrt{cV}\sqrt{\int_{\omega_1}^{\omega_2} \omega^2 H^2 \ d\omega} \ , \tag{26.32}$$

where H is the frequency response yielding the displacement of the sprung mass.

In a similar way the r.m.s. value of the variable component of the vertical road-tire force on the ground is

$$F_{z_{rms}} = \sqrt{\int_{\omega_1}^{\omega_2} H_F^2 S \ d\omega} = \sqrt{cV}\sqrt{\int_{\omega_1}^{\omega_2} \frac{H_F^2}{\omega^2} \ d\omega} \ , \tag{26.33}$$

where H_F is the frequency response yielding the variable component of the force.

The r.m.s. values of the acceleration and of the dynamic component of force F_z are easily computed for different values of the damping coefficient of the shock absorber. The plot of the former versus the latter allows some interesting conclusions to be drawn on the choice of the value of the damping (Fig. 26.17).

It must be remembered that ratios $a_{rms}/\sqrt{cV}$ and $F_{z_{rms}}/\sqrt{cV}$ are independent from c, that is from the characteristics of the road, but not completely independent from the speed. Actually they would be so if the integrals where computed in a frequency range from 0 to infinity, but the speed is included in the integration limits here defined.

Remark 26.5 *The conditions leading to optimum comfort (in the sense of minimum acceleration) and to optimum handling (in the sense of minimum force*

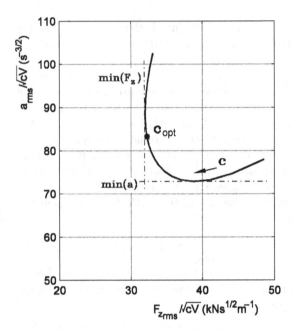

FIGURE 26.17. Ratio $a_{rms}/\sqrt{cV}$ versus $F_{z_{rms}}/\sqrt{cV}$ for the quarter car model with two degrees of freedom with the same data as in Fig. 26.14b ($m_s = 238$ kg; $m_u = 38$kg; $K = 15.7$ kN/m; $P = 135$ kN/m). Computation referred to a speed of 30 m/s

variations) are readily identified: The first is obtained with a damping lower than the optimum damping defined above, while the second for a damping value that is higher. This result is in better accordance with experimental results than the previous one.

As already stated, the presence of the tires has a negligible effect on the frequency response at low frequency, while at higher frequencies their stiffness must be accounted for. A comparison between the results obtained using the quarter car models with one and two degrees of freedom is shown in Fig. 26.18.

26.4.3 International Roughness Index

As already stated, the International Roughness Index (IRI) is defined with reference to a particular quarter car with two degrees of freedom moving at a specified speed. The data of the quarter car, often defined as *golden car*, are:

$$\frac{K}{m_s} = 63,3 \ \text{s}^{-2} \ , \quad \frac{P}{m_s} = 653 \ \text{s}^{-2} \ , \quad \frac{m_u}{m_s} = 0,15 \ , \quad \frac{c}{m_s} = 6 \ \text{s}^{-1} \ .$$

The value of the optimum damping is

$$\frac{c_{opt}}{m_s} = 6,147 \ \text{s}^{-1}$$

and thus the model has a damping that is close to optimal.

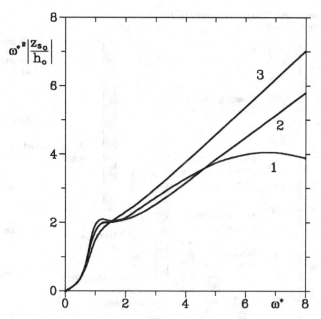

FIGURE 26.18. Acceleration of the sprung mass as a function of the frequency for a unit displacement input. Comparison between the quarter car model with one and two degrees of freedom (in the latter case $P = 4K$, $m_s = 10m_u$). 1): 2 degrees of freedom; 2): 1 degree of freedom, damping defined by Eq. (26.29); 3): 1 degree of freedom, damping defined by Eq. (26.12).

The reference speed is 80 km/h.

To define the Roughness Index of a given road profile, the motion of the quarter car is simulated and the cumulative travel of the sprung mass with respect to the unsprung mass is calculated over time. The index is the total value of the travel divided by the distance travelled by the vehicle

$$\text{IRI} = \frac{1}{VT} \int_0^T |\dot{z}_s - \dot{z}_s| \, dt \, . \tag{26.34}$$

The index so defined is a non-dimensional quantity, but one that is often measured in non-consistent units, [m/km] or [in/mi]. A correlation between the road characteristics and the roughness index is shown in Fig. 26.19.

Remark 26.6 *The Roughness Index may also be interpreted as the average value of the absolute value of the relative speed of the two masses divided by the vehicle speed.*

The use of an index to define the quality of the road surface, one based on the ratio between the total motion of the suspension and the distance travelled, dates back to the 1940s. It is used at present by many international organizations. Since 1982 the World Bank has used it to compare of road conditions in various

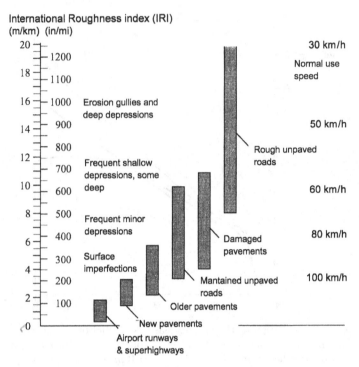

FIGURE 26.19. Correlation between the road characteristics and the roughness index

Countries. It has been shown that a good correlation exists between the index and both the vertical acceleration and the variation of the force on the ground; this property allows the comfort and the performance on a given road to be understood.

26.4.4   Quarter car with secondary suspension (three degrees of freedom)

In many vehicles the suspensions are not assembled directly to the body, but are mounted on a secondary frame, one that often carries other elements with a non-negligible mass as well. This auxiliary frame is connected to the chassis with a secondary suspension made by elastomeric mounts. A quarter car of this type is shown in Fig. 26.7c).

Its equation of motion is

$$
\begin{bmatrix} m_s & 0 & 0 \\ 0 & m_t & 0 \\ 0 & 0 & m_u \end{bmatrix} \begin{Bmatrix} \ddot{z}_s \\ \ddot{z}_t \\ \ddot{z}_u \end{Bmatrix} + \begin{bmatrix} c & -c & 0 \\ -c & c+c_t & -c_t \\ 0 & -c_t & c_t+c_p \end{bmatrix} \begin{Bmatrix} \dot{z}_s \\ \dot{z}_t \\ \dot{z}_u \end{Bmatrix} + \quad (26.35)
$$

$$+ \begin{bmatrix} K & -K & 0 \\ -K & K + K_t & -K_t \\ 0 & -K_t & K_t + P \end{bmatrix} \begin{Bmatrix} z_s \\ z_t \\ z_u \end{Bmatrix} = \begin{Bmatrix} 0 \\ 0 \\ c_p \dot{h} + Ph \end{Bmatrix}.$$

The response to a harmonic excitation can be computed as with previous models.

Example 26.4 *Repeat the computations of the previous example assuming that a secondary suspension is located between the sprung and unsprung masses. The data are: sprung mass $m_s = 250$ kg; mass of the auxiliary frame $m_t = 10$ kg; unsprung mass $m_u = 25$ kg; stiffness of the spring $K = 25$ kN/m; stiffness of the spring of the auxiliary suspension $K_t = 100$ kN/m; stiffness of the tire $k_t = 100$ kN/m; damping coefficient of the damper $c = 2,150$ Ns/m; damping coefficient of the auxiliary suspension $c_t = 5,000$ Ns/m.*

The values of the natural frequencies are

$$\begin{cases} \omega_1 = 8.30 \; rad/s \; = 1.32 \; Hz \\ \omega_2 = 59.68 \; rad/s \; = 9.50 \; Hz \\ \omega_3 = 130.28 \; rad/s \; = 20.73 \; Hz \end{cases}$$

The dynamic compliance $H(\omega)$ and the inertance $\omega^2 H$ are plotted in Fig. 26.20, along with the power spectral density of the acceleration of the sprung mass when the vehicle travels at 30 m/s on a road defined as in-between classes B and C ISO 8606: 1995.

The r.m.s. value of the acceleration is

$$a_{rms} = 1.21 \; m/s^2 \; = \; 0.12 \; g.$$

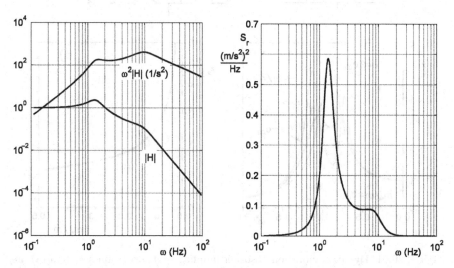

FIGURE 26.20. Dynamic compliance and inertance of the quarter car model with three degrees of freedom. Power spectral density of the acceleration due to motion on a road at a speed of 30 m/s

The auxiliary suspension improves comfort slightly. By comparing the plots, however, it is clear that the improvement is concentrated in the medium-high frequency range. If the comparison were done at a higher speed, in a condition in which high frequency excitation were more significant, the improvement would be larger.

26.4.5   Quarter-car model with dynamic vibration absorber

A dynamic vibration absorber essentially consists of a mass connected to the system through a spring and possibly a damper (Fig. 26.21a). If properly tuned, it can reduce the amplitude substantially at one of the resonances of the original system, but it introduces an additional resonance whose peak amplitude is controlled by the value of its damping c_d.

The frequency response of the system of Fig. 26.21a is shown in Fig. 26.21c for 3 different values of damping: If c_d is low, two resonance peaks are present,

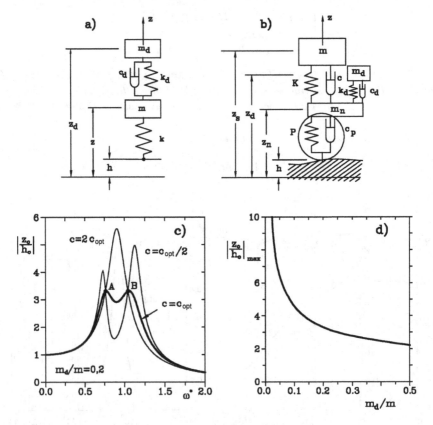

FIGURE 26.21. Dynamic vibration absorber, applied to a spring-mass system (a) and to a quarter car model (b). Frequency response of the first of the two systems for different values of c_d and with $m_d/m = 0.2$ (c) and value of the peak amplitude with optimum damping as a function of the mass ratio m_d/m (d)

while if c_d is high, there is only one peak. If the damping tends to zero the two peaks have an infinite height, while if it tends to infinity the system reduces to an undamped system with a single degree of freedom and thus a single peak with infinite height. It is possible to demonstrate that the stiffness k_d which reduces the amplitude of the motion of mass m to a minimum is[10]

$$(K_d)_{opt} = K \frac{mm_d}{(m + m_d)^2} . \tag{26.36}$$

The value $(c_d)_{opt}$ of the damping necessary to obtain such a minimum and the peak amplitude are respectively

$$(c_d)_{opt} = \frac{m_d}{m + m_d} \sqrt{K \frac{3mm_d}{2(m + m_d)}} , \qquad \left| \frac{z_0}{h_0} \right|_{max} = \sqrt{1 + \frac{2m}{m_d}} . \tag{26.37}$$

Sometimes the vibration absorber may have no elastic element or may be provided with a damper whose behavior is modelled better by dry friction than by viscous damping.

Dynamic vibration absorbers are sometimes used in motor vehicle suspensions, as in the quarter car model of Fig. 26.21b, in which a standard shock absorber is also represented. The equation of motion of the system is

$$\begin{bmatrix} m_s & 0 & 0 \\ 0 & m_d & 0 \\ 0 & 0 & m_u \end{bmatrix} \begin{Bmatrix} \ddot{z}_s \\ \ddot{z}_d \\ \ddot{z}_u \end{Bmatrix} + \begin{bmatrix} c & 0 & -c \\ 0 & c_d & -c_d \\ -c & -c_d & c + c_d + c_p \end{bmatrix} \begin{Bmatrix} \dot{z}_s \\ \dot{z}_d \\ \dot{z}_u \end{Bmatrix} +$$

$$+ \begin{bmatrix} K & 0 & -K \\ 0 & K_d & -K_d \\ -K & -K_d & K + K_d + P \end{bmatrix} \begin{Bmatrix} z_s \\ z_d \\ z_u \end{Bmatrix} = \begin{Bmatrix} 0 \\ 0 \\ c_p \dot{h} + Ph \end{Bmatrix} . \tag{26.38}$$

The various frequency responses may be obtained immediately as with the other quarter car models. Some results are reported in Fig. 26.22. Because no attempt has been made to optimize the suspension, the figure has only a qualitative interest.

Curve 1 deals with a conventional quarter car model as studied in the previous section. Curves 2 and 3 refer to a system in which a vibration absorber is applied to the unsprung mass, tuned on the first and the second natural frequency ($\sqrt{k_d/m_d} = 0.89$ and $\sqrt{k_d/m_d} = 7.09$). The mass of the vibration absorber is $1/20$ of the sprung mass and the damping c_d is $2\sqrt{k_d m_d}$.

To add a vibration absorber to a conventional suspension changes its performance only slightly, both in terms of acceleration of the sprung mass and of forces on the ground. But vibration absorbers are interesting because of the possibility of using them instead of conventional shock absorbers, as in the case shown by curve 4.

---

[10] J.P. Den Hartog, *Mechanical vibrations*, McGraw Hill, New York, 1956.

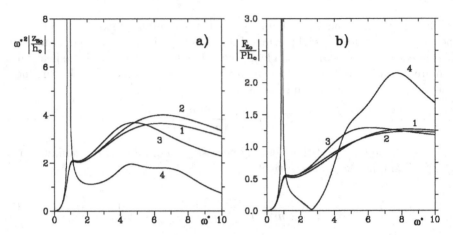

FIGURE 26.22. Quarter car model with dynamic vibration absorber: Non-dimensional amplitude of the acceleration (a) and of the dynamic component of the force F_z (b) as functions of the non-dimensional frequency ω^* ($P = 4K$; $m_s = 10m_u$; $c = \sqrt{6mK/8}$). Line 1: Quarter car without vibration absorber; line 2: $m_d = 0.05m_s$, $\sqrt{k_d/m_d} = 0.89$, $c_d = 2\sqrt{k_d m_d}$; line 3: $m_d = 0.05m_s$, $\sqrt{k_d/m_d} = 7.09$, $c_d = 2\sqrt{k_d m_d}$; line 4: $m_d = 0.05m_s$, $c = 0$, $\sqrt{k_s/m_s} = 4.8$, $c_s = 0.8\sqrt{k_s m_s}$

In the case shown, in which the values of the parameters were obtained by trial and error without a true optimization, the acceleration of the sprung mass is quite low in the entire frequency range, except for a strong resonance peak at low frequency.

Remark 26.7 *The height of the peak is obviously limited, however, because some damping is present, and in practice is further limited by the other forms of damping present in the actual system, such as that due to the tire.*

If the stiffness of the springs K is low, the peak occurs at very low frequency, where its importance may be marginal, and the advantages of the vibration absorber, primarily linked to lower cost and complexity of the system due to elimination of the need for an element mounted between the body and the wheel, add to its excellent suspension performance.

Dynamic vibration absorbers, used instead of conventional shock absorbers, proved advantageous on several low cost small cars with soft suspensions; they may also, however, be added to conventional luxury cars to further increase ride comfort.

26.4.6 Quarter car with many degrees of freedom to study the suspension-tire interaction

In the model of Fig. 26.7c, an additional degree of freedom has been included to account for the compliance of the tire. To proceed in a more comprehensive way it is possible to use the component mode synthesis approach, which is theoretically applicable only if the tire is a linear elastic system, or at least a lightly damped but nonetheless linear system.

The part of the system that may be considered a substructure can be identified. When the system is discretized and the generalized coordinates are the displacements of a certain number of points (the nodes), it is possible to subdivide the nodes into two groups: the connection nodes, that are common to the substructure and to other parts of the system, and the internal nodes. The vector of the generalized coordinates and the stiffness matrix of the substructure may be accordingly partitioned

$$\mathbf{q} = \left\{ \begin{array}{c} \mathbf{q}_1 \\ \mathbf{q}_2 \end{array} \right\}, \qquad \mathbf{K} = \left[\begin{array}{cc} \mathbf{K}_{11} & \mathbf{K}_{12} \\ \mathbf{K}_{21} & \mathbf{K}_{22} \end{array} \right], \qquad (26.39)$$

where subscripts 1 and 2 refer respectively to the boundary and the internal degrees of freedom. The other matrices (mass and damping matrices) may be partitioned in the same way.

In the present case, if the tire is a substructure, the connection nodes are located on the wheel rim and the internal nodes are all others. If the rim is a rigid body and, as in the case of the quarter car model, only the vertical displacement is considered, the only generalized coordinate that is common to the tire and the other parts of the system is displacement z_u ($\mathbf{q}_1$ has just one element).

Consider the tire in its deformed configuration under the static forces due to the load applied to the suspension and linearize its behavior about this configuration. Neglecting the forces applied to the internal nodes in the static configuration, vector $\mathbf{q}_2$ is

$$\mathbf{q}_2 = \mathbf{K}_{22}^{-1}\mathbf{K}_{21}\mathbf{q}_1 . \qquad (26.40)$$

To express the dynamic deflected configuration it is ideally possible to lock the boundary nodes (in this case by constraining the rim of the wheel) and perform dynamic analysis. The natural frequencies and the mode shapes of the tire are then obtained by solving the eigenproblem related to matrices $\mathbf{K}_{22}$ and $\mathbf{M}_{22}$:

$$\det\left(-\omega^2\mathbf{M}_{22} + \mathbf{K}_{22}\right) = 0.$$

Once the eigenproblem has been solved, it is possible to use the eigenvector matrix $\boldsymbol{\Phi}$ to perform the modal transformation

$$\mathbf{q}_2 = \boldsymbol{\Phi}\boldsymbol{\eta}_2 .$$

The generalized coordinates of the substructure can thus be expressed as

$$\left\{ \begin{array}{c} \mathbf{q}_1 \\ \mathbf{q}_2 \end{array} \right\} = \left\{ \begin{array}{c} \mathbf{q}_1 \\ -\mathbf{K}_{22}^{-2}\mathbf{K}_{21}\mathbf{q}_1 + \boldsymbol{\Phi}\boldsymbol{\eta}_2 \end{array} \right\} =$$
$$= \left[\begin{array}{cc} \mathbf{I} & \mathbf{0} \\ -\mathbf{K}_{22}^{-1}\mathbf{K}_{21} & \boldsymbol{\Phi} \end{array} \right] \left\{ \begin{array}{c} \mathbf{q}_1 \\ \boldsymbol{\eta}_2 \end{array} \right\} = \boldsymbol{\Psi} \left\{ \begin{array}{c} \mathbf{q}_1 \\ \boldsymbol{\eta}_2 \end{array} \right\}.$$

$$(26.41)$$

Equation (26.41) is a coordinate transform, allowing the deformation of the internal part of the substructure to be expressed in terms of constrained and internal modes. Matrix $\boldsymbol{\Psi}$ expressing this transformation can be used to compute new mass, stiffness and, if needed, damping matrices and a force vector

$$\mathbf{K}^* = \boldsymbol{\Psi}^T \mathbf{K} \boldsymbol{\Psi} \ , \quad \mathbf{M}^* = \boldsymbol{\Psi}^T \mathbf{M} \boldsymbol{\Psi} \ , \quad \mathbf{C}^* = \boldsymbol{\Psi}^T \mathbf{C} \boldsymbol{\Psi} \ , \quad \mathbf{f}^* = \boldsymbol{\Psi}^T \mathbf{f} \ . \quad (26.42)$$

If the constrained coordinates are m (in the present case $m = 1$) and the internal coordinates are n and only k modes of the constrained substructure are considered ($k < n$), the size of the original $\mathbf{M}$, $\mathbf{K}$, etc. matrices is $m + n$, while that of matrices $\mathbf{M}^*$, $\mathbf{K}^*$, etc. is $m + k$.

Remark 26.8 *If all internal modes are considered ($k = n$) the method does not introduce errors, but there is no simplification. Both approximation and simplification increase while decreasing the number of modes considered.*

The substructure so obtained can be easily assembled to the other parts of the system. If, for instance, only one boundary degree of freedom (the vertical displacement of the unsprung mass) and only one vibration mode of the tire are considered, the quarter car model has three degrees of freedom, two 'physical' ones (displacements of the sprung and unsprung masses) plus a modal one.

Example 26.5 *Consider the quarter car model of the previous examples, taking into account the inertia of the tire as well. A realistic model of the tire not being available, consider it as made of a number of rigid rings, the first being attached to the rim and the last connected to the ground, connected to each other by linear springs and dampers (Fig. 26.23a, where the rigid rings are 4). The dynamic model of the quarter car is shown in Fig. 26.23b.*

Assume the following data: sprung mass $m_s = 250$ kg; unsprung mass $m_u = 23$ kg; masses of the 4 rings modelling the tire $m_i = 1$ kg ($i = 0, ..., 3$), stiffness of the suspension spring $K = 25$ kN/m; stiffness of the springs simulating the tire $k_i = 300$ kN/m ($i = 0,..., 3$); damping coefficient of the shock absorber $c = 2\,150$ Ns/m; damping coefficient of the dampers simulating the tire $c_i = 100$ Ns/m ($i = 0, ..., 3$).

The values of k_i were chosen so that, in static conditions, the stiffness of the tire (that is, the stiffness of the three springs in series) is the same as in the previous examples. A certain damping of the tire must be introduced in the present model; otherwise, the response tends to infinity at the resonance of the latter. The value chosen is, however, low enough not to influence the results at other frequencies, and it allows the tire to be studied with the assumption of small damping.

The tire is a system with three degrees of freedom; its mass, stiffness and damping matrices and the forcing vector due to the motion of the contact point are

$$\mathbf{M} = m_0 \begin{bmatrix} 1 & 0 & 0 \\ 0 & 1 & 0 \\ 0 & 0 & 1 \end{bmatrix} , \quad \mathbf{K} = k_0 \begin{bmatrix} 1 & -1 & 0 \\ -1 & 2 & -1 \\ 0 & -1 & 2 \end{bmatrix} ,$$

$$\mathbf{C} = c_0 \begin{bmatrix} 1 & -1 & 0 \\ -1 & 2 & -1 \\ 0 & -1 & 2 \end{bmatrix} , \quad \mathbf{f} = h \left\{ \begin{array}{c} 0 \\ 0 \\ i\omega c_0 + k_0 \end{array} \right\} .$$

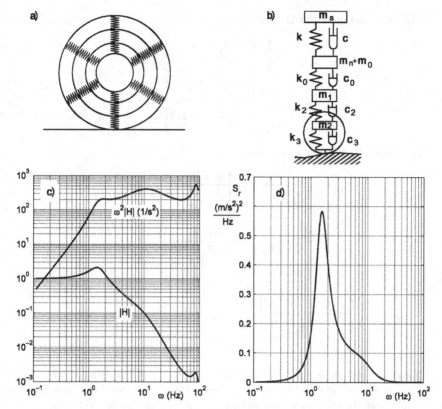

FIGURE 26.23. a): Simplified dynamic model of a tire made of 4 rigid rings connected to each other by springs. b): Dynamic model of a quarter car with the tire simulated as in a). c) and d): Response of the quarter car model with three degrees of freedom based on the model in b)

The first degree of freedom coincides with the vertical displacement of the unsprung mass and is thus a constrained degree of freedom; the other two are internal degrees of freedom of the tire. The stiffness matrix must then be partitioned as

$$K_{11} = k_0 \ , \ K_{12} = \begin{bmatrix} -k_0 & 0 \end{bmatrix} \ , \ K_{21} = \begin{bmatrix} -k_0 \\ 0 \end{bmatrix} \ , \ K_{22} = k_0 \begin{bmatrix} 2 & -1 \\ -1 & 2 \end{bmatrix} .$$

The other matrices are partitioned in the same way. Modal analysis of the internal modes yields the following values for the natural frequencies and the eigenvectors

$$\begin{cases} \omega_1 = 457 \ rad/s = 87 \ Hz \\ \omega_1 = 949 \ rad/s = 151 \ Hz \end{cases} \qquad \Phi = \begin{bmatrix} \dfrac{\sqrt{2}}{2} & \dfrac{-\sqrt{2}}{2} \\ \dfrac{\sqrt{2}}{2} & \dfrac{\sqrt{2}}{2} \end{bmatrix} .$$

The transformation matrix $\boldsymbol{\Psi}$ is

$$\boldsymbol{\Psi} = \begin{bmatrix} 1 & 0 & 0 \\ \dfrac{2}{3} & \dfrac{\sqrt{2}}{2} & \dfrac{-\sqrt{2}}{2} \\ \dfrac{1}{3} & \dfrac{\sqrt{2}}{2} & \dfrac{\sqrt{2}}{2} \end{bmatrix}.$$

The transformed matrices are then

$$\mathbf{M}^* = m_0 \left[\begin{array}{c|cc} 1.556 & 0.7071 & -0.2357 \\ \hline 0.7071 & 1 & 0 \\ -0.2357 & 0 & 1 \end{array} \right],$$

$$K^* = k_0 \left[\begin{array}{c|cc} 0.3333 & 0 & 0 \\ \hline 0 & 1 & 0 \\ 0 & 0 & 3 \end{array} \right],$$

$$\mathbf{C}^* = c_0 \left[\begin{array}{c|cc} 0.3333 & 0 & 0 \\ \hline 0 & 1 & 0 \\ 0 & 0 & 3 \end{array} \right], \qquad \mathbf{f}^* = h\,(iwc_0 + k_0) \left\{ \begin{array}{c} \dfrac{1}{3} \\ -\sqrt{2} \\ \dfrac{2}{\sqrt{2}} \\ 2 \end{array} \right\}.$$

If a model with three degrees of freedom is required, just one of the internal modes of the tire is needed. The third row and column of the transformed matrices may then be cancelled and the tire can be assembled to the quarter car model, obtaining

$$\begin{bmatrix} m_s & 0 & 0 \\ 0 & m_u + 1.556 m_0 & 0.7071 m_0 \\ 0 & 0.7071 m_0 & m_0 \end{bmatrix} \left\{ \begin{array}{c} \ddot{z}_s \\ \ddot{z}_u \\ \ddot{\eta} \end{array} \right\} +$$

$$+ \begin{bmatrix} c & -c & 0 \\ -c & c + 0.3333 c_0 & 0 \\ 0 & 0 & c_0 \end{bmatrix} \left\{ \begin{array}{c} \dot{z}_s \\ \dot{z}_u \\ \dot{\eta} \end{array} \right\} +$$

$$+ \begin{bmatrix} K & -K & 0 \\ -K & K + 0.3333 k_0 & 0 \\ 0 & 0 & k_0 \end{bmatrix} \left\{ \begin{array}{c} z_s \\ z_u \\ \eta \end{array} \right\} = h\,(iwc_0 + k_0) \left\{ \begin{array}{c} 0 \\ 1 \\ \dfrac{3}{-\sqrt{2}} \\ 2 \end{array} \right\}.$$

The values of the natural frequencies are

$$\begin{cases} \omega_1 = 8.93 \ rad/s = 1.42 \ Hz \\ \omega_2 = 72.97 \ rad/s = 11.61 \ Hz \\ \omega_3 = 553.74 \ rad/s = 88.13 \ Hz \ . \end{cases}$$

The dynamic compliance $H(\omega)$ and the inertance $\omega^2 H$ are reported in Fig. 26.23 together with the power spectral density of the acceleration of the sprung mass.

The r.m.s. value of the acceleration is

$$a_{rms} = 1.36 \ m/s^2 \ = \ 0.14 \ g.$$

The inertia of the tire has no major effect on the results, except in a narrow frequency range about its natural frequency, where a small peak can be seen in the frequency responses. The r.m.s. value of the acceleration is slightly higher and, because of the presence of high frequency components, an auxiliary suspension may be useful. Because of the approximate tire model here used, this example is simply an indication of how the component modes synthesis approach can be used.

26.4.7   Effect of the suspension kinematics

In the previous cases the motion of the unsprung mass is only a vertical translation, as if the suspension kinematism were a prismatic guide with axis parallel to the body-fixed z-axis. This model for an independent suspension is, however, quite rough, because no actual suspension is made with a prismatic guide. Each type of suspension has its own kinematics, or better elasto-kinematics because the various linkages are rigid only as a first approximation.

Usually the deviations of the trajectory of the unsprung mass from a straight line parallel to the z-axis are considered shortcomings of the guiding kinematic arrangement, as if straight motion were the ideal situation. But it is actually advisable, on the contrary, that when the wheel gets a shock in the horizontal direction the suspension allows it to move backwards, reducing the excitation in the x-direction transferred to the vehicle body. Moreover, as seen in Part I and in a later section, only by accepting that the wheel does not move exactly in the z-direction is it possible to counteract the *dive* effects that occur when braking, and the *lift* (or *squat*) effects found when driving.

The first of these effects, the ability to absorb horizontal shocks, may be obtained in two ways: by using a kinematic arrangement to produce a suitable trajectory, as in the case of trailing arms (note that the opposite arrangements, which could be defined as leading arms, cause a forward displacement of the wheel when the latter moves upwards and for this reason are seldom used today); or by giving a suitable horizontal compliance to the suspension.

In the first case, it is still possible to use the quarter car model, while in the second the model must include a further degree of freedom for each wheel, namely horizontal displacement.

While a knowledge of the exact elasto-kinematics of the suspension is essential for stating the position of the wheel with respect to the ground and thus for assessing contact forces, its actual effect on the inertia forces acting on the elements of the suspension and on the forces due to the spring and the shock absorber is limited. Thus it is possible to neglect the compliance of the guiding elements and proceed to a first approximation study related to comfort by using models based on the quarter car approach.

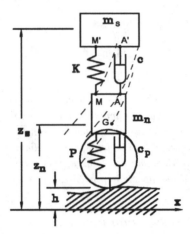

FIGURE 26.24. Quarter car model with two degrees of freedom. The dashed lines are the trajectories of points G, A and M in the xz plane (functions f_G, f_A and f_M)

It is possible to define the trajectories of points G, M and A (Fig. 26.24), i.e. of the center of mass of the unsprung mass and the attachment points of the spring and of the shock absorber in a reference frame $x_1 y_1 z_1$ fixed to the sprung mass. The trajectory of a generic point P of the unsprung mass is

$$\begin{cases} f_{1P}(x_1, y_1, z_1) = 0 \\ f_{2P}(x_1, y_1, z_1) = 0 \ . \end{cases} \qquad (26.43)$$

The motion of the unsprung mass occurs mostly in the z_1 direction, and thus it is expedient to transform Equations (26.43) by solving them in x_1 and y_1. The coordinates of point P are then linked to each other by the relationships

$$\begin{cases} x_{1P} = f_P(z_{1P}) \\ y_{1P} = g_P(z_{1P}) \ . \end{cases} \qquad (26.44)$$

Using geometrical considerations it is possible to obtain a third equation expressing coordinate z_{1P} of point P as a function of the coordinate z_{1G} of the center of mass of the unsprung mass

$$z_{1P} = h_P(z_{1G}) \ . \qquad (26.45)$$

The inertial coordinate z_u of the center of mass of the unsprung mass is

$$z_u = z_s + z_{1G} \ . \qquad (26.46)$$

The kinematics of the suspension is then completely defined.

To write a linearized equation of motion for studying small oscillations about a reference position (for instance, that of static equilibrium), the equations expressing the trajectory of the generic point P may be linearized as

$$
\begin{cases}
x_{1P} = x_{1P0} + \left(\frac{df_P}{dz_{1P}}\right)_0 z_{1P} \\
y_{1P} = y_{1P0} + \left(\frac{dg_P}{dz_{1P}}\right)_0 z_{1P} \\
z_{1P} = z_{1P0} + \left(\frac{dh_P}{dz_{1G}}\right)_0 z_{1G} \ ,
\end{cases}
\tag{26.47}
$$

that is

$$
\begin{cases}
x_{1P} = x_{1P0} + \left(\frac{df_P}{dz_{1P}}\right)_0 z_{1P0} + \left(\frac{df_P}{dz_{1P}}\right)_0 \left(\frac{dh_P}{dz_{1G}}\right)_0 z_{1G} \\
y_{1P} = y_{1P0} + \left(\frac{dg_P}{dz_{1P}}\right)_0 z_{1P0} + \left(\frac{dg_P}{dz_{1P}}\right)_0 \left(\frac{dh_P}{dz_{1G}}\right)_0 z_{1G} \\
z_{1P} = z_{1P0} + \left(\frac{dh_P}{dz_{1G}}\right)_0 z_{1G} \ .
\end{cases}
\tag{26.48}
$$

The velocity of point P can be expressed in $x_1 y_1 z_1$ reference frame as

$$
V_P = \left\{
\begin{array}{c}
\left(\frac{df_P}{dz_{1P}}\right)_0 \left(\frac{dh_P}{dz_{1G}}\right)_0 \dot{z}_{1G} \\
\left(\frac{dg_P}{dz_{1P}}\right)_0 \left(\frac{dh_P}{dz_{1G}}\right)_0 \dot{z}_{1G} \\
\left(\frac{dh_P}{dz_{1G}}\right)_0 \dot{z}_{1G}
\end{array}
\right\} ,
\tag{26.49}
$$

where

$$
\dot{z}_{1G} = \dot{z}_u - \dot{z}_s \ .
\tag{26.50}
$$

In the case of point G, function h_G and its derivative are

$$
h_G(z_{1G}) = z_{1G} \ , \qquad \frac{dh_G}{dz_{1G}} = 1
\tag{26.51}
$$

and thus the velocity of the unsprung mass is

$$
V_G = \left\{
\begin{array}{c}
\left(\frac{df_G}{dz_{1G}}\right)_0 (\dot{z}_u - \dot{z}_s) \\
\left(\frac{dg_G}{dz_{1G}}\right)_0 (\dot{z}_u - \dot{z}_s) \\
\dot{z}_u
\end{array}
\right\} .
\tag{26.52}
$$

The translational kinetic energy of the quarter car with two degrees of freedom is then

$$
\mathcal{T} = \frac{1}{2} m_s \dot{z}_s^2 + \frac{1}{2} m_u \dot{z}_u^2 + \frac{1}{2} m_u \left[\left(\frac{df_G}{dz_{1G}}\right)^2 + \left(\frac{dg_G}{dz_{1G}}\right)^2 \right] (\dot{z}_u - \dot{z}_s)^2 \ ,
\tag{26.53}
$$

i.e.:

$$
\mathcal{T} = \frac{1}{2} (m_s + m_u \beta) \dot{z}_s^2 + \frac{1}{2} m_u (1 + \beta) \dot{z}_u^2 + m_u \beta \dot{z}_u \dot{z}_s \ ,
\tag{26.54}
$$

where β is a constant whose value is

$$\beta = \left(\frac{df_G}{dz_{1_G}}\right)_0^2 + \left(\frac{dg_G}{dz_{1_G}}\right)_0^2 . \tag{26.55}$$

The mass matrix can be immediately obtained from the kinetic energy

$$\mathbf{M} = \begin{bmatrix} m_s + m_u\beta & m_u\beta \\ m_u\beta & m_u\left(1+\beta\right) \end{bmatrix} . \tag{26.56}$$

Distance MM' must be computed to obtain the potential energy of the spring. The coordinates of point M' are constant in the $x_1y_1z_1$ frame. They can be written as $x_{0_{M'}}$, $y_{0_{M'}}$, $z_{0_{M'}}$. The potential energy of the spring is then

$$\mathcal{U} = \frac{1}{2}K\left[\left(x_M - x_{M'}\right)^2 + \left(y_M - y_{M'}\right)^2 + \left(z_M - z_{M'}\right)^2\right] , \tag{26.57}$$

where the system has been assumed to behave linearly about its static equilibrium position. Only the quadratic terms of the potential energy enter the stiffness matrix. The linear terms actually produce constant terms (generalized forces) in the equation of motion that do not affect the dynamic behavior about the equilibrium position, and these constant terms are arbitrary. Taking into account only the quadratic terms, the potential energy reduces to

$$\mathcal{U} = \frac{1}{2}K\gamma z_{1_G}^2 = \frac{1}{2}K\gamma\left(z_u - z_s\right) , \tag{26.58}$$

where

$$\gamma = \left(\frac{dh_M}{dz_{1_G}}\right)_0^2\left[1 + \left(\frac{df_M}{dz_{1_M}}\right)_0^2 + \left(\frac{dg_M}{dz_{1_M}}\right)_0^2\right] . \tag{26.59}$$

Taking into account also the deformation potential energy of the tire

$$\mathcal{U}_P = \frac{1}{2}Pz_u^2$$

(after neglecting constant and linear terms), the stiffness matrix of the system is

$$\mathbf{K} = \begin{bmatrix} K\gamma & -K\gamma \\ -K\gamma & K\gamma + P \end{bmatrix} . \tag{26.60}$$

By substituting subscript M with subscript A, Eq. (26.49) yields directly the relative velocity of point A with respect to the sprung mass, i.e. to point A'. However, what matters for the computation of the forces due to the shock absorber is not the relative velocity but only its component in the direction AA'. Distance AA' is:

$$\overline{AA}' = \sqrt{\left(x_A - x_{A'}\right)^2 + \left(y_A - y_{A'}\right)^2 + \left(z_A - z_{A'}\right)^2} . \tag{26.61}$$

The Raleigh dissipation function of the shock absorber is then

$$\mathcal{F} = \frac{1}{2}c\left(\frac{d\overline{AA}'}{dt}\right)^2 .$$

(26.62)

The dissipation function can be simplified by linearizing it about the equilibrium position, as

$$\mathcal{F} = \frac{1}{2}c\delta\left(\dot{z}_u - \dot{z}_s\right)^2 ,$$

(26.63)

where

$$\delta = \left(\frac{dh_A}{dz_{1_G}}\right)_0^2 \frac{\left[(x_{A_0} - x_{A_0'})\left(\frac{df_A}{dz_{1_A}}\right)_0 + (y_{A_0} - y_{A_0'})\left(\frac{dg_A}{dz_{1_A}}\right)_0 + (z_{A_0} - z_{A_0'})\right]^2}{(x_{A_0} - x_{A_0'})^2 + (y_{A_0} - y_{A_0'})^2 + (z_{A_0} - z_{A_0'})^2} ,$$

(26.64)

By also inserting the term due to the damping of the tire into the expression of the dissipation function, the damping matrix of the system is obtained

$$\mathbf{C} = \begin{bmatrix} c\delta & -c\delta \\ -c\delta & c\delta + c_P \end{bmatrix} .$$

(26.65)

As far as the elastic and damping terms are concerned, the linearized equation of motion is still Eq. (26.22), except for the values of the stiffness of the spring or the damping coefficient of the shock absorber which are 'reduced' through coefficients γ and δ. The mass matrix is, on the other hand, different, because it is not diagonal. An inertial coupling proportional to the value of the unsprung mass is then present. It will cause a larger motion of the sprung mass at frequency ranges typical of the motion of the unsprung mass.

In any case, the quarter car model assumes that the sprung mass moves along the z direction, an approximation that is increasingly unrealistic with increasing coupling of horizontal and vertical motion of the unsprung mass due to the kinematics of the suspension. In particular, a motion of the sprung mass in the x direction due to a motion in the z direction may be detrimental to comfort, and cannot be studied using such simple models.

Example 26.6 *Consider the quarter car with two degrees of freedom shown in Fig. 26.25. The trailing arm suspension is hinged about an axis parallel to y-axis of the vehicle. The data are: sprung mass $m_s = 250$ kg; unsprung mass $m_u = 25$ kg; stiffness of the spring $K = 700$ kN/m; damping coefficient $c = 19.35$ kNs/m; Data of the tire: $P = 100$ kN/m; $c_P = 0$. Geometrical data: $l = 500$ mm, $l_A = 300$ mm, $l_M = 100$ mm; position of C: $\begin{bmatrix} 200 & 0 & -300 \end{bmatrix}^T$ mm; position of M': $\begin{bmatrix} -200 & 0 & -360 \end{bmatrix}^T$ mm; position of A': $\begin{bmatrix} 200 & 0 & 0 \end{bmatrix}^T$ mm.*

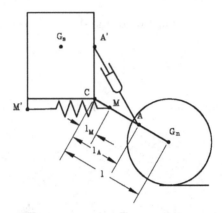

FIGURE 26.25. Trailing arm quarter car: geometrical definitions

The functions f_1 and f_2 defining the trajectory of a point P of the suspension are

$$\begin{cases} (x_1 - x_0)^2 + (z_1 - z_0)^2 - l_P^2 = 0 \\ y_1 - y_0 = 0 , \end{cases}$$

where x_0, y_0 and z_0 are the coordinates of point C. The corresponding functions f_P and g_P are

$$\begin{cases} f_P = x_0 \pm \sqrt{l_P^2 - (z_{1P} - z_0)^2} \\ g_P = y_0 . \end{cases}$$

The double sign does not give problems, because it is determined by the geometry of the system: Sign $(+)$ must be used if point P is forward of the axis of the hinge. Because the suspension lies in a plane parallel to the xz plane, coordinate y of all points can be assumed as zero $(y_0 = 0)$.
The derivative $\left(\frac{df}{dz}\right)_0$ is then

$$\left(\frac{df_P}{dz_{1P}}\right)_0 = \left(\frac{-(z_{1P} - z_0)}{\sqrt{l_P^2 - (z_{1P} - z_0)^2}}\right)_0$$

The static equilibrium position is defined by

$$\begin{cases} x_{1G_0} = x_0 + l \cos(\theta_0) = 633 \ mm \\ z_{1G_0} = z_0 - l \sin(\theta_0) = -550 \ mm \end{cases}$$

and thus

$$\left(\frac{df_G}{dz_{1G}}\right)_0 = 0.577 , \qquad \beta = 0.333 .$$

The mass matrix is then

$$\mathbf{M} = \begin{bmatrix} 258,33 & 8,33 \\ 8,33 & 33,33 \end{bmatrix} .$$

By remembering that triangles CMH' are CGH similar, the coordinate z_1 of point M is obtained:

$$z_{1M} = z_0 - \frac{l_M}{l}\sqrt{l^2 - (z_{1G} - z_0)^2}$$

and then

$$\left(\frac{dh_M}{dz_{1_G}}\right)_0 = -\frac{l_M}{l}\left(\frac{(z_{1G} - z_0)}{\sqrt{l^2 - (z_{1G} - z_0)^2}}\right)_0 .$$

Thus it follows that

$$\left(\frac{df_M}{dz_{1P}}\right)_0 = 1.732 \ , \qquad \left(\frac{dh_M}{dz_{1_G}}\right)_0 = 0.116 \ , \qquad \gamma = 0.0364 \ .$$

The coordinate of point A is obtained in a similar way

$$z_{1A} = z_0 + \frac{l_A}{l}(z_{1G} - z_0)$$

and thus

$$\left(\frac{dh_A}{dz_{1_G}}\right)_0 = \frac{l_A}{l} = 0.6 \ .$$

Finally, it follows that

$$\left(\frac{df_M}{dz_{1G}}\right)_0 = 0.577 \ , \qquad \delta = 0.111 \ .$$

Note that the values of K and c were chosen in such a way that the reduced values $K\gamma$ and $c\delta$ were practically identical to those of example 26.3.

The dynamic compliance $H(\omega)$ and the inertance $\omega^2 H$ are plotted in Fig. 26.26 together with the power spectral density of the acceleration of the sprung mass.

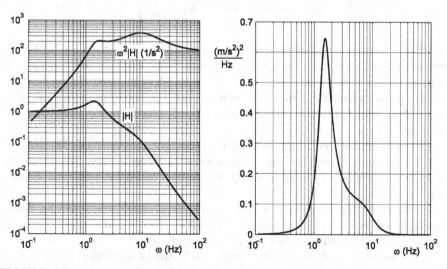

FIGURE 26.26. Dynamic compliance and inertance of the trailing arm quarter car model. Power spectral density of the acceleration due to motion on a road at a speed of 30 m/s

The r.m.s. value of the acceleration is

$$a_{rms} = 1.34 \ m/s^2 \ = \ 0.14 \ g.$$

By comparing the plot with that obtained for the simple two degrees of freedom quarter car model, it will be noted that the peak of the response is higher, and also that the response is in general higher at frequencies close to the resonances of both sprung and unsprung masses. At higher frequencies the response is lower, primarily because the inertia is in some way increased by the coupling of longitudinal and vertical motion, causing the suspension to behave as if it were softer. The reduction of the r.m.s. value of the acceleration is, however, minimal.

26.5   HEAVE AND PITCH MOTION

26.5.1   *Simplified models with rigid tires*

The heave motion of the vehicle is strictly coupled with pitch motion. The simplest model for studying the heave-pitch coupling is shown in Fig. 26.27a. Its equation of motion is that of a beam on two elastic and damped supports

$$
\begin{bmatrix} m_S & 0 \\ 0 & J_y \end{bmatrix} \left\{ \begin{array}{c} \ddot{Z}_s \\ \ddot{\theta} \end{array} \right\} + \begin{bmatrix} c_1 + c_2 & -ac_1 + bc_2 \\ -ac_1 + bc_2 & a^2 c_1 + b^2 c_2 \end{bmatrix} \left\{ \begin{array}{c} \dot{Z}_s \\ \dot{\theta} \end{array} \right\} +
$$

$$
+ \begin{bmatrix} K_1 + K_2 & -aK_1 + bK_2 \\ -aK_1 + bK_2 & a^2 K_1 + b^2 K_2 \end{bmatrix} \left\{ \begin{array}{c} Z_s \\ \theta \end{array} \right\} = \qquad (26.66)
$$

$$
= \left\{ \begin{array}{c} c_1 \dot{h}_A + c_2 \dot{h}_B + K_1 h_A + K_2 h_B \\ -ac_1 \dot{h}_A + bc_2 \dot{h}_B - aK_1 h_A + bK_2 h_B \end{array} \right\}.
$$

The overturning moment due to weight (term $-m_S g h$ to be added in position 22 in the stiffness matrix) is not included in Eq. (26.66), because no assumption has been made on the height of the pitch center over the road plane. Nor is any aerodynamic term is introduced into the equation of motion. The longitudinal position of the springs and the shock absorbers has been assumed to be the same.

The forcing functions were written in a form that considers only the vertical motion of points A and B, neglecting horizontal forces at the ground-wheels interface and the possible coupling between vertical and horizontal motions due to suspensions.

If mass m_s and moment of inertia J_y are those of the whole sprung mass, the stiffnesses K_i and the damping coefficients c_i are those of a whole axle and are then twice those of a single spring or shock absorber.

The compliance of the tires was neglected in the beam model shown in Fig. 26.27a. It may thus be considered an evolution of the quarter car with a single degree of freedom. In some cases it can be reduced to a pair of quarter cars,

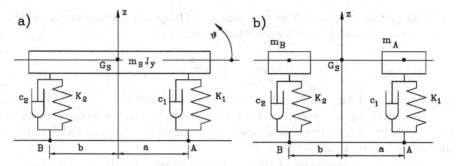

FIGURE 26.27. Beam models for heave and pitch motions

as shown in Fig. 26.27b. To compare the two models, it is possible to use the coordinates Z_A and Z_B instead of Z_s and θ to describe the motion of the beam. The coordinate transformation may be expressed as:

$$\left\{ \begin{array}{c} Z_s \\ \theta \end{array} \right\} = \frac{1}{l} \left[\begin{array}{cc} b & a \\ -1 & 1 \end{array} \right] \left\{ \begin{array}{c} Z_A \\ Z_B \end{array} \right\} . \tag{26.67}$$

The mass matrix to be included in the equation of motion when using the new coordinates is

$$\mathbf{M}' = \mathbf{T}^T \mathbf{M} \mathbf{T},$$

where $\mathbf{T}$ is the transformation matrix defined by Eq. (26.67). All other matrices can be obtained in the same way. Eq. (26.66) then becomes

$$\frac{m_S}{l^2} \left[\begin{array}{cc} b^2 + r_y^2 & ab - r_y^2 \\ ab - r_y^2 & a^2 + r_y^2 \end{array} \right] \left\{ \begin{array}{c} \ddot{Z}_A \\ \ddot{Z}_B \end{array} \right\} + \left[\begin{array}{cc} c_1 & 0 \\ 0 & c_2 \end{array} \right] \left\{ \begin{array}{c} \dot{Z}_A \\ \dot{Z}_B \end{array} \right\} +$$

$$+ \left[\begin{array}{cc} K_1 & 0 \\ 0 & K_2 \end{array} \right] \left\{ \begin{array}{c} Z_A \\ Z_B \end{array} \right\} = \left\{ \begin{array}{c} c_1 \dot{h}_A + K_1 h_A \\ c_2 \dot{h}_B + K_2 h_B \end{array} \right\} , \tag{26.68}$$

where r_y is the radius of gyration of the sprung mass about the y-axis.

A *dynamic index* I_d of the sprung mass can thus be defined as

$$I_d = \frac{r_y^2}{ab} . \tag{26.69}$$

If I_d is equal to unity, i.e. if

$$J_y = m_S ab ,$$

that is

$$r_y^2 = ab ,$$

the two equations uncouple from each other, yielding the two equations of motion of two separate quarter cars with sprung masses

$$m_S \frac{b}{l} \quad \text{and} \quad m_S \frac{a}{l} ,$$

and the model of Fig. 26.27a reduces to that of Fig. 26.27b.

This condition is usually not verified in practice. The tendency to increase the wheelbase for stability reasons leads to values of the dynamic index usually smaller than 1, even smaller than 0.8.

The natural frequencies of the undamped system may be computed using the homogeneous equation associated with Eq. (26.66) or (26.68), after cancelling the damping term. If the solution

$$\left\{ \begin{array}{c} Z_A \\ Z_B \end{array} \right\} = \left\{ \begin{array}{c} Z_{A_0} \\ Z_{B_0} \end{array} \right\} e^{i\omega t} , \tag{26.70}$$

is introduced in the second of the mentioned equations, the characteristic equation

$$\det \left[-\omega^2 \frac{m_S}{l^2} \left[\begin{array}{cc} b^2 + r_y^2 & ab - r_y^2 \\ ab - r_y^2 & a^2 + r_y^2 \end{array} \right] + \left[\begin{array}{cc} K_1 & 0 \\ 0 & K_2 \end{array} \right] \right] = 0 \tag{26.71}$$

is obtained.

The natural frequencies are then the roots of equation

$$\omega^4 - \omega^2 \frac{K_1(r_y^2 + a^2) + K_2(r_y^2 + b^2)}{m_S r_y^2} + K_1 K_2 \frac{l^2}{m_S^2 r_y^2} = 0 , \tag{26.72}$$

that yields

$$\omega_i = \frac{\sqrt{\left(b^2 + r_y^2\right) K_2 + \left(a^2 + r_y^2\right) K_1 \pm \Delta}}{r_y \sqrt{2 m_S}} , \tag{26.73}$$

where

$$\Delta = \sqrt{\left(b^2 + r_y^2\right)^2 K_2^2 + 2 K_1 K_2 \left[\left(ab - r_y^2\right)^2 - r_y^2 l^2\right] + \left(a^2 + r_y^2\right)^2 K_1^2} .$$

The corresponding eigenvectors are

$$\mathbf{q}_i = \left\{ \begin{array}{c} \dfrac{\left(b^2 + r_y^2\right) K_2 - \left(a^2 + r_y^2\right) K_1 \mp \Delta}{2 K_1 \left(ab - r_y^2\right)} \\ 1 \end{array} \right\} .$$

The solution with (+) sign yields two positive values, that generally are not equal to each other. The motion of the beam is neither a rotation about its center of mass (pitch) nor a translational motion in the z direction (heave), but the very fact that the displacements at the front and rear axles have the same

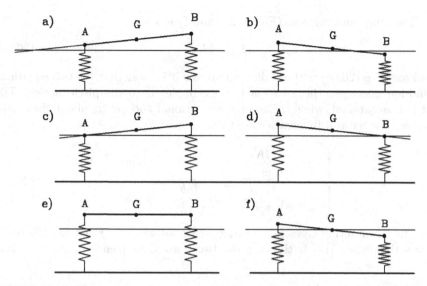

FIGURE 26.28. Heave and pitching motions. (a) and (b): General case; primarily heave (a) and primarily pitching (b) motions. (c) and (d): Case with $I_d = 1$. (e) and (f): Case with $aK_1 = bK_2$.

sign means that the node (the point with zero displacement) lies outside of the wheelbase, and thus the motion is primarily translational (heave, Fig. 26.28a). The solution with $(-)$ sign (Fig. 26.28b) yields a positive and a negative value: the displacements of the front and rear axles are one positive and one negative and the node is within the wheelbase. The motion is primarily rotational, even if not about the center of mass, and it is primarily a pitching motion.

If the dynamic index I_d has a unit value (Fig. 26.28c e d), it follows that

$$\Delta = \left(b^2 + r_y^2\right) K_2 - \left(a^2 + r_y^2\right) K_1 = l \left(bK_2 - aK_1\right)$$

and then

$$\omega_1 = \sqrt{\frac{lK_2}{am_S}} \ , \quad \omega_2 = \sqrt{\frac{lK_1}{bm_S}} . \tag{26.74}$$

As previously stated, the natural frequencies in this case are those of the two separate quarter cars of Fig. 26.27b. The corresponding eigenvectors are

$$\mathbf{q}_1 = \left\{ \begin{array}{c} 0 \\ 1 \end{array} \right\} , \quad \mathbf{q}_2 = \left\{ \begin{array}{c} 1 \\ 0 \end{array} \right\} ,$$

and the free oscillations of the sprung mass are rotations about the points where the suspensions are connected to the body. It is impossible to identify a heave and a pitch mode; it is more accurate to speak about a front-axle and a rear-axle mode. The limiting case is where the node internal to the wheelbase and that external to it tend to the ends of the wheelbase.

The other limiting case (Fig. 26.28e and f) is when

$$aK_1 = bK_2 . \tag{26.75}$$

From Eq. (26.66) without damping terms it is clear that the two equations of motion uncouple: the heave motion uncouples from the pitch motion. The first is translational, while the second is rotational and occurs about the center of mass. The natural frequencies are then

$$\begin{cases} \omega_1 = \sqrt{\dfrac{lK_1}{bm_S}} & \text{bounce,} \\[2ex] \omega_2 = \sqrt{\dfrac{laK_1}{r_y^2 m_S}} = \omega_1 \sqrt{\dfrac{ab}{r_y^2}} & \text{pitch.} \end{cases} \tag{26.76}$$

The two limiting cases may also occur simultaneously. From Eq. (26.76) it follows that when this is the case the two natural frequencies have the same value.

Remark 26.9 *This solves an apparent inconsistency; if*

$$aK_1 = bK_2 ,$$

the centers of rotation are one in the centre of mass (pitch mode) and one at infinity (heave mode), while when the dynamic index has an unit value they are at the end of the wheelbase. When both conditions occur simultaneously, the two natural frequencies coincide; in this case, any linear combination of the eigenvectors is itself an eigenvector. Thus in the case of a rigid beam, any point of the beam (or better, of the straight line constituting the beam axis) may be considered as a center of rotation.

Example 26.7 *Consider a vehicle having the following characteristic: sprung mass $m_s = 1,080$ kg; pitching moment of inertia $J_y = 1,480$ kg m^2; stiffness of the suspensions (referred to the axles) $K_1 = 45$ kN/m; $K_2 = 38$ kN/m; $a = 1.064$ m; $b = 1.596$ m, ($l = 2.66$ m). Study the pitching oscillations of the vehicle, using a beam model.*
The pitching radius of gyration and the dynamic index are

$$r_y = \sqrt{\frac{J_y}{m}} = 1.17 \ m \ , \qquad I_d = \frac{r_y^2}{ab} = 0.807 \ . \tag{26.77}$$

The sprung mass may be subdivided into two masses, one at the front axle and one at the rear

$$m_1 = \frac{bm_s}{l} = 648 \ kg \ , \qquad m_2 = \frac{am_s}{l} = 432 \ kg . \tag{26.78}$$

The two natural frequencies of the independent quarter car models are

$$\omega_1 = 1.33 \ Hz \ , \qquad \omega_2 = 1.49 \ Hz . \tag{26.79}$$

Because the dynamic index is different from 1, the approximation so obtained is a rough one. By solving the characteristic equation, the correct natural frequencies are obtained

$$\omega_1 = 1.36 \; Hz \,, \qquad \omega_2 = 1.62 \; Hz \,. \tag{26.80}$$

The corresponding eigenvectors, normalized so that the largest element has a unit value, are

$$\mathbf{q}_1 = \left\{ \begin{array}{c} 1 \\ -0.32 \end{array} \right\} \,, \qquad \mathbf{q}_2 = \left\{ \begin{array}{c} 0.44 \\ 1 \end{array} \right\} \,,$$

The node of the first mode lies within the wheelbase and, because it is 2,015 m from the front axle, is behind the center of mass. This is essentially a pitching mode. The node of the second mode, a heave mode, is in front of the vehicle, 2,09 m from the front axle.

26.5.2   Pitch center

No guiding linkage is considered in the model of Fig. 26.27a, but it is implicitly assumed that the connection points of the suspension to the body may move only in a vertical direction. The wheelbase of the vehicle is then not affected either by bounce or pitch motion. Moreover, it is assumed that the inertia of the unsprung masses does not affect the motion of the body.

It is, in any caser, possible to find a point along the x-axis such that a vertical force applied to it produces a vertical motion but no pitching. This point is the pitch center.

To define the position of the pitch center, a static force F can be applied to the body in a vertical direction at in a point on the x-axis at a generic distance d from the center of mass. Equation (26.66) becomes

$$\left[\begin{array}{cc} K_1 + K_2 & -aK_1 + bK_2 \\ -aK_1 + bK_2 & a^2 K_1 + b^2 K_2 \end{array} \right] \left\{ \begin{array}{c} Z_s \\ \theta \end{array} \right\} = F \left\{ \begin{array}{c} 1 \\ d \end{array} \right\} \,. \tag{26.81}$$

Solving for the pitch angle θ, it follows that

$$\theta = \frac{aK_1 - bK_2 - d\,(K_1 + K_2)}{K_1 K_2 l^2} \,. \tag{26.82}$$

Equating the numerator to zero, it follows that if

$$d = \frac{aK_1 - bK_2}{K_1 + K_2} \tag{26.83}$$

angle θ vanishes. This value of d is the distance in the x direction of the pitch center from the center of mass; it is positive if the pitch center is forward of the mass center. In the majority of cases, d is positive. If Eq. (26.75) holds, $d = 0$ and the mass center is above or below the mass center.

The presence of kinematic guides for suspensions does not change things: Eq. (26.66) still holds, even if the meaning of the terms may vary. Consider, for

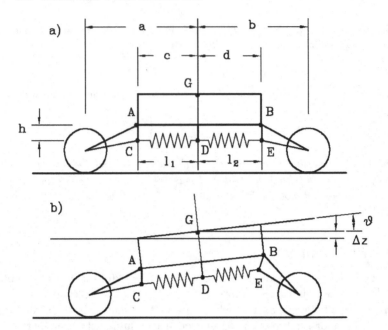

FIGURE 26.29. Vehicle with longitudinal swing arm suspensions. a): sketch of the system; b) position during bounce and pitch motion

instance, a vehicle with longitudinal swing arm suspensions and springs located below the floor. Linearize the equations of motion about a reference position (Fig. 26.29).

Assume that the suspended mass moves vertically by a distance Δz and rotates in pitch by an angle θ. Points A and B move vertically by

$$\begin{cases} \Delta z_A = \Delta z - c\theta \\ \Delta z_B = \Delta z + d\theta \ . \end{cases} \tag{26.84}$$

By linearizing the system about the horizontal position, rotations ϕ_i of the two swing arms are

$$\begin{cases} \Delta \phi_1 = \dfrac{\Delta z_A}{a - c} \Delta x_A = \dfrac{\Delta z - c\theta}{a - c} \\[4mm] \Delta \phi_2 = -\dfrac{\Delta z_B}{b - d} \Delta x_A = -\dfrac{\Delta z + d\theta}{b - d} \ . \end{cases} \tag{26.85}$$

The springs stretch by

$$\begin{cases} \Delta l_1 = -\Delta \phi_1 h = -h \dfrac{\Delta z - c\theta}{a - c} \\[4mm] \Delta l_2 = \Delta \phi_2 h = -h \dfrac{\Delta z + d\theta}{b - d} \ . \end{cases} \tag{26.86}$$

The change in elastic potential energy is then

$$\Delta\mathcal{U} = \tfrac{1}{2}K_1\Delta l_1^2 + \tfrac{1}{2}K_2\Delta l_2^2 =$$

$$= \frac{1}{2}K_1\left(h\frac{\Delta x - c\theta}{a-c}\right)^2 + \frac{1}{2}K_2\left(h\frac{\Delta x + d\theta}{b-d}\right)^2 . \tag{26.87}$$

By differentiating the potential energy with respect to the generalized displacements, the following relationships linking the vertical force and the pitching moment applied to the vehicle body with the generalized displacements emerge

$$\left\{\begin{array}{c} F_z \\ M_y \end{array}\right\} = \left[\begin{array}{cc} K_1^* + K_2^* & -cK_1^* + dK_2^* \\ -cK_1^* + dK_2^* & c^2K_1^* + d^2K_2^* \end{array}\right]\left\{\begin{array}{c} \Delta z_s \\ \theta \end{array}\right\}, \tag{26.88}$$

where:

$$K_1^* = K_1\left(\frac{h}{a-c}\right)^2 , \quad K_2^* = \left(\frac{h}{b-d}\right)^2 .$$

As clearly seen, the structure of the stiffness matrix is identical to the general case, even if it includes terms that are typical of the particular type of suspension. The damping matrix may be obtained along the same lines.

The suspension type also affects the mass matrix to be introduced into Eq. (26.66), because heave and pitch motions also cause some movements of the unsprung masses, causing their inertial parameters to enter the mass matrix as well.

The height of the pitch center becomes important when the wheels exert longitudinal forces, because the coupling between driving (or braking) and pitching depend on it. The *antidive* and *antilift* (or *antisquat*) characteristics of the suspensions also depend on the height of the pitch center.

If the wheels do not exert longitudinal forces, the pitch center is assumed to lie roughly at the height of the centers of the wheels, which amounts to assuming that the wheels travel at constant speed even when the body oscillates in heave or pitch[11].

26.5.3 Empirical rules for the design of suspensions

As already stated, the bounce and pitch dynamics of the suspended mass are strictly related to each other. Some empirical criteria for the choice of the relevant parameters are here reported: They date back to the 1930s and were introduced by Maurice Olley[12].

- The vertical stiffness of the front suspension must be about 30% lower than that of the rear suspension;

[11]Milliken W.F., Milliken D.L., *Chassis Design*, Professional Engineering Publishing, Bury St. Edmunds, 2002.
[12]T.D. Gillespie, *Fundamentals of vehicle dynamics*, SAE, Warrendale, 1992.

- The pitch and bounce frequencies must be close to each other; the bounce frequency should be less than 1.2 times the pitch frequency;

- Neither frequency should be greater than 1.3 Hz;

- The roll frequency should be approximately equal to the bounce and pitch frequency.

The first rule states that the natural frequency of the rear suspension is higher than that of the front, at least if the weight distribution is not such that the rear wheels are far more loaded than those in front . The importance of having a lower natural frequency for the front suspension may be explained by observing that any road input reaches the front suspension first and then, only after a certain time, the rear one. If the natural frequency of the latter is higher, when the vehicle rides over a bump the rear part quickly "catches up" to the motion of the front and, after the first oscillation, the body of the vehicle moves in bounce rather than pitch, a favorable factor for ride comfort. Then the rear part of the vehicle should lead the motion, but by that time damping has caused the amplitude to decrease.

The second rule is easily fulfilled in modern cars. The problem here may be that of having the pitch frequency much higher than the bounce one, and higher than 1.3 Hz (third rule), as may happen when the dynamic index is smaller than unity (vehicle with long wheelbase and small front/rear overhang). Generally speaking, a dynamic index close to unity is considered a desirable condition for good ride properties, while a complete bounce-ride uncoupling as occurs when $aK_1 = bK_2$ is considered a nuisance. Coupling between bounce and pitching is good as it tends to avoid strong pitch oscillations.

The fourth rule has nothing to do with pitch motion, and will be discussed later.

Example 26.8 *Check whether the vehicle studied in the previous example complies with the criteria defined by Olley. Study the response of the vehicle when crossing a road irregularity at a speed of 100 km/h = 27.8 m/s by using an impulsive model, assuming that the impulse given by the irregularity first to the front axle and then to the rear axle has a unit value.*

To study the motion of the body after crossing the irregularity, assume that both suspensions are damped with a damping coefficient equal to the optimum value computed using a quarter car model with a single degree of freedom.

The natural frequencies of the suspensions, computed using the model with two independent quarter cars, are 1.33 (front axle) and 1.49 Hz (rear axle). The second is higher than the first by about 12%. By considering that the natural frequencies are proportional to the square root of the stiffness, this corresponds to a stiffness of the rear axle 24% greater than that of the front, a value not much different from the suggested 30%.

Because the dynamic index has no unit value ($I_d = 0.807$), the model made by two quarter cars is not accurate. If the system is considered as a coupled system, the *frequencies for bounce and pitch motions are 1,36 and 1,62 Hz, which does not coincide with those previously computed (the first is not much different, while the second is greater by about 8%). The first is smaller than 1.2 times the second (actually smaller than that) and the frequencies are relatively similar. However, the natural frequency in pitch is greater than 1,3 Hz, and is higher than what has been suggested, even if not by much.*

The values of the damping of the shock absorbers, computed using the quarter car model with a single degree of freedom, are

$$c_1 = \sqrt{\frac{K_1 m_1}{2}} = 3,820 \ Ns/m \ , \qquad c_2 = \sqrt{\frac{K_2 m_2}{2}} = 2,865 \ Ns/m \ . \qquad (26.89)$$

The delay between the instant the front axle is excited and that when the rear axle is on the irregularity is, at 100 km/h,

$$\tau = \frac{l}{V} = 0.096 \ s \ . \qquad (26.90)$$

To compute the response of the model made by two independent quarter car models to a unit impulse, it is enough to compute the free responses of the two systems with the initial conditions due to the impulse. The front suspension will start with the following initial conditions

$$(z_A)_0 = 0 \ , \qquad (\dot{z}_A)_0 = \frac{I}{m_A} \ for \ t = 0 \ . \qquad (26.91)$$

In the same way, the initial conditions for the rear suspension are

$$(z_B)_0 = 0 \ , \qquad (\dot{z}_B)_0 = \frac{I}{m_B} \ for \ t = \tau \ . \qquad (26.92)$$

The result is reported in Fig. 26.30a in terms of time histories. The vertical displacement at the center of mass and the pitch angle may be computed through Eq. (26.67) from the displacements of the points where the suspensions are attached. The result is plotted in Fig. 26.30b, dashed curves. Because the natural frequency of the rear axle is greater than that of the front axle, the two masses move synchronously after a single oscillation is completed: pitch motions extinguish faster than bounce.

To avoid the approximations due to the model with two independent quarter cars, it is possible to numerically integrate the system's equations of free motion (homogeneous equation associated to Eq. (26.66)) in two distinct intervals of time, between $t = 0$ and $t = \tau$ and after $t = \tau$. In the first interval the initial conditions are those following the first impulse.

$$\left\{ \begin{matrix} z_G \\ \theta \end{matrix} \right\}_0 = \left\{ \begin{matrix} 0 \\ 0 \end{matrix} \right\} \ , \qquad \left\{ \begin{matrix} \dot{z}_G \\ \dot{\theta} \end{matrix} \right\}_0 = \left\{ \begin{matrix} \frac{I}{m} \\ -\frac{aI}{J} \end{matrix} \right\} \ for \ t = \tau \ . \qquad (26.93)$$

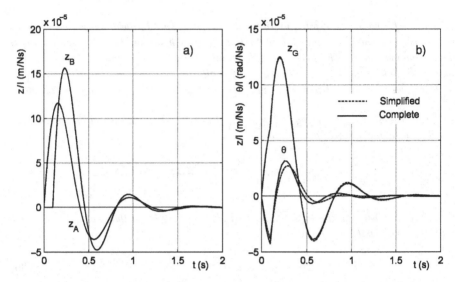

FIGURE 26.30. Bounce and pitch response for a vehicle crossing an obstacle with the front and the the rear axle at 100 km/h. The obstacle causes a unit vertical impulse. Displacements of points A and B (a) and of the center of mass and rotation (b) computed using the simplified (Fig. 26.27b) and complete (Fig. 26.27a) model

For the second interval, that following the second impulse, the initial conditions are

$$\left\{ \begin{array}{c} z_G \\ \theta \end{array} \right\}_0 = \left\{ \begin{array}{c} z_G \\ \theta \end{array} \right\}_1 , \quad \left\{ \begin{array}{c} \dot{z}_G \\ \dot{\theta} \end{array} \right\}_0 = \left\{ \begin{array}{c} \dot{z}_G \\ \dot{\theta} \end{array} \right\}_1 + \left\{ \begin{array}{c} \frac{I}{m} \\ \frac{bI}{J} \end{array} \right\} \quad for \; t = \tau \; , \quad (26.94)$$

where subscript 1 designates the condition at the end of the first part of the integration, just before receiving the second impulse. The result is shown in Fig. 26.30b, full lines. The errors due to the model made by two independent quarter cars are small, even if the dynamic index is smaller than one.

26.5.4 Frequency response of the model with two degrees of freedom

If, when using the quarter car model road, roughness excites only bounce motions, in a complete vehicle it excites both bounce and pitching motions, as already seen. Neglecting roll, it is possible to use a model of the type seen in Fig. 26.27a, assuming that laws $h_A(t)$ and $h_B(t)$ are identical, except for the fact that the second is delayed with respect to the first by time

$$\tau = \frac{l}{V}, \quad (26.95)$$

needed to travel a distance equal to the wheelbase. To compute the frequency response of the vehicle, the forcing functions to be introduced into equation

(26.68) are then

$$\begin{cases} h_A = h_o \sin(\omega t) \\ h_B = h_o \sin[\omega(t + \tau)] = h_o[\sin(\omega t)\cos(\omega \tau) + \cos(\omega t)\sin(\omega \tau)] \,. \end{cases} \quad (26.96)$$

Term $\omega \tau$ is proportional to the ratio between the wheelbase l of the vehicle and the wavelength of road irregularities λ. The frequency $\bar{\omega}$ related to space instead of time is linked with the wavelength by the relationship

$$\lambda = \frac{2\pi}{\bar{\omega}} \quad (26.97)$$

and to the frequency related to time by the relationship

$$\bar{\omega} = \frac{\omega}{V} \,. \quad (26.98)$$

It follows then

$$\omega \tau = \frac{\omega l}{V} = 2\pi \frac{l}{\lambda} \,. \quad (26.99)$$

At low frequency, the excitation at the two axles occurs almost in phase, with the result that pitch motions are little excited. In a similar way, if l is a whole multiple of the wavelength λ (the wavelength is equal to the wheelbase or to one of its whole sub-multiples), $\omega \tau$ is a whole multiple of 2π and then $\cos(\omega \tau) = 1$ and $\sin(\omega \tau) = 0$. The two axles are excited in phase: if the equations of motion were uncoupled and the center of mass were at mid-wheelbase, only bounce motion would be excited and no pitching would occur. Although this is not exactly true due to coupling, the result is that the vehicle pitches much less than it bounces.

If, on the contrary, l is an odd multiple of $\lambda/2$ (the wavelength of the irregularities is twice the wheelbase or is a whole multiple of twice the wheelbase), it follows that $\cos(\omega \tau) = -1$ and $\sin(\omega \tau) = 0$, and the two axles are excited with 180° phasing. In this case, if the center of mass were at mid-wheelbase and the system uncoupled, no bouncing would occur and the vehicle would only pitch. This consideration holds qualitatively for actual cases.

This phenomenon, usually referred to as wheelbase filtering, introduces a dependence between the response of the system and speed. If, for instance, the wheelbase is 2 m and the speed is 20 m/s, the delay τ is 0,1 s. The maximum pitch response, with a very low bounce, occurs when the irregularities have a wavelength equal to twice the wheelbase or one of the odd submultiples of twice the wheelbase, that is 4, 4/3, 4/5, ... m. At 20 m/s, the corresponding frequencies at which bounce motions are minimal are 5, 15, 25, ... Hz. In the same way, the maximum bounce motions with little pitching occur at wavelengths equal to the wheelbase and its whole submultiples, 2, 1, 0,5, ... m. At a speed of 20 m/s, the corresponding frequencies are 10, 20, 30, ... Hz. Moreover, little pitch excitation occurs at very low frequency, as already stated, and as a consequence pitch excitation is minimal in highway driving.

The situation may be different for industrial vehicles owing to the larger wheelbase and lower speed coupled with high spring stiffness: wheelbase filtering may lead to strong pitch response, accompanied by low bounce. The effect is further worsened by the fact that in tall vehicles pitch excitation causes longitudinal oscillation in points above the center of mass that may prove quite inconvenient.

In general, the expression of the excitation vector is

$$
h_0 \left\{ \begin{array}{c} \left[K_1 + K_2 \cos(\omega\tau) - c_2\omega\sin(\omega\tau) \right] \sin(\omega t) + \\ + \left[c_1\omega + c_2\omega\cos(\omega\tau) + K_2\sin(\omega\tau) \right] \cos(\omega t) \\ \left[-aK_1 + bK_2\cos(\omega\tau) - bc_2\omega\sin(\omega\tau) \right] \sin(\omega t) + \\ + \left[-ac_1\omega + bc_2\omega\cos(\omega\tau) + bK_2\sin(\omega\tau) \right] \cos(\omega t) \end{array} \right\} . \tag{26.100}
$$

The equation of motion (26.66) for vertical and pitch oscillations can then be written as

$$
\begin{bmatrix} m_S & 0 \\ 0 & J_y \end{bmatrix} \left\{ \begin{array}{c} \ddot{Z}_s \\ \ddot{\theta} \end{array} \right\} + \begin{bmatrix} c_1 + c_2 & -ac_1 + bc_2 \\ -ac_1 + bc_2 & a^2c_1 + b^2c_2 \end{bmatrix} \left\{ \begin{array}{c} \dot{Z}_s \\ \dot{\theta} \end{array} \right\} +
$$

$$
+ \begin{bmatrix} K_1 + K_2 & -aK_1 + bK_2 \\ -aK_1 + bK_2 & a^2K_1 + b^2K_2 \end{bmatrix} \left\{ \begin{array}{c} Z_s \\ \theta \end{array} \right\} = \tag{26.101}
$$

$$
= h_0 \left\{ \begin{array}{c} f_1(\omega\tau)\sin(\omega t) + g_1(\omega\tau)\cos(\omega t) \\ f_2(\omega\tau)\sin(\omega t) + g_2(\omega\tau)\cos(\omega t) \end{array} \right\} ,
$$

where:

$$
\begin{aligned}
f_1(\omega\tau) &= K_1 + K_2\cos(\omega\tau) - c_2\omega\sin(\omega\tau) , \\
f_2(\omega\tau) &= -aK_1 + bK_2\cos(\omega\tau) - bc_2\omega\sin(\omega\tau) , \\
g_1(\omega\tau) &= c_1\omega + c_2\omega\cos(\omega\tau) + K_2\sin(\omega\tau) , \\
g_2(\omega\tau) &= -ac_1\omega + bc_2\omega\cos(\omega\tau) + bK_2\sin(\omega\tau) .
\end{aligned} \tag{26.102}
$$

Functions $f_i(\omega\tau)$ and $g_i(\omega\tau)$ may be considered as filters that, applied to the sine and cosine components of the excitation due to the road profile, yield the bounce and pitch excitation. However, because of coupling between the equations of motion, all terms of the excitation contribute to both bouncing and pitching.

To obtain a first approximation evaluation of the effect of wheelbase filtering, assume that the equations of motion are uncoupled ($aK_1 = bK_2$ and $bc_2 = ac_1$) and that the center of mass is at mid wheelbase ($a = b$). To comply with both these conditions the front and rear suspensions must have the same elastic and damping characteristics ($K_1 = K_2$ and $c_2 = c_1$).

The two equations of motion uncouple, reducing to

$$
m_S\ddot{Z}_s + 2c_1\dot{Z}_s + 2K_1Z_s = h_0\left[f_1(\omega\tau)\sin(\omega t) + g_1(\omega\tau)\cos(\omega t) \right], \tag{26.103}
$$

where

$$
\begin{aligned}
f_1(\omega\tau) &= K_1\left[1 + \cos(\omega\tau) \right] - c_1\omega\sin(\omega\tau) , \\
g_1(\omega\tau) &= c_1\omega\left[1 + \cos(\omega\tau) \right] + K_1\sin(\omega\tau) ,
\end{aligned} \tag{26.104}
$$

for vertical motions, and

$$J_y\ddot{\theta} + 2a^2c_1\dot{\theta} + 2a^2K_1\theta = h_0\left[f_2(\omega\tau)\sin(\omega t) + g_2(\omega\tau)\cos(\omega t)\right], \qquad (26.105)$$

where

$$\begin{aligned} f_2(\omega\tau) &= aK_1\left[-1 + \cos(\omega\tau)\right] - ac_1\omega\sin(\omega\tau) , \\ g_2(\omega\tau) &= ac_1\omega\left[-1 + \cos(\omega\tau)\right] + aK_1\sin(\omega\tau) , \end{aligned} \qquad (26.106)$$

for pitching motions.

If the phasing between bounce and pitch motion is not to be computed, it is useless to obtain the sine and cosine components of the response separately: what matters is solely its amplitude. The amplitude of the excitation for bouncing motions is

$$h_0\sqrt{2\left(K_1^2 + c_1^2\omega^2\right)}\sqrt{1 + \cos(\omega\tau)} . \qquad (26.107)$$

The corresponding frequency response is then

$$\left|\frac{Z_{s_0}}{h_0}\right| = \sqrt{\frac{4\left(K_1^2 + c_1^2\omega^2\right)}{\left(2K_1 - m\omega^2\right)^2 + 4c_1^2\omega^2}}\sqrt{\frac{1 + \cos(\omega\tau)}{2}} . \qquad (26.108)$$

The first square root is nothing else than the amplification factor of a quarter car with a single degree of freedom with mass $m/2$, stiffness K_1 and damping c_1. The second term gives the wheelbase filtering effect for vertical motions. Function

$$\sqrt{\frac{1 + \cos(\omega\tau)}{2}}$$

is plotted in Fig. 26.31a versus the frequency, together with the frequency response for the acceleration (inertance) of the quarter car model and their product. The values of the speed and the wheelbase used to plot the figure are 30 m/s and 2.16 m respectively.

In a similar way, the amplitude of the excitation entering the second equation, that for pitching motions, is

$$h_0\sqrt{2a^2\left(K_1^2 + c_1^2\omega^2\right)}\sqrt{1 - \cos(\omega\tau)} . \qquad (26.109)$$

The frequency response for pitch motion is

$$\left|\frac{\theta_0}{h_0}\right| = \sqrt{\frac{4a^2\left(K_1^2 + c_1^2\omega^2\right)}{\left(2a^2K_1 - J_y\omega^2\right)^2 + 4a^2c_1^2\omega^2}}\sqrt{\frac{1 - \cos(\omega\tau)}{2}} , \qquad (26.110)$$

that is, introducing the dynamic index I_d,

$$\left|\frac{\theta_0}{h_0}\right| = \sqrt{\frac{4\left(K_1^2 + c_1^2\omega^2\right)}{\left(2K_1 - mI_d\omega^2\right)^2 + 4c_1^2\omega^2}}\sqrt{\frac{1 - \cos(\omega\tau)}{2}} . \qquad (26.111)$$

If the dynamic index has a unit value, the first square root coincides with that seen for vertical motions, that is, it coincides with the amplification factor

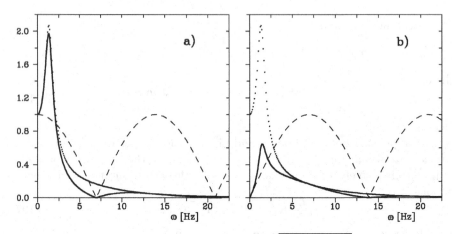

FIGURE 26.31. Wheelbase filtering. (a): Function $\sqrt{2[1 + \cos(\omega\tau)]}$ (dashed line), inertance of the quarter car model (dotted line) and product of the two (full line); (b): Same as in (a), but for function $\sqrt{2[1 - \cos(\omega\tau)]}$. $V = 30$ m/s; $l = 2.16$ m.

of a quarter car with a single degree of freedom with mass $m/2$, stiffness K_1 and damping c_1; reference must otherwise be made directly to the pitching oscillations of the beam constituting the model of the vehicle. The second term yields the wheelbase filtering effect for pitching motions. Function

$$\sqrt{\frac{1 - \cos(\omega\tau)}{2}}$$

is plotted in Fig. 26.31b versus the frequency, together with the inertance of the quarter car and their product, using the same values of V and l, as in Fig. 26.31a.

Remark 26.10 *The subjective feeling of riding comfort is also affected by the position of the passengers; when they are close to the centre of mass, pitching oscillations are slight, but they may be a nuisance in points located a greater distance from it. Bounce - pitch coupling due to suspensions may severely reduce riding comfort.*

Example 26.9 *Compute functions $f_i(\omega\tau)$ and $g_i(\omega\tau)$ for the vehicle of the previous examples at a speed of 100 km/h = 27.8 m/s and plot the frequency responses for bounce and pitch oscillations.*

The computation will be performed initially using a first approximation model (uncoupling between bounce and pitch and unit dynamic index), and then factoring in the actual value of the parameters.

To uncouple the equations, the actual values of a and b are substituted by $l/2$ and those of K_1, K_2, c_1 and c_2 by the mean values of the stiffnesses and damping coefficients. Morover, J_y is assumed to be equal to mab.

The results are plotted in Fig. 26.32a and b. As expected, at vanishing frequency and when l is equal to a whole multiple of λ ($\omega = 0$, $\omega = 10.44$ Hz, ...) the pitching

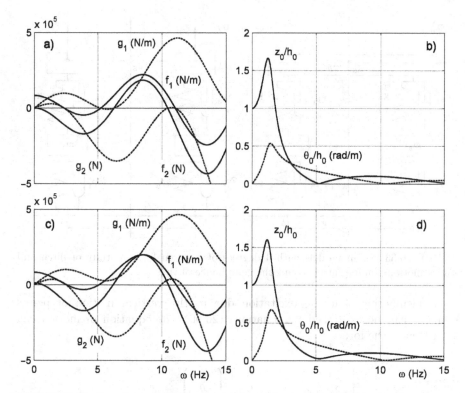

FIGURE 26.32. Functions $f_i(\omega\tau)$ and $g_i(\omega\tau)$ and frequency responses for bounce and pitch motions for the vehicle of the previous examples at 100 km/h. (a) and (b): simplified uncoupled model. (c) and (d): actual value of the parameters

response vanishes and only bounce is present. If l is an odd multiple of $\lambda/2$ ($\omega = 5.22$ Hz, $\omega = 15.66$ Hz, ...) the bounce response vanishes and only pitch is present.

The results obtained from the actual values of the parameters are shown in Fig. 26.32c and d. As is clear from the figure, the results differ from those obtained by uncoupling the equations, but the difference is not large. In particular, the bounce response never vanishes, even if at about 5 and 15 Hz it becomes quite small.

It must be noted that the model with stiff tires used here should not be used for frequencies higher than $4 - 6$ Hz.

26.5.5   Effect of tire compliance

If the compliance of tires is accounted for, the model must contain also the unsprung masses. The minimum number of degrees of freedom needed to study bounce and pitch motions is four (Fig 26.33a). If the dynamic index has a unit value, the model of Fig 26.33a may be substituted by that of Fig 26.33b.

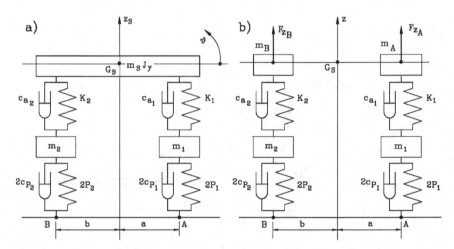

FIGURE 26.33. Beam models with 4 degrees of freedom for the study of pitch and bounce motions taking into account the compliance of tires

Remembering that the excitation due to the vertical motion of points A and B can be expressed by equations (26.96), the equation of motion can be written in the form

$$
\begin{bmatrix} m_S & 0 & 0 & 0 \\ 0 & J_y^* & 0 & 0 \\ 0 & 0 & m_1 & 0 \\ 0 & 0 & 0 & m_2 \end{bmatrix} \begin{Bmatrix} \ddot{Z}_s \\ \ddot{\theta} \\ \ddot{Z}_1 \\ \ddot{Z}_2 \end{Bmatrix} +
$$

$$
+ \begin{bmatrix} c_1 + c_2 & -ac_1 + bc_2 & -c_1 & -c_2 \\ & a^2c_1 + b^2c_2 & ac_1 & -bc_2 \\ & & c_1 + 2c_{p_1} & 0 \\ \text{symm.} & & & c_2 + 2c_{p_2} \end{bmatrix} \begin{Bmatrix} \dot{Z}_s \\ \dot{\theta} \\ \dot{Z}_1 \\ \dot{Z}_2 \end{Bmatrix} + \quad (26.112)
$$

$$
+ \begin{bmatrix} K_1 + K_2 & -aK_1 + bK_2 & -K_1 & -K_2 \\ & a^2 K_1 + b^2 K_2 & aK_1 & -bK_2 \\ & & K_1 + 2P_1 & 0 \\ \text{symm.} & & & K_2 + 2P_2 \end{bmatrix} \begin{Bmatrix} Z_s \\ \theta \\ Z_1 \\ Z_2 \end{Bmatrix} =
$$

$$
= h_0 \begin{Bmatrix} 0 \\ 0 \\ 2P_1 \sin(\omega t) + 2\omega c_{p_1} \cos(\omega t) \\ 2f(\omega\tau)\sin(\omega t) + 2g(\omega\tau)\cos(\omega t) \end{Bmatrix} ,
$$

where m_1 and m_2 are the unsprung masses of the two axles.

$$
\begin{aligned} f(\omega\tau) &= P_2 \cos(\omega\tau) - c_{p_2}\omega \sin(\omega\tau) , \\ g(\omega\tau) &= c_{p_2}\omega \cos(\omega\tau) + P_2 \sin(\omega\tau) . \end{aligned} \quad (26.113)
$$

Example 26.10 *Compute the frequency responses for bounce and pitch motion using the model with 4 degrees of freedom. Compare the results with those obtained from the model with two degrees of freedom.*

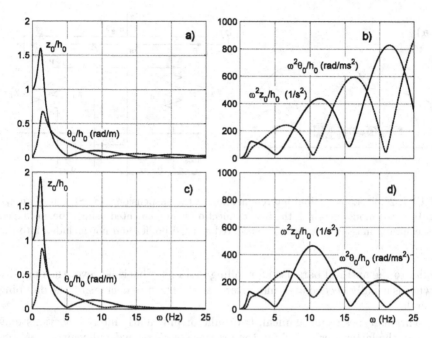

FIGURE 26.34. Frequency responses for displacements and accelerations in bounce and pitch motions for the vehicle of the previous examples at 100 km/h. (a) and (b): model with two degrees of freedom; (c) and (d): model with four degrees of freedom

Data of the unsprung masses: $m_{n1} = m_{n2} = 65$ *kg,* $P_1 = P_2 = 125$ *kN/m,* $c_{p1} = c_{p2} = 0$.

The results obtained using the model with 2 degrees of freedom are reported in Fig. 26.34a and b, while those obtained using the model of 4 degrees of freedom are reported in Fig. 26.34c and d. The values of the natural frequencies of the undamped system are 1,25, 1,51, 10,61 e 10,74 Hz, while those of the simplified model are 1,36 e 1,62 Hz .

At low frequency the results obtained from the two models are similar, while at frequencies higher than those of the unsprung masses the filtering effect of the tires reduces the amplitude of the response.

26.5.6 Interconnected suspensions

If the value of the pitch natural frequency is too high when compared with that of the bounce motions, ride comfort may be affected. To control the natural frequencies of pitch and bounce independently, without changing the wheel positions and the inertial properties of the body, the suspensions can be interconnected. Pitching frequencies can be raised without increasing those in bounce if the front and rear wheels are connected by a spring opposing pitching motions,

a)

b)

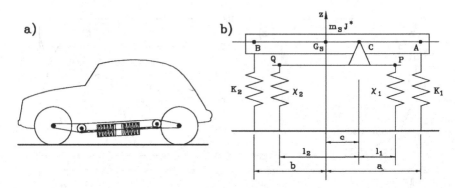

FIGURE 26.35. Longitudinal interconnection of the suspensions. (a) Sketch of an application; (b) model in which the interconnection is implemented using a beam hinged to the sprung mass. The tires are considered as rigid bodies and not included into the model

similar to the anti-roll bars used for rolling motions. This is however the opposite of what is usually needed and, moreover, has the effect of decreasing the damping of pitch.

Various types of mechanical, hydraulic or pneumatic interconnections may be used, the latter particularly in the presence of air or hydraulic springs. A mechanical solution is shown in Fig. 26.35a: The vehicle is based on longitudinal swing arm suspensions, with springs located longitudinally under the sprung mass. The springs are connected to a further element, itself elastically connected to the vehicle body. The system is functionally similar (even if simpler) to the model shown in Fig. 26.35b, in which the intermediate element is a beam, hinged to the vehicle body and connected to the unsprung masses through springs. The tires are considered here as rigid bodies and have not been included in the model.

If the beam and springs with stiffness χ_1 and χ_2 were not included, the equation of motion would have been Eq. (26.66), without the damping matrix, as in the figure, if damping is neglected. If the inertia of the beam is neglected, no further degree of freedom is needed, because the position of the beam is determined by the displacement z and the rotation θ of the sprung mass. The stiffness matrix may be obtained simply by adding the potential energy of springs χ_1 and χ_2 and performing the relevant derivatives.

The positions of points P and Q are simply

$$\begin{cases} z_P = z + c\theta - l_1\gamma \\ z_Q = z + c\theta + l_2\gamma , \end{cases} \tag{26.114}$$

where γ is the angle between line PQ and the horizontal and all relevant angles are assumed to be small.

The potential energy due to the two added springs is

$$2\mathcal{U} = \chi_1 z_P^2 + \chi_2 z_Q^2 = (\chi_1 + \chi_2)\left(z^2 + c^2\theta^2 + 2c\theta z\right) +$$

$$+\gamma^2(l_1^2\chi_1 + l_2^2\chi_2) + 2\gamma(z + c\theta)(l_1\chi_1 - l_2\chi_2) . \tag{26.115}$$

The value of γ can be easily computed by stating

$$\frac{\partial \mathcal{U}}{\partial \gamma} = 0 \,,$$

which yields

$$\gamma = -(z + c\theta)\frac{l_1\chi_1 - l_2\chi_2}{l_1^2\chi_1 + l_2^2\chi_2} \,. \tag{26.116}$$

The final expression for the potential energy is then

$$\mathcal{U} = \frac{1}{2}(z + c\theta)^2 \frac{\chi_1\chi_2(l_1 + l_2)^2}{l_1^2\chi_1 + l_2^2\chi_2} \,. \tag{26.117}$$

By performing the relevant derivatives, the stiffness matrix becomes

$$\mathbf{K} = \begin{bmatrix} K_1 + K_2 + \chi & -aK_1 + bK_2 + \chi c \\ -aK_1 + bK_2 + \chi c & a^2 K_1 + b^2 K_2 + \chi c^2 \end{bmatrix} \,, \tag{26.118}$$

where

$$\chi = \frac{\chi_1\chi_2(l_1 + l_2)^2}{l_1^2\chi_1 + l_2^2\chi_2} \,.$$

From Eq. (26.118) it is clear that the terms due to the interconnection between front and rear suspensions affect in a different way the various elements of the stiffness matrix and allow to modify independently the values of the bounce and pitch natural frequencies, possibly lowering the latter without affecting the former.

26.6   ROLL MOTION

26.6.1   Model with a single degree of freedom

As already stated, roll is coupled with handling and not with ride comfort. However it is also true that rolling can affect strongly the subjective feeling of riding comfort.

The simplest model for studying roll motion is a rigid body, simulating the sprung mass, free to rotate about the roll axis, constrained to the ground by a set of springs and damper with a stiffness and a damping coefficient equal to those of the suspensions (Fig. 26.36). If J_x, m_s, χ_i and Γ_i are respectively the moment of inertia about the roll axis, the sprung mass, the torsional stiffness and the damping coefficient of the ith suspension, the equation of motion is

$$J_x\ddot{\phi} + (\Gamma_1 + \Gamma_2)\dot{\phi} + (\chi_1 + \chi_2)\phi - m_s g h_G \sin(\phi) = \tag{26.119}$$

$$= \Gamma_1\dot{\alpha}_{t_1} + \Gamma_2\dot{\alpha}_{t_2} + \chi_1\alpha_{t_1} + \chi_2\alpha_{t_2} \,,$$

where the forcing functions are those due to the transversal inclination of the road α_{t_i} at the ith suspension.

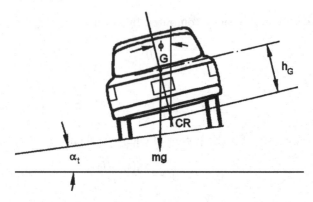

FIGURE 26.36. Model with a single degree of freedom for the study of roll motion. Cross section in a plane containing the center of mass G of the sprung mass. The roll axis goes through point CR

The inertia of the unsprung masses and the compliance of the tires are not included in such a simple model, which is formally identical to the quarter car with a single degree of freedom.

For small values of the roll angle, the model may be linearized, stating $\sin(\phi) \approx \phi$. The roll natural frequency is then

$$\omega_{roll} = \sqrt{\frac{\chi_1 + \chi_2 - m_s g h_G}{J_x}} . \tag{26.120}$$

The optimum damping value may be obtained from Eq. (26.12):

$$\Gamma_{opt} = \sqrt{\frac{J_x(\chi_1 + \chi_2 - m_s g h_G)}{2}} . \tag{26.121}$$

This condition is generally not satisfied, particularly if the vehicle has anti-roll bars. The torsional damping of the suspensions is supplied by the same shock absorbers normally designed to optimize vertical motion; the roll damping they supply is usually lower than needed. The increase in stiffness due to anti-roll bars is not accompanied by an increase in damping. The effect is causes a decrease of the damping ratio, together with an increase of the natural frequency.

The stiffer the suspension in torsion, the more underdamped the roll behavior, if the increase in stiffness is due to anti-roll bars. Although reducing rolling in stationary conditions, they may increase it in dynamic conditions. An overelongation in the step response, as when roll is due to a moment abruptly applied (steering step input, wind gusts or other similar instances), may then result. A large roll in dynamic conditions my cause rollover.

The stationary value of the roll angle on a road with transversal slope $\alpha_t = \alpha_{t_1} = \alpha_{t_2}$ is

$$\phi = \alpha_t \frac{\chi_1 + \chi_2}{\chi_1 + \chi_2 - m_s g h_G} . \tag{26.122}$$

The importance of a center of mass not too high on the roll axis and stiff suspensions (in roll) is then clear. The last condition contradicts the need for a small roll in dynamic conditions.

Example 26.11 *Consider the vehicle studied in Example 26.7, and assume that the moment of inertia J_x is equal to 388.8 kg m^2 and the sprung mass is 1,080 kg. Compute the time history of the roll angle when the vehicle encounters a ramp leading from a horizontal road to a transversal slope $\alpha_t = 5°$ in a distance of 10 m at a speed of 30 m/s.*

Other data: stiffnesses of the axles $K_1 = 45$ kN/m, $K_2 = 38$ kN/m, damping of the axles $c_1 = 3,820$ Ns/m, $c_2 = 2,865$ Ns/m, distance of the springs and dampers from the symmetry plane $d = 0.5$ m, wheelbase $l = 2.66$ m. Repeat the computation, adding an anti-roll bar at the front axle, with a stiffness $\chi_b = 4,000$ Nm/rad.

Computation without anti-roll bar.

The stiffnesses and damping coefficients of the axles can be computed using formulae of the type

$$\chi_i = K_i d_i^2 . \tag{26.123}$$

It then follows that: $\chi_1 = 11.25$ kNm/rad, $\chi_2 = 9.5$ kNm/rad, $\Gamma_1 = 955$ Nms/rad, $\Gamma_2 = 716$ Nms/rad. The total roll damping $\Gamma_1 + \Gamma_2 = 1.671$ Nms/rad is then smaller than the optimum damping computed by neglecting the gravitational effect (2.008 Nms/rad), while only slightly smaller than that computed by taking it into account (1.733 Nsm/rad).

The roll natural frequencies are then $\omega = 1.00$ Hz (gravitational effect included) or $\omega = 1.16$ Hz.

The linearized, state space equation

$$
\left\{ \begin{array}{c} \dot{v}_\phi \\ \dot{\phi} \end{array} \right\} = \left[\begin{array}{cc} -\frac{(\Gamma_1 + \Gamma_2)}{J_x} & -\frac{\chi}{J_x} \\ 1 & 0 \end{array} \right] \left\{ \begin{array}{c} v_\phi \\ \phi \end{array} \right\} + \frac{1}{J_x} \left[\begin{array}{cccc} \Gamma_1 & \Gamma_2 & \chi_1 & \chi_2 \\ 0 & 0 & 0 & 0 \end{array} \right] \left\{ \begin{array}{c} \dot{\alpha}_{t_1} \\ \dot{\alpha}_{t_2} \\ \alpha_{t_1} \\ \alpha_{t_2} \end{array} \right\}
\tag{26.124}
$$

where $v_\phi = \dot{\phi}$, can be written and then solved numerically to compute the time history of the response.

The input to the system is given by angles α_i and their derivatives. If the excitation is due to motion at a speed V on a ramp having a length l_r leading linearly to a transversal slope α_t, assuming that at time $t = 0$ the front axle meets the ramp, it follows that

$$
\alpha_{t_1} = \left\{ \begin{array}{ll} 0 & for\ t \le 0 \\ \alpha_t t \frac{V}{l_r} & for\ 0 < t < \frac{l_r}{V} \\ \alpha_t & for\ t \ge \frac{l_r}{V} \end{array} \right. \qquad
\alpha_{t_2} = \left\{ \begin{array}{ll} 0 & for\ t \le \frac{l}{V} \\ \alpha_t \left(t - \frac{l}{V} \right) \frac{V}{l_r} & for\ \frac{l}{V} < t < \frac{l_r + l}{V} \\ \alpha_t & for\ t \ge \frac{l_r + l}{V} \end{array} \right.
\tag{26.125}
$$

$$
\dot{\alpha}_{t_1} = \left\{ \begin{array}{ll} 0 & for\ t \le 0 \\ \alpha_t \frac{V}{l_r} & for\ 0 < t < \frac{l_r}{V} \\ \alpha_t & for\ t \ge \frac{l_r}{V} \end{array} \right. \qquad
\dot{\alpha}_{t_2} = \left\{ \begin{array}{ll} 0 & for\ t \le \frac{l}{V} \\ \alpha_t \frac{V}{l_r} & for\ \frac{l}{V} < t < \frac{l_r + l}{V} \\ \alpha_t & for\ t \ge \frac{l_r + l}{V} \end{array} \right.
\tag{26.126}
$$

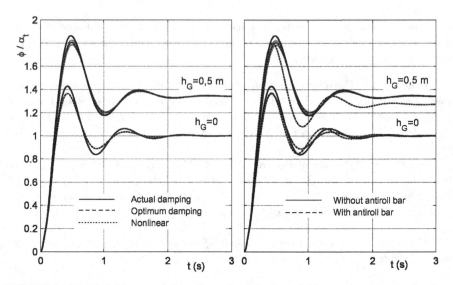

FIGURE 26.37. Time history of the roll angle when the vehicle manages a ramp leading from level road to a transversal slope α_t. Model with 1 degree of freedom

Computation with anti-roll bar.

The computation is then repeated after adding the stiffness of the anti-roll bar to that of the suspensions. The roll natural frequency is now $\omega = 1.13$ Hz.

The results are reported in non-dimensional form in Fig. 26.37. Results obtained with both the actual and the optimum damping are reported for the case without anti-roll bar. In this case, the actual damping is smaller than the optimum and the difference between the two results is small.

As expected, the steady state inclination of the body coincides with the transversal slope of the road, if the effect of the weight is neglected. If weight is accounted for, the final inclination of the body is greater. The nonlinear model has been integrated numerically to check whether the results obtained are realistic: owing to the low values of the angles, the difference between the linearized and the nonlinear results is negligible.

If an anti-roll bar is present the whole curve decreases, when the effect of weight is taken into account. If the latter is neglected, the effect of the anti-roll bar is minimal.

26.6.2   Model with many degrees of freedom

A simple model with three degrees of freedom may be used to take rolling of unsprung masses and compliance of the tires into account. The unsprung masses are modelled as rigid bodies free to rotate about the roll axis of the vehicle. It is clear that this model is a rough approximation, particularly if independent suspensions are used. This further approximation, however, does not worsen matters, because the largest errors come from studying roll motion without taking into account that they are coupled with handling motions.

The linearized equation for the study of roll motion is

$$
\begin{bmatrix} J_x & 0 & 0 \\ 0 & J_{x_1} & 0 \\ 0 & 0 & J_{x_1} \end{bmatrix} \begin{Bmatrix} \ddot{\phi} \\ \ddot{\phi}_1 \\ \ddot{\phi}_2 \end{Bmatrix} + \begin{bmatrix} \Gamma_1 + \Gamma_2 & -\Gamma_1 & -\Gamma_2 \\ -\Gamma_1 & \Gamma_1 + \Gamma_{p1} & 0 \\ -\Gamma_2 & 0 & \Gamma_2 + \Gamma_{p2} \end{bmatrix} \times
$$

$$
\times \begin{Bmatrix} \dot{\phi} \\ \dot{\phi}_1 \\ \dot{\phi}_2 \end{Bmatrix} + \begin{bmatrix} \chi_1 + \chi_2 - m_s g h_G & -\chi_1 & -\chi_2 \\ -\chi_1 & \chi_1 + \chi_{p1} & 0 \\ -\chi_2 & 0 & \chi_2 + \chi_{p2} \end{bmatrix} \begin{Bmatrix} \phi \\ \phi_1 \\ \phi_2 \end{Bmatrix} =
$$

$$
= \begin{Bmatrix} 0 \\ \Gamma_{p1} \dot{\alpha}_{t_1} + \chi_{p1} \alpha_{t_1} \\ \Gamma_{p2} \dot{\alpha}_{t_2} + \chi_{p2} \alpha_{t_2} \end{Bmatrix} ,
$$

$$(26.127)$$

where χ_i, χ_{p_i}, Γ_i, Γ_{p_i} are the stiffness and the damping of the suspensions and of the tires. The excitation is given by the transversal slope of the road α_{t_1} and α_{t_2} at the front and rear axles.

Equation (26.127) can be solved numerically and allows the natural frequencies of roll oscillations to be computed.

To drastically simplify the model, the moment of inertia of the unsprung masses and the damping of the tires can be neglected. The equations of motion are thus a set of a second order equations plus two first order ones. The state space equation so obtained is of the fourth order:

$$
\begin{Bmatrix} \dot{v}_\phi \\ \dot{\phi} \\ \dot{\phi}_1 \\ \dot{\phi}_2 \end{Bmatrix} = \begin{bmatrix} 0 & m_s g h_G / J_x & -\chi_{p1}/J_x & -\chi_{p2}/J_x \\ 1 & 0 & 0 & 0 \\ 1 & \chi_1/\Gamma_1 & -(\chi_1 + \chi_{p1})/\Gamma_1 & 0 \\ 1 & \chi_2/\Gamma_2 & 0 & -(\chi_2 + \chi_{p2})/\Gamma_2 \end{bmatrix} \times
$$

$$
\times \begin{Bmatrix} v_\phi \\ \phi \\ \phi_1 \\ \phi_2 \end{Bmatrix} + \begin{bmatrix} \chi_{p1}/J_x & \chi_{p2}/J_x \\ 0 & 0 \\ \chi_{p1}/\Gamma_1 & 0 \\ 0 & \chi_{p2}/\Gamma_2 \end{bmatrix} \begin{Bmatrix} \alpha_{t_1} \\ \alpha_{t_2} \end{Bmatrix} .
$$

$$(26.128)$$

Example 26.12 *Consider the vehicle of the previous example, assuming that the stiffness of the tires is 125 kN/m. Assuming a value of 1.48 m for the track, the torsional stiffness for the unsprung masses is $\chi_1 = \chi_2 = 136.9$ kNm/rad. The results are reported in Fig. 26.38 in nondimensional form.*

As is clear from the figure, the effect of the compliance of the tire is not large.

26.7 EFFECT OF NONLINEARITIES

26.7.1 Shock absorbers

As previously stated, shock absorbers are far from being linear viscous dampers. In fact, most automotive shock absorbers are unsymmetrical, with a damping

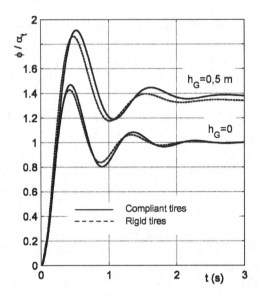

FIGURE 26.38. Time history of the roll angle when the vehicle manages a ramp leading from level road to a transversal slope α_t. Model with 3 degrees of freedom

which is larger in the rebound stroke. Apart from the nonlinearities in the behavior of the shock absorbers and those due to the geometry of the suspension, along with asymmetries purposely built in, other unwanted nonlinear effects, such as dry friction, are often present. Particular care must be devoted to the effects of lateral loads in McPherson suspensions, due to a more or less pronounced dependence of the characteristics on temperature and cavitation. The latter phenomenon is primarily felt at high temperature, and consists in the vaporization of the fluid or the expansion of the gasses dissolved in it.

Moreover, even in cases where shock absorbers are assumed to act in the same direction as other forces, some deviations may occur in practice, introducing further nonlinearities that should be accounted for.

By neglecting the inertia of moving elements and temperature variations, the force exerted by a shock absorber may be considered as a function of both relative displacement and relative velocity of its endpoints:

$$F = F(z, \dot{z}) . \tag{26.129}$$

The experimental results are often reported in the form of a force-displacement plot (Fig. 26.39a). If the force were proportional to velocity (viscous damping) the plot obtained in harmonic motion conditions would be an ellipse, with a ratio between its axes proportional to the frequency. If the characteristics were linear but unsymmetrical (i.e. bilinear) the plot would be made by two semi-ellipses, one above (the smallest) and one below (the largest) the abscissa's axis.

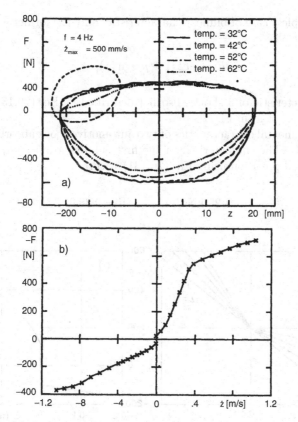

FIGURE 26.39. (a): Force-displacement experimental plot for a shock absorber at various temperatures. Note the anomaly, likely due to cavitation, in the second quadrant. (b): Force-speed plot for the same shock absorber, obtained in slightly different conditions

The force-speed plot (characteristic plot) of the same shock absorber is reported in Fig. 26.39b. If no cavitation occurs, the force depends only on the speed, i.e. the intersections of the surface (26.129) with planes with $\dot{z}$ constant are horizontal straight lines, and the characteristic diagram is unique. In this case, force F depends only on $\dot{z}$ and may be written as the sum of a linear characteristic (viscous damping), an odd function $f_o(\dot{z})$ and an even function $f_e(\dot{z})$ of the speed $\dot{z}$:[13]

$$F = -c\,\dot{z} - f_e(\dot{z}) - f_o(\dot{z}) \ . \tag{26.130}$$

The two functions are, respectively, the *deviation from symmetry* and the *deviation from linearity*.

---

[13]G. Genta, P. Campanile, *An Approximated Approach to the Study of Motor Vehicle Suspensions with Nonlinear Shock Absorbers*, Meccanica, Vol. 24, 1989, pp. 47-57.

In the simplest case of bilinear characteristic, only the former is present and the characteristic is

$$F = -c\dot{z}\left[1 + \mu \, \mathrm{sgn}\,(\dot{z})\right] . \qquad (26.131)$$

The characteristic of a shock absorber described by Eq. (26.131) is plotted for various values of μ, in Fig. 26.40a.

The experimental characteristics of two automotive shock absorbers are plotted in Fig. 26.41. The characteristic of the first is bilinear, and may be approximated with good precision using Eq. (26.131) with:

$$c = 3.25 \text{ kNs/m} , \qquad \mu = 0.3846 .$$

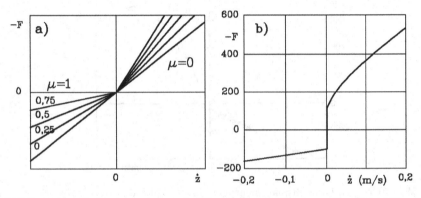

FIGURE 26.40. a): Characteristics of bi-linear shock absorbers (Eq. (26.131)), with various values of μ. b) Effect of dry friction on the characteristics of a nonlinear nonsymmetric shock absorber

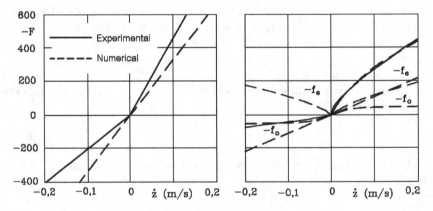

FIGURE 26.41. Characteristics of two shock absorbers, one bi-linear and the other one nonlinear

The characteristic of the second is more complicated and can be expressed by Eq. (26.130) with

$$c = 1000 \text{ kNs/m} , \qquad f_o = 268.6\sqrt{|\dot{z}|} + 350|\dot{z}| ,$$
$$f_e = 268.6\sqrt{|\dot{z}|}\text{sgn}(\dot{z}) - 350\dot{z} ,$$

where $\dot{z}$ is in m/s.

A step function centered in the origin can be added to $f_o(\dot{z})$ to include dry friction in the model. The characteristics of Fig. 26.41b with dry friction added is shown in Fig. 26.40b.

Consider a quarter car model with a single degree of freedom with a non-linear shock absorber whose characteristic is expressed by Eq. (26.130). The equation of motion is

$$m\ddot{z} + c\dot{z} + Kz + f_e\left(\dot{z} - \dot{h}\right) + f_o\left(\dot{z} - \dot{h}\right) = c\dot{h} + Kh . \qquad (26.132)$$

In this case, it is easier to write the equation in terms of relative displacement

$$z_r = z - h,$$

instead of the displacement z, obtaining

$$m\ddot{z}_r + c\dot{z}_r + Kz_r + f_e\dot{z}_r + f_o\dot{z}_r = -m\ddot{h} . \qquad (26.133)$$

Because the system is nonlinear, the response will be a generic periodic but non-harmonic law even if the forcing function is harmonic. It may, at any rate, be expressed by a Fourier series

$$z_r = z_0 + \sum_{i=0}^{\infty} z_i \sin(i\omega t + \phi_i) , \qquad (26.134)$$

where all harmonics, including that of order 0, may be present, because the nonlinear function contains both even and odd terms. If the nonlinearities are not too strong, a first approximation solution may be obtained by truncating the series after the term with $i = 1$. Working in phase with the response and not with the excitation, it is possible to write

$$\begin{cases} h = h_0 \sin(\omega t - \phi) \\ z_r = z_0 + z_1 \sin(\omega t) . \end{cases} \qquad (26.135)$$

By introducing solution (26.135) into Eq. (26.133), the latter transforms into the algebraic equation

$$\mathcal{F}(t) = 0 , \qquad (26.136)$$

where

$$\begin{aligned} \mathcal{F}(t) = {} & z_1\left[(K - m\omega^2)\sin(\omega t) + c\omega\cos(\omega t)\right] + \\ & + z_0 K + f_e(\omega z_1 \cos(\omega t)) + f_o(\omega z_1 \cos(\omega t)) + \qquad (26.137) \\ & - \omega^2 m h_0 \left[\cos(\phi)\sin(\omega t) - \sin(\phi)\cos(\omega t)\right] . \end{aligned}$$

An approximated solution for z_0 and z_1 may be obtained by stating that Eq. (26.136) holds as an average for a whole period, instead of holding in each instant. This may be formalized by stating that the integral of the virtual work

$$\mathcal{F}(t)\delta z_r = \mathcal{F}(t)\left[\delta z_0 + \delta z_1 \sin{(\omega t)}\right] \tag{26.138}$$

for a period vanishes.

Because the virtual displacements δz_0 and δz_1 are arbitrary, this amounts to stating

$$\begin{cases} \int_0^T \mathcal{F}(t)dt = 0 \\ \int_0^T \mathcal{F}(t) \sin{(\omega t)} \, dt = 0. \end{cases} \tag{26.139}$$

Since the integrals over a period of sine and cosine functions and of all odd functions of any trigonometric function vanish, the first equation yields

$$z_0 KT + \int_0^T f_e\left(\omega z_1 \cos{(\omega t)}\right) dt = 0 \ . \tag{26.140}$$

It follows then that

$$z_0 = -\frac{1}{2\pi K} \int_0^{2\pi} f_e\left(\omega z_1 \cos{(\omega t)}\right) d\left(\omega t\right) \ . \tag{26.141}$$

In the case of the bi-linear shock absorber, from Eq. (26.131) it follows

$$f_e = c\dot{z}\mu \ \mathrm{sgn}\left(\dot{z}\right) = c\omega z_1 \mu \left|\cos{(\omega t)}\right| \tag{26.142}$$

and then

$$z_0 = -\frac{c\omega z_1 \mu}{2\pi K} \int_0^{2\pi} \left|\cos{(\omega t)}\right| d\left(\omega t\right) = -\frac{2c\omega z_1 \mu}{\pi K} \ . \tag{26.143}$$

From simple symmetry considerations, it follows that

$$\int_0^T f_e\left(\omega z_1 \cos{(\omega t)}\right) \sin{(\omega t)} \, dt = 0$$
$$\int_0^T f_o\left(\omega z_1 \cos{(\omega t)}\right) \sin{(\omega t)} \, dt = 0$$

and then the second Eq. (26.131) yields

$$z_1\left(K - m\omega^2\right) = \omega^2 m h_0 \cos{(\phi)} \ . \tag{26.144}$$

The phasing between the forcing function and the harmonic component of the response can be computed by stating that the energy dissipated in a cycle by the damper is equal to the energy supplied by the forcing function

$$\int_0^T \left[c\ \dot{z} + f_e\left(\dot{z}\right) + f_o\left(\dot{z}\right)\right] \dot{z}dt = \int_0^T -m\ddot{h}\dot{z}dt \ , \tag{26.145}$$

that is

$$\int_0^T \left[cz_1\omega\cos^2(\omega t) + f_e(\omega z_1\cos(\omega t))\cos(\omega t) + \tag{26.146}\right.$$

$$\left. + f_o(\omega z_1\cos(\omega t))\cos(\omega t)\right] dt = -mh_0\omega^2\int_0^T \cos^2(\omega t)\sin(\phi)\,dt.$$

Because

$$\int_0^T f_e(\omega z_1\cos(\omega t))\cos(\omega t)\,dt = 0 , \tag{26.147}$$

it follows that

$$cz_1\omega + \frac{1}{\pi}\int_0^{2\pi} f_o(\omega z_1\cos(\omega t))\cos(\omega t)\,d(\omega t) = -mh_0\omega^2\sin(\phi) . \tag{26.148}$$

Equations (26.144) and (26.148) allow the two remaining unknowns, z_1 and ϕ to be computed By adding the squares of the two equations, it follows that

$$z_1^2\left(K - m\omega^2\right)^2 + \left[cwz_1 + \frac{1}{\pi^2}\int_0^{2\pi} f_o(\omega z_1\cos(\omega t))\cos(\omega t)\,d(\omega t)\right]^2 = \omega^4 m^2 h_0^2 . \tag{26.149}$$

Once function $f_o(\dot{z})$ has been stated, this equation allows the amplitude of the motion z_1 to be computed.

By dividing Eq. (26.148) by Eq. (26.144) it follows that

$$\phi = \text{artg}\left[-\frac{\pi cz_1\omega + \int_0^{2\pi} f_o(\omega z_1\cos(\omega t))\cos(\omega t)\,d(\omega t)}{\pi z_1\left(K - m\omega^2\right)}\right] . \tag{26.150}$$

It is then possible to demonstrate that the even function (deviation from symmetry) causes a displacement of the center of oscillation from the static equilibrium position, but has little effect on the dynamic response of the system. If the deviation from symmetry is neglected, the characteristics of the shock absorber can be linearized in the origin, and it is possible to use the equivalent linear viscous damping to study the small oscillations of the system. This explains why linearized models may be used even when the effect of nonlinearities seems to be important. This holds true even for small oscillations.

Example 26.13 *Consider a quarter car with two degrees of freedom with the parameters typical of the suspension of a small car:* $m_s = 240$ *kg,* $m_u = 25$ *kg,* $K = 20,8$ *kN/m,* $P = 125$ *kN/m. Assume that the shock absorber is nonlinear and asymmetrical, and that its characteristics may be modeled using Eq. (26.131) with*

$$c = 1.8 \ kNs/m , \qquad \mu = 0.65 .$$

Moreover, dry friction is also present. It may be modeled using the following odd function

$$f_o = 60 \ sign\,(\dot{z}) \ N .$$

Compute the response to harmonic excitation with amplitude $h_0 = 100$ mm by numerically integrating the equation of motion, and compare this result with the linearized solution and with the approximated solution of the nonlinear equation. Repeat the computation for an amplitude of the forcing function of 10 mm.

The equation of motion is Eq. (26.22), to which the nonlinear terms are added. However, to simplify the equation, it is possible to substitute coordinates

$$\begin{cases} z_1 = z_u - h \\ z_2 = z_s - z_u \ . \end{cases} \qquad (26.151)$$

to variables z_s and z_u.

Neglecting the damping of the tire, the equation of motion becomes

$$\begin{bmatrix} m_T & m_s \\ m_s & m_s \end{bmatrix} \begin{Bmatrix} \ddot{z}_1 \\ \ddot{z}_2 \end{Bmatrix} + \begin{bmatrix} 0 & 0 \\ 0 & c \end{bmatrix} \begin{Bmatrix} \dot{z}_1 \\ \dot{z}_2 \end{Bmatrix} +$$

$$+ \begin{bmatrix} P & 0 \\ 0 & K \end{bmatrix} \begin{Bmatrix} z_1 \\ z_2 \end{Bmatrix} + \begin{Bmatrix} 0 \\ f_e\,(\dot{z}_2) + f_o\,(\dot{z}_2) \end{Bmatrix} = \begin{Bmatrix} m_T \ddot{h} \\ m_s \ddot{h} \end{Bmatrix} , \qquad (26.152)$$

where

$$m_T = m_s + m_u \ , \qquad f_e = -c\mu\dot{z}\ \mathrm{sgn}\,(\dot{z}) \qquad (26.153)$$

and f_o is given by the above mentioned expression

A solution of the type of Eq. (26.135) is

$$\begin{cases} h = h_0 \sin{(\omega t - \phi)} = h_0 \left[\sin{(\omega t)} \cos{(\phi)} - \cos{(\omega t)} \sin{(\phi)} \right] \\ z_2 = z_{20} + z_{21} \sin{(\omega t)} \\ z_1 = z_{10} + z_{11s} \sin{(\omega t)} + z_{11c} \cos{(\omega t)} \ . \end{cases} \qquad (26.154)$$

By introducing this solution into the first equation of motion, which is linear, and remembering that the damping of the tire has been neglected, it follows that

$$\begin{cases} z_{10} = 0 \\ z_{11s} = \omega^2 \dfrac{-m_T h_0 \cos{(\omega t)} + m_s z_{21}}{P - \omega^2 m_T} \\ z_{11c} = \omega^2 \dfrac{m_T h_0 \sin{(\omega t)}}{P - \omega^2 m_T} \ . \end{cases} \qquad (26.155)$$

By introducing the values of the unknowns so obtained into the second equation of motion, an equation formally identical to that of a quarter car with a single degree of freedom (Equations (26.136) and (26.137)) is obtained, once

$$z_{20} \ , \qquad m_s \dfrac{P - \omega^2 m_u}{P - \omega^2 m_T} \ , \qquad h_0 \dfrac{P}{P - \omega^2 m_T} \qquad (26.156)$$

are substituted for z_1, m_s and h_0.

The results for a forcing function with an amplitude of 100 mm are reported in Fig. 26.42a. It is clear that the amplitude of the motion of both the sprung and unsprung masses (in terms of z_s and z_u and not of z_1 and z_2) obtained using numerical integration and the approximated nonlinear computations are close to each other. Moreover, the

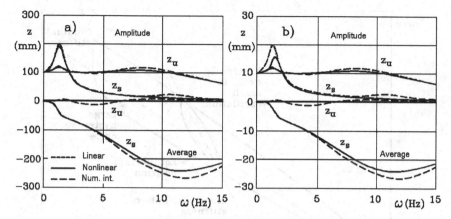

FIGURE 26.42. Response to a harmonic forcing function with amplitudes of 100 mm (a) and 10 mm (b) as a function of frequency for a quarter car model with two degrees of freedom provided with a nonlinear shock absorber

amplitude of the motion almost coincides with that obtained from the linearized model, with the difference that in this case there is a displacement of the central position of the oscillation.

The results obtained for an amplitude of the forcing function of 10 mm are shown in Fig. 26.42b. In this case, there is some difference between the linearized and the nonlinear solution at low frequency, due to dry friction that locks the suspension in this condition. In general, however, the accuracy of the linearized model is confirmed.

26.7.2 Springs

Dry friction in leaf springs introduces hysteresis and an apparent increase of stiffness in low amplitude motion. A qualitative force-deflection characteristics of a leaf spring is shown in Fig. 26.43: The hysteresis cycle is readily visible. The overall elastic behavior is practically linear, with a hysteresis cycle occurring about the straight line representing the average stiffness. If small amplitude oscillations occur about the equilibrium position, the apparent stiffness is strongly dependent on the amplitude, with a value tending to infinity when the amplitude tends to zero. This behavior is typical of dry friction that causes the spring to lock when very small movements are required. The stiffness for the small oscillations typical of ride behavior can then be much larger than the overall stiffness of the spring.

The presence of dry friction makes linear models inapplicable, or at least makes their results inaccurate, and causes a deterioration of the ride qualities of the suspension.

Other nonlinearities may be introduced by nonlinear springs, which are sometimes used for industrial vehicles in order to avoid large variations of the natural frequencies with the load. Air springs are also widely used on industrial vehicles, and their characteristics are strongly nonlinear. However, nonlinearities

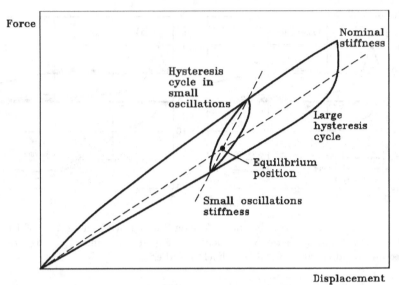

FIGURE 26.43. Load-deflection characteristics of a leaf spring exhibiting hysteretic behavior. The hysteresis cycle for small displacements about the equilibrium position is shown

of the behavior of the springs can be dealt with in the same way as those due to the kinematics of the suspensions and, in the motion about any equilibrium condition, a linearized study holds with good approximation.

26.8   CONCLUDING REMARKS ON RIDE COMFORT

The linearized study of suspension motions, based primarily on quarter-car models, shows that the value of shock absorber damping for optimizing comfort is the same as that for reducing the dynamic component of the force on the ground to a minimum, and hence optimizing handling. However, some results obtained considering the root mean square value of the acceleration and the dynamic component of the force show that, even when using a simplified linearized model, the value optimizing comfort is lower than that optimizing handling.

The last statement is also confirmed by other considerations. Firstly, the reduction of the force is not the only goal in handling optimization. The displacement of the sprung with respect to the unsprung masses is also important. Every type of suspension has some deviations from a perfect kinematic guide, thus causing the wheels to be set in a position different from the nominal (e.g., changes of the camber angles, roll steer etc.); this negatively affects the handling characteristics of the vehicle. The larger the displacement of the sprung mass, the worse the problem.

Operating in the same way as for minimizing the acceleration, it can be shown that the value of the damping minimizing displacement is

$$c = \sqrt{\frac{m(P+K)(P+2K)}{2P}} \, , \tag{26.157}$$

which is higher than the optimum value computed above.

This also suggests an increase in the stiffness of the suspensions and goes against the criterion of "the softer the better" deriving from consideration of the vertical acceleration alone.

Another point is linked to roll oscillations. The damping of the shock absorbers is usually chosen with bounce in mind; this causes rolling motions in most cases to be excessively underdamped. When anti-roll bars are used, the situation becomes worse: By increasing roll stiffness without increasing the corresponding damping, they cause a more marked underdamped behavior and a decrease of the dynamic stability of roll motions. This not only increases the amplitude of rolling motions and the dynamic load transfer, while lowering the roll angle in steady state conditions, but also makes rollover in dynamic conditions easier.

The increase of shock absorber damping beyond the value defined above as optimum is effective in reducing these effects, which affect handling more than comfort.

On the other hand, the need for reducing jerk to increase comfort goes in the opposite direction. The value of damping minimizing jerk is lower than that minimizing acceleration, which leads to better comfort when damping is decreased.

The effect of the stiffness of springs on comfort is in a way contradictory: On one hand, as already stated, the need of reducing vertical accelerations suggests that stiffness be reduced as much as possible, but this would lead to very low natural frequencies which may, in turn, cause motion sickness and similar effects.

The compliance of the frame or of other parts of the vehicle may also affect riding comfort. The effect of the compliance of those elements that, in simplified models, are assumed to be stiff, is at any rate smaller than the effect the flexibility of the same elements has on handling. While, as already stated, the compliance of the body in bending in the xy plane and above all the torsional compliance about the z axis may have a strong effect (usually reducing performance) on handling, bending compliance in the xz plane may affect comfort, though not always in a negative way.

The local compliance of the body and the frame may strongly affect acoustic and vibrational comfort when it leads to natural frequencies that can be excited by the forcing functions that are always present on a vehicle. A typical example is the compliance of the supporting structure of the engine ancillaries (alternator, air conditioning compressor, etc.) that may cause resonances of the system made by the same elements, the supporting brackets, the belts and other elements connected to them. Because such a system is located close to, or even directly on, the engine, which is a strong source of excitations at various frequencies, many resonant vibrations are possible.

As usual, when local resonances are possible, there are many different cures:

- Increasing the damping of the system, usually by adding damping material, to reduce the amplitude of vibration below acceptable limits. This is the simplest cure, one that may induce a non-negligible increase of weight and often gives only marginal improvements. A typical example is the application of damping paints or sheet metal covered by damping material on the floor or the firewall.

- Increasing the stiffness of the structure, so as to increase the natural frequencies and thus move the resonance to frequencies at which there is little excitation. The opposite method, reducing the stiffness to decrease the natural frequencies, is usually not applicable in the automotive field, because it would cause excessive deformations and would, at any rate, induce several low frequency resonances. This cure usually causes a weight penalty as well and to increase the stiffness of a vibrating system without also increasing the damping makes the system more underdamped, with the consequence of increasing the amplitude of vibration in case a resonance occurs.

- Reducing the amplitude of the excitation at the source. Although this is the most effective cure, it is seldom applicable. To reduce the vibration caused by imbalance of rotating elements the best procedure is to improve balancing, but the prescribed balancing grade is usually chosen compatibly with constraints such as construction techniques and costs. Moreover, wear can reduce balance over time and the compliance of rotating elements may make it difficult to obtain good balancing in all operating conditions.

- Preventing vibration transmitted from the source to the resonant element. This often requires design changes or innovative concepts. For instance, the transmission of vibration from the engine to the passenger compartment is drastically reduced, with improvements in acoustic and vibrational comfort, by substituting the standard rigid linkage for gearbox control with a device bases on flexible cables.

- Adding dampers close to the zones affected by vibration. The use of dynamic vibration absorbers is widespread in automotive technology, both on the chassis and in the engine (crankshaft dampers, etc.). Because the components of the vehicle are excited by a number of frequencies in a wide range, damped vibration absorbers (i.e. containing dissipative elements) are usually used instead of purely dynamic absorbers. The example given for the quarter car with dynamic vibration absorber can be extended to other cases.

27

CONTROL OF THE CHASSIS AND 'BY WIRE' SYSTEMS

27.1 MOTOR VEHICLE CONTROL

As already stated, a road vehicle on pneumatic tires cannot maintain a given trajectory under the effect of external perturbations unless managed by some control device, which is usually a human driver. Its stability solely involves such state variables as the sideslip angle β and the yaw velocity r.

In the case of two-wheeled vehicles the capsize motion is intrinsically unstable forcing the driver not only to control the trajectory but stabilize the vehicle.

A possible scheme of the vehicle-driver system is shown in Fig. 27.1. The driver is assumed to be able to detect the yaw angle ψ, the angular and linear accelerations $\dot{\beta}$, $\dot{r}$, dV/dt, V^2/R and to be able to assess his position on the road (X and Y). Moreover, the driver receives other information from the vehicle, such as forces, moments, noise, vibrations, etc. that allow him to assess, largely unconsciously, the conditions of the vehicle and the road-wheel interactions.

27.1.1 Conventional vehicles

In all classical vehicles of the second half of the twentieth century up to the 1990s, the driver had to perform all control and monitoring tasks. The only assistance came from devices like power steering or power brakes that amplified the force the driver exerted on the controls. In this situation, the human controller is fully inserted in the control loop or, as usually said, the systems include a *human in the loop.*

G. Genta, L. Morello, *The Automotive Chassis, Volume 2: System Design,*
Mechanical Engineering Series,
© Springer Science+Business Media B.V. 2009

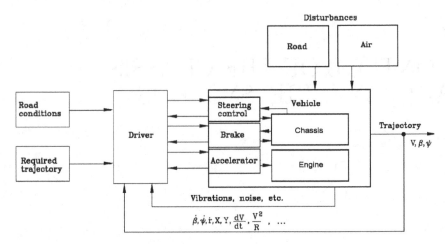

FIGURE 27.1. Simplified scheme of the vehicle-driver system.

Actually the driver must control high level functions (choice of the trajectory, decisions about speed and driving style, about manoeuvres like overtaking, etc.) and intermediate level functions (reacting to perturbations coming from the air and the road, following the chosen trajectory, etc.). Only stability at the lowest level, involving the sideslip angle and the yaw velocity, is provided by the dynamic behavior of the vehicle. As already stated, in motorbikes the driver must also act as a stabilizer against capsizing.

In particular:

- Direction control is implemented by applying a torque to the steering wheel that is then transmitted through a mechanical system (steering box, steering arms, various linkages) to the steering wheels, which are always the front wheels. The torque exerted by the driver may be increased by an hydro-pneumatic or electromechanical system (power steering) that nonetheless never replaces the driver by exerting the whole moment. The required sensitivity is provided by the torque the steering system exerts on the driver through the aligning torque and the contact forces at the wheel-road interface. These, in turn, depend upon the geometry of the steering system (caster angle, toe in, offsets, etc.).

- The control of the power supplied by the engine is managed through the accelerator pedal, operating directly through a mechanical leverage. Sensitivity is supplied by the elastic reaction of a spring that reacts to the motion of the pedal. The driver must control the power accurately enough so that the maximum force the wheel can exert on the ground is not exceeded.

- Engine control is accompanied by control of the gearbox and the clutch, which operate through the clutch pedal and the gear lever. These controls are often automatic.

- Braking control is performed by applying a force on the brake pedal that is then transmitted through a system (usually hydraulic, but pneumatic in industrial vehicles) to the brakes located in all wheels. Here the force exerted by the driver can also be augmented by a hydro-pneumatic device (power braking). In all cases, sensitivity is granted by the fact that the force exerted by the driver is proportional (or at least depends in an almost linear way) to the braking torque and then to the braking force. The driver must control the braking force so that the wheels do not lock.

These basic controls are accompanied by many secondary controls, such as those of the lighting systems, window cleaning and defrosting, parking brake etc. Although not directly used to control the motion of the vehicle, these are extremely important for driving safety. The basic controls are standardized on all vehicles, with some difference in special vehicles, and are subjected to detailed standards. In the case of particular arrangements, to be used by persons with disabilities of various kinds that do not allow them to operate conventional controls directly, a non-conventional *user interface* is provided, designed as needed for each particular installation. The transmission of commands, however, remains the same: for instance, the accelerator control may be brought to the steering wheel with a ring coaxial to the wheel that can be moved axially. This, in turn, operates the conventional accelerator control through levers.

The situation with two-wheeled vehicles is essentially the same, the only difference being that the driver can change the inertial and geometrical characteristics of the vehicle, using these changes as control inputs: for instance, he can move the center of mass sideways or change the aerodynamic characteristics. The controls are obviously different with the front and rear brakes often operating independently.

27.1.2 Automatic and intelligent vehicles

The possibility of introducing automatic control devices in road vehicles has led in recent years to many studies aimed at designing vehicles able to perform automatically a number of those control functions that at present are entrusted to the driver, with the long term goal of building road vehicles able to perform all their functions automatically, essentially transforming the driver into a passenger. This goal is still distant and, as with predictions of so-called strong artificial intelligence, there are doubts as to its achievability, at least with today's technologies and those likely to be developed in the foreseeable future.

However, while the goal of building a fully automatic road vehicle may be a long way off, many partial applications are already available or are about to be realized.

One source of inspiration is what has been done in the field of aeronautics. Since World War II, devices able to keep an aircraft at a given attitude and on a prescribed course, allowing the pilot to leave the controls for a more or less prolonged time, have entered common use. Such devices do not need to sense external conditions and adapt to them; they are simple regulators, that only need to maintain the predetermined motion conditions. Devices of this kind have only a limited use in road vehicles (for instance, *cruise control* devices) because vehicles must continuously adapt their motion to the road and traffic conditions.

Military aircraft are increasingly built on configurations that reduce intrinsic stability or are even unstable, with the goal of improving manoeuvrability; the task of stabilizing the aircraft is given to suitable control devices. Moreover, the senses of the pilot have been enhanced by supplying additional information through the control devices, such as devices that shake the control stick when stall conditions approach. Artificial stability may prove interesting in the vehicular field as well, not so much for improving manoeuvrability as for allowing the use of configurations that are advantageous but reduce stability.

Devices providing an artificial sensibility, often referred to as *haptic*, are those that provide a reaction force through *by wire* controls that is similar to the reaction that conventional mechanical controls would supply. They may also add further information, like the devices that cause the accelerator or the brake pedal to shake when getting close to slip conditions in traction or braking. Such devices are intrinsically necessary when controls are made automatic. They are at present under study and in same cases already on the market.

Nowadays in the aeronautical field commands are no longer transmitted by mechanical (rods, cables, etc.) or hydraulic devices but by electric systems (*fly by wire*), the only exception being small and low cost aircraft. There are two main advantages: first, freedom in architectural and layout is greatly increased (it is much easier to route electric cables than mechanical controls), resulting in a mass reduction. Second, it is much easier to integrate control systems, which are mostly electronic, in *by wire* than in conventional architectures. In the most modern aircraft, the pilot interacts with a computer that in turn actuates the control surfaces through *by wire* devices.

A similar evolution is also underway in the automotive industry. Here the term *steer by wire* is used for the steer control, *brake by wire* for the braking function and *drive by wire* for the accelerator control. The generic term for these systems is *X by wire*, where the generic *X* stands for the various controls. The advantages are similar to those in the aeronautical field, with the added bonus of allowing the use of different user interfaces that can, for example, be designed specifically for disabled persons and even adapted for individual cases.

However, the transfer from the aeronautic *fly by wire* to the automotive *X by wire* is not simple. A first difference between the two fields is linked to cost, or better, to the ratio reliability/cost. The total cost of an aircraft is greater than the cost of a motor vehicle by orders of magnitude, allowing the use of control systems and components much more expensive than those that may be used in vehicles. Something similar can be said for the low cost segment of the aeronautical market: *fly by wire* systems are still not used in light and ultralight aviation.

The scale of production may mitigate this problem: development costs are subdivided, in the automotive market, into a much greater (even by orders of magnitude) number of machines than in the aeronautical market. Reliability is strictly linked to costs: when dealing with functions that are vital for safety, like steering or braking, the need for extremely high reliability leads to high costs, because the required safety is obtained through redundancy of sensors, actuators, control units and communication lines as well as high quality components. Electronic and computer based devices have been available for motor vehicles for several years in non-vital functions and, more often, in gadgets performing tasks that are of little practical use.

However, it is not just a matter of cost: motor vehicles are designed for general use; their mission analysis is less determinate than that of aircraft, and they must be able to work in conditions far from those for which they have been designed, with a less stringent respect for maintenance schedules. This makes technology transfer from aeronautics to automotive industry even more difficult, particularly where complex and even critical technologies are concerned.

One field where technology transfer may be facilitated is that of racing cars, and in particular Formula 1 racers, because these vehicles must be optimized with a limited number of parameters in mind, accrue higher costs and are used in controlled conditions. However their design specifications are strictly linked to racing regulations, which at present (2008) do not allow the use of automatic and control devices.

At present, the fields in which control devices are more common or are at least being actively studied are:

- Engine control systems. All modern automotive internal combustion engines are provided with one or more electronic control units (ECU) that control its main functions. The motor control may be conventional or *by wire*, but in the latter case there is no problem in supplying the driver with adequate sensory inputs. Because these systems are studied in conjunction with the engine and not with the chassis, they will not be dealt with here.

- Longitudinal slip control in traction, (ASR, Anti Spin Regulator[1]). These are systems that detect the beginning of driving wheel skid and reduce the power supplied by the engine. Theoretically, they should measure the longitudinal slip of the tires, but in practice they measure the acceleration of the driving wheels.

- Longitudinal slip control in braking, (ABS, Antilock Braking System). They are systems that detect the beginning of wheel skid and reduce the braking torques. They, too, should measure the longitudinal slip of the tires, but actually measure the deceleration of the wheels.

---

[1] The acronyms here mentioned are often trade marks of a particular manufacturer, even if many of them have entered technical jargon to designate a variety of similar devices.

- Vehicle dynamics control systems, (VDC, Vehicle Dynamic Control, ESP, Enhanced Stability Program, DSC, Dynamics Stability Control). The goal of these systems is to improve the dynamic response of the vehicle. They often act by differentially braking (and sometimes differentially driving) the wheels of the same axle to produce a yaw torque. The driver controls the trajectory normally through the steering wheel, while the control device tries to counteract the difference between the behavior required and that actually obtained by applying yaw torques.

- Suspension control systems. Many different types of controlled, semi-active and active suspensions have been and are being developed. These can simply adapt the suspension characteristics to the type and conditions of the road or, in the most advanced cases, completely substitute an active system for the conventional suspension.

- Electric Power Steering (EPS). Strictly speaking, EPS should not be considered a control system any more than conventional power steering, but electric actuation allows steering control functions to be added. EPS, then, may be considered as a first step towards *steer by wire*.

- Electric braking. A wide span of functions are available through electric braking, from simple electric power braking with an electric actuator on the master cylinder of a conventional hydraulic system (which should not be listed here) to a true *brake by wire* system, with the electric actuators at the wheels.

- Servo controlled gearbox and clutch. These systems provide automatic gearbox functions by controlling a more or less conventional manual transmission using suitable actuators, with all the advantages of classic automatic transmissions but with a much more efficient mechanical transmission without a torque converter.

- Finally, the parking brake must be counted among the secondary controls that may be made automatic. The advantages are that it is possible to ensure that the brake is applied every time the driver leaves the vehicle, without the possibility of forgetting it, while reducing the effort needed to engage and disengage the brake. Electric parking brake yields a larger freedom to the designer of the interior of the vehicle.

The components of some of these systems and the main control strategies have already been described in Part I.

All the mentioned systems allow the tasks of the driver to be simplified and safety increased, assuming that they meet reliability standards. The driver is still in the control loop, but his work is made simpler by avoiding low level control tasks so that he can concentrate on high level decisions.

Strong research activity is now devoted to going beyond this approach by making it possible to perform higher level functions automatically, as in systems able to recognize and follow the road automatically using video cameras that identify the outer edges of the road and the lines delimiting the lanes. Other examples include systems able to regulate the speed, keeping a constant distance from the preceding vehicle, and anti-collision systems based on obstacle recognition.

There is no doubt that systems of this kind are feasible once the critical technologies have been developed at acceptable costs and with the required reliability. However, systems that can do completely without the presence of a human in the loop are beyond present and predictable technology.

27.2   MODELS FOR THE VEHICLE-DRIVER SYSTEM

Before embarking on the study of automatic systems aimed at controlling the vehicle, it is advisable to study the vehicle driver system in conventional vehicles, where the human driver is fully integrated in the control loop. Such a study has two primary goals:

- To build a mathematical model of a human driver that can be integrated into the mathematical model of the vehicle in simulations. It is not necessarily true that a system made of two subsystems that are both stable is itself stable. The study of stability, therefore, should take into account the behavior of the driver even if the vehicle is intrinsically stable. Moreover, in the case of motorcycles, the intrinsic instability of the system makes it necessary to introduce a driver model, at least as a roll stabilizer, to allow the dynamic behavior of the system to be numerically simulated.

- To supply guidelines for the design of automatic control systems. Automatic controllers are often inspired by the behavior of the human controller, if for no other reasons than that it is the only available model. Moreover, the performance of human controllers is better than that expected from automatic devices. Automatic control systems must interact with a human controller and supply the latter with information and sensory inputs that are not much different from those he is used to.

It is clear that stability of the vehicle-driver system is mandatory, but it is not sufficient to assess the required handling and comfort characteristics of the vehicle. The greater the stability with free and locked controls of the vehicle itself, the fewer the corrections the driver has to introduce to obtain the required trajectory. A vehicle that is stable in β and r requires from the driver only those inputs needed to follow the required trajectory, but not those needed to stabilize the motion on it.

On the other hand, a vehicle that is too stable may lack the manoeuvrability needed to cope with emergency conditions or simply to allow sport driving.

The amount of stability must be assessed in each case, taking into account the type of vehicle, the market target, the traditions and image of the manufacturer.

Usually stability, handling and comfort characteristics of a vehicle are assessed on the basis of prolonged road testing performed by skilled test drivers. This approach has the drawback of being in a way subjective, and above all of focusing on the global characteristics of the vehicle, without giving detailed suggestions on causal relationships between the construction parameters of the vehicle and its behavior. It also demands that long and costly road tests be performed and, above all, forces evaluation of the performance of the vehicle to be postponed to a stage in which prototypes are available.

The availability of mathematical models for the driver-vehicle interaction has a number of advantages that are too obvious for a detailed discussion. The difficulty of translating concepts like comfort and user friendliness into mathematical functions is a serious obstacle in this study, making experimental and numerical approaches likely to remain complementary.

A model able to simulate the behavior of the driver must be built for the study of man-machine interactions. The difficulties encountered in such a task are so large that many different approaches have been attempted. Up to now there is no standard driver model that can be applied.

The first systematic studies were performed in the aeronautical field[2], but beginning in the 1970s, a large number of models specialized for the vehicular field have been published. A quick bibliographic scan identifies more than sixty models published in less than 25 years. These span from simple constant-parameter single-input single-output linear models to multi-variable, nonlinear, adaptive models or models based on fuzzy logic and/or neural networks.

As always, the complexity of the model must be chosen in a way that is consistent with the aims of the study and the availability of significant input data.

27.2.1 Simple linearized driver model for handling

As previously stated, the driver may be thought as a controller receiving a number of inputs from the vehicle and the environment and outputting a few control signals to the vehicle. Under manual control, the driver performs the tasks of the sensors, the controller, the actuators and the source of control power, even if his control actions may be assisted by devices such as power steering or braking.

In building a simple driver model, a small number of the inputs the driver receives is selected and simple control algorithms are chosen to link them with the outputs. The latter are usually only the steering angle δ and the position of the accelerator/brake pedals. Only the former is considered if the driver model is used in connection with a constant speed handling model.

---

[2]See, for instance, D.T. McRuer, E.S.Crendel, *Dynamic response of human operators*, WADC T.R. 56-524, Oct. 1957.

The simplest driver model is a proportional linear tracking system reacting to the error $\psi - \psi_0$, where ψ_0 is the desired yaw angle, with a control action in terms of steering angle δ proportional to the error. Because the controller has a delay τ, this means

$$\delta(t + \tau) = -K_g[\psi(t) - \psi_0(t)] , \qquad (27.1)$$

where K_g is the proportional gain of the controller. By developing function $\delta(t + \tau)$ in Taylor series about time t and truncating the series after the linear term, it follows that

$$\tau\dot{\delta}(t) + \delta(t) = -K_g[\psi(t) - \psi_0(t)] , \qquad (27.2)$$

Equation (27.2) is only an approximation, yielding results that are increasingly inadequate with increasing delay τ. In the present case, it is possible to assume that the values of the delay range between 0,08 s for a professional driver to more than 0,25 s for an occasional driver. Consequently, Eq. (27.2) may lead to non-negligible errors.

The transmission ratio of the steering system must be introduced into the gain K_g, because δ is the steering angle at the wheels and not at the steering wheels.

The simplest handling model that may be coupled to the driver model is a rigid body model that, assuming that the vehicle is neutral steer, reduces to a first order system (Eq.(25.108a)):[3]

$$J_z\dot{r} = N_r r + N_\delta \delta + M_{z_e}. \qquad (27.3)$$

Remembering that

$$r = \dot{\psi} ,$$

the dynamic equation of the controlled system in the state space is

$$\left\{ \begin{array}{c} \dot{r} \\ \dot{\delta} \\ \dot{\psi} \end{array} \right\} = \mathbf{A} \left\{ \begin{array}{c} r \\ \delta \\ \psi \end{array} \right\} + \mathbf{B}_c \psi_0 + \mathbf{B}_d M_{z_e}, \qquad (27.4)$$

where the dynamic matrix and the control and disturbances input gain matrices are

$$\mathbf{A} = \begin{bmatrix} \dfrac{N_r}{J_z} & \dfrac{N_\delta}{J_z} & 0 \\ 0 & -\dfrac{1}{\tau} & -\dfrac{K_g}{\tau} \\ 1 & 0 & 0 \end{bmatrix} , \mathbf{B}_c = \begin{bmatrix} 0 \\ \dfrac{K_g}{\tau} \\ 0 \end{bmatrix} , \mathbf{B}_d = \begin{bmatrix} \dfrac{1}{J_z} \\ 0 \\ 0 \end{bmatrix} .$$

If the delay vanishes, the vehicle-driver system reduces to a second order system

$$J_z\ddot{\psi} - N_r\dot{\psi} + N_\delta K_g\psi = N_\delta K_g\psi_0(t) + M_{z_e}. \qquad (27.5)$$

[3]P.G. Perotto, *Sistemi di automazione, Vol.I, Servosistemi*, UTET, Torino, 1970.

Because N_r is always negative, while product $N_\delta K_g$ is always positive, the system is always stable, both statically and dynamically. Its behavior is not oscillatory if

$$|N_r| > 2\sqrt{J_z N_\delta K_g} \,, \qquad (27.6)$$

i.e.

$$K_g < \frac{N_r^2}{4 J_z N_\delta} \,. \qquad (27.7)$$

If the derivatives of stability are computed considering the cornering forces of the tires alone, such a condition becomes

$$K_g < \frac{a l^2 C_1}{4 J_z} \frac{1}{V^2} \,. \qquad (27.8)$$

If the delay τ of the driver is accounted for, the stability of the system can be studied by searching for the eigenvalues of the dynamic matrix. The characteristic equation is

$$s^3 + \left(\frac{1}{\tau} - \frac{N_r}{J_z}\right) s^2 - \frac{N_r}{\tau J_z} s + \frac{N_\delta K_g}{\tau J_z} = 0 \,. \qquad (27.9)$$

From the Routh-Hurwitz criterion, it follows that the real parts of the solutions of the cubic equation

$$a s^3 + b s^2 + c s + d = 0$$

are all negative if

$$a > 0 \,, \qquad b > 0 \,, \qquad \det \begin{bmatrix} b & a \\ d & c \end{bmatrix} = bc - ad > 0 \,,$$

$$\det \begin{bmatrix} b & a & 0 \\ d & c & b \\ 0 & 0 & d \end{bmatrix} = d\,(bc - ad) > 0 \,.$$

In the present case, the first two conditions are always satisfied (N_r is always negative) The last condition is always satisfied provided that the third one is, because $d > 0$. The condition for stability is then the third condition

$$\tau \left(1 - \frac{J_z N_\delta K_g}{N_r^2}\right) > \frac{J_z}{N_r} \,. \qquad (27.10)$$

Because the term at the right side is negative, the system is always stable if the term in brackets is positive, i.e. when

$$K_g < \frac{N_r^2}{J_z N_\delta} \,. \qquad (27.11)$$

If such a condition is not met, the system is stable if

$$\tau < \frac{J_z N_r}{N_r^2 - J_z N_\delta K_g} \, . \tag{27.12}$$

To explicitly express the dependence of the stability of the vehicle-driver system on speed, and remembering that N_r is proportional (at least as a first approximation) to $1/V$, it is possible to introduce parameter $V N_r$, the reduced delay τ' and the reduced gain K_g' defined as

$$K_g' = K_g \frac{J_z N_\delta}{V^2 N_r^2}, \qquad \tau' = \tau \frac{V \, |N_r|}{J_z}.$$

The first condition for stability then becomes

$$K_g' < \frac{1}{V^2} \, . \tag{27.13}$$

If such a condition is not met, the system is stable if

$$\tau' < \frac{V}{K_g' V^2 - 1} \, .$$

For the system to be stable, the driver must react quickly (with minimal delay) and gradually (small value of the gain). These requests become more demanding with increasing speed. If the delay τ (or better τ') is given, a single condition for stability in terms of K_g (or better K_g') can be written

$$K_g' < \frac{\tau' + V}{\tau' V^2} \, . \tag{27.14}$$

Condition (27.14) is plotted in Fig. 27.2.

If only the cornering forces of the tire are considered in computing the derivatives of stability, the expressions of K_g' and τ' reduce to

$$K_g' = K_g \frac{J_z}{al^2 C_1}, \qquad \tau' = \tau \frac{al C_1}{J_z} \, . \tag{27.15}$$

To avoid the approximations of Eq. (27.2), it is still possible to obtain the response of the controlled system by numerically integrating the equations of motion. By introducing the steering angle at time t

$$\delta(t) = -K_g [\psi(t - \tau) - \psi_0(t - \tau)] \tag{27.16}$$

into the equation of motion (25.108) of a neutral steer vehicle, it follows that

$$\left\{ \begin{matrix} \dot{r} \\ \dot{\psi} \end{matrix} \right\} = \begin{bmatrix} \dfrac{N_r}{J_z} & 0 \\ 1 & 0 \end{bmatrix} \left\{ \begin{matrix} r \\ \psi \end{matrix} \right\} + \tag{27.17}$$

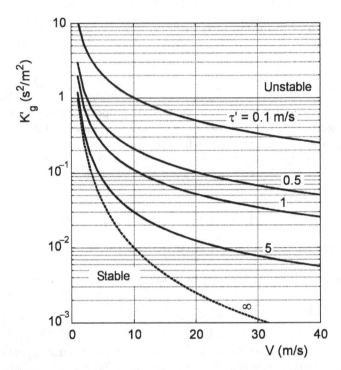

FIGURE 27.2. Values of the gain K'_g allowing a stable working of the vehicle-driver system versus the speed, for different values of the delay τ'.

$$+ \begin{bmatrix} \dfrac{K_g N_\delta}{J} & \dfrac{1}{J_z} \\[2mm] 0 & 0 \end{bmatrix} \left\{ \begin{array}{c} -[\psi(t-\tau) - \psi_0(t-\tau)] \\ M_{z_e}(t) \end{array} \right\}.$$

Example 27.1 *Consider a neutral vehicle with the following characteristics: $J_z = 1,428\ kg\ Nm^2$, $a = 1.3\ m$, $b = 1.35\ m$, $C_1 = 50\ kN/rad$. A linearized drive modelled by Eq. (27.2) steers it along a standard ISO lane change manoeuvre at 80 km/h. Assume a delay $\tau = 0.1\ s$ and a value of the gain such that the system is stable but not overly oscillatory.*

Plot the root locus of the vehicle-driver system for various values of the speed and compute the trajectory obtained through numerical integration of Equations (27.2) and (27.17). Repeat the computations for a delay of 0.3 s.

The ISO lane change manoeuvre, intended to simulate overtaking, was described in Part III. It requires the vehicle to travel for 15 m in the original lane, to change lane with a lateral displacement of 3.5 m in 30 m, to stay in this lane for 25 m and to return to the original lane in 25 m. The manoeuvre must be performed at 80 km/h. The lane changes may be performed using any trajectory, provided that none of the cones delimiting the three straight lanes are touched. The width of these lanes are, in meters, 1.1 B + 0.25 for the first lane, 1.2 B + 0.25 for the second and 1.3 B + 0.25 for the third, where B is the width of the vehicle. If the vehicle is 1.56 m wide, the three lanes are then 1.966, 2.122 and 2.278 m wide, leaving a margin of 0.203, 0.281 and 0.359 m on both sides of the theoretical trajectory.

The actual lane changes are left to the driver. In the present simulation, a cosine function is used, which has the advantage of being simple and the drawback of yielding a discontinuity of curvature at each transition with a straight path.

The trajectory and angle ψ_0 are then

$$
\begin{cases}
Y = 0 & \text{for } X < 15 \\
Y = \dfrac{3,5}{2}\left\{1 - \cos\left[\dfrac{\pi}{30}(X - 15)\right]\right\} & \text{for } 15 \leq X < 45 \\
Y = 3,5 & \text{for } 45 \leq X < 70 \\
Y = \dfrac{3,5}{2}\left\{1 + \cos\left[\dfrac{\pi}{25}(X - 70)\right]\right\} & \text{for } 70 \leq X < 95 \\
Y = 0 & \text{for } 95 \leq X < 125
\end{cases}
$$

$$
\begin{cases}
\psi_0 = 0 & \text{for } X < 15 \\
\psi_0 = \arctan\left\{\dfrac{3,5\pi}{60}\sin\left[\dfrac{\pi}{30}(X - 15)\right]\right\} & \text{for } 15 \leq X < 45 \\
\psi_0 = 0 & \text{for } 45 \leq X < 70 \\
\psi_0 = -\arctan\left\{\dfrac{3,5\pi}{50}\sin\left[\dfrac{\pi}{25}(X - 70)\right]\right\} & \text{for } 70 \leq X < 95 \\
\psi_0 = 0 & \text{for } 95 \leq X < 125
\end{cases}
$$

The value of τ' corresponding to $\tau = 0.1$ s, computed using Eq. (27.15) is $\tau' = 12.06$ m/s. The maximum value of K'_g to obtain stability at 80 km/h is 0.0058 s^2/m^2, corresponding to a gain of $K_g = 1.84$. To guarantee stability and minimal oscillations, a value equal to 20% of the maximum allowable is assumed: $K_g = 0.368$.

The roots locus (at varying speed) obtained for those values of the gain is plotted in Fig. 27.3a. The results of the numerical simulation are reported in Fig. 27.3b.

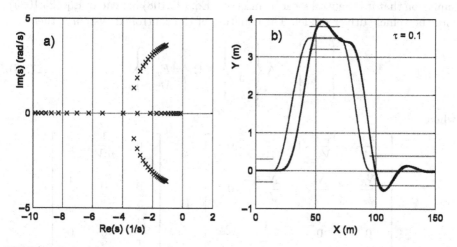

FIGURE 27.3. (a): Root locus of the vehicle-driver system with a delay of 0.1 s. (b): Trajectory obtained during an ISO lane change test computed while taking into account the delay without approximations (full line) and using the approximation of Eq. (27.2).

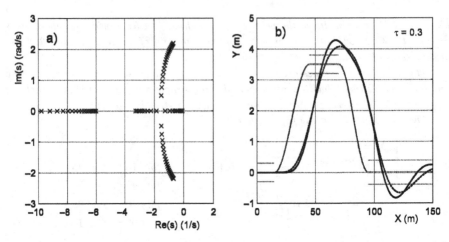

FIGURE 27.4. (a): Root locus of the vehicle-driver system with a delay of 0.3 s. (b): Trajectory obtained during an ISO lane change test computed while taking into account the delay without approximations (full line) and using the approximation of Eq. (27.2).

The computation was repeated with a delay $\tau = 0.3$ s, corresponding to $\tau' = 36.18$ m/s. The maximum value of K'_g to obtain stability at 80 km/h is 0.0033 s^2/m^2, corresponding to a gain of $K_g = 1.04$. Again, to have stability and minimal oscillations, a value equal to 20% of the maximum is assumed: $K_g = 0.21$. The trajectory is shown in Fig. 27.4b. The trajectory follows that required with much delay and strong oscillations. Note that the approximation of Eq. (27.2) leads in this case to non-negligible errors.

In both cases, the driver is unable to perform the lane change manoeuvre without hitting the cones.

To consider a more realistic vehicle behavior, it is possible to remove the assumption that it is neutral steer by inserting Eq. (25.108) instead of Eq. (25.108a) into the vehicle driver model. The equation of the controlled system is then

$$
\left\{
\begin{array}{c}
\dot{\beta} \\
\dot{r} \\
\dot{\delta} \\
\dot{\psi}
\end{array}
\right\}
= \mathbf{A}
\left\{
\begin{array}{c}
\beta \\
r \\
\delta \\
\psi
\end{array}
\right\}
+ \mathbf{B}
\left\{
\begin{array}{c}
\psi_0 \\
F_{y_e} \\
M_{z_e}
\end{array}
\right\},
\tag{27.18}
$$

where

$$
\mathbf{A} =
\begin{bmatrix}
\dfrac{Y_\beta}{mV} & \dfrac{Y_r}{mV} - 1 & \dfrac{Y_\delta}{mV} & 0 \\[2mm]
\dfrac{N_\beta}{J_z} & \dfrac{N_r}{J_z} & \dfrac{N_\delta}{J_z} & 0 \\[2mm]
0 & 0 & -\dfrac{1}{\tau} & -\dfrac{K_g}{\tau} \\[2mm]
0 & 1 & 0 & 0
\end{bmatrix}
, \quad
\mathbf{B} =
\begin{bmatrix}
0 & \dfrac{1}{mV} & 0 \\[2mm]
0 & 0 & \dfrac{1}{J_z} \\[2mm]
\dfrac{K_g}{\tau} & 0 & 0 \\[2mm]
0 & 0 & 0
\end{bmatrix}
.
$$

Remark 27.1 *It is, however, uncertain whether it is worthwhile to introduce a more realistic vehicle model when the driver is modelled in such a rough way. An approach of this type is too simple to produce realistic results. The lack of a predictive action and the assumption that the driver reacts only to the yaw angle leads to a reduced overall stability.*

Remark 27.2 *The only interesting result of this model, although quite obvious, is that, to avoid instability, driving must be quick, i.e. with a limited delay, and smooth, i.e. with a low gain.*

27.2.2   More realistic models of linearized driver

The delay is usually assumed to be the sum of three different delays: the first due to the reaction time, i.e. the time needed for the driver to elaborate the information coming from the vehicle and the environment; a neuromuscular delay, the time needed for the command to reach the muscles involved in the control action; and an actuation delay, due to the time needed to actually perform the control action. Some models consider the three delays in distinct ways, and also factor in any predictive action the human operator can perform. Experience shows that such predictive action, which can be much improved with training, is of paramount importance in actual driving conditions. A simple open loop transfer function of a linearized driver is

$$\frac{y(s)}{u(s)} = K_g \frac{(1 + T_L s)e^{-\tau s}}{1 + T_D s} , \qquad (27.19)$$

where y, u, K_g, T_L, τ, T_D are respectively the output and the input of the driver, the gain, the prediction time, the reaction time and the neuromuscular delay.

In many cases, the prediction time is neglected and all the delays are added together in a single delay τ, yielding a simpler open loop transfer function

$$\frac{y(s)}{u(s)} = K_g e^{-\tau s} . \qquad (27.20)$$

By expressing the exponential as a power series and truncating it after three or two terms, its expression reduces to

$$\frac{y(s)}{u(s)} \approx K_g \frac{1}{1 + \tau s + \frac{1}{2}\tau^2 s^2} \approx K_g \frac{1}{1 + \tau s} . \qquad (27.21)$$

The last of these expressions corresponds, in the time domain, to the already seen expression

$$\tau \dot{y}(t) + y(t) = K_g u(t) . \qquad (27.22)$$

The choice of which inputs to consider is a delicate one. The results are different if a quantity linked to the position, such as coordinates X and Y or, better, the deviation from the required trajectory, or the yaw angle ψ is used.

It is a common experience that a snaking trajectory is obtained when the driver uses a reference a point close to the front end of the vehicle, as when driving in the fog looking at the curb, while the oscillations disappear when the reference point is far in front of the vehicle.

A simple way to incorporate a kind of predictive behavior into the model is that of using as error not the difference between the desired and the actual value of the yaw angle, but the distance d between a point on the vehicle x-axis at a given distance L in front of the vehicle and the required trajectory (distance d in Fig. 27.5a).

With simple computations and assuming that angle $\psi - \psi_1$ is small, such a distance can be approximated as

$$d = L\left(\psi - \psi_1 + \frac{y}{L}\right) , \tag{27.23}$$

where y is the lateral displacement of the vehicle, i.e. the integral of the lateral velocity v. If the speed of the vehicle is constant, with the usual linearization, it coincides with the integral of β, multiplied by V. Angle ψ_1 is the angle between the X-axis and a line passing through two points of the trajectory at a distance L; it may be easily computed from the shape of the trajectory.

By using the linearized expression for the delay seen above, the equation expressing the time domain model of the driver is then

$$\tau\dot{\delta}(t) + \delta(t) = -K_g\left[\psi(t) - \psi_1(t) + \frac{y(t)}{L}\right] . \tag{27.24}$$

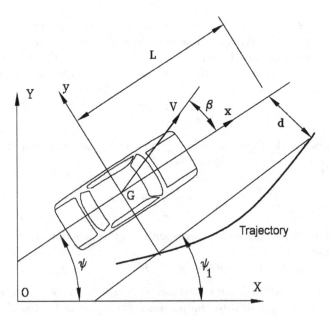

FIGURE 27.5. Definition of distance d.

By introducing Eq. ((27.24) into the simplest mathematical open-loop model of linearized vehicle, remembering that

$$\dot{y} = V\beta\,,$$

and operating as for the previous model, the state equation for the vehicle-driver system is

$$\left\{\begin{array}{c} \dot{\beta} \\ \dot{r} \\ \dot{\delta} \\ \dot{\psi} \\ \dot{y} \end{array}\right\} = \mathbf{A} \left\{\begin{array}{c} \beta \\ r \\ \delta \\ \psi \\ y \end{array}\right\} + \mathbf{B}_c\psi_1 + \mathbf{B}_c \left\{\begin{array}{c} F_{y_e} \\ M_{z_e} \end{array}\right\}, \qquad (27.25)$$

where

$$\mathbf{A} = \begin{bmatrix} \dfrac{Y_\beta}{mV} & \dfrac{Y_r}{mV} - 1 & \dfrac{Y_\delta}{mV} & 0 & 0 \\[2mm] \dfrac{N_\beta}{J_z} & \dfrac{N_r}{J_z} & \dfrac{N_\delta}{J_z} & 0 & 0 \\[2mm] 0 & 0 & -\dfrac{1}{\tau} & -\dfrac{K_g}{\tau} & -\dfrac{K_g}{L\tau} \\[2mm] 0 & 1 & 0 & 0 & 0 \\[1mm] V & 0 & 0 & 0 & 0 \end{bmatrix}, \quad \mathbf{B}_d = \begin{bmatrix} \dfrac{1}{mV} & 0 \\[2mm] 0 & \dfrac{1}{J_z} \\[2mm] 0 & 0 \\[2mm] 0 & 0 \\[1mm] 0 & 0 \end{bmatrix}$$

and

$$\mathbf{B}_c = \begin{bmatrix} 0 & 0 & \dfrac{K_g}{\tau} & 0 & 0 \end{bmatrix}^T.$$

The errors linked with Eq. (27.24) can be avoided by numerically integrating the equations of motion. Neglecting external disturbances and writing the steering angle at time t

$$\delta(t) = -K_g \left[\psi(t-\tau) - \psi_1(t-\tau) + \frac{y(t-\tau)}{L} \right], \qquad (27.26)$$

the equation of motion of the vehicle-driver system is

$$\left\{\begin{array}{c} \dot{\beta} \\ \dot{r} \\ \dot{\psi} \\ \dot{y} \end{array}\right\} = \begin{bmatrix} \dfrac{Y_\beta}{mV} & \dfrac{Y_r}{mV} - 1 & 0 & 0 \\[2mm] \dfrac{N_\beta}{J_z} & \dfrac{N_r}{J_z} & 0 & 0 \\[2mm] 0 & 1 & 0 & 0 \\[1mm] V & 0 & 0 & 0 \end{bmatrix} \left\{\begin{array}{c} \beta \\ r \\ \psi \\ y \end{array}\right\} + \qquad (27.27)$$

$$-K_g \left\{ \begin{array}{c} \dfrac{Y_\delta}{mV} \\[2mm] \dfrac{N_\delta}{J_z} \\[2mm] 0 \\[2mm] 0 \end{array} \right\} \left[\psi(t-\tau) - \psi_1(t-\tau) + \dfrac{y(t-\tau)}{L} \right] .$$

Example 27.2 *Repeat the simulation of the previous example using the driver model of Eq. (27.24).*

The values of the time delay, the gain and the prediction distance L are assumed to be, respectively, 0.20 s, 0.25 and 30 m.

The trajectory is shown in Fig. 27.6b. The driver model is now successful in performing the required manoeuvre. The tendency to oscillate about the required trajectory is much reduced, and the driver is successful in anticipating the required correction.

Remark 27.3 *Both models here described are a drastic oversimplification of actual human behavior, but the second performs satisfactorily in many instances. In particular, the results obtained in the previous examples were based on a trial and error definition of the characteristics of the driver that satisfied the requirements of the particular case studied. To obtain acceptable results in different manoeuvres, a further adaptation of driver parameters would be needed. To perform simulations of more practical value, more sophisticated models are needed.*

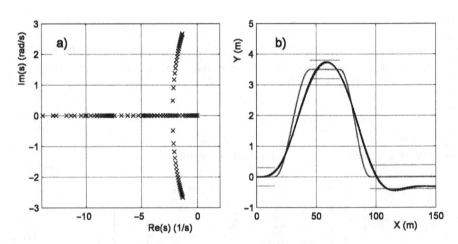

FIGURE 27.6. (a): Root locus of the vehicle-driver system with a delay of 0.2 s. (b): Trajectory obtained during an ISO lane change test computed while taking into account the delay without approximations (full line) and using the approximation of Eq. (27.2).

27.2.3 Longitudinal control

Apart from acting on the steering wheel to maintain course, the driver needs to regulate the vehicle speed by acting on the accelerator pedal and, occasionally, the brakes. If traffic is not intense and the road is largely straight or contains bends with a large radius, it is possible to use a simple regulator maintaining a constant speed at a value set by the driver. These regulators are usually referred to as *cruise control systems*. They are now found in most cars, at least as an option, but they are practically useful in particular conditions: they were first applied in the United States, where they are usable primarily on interstate highways, while in Europe they can be used on highways only when traffic is particularly low.

When traffic is intense, and in particular when there is a line of cars, the driver has to control the accelerator pedal, and occasionally the brakes as well, to maintain a constant distance from the vehicle travelling in the same lane in front of him. In these conditions, long lines of cars can form on highways, with each driver trying to regulate the speed of his vehicle so that the distance from the previous vehicle is constant. Many *anti-collision* systems, imitating this behavior, have been developed.

The simplest model of driver, or of anti-collision system, that follows this approach is a linear controller based on measuring the distance from the previous vehicle. Let V_i and d_i be the speed of the ith vehicle and the distance between the ith and the $(i-1)$th vehicle, whose speed is V_{i-1}.

The derivative of the distance with respect to time is obviously linked to the speed by the relationship

$$\frac{d}{dt}(d_i) = V_{i-1} - V_i \ . \tag{27.28}$$

The driver of the ith vehicle tries to maintain the distance at a fixed value by accelerating when the distance increases and decelerating while it decreases, but this action is applied with a certain delay. A linear model for this action is the following

$$\frac{dV_i(t+\tau)}{dt} = K\frac{d}{dt}(d_i) = K(V_{i-1} - V_i) \ , \tag{27.29}$$

where K is the gain and τ the delay time.

By using the series for the function $V_i(t+\tau)$ truncated after the second term, Eq. (27.29) reduces to

$$\tau\ddot{V}_i + \dot{V}_i + KV_i = KV_{i-1} \ , \tag{27.30}$$

which is formally identical to the equation of motion of a second-order system, with mass τ, unit damping and stiffness K. If

$$K > \frac{1}{4\tau} \tag{27.31}$$

the system has an oscillatory behavior with natural frequency[4] and damping ratio

$$\omega_n = \sqrt{\frac{K}{\tau}} \ , \quad \zeta = \frac{1}{2\sqrt{K\tau}}. \tag{27.32}$$

Equation (27.30) can be used to predict the behavior of the vehicle-driver system that follows another vehicle travelling at constant speed or, at least, at a speed having a known time history (the law $V_{i-1}(t)$ is known). The behavior must be quick and smooth (both τ and K must be small), so that no oscillations are induced.

The delay here is likely larger than in the previous cases linked with handling, because it is much more difficult to perceive when the distance from the vehicle in front of us changes than to detect changes of trajectory.

The same equation may be used to study the behavior of a line of vehicles. Consider n vehicles ($i = 1, ..., n$) in a line behind a first vehicle ($i = 0$). Let τ_i and K_i be the delay and the gain of the ith vehicle; the n equations of the type of Eq. (27.30) are

$$\begin{bmatrix} \tau_1 & 0 & \dots & 0 \\ 0 & \tau_2 & \dots & 0 \\ \dots & \dots & \dots & \dots \\ 0 & 0 & \dots & \tau_n \end{bmatrix} \begin{Bmatrix} \ddot{V}_1 \\ \ddot{V}_2 \\ \dots \\ \ddot{V}_n \end{Bmatrix} + \mathbf{I} \begin{Bmatrix} \dot{V}_1 \\ \dot{V}_2 \\ \dots \\ \dot{V}_n \end{Bmatrix} + \tag{27.33}$$

$$+ \begin{bmatrix} K_1 & 0 & \dots & 0 \\ -K_2 & K_2 & \dots & 0 \\ \dots & \dots & \dots & \dots \\ 0 & 0 & \dots & K_n \end{bmatrix} \begin{Bmatrix} V_1 \\ V_2 \\ \dots \\ V_n \end{Bmatrix} = \begin{Bmatrix} K_1 V_0 \\ 0 \\ \dots \\ 0 \end{Bmatrix}.$$

As can be seen, the equation describing the behavior of a line of vehicles is then similar, although not identical, to that governing a system made of a set of masses linked to each other by springs and dampers. The stiffness matrix is not symmetrical because the behavior of each vehicle depends on those preceding it, but not those following it.

The eigenvalues can be immediately computed, because the characteristic equation is

$$\prod_{i=1,n} \left(\tau_i s^2 + s + K_i \right) = 0, \tag{27.34}$$

whose $2n$ solutions are

$$s = -\frac{1}{2\tau_i} \pm i \frac{\sqrt{1 - 4\tau_i K_i}}{2\tau_i} \qquad \text{per} \ \ i = 1, ..., n \tag{27.35}$$

and thus coincide with those of n independent systems described by Eq. (27.30), with the various values of the parameters.

---

[4]The natural frequency is referred, as usual, to the undamped system. If the damping ratio ζ is larger than one, the damped system will not have free oscillations.

The frequency response of the ith vehicle with respect to the first one is

$$\frac{(V_0)_i}{(V_0)_0} = \frac{\prod_{k=1,i} K_k}{\prod_{k=1,i} \sqrt{(K_k - \tau_k \omega^2)^2 + \omega^2}} .$$ (27.36)

Remark 27.4 *The present model does not consider the characteristics of the vehicle, and in particular the possibility that its acceleration is not sufficient to get close to the preceding one in the predicted way. If a limitation to the acceleration of the vehicle is introduced, the model is no longer linear, and the approach followed here no longer holds.*

Example 27.3 *Consider a line of vehicles controlled by identical drivers, with a gain $K = 1.6$ 1/s and delay $\tau = 0.6$ s. Plot the frequency response of the various vehicles to an harmonic variation of the speed of the first one.*

The frequency response is plotted in Fig. 27.7 for some values of i. The line of vehicles has a resonant frequency and may be expected to oscillate.

This linearized model is obviously too rough to give quantitative indications, but qualitatively explains the oscillatory behavior of long lines of vehicles and also the fact that in the case of dense highway traffic periodic stops can be experienced without any obvious explanation. Clearly, when the oscillations become too large the linearized model loses validity: The acceleration required for some vehicles may be too large or the distance too small, forcing some vehicles to stop.

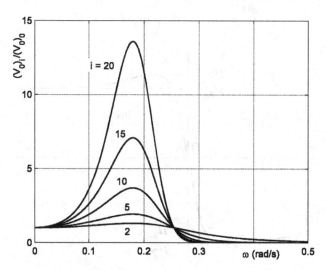

FIGURE 27.7. Frequency response of the velocity of the i-th vehicle in a line using as input the velocity variations of the first one. $K = 1.6$ 1/s; $\tau = 0.6$ s.

27.3 ANTILOCK (ABS) AND ANTISPIN (ASR) SYSTEMS

27.3.1 Basic principles

The force a tire can exert in the longitudinal direction at the wheel-road contact is limited by the available traction. As previously stated when dealing with the tire, to exert a longitudinal force the tire must work with a longitudinal slip, i.e. its angular velocity must be different (smaller when braking, larger when driving) from that characterizing pure rolling. Longitudinal slip was defined as

$$\sigma = \frac{v}{V} \, ,$$

where v is the average slip velocity of the tire on the ground. The longitudinal force, or better the longitudinal force coefficient

$$\mu_x = \frac{F_x}{F_z}$$

is linked to the slip by a nonlinear relationship of the type shown in Fig. 27.8, where both force and slip are assumed to be negative in braking and positive in driving.

The dashed lines indicate unstable working of the tire: once the traction has reached its peak value, not only does the force decrease, but the wheel tends to lock if braking, or to accelerate until a full slip is reached if driving. The equation of motion of a free wheeling wheel in braking is

$$J_r \frac{d\Omega}{dt} = M_m - M_b = |\mu_x| F_z r_l - M_b \, , \qquad (27.37)$$

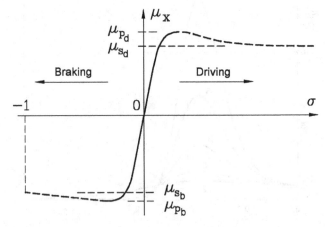

FIGURE 27.8. Longitudinal force coefficient of a tire as a function of the longitudinal slip.

where the driving torque acting on the wheel M_m is due to the longitudinal force at the wheel-road contact $|\mu_x| F_z$ multiplied by the loaded radius r_l. If $|\mu_x|$ decreases at increasing slip, any slowing down of the wheel, with an accompanying increase in slip, will cause a decrease of the longitudinal force and then a further slowing down of the wheel and a further increase of the slip. If the driver does not reduce the braking torque by releasing the pressure on the brakes, the wheels lock. In a similar way, a driving wheel accelerates until it spins freely.

Note that the peak, and the following decrease of traction, are more pronounced in case of high performance tires: racing tires can show values of μ_{xp} much higher than one (even up to 1,8 - 2), while the value at complete slip μ_{xs} remains not much greater than 1. The tendency for the wheel to lock or go into uncontrolled spin is then much greater.

The problems linked with the slipping of a pneumatic tire are, however, not linked solely to the ensuing decrease of the longitudinal force, but deeply affect handling and the very stability of the vehicle. When the traction used in the longitudinal direction increases, the ability of the tire to supply cornering forces decreases and, when the limit traction conditions are reached, the force the tire exerts on the ground has the same direction as the relative velocity between the thread band and the ground. In these conditions, the cornering stiffness of the tire practically vanishes and the wheel loses its ability to supply cornering forces. If the rear wheels lock, the vehicle becomes unstable, while locking of the front wheels causes the vehicle to be uncontrollable, in the sense that it can move only on a straight trajectory.

In traditional vehicles it is the driver who, with his ability to understand the conditions of the vehicle, controls braking forces so that maximum traction conditions are never exceeded. The usefulness of devices that can help the driver in this task is obvious. They can keep the slip of the wheels under control and prevent maximum traction conditions from being exceeded. Actually, it is difficult to measure the longitudinal slip of the tires, because it would require the angular velocity of the wheels to be measured (a simple matter) as well as the speed of the vehicle, a complicated task, and one that requires instrumentation of a kind not usually provided on vehicles.[5]

The first devices of this kind to be used were the anti-lock systems (ABS) used on aircraft for braking during landings and aborted takeoffs. Such devices only measured the speed of the wheel and computed its derivative: a quick decrease of speed shows that locking is about to occur. Subsequently, the device directly reduces the pressure in the hydraulic braking system and then reduces the braking torque. The wheel then returns to a speed close to that of pure rolling and the system ceases to intervene. Braking then recurs, making it likely that the locking conditions repeat, with an ensuing new intervention of the ABS

---

[5]Normal speedometers, always present on motor vehicles, estimate the velocity of the vehicle from that of an element connected with the output shaft of the gearbox, assuming that the wheels are in pure rolling conditions. Regulations state that speedometers may have only positive errors, i.e. the speed they show must never be less than the actual.

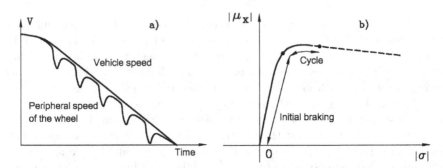

FIGURE 27.9. Working of an anti-lock system (ABS). (a) Time history of the speed of the vehicle and of the peripheral velocity of the wheel during braking with ABS. (b) Zone of the curve $\mu_x(\sigma)$ where the ABS keeps the longitudinal force coefficient.

device. Brakes then act as before, with frequent intervention by the ABS, and the longitudinal force coefficient remains close to its maximum value (Fig. 27.9).

Controlling braking in this way does not allow a traction as good as that at the maximum of the curve $\mu_x(\sigma)$ to be obtained, but it does prevent the wheel from locking and allows a fair lateral force and cornering stiffness to be maintained.

In the automotive field, ABS were initially applied to trailers, owing to the difficulties the driver has in understanding trailer braking conditions. They were then applied to industrial vehicles and luxury cars. Most cars sold today are equipped with ABS.

The control logic of anti-lock systems evolved, even if it is still based mostly on the measurement of the speed of the wheels and on the computation of their deceleration. However, while this method is adequate for free wheels, things are more complex for driving wheels. Here, the inertia of the engine and the transmission must be accounted for along with the inertia of the wheel, and the value of J_r in Eq. (27.37), which depends on the transmission ratio, is much higher. While in the highest gear the apparent increase of the inertia of the wheel may be of 200% or 300%, in the lowest gear the inertia may increase by one or two orders of magnitude. The deceleration of the wheel in these conditions may also depend on both the braking torque of the engine and the position of the accelerator. It may be low enough to prevent the increase of the longitudinal slip from being detected in wheel acceleration measurements.

It is important to evaluate the slip but, as already stated, a measurement of the speed of the vehicle that does not depend on the speed of the wheels is needed, requiring complex and costly equipment. Different strategies are possible, and sometimes more than one can apply. A reference velocity can be defined by elaborating the speeds of all wheels and possibly the longitudinal deceleration of the vehicle as well. By averaging the speed of the wheels, it is possible to obtain a reliable value for the reference velocity, until a quick deceleration of the wheels shows that the longitudinal slip has begun to increase.

By integrating the acceleration of the vehicle in time it is possible to update the values of the speed to obtain a better estimate of the slip. At this point it is possible to use different definitions of the reference speed and different algorithms to compute it, possibly using different definitions for the various axles as well, so that the required performance can be obtained. Suitable corrections may be introduced to take different factors into account, such as the roughness of the road, which is detected from oscillations in the wheel deceleration. It is possible that in the near future devices allowing the vehicle speed to be obtained directly will be available, devices based, for instance, on GPS. By estimating, or even better, by measuring the vehicle speed and then the longitudinal slip, the control system can not only prevent the wheels from locking, but may also try to maintain the slip in the zone where the braking force is highest.

27.3.2 Control strategies for ABS

How the anti-lock system acts on the various wheels has a large effect on stability. Older systems were simple *two channel* devices, i.e. they jointly controlled the two wheels of the same axle. The control of an axle may be performed following two distinct strategies, usually referred to as *select high* and *select low*. In the former case, if the wheels are in different conditions, the wheel governing the behavior of the system is the one in the best condition. In other words, the ABS device allows the wheel in the worst condition to slip, reducing the pressure in the braking system only when the wheel in the best situation begins to slip. The second strategy, on the other hand, begins to reduce the braking pressure when the wheel in the worst condition encounters a critical situation.

The latter strategy guarantees that the two wheels exert the same longitudinal force, thus preventing yawing moments. But it decreases the total amount of braking force to what the axle could exert when all its wheels are in the condition of the wheel that initially slips. Another advantage is that it guarantees a high value for the cornering stiffness of the axle.

Select high strategy, on the other hand, allows the ability of the wheel in the best condition to exert a high force to be exploited, while the other wheel works in conditions close to slipping. The braking force is much higher, but the axle produces a yawing moment that may be quite strong. The ability of the axle to produce cornering forces is compromised, because the cornering stiffness of the wheel in the worst condition vanishes, while that of the other wheel is reduced due to the strong longitudinal force.

A reasonable global strategy is to use *select low* at the rear wheels, which do not, in any case, exert large braking forces, and *select high* at the front wheels which produce most of the braking force. The total braking force is thus close to the highest possible, the rear axle has good cornering stiffness, while that of the front axle decreases, with the result that the vehicle is more understeer and thus mode stable. The drawback is the generation of a yawing moment that compels the driver to control the trajectory by acting on the steering wheel, with a steering angle that, owing to the decrease of the ability of the front wheels to generate lateral forces, may be large.

Performance is increased by using two separate ABS systems at the front axle, so that the wheel in worst conditions does not lock, with an increase of the overall braking force, decrease of the yawing moment and, above all, lower reduction of the corner stiffness of the front wheels. The vehicle remains more manoeuvrable, and it is easier for the driver to counteract the yawing moment.

The rear axle may remain controlled by a single ABS device with a *select low* logic. The system then has three channels, or it may have two distinct ABS devices. In the latter case (four channel ABS), the *select low* strategy may be implemented on the ECU, keeping open the possibility of modifying the strategy depending on the values of a number of parameters.

The ground is usually not uniform, so the traction may change from wheel to wheel. However, the average traction (in time) is the same usually for all wheels. In some cases, the road under the wheels on one side has characteristics that are not the same as those at the other side, such as when the wheels near the curb are on wet road or, even worse, on snow or ice, while those toward the center of the road are on a clean and dry surface. These conditions are usually referred to as *μ-split*. In this case, a *select low* strategy at the front axle would lead to very low values of the braking force, while a *select high* strategy may lead to very high yawing moments. The use of an accelerometer measuring yawing accelerations or an instrument that can measure the yaw angular velocity allows intermediate strategies to be implemented. The braking force on the front wheel that is in better conditions can be limited when the yaw angular acceleration increases, so that the stability of the vehicle can be maintained without overly penalizing the braking force. A strategy of this type depends largely on the characteristics of the vehicle, and particularly on its moment of inertia about the yaw axis and its geometrical characteristics (wheelbase, track). The larger the yaw moment of inertia, the larger the allowable yaw moments.

The devices used to implement an ABS system were described in Part I. Here we must just remember that although an ABS device is conceptually simple (to reduce the braking torque it is sufficient to use an electrovalve discharging a quantity of high pressure fluid from the hydraulic system, thus reducing the pressure in the cylinders of the brakes), in practice a simple ABS of this kind cannot be used, because after a number of interventions the brake pedal would sink owing to the discharge of fluid. This problem may be solved at the cost of greater system complexity, using a pump actuated by an electric motor that takes the discharged fluid and reintroduces it into the high pressure part of the circuit. There are alternatives that avoid adding an electric motor with its control devices to a system that is already complex, but a pump allowing the system to be put under pressure with no intervention by the driver on the brake pedal is required for other functions like traction control.

The ABS system interferes with other devices usually included in the braking system, such as the pressure proportioning valve. The function of the latter can actually be integrated into the ABS system, obtaining what is often referred to as EBD (Electronic Brake Distributor). However, the strategy of an ABS system and a pressure proportioning valve are radically different. The first must step in when locking conditions are approached, while the second must always function, so that the braking torques at the rear axle are reduced when the weight acting on it is reduced because of longitudinal load shift. Conceptually, the slip of the front and rear wheels must be continuously monitored so that the longitudinal slip at the front is larger than that at the rear. All the difficulties seen for measurement of the slip while dealing with ABS systems are present, with the simplification that what matters here is not an absolute measurement but simply the measurement in the difference of slip between the axles. The measurement of the longitudinal acceleration, and then of the load shift, may be very useful.

27.3.3 Traction control systems (TCS, ASR)

The problem of preventing driving wheels from slipping is similar to that of braking wheels, even if it is usually less severe and occurs only in the case of powerful vehicles or in conditions of poor traction. Wheel slipping in this case has two effects: a decrease of the force exerted in the longitudinal direction and a loss of the ability to exert transversal forces, with effects on stability and driveability. The latter are considered less severe in front wheel drive vehicles, which become less controllable, than in rear wheels drive vehicles, which become less stable.

The systems that control the slipping of driving wheels are usually referred to as Traction Control Systems (TCS) or Anti Spin Regulators (ASR). The sensors are the same as for the anti-lock system: they measure the angular velocity of the wheels so that their acceleration can be computed. When a wheel begins to spin, it is possible to react in two ways: either by reducing the power supplied by the engine or operating the brakes. The second strategy is usually quicker, but the actual implementation follows a mixed strategy: The brakes are first used to slow the wheel that has begun to spin, after which the power of the engine is reduced.

The two strategies have different effects and are used to solve different problems. If the vehicle is in symmetrical conditions, i.e., if the right and left wheels are in the same conditions, it is useless to use brakes and not advisable, because both wheels of the same axles should be braked. The result is that the transmission is much more stressed than usual, at least until the power from the engine is reduced. In this case, the reduction of the driving torque prevents both wheels from slipping.

The advantage is not so much improved performance, because the driving force the wheels can exert when not slipping is not much greater than they would exert when slipping (except in the case of high performance tires), but the possibility of exerting side forces. In particular, in the case of rear wheel drive vehicles, the vehicle remains stable, while in that of front wheel drive, the vehicle remains manoeuvrable.

The reduction of power may then be realized by acting on the motor control system.

In case of asymmetric conditions (μ-*split*), this strategy would penalize performances excessively, because it would apply a sort of *select low* strategy: both wheels would exert a force equal to what the wheel in the worst conditions can exert.

On the other hand, by braking the slipping wheel, the differential gear subdividing the driving torque between the driving wheels allows the wheel in better condition to transfer a torque equal to the sum of the braking torque applied on the other wheel plus the torque the latter is able to transfer. By acting on the brakes an increase in performance is obtained, but has a small effect on stability or manoeuvrability. TCS systems can thus be used as an alternative to controlled slip differential because, by applying a braking torque on the wheel that would slip, the system allows a certain driving torque to be transferred to the other wheel even in case where the differential is of the simplest type.

While ABS systems act on the braking system to reduce the pressure exerted by the driver, possibly assisted by the power brake, the TCS must use a pump to put the hydraulic system under pressure, independent of the force exerted by the driver on the brake pedal. However, in many cases such a pump is already included in the ABS system, so the complexity introduced is not great.

To combine the requirements on performance with those on stability, TCS systems must act on both the engine and the brakes, even at the expense of added complexity. The two strategies can be mixed following a logic that is based on many parameters, apart from the wheel slip and the acceleration of the vehicle: for instance, at low speed it is possible to give priority to acceleration performance through a strategy based on the use of brakes, while at high speed it is possible to give priority to stability by acting on the engine.

TCS systems allow drivers with limited ability to drive difficult vehicles, such as rear wheel drive high powered cars, even in critical road conditions.

27.3.4   Electric braking

Electric braking may be performed in two radically different ways: by electrically actuating the pump of a conventional hydraulic (pneumatic in industrial vehicles) system, or by substituting an electric braking system for the hydraulic one. This second strategy may in turn be implemented in two different ways, by replacing the hydraulic system with an electromechanical one (including, for instance, an electric motor in each wheel that operates the calipers using a ball screw) or by putting a pump operated by an electric motor in each wheel, sending pressurized liquid to a more or less conventional caliper.

Even if all these approaches allow ABS and possibly TCS functions to be integrated more easily into the braking system, only the total replacement of the hydraulic system with an electromechanical one where the electric control reaches each wheel directly (possibly using a fully electromechanical actuator) allows the full performance predicted for *by wire* devices to be attained, but at the cost of many reliability problems.

27.4 HANDLING CONTROL

27.4.1 General considerations

As seen in Part I, the lateral dynamics of the vehicle is controlled by the driver, who creates a yawing moment by setting the steering wheels at a certain steering angle; this torque causes a yaw rotation that puts all wheels at a certain yaw angle and consequently produces a lateral force that alters the trajectory. The yawing torque may be produced in many ways: acting on the front steering wheels (setting them at a sideslip angle and then creating a side force that, being applied in front of the center of mass, produces a yawing moment), acting on all wheels that are steering (setting them at a sideslip angle in the opposite direction and then producing a yawing moment), and also by creating a yawing torque directly through differential braking or traction on the right and left wheels. It is also possible to act on all the wheels, setting them at a steering angle in the same direction and thus producing directly the side forces needed to alter the trajectory without any yaw rotation or sideslip angle of the vehicle.

Traditional vehicles are controlled using only the first of these strategies: manual direction control through front wheel steering. If the road conditions are good and the required manoeuvre is not too severe, the simplified models seen in Chapter 25 allow the response of the vehicle to be predicted fairly well. The average driver is able to maintain control without difficulty.

In particular, the dynamic analogy of the vehicle with a mass, spring, damper system (equations (25.110) and (25.111)) is fairly accurate and the response to a steering command $\delta(t)$ is that of a second order system excited by a linear combination of laws $\delta(t)$ and $\dot{\delta}(t)$. The response in terms of yaw velocity and lateral acceleration is expressed by Equations (25.117) and (25.115), and that in term of sideslip angle $\beta(t)$ is expressed by Eq. (25.116).

Even if the response is that of a non-minimum phase system, in which the sideslip angle may initially take values of opposite sign with respect to the steady state value, the sideslip angle can be felt by the driver only to a limited extent and does not create confusion or dangerous situations.

It is interesting to study the response of the vehicle if the control strategy is based on rear wheel steering. The non-stationary response can be computed from previously seen equations by stating $K_1 = 0$ and $K_2 = -1$ into Equations (25.183) and (25.184). By using the simplified expressions of the derivatives of stability, it follows that

$$\frac{r_0}{\delta_0} = C_2 \frac{mbVs + C_1 l}{\Delta} \,, \tag{27.38}$$

$$\frac{a_{y0}}{\delta_0} = C_2 \frac{-VJ_z s^2 - C_1 las + VC_1 l}{\Delta} \,, \tag{27.39}$$

$$\frac{\beta_0}{\delta_0} = C_2 \frac{-\left(mbV^2 + C_1 la\right)s + J_z V}{V\Delta} \,, \tag{27.40}$$

where Δ is still expressed by Eq. (25.168).

Note that the steady state response in terms of yaw velocity r and lateral acceleration is that already seen for front wheel steering, while that in terms of β is similar except for the term C_2 substituting for C_1. The steady state response is little changed if the vehicle is steered by operating the rear instead of the front wheels.

In non-stationary conditions, things are quite different. By setting the numerator of the first and the third transfer functions to zero, one can see that the responses in terms of r and β have only one real positive zero and thus cause no problem. The zeros of the transfer function a_{y_0}/δ_0 are

$$s = C_2 \frac{-C_1 l a \pm \sqrt{C_1^2 l^2 a^2 + 4 J_z V^2 C_1 l}}{2 V J_z}. \tag{27.41}$$

The two zeros are real, one positive and one negative. The response is then that of a non-minimum phase system at all speeds. The vehicle initially accelerates laterally in a direction opposite to that in which it will accelerate in steady state, disorienting the driver.

It has been proven that it is still possible to drive the vehicle under these conditions, provided that the driver is suitably trained, but driving using the rear wheels is much more difficult. This solution is used only on very slow vehicles (earth-moving machines, dumpers, etc.), because the zero in the right half plane of the roots locus tends to move towards the origin when the speed tends to zero.

The above mentioned considerations are based on linearized models. The experience of the average driver is based on driving conditions in which the behavior of the vehicle is essentially linear. If the limit conditions are approached, either because road conditions are poor or because high performance is required, the nonlinearity of the system, due both to the tires (and possibly to aerodynamic actions), and the geometry of the system, starts to become important. The vehicle may start to behave differently from what the driver expects.

It is difficult for the driver to assess the traction available due to road conditions and the sideslip angle of the tires, so when the wheels start slipping they do so abruptly. The feeling the driver has in normal conditions is that of kinematic driving (the wheels seem not to be at a sideslip angle and the trajectory seems to be defined by the position of the steering wheels in a geometrical way). When the sideslip angles increase to values that can no longer be neglected, the driver feel he has lost control of the vehicle. And indeed he has, for the average driver is unable to control the vehicle when the sideslip angles are large.

Actually, the behavior of the vehicle may change considerably when the sideslip angles take values beyond the linearity range. From the viewpoint of theoretical study, it is possible to compute the steady state working conditions using nonlinear models and then to linearize the equations about those conditions. This allows us to study, for instance, the stability or manoeuvrability under these conditions (Chapter 25). The aim of a study of this kind should be to reduce the difference between handling in limit and in linearized conditions so the driver is able to control the vehicle even outside the linearity range.

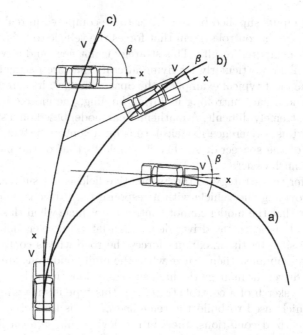

FIGURE 27.10. Response to a steering input. Curve a): conditions far from the limit. Curve b): conditions close to the limit. Curve c): conditions beyond the limit.

Consider, for instance, a steering input, aimed at putting the vehicle on a curved trajectory (Fig. 27.10). If road conditions are good, the vehicle moves on its trajectory in conditions that are far from limit conditions. The sideslip angles are small and the behavior is essentially that of a linear system (curve a). If road conditions are worse (or if the driver steers the tires so as to develop side forces that approach the limit) the sideslip angles become larger and the tires work in nonlinear conditions (curve b). Control is still possible, but the driver must have a proficiency beyond that of an average driver. Finally, if the road conditions are worse still, the vehicle may rotate about the yaw axis, but the wheels slip laterally, so that the vehicle can no longer follow the required trajectory (curve c). The sideslip angles increase in an unbounded way and the yaw rotation becomes uncontrolled. An alternative outcome is that the vehicle cannot rotate due to lateral slipping of the steering wheels. It may then go out of its trajectory with limited, or even small, values of the sideslip angles.

27.4.2 Control using a reference model

It is impossible for the vehicle to maintain a behavior corresponding to what the driver is used to based on his experience with the linearized response when he approaches conditions close to the limit. However, it is possible to introduce control strategies giving the vehicle an apparent behavior the user can find predictable. This can be done by building a reference model allowing the response to the control inputs supplied by the driver to be computed in real time, and by implementing it on a control system that forces the vehicle to behave as close as possible to the computed results. This strategy is not new, and is widely used in the aerospace field, particularly for flying aircraft that are particularly difficult to control. The most typical example is the Space Shuttle[6]: its aerodynamic configuration is such that controlling it during landing, the most critical phase of the flight, is extremely difficult. A mathematical model based on a standard civil aircraft is used as a reference model; it runs on a control system that corrects the behavior of the spacecraft so that it simulates that of the model, making maneuvering much easier.

To transfer this strategy to the automotive field is not simple, but the advantage of providing the vehicle with a response the driver is used to is clear. It is also clear that the model cannot control the vehicle when this is physically impossible, such as when the driver demands a lateral acceleration higher than that made possible by the maximum forces the road-wheels contact can exert. The control system must then realize when the limit conditions are approaching and warn the driver, or manage the high slip conditions in the best possible way.

A possible sketch of a control strategy of this type is shown in Fig 27.11.

If the vehicle used to build the reference model is the actual vehicle used under the linearized conditions, the control device will perform little work in conditions far from the limit, and only when the behavior deviates from the linearized behavior will it be asked to correct the response. If, on the other hand, this strategy is used to induce a vehicle behavior different from standard operations, the control device will have to act in all driving conditions.

Only the steering angle δ is sensed in the sketch of Fig 27.11, but it is possible to use devices that are much more complex, measuring the driving or braking forces (at minimum, sensing the position of the accelerator pedal and the pressure in the braking system), because directional control is much influenced

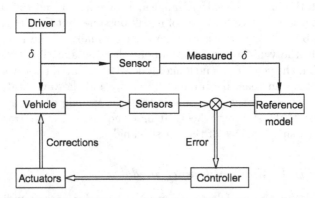

FIGURE 27.11. Sketch of a possible strategy based on a reference model to implement lateral control.

[6]D. Karnopp, *Vehicle Stability*, Dekker, New York, 2004

by longitudinal forces exerted by the tires. The sensors for the actual behavior of the vehicle may be of a different type, including not only a yaw velocity and a lateral acceleration sensor, but also other sensors allowing the slip of the various wheels to be computed as in ABS systems. Actuators may also be of a different type and corrections may be exerted using different strategies.

27.4.3 VDC systems

Acronyms for the devices implementing lateral dynamics control are many, and they are often trademarks of the various manufacturers, designating systems working with different strategies. In this chapter they will be referred to under the general name of Vehicle Dynamics Control (VDC).

The simplest way to control vehicle dynamics is to leave full control of the steering to the driver, performing the control action by differentially braking the wheels using the existing braking and traction control system, assuming the vehicle already has ABS and TCS devices.

Assume, for instance, that the control system detects a yaw speed that is higher than that computed by the reference mathematical model for the measured steering angle and speed. A situation of this kind, similar to that shown in Fig 27.10, curves b or c, may be defined as an excess of oversteer and corresponds to an excessive sideslip angle of the rear wheels. It may occur for various reasons, such as an underinflated rear tire, low traction at the rear wheels, a center of mass located far to the rear, driving forces created by a rear wheel drive vehicle, or many other possibilities. To reduce the yaw angular velocity it is possible to brake the wheels outside the curved trajectory. In this case, it is expedient to brake the front wheel, both because it is likely that the front axle still has traction available and because by doing so the vehicle becomes more understeer. In a similar way, if the yaw velocity is lower than the computed value, it is possible to brake the rear wheel inside the turn (Fig 27.12).

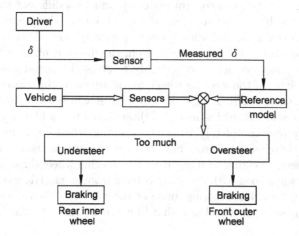

FIGURE 27.12. Differential braking used to correct directional behavior.

Instead of using differential braking, it is possible to have a steering front axle controlled by the VDC system. This strategy is different from that assumed in the section on 4WS where the rear steering was controlled directly by the driver along with the front steering. In the present case, the driver controls the vehicle using the front steering only, while the rear steering is used to perform corrections aimed at forcing the vehicle to follow the behavior of the reference model.

Differential braking has the advantage of acting more quickly than rear steering, because the latter requires not only that the wheels be set at a sideslip angle, something that cannot occur instantly, but also that the wheels start exerting cornering forces, something that requires the vehicle to travel a distance equal to the relaxation length. Nor is the action of the brakes is instantaneous, but its characteristic time is lower.

This drawback of controlling by steering particular wheels is not typical of rear steering, but is present whenever tires exert cornering forces. Another drawback of this strategy is that any steering control performed on the rear wheels involves a non-minimum phase system. The control action initially produces a lateral acceleration in a direction that is opposite to that occurring in steady state. Although this may in itself be an advantage (non-minimum phase systems are often quicker than standard systems), it may be felt as a drawback by the driver, who may prefer a vehicle without VDC.

Instead of using rear steering, or in addition to it, the VDC system may control front steering, removing it in part or totally from the direct control of the driver. While the strategies seen above are additional to the usual ways of controlling the vehicle's course, this method does offer an alternative. If the control system uses a reference model to perform driver commands, the latter may not even realize that the control system is overriding his command. He may have the feeling that the vehicle behaves as expected, which is true: The goal of the driver is to keep the vehicle on the required trajectory, and it does not matter whether to do so the steering wheels are set to the angle the driver sets on the steering wheel or the control system produces a different steering angle, possibly while the brakes act differentially and the rear wheels steer.

If the vehicle is provided with a *steer by wire* system, there is no difficulty in removing, in part or totally, steering from the direct control of the driver: It is enough that the steer actuator not only receives the signal from the sensor measuring the angle of the steering wheel, but is driven by the controller that acts according to a more complex logic. However, as it will be seen later, at present (2008) there are still difficulties, included those linked to standards, in uncoupling the steering system completely from the steering wheel. It is possible to use a partially mechanical solution, where the steering mechanism is connected to a differential gear, receiving an input from the steering wheel through the shaft in a conventional way, along with the input coming from an electric motor controlled by the VDC system. The steering angle of the wheels is then a combination of the angle of the steering wheel and that imposed by the actuator.

Apart from operating the brakes, the rear steering and the front steering, the VDC system may also operate through the suspensions. If an active anti-roll bar is used, it is possible to shift the load transfer from the front to the rear axle, or vice versa, making the vehicle more or less oversteering. This strategy will be dealt with later when we discuss active suspensions.

Devices of the kind seen above not only affect the driveability of a vehicle and the feeling the driver gets from it, but may also change its intrinsic stability. While the steering on all wheels, implemented by controlling the steering boxes of the two axles using the steering control, has no effect on the stability of the vehicle (although it may affect the stability of the vehicle driver system), a device in which a closed loop controller oversees the lateral behavior (like that sketched in Fig 27.11 or in Fig 27.12) may alter it to some extent.

In general, a VDC system may make the vehicle more stable, by making it less subject to external disturbances, as is predictable for a closed-loop device. If a strategy based on a reference model is applied, the stability of the vehicle should be brought to coincide with that of the model, but this must be accurately studied at the design stage. It must be noted that if the VDC system is designed to operate in conditions in which the vehicle has a nonlinear behavior (in conditions far from the limit, where the vehicle is essentially linear, the usefulness of VDC is debatable) both the system and the controller are nonlinear, complicating the study of stability. This also applies to cases where the goal of the control device is to make the global behavior of the controlled system (as seen by the driver) as linear as possible.

Finally, any system controlling the handling of a vehicle has limitations. The vehicle may remain on the trajectory the driver sets, thanks to the control system, only until the maximum cornering forces are reached. From this viewpoint, a device that changes the behavior of the system so that it remains close to the behavior in conditions far from critical may be dangerous, because it prevents the driver from realizing that he is dangerously approaching critical conditions.

If the control system must warn the driver of the approaching of limit conditions, the controller must be designed so that it changes its control logic when the limit conditions are reached. If, notwithstanding the intervention of the controller, the vehicle cannot be kept on the desired trajectory, the device must perform an emergency manoeuvre, quite a difficult thing because also the driver may be trying to do something similar.

It is unlikely that the control system should have authority to override the driver, because the latter has information the control system ignores (if nothing else, information on the road conditions) and, if the driver is proficient, he may be able to manage the emergency situation better than any control device. This consideration may not apply to average drivers, who may panic and react in a completely wrong way.

Another reason to limit the authority of the controller is to give the driver some margin to react to a malfunctioning of the device. This consideration may be important for present applications, while in the future the authority of the system may be increased as more experience on these systems is gained. However, it is perhaps unrealistic to believe that in the future stability control systems that cannot be a cause of accidents a particularly good human driver could prevent will be built. We must probably reason in statistical terms: A system is useful when the number of accidents it may prevent is larger than the number of accidents it may cause. This criterion is, however, difficult to implement, because it is impossible to say how many accidents a vehicle dynamics control system may prevent.

A further point is that the authority of the VDC system may need to be limited to prevent the driver from feeling that he cannot master his vehicle. An expert driver may appreciate the feeling of being in a complete control and may even enjoy the nonlinear behavior in conditions close to the limit. For this reason the driver can be given the option to switch off the VDC system, as he chooses. The vehicle may have completely different behavior in the two cases, satisfying customers who might otherwise buy two different vehicles, one for everyday use and one for sport driving.

Finally it would be a mistake to think that modifying the handling of a vehicle using a control system makes efforts aimed at designing well-handling vehicles useless. As already stated, it is better to avoid giving too much authority to the control system for safety reasons, and a failure of the control system that makes the vehicle outright unstable must be avoided. Moreover, it must be added that the control system is unable to generate arbitrarily high control forces, leading to limitations to its capabilities.

27.4.4 Simplified VDC with yaw velocity control.

Consider a device controlling the yaw velocity of the vehicle, implemented by steering the front wheels. The steering angle of the front wheels is then a linear combination, one that can be implemented using a differential gear that has as inputs the steering wheel and the controlled actuator, of angles δ_g expressed by the driver and δ_c imposed by the control system. A possible implementation of this system is shown in Section 6.1.2 of Part I.

Assuming that the coefficients of the linear combination have a unit value, it follows that

$$\delta = \delta_g + \delta_c \ . \tag{27.42}$$

Let the reference model be the two degrees of freedom model studied in Chapter 25:

$$\left\{ \begin{array}{c} \dot{v}_r \\ \dot{r}_r \end{array} \right\} = \mathbf{A}_r \left\{ \begin{array}{c} v_r \\ r_r \end{array} \right\} + \mathbf{B}_r \delta_g \ , \tag{27.43}$$

where the dynamic matrix at constant speed and the input gain matrix are

$$\mathbf{A}_r = \begin{bmatrix} \dfrac{Y_{\beta r}}{mV} & \dfrac{Y_{rr}}{m} - V \\[4mm] \dfrac{N_{\beta r}}{J_z V} & \dfrac{N_{rr}}{J_z} \end{bmatrix} , \tag{27.44}$$

$$\mathbf{B}_r = \left\{ \begin{array}{c} \dfrac{Y_{\delta r}}{m} \\[2ex] \dfrac{N_{\delta r}}{J_z} \end{array} \right\}. \tag{27.45}$$

The parameters of the reference model may be chosen with different criteria. For instance, $Y_{\beta r}$, Y_{rr}, ... may be chosen to be equal to the values Y_β, Y_r, ... as computed for the actual vehicle. In this way the control system will be almost inactive when the vehicle behavior is similar to the linearized model, and the driver will have the impression that the behavior of the vehicle is what he is used to in conditions of smooth driving.

Another strategy may be to modify the behavior of the vehicle in all conditions, giving it, for instance, a neutral steer response. In this case

$$N_{\beta r} = 0, \quad Y_{rr} = 0 \tag{27.46}$$

while the derivatives of stability may be

$$Y_{\beta r} = -\frac{l}{b} C_1, \qquad N_{rr} = -\frac{al}{V} C_1,$$

$$Y_{\delta r} = C_1, \qquad N_{\delta r} = a C_1. \tag{27.47}$$

If only the yaw velocity r is to be controlled, it is possible to define an error

$$e = r - r_r. \tag{27.48}$$

As an example, a purely proportional controller may be used: It generates a steering angle δ_c proportional to the error through a gain k of the controller, yielding

$$\delta_c = -ke = -k(r - r_r). \tag{27.49}$$

Although a control of this kind is easy to implement (only r must be measured and r_r computed), its performance would be poor, particularly in terms of control robustness.

An example based on the *sliding mode* technique is reported in *Vehicle Stability* by Dean Karnopp.[7]

Example 27.4 *Consider the vehicle studied in Example 26-4 and apply to it a VDC system of the type shown above with the aim of making it neutral steer, using a simple proportional control. Compute the gains of the vehicle in steady state conditions and simulate a steering step manoeuvre to put it on a circular trajectory with a radius of 200 m at a speed of 100 km/h = 27.78 m/s.*

The behavior of the system may be modelled using an equation of the type

$$\dot{\mathbf{z}} = \mathbf{A}\mathbf{z} + \mathbf{B}\mathbf{u} + \mathbf{d}, \tag{27.50}$$

---

[7] D. Karnopp, *Vehicle Stability*, Dekker, New York, 2004.

where $\mathbf{z}$ is the state vector, the input $\mathbf{u}$ can be assumed as

$$\delta = \delta_g + \delta_c$$

and $\mathbf{d}$ is a vector containing the external disturbances and those due to the deviation from the linear behavior implicit in matrices $\mathbf{A}$ and $\mathbf{B}$.

The order n of the model used for the vehicle may be different from the order of the reference model, which is equal to 2, but in the present case also the former is assumed to be a model with two degrees of freedom. Thus assume $n = 2$. Neglecting disturbances and nonlinearities, $\mathbf{d} = 0$

If the output of the system is only the yaw velocity, the output equation reduces to

$$r = \mathbf{C}\mathbf{z}, \tag{27.51}$$

where the output gain matrix $\mathbf{C}$ is a row matrix which, if r is one of the states of the system, has all elements equal to zero and one equal to 1. In the present case, $\mathbf{C}$ has just 2 elements.

The state equation for the system made by the vehicle and the reference model is

$$\begin{cases} \dot{\mathbf{z}} = \mathbf{A}\mathbf{z} + \mathbf{B}\,(\delta_g + \delta_c) + \mathbf{d} \\ \dot{\mathbf{z}}_r = \mathbf{A}_r\mathbf{z}_r + \mathbf{B}_r\delta_g \end{cases} \tag{27.52}$$

Remembering equations (27.49) and (27.51) and neglecting $\mathbf{d}$, Eq. (27.52) becomes

$$\left\{ \begin{array}{c} \dot{\mathbf{z}} \\ \dot{\mathbf{z}}_r \end{array} \right\} = \left[\begin{array}{cc} \mathbf{A} & 0 \\ 0 & \mathbf{A}_r \end{array} \right] \left\{ \begin{array}{c} \mathbf{z} \\ \mathbf{z}_r \end{array} \right\} + \left[\begin{array}{c} \mathbf{B} \\ \mathbf{B}_r \end{array} \right] \delta_g - k \left[\begin{array}{c} \mathbf{B} \\ 0 \end{array} \right] (\mathbf{C}\mathbf{z} - \mathbf{C}_r\mathbf{z}_r), \tag{27.53}$$

where $\mathbf{C}_r$ is a matrix similar to $\mathbf{C}$, but for the reference model. In the present case, the two matrices are the same.

With simple computations, it follows that

$$\left\{ \begin{array}{c} \dot{\mathbf{z}} \\ \dot{\mathbf{z}}_r \end{array} \right\} = \left[\begin{array}{cc} \mathbf{A} - k\mathbf{B}\mathbf{C} & k\mathbf{B}\mathbf{C}_r \\ 0 & \mathbf{A}_r \end{array} \right] \left\{ \begin{array}{c} \mathbf{z} \\ \mathbf{z}_r \end{array} \right\} + \left[\begin{array}{c} \mathbf{B} \\ \mathbf{B}_r \end{array} \right] \delta_g . \tag{27.54}$$

Assume a unit value for k (note that k in this case is not non-dimensional, but has the dimension of a time. It must then be said that $k = 1s$). Using symbols $\mathbf{A}_{cl}$ and $\mathbf{B}_{cl}$ for the matrices in Eq. (27.54), the steady state response is

$$\left\{ \begin{array}{c} \dot{\mathbf{z}} \\ \dot{\mathbf{z}}_r \end{array} \right\}_{ss} = -\mathbf{A}_{cl}^{-1}\mathbf{B}_{cl}\delta_g. \tag{27.55}$$

The trajectory curvature gain and the sideslip angle gain are

$$\frac{1}{R\delta_g} = -\frac{1}{V} \left[\begin{array}{cccc} 0 & 1 & 0 & 0 \end{array} \right] \mathbf{A}_{cl}^{-1}\mathbf{B}_{cl}. \tag{27.56}$$

$$\frac{\beta}{\delta_g} = -\frac{1}{V} \left[\begin{array}{cccc} 1 & 0 & 0 & 0 \end{array} \right] \mathbf{A}_{cl}^{-1}\mathbf{B}_{cl}.$$

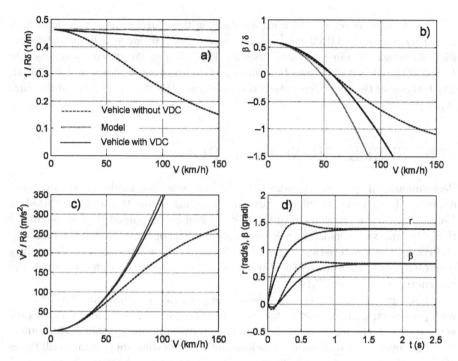

FIGURE 27.13. Trajectory curvature gain (a), sideslip angle gain (b) and lateral acceleration gain (c) as functions of the speed for the vehicle of example 25.4 for the reference model (neutral steer) and for the vehicle with VDC. In (d) the response to a step steering input to insert the vehicle on a trajectory with 200 m radius is shown.

The trajectory curvature gain, the sideslip angle gain and the lateral acceleration gain are plotted as functions of the speed in Fig 27.13. The reference model is that of a neutral steer vehicle with the same geometry and front tires.

From the trajectory curvature gain it is clear that the vehicle has an almost neutral steer behavior, but a proportional control cannot completely compensate for stationary conditions, as expected. To produce an exactly neutral behavior at least a proportional integrative (PI) control must be used.

At a speed of 100 km/h the original vehicle, strongly understeer, has complex conjugate poles. From the model with two degrees of freedom it follows that

$$s_{1,2} = -5.1915 \pm 4.5730i.$$

The poles of the vehicle with VDC are obviously 4, because the model has 4 states, and their values are

$$s_{1,..4} = \begin{bmatrix} -3.7585 & -4.8822 & -7.7146 & -48.8114 \end{bmatrix}.$$

Even if a much simplified control has been used, and no optimization of the gain has been attempted, the vehicle is stable.

To obtain a trajectory with a radius of 200 m the vehicle without VDC must have a steering angle $\delta_g = 0.0202$ rad $= 1.159°$, values that coincide with those of Example 25.4. To obtain the same curvature with the VDC system, the steering angle must be $\delta_g = 0.0115$ rad $= 0.658°$. The time history of the response is shown in Fig 27.13d). The behavior of the understeer vehicle is oscillatory, while that of the controlled vehicle is not.

27.5 SUSPENSIONS CONTROL

Performance of passive suspensions (i.e. suspensions made of springs and dampers) both in terms of comfort and handling is limited and, as stated in Chapter 26, any improvement of the first is often accompanied by a worsening of the latter and vice versa. The role of damping in the optimization of the handling and comfort characteristics is shown in Fig. 26.17. The figure is here shown in a qualitative way, with its scales starting at zero to show the range in which the parameters span (Fig. 27.14).

The curve is the lower envelope of the performance that may be obtained with passive suspensions, while the zone between the horizontal and vertical tangents shown in the figure is the locus of the points defining optimal performance.

Moreover, as stated earlier, the kinematic errors that are present in all types of suspensions make it impossible to use very soft springs, because they will lead to larger suspension movements and thus to larger unwanted motions of the unsprung mass (changes of track, steering and camber angles, etc., with heave and roll motion).

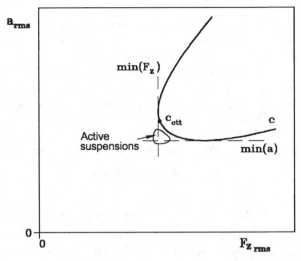

FIGURE 27.14. r.m.s. value of the acceleration of the sprung mass versus the r.m.s. value of the oscillating component of the force on the ground for a quarter car excited by a white noise in velocity. Lower envelope for passive suspensions and zone for active suspensions.

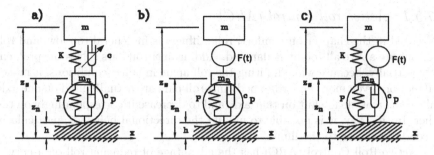

FIGURE 27.15. Quarter car with semiactive (a) suspension (the damping coefficient c is variable in a controlled way) and active suspensions with a controlled actuator supplying all the force exerted between the sprung and unsprung masses with no other mechanical device (b) or with the help of a spring (c).

To improve the situation, suspensions controlled by suitable devices have been studied and sometimes applied in mass produced vehicles. As shown in Section 5.3 in Part I, it is possible to distinguish between *semi-active* suspensions (Fig. 27.15a) and *active suspensions* (Fig. 27.15 b and c) depending on whether the control system simply controls the parameters of the system (usually damping) with limited energy requirements, or directly controls the force exerted by the suspension on the sprung mass, using actuators of a different type that involve a relatively large power consumption.

The performance of semi-active suspensions lies in Fig. 27.14 on the same curve shown for passive suspensions, with the difference that while the latter are represented by a single point, the former may move on the curve when the value of the damping coefficient is changed. It is also possible to imagine semi-active suspensions where the stiffness of the spring can also be changed, as in the system in which the attitude of the vehicle is controlled, shown in the referenced section of Part I. However, devices of this kind usually require much more power than that needed to control damping.

In the case of active suspensions, it is possible to have performance represented by points lying below the envelope, in the zone shown in the plot. In the figure such a zone is very small, because it has been assumed that the stiffness of the suspension is the same as that of the passive suspension, but it may be much larger if a fully active suspension is used.

Another important parameter is the maximum frequency at which the active system operates. The practical difficulties and the power needed in an active system increase with increasing frequency. If the suspension must control the attitude of the vehicle and the frequency at which it works is low, on the order of the frequency of the sprung mass, the system is relatively straightforward. If the control must operate at frequencies on the order of the unsprung masses, or even higher, the systems becomes more complicated.

27.5.1   Active roll control (ARC)

As stated in the chapter on comfort, roll stiffness is in general too low, and roll motions are above all too little damped. Adding anti-roll bars solves the problem only partially, because while limiting the roll angle in static conditions, it makes roll motions even more underdamped. anti-roll bars are actually used on one axle only, to increase load shift on that axle and to decrease it correspondingly on the other. In this way, it is possible to modify the directional behavior, particularly in conditions close to the limit.

Active Roll Control (ARC) has the advantage of reducing roll on a curved path, while increasing roll damping or modifying handling characteristics, depending on the control strategies used, without the drawbacks involved in an increase of the roll stiffness of the vehicle.

First consider roll in quasi-static conditions. If the center of mass is above the roll center, as is almost always the case, the vehicle rolls toward the outside of the path. This is inadvisable both for stability (because the center of mass moves outwards, reducing the rollover factor – making rollover easier – and increasing load shift) and for comfort. Moreover, because of the rolling condition, the wheels take a camber angle that results in camber forces usually directed against the sideslip forces that bend the trajectory. The last effect depends strongly on the type of suspensions used and particularly on their camber recovery.

Even if suspensions were infinitely stiff, the subjective feeling of the passengers when rounding a turn would be that their bodies were inclined outward, because the local vertical direction is inclined with respect to the true vertical. The actual inclination of the vehicle body adds to this effect with results that may be detrimental to comfort, particularly if the lateral acceleration is high.

Roll may be controlled by an actuator exerting a torque between the body and one or more axles. The simplest scheme is that of an active anti-roll bar. Conceptually, imagine cutting the anti-roll bar of an axle, for instance, in the middle, and inserting a rotational actuator exerting a torque. An example of a device of this kind is discussed in section 6.3.4 of Part I.

With the torsional stiffness of the axle unchanged, the vehicle behaves passively if the actuator does not work or is locked. If the actuator exerts a torque to rotate the body back to a position perpendicular to the ground, the effect is an apparent increase of the roll stiffness because the angle is reduced. But the actual stiffness does not change nor does underdamping of the suspension increase, as it would if the stiffness actually had increased.

A first strategy is to cancel the roll angle, obtaining an apparently infinite stiffness. The feeling of lateral tilting of the vehicle is reduced but not cancelled, while above all the changes of track, steering, camber, etc. due to roll vanish. Because kinematic errors related to suspensions linked with roll (or better with the static component of roll) are completely cancelled, handling is improved.

From the viewpoint of mathematical modelling, rigid-body models become more precise, because the assumption that roll angle is vanishingly small holds exactly. Often, while applying this strategy, roll is compensated only up to a certain lateral acceleration. Some rolling toward the outside of the path, although smaller than usual, is accepted. This may be necessary, because the torque the actuator must exert increases with the lateral acceleration and may become quite large. Operating in this way, the roll angle can warn the driver that critical conditions are approaching, but it is doubtful whether the geometric roll angle may be separated from the apparent roll angle due to the inclination of the local vertical.

Another strategy is to overcompensate for rolling by inclining the vehicle toward the inside of the trajectory. The limit condition is defined by aligning axis z with the local vertical. This option will be discussed later in detail, in relation to tilting vehicles.

As seen when dealing with two-wheeled vehicles, the roll angle is

$$\phi = \arctan\left(\frac{V^2}{Rg}\right) . \tag{27.57}$$

There are advantages in both comfort and handling. If the control is precise and quick enough, passengers do not feel any centrifugal acceleration. In rail transportation the diffusion of tilting trains owes mostly to comfort, even if the larger radii of railways when compared to roads cause lower lateral accelerations than in motor vehicles. On the other hand, the great sensitivity of the organs of human equilibrium in detecting lateral acceleration allows us to detect a tilt of less than one degree, and an angle of just a few degrees causes a strong discomfort.

In terms of handling, if roll compensates exactly for lateral acceleration, load shift is exactly zero and all its effects are cancelled. On the other hand, the effects of kinematic suspensions errors do not vanish and may be larger, although of opposite sign, than those typical of conventional vehicles. If the suspensions keep the midplanes of the wheels parallel to the xz plane, i.e. if

$$\frac{\partial \gamma}{\partial \phi} = 1 ,$$

the position of the wheels on the ground is similar to that of the wheels of motorcycles. The strong camber forces then add to the side forces due to sideslip.

The roll angle needed to compensate for the lateral acceleration may be large. For instance, at an acceleration of 0.2 g the angle is 11° while if the acceleration is 0.5 g the angle is 27°. While angles larger than 40° are needed to compensate for the lateral acceleration in high performance cars, it is impossible to use this strategy in racing cars because angles of 70° or more would be required. The steady state torque the actuator must exert may be small even in the case of large lateral accelerations. The actuator must maintain an unstable equilibrium position If it is maintained with precision, reacting quickly to deviations, the moment the actuator must exert is theoretically very small. It is, however, impossible that the equilibrium condition be followed instant by instant with a precision sufficient to cancel all rollover moments, and the actuator may be called to exert a high torque.

Many studies on tilting vehicles based on active roll control have recently been undertaken. These were followed by prototypes and even some small scale production. The vehicles are mostly narrow, often designed for city use, and can be considered a synthesis of motorcycle and car. If the vehicle is very narrow it is possible to reach inclination angles large enough to compensate for lateral acceleration without the body of the vehicle touching the ground at the inside of the path. Many of these *tilting body vehicles* have three wheels.

The simplest model for roll control is the model with a single degree of freedom described in Section 26.6.1 (Fig.26.36). It is quite a rough model because it neglects the compliance of the tires and above all because it is impossible to study roll dynamics separately from lateral dynamics. The equation of motion of a passive vehicle is Eq. (26.119), repeated here

$$J_x \ddot{\phi} + (\Gamma_1 + \Gamma_2) \dot{\phi} + (\chi_1 + \chi_2) \phi - m_s g h_G \sin(\phi) =$$

$$= \Gamma_1 \dot{\alpha}_{t_1} + \Gamma_2 \dot{\alpha}_{t_2} + \chi_1 \alpha_{t_1} + \chi_2 \alpha_{t_2} ,$$

where the forcing functions are due to the lateral slope of the road α_{t_i} at the i-th suspension. J_x is the moment of inertia about the roll axis of the sprung mass alone.

If two active anti-roll bars exerting a torque M_{a_i} ($i = 1, 2$) are added and if a generic moment M_e is included into the model, the equation of motion becomes

$$J_x \ddot{\phi} + (\Gamma_1 + \Gamma_2) \dot{\phi} + (\chi_1 + \chi_2) \phi - m_s g h_G \sin(\phi) = \qquad (27.58)$$

$$= M_{a_1} + M_{a_2} + M_e + \Gamma_1 \dot{\alpha}_{t_1} + \Gamma_2 \dot{\alpha}_{t_2} + \chi_1 \alpha_{t_1} + \chi_2 \alpha_{t_2} .$$

If moment M_e is due to the centrifugal force on a path with radius R, its value is

$$M_e = \frac{m_s h_G V^2}{R} \cos(\phi) . \qquad (27.59)$$

Assuming that the moment exerted by the anti-roll bar coincides with that due to the actuator, reduced to the bar through suitable transmission ratios, the stiffnesses χ_i and damping coefficients Γ_i are those of the passive suspensions. If the actuator is controlled by an ideal proportional-derivative (PD) system that uses the roll angle as error, the moments are

$$M_{a_i} = -k_{pi} \phi - k_{di} \dot{\phi} ,$$

where k_{pi} and k_{di} are the i-th proportional and derivative gains.

The system behaves as a passive system, with stiffness and damping increased by the gains

$$J_x \ddot{\phi} + (\Gamma_1 + \Gamma_2 + k_{d1} + k_{d2}) \dot{\phi} + (\chi_1 + \chi_2 + k_{p1} + k_{p2}) \phi + \qquad (27.60)$$

$$-m_s g h_G \sin(\phi) = M_e + \Gamma_1 \dot{\alpha}_{t_1} + \Gamma_2 \dot{\alpha}_{t_2} + \chi_1 \alpha_{t_1} + \chi_2 \alpha_{t_2} .$$

Assuming that the roll angle is small, the steady state roll angle in a bend is

$$\phi = \frac{m_s h_G V^2}{R\left(\chi_1 + \chi_2 + k_{p1} + k_{p2} - m_s g h_G\right)} .$$ (27.61)

As expected, a PI control is unable to compensate for steady state roll, even if it is possible to reduce it by increasing the proportional gains. There are, however, limitations on the values of the gains, mainly for stability reasons.

The steady state torque the actuators must exert is

$$M_{a_i} = -k_{pi}\phi ,$$

and it grows with V^2.

The load shift on the i-th axle, with track t_i, is

$$\Delta F_{z_i} = \frac{m_s h_G V^2 \left(\chi_i + k_{pi}\right)}{R t_i \left(\chi_1 + \chi_2 + k_{p1} + k_{p2} - m_s g h_G\right)} .$$ (27.62)

To compensate for the steady state roll it is possible to use a proportional-integrative-derivative (PID) control

$$M_{a_i} = -k_{pi}\phi - k_{di}\dot{\phi} - k_{ii}\int \phi dt .$$

By introducing the auxiliary states v_ϕ and i_ϕ, respectively the derivative and the integral of ϕ, the state-space model of the system is

$$\left\{ \begin{array}{c} \dot{v}_\phi \\ \dot{\phi} \\ \dot{i}_\phi \end{array} \right\} = \left[\begin{array}{ccc} -\frac{K}{J_x} & -\frac{C}{J_x} & -\frac{D}{J_x} \\ 1 & 0 & 0 \\ 0 & 1 & 0 \end{array} \right] \left\{ \begin{array}{c} v_\phi \\ \phi \\ i_\phi \end{array} \right\} + \qquad (27.63)$$

$$+ \frac{1}{J_x} \left\{ \begin{array}{c} m_s g h_G \sin{(\phi)} + M_e + \Gamma_1 \dot{\alpha}_{t_1} + \Gamma_2 \dot{\alpha}_{t_2} + \chi_1 \alpha_{t_1} + \chi_2 \alpha_{t_2} \\ 0 \\ 0 \end{array} \right\} ,$$

where

$$\begin{aligned} K &= \chi_1 + \chi_2 + k_{p1} + k_{p2}, \\ C &= \Gamma_1 + \Gamma_2 + k_{d1} + k_{d2}, \\ D &= k_{i1} + k_{i2}. \end{aligned} \qquad (27.64)$$

Example 27.5 *Consider the vehicle of example 26.11. The data entering the simplified roll model are $J_x = 388.8$ kg m^2, $m_s = 1,080$ kg, $\chi_1 = 11.25$ kNm/rad, $\chi_2 = 9.5$ kNm/rad, $\Gamma_1 = 955$ Nms/rad, $\Gamma_2 = 716$ Nms/rad, $h_G = 0.5$ m. Compute the time history of the roll angle and of the load shift of the vehicle without active systems after a step steering input to insert it on a curve with a radius of 200 m at a speed of 100 km/h = 27.7 m/s.*

Repeat the computation for a vehicle with an active anti-roll bar at the rear axle, with a PD controller with $k_{pi} = 10$ kNm/rad, $k_{di} = 3$ kNms/rad.

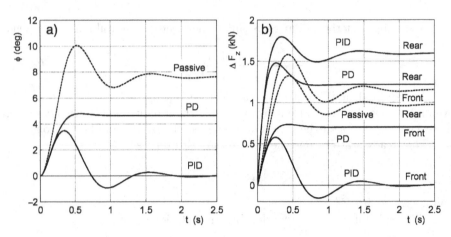

FIGURE 27.16. Time history of the roll angle (a) and of the load shift on both axles (b) for the three cases of a passive vehicle and a vehicle with an active anti-roll bar at the rear axle with PD and PID controller.

Add then an integral control with $k_{ii} = 100$ kNm/srad. For computing load shift, assume a track of 1.3 m on both axles.

Because the system is nonlinear, the equation of motion was integrated numerically. The results are reported in Fig.27.16(a) and (b) for the roll angle and the load shift.

Note that the load shift is larger at the front angle in the case of the passive vehicle, while the active anti-roll bar displaces it at the rear axle. The high value of the derivative gain leads to an almost non-oscillatory behavior of the PD system, while oscillations are caused by the integrative control in the case of PID.

The example is only an indication, because the model is quite rough (even a step steering input causes the lateral acceleration to grow more gradually) and because the values of the gains were assumed arbitrarily. In an actual case it would not be advisable to move the load shift to the rear axle (except for correcting understeer), because in this way the oversteer characteristics of the vehicle would increase.

Remark 27.5 *Up to now it has been assumed that the controller has used the absolute roll angle as a reference, something that requires the use of an artificial horizon like those used on aircraft. If the roll angle is measured with reference to the position of the axle, the vehicle body tends to follow the transversal slope of the road, while if the roll angle is measured with reference to the local vertical, a measurement easily performed using an accelerometer, the vehicle tends to tilt toward the inside of the bend, as already shown for tilting vehicles.*

Example 27.6 *Repeat the study of the previous example, using as a reference the roll angle measured from the local vertical. Owing to the type of input assumed, during the entire manoeuvre the local vertical makes an angle*

$$\phi_r = -artg\left(\frac{V^2}{Rg}\right) = -21.36°$$

with the perpendicular to the ground.

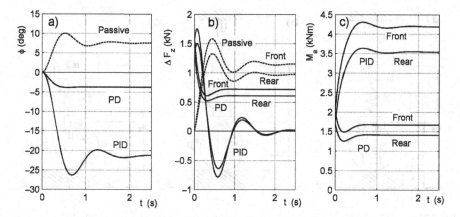

FIGURE 27.17. Time history of the roll angle (a), of the load shift on both axles (b) and of the control torques (c) for the three cases of a passive vehicle and a vehicle with an active anti-roll bar at both the rear and front axles with PD and PID controller.

Because ϕ_r is constant, the moment exerted by the actuators is

$$M_{a_i} = -k_{pi}\left(\phi - \phi_r\right) - k_{di}\dot{\phi} - k_{ii}\int\left(\phi - \phi_r\right)dt =$$

$$= -k_{pi}\phi - k_{di}\dot{\phi} - k_{ii}\int\phi dt + k_{pi}\phi_r + k_{ii}\phi_r t \ .$$

The results of the numerical integration are shown in Fig.27.17(a), (b) and (c) for the roll angle, the load transfer and the torque exerted by the actuators. To avoid too large a load shift on a single axle, two active anti-roll bars are used, with gains distributed between the axles in the same ratio as the stiffness of the suspensions. The sum of the gains on the two axles is, however, the same.

Note that the device with a PID controller succeeds in keeping the vehicle inclined toward the inside of the path so as to compensate for load transfer, even if the actuators must exert large torques. This is due to the fact that they must balance the torques due to the suspension springs, which try to keep the vehicle upright: if the control system is used to keep the vehicle inclined in the curve like a motorcycle, no passive stiffness must be put in series to the actuator. If the suspension must be passively stable with non-working actuators, the suspension springs must be put in series and not in parallel to the actuators.

27.5.2 Heave control

Quarter car with ideal *skyhook*

Consider a quarter car with two degrees of freedom, like that shown in Fig. 26.7b. A damper located not between the two masses but between the sprung mass and

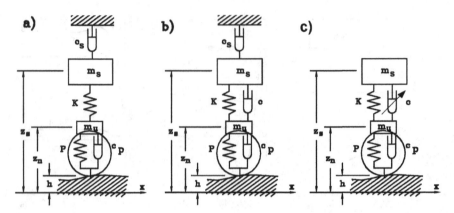

FIGURE 27.18. Quarter car model with two degrees of freedom with skyhook. a) and b): Ideal skyhook, on a quarter car without and with shock absorber between the two masses. c): Practical implementation, in which the ideal skyhook is approximated by a controlled damper located between the two masses (semi-active solution). If the controlled damper is substituted by an actuator, an active solution is obtained.

a fixed point would be needed to damp the motion of the vehicle body in an optimal way. A model with this configuration, usually referred to as *skyhook*[8], is shown in Fig. 27.18 a) or b). The point to which the damper is attached is fixed in an inertial frame; the damper substitutes for the conventional shock absorber in the first scheme, while it is added to it in the second.

With reference to Fig. 27.18 a), the equation of motion is

$$
\begin{bmatrix} m_s & 0 \\ 0 & m_u \end{bmatrix} \begin{Bmatrix} \ddot{z}_s \\ \ddot{z}_u \end{Bmatrix} + \begin{bmatrix} c_s & 0 \\ 0 & c_p \end{bmatrix} \begin{Bmatrix} \dot{z}_s \\ \dot{z}_u \end{Bmatrix} +
$$
$$
+ \begin{bmatrix} K & -K \\ -K & K+P \end{bmatrix} \begin{Bmatrix} z_s \\ z_u \end{Bmatrix} = \begin{Bmatrix} 0 \\ c_p \dot{h} + Ph \end{Bmatrix} ,
$$
(27.65)

where z_s and z_u are the displacements with respect to the static equilibrium position and are measured in an inertial reference frame.

By introducing the non-dimensional ratios

$$
a = \frac{m_u}{m_s} , \quad b = \frac{P}{K} , \quad \zeta_s = \frac{c_s}{2\sqrt{m_s K}} , \quad \zeta_p = \frac{c_p}{2\sqrt{m_s K}} ,
$$
(27.66)

---

[8]The term *skyhook* as used in automotive technology must not be confused with the same term used in aerospace, indicating a long (and hypothetical) cable system attached to a planet (e.g. the Earth), extending beyond synchronous orbit and rotating at the same speed as the planet.

the equation of motion may be written as

$$m_s \begin{bmatrix} 1 & 0 \\ 0 & a \end{bmatrix} \begin{Bmatrix} \ddot{z}_s \\ \ddot{z}_u \end{Bmatrix} + 2\sqrt{m_s K} \begin{bmatrix} \zeta_s & 0 \\ 0 & \zeta_p \end{bmatrix} \begin{Bmatrix} \dot{z}_s \\ \dot{z}_u \end{Bmatrix} +$$

$$+ K \begin{bmatrix} 1 & -1 \\ -1 & 1+b \end{bmatrix} \begin{Bmatrix} z_s \\ z_u \end{Bmatrix} = \begin{Bmatrix} 0 \\ c_p \dot{h} + Ph \end{Bmatrix}. \tag{27.67}$$

By neglecting the damping of the tire c_p, which is usually quite small, the amplification factor of the sprung and unsprung masses are

$$\frac{|z_{s_0}|}{|h_0|} = PK\sqrt{\frac{1}{f^2(\omega) + c_s^2 \omega^2 g^2(\omega)}} \tag{27.68}$$

where

$$\begin{cases} f(\omega) = m_s m_u \omega^4 - [Pm_s + K(m_s + m_u)]\omega^2 + KP \\ g(\omega) = m_u \omega^2 - K - P. \end{cases}$$

If the shock absorber between the two masses is still present (Fig. 27.18 b), the equation of motion is

$$\begin{bmatrix} m_s & 0 \\ 0 & m_u \end{bmatrix} \begin{Bmatrix} \ddot{z}_s \\ \ddot{z}_u \end{Bmatrix} + \begin{bmatrix} c+c_s & -c \\ -c & c+c_p \end{bmatrix} \begin{Bmatrix} \dot{z}_s \\ \dot{z}_u \end{Bmatrix} +$$

$$+ \begin{bmatrix} K & -K \\ -K & K+P \end{bmatrix} \begin{Bmatrix} z_s \\ z_u \end{Bmatrix} = \begin{Bmatrix} 0 \\ c_p \dot{h} + Ph \end{Bmatrix}, \tag{27.69}$$

i.e.

$$m_s \begin{bmatrix} 1 & 0 \\ 0 & a \end{bmatrix} \begin{Bmatrix} \ddot{z}_s \\ \ddot{z}_u \end{Bmatrix} + 2\sqrt{m_s K} \begin{bmatrix} \zeta_s+\zeta & -\zeta \\ -\zeta & \zeta_p+\zeta \end{bmatrix} \begin{Bmatrix} \dot{z}_s \\ \dot{z}_u \end{Bmatrix} +$$

$$+ K \begin{bmatrix} 1 & -1 \\ -1 & 1+b \end{bmatrix} \begin{Bmatrix} z_s \\ z_u \end{Bmatrix} = \begin{Bmatrix} 0 \\ c_p \dot{h} + Ph \end{Bmatrix}. \tag{27.70}$$

The frequency response of a quarter car with skyhook is plotted in non-dimensional form in Fig. 27.19. The figure is similar to Fig. 26.12a and c, and was obtaining using the same values of the non-dimensional parameters $b = P/K = 4$, $a = m_u/m_s = 0.1$. The response of the quarter car with skyhook with $\zeta_s = 1$ is compared with that of a conventional quarter car with optimum damping (Eq. (26.29)) and to a quarter car with skyhook plus a damper between the two masses with damping equal to $1/3$ of the optimum value.

The amplification factor of the unsprung mass is shown in Fig. 27.19a, while the inertance is reported in Fig. 27.19b. The presence of the skyhook greatly reduces the displacement and the acceleration at low frequency, while causing a very high, although not infinite, resonance peak at the resonant frequency of the

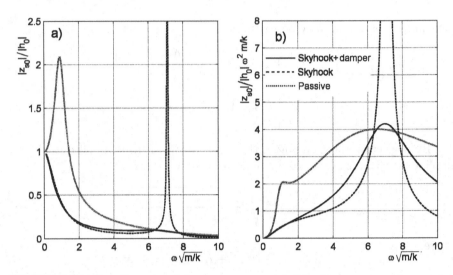

FIGURE 27.19. Non-dimensional frequency response of the sprung mass of a quarter car with two degrees of freedom. Non-dimensional parameters: $b = P/K = 4$ and $a = m_u/m_s = 0.1$. The response of a passive system with optimum damping ($\zeta = 0.433$) is compared with that of a system with skyhook ($\zeta_s = 1$) and with skyhook with the same value of ζ_s plus a damper between the two masses with damping equal to $1/3$ of the optimum value, ($\zeta = 0.1443$).

unsprung mass. The peak is fairly narrow and in theory can easily be reduced, because when c_s tends to infinity the response tends to 0. It is, however, impossible to increase the damping of the skyhook indefinitely, with the result that the peak cannot be eliminated with this configuration. If a damper, even a small one, is added between the two masses, the resonance peak in the medium frequency range disappears without greatly changing the response at low frequency.

A skyhook damper of this kind is quite effective in controlling the motion of the sprung mass, but the control of the unsprung mass is unsatisfactory. The amplification factors of the displacement and the acceleration of the unsprung mass are shown in Fig. 27.20 a and b: The skyhook has practically no effect on the motion of the unsprung mass at frequencies close to its resonance. However, the presence of a damper between the two masses strongly reduces the displacement and the acceleration of the unsprung mass. In the figure, the damping of the conventional damper is quite low, being about one third of the optimum; if it is increased the height of the resonance peak decreases, disappearing when the optimum value is reached. In these conditions there is a small increase of the response at low frequency.

The ideal skyhook is, then, an ideal solution to control the low frequency motions of the sprung mass, but it is only a reference solution, because it cannot be implemented in practice.

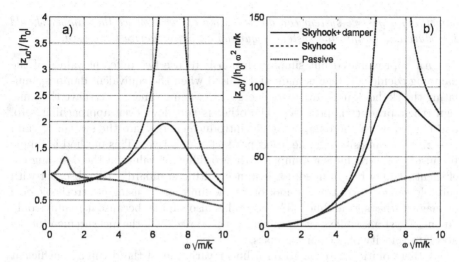

FIGURE 27.20. Non-dimensional frequency response of the unsprung mass of the same quarter car with two degrees of freedom studied in Fig. 27.19.

Semi-active quarter car with 'real world' skyhook

The fixed point where the skyhook damper is attached does not exist in the real world. This strategy must therefore be implemented using a device located between the sprung and the unsprung masses. The semi-active solution, based on a damper with controllable damping coefficient, is shown in Fig. 27.18c and is described in Section 6.3.2. of Part I.

The controlled damper must supply a force

$$F = -c_s \dot{z}_s - c \left(\dot{z}_s - \dot{z}_u \right), \qquad (27.71)$$

that is

$$F = - \left(c_s \frac{\dot{z}_s}{\dot{z}_s - \dot{z}_u} + c \right) \left(\dot{z}_s - \dot{z}_u \right). \qquad (27.72)$$

Theoretically, it should be possible to implement a device able to simulate the skyhook simply by modulating the damping coefficient of the damper so that it is, in each instant, equal to

$$c_{eq} = c_s \frac{\dot{z}_s}{\dot{z}_s - \dot{z}_u} + c. \qquad (27.73)$$

Actually, even operating in this way, only an approximation of the ideal skyhook can be obtained because, even if the forces it exerts on the sprung mass are those of the ideal device, the forces exerted on the unsprung mass are different.

Remark 27.6 *Equation (27.73) cannot be implemented by a passive device. When the equivalent damping coefficient is positive, the device dissipates energy,* *something a passive system can do, but when c_{eq} is negative the damper should introduce energy into the system, requiring an active device to be used.*

An approximated but simple solution is to use two different values of the damping coefficient. One is high, and is used when the equivalent damper simulating the skyhook must dissipate energy, i.e. when $\dot{z}_s$ and $\dot{z}_s - \dot{z}_u$ have the same sign (their product is positive). The other is very low, even approaching zero, and is used when the damper should introduce energy into the system ($\dot{z}_s$ and $\dot{z}_s - \dot{z}_u$ have opposite sign, i.e. their product is negative). This method is simple because it only requires a damper with two different values of the damping coefficient. It may be obtained, for instance, with a standard shock absorber with suitable valves to control the motion of the fluid, or by using electrorheological or magnetorheological fluids. The control is also simple, because it requires only an on-off system, while greater difficulties can arise from the measurement of the absolute velocity of the sprung mass.

Other solutions are based on a linear variation of the damping coefficient with ratio $\dot{z}_s / (\dot{z}_s - \dot{z}_u)$. The damping coefficient is nonetheless set to zero or to a very small positive value when it should be negative.

An approach of this kind leads to a nonlinear behavior of the system, making it impossible to make a general comparison between the behavior of this device and that of other suspensions.

Active quarter car with 'real world' skyhook

An active system, able to transfer energy to the system, is needed to follow the law (27.73). As usually stated, a device operating on four quadrants must be used. This expression comes from the force-velocity plot of the damper: All the conditions in which a passive system can operate lie in the second and fourth quadrants, i.e. in the quadrants where force and velocity have opposite signs (if the force is that exerted by the damper to one of its end points and the velocity is that of the same point while the other is constrained). An active system may also exert forces with the same sign of the velocity, in which case it works in all quadrants.

Consider for instance the quarter car of Fig. 26.7c, where the damper with controllable damping is substituted by an actuator operating on four quadrants. By neglecting the damping of the tires, the equation of motion is

$$\begin{bmatrix} m_s & 0 \\ 0 & m_u \end{bmatrix} \left\{ \begin{array}{c} \ddot{z}_s \\ \ddot{z}_u \end{array} \right\} + \begin{bmatrix} K & -K \\ -K & K+P \end{bmatrix} \left\{ \begin{array}{c} z_s \\ z_u \end{array} \right\} = \left\{ \begin{array}{c} F \\ -F \end{array} \right\} + \left\{ \begin{array}{c} 0 \\ Ph \end{array} \right\},$$
(27.74)

where F is the force exerted by the actuator on the sprung mass, that is

$$m_s \begin{bmatrix} 1 & 0 \\ 0 & a \end{bmatrix} \left\{ \begin{array}{c} \ddot{z}_s \\ \ddot{z}_u \end{array} \right\} + K \begin{bmatrix} 1 & -1 \\ -1 & 1+b \end{bmatrix} \left\{ \begin{array}{c} z_s \\ z_u \end{array} \right\} = \left\{ \begin{array}{c} F \\ -F \end{array} \right\} + \left\{ \begin{array}{c} 0 \\ Ph \end{array} \right\}.$$
(27.75)

To simulate the skyhook, such a force must be (Eq. (27.73)):

$$F = -c_s \dot{z}_s - c\left(\dot{z}_s - \dot{z}_u\right) = -\left(c_s + c\right)\dot{z}_s + c\dot{z}_u \ . \qquad (27.76)$$

The system is equivalent to an ideal proportional control on the states (velocities $\dot{z}_s$ and $\dot{z}_u$ are two of the states of the system). Clearly, this is an idealized system, both because the gains are considered as constants and because in practice it is impossible to measure the absolute velocity of the sprung mass directly. A system of this kind may be approximated using a state observer.

By introducing Eq. 27.76 into Eq. 27.75, it follows

$$m_s \begin{bmatrix} 1 & 0 \\ 0 & a \end{bmatrix} \begin{Bmatrix} \ddot{z}_s \\ \ddot{z}_u \end{Bmatrix} + 2\sqrt{Km_s} \begin{bmatrix} \zeta_s + \zeta & -\zeta \\ -(\zeta_s + \zeta) & \zeta \end{bmatrix} \begin{Bmatrix} z_s \\ z_u \end{Bmatrix} + \qquad (27.77)$$

$$+K \begin{bmatrix} 1 & -1 \\ -1 & 1+b \end{bmatrix} \begin{Bmatrix} z_s \\ z_u \end{Bmatrix} = \begin{Bmatrix} 0 \\ Ph \end{Bmatrix} .$$

As predicted, the damping matrix is not symmetrical, because there is no Raleigh dissipation function able to express the damping of the system.

Consider the quarter car studied in figures 27.19 and 27.20. Because the system is active, its stability must first be checked. The poles of the system are

$$s = \begin{cases} -1.5732 \pm 6.4207\ i \\ -1.3519, \\ -0.6771. \end{cases}$$

All poles have a negative real part, hence the system is stable. One of the poles has a non-vanishing imaginary part, and the behavior of the system is oscillatory.

The frequency response of the sprung mass (in terms of displacement and acceleration) is shown in Fig. 27.21. The response of the quarter car with a skyhook of this kind is not essentially different from that with an ideal skyhook.

The frequency response of the unsprung mass is shown in Fig. 27.22. The response of the quarter car is in this case much different from that with an ideal skyhook and, more significantly, is unsatisfactory. To solve this problem a larger value of ζ is needed. Note that damping between the two masses may be supplied by a passive damper, leaving the active system with only the task of simulating the skyhook.

Quarter car with *groundhook*

The very concept of skyhook was introduced to minimize the vertical accelerations of the sprung mass, with no consideration about the motion of the unsprung mass. It is not surprising, then, that the motion of the unsprung mass is too large and the corresponding variations of the force on the ground are unacceptable. As said in Section 7.4.2, the variable component of the vertical force on the ground F_z may be approximated by neglecting the damping of the tire

$$F_z = -P(z_u - h) \ .$$

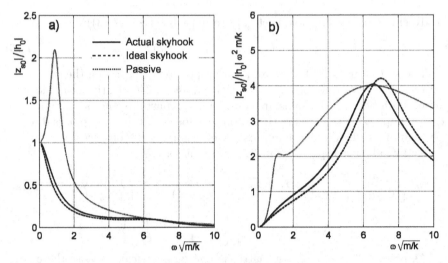

FIGURE 27.21. Non-dimensional frequency response of the sprung mass of a quarter car with two degrees of freedom with $b = P/K = 4$ and $a = m_u/m_s = 0.1$. The response of a passive quarter car with optimum damping ($\zeta = 0.433$) is compared with one with an ideal skyhook ($\zeta_s = 1$, ζ equal to 1/3 of the optimum value, $\zeta = 0.1443$) and with an actual skyhook with the same values of the parameters.

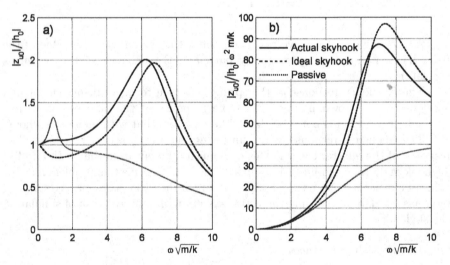

FIGURE 27.22. Frequency response of the unsprung mass of the same quarter car of Fig. 27.21.

The frequency response in terms of tire-ground force of the quarter car of Fig. 27.21 is shown in Fig. 27.23. As can be clearly seen, there is a strong improvement at low frequency, but at high frequency things are much worse.

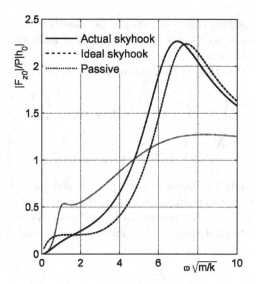

FIGURE 27.23. Frequency response in terms of force on the ground of the same quarter car of Fig. 27.21.

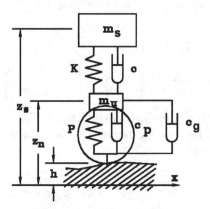

FIGURE 27.24. Quarter car with two degrees of freedom with a damper ideally located between the unsprung mass and the contact point with the ground (*groundhook*).

To minimize the variable component of the force on the ground either the stiffness of the tire or the deformation $(z_u - h)$ must be reduced. To reduce the latter a damper between the unsprung mass and the contact point with the ground may be ideally introduced (Fig. 27.24). Using a designation similar to the skyhook, this approach is usually referred to as *groundhook*.

It is clear from the figure that introducing a groundhook is equivalent to increasing the damping of the tire. Actually, it impossible to do so in an actual situation, and not only for practical reasons. Because the tire rotates, an increase of c_p would cause an unacceptable increase of rolling resistance, accompanied by a strong heating of the tire.

The equation of motion of the system of Fig. 27.24 is

$$\begin{bmatrix} m_s & 0 \\ 0 & m_u \end{bmatrix} \begin{Bmatrix} \ddot{z}_s \\ \ddot{z}_u \end{Bmatrix} + \begin{bmatrix} c & -c \\ -c & c+c_p+c_g \end{bmatrix} \begin{Bmatrix} \dot{z}_s \\ \dot{z}_u \end{Bmatrix} + \\ + \begin{bmatrix} K & -K \\ -K & K+P \end{bmatrix} \begin{Bmatrix} z_s \\ z_u \end{Bmatrix} = \begin{Bmatrix} 0 \\ (c_p+c_g)\dot{h}+Ph \end{Bmatrix}. \tag{27.78}$$

Here it is not only the homogeneous part of the equation that is affected: The forcing function is also changed. Again, this approach is ideal and in practice may be realized using only the scheme of Fig. 27.18 c), where an active or semi-active device is located between the two masses. If an active system is used, the force exerted on the unsprung mass is

$$F = -c_g \dot{z}_u . \tag{27.79}$$

In this way, not only is an equal and opposite force exerted on the sprung mass, but the effect linked with the vertical motion of the contact point on the ground is lost. The equation of motion is not Eq. (27.78), but becomes

$$\begin{bmatrix} m_s & 0 \\ 0 & m_u \end{bmatrix} \begin{Bmatrix} \ddot{z}_s \\ \ddot{z}_u \end{Bmatrix} + \begin{bmatrix} c & -c-c_g \\ -c & c+c_p+c_g \end{bmatrix} \begin{Bmatrix} \dot{z}_s \\ \dot{z}_u \end{Bmatrix} + \\ + \begin{bmatrix} K & -K \\ -K & K+P \end{bmatrix} \begin{Bmatrix} z_s \\ z_u \end{Bmatrix} = \begin{Bmatrix} 0 \\ c_p\dot{h}+Ph \end{Bmatrix}. \tag{27.80}$$

Note as well that the damping matrix is not symmetrical.

A combination of the two strategies may be implemented introducing the following damping matrix

$$\mathbf{C} = 2\sqrt{Km_s} \begin{bmatrix} \zeta_s+\zeta & -(\zeta_g+\zeta) \\ -(\zeta_s+\zeta) & \zeta_g+\zeta+\zeta_p \end{bmatrix}. \tag{27.81}$$

Because $\mathbf{C}$ is not symmetrical, it can be subdivided into a symmetric and a skew-symmetric matrix or, a more expedient procedure, in a matrix corresponding to a passive damper with constant damping coefficient and a matrix corresponding to an active system.

By introducing the mean and the deviatoric values of the damping of the skyhook and the groundhook

$$\zeta_m = \frac{\zeta_s+\zeta_g}{2}, \quad \zeta_d = \frac{\zeta_s-\zeta_g}{2}, \tag{27.82}$$

it follows that

$$\mathbf{C} = 2\sqrt{Km_s} \begin{bmatrix} \zeta_0 & -\zeta_0 \\ -\zeta_0 & \zeta_0+\zeta_p \end{bmatrix} + 2\sqrt{Km_s} \begin{bmatrix} \zeta_d & \zeta_d \\ -\zeta_d & -\zeta_d \end{bmatrix}, \tag{27.83}$$

where the passive damping is

$$\zeta_0 = \zeta + \zeta_m = \zeta + \frac{\zeta_s + \zeta_g}{2}. \qquad (27.84)$$

The dynamic behavior of the quarter car is then determined by a single parameter, ζ_d, that states the entity of force

$$F = -2\sqrt{Km_s}\zeta_d\left(\dot{z}_s + \dot{z}_u\right) \qquad (27.85)$$

the actuator exerts on the sprung mass. A positive value of ζ_d shows that the skyhook effect dominates, while a negative value shows that the system is primarily a groundhook. For instance, the suspension with skyhook of Fig. 27.21 is equivalent to a suspension with $\zeta_0 = 0.6443$ and $\zeta_d = 0.5$.

To compare the two control strategies, consider the same quarter car of Fig. 27.21, choosing a value $\zeta_0 = 0.433$, corresponding to the optimal value of the passive suspension, and two values of ζ_d, equal to 0.4 and -0.4 (Fig. 27.25). In the first case, corresponding to a skyhook, the performance of the active suspension is better than that of a passive system for the sprung mass, in a low frequency range. In the second case (groundhook), on the other hand, performance is improved at high frequencies, particularly if the unsprung mass is considered.

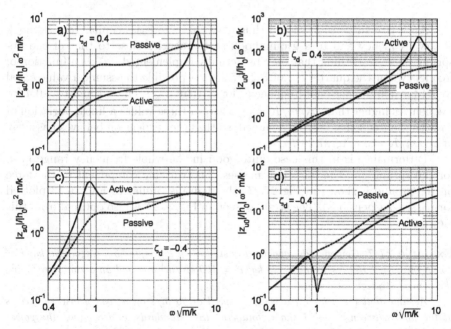

FIGURE 27.25. Non-dimensional frequency response of the sprung (a,c) and unsprung (b, d) mass of a quarter car with two degrees of freedom with $b = P/K = 4$ and $a = m_u/m_s = 0.1$. The response of the passive quarter car with optimum damping ($\zeta = 0.433$) is compared with that with skyhook ($\zeta_d = 0.4$, a,b) and groundhook ($\zeta_d = -0.4$, c,d).

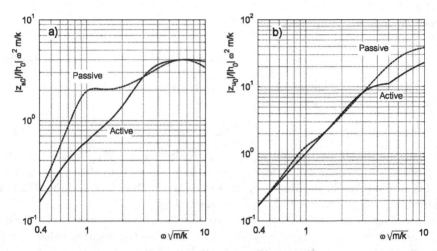

FIGURE 27.26. Non-dimensional frequency response of the sprung (a) and unsprung (b) masses of a quarter car with two degrees of freedom with $b = P/K = 4$ and $a = m_u/m_s = 0.1$. The response of a passive suspension with optimum damping ($\zeta = 0.433$) is compared with that of a quarter car with a value of ζ_d variable with the frequency between 0.4 and -0.4.

An active suspension should thus have characteristics similar to a skyhook at low frequency and to a groundhook at frequencies close to that of the unsprung mass, something that may be actually implemented if ζ_d is a function of frequency. For the quarter car of Fig. 27.25 it is possible to assume a value equal to 0.4 for non-dimensional frequencies lower than $\omega\sqrt{m/K} = 1$ and to -0.4 for frequencies higher than $\omega\sqrt{m/K} = 5$. Between these values a linear variation of ζ_d with the frequency may be assumed, so that the suspension behaves passively at $\omega\sqrt{m/K} = 3$ (Fig. 27.26).

Performance is in this case quite good in the whole frequency range both for the sprung and the unsprung masses. The variable component of the force on the ground is shown in Fig. 27.27 (the figure is similar to Fig. 26.13 plotted for passive suspensions). The strategy seen in Fig. 27.26 is also optimal from this viewpoint.

Example 27.7 *Consider the quarter car studied in Example 26.2 ($m_s = 250$ kg; $K = 25$ kN/m; $c = 2,150$ Ns/m $m_u = 25$ kg; $P = 100$ kN/m) and add an actuator realizing the control strategy seen above.*

Assume that the vehicle travels at a speed of 30 m/s on a road whose surface is at the limit between zones B and C following ISO standards, and compute the power spectral density of the acceleration of the sprung mass and the related r.m.s. value.

Compare the results in terms of comfort with those seen for the passive suspension.

The natural frequency of the sprung mass $\sqrt{K/m}$ is equal to 10 rad/s = 1.59 Hz, in which case the deviatoric damping of the active suspensions takes a value $\zeta_d = 0.4$ up to 1.59 Hz and $\zeta_d = -0.4$ above 7.96 Hz. Between these values it varies linearly.

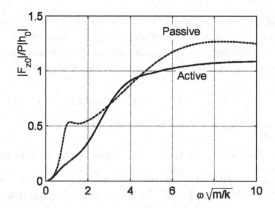

FIGURE 27.27. Variable component of the force in z direction on the ground as a function of frequency for the quarter car of Fig. 27.26.

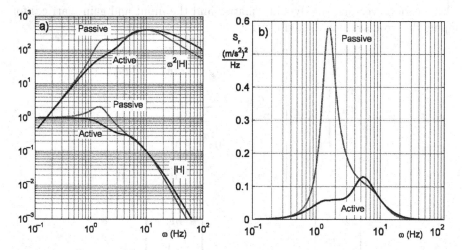

FIGURE 27.28. Dynamic compliance and inertance of the passive and active quarter car with two degrees of freedom (a). Power spectral density of the acceleration of the sprung mass while travelling on a road between zones B and C of ISO standards at a speed of 30 m/s (b).

Frequency response. *The dynamic compliance $H(\omega)$, i.e. the ratio between the displacement of the sprung mass and that of the supporting point on the ground, and the inertance, that is the ratio between the acceleration of the sprung mass and the displacement of the same point, $\omega^2 H$, are plotted in Fig. 27.28a) for both the passive and active suspension. The ability of the active suspension to filter out the perturbations reaching the sprung mass at its natural frequency is clear.*

Response to road excitation. *The power spectral density of the acceleration of the sprung mass may be computed as seen in the previous examples. The result is shown in Fig. 27.28 b). The resonance peak is completely cancelled, while at frequencies close* to the resonance of the unsprung mass the active suspension is slightly worse than the passive. This is due to the groundhook effect, which has been introduced to reduce the variable component of the force on the ground and thus to improve handling.

Comfort is, however, much improved, because the r.m.s. value of the acceleration is

$$a_{rms} = 1.11 \ m/s^2 \ = \ 0.113 \ g$$

about 20% less than in the case of the passive suspension

Quarter car controlled following the acceleration of the sprung mass

Consider a quarter car with two degrees of freedom and put an actuator in parallel to the spring and the shock absorber. By neglecting, as usual, the damping of the tire, the state space equation of the controlled system is

$$\dot{\mathbf{z}} = \mathbf{A}\mathbf{z} + \mathbf{B}_d\mathbf{u}_d + \mathbf{B}_c\mathbf{u}_c, \tag{27.86}$$

where various vectors, the dynamic matrix and the input and gain matrices for inputs due to disturbances and control are

$$\mathbf{z} = \left\{ \begin{array}{c} v_s \\ v_u \\ z_s \\ z_u \end{array} \right\}, \mathbf{A} = \left[\begin{array}{cccc} -\dfrac{c}{m_s} & \dfrac{c}{m_s} & -\dfrac{K}{m_s} & \dfrac{K}{m_s} \\[2mm] \dfrac{c}{m_u} & -\dfrac{c}{m_u} & \dfrac{K}{m_u} & -\dfrac{K+P}{m_u} \\[2mm] 1 & 0 & 0 & 0 \\[2mm] 0 & 1 & 0 & 0 \end{array} \right],$$

$$\mathbf{B}_d = \left\{ \begin{array}{c} 0 \\ \dfrac{P}{m_u} \\ 0 \\ 0 \end{array} \right\}, \ \mathbf{u}_d = h \ , \ \mathbf{B}_c = \left\{ \begin{array}{c} \dfrac{1}{m_s} \\ -\dfrac{1}{m_u} \\ 0 \\ 0 \end{array} \right\}, \ \mathbf{u}_c = F \ .$$

Consider the acceleration measured by an accelerometer on the sprung mass as output of the system. Assuming that the sensor is ideal, its output is

$$y = \mathbf{C}\mathbf{z} \ , \tag{27.87}$$

where the output gain matrix is

$$\mathbf{C} = \left[\begin{array}{cccc} -\dfrac{c}{m_s} & \dfrac{c}{m_s} & -\dfrac{K}{m_s} & \dfrac{K}{m_s} \end{array} \right].$$

Assuming a simple proportional control law, the control loop can be closed simply by stating that the force exerted by the actuator is proportional to the acceleration of the sprung mass through the gain G:

$$F = -Gy. \tag{27.88}$$

The dynamic equation of the system is

$$\dot{\mathbf{z}} = (\mathbf{A} - \mathbf{B}_c G \mathbf{C})\,\mathbf{z} + \mathbf{B}_d h. \tag{27.89}$$

The transfer function yielding the four components of the state vector is

$$\frac{1}{h}\mathbf{z} = (s\mathbf{I} - \mathbf{A} - \mathbf{B}_c G \mathbf{C})^{-1}\mathbf{B}_d. \tag{27.90}$$

The transfer function for the control force is

$$\frac{F}{h} = G\mathbf{C}\,(s\mathbf{I} - \mathbf{A} - \mathbf{B}_c G \mathbf{C})^{-1}\mathbf{B}_d. \tag{27.91}$$

If

$$G = m_s,$$

the first row of the closed loop dynamic matrix $\mathbf{A} - \mathbf{B}_c G \mathbf{C}$ vanishes, allowing the suspension to filter out the road irregularities completely. Such a solution, apart from problems linked with its practical feasibility, is also impossible from a theoretical viewpoint, because it would optimize the acceleration of the sprung mass but would lead to a strong increase of the dynamic component of the force on the ground.

Example 27.8 *Repeat the previous example, using a quarter car controlled on the acceleration of the sprung mass. Look for a gain of the control system that reduces the r.m.s. value of the acceleration of the sprung mass with reference to the conditions seen in the previous example (speed of 30 m/s on a road of the same type) without overly penalizing performance in term of forces on the ground or leading to high control forces.*

First, the r.m.s. values of the acceleration, the force on the ground and the control force are computed for different values of the gain. The computation is fairly long but not complicated: The transfer functions can be computed for each value of the gain through Equations (27.90) and (27.91).

Remembering that $s = i\omega$, the frequency responses for the power spectral density of the response can be computed. The required r.m.s. values are obtained by integrating the latter.

The values so computed are plotted in Fig. 27.29a. A range for the gain between 0 (passive system) and 300 Ns^2/m, was chosen. It includes the value $G = 250\ Ns^2/m$, at which the response in terms of acceleration of the sprung mass vanishes.

The r.m.s. values of the force on the ground and the acceleration were made non-dimensional by dividing them by the values of the passive system, while those of the control force were divided by the value obtained for a gain $G = 300\ Ns^2/m$. As predictable,

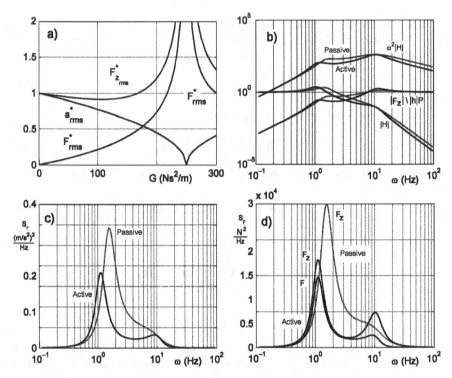

FIGURE 27.29. (a): Non-dimensional r.m.s. values of the acceleration of the sprung mass, the force on the ground and the control force at varying gain. (b): Frequency responses of the displacement and acceleration of the sprung mass and of the variable component of the force on the ground. (c) e (d): Power spectral density of the acceleration of the sprung mass and of the forces.

the acceleration decreases with increasing G, but the force has a shallow minimum to increase again, with a strong increase of the control force. A good compromise value may be

$$G = 120 \ Ns^2/m.$$

With this gain

$$a_{rms} = 0.964 \ m/s^2 \ = \ 0.098 \ g, \quad F_{z_{rms}} = 356 \ N \ , \quad F_{rms} = 223 \ N.$$

With respect to the passive solution, the acceleration is reduced by 28% and the force on the ground by little more than 8%.

The frequency responses of the displacement and the acceleration of the sprung mass and of the variable component of the force on the ground are plotted in Fig. 27.29b, while the power spectral density of the acceleration of the sprung mass and of the forces are reported in Fig. 27.29c and d.

The results show a strong improvement of comfort performance, even if of a different kind than that plotted in the previous example. In that case, the improvement was entirely at low frequency where the peak disappeared, while at high frequency performance was similar, if not worse, than that of the passive suspension.

In this case, the improvement is distributed over the entire frequency range. The handling performance seems to be essentially unchanged, at least in terms of the force on the ground. This could be expected, owing to the type of control used.

Final considerations

Some control strategies for active suspensions were considered in the previous sections. It must be first stressed that this is only a theoretical study, because the strategies that may actually be implemented depend on the actual characteristics of the actuators, sensors and power systems used, and in particular on their limitations in terms of manageable power and frequency response. Moreover, all solutions considered are based on a suspension including a passive spring, while the active or semi-active systems introduce a damping force only.

This solution, often considered because it allows relatively small forces to be controlled, clearly has a basic limitation: The elastic characteristics of the suspension are those of the passive device from which it derives. The zone labelled as active suspensions in Fig. 27.14 refers to solutions of this kind. If the stiffness of the suspension can also be controlled or, even better, if the force exerted between the sprung and the unsprung mass is completely controllable, performance can be optimized in a much wider range.

The techniques used are often based on optimal control, with observers used to estimate the quantities that cannot be measured directly, but the greater difficulties are found not so much in the definition and implementation of the control algorithms, but in the implementation and in the optimization of the actuation and power systems.

The models shown above are based on the quarter car model and assume a decentralized control in which each wheel (or as is often said, each corner of the vehicle) acts in an independent way. The roll and pitch characteristics of the vehicle come from those of single suspensions, as was also seen for the passive suspensions. A decentralized control is often implemented, and the interactions between the suspensions are considered as non-modelled dynamics, as well as other phenomena that are neglected. As seen in Section 27.5.1, roll motions may be controlled independently from heave motions by using active anti-roll bars. In the same way it is possible to control pitch motions by connecting the control of the front and rear axis suspensions in what may be defined as a centralized control, even if solutions of this kind are in general not yet applied.

27.6 *By wire* systems

The active systems used on motor vehicles are usually based on hydraulic components or, particularly on industrial vehicles, on pneumatic devices. Control of the actuators is performed by electrovalves and the actuation power is supplied by pumps, which are usually powered by the engine or by electric motors. Traditionally, the same holds for systems helping the driver, such as power steering or power brakes.

The development of electromechanical components and systems has led to the consideration of alternative solutions in which electric actuators are controlled directly without the need for electrovalves, pumps and other hydraulic or pneumatic devices. The tendency to replace hydraulic and pneumatic devices with electric systems is widespread and in many fields *more-electric* or even *all-electric* systems are considered. The advantages are many, among them the following:

- interfacing electric devices to control systems is easier than hydraulic devices;

- the transmission of power and command signals requires electric cables, which allow much greater freedom of layout than hydraulic pipes or mechanical transmissions as used with mechanical controls;

- electric devices are less affected by environmental conditions, above all temperature, than hydraulic systems. In particular, in the latter the viscosity of hydraulic fluids is strongly dependent on temperature;

- the fluids used in hydraulic devices require anti-pollution measures during construction, maintenance and ultimately disposal of the vehicle that are not required in electric devices;

The drawbacks of electric devices that have up to now hampered their diffusion are:

- electric actuators are in general heavier and often more bulky than the corresponding hydraulic actuators;

- the cost of high performance magnetic materials (in particular rare earth magnets) is still high for automotive applications;

- electric systems are in general 'stiffer' than hydraulic systems requiring a more precise control. They may cause more noise and vibration.

Progress in the fields of magnetic materials (above all permanent magnets), control systems and power amplifiers is gradually reducing these drawbacks allowing us to predict that the implementation of hydraulic and pneumatic systems will diminish. The cost of permanent magnets with high energy density is also decreasing, thanks to a liberalization of the markets due to the expiry of many patents.

An apparently marginal problem that is nonetheless hampering the introduction of *by wire* devices is the voltage of the on-board electric system. The increase in the power of the electric devices on motor vehicles makes it convenient to increase the voltage from the traditional (on cars) 12 V to at least

24 V, as on industrial vehicles, or even to 36 or 48 V. In this way, the current needed by the various devices would be reduced, with advantages in cost and weight. This is not a marginal change, as it might seem, because it would require the simultaneous redesign, production and marketing of a large number of new electric components (batteries, bulbs, switches, electric motors, etc.).

Another problem under intense study is electromagnetic compatibility. The electromagnetic environment in which automotive electromechanical and electronic devices must operate is very dirty, which may induce malfunctionings of different kinds. These must be by all means prevented when electric systems are entrusted functions that are vital for the safety of the vehicle.

By wire systems are now one of the most actively researched fields in the automotive industry, with different solutions under study. They will probably enter mass production in the relatively near future. The following sections will deal briefly with some applications already mentioned in this text. Note that the problems to be solved before vehicles completely controlled *by wire* may be marketed are not only technical (design, production, marketing, etc.) but also legal, regulatory and standards-based.

27.6.1   Steer by wire

Electric steering systems like Electric Power Steering (EPS) are in common use, primarily in cars in the low or medium market segment. Their application does not imply substitution for the mechanical steering system, but simply the presence of an electric actuator in parallel with the manual steering system. The electric motor may act directly, exerting a torque on the steering wheel shaft, or it may exert a force on the rack. The torque exerted by the power steering is proportional to that applied by the driver, following strategies that may depend on many parameters. The steering control may be made 'harder' with increasing speed, to compensate for the decrease of steering torque typical of many cars and to induce the driver to act on the wheel with more care.

As already stated in our discussion of handling control devices, the steering actuator may act independently of the command given by the driver, as when using a differential gear having as inputs the steering wheel shaft and an actuator controlled independently. In this case, direct control remains, but the authority of the control system is greater.

In a true *by wire* system, there is no direct control link and the steering wheel is connected solely to a rotation sensor (potentiometer, encoder...) or a torque sensor supplying the value of the angle or the moment to the system that controls the steering actuator. A system of this kind is much more flexible, allowing different control strategies to be used, such as a variable ratio between the rotation of the steering wheel and the steering rotation of the wheels. This allows the command from the driver to interact in more complex way with the command from the control system. Because there is no direct link between the wheels and the steering wheel, there must be an actuator exerting a torque on the steering wheel to supply the driver with information on the working conditions of the wheels (haptic controls).

27.6.2   Brake by wire

Electric power brakes may be simple devices in which an electric motor actuates a pump amplifying the command given by the driver through the master cylinder connected with the brake pedal. In more complicated systems no pump is connected with the brake pedal, with the latter simply supplying a position or force signal that, through a control system, acts on the actuators at the brakes.

The actuator may be a single pump supplying high pressure fluid to a more or less conventional braking system, or an electric actuator located in each wheel. In the latter case, the actuator may pressurize a fluid acting on the pistons of the caliper or may directly actuate the caliper through a ball screw or a mechanical system of other kinds. The choice among the various solutions must take into account the mass, the cost, the reliability of the system, the need for maintenance and adjustments, and the possibility of self-adjusting.

Clearly, an actuator in each wheel allows functions like ABS, TCS, VDC, etc. to be performed without the need of valves discharging the pressure from the branch of the system in each wheel or of pumps that recover the fluid. The brake in each wheel is controlled independently following the commands from the driver and the various control devices.

27.6.3   Electromechanical suspensions

Semi-active suspensions already use systems that are at least partially electromechanical, such as the shock absorbers based on electrorheological of magnetorheological fluids. Electric actuators may replace hydraulic actuators in active suspensions (for instance in ARC systems) and above all, electromechanical eddy current dampers may replace classical shock absorbers. In particular, eddy current dampers may work as passive, uncontrolled components (in this case they may behave as almost perfect viscous dampers, without the drawback of the presence of the fluid and with a greater stability in changing environmental conditions), they may be inserted in an electric circuit containing controlled elements or they may work in a fully active way. At present, their mass is comparable to that of similar hydraulic devices and their cost, although still higher, is quickly decreasing.

27.6.4   Other by wire controls

As already stated, all functions of the vehicle may be controlled by electromechanical actuators. Because control of the engine is the simplest, conversion, *by wire* accelerators are now common. The clutch and the gearbox may also be controlled by electric devices. Apart from the advantage of avoiding mechanical linkages between the gearbox and the passenger compartment, which may transmit vibration, this solution allows for the building of fully automated gearboxes. Secondary controls, such as the parking brake, may also be replaced by electric devices, with the advantages of automation (in the case of the parking brake, it is possible to guarantee that the brake is applied when the vehicle is stopped and released when it moves), thus allowing a greater freedom in command placement and design of the user interface, as well as a much simpler design of the control transmission.

Part V

MATHEMATICAL MODELLING

INTRODUCTION TO PART V

For decades the design of motor vehicles was based on a very close interaction between simplified analytical methods , sometimes deriving from empirical studies, and experimental verification, primarily performed by experienced test drivers on tracks simulating the various operating conditions. This approach was justified by the great complexity of the vehicle, seen as a system, and by the number of requirements, often in contradiction with each other, it has to satisfy.

Starting in the 1960s, an approach based on greater use of mathematical models was introduced. Modeling was used not only to verify the ability of the components to withstand the various predictable load conditions (this had begun in civil engineering and made its first appearance as early as the XVII Century, spreading into the XIX Century primarily through new demands imposed by the construction of railways and rolling stock), but also to simulate in detail the performance at a system level. There is no doubt that grater use of mathematical modelling was initially dictated by the needs of the aerospace industry, in which the behavior an aircraft or a spacecraft would demonstrate in flight had to be known in detail before performing any full scale experiment.

In the same 1960s, these new demands for a wider use of mathematical methods in design were met by the availability of new computational instruments with a power up to then unheard of. The constant increase of the power of computers, together with an equally striking reduction in their cost, and the implementation of software allowing the use of these powerful tools by a wider number of technicians, not only in Universities and Research centers, but also in Industry, allowed these methods to spread from the fields in which they originated to other industries where the need for predicting the behavior of machines mathematically was not so compelling.

One of the fields that was significantly transformed by this new approach was the automotive industry. It obtained advantages, linked largely to the economy of the production process. A major reason for this is the scale of production: when thousands (millions) of essentially identical machines, or at least machines that share many components, have to be produced, it is possible to accept very high design costs that are subdivided by the large number of produced units. Even if a more accurate design causes only small savings per unit, the cumulative effect may be large.

Numerical simulation allows the effects of complex phenomena to be studied, even if they are nonlinear, before building and experiments begin on prototypes, something that was completely beyond the scope of traditional analytical methods based on much simplified and linearized models. If these traditional methods are still useful and probably necessary instruments, at least in a qualitative way, physical phenomena that may be very complex demand numerical simulation as the only way able to supply results quantitatively adequate to validate design choices and to optimize a design, at least initially, before resorting to costly experimental techniques.

Part V of this book will be devoted to the mathematical modelling of motor vehicles, particularly those parts for which the modelling of the chassis and its components is essential. The models already seen in the preceding sections will be further developed, increasing both their complexity and their ability to simulate in detail the behavior of the vehicle. It must be remembered that very complex models must be implemented using complicated codes that are usually produced by specialized companies and are used as design tools.

The aim of these chapters is, then, not to supply the knowledge allowing designers to prepare their own codes for performing dynamic simulations, but to help the users of commercial codes to gain the knowledge needed to prepare the models and interpret the results correctly. Any mathematical model is based on a simplification of the physical world that in come cases may be rough. Its use requires a thorough knowledge of the mechanisms and assumptions on which it is based.

It may be said that the correct use of simulation codes requires that users have a knowledge of the mathematical methods and the schematization of the actual phenomena only slightly less than that needed to design and implement the code.

After a chapter devoted to mathematical models and numerical simulation in general, a chapter dealing with multibody models of motor vehicles, i.e. models in which the whole vehicle is simulated by a number of rigid bodies connected together by springs and dampers, will follow.

Another chapter will be devoted to the detailed modelling of the driveline. The section will be completed by a chapter on the modelling of tilting body vehicles. This last chapter is justified by the fact that such vehicles, characterized by the ability to keep the local vertical in the symmetry plane by inclining the whole body under the action of suitable control devices, are at present being tested by many manufacturers. Their mathematical modelling is peculiar, because the much higher roll angle at which they work precludes the possibility of introducing simplifications usually accepted in vehicle dynamics.

28

MATHEMATICAL MODELS FOR THE VEHICLE

An increasingly competitive automotive market offers its products to increasingly demanding customers. Numerous standards and rules, primarily regarding safety and environmental impact, are issued by regulating bodies and governments, making today's vehicles more and more complex.

The specifications vehicles must comply with often contrast with each other. The time between the conception of a vehicle and its entering the market, the so called *time to market*, is an essential factor for its commercial success. The traditional approach, based on the construction of prototypes, subsequent experimentation and modification, is no longer adequate.

When design changes are introduced during the development of any machine, they should be implemented as early as possible. The later such changes are made, the smaller the advantages, both in terms of economics but also more generally, while related costs are higher (Fig. 28.1). In the early stages of development, when the vehicle is merely an idea or the result of a preliminary study, major changes may be introduced at low cost, but when the design proceeds toward the construction of prototypes or the product launch, the freedom of designers is reduced and the costs linked to changes, not only to the design but also to prototypes and above all to production equipment, soar.

This consideration alone can explain the increasing significance of activities like the construction of mathematical models, virtual experimentation and simulation in the motor vehicle industry.

The costs linked with the construction of prototypes and physical experimentation are continuously growing, while the increase of computational power causes a decrease of computational times and thus of the costs linked with the construction of mathematical models and the ensuing numerical experimentation.

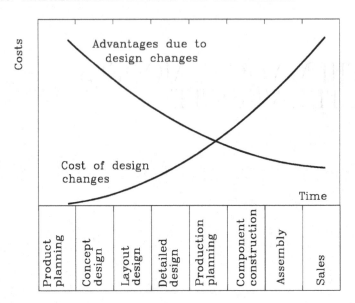

FIGURE 28.1. Qualitative trend of costs and advantages of design changes in the various stages of the design and construction of a machine.

The importance of being able to predict the behavior of the system and its components before the construction of prototypes is therefore increasing in automotive technology. The goal, at present still far from being achieved, is to reduce the importance of physical testing to a simple verification of what virtual testing has ascertained.

28.1   MATHEMATICAL MODELS FOR DESIGN

The computational predictions of the characteristics and the performance of a physical system are based on the construction of a *mathematical model*[1], constructed from a number of equations, whose behavior is similar to that of the physical system it replaces. In the case of dynamic models, such as those used to predict the performance of motor vehicles, the model is usually built from a number of ordinary differential equations[2] (ODE).

The complexity of the model depends on many factors, that represent the first choice the analyst has to make. The model must be complex enough to allow

---

[1] Note that simulation is not always based on a mathematical model in a strict sense. In the case of analog computers, the model was made from an electric circuit whose behavior simulated that of the physical system. Simulation on digital computers is based on actual mathematical models.

[2] A dynamic model, or a dynamic system, is a model expressed by one or more differential equations containing derivatives with respect to time.

a realistic simulation of the system's characteristics of interest, but no more. The more complex the model, the more data it requires, and the more complicated are the solution and interpretation of results. Today it is possible to built very complex models, but overly complex models yield results from which it is difficult to extract useful insights into the behavior of the system.

Before building the model, the analyst must be certain what he wants to obtain from it. If the goal is a good physical understanding of the underlying phenomena, without the need for numerically precise results, simple models are best. Skilled analysts have simulated even complex phenomena with precision using models with a single degree of freedom. If, on the contrary, the aim is precise quantitative results, even at the price of more difficult interpretation, the use of complex models becomes unavoidable.

Finally, it is important to take into account the available data at the stage reached by the project: Early in the definition phase, when most data are still not available, it can be useless to use complex models, into which more or less arbitrary estimates of the numerical values must be introduced. Simplified, or synthetic, models are the most suitable for a preliminary analysis. As the design is gradually defined, new features may be introduced into the model, reaching comprehensive and complex models for the final simulations.

Such complex models, useful for simulating many characteristics of the vehicle, may be considered as true *virtual prototypes*. Virtual reality techniques allow these models to yield a large quantity of information, not only on performance and the dynamic behavior of the vehicle, but also on the space taken by the various components, the adequacy of details and the esthetic qualities of the vehicle, that is comparable to what was once obtainable only from physical prototypes.

The models of a given vehicle often evolve initially toward a greater complexity, from synthetic models to virtual prototypes, to return later to simpler models. Models are useful not only to the designer in defining the vehicle and its components, but also to the test engineer in interpreting the results of testing and performing all adjustments. Simplified models can be used on the test track to allow the test engineer to understand the effect of adjustments and reduce the number of tests required, provided they are simple enough to give an immediate idea of the effect of the relevant parameters. Here the final goal is to adjust the virtual prototype on the computer, transferring the results to the physical vehicle and hoping that at the end of this process only a few validation tests are required.

Simplified models that can be integrated in real time on relatively low power hardware are also useful in control systems. A mathematical model of the controlled system (*plant*, in control jargon) may constitute an *observer* (always in the sense of the term used in control theory) and be a part of the control architecture.

The analyst has the duty not only of building, implementing and using the models correctly, but also of updating and maintaining them. The need to build a mathematical model of some complexity is often felt at a certain stage of the design process, but the model is then used much less than necessary, and above all is not updated with subsequent design changes, with the result that it becomes completely useless or needs updating when the need for it again arises.

There are usually two different approaches to mathematical modelling: models made by equations describing the physics of the relevant phenomena, − these may be defined as *analytical models* − and empirical models, often called *black box* models.

In analytical models the equations approximating the behavior of the various parts of the system, along with the required approximations and simplifications, are written. Even if no real world spring behaves exactly like the linear spring, producing a force proportional to the relative displacement of its ends through a constant called stiffness, and even if no device dissipating energy is a true linear damper, the dynamics of a mass-spring-damper system can be described, often to a very good approximation, by the usual ordinary differential equation (ODE)

$$m\ddot{x} + c\dot{x} + kx = f(t) \ . \tag{28.1}$$

The behavior of a tire, on the other hand, is so complex that writing equations to describe it beginning with the physical and geometrical characteristics of its structure is forbiddingly difficult. The *magic formula* is a typical example of the empirical, black box, model. The behavior of the tire is studied experimentally after which a mathematical expression able to describe it is sought, identifying the various parameters from the experimental data. While each of the parameters m, c and k included in the equation of motion of the mass-spring-damper system refers to one of the parts of the system and has a true physical meaning, the many coefficients a_i, b_i and c_i appearing in the expressions for A, B, C, etc. in the magic formula have no physical meaning, and refer to the system as a whole.

Among the many ways to build black box models, that based on neural networks must be mentioned[3]. Such networks can simulate complex and highly nonlinear systems, adapting their parameters (the *weights* of the network) to produce an output with a relationship to the input that simulates the input-output relationship of the actual system.

Actually, the difference between analytical and black-box models is not as clear cut as it may seem. The complexity of the system is often such that it is difficult to write equations precisely describing the behavior of its parts, while the values of the parameters cannot always be known with the required precision. In such cases the model is built by writing equations approximating the general pattern of the response of the system, with the parameters identified to make the response of the model as close as possible to that of the actual system. In this case, the identified parameters lose a good deal of their physical meaning related to the various parts of the system they are conceptually linked to and become global parameters of the system.

---

[3]Strictly speaking, neural networks are not sets of equations and thus do not belong to the mathematical models here described. However, at present neural networks are usually simulated on digital computers, in which case their model is created by a set of equations.

In the following, primarily analytical models will be described and an attempt will be made to link the various parameters to the components of the system.

28.2 CONTINUOUS AND DISCRETIZED MODELS

The objects constituting our real world are all more or less compliant and are quite well modelled as continuous systems. A compliant body is usually modelled as a continuum or, if its behavior can be considered linear and damping is neglected, as a linear elastic continuum. It is clear that the elastic continuum is only a model, because no actual body is such at an atomic scale, but for most objects studied by structural dynamics the continuum model is more than adequate.

An elastic body may then be thought of as consisting of an infinity of points. The configuration at any time t can be obtained from the initial configuration once a vector function expressing the displacements of all points is known (Fig. 28.2). The displacement of a point is a vector, with a number of components equal to the number of dimensions of the reference frame. The components of this vector are usually taken as the degrees of freedom of each point, and thus the number of degrees of freedom of a deformable body is infinite. The corresponding generalized coordinates can be manipulated as functions of space and time coordinates, usually continuous and differentiable up to a suitable order, while the characteristics of the material are defined by functions of the coordinates in the whole part of space occupied by the continuum. In general, these functions need not be continuous.

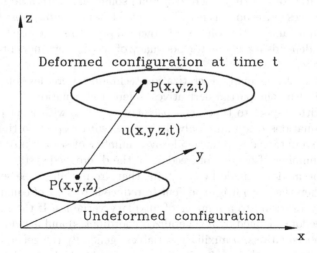

FIGURE 28.2. Deformation of an elastic continuum; reference frame and displacement vector.

The theory of continuous functions is the natural tool for dealing with deformable continua. Function $\vec{u}(x, y, z, t)$ describing the displacement of the points of the body is differentiable with respect to time at least twice; the first derivative gives the displacement velocity and the second the acceleration. Usually, however, higher-order derivatives also exist.

Assuming that the forces acting on the body are expressed by function $\vec{f}(x, y, z, t)$, the equation of motion can generally be written as

$$D[\vec{u}(x, y, z, \dot{x}, \dot{y}, \dot{z}, t)] = \vec{f}(x, y, z, t) , \qquad (28.2)$$

where the differential operator D completely describes the behavior of the body. Eq. (28.2) is a partial derivatives differential equation, in general nonlinear, containing derivatives with respect to both time and space coordinates and velocities.

If the system is linear, Eq. (28.2) is also linear and does not contain the velocities $\dot{x}$, $\dot{y}$, $\dot{z}$ if the system is conservative. To such an equation other equations expressing the boundary conditions and the initial conditions must be added.

The actual form of the differential operator may be obtained by resorting directly to the dynamic equilibrium equations or by writing the kinetic and potential energies and using Lagrange equations. The boundary conditions usually follow from geometrical considerations.

Equation (28.2) may be solved in closed form only in a small number of cases, owing to the difficulties deriving from the differential equation and especially from the boundary conditions.

For complex systems the only feasible approach is the discretization of the continuum and then the application of the methods used for discrete systems. The replacement of a continuous system, characterized by an infinite number of degrees of freedom, with a discrete system, sometimes with a large but finite number of degrees of freedom, is usually referred to as *discretization*. This step is of primary importance in the solution of practical problems, because the accuracy of the results depends largely on the adequacy of the discrete model to represent the actual system.

In recent centuries many discretization techniques were developed with the aim of substituting the partial derivatives differential equation of motion (with derivatives with respect to time and space coordinates) with a set of ordinary differential equations containing only derivatives with respect to time. The set of equations so obtained is often made by a number of second-order equations equal to the number of degrees of freedom of the discretized system.

When the model is made by equations that are not all of the second-order (and often when they are) it is expedient to reduce it to a set of equations of the first order, by resorting to a number of auxiliary variables. If the model has n degrees of freedom (defined by n generalized coordinates) and is made of a set of n second-order equations, n auxiliary variables (generally the generalized velocities) are needed, and the resulting model is made by $2n$ first-order equations. The $2n$ variables (n generalized coordinates and n generalized velocities) are the *state variables* of the system. The vector containing the state variables is called the *state vector* and is usually indicated by $\mathbf{z}$.

If the set of $2n$ first-order equations is solved in the derivatives of the state variables, or it is solved in monic form, it has the form

$$\dot{\mathbf{z}} = f(\mathbf{z}, t) \, . \qquad (28.3)$$

The simplest way to discretize a model is by concentrating its inertial characteristics in a certain number of rigid bodies, or even material points, with its elastic and damping properties in massless springs and dampers. The models seen in the previous chapters to analyze the dynamic behavior of motor vehicles belong mostly to this type, as for instance the spring-mass-damper model expressed by Eq. (28.1).

Because a point has 3 degrees of freedom in three-dimensional space, the most obvious choice is to use as generalized coordinates the 3 coordinates of the point referred to an inertial frame. A rigid body has, in a three-dimensional space, 6 degrees of freedom. Thus a reasonable choice for the generalized coordinates is to use the three components of the displacement of one of its points, usually the center of mass, and 3 rotations. While the displacement is a vector, so that there is no difficulty in choosing the 3 coordinates related to displacement, things are much more complicated for the coordinates linked to rotations and, as will be seen later, different choices are possible.

The choice of the generalized coordinates and the equations of motion of a rigid body will be described in detail in Appendix A.

If it becomes impossible to neglect the compliance of some parts of the system, it is possible to resort to the Finite Element Method (FEM). The body is subdivided into a number of regions, the finite elements, so-called to specify their difference from the infinitesimal regions of space used to write the equations of motion of continuous systems. The inflected shape of each region is approximated by the linear combination of a number of functions of space coordinates through parameters that are taken as the generalized coordinates of the element. Usually such functions of space coordinates (the shape functions) are very simple and the generalized coordinates have a physical meaning, such as generalized displacements of some points of the element, called nodes. The analysis then proceeds by writing a set of differential equations of the type typical of discrete systems.

28.3   ANALYTICAL AND NUMERICAL MODELS

Once the model has been discretized and the equations of motion written, there is no difficulty in studying the response to any input, assuming the initial conditions are stated. A general approach is to numerically integrate the ordinary differential equation constituting the model, using any of the many available numerical integration algorithms. In this way, the *time history* of the generalized coordinates (or of the state variables) is obtained from any given time history of the inputs (or of the forcing functions).

This approach, usually referred to as simulation or numerical experimentation, is equivalent to physical experimentation, where the system is subjected to given conditions and its response measured.

This method is broadly applicable, because it

- may be used on models of any type and complexity, and

- allows the response to any type of input to be computed.

Its limitations are also clear:

- it doesn't allow the general behavior of the system to be known, but only its response to given experimental conditions,

- it may require long computation time (and thus high costs) if the model is complex, or has characteristics that make performing the numerical integration difficult, and

- it allows the effects of changes of the values of the parameters to be predicted only at the cost of a number of different simulations.

If the model can be reduced to a set of linear differential equations with constant coefficients, it is possible to obtain a general solution of the equations of motion. The free behavior of the system can be studied independently from its forced behavior, and it is possible to use mathematical instruments such as Laplace or Fourier transforms to obtain solutions in the *frequency domain* or in the *Laplace domain*. These solutions are often much more expedient than solutions in the *time domain* that, as stated above, are in general the only type of solution available for nonlinear systems.

The possibility of obtaining general results makes it convenient to start the study by writing a linear model through suitable linearization techniques. Only after a good insight into the behavior of the linearized models is obtained will the study of the nonlinear model be undertaken. When dealing with nonlinear systems it is often expedient to begin with simplified methods, such as harmonic balance, or to look for series solutions before starting to integrate the equations numerically.

29

MULTIBODY MODELLING

A vehicle on elastic suspensions may be modelled as a system made by a certain number of rigid bodies connected with each other by mechanisms of various kinds and by a set of massless springs and dampers simulating the suspensions. A vehicle with four wheels can be modelled as a system with 10 degrees of freedom, six for the body and one for each wheel. This holds for any type of suspension, if the motion of the wheels due to the compliance of the system constraining the motion of the suspensions (longitudinal and transversal compliance of the suspensions) is neglected. The wheels of each axle may be suspended separately (independent suspensions) or together (solid axle suspensions), but the total number of degrees of freedom is the same (Fig. 29.1). Additional degrees of freedom, such as the rotation of the wheels about their axis or about the kingpin, can be inserted into the model to allow the longitudinal slip or the compliance of the steering system to be taken into account.

The multibody approach can be pushed much further, by modelling, for instance, each of the links of the suspensions as a rigid body. To model a short-long arms (SLA) suspension it is possible to resort to three rigid bodies, simulating the lower and upper triangles and the strut, plus a further rigid body simulating the steering bar. While modelling the system in greater detail, the number of rigid bodies included in the model increases. However, if the compliance of the various elements is neglected (i.e. if these bodies are rigid bodies), the number of degrees of freedom does not increase along with the number of bodies: an SLA suspension always has a single degree of freedom, even if it is made up of a number of rigid bodies simulating its various elements.

The mathematical model of a multibody system is thus made up of the equations of motion of the various elements, which in tri-dimensional space are $6n$, if n is the number of the rigid bodies, plus a suitable number of constraint equations.

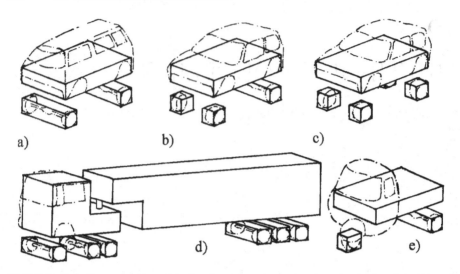

FIGURE 29.1. Example of models for the dynamic study of road vehicles. (a), (b) and (c): Vehicle with two axles, 10 d.o.f.; (d): Articulated truck with 6 axles, 21 d.o.f.; (e): Vehicle with 3 wheels; 9 d.o.f.

Consider for instance the articulated truck of Fig. 29.1d. The rigid bodies are 8 (tractor, trailer and 6 rigid axles) and thus the equations of motion are 48 second order differential equations. The constraint equations are 27: 3 equations for the constraint between tractor and trailer (these state that the coordinates of the center of the hitch, assumed to be a spherical hinge, are the same if this point is seen as belonging to the tractor or to the trailer) and 4 equations for each axle, leaving to each one of them just two, out of its 6 degrees of freedom. The 27 constraint equations are algebraic equations containing only the generalized coordinates but not the velocities (holonomic constraints).

By using the 27 constraint equations to eliminate 27 of the generalized coordinates, a set of 21 equations in the 21 independent generalized coordinates is obtained. It must be emphasized that in the 48 equations of motion originally written, the forces the various bodies exchange at the constraints are included; these forces are then eliminated when the constraint equations are introduced. The 27 equations so eliminated can be used to compute the constraint forces.

This approach is the broadest, and is usually implemented in *general purpose* multibody computer codes.

It is possible to resort to a simpler approach in the case of the multibody models used in motor vehicle dynamics, because the internal constraints are holonomic and the system is branched. One of the bodies may be chosen as *main body*, to which a number of secondary, *first level* bodies are attached. Other secondary bodies, considered as *second level* bodies, are then attached, and so on. Secondary bodies have only the degrees of freedom allowed by the constraints. In this way, the minimum number of equations needed for the study are directly obtained. Such equations are all differential equations, usually of the second

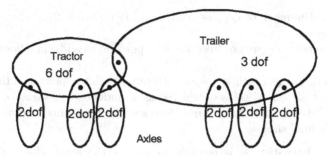

FIGURE 29.2. Model of the articulated truck with 6 axles shown as a branched model: the tractor is the main body, the axles of the tractor and the trailer are first level secondary bodies and the axles of the trailer are second level secondary bodies. The constraint between tractor and trailer is a spherical hinge constraining 3 degrees of freedom, while those constraining the axles to the sprung masses lock 4 degrees of freedom each.

order. The forces exchanged at the constraints between the bodies do not appear explicitly in the equations and need not be computed in the dynamic study of the system as a whole. The model of the articulated truck with 6 axles shown in Fig. 29.1d obtained in this way is sketched in Fig. 29.2: the tractor is considered as the main body, the axles of the tractor and the trailer are first level secondary bodies, and the axles of the trailer are second level secondary bodies.

29.1   ISOLATED VEHICLE

The model for an isolated vehicle can thus be easily built through the following steps:

- choice of the generalized coordinates;

- computation of the expressions for the kinetic and potential energies, the dissipation function and the virtual work of external forces (road-wheels forces and aerodynamic forces);

- writing the equations of motion through Lagrange equations.

The basic degrees of freedom of the model are:

- six degrees of freedom for the sprung mass (usually three components of the displacement define the position of the vehicle and three rotations define its orientation in space);

- two degrees of freedom for each rigid axle;

- one degree of freedom for each independent suspension.

The total number of degrees of freedom is then $6 + 2m$, where m is the number of axles.

To these basic degrees of freedom, it is possible to add the following:

- The rotation χ of each wheel, or better its angular velocity $\dot{\chi}$, if the angle of rotation of each wheel is considered as an independent variable. It is then possible to compute longitudinal forces at the road-wheel contact from the longitudinal slip.

  As an alternative, it is possible to neglect the longitudinal slip and to compute the wheel rotations from the space covered by the vehicle or, better, to compute their velocity from the velocity of the vehicle, but in this case the wheel rotations are not degrees of freedom of the system.

- The steering angle δ of the steering wheels. It may be considered as:

  - a given quantity, or better, an input, if the motion is studied with locked controls and the compliance of the steering system is neglected;

  - a known function of a single variable, the angle at the steering wheel δ_v, if the motion is studied with free controls but the compliance of the steering system is neglected;

  - a variable, linked by a an equation expressing the compliance of the steering system, to the angle at the steering wheel δ_v that is a given quantity, or better, an input, if the motion is studied with locked controls and the compliance of the steering system is accounted for;

  - an independent variable, to which a further independent variable, the angle at the steering wheel δ_v is added, if the motion is studied with free controls and the compliance of the steering system is accounted for.

A motor vehicle with four wheels can then be described by a model with 10 degrees of freedom (Fig. 29.1) in the simplest case (locked controls, neglecting longitudinal slip and the compliance of the steering system), that become 14 if the slip of the wheels is considered, 15 if the study is performed with free controls and rigid steering, or 16, 17 or 19 depending on how the steering system is modelled and, in the latter case, on how many wheels steer.

Once the kinematics of the suspensions has been defined, it is possible to write the equations of motion. An approach also used in commercial codes is to introduce the elasto-kinematic characteristics of the suspensions directly (often the kinematic characteristics alone, because the suspensions are assumed to be made of rigid bodies) to define the kinematics of the system. The characteristics of the tires, including the cornering forces, the aligning torques, and the relationships linking the longitudinal slip with the longitudinal forces (in case the longitudinal slip is not neglected) can be expressed using the magic formula.

The ten (or more, depending on the model used) equations can thus be written. They are quite complicated nonlinear equations, difficult to write in explicit form. They will not be shown here.

The solution of such a set of equations can be undertaken only by numerically integrating the equations in time starting from a given set of initial conditions and specifying the time history of the various inputs (steering angle, if the manoeuvre with locked controls, or the torque acting on the steering wheel for the motion with free controls). As an alternative, it is possible to use a model of the driver to simulate the behavior of the vehicle-driver system. Suspensions are modelled by introducing their elasto-kinematic characteristics directly.

Several commercial computer codes operating in this way exist. One of the most common is Carsim®, based on a model with 14 degrees of freedom.

A more complex alternative is the use of one of the standard multibody general purpose codes to simulate the suspensions in detail, taking into account the exact kinematics of the system and again simulating the tires using the magic formula. An example of a commercial code operating along these lines is ADAMS-Car®, based on the general purpose code ADAMS®.

Codes of the latter type draw upon a much larger number of equations, because they do not use just the minimum number of generalized coordinates, but are based on the explicit equations of motion of the various parts and on the relevant constraint equations.

The two approaches are equivalent to the user, because in both cases it is necessary to resort to the numerical integration of the equations of motion, simulating the dynamic behavior of the vehicle. The only actual difference for the user is that in the first case the behavior of the suspensions is introduced in synthetic form, computing their kinematic characteristics separately or measuring experimentally and then introducing them into the computations, while in the second case the geometry of the suspensions is directly introduced in analytic form.

29.2 LINEARIZED MODEL FOR THE ISOLATED VEHICLE

29.2.1 Basic assumptions

While in the case of the nonlinear model the exact geometry or the elasto-kinematic characteristics of the suspensions must be introduced, in the case of linearization it is possible to write a model that is fairly precise and quite general, while allowing analytical solutions to be obtained. In this case it is worthwhile to write the equations of motion in explicit form, so that it is possible to obtain general results and closed form solutions. In particular, it will be possible to obtain solutions in the frequency domain and to perform stability studies.

The model is based on the following additional assumptions:

- Reference is made to a certain configuration of the vehicle. It may be the static equilibrium condition with the vehicle at standstill or travelling at a constant speed. If the shape of the vehicle is such that it produces little aerodynamic lift and pitching moment, the first choice may be the best, because it allows the motion at constant speed to be studied. However, if aerodynamic forces are important, as in racing cars, the configuration of the suspensions may change considerably at varying speed, so much so that linearization is not possible, and reference must then be made to the equilibrium configuration at the given speed. The linearization of the model allows us to resort to the superimposition of effects and then to neglect static forces (weight, aerodynamic lift in the reference conditions, etc.) in the dynamic study.

- The kinematics of suspensions is linearized around the reference position.

- Pitch and roll angles are small enough to linearize their trigonometric functions. Also, the displacements in Z direction and all linear and angular velocities, with the exception of the forward speed and the rotation speed of wheels, are considered as small quantities.

Two other assumptions, needed to further simplify the equations of motion, are then added:

- The vertical plane xz through the center of mass is a symmetry plane for the vehicle and its parts.

- Sideslip angles of the wheels are small enough to linearize the cornering forces and the aligning torque

29.2.2 Sprung mass

Let the reference frame of the sprung mass be $x_s y_s z_s$, with axis x_s parallel to the roll axis and axis z_s lying in the plane of symmetry. Axis x^* coincides with the projection of the roll axis on the ground. A plane perpendicular to the road and to axis x^* containing the centre of mass G of the vehicle in the reference position is defined (section B-B in Fig. 29.3). The roll axis intersects such a plane in H; O is the point on the ground vertically under H (it may be located above H, as the roll axis may lie below the ground in particular cases). Two further reference frames are defined. They are:
— xyz, fixed to the sprung mass, with origin in point H, x-axis coinciding with the roll axis, z axis laying in the symmetry plane of the sprung mass;
— $x^* y^* z^*$ with origin in point O; x^*-axis coincides with the projection on the ground of the roll axis and z^* axis is perpendicular to the road.
Instead of using the coordinates of the centre of mass G_s of the sprung mass to define the generalized coordinates for the translational degrees of freedom, the

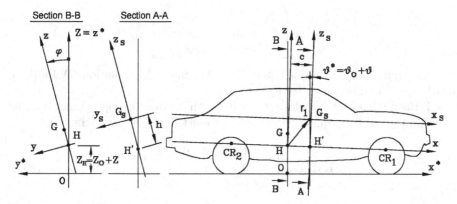

FIGURE 29.3. Reference frames for the sprung mass and definition of points H and O.

coordinates X_H, Y_H and Z_H of point H in the inertial frame $OX_iY_iZ_i$ will be used. In the following, to simplify the notation it will be $X = X_H$ and $Y = Y_H$. Operating in this way, if the roll and pitch motions are locked, the frame $x^*y^*z^*$ coincides with frame xyz defined in Fig. 25.15 and the model reduces to that of a rigid vehicle.

Coordinate Z_H can be considered as the sum of a constant value Z_0 corresponding to a reference position and a displacement Z:

$$Z_H = Z_0 + Z . \tag{29.1}$$

The generalized coordinates for translations of the sprung mass are then X, Y and Z, with Z considered as a small displacement with respect to the reference position.

The generalized coordinates for rotations are three Tait-Bryan angles (Appendix A, Fig. A.5): the yaw angle ψ, the pitch angle, here considered as the sum of a constant value θ_0 related to the reference position and a pitch generalized coordinate θ, and the roll angle ϕ. Angle θ_0 is the inclination on the horizontal direction of the roll axis in the reference position and will be considered as a small angle. All generalized coordinates and velocities, except v_x, are then small quantities.

Velocity v_x may be confused with the velocity V along the vehicle path. This is made possible by the smallness of the pitch angle $\theta_0 + \theta$ and of the sideslip angle β.

The rotation matrix allowing us to pass from the frame $Gxyz$ fixed to the body to the inertial frame $X_iY_iZ_i$ is:

$$\mathbf{R} = \mathbf{R}_1\mathbf{R}_2\mathbf{R}_3 , \tag{29.2}$$

where:

$$\mathbf{R}_1 = \begin{bmatrix} \cos(\psi) & -\sin(\psi) & 0 \\ \sin(\psi) & \cos(\psi) & 0 \\ 0 & 0 & 1 \end{bmatrix} , \quad \mathbf{R}_2 = \begin{bmatrix} \cos(\theta_0 + \theta) & 0 & \sin(\theta_0 + \theta) \\ 0 & 1 & 0 \\ -\sin(\theta_0 + \theta) & 0 & \cos(\theta_0 + \theta) \end{bmatrix} ,$$

$$\mathbf{R}_3 = \begin{bmatrix} 1 & 0 & 0 \\ 0 & \cos(\phi) & -\sin(\phi) \\ 0 & \sin(\phi) & \cos(\phi) \end{bmatrix} .$$

Its explicit expression is reported in Appendix A (Equation (A.106)), in which $\theta_0 + \theta$ is substituted for θ).

If the pitch and roll angles are small, i.e. their cosine is approximately equal to 1 and their sine is equal to the angle, product $\mathbf{R}_2\mathbf{R}_3$ is, approximately:

$$\mathbf{R}_2\mathbf{R}_3 \approx \begin{bmatrix} 1 & 0 & \theta_0 + \theta \\ 0 & 1 & -\phi \\ -\theta_0 - \theta & \phi & 1 \end{bmatrix} . \tag{29.3}$$

Because the generalized forces are written in the body-fixed frame, it is expedient to write the kinetic energy in term of the components v_x, v_y and v_z of the velocity written in the $x^*y^*z^*$ frame and the components Ω_x, Ω_y and Ω_z of the angular velocity in frame $Gxyz$.

The components of the velocity and angular velocity so obtained are not the derivatives of true coordinates, but are linked with the derivatives of the coordinates by the six kinematic equations

$$\mathbf{V} = \begin{Bmatrix} v_x \\ v_y \\ v_z \end{Bmatrix} = \mathbf{R}_1^T \begin{Bmatrix} \dot{X} \\ \dot{Y} \\ \dot{Z} \end{Bmatrix} , \tag{29.4}$$

$$\begin{Bmatrix} \Omega_x \\ \Omega_y \\ \Omega_z \end{Bmatrix} = \begin{bmatrix} 1 & 0 & -\sin(\theta) \\ 0 & \cos(\phi) & \sin(\phi)\cos(\theta) \\ 0 & -\sin(\phi) & \cos(\theta)\cos(\phi) \end{bmatrix} \begin{Bmatrix} \dot{\phi} \\ \dot{\theta} \\ \dot{\psi} \end{Bmatrix} . \tag{29.5}$$

The third Eq. (A.109) is justified because Z differs from Z_H by a constant. The vector of the generalized coordinates is then

$$\mathbf{q} = \begin{bmatrix} X & Y & Z & \phi & \theta & \psi \end{bmatrix}^T . \tag{29.6}$$

Let the generalized velocities for translational degrees of freedom be the components of the velocity in the $x^*y^*z^*$ frame. For the rotational degrees of freedom, on the other hand, the derivatives $\dot{\phi}$, $\dot{\theta}$ and $\dot{\psi}$ of coordinates ϕ, θ and ψ, which in the following will be indicated as v_ϕ, v_θ and v_ψ, will be used instead of the components Ω_x, Ω_y and Ω_z of the angular velocity, as shown in Appendix A. This choice is due to the fact that the yawing moments are more easily expressed when considering an axis perpendicular to the road, and also because the linearization of the model allows us to proceed in this way without difficulties.

The generalized velocities are then

$$\mathbf{w} = \begin{bmatrix} v_x & v_y & v_z & v_\phi & v_\theta & v_\psi \end{bmatrix}^T . \tag{29.7}$$

The relationship linking the generalized velocities to the derivatives of the generalized coordinates may then be written as

$$\mathbf{w} = \mathbf{A}^T \dot{\mathbf{q}} , \qquad (29.8)$$

where matrix $\mathbf{A}^1$ is:

$$\mathbf{A} = \begin{bmatrix} \mathbf{R}_1 & \mathbf{0}_{3\times3} \\ \mathbf{0}_{3\times3} & \mathbf{I}_{3\times3} \end{bmatrix} . \qquad (29.9)$$

The inverse transformation is Eq. (A.85):

$$\dot{\mathbf{q}} = \mathbf{B}\mathbf{w} ,$$

where^2 $\mathbf{B} = \mathbf{A}^{-T}$ is the inverse of the transpose of $\mathbf{A}$. In this case, $\mathbf{A}$ is a rotation matrix, and then

$$\mathbf{A}^{-1} = \mathbf{A}^T ; \qquad \mathbf{B} = \mathbf{A} . \qquad (29.10)$$

If $\mathbf{r}_1$ is the vector defining the position of the center of mass of the sprung mass G_S with respect to point H, the position of the former in the inertial frame is

$$(\overline{G_S - O'}) = (\overline{H - O'}) + \mathbf{R}\mathbf{r}_1. \qquad (29.11)$$

Assume that the vehicle body has a symmetry plane and that this plane coincides with plane xz in Fig. 29.3. In the reference position, points G and G_S then belong to the symmetry plane and the second component of vector $\mathbf{r}_1$ vanishes. The coordinates of the center of the sprung mass are c, 0 and h in the xyz frame, and thus the expression of vector $\mathbf{r}_1$ is:

$$\mathbf{r}_1 = \begin{bmatrix} c & 0 & h \end{bmatrix}^T . \qquad (29.12)$$

Because $\mathbf{r}_1$ is constant, the velocity of point G_S is:

$$\mathbf{V}_{G_S} = \begin{bmatrix} \dot{X} & \dot{Y} & \dot{Z} \end{bmatrix}^T + \dot{\mathbf{R}}\mathbf{r}_1 . \qquad (29.13)$$

i.e.,

$$\mathbf{V}_{G_S} = \mathbf{R}_1 \mathbf{V} + \dot{\mathbf{R}}\mathbf{r}_1 . \qquad (29.14)$$

and then the translational kinetic energy of the sprung mass is

$$\mathcal{T}_t = \frac{1}{2}m \left(\mathbf{V}^T\mathbf{V} + \mathbf{r}_1{}^T\dot{\mathbf{R}}^T\dot{\mathbf{R}}\mathbf{r}_1 + 2\mathbf{V}^T\mathbf{R}_1^T\dot{\mathbf{R}}\mathbf{r}_1 \right) . \qquad (29.15)$$

Because xz is a symmetry plane for the sprung mass, its inertia tensor is

$$\mathbf{J}_s = \begin{bmatrix} J_{x_s} & 0 & -J_{xz_s} \\ 0 & J_{y_s} & 0 \\ -J_{xz_s} & 0 & J_{z_s} \end{bmatrix} . \qquad (29.16)$$

---

^{1}Matrix $\mathbf{A}$ here defined has nothing to do with the dynamic matrix of the system, which is also indicated as $\mathbf{A}$.

^{2}Matrix $\mathbf{B}$ here used must not be confused with the input gain matrix, which is also usually indicated as $\mathbf{B}$.

The rotational kinetic energy of the sprung mass is then

$$T_w = \frac{1}{2}\mathbf{\Omega}^T \mathbf{J}_s \mathbf{\Omega}. \tag{29.17}$$

Performing the relevant computations, expressing the components of the angular velocity as functions of the derivatives of the coordinates and neglecting the terms containing powers higher than 2 of small quantities, it follows that

$$
\begin{aligned}
T_s &= \tfrac{1}{2}m_s \left(v_x{}^2 + v_y{}^2 + v_z{}^2\right) + \tfrac{1}{2}\left(m_s h^2 + J_{x_s}\right)^2 \dot{\phi}^2 + \\
&+ \tfrac{1}{2}\left[m_s\left(h^2 + c^2\right) + J_{y_s}\right]\dot{\theta}^2 + \tfrac{1}{2}\left(m_s c^2 + J_{z_s}\right)\dot{\psi}^2 - \left(m_s ch + J_{xz_s}\right)\dot{\psi}\,\dot{\phi} + \\
&- m_s v_x \left[\left(c\theta_0 - h\right)\dot{\theta} - h\dot{\psi}\phi + c\dot{\theta}\theta\right] - m_s v_y \left(h\dot{\phi} - c\dot{\psi}\right) - m_s c v_z\dot{\theta}.
\end{aligned} \tag{29.18}
$$

The height of the center of mass of the sprung mass on the road is

$$Z_G = Z_0 + Z + \mathbf{e}_3^T \mathbf{R}\mathbf{r}_1 , \tag{29.19}$$

where:

$$\mathbf{e}_3 = \begin{bmatrix} 0 & 0 & 1 \end{bmatrix}^T$$

is the unit vector of axis Z.

The potential energy of the sprung mass is simply its gravitational potential energy; its expression is:

$$\mathcal{U}_s = m_s g\left(Z_0 + Z\right) + m_s g \mathbf{e}_3^T \mathbf{R}\mathbf{r}_1 , \tag{29.20}$$

or, performing the relevant computations

$$\mathcal{U}_s = m_s g\left(Z_0 + Z\right) + m_s g\left[-c\sin\left(\theta_0 + \theta\right) + h\cos\left(\theta_0 + \theta\right)\cos\left(\phi\right)\right] . \tag{29.21}$$

Because the model is linearized, the trigonometric functions of small angles may be substituted by their series, truncated after the quadratic term

$$\sin\left(\theta_0 + \theta\right) \approx \theta_0 + \theta , \qquad\qquad \cos\left(\phi\right) \approx 1 - \frac{\phi^2}{2} ,$$

$$\cos\left(\theta_0 + \theta\right) \approx 1 - \frac{\left(\theta_0 + \theta\right)^2}{2} = 1 - \frac{\theta_0^2}{2} - \frac{\theta^2}{2} - \theta_0\theta .$$

Neglecting the constant term, it follows that

$$\mathcal{U}_s = m_s g\left[Z - \left(c + h\theta_0\right)\theta - h\frac{\theta^2}{2} - h\frac{\phi^2}{2}\right] . \tag{29.22}$$

29.2.3 General solid axle suspension

Geometry of the suspension

A solid axle suspension can be modelled as a secondary rigid body having two degrees of freedom with respect to the main body.

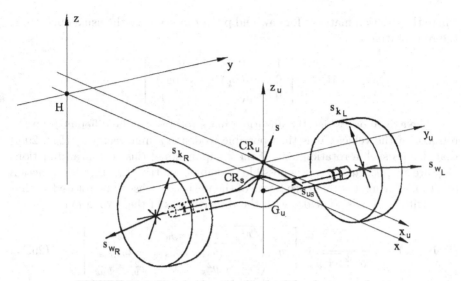

FIGURE 29.4. Sketch of an idealized solid axle suspension.

The geometry of the suspension may be simplified by assuming that it is possible to identify a roll center CR in the motion about the reference position. This is a point belonging to the roll axis x and to a plane perpendicular to the ground passing through the centers of the wheels. In the heave motion, point CR belonging to the sprung mass, indicated as CR_s, will not coincide with the corresponding point CR_u belonging to the axle. Assume that the latter moves along a trajectory belonging to the xz plane fixed to the vehicle body. For small displacements of the body, substitute the trajectory with its tangent in CR_s.(Fig. 29.4).

Let the position of CR_s in the frame fixed to the sprung mass be

$$\overline{(CR_s - H)} = \begin{bmatrix} 0 & 0 & x_u \end{bmatrix}^T . \tag{29.23}$$

If the unit vector tangent to the trajectory of CR_u in CR_s is

$$\mathbf{s} = \begin{bmatrix} s_x & 0 & s_z \end{bmatrix}^T \tag{29.24}$$

and ζ is the distance between these two points, the position of CR_u in the frame fixed to the sprung mass is

$$\overline{(CR_u - H)} = \begin{bmatrix} x_u + \zeta s_x & 0 & \zeta s_z \end{bmatrix}^T \approx \begin{bmatrix} x_u + \zeta s_x & 0 & \zeta \end{bmatrix}^T . \tag{29.25}$$

The second of the two expressions is justified by the fact that the angle between vector $\mathbf{s}$ and axis z is small.

If the roll rotation of the unsprung mass occurs about an axis parallel to the roll axis x, its rotation matrix is

$$\mathbf{R}_k = \mathbf{R}_1 \mathbf{R}_2 \mathbf{R}_{3k} , \tag{29.26}$$

where the rotation matrices for yaw and pitch rotations are the usual ones, while the roll rotation is

$$\mathbf{R}_{3k} = \begin{bmatrix} 1 & 0 & 0 \\ 0 & \cos(\phi_k) & -\sin(\phi_k) \\ 0 & \sin(\phi_k) & \cos(\phi_k) \end{bmatrix} .$$

The axis about which the unsprung mass rotates may be different from the roll axis of the vehicle. It is then possible to define a unit vector $\mathbf{s}_k$ (Fig. 29.4) that defines such a rotation axis. If the components of this vector, all functions of ζ, are x_{us}, y_{us} and z_{us}, it is possible to define a matrix $\mathbf{R}_{us}$, that is a function of ζ too, allowing the reference frame of the unsprung mass to be rotated so that its longitudinal axis coincides with the rotation axis of the sprung mass

$$\mathbf{R}_{us}(\zeta) = -\frac{1}{\sqrt{x_{us}^2 + y_{us}^2}} \begin{bmatrix} x_{us}\sqrt{x_{us}^2 + y_{us}^2} & -y_{us} & -z_{us}x_{us} \\ y_{us}\sqrt{x_{us}^2 + y_{us}^2} & x_{us} & -z_{us}y_{us} \\ z_{us}\sqrt{x_{us}^2 + y_{us}^2} & 0 & x_{us}^2 + y_{us}^2 \end{bmatrix} . \tag{29.27}$$

If the deviation of the roll axis of the unsprung mass from the longitudinal direction is small, vector $\mathbf{s}_k$ is contained in the symmetry plane, and the rotation matrix reduces to

$$\mathbf{R}_{us}(\zeta) \approx \begin{bmatrix} 1 & 0 & -z_{us} \\ 0 & 1 & 0 \\ z_{us} & 0 & 1 \end{bmatrix} . \tag{29.28}$$

Translational kinetic energy

The rotation matrix of the unsprung mass is then

$$\mathbf{R}_k = \mathbf{R}_1 \mathbf{R}_2 \mathbf{R}_{us} \mathbf{R}_{3k} . \tag{29.29}$$

Once linearized, the product of matrices $\mathbf{R}_2 \mathbf{R}_{us} \mathbf{R}_{3k}$ is

$$\mathbf{R}_2 \mathbf{R}_{us} \mathbf{R}_{3k} \approx \begin{bmatrix} 1 & 0 & \theta + \theta_0 - z_{us} \\ 0 & 1 & -\phi_k \\ -\theta - \theta_0 + z_{us} & \phi_k & 1 \end{bmatrix} . \tag{29.30}$$

Let the coordinates of the center of mass G_u of the unsprung mass in the reference position in the reference frame of the vehicle be x_{Gu}, y_{Gu} and z_{Gu} ($y_{Gu} = 0$ for symmetry reasons). Its position in the inertial frame is

$$(\overline{G_u - O'}) = (\overline{H - O'}) + \mathbf{R}_1 \mathbf{r} , \tag{29.31}$$

where

$$\mathbf{r} = \mathbf{R}_2 \left[\mathbf{R}_3 \left\{ \begin{array}{c} x_u + \zeta s_x \\ 0 \\ \zeta \end{array} \right\} + \mathbf{R}_{us} \mathbf{R}_{3N} \left\{ \begin{array}{c} x_{Gu} - x_u \\ 0 \\ z_{Gu} \end{array} \right\} \right] , \tag{29.32}$$

that is, linearizing,

$$
\mathbf{r} = \left\{
\begin{array}{c}
x_{Gu} - z_{us}z_{Gu} + \theta_0 z_{Gu} + \theta z_{Gu} \\
-\phi_k z_{Gu} \\
\zeta + z_{Gu} + z_{us}(x_{Gu} - x_u) - x_{Gu}(\theta + \theta_0)
\end{array}
\right\} .
\tag{29.33}
$$

The height of the center of mass of the kth suspension from the ground may be considered as the sum of a constant value related to the reference position, plus a displacement of the same order of the other small quantities (like Z):

$$
Z_{G_u} = Z_{0k} + Z_k = Z_0 + Z + \mathbf{e}_3^T \mathbf{R}_1 \mathbf{r} .
\tag{29.34}
$$

Performing the relevant computations and linearizing the trigonometric functions of small angles, it follows that

$$
Z_{0k} + Z_k = Z_0 + Z - x_{Gu}(\theta_0 + \theta) + \zeta + z_{Gu} + z_{us}(x_{Gu} - x_u) .
\tag{29.35}
$$

In the reference position, its value is

$$
Z_{0k} = Z_0 - x_{Gu}\theta_0 + z_{Gu} + z_{us}(x_{Gu} - x_u)
\tag{29.36}
$$

and then the relationship linking ζ to Z_k is simply

$$
\zeta = Z_k - Z + x_{Gu}\theta .
\tag{29.37}
$$

The component variable in time Z_k of the Z coordinate of the center of mass of the kth suspension will be assumed to be the generalized coordinate for the vertical displacement of the unsprung mass.

Introducing the linearized value of ζ into the expression for $\mathbf{r}$, it follows that

$$
\mathbf{r} = \left\{
\begin{array}{c}
x_{Gu} - z_{us}z_{Gu} + \theta_0 z_{Gu} + \theta z_{Gu} \\
-\phi_k z_{Gu} \\
z_{Gu} + z_{us}(x_{Gu} - x_u) - x_{Gu}\theta_0 + Z_k - Z
\end{array}
\right\} .
\tag{29.38}
$$

Its derivative with respect to time, once linearized, is

$$
\dot{\mathbf{r}} = \left\{
\begin{array}{c}
\dot{\theta} z_{Gu} \\
-\dot{\phi}_k z_{Gu} \\
\dot{Z}_k - \dot{Z}
\end{array}
\right\} .
\tag{29.39}
$$

The velocity of the center of mass of the unsprung mass is then

$$
\mathbf{V}_{G_u} = \begin{bmatrix} \dot{X} & \dot{Y} & \dot{Z} \end{bmatrix}^T + \mathbf{R}_1 \dot{\mathbf{r}} + \dot{\mathbf{R}}_1 \mathbf{r} .
\tag{29.40}
$$

The velocity $\mathbf{V}_{G_u}$ in the $x^*y^*z^*$ frame is obtained by premultiplying this expression by $\mathbf{R}_1^T$. Remembering that

$$
\mathbf{R}_1^T \mathbf{R}_1 = \mathbf{I} , \quad \mathbf{R}_1^T \dot{\mathbf{R}}_1 = \dot{\psi}\mathbf{S} = \dot{\psi}\begin{bmatrix} 0 & -1 & 0 \\ 1 & 0 & 0 \\ 0 & 0 & 0 \end{bmatrix} ,
\tag{29.41}
$$

it follows that

$$\mathbf{V}_{G_u} = \mathbf{V} + \dot{\mathbf{r}} + \dot{\psi}\mathbf{S}\mathbf{r} \ . \tag{29.42}$$

The translational kinetic energy of the unsprung mass is then

$$\mathcal{T}_t = \frac{1}{2}m_u \left(\mathbf{V}^T\mathbf{V} + \dot{\mathbf{r}}^T\dot{\mathbf{r}} + \dot{\psi}^2\mathbf{r}^T\mathbf{S}^T\mathbf{S}\mathbf{r} + 2\mathbf{V}^T\dot{\mathbf{r}} + 2\dot{\psi}\mathbf{V}^T\mathbf{S}\mathbf{r} + 2\dot{\psi}\dot{\mathbf{r}}^T\mathbf{S}\mathbf{r} \right) \ , \tag{29.43}$$

that is, by indicating as r_x, r_y, r_z the components of vector $\mathbf{r}$,

$$\mathcal{T}_t = \tfrac{1}{2}m_u \left\{ v_x^2 + v_y^2 + v_z^2 + \dot{r}_x^2 + \dot{r}_y^2 + \dot{r}_z^2 + \dot{\psi}^2\left(r_x^2 + r_y^2 \right) + \right.$$
$$\left. +2\left(v_x\dot{r}_x + v_y\dot{r}_y + v_z\dot{r}_z \right) + 2\dot{\psi}\left(-v_x r_y + v_y r_x - \dot{r}_x r_y + \dot{r}_y r_x \right) \right\}. \tag{29.44}$$

Only the constant and linear terms of $\mathbf{r}$ are present in all terms, except for the term in $v_x\dot{r}_x$. Only the expression of $\dot{r}_x$ containing also the quadratic terms

$$\dot{r}_x = \beta_1\dot{\theta} + x_{Gu}\dot{\theta}\theta + \beta_3\left(v_k - v_z \right) + \theta\left(\dot{Z}_k - \dot{Z} \right) + \dot{\theta}\left(Z_k - Z \right), \tag{29.45}$$

where

$$\beta_1 = z_{Gu} + z_{us}\left(x_{Gu} - x_u \right) + x_u s_x \ , \quad \beta_3 = \theta_0 + s_x \tag{29.46}$$

and $v_k = \dot{Z}_k$ is the velocity of the kth unsprung mass, needs to be written explicitly.

The following simplified expression is so obtained

$$\mathcal{T}_t = \tfrac{1}{2}m_u \left\{ v_x^2 + v_y^2 + v_k^2 + \dot{\phi}_k^2 z_{Gu}^2 + \left(z_{Gu}^2 + x_{Gu}^2 \right)\dot{\theta}^2 + x_{Gu}^2\dot{\psi}^2 + \right.$$
$$-2\dot{Z}_k\dot{\theta}x_{Gu} + 2v_x\left[\left(\dot{Z}_k - \dot{Z} \right)\beta_3 + \dot{\theta}\beta_1 + \dot{\theta}\left(Z_u - Z \right) + \right.$$
$$\left. +\theta\left(\dot{Z}_k - \dot{Z} \right) + x_{Gu}\theta\dot{\theta} + \dot{\psi}\phi_k z_{Gu} \right] +$$
$$\left. +2v_y\left[-\dot{\phi}_k z_{Gu} + \dot{\psi}x_{Gu} \right] + 2\dot{\psi}\dot{\phi}_k z_{Gu}x_{Gu} \right\}. \tag{29.47}$$

Angular velocity of the wheels

If the axle did not rotate with respect to the body about its longitudinal axis, its absolute angular velocity about its longitudinal axis, as expressed in its own reference frame, is

$$\Omega_k = \mathbf{R}_{us}^T\Omega = \left\{ \begin{array}{c} \Omega_x + z_{us}\Omega_z \\ \Omega_y \\ \Omega_z - z_{us}\Omega_x \end{array} \right\} . \tag{29.48}$$

In reality, the unsprung mass is free to rotate about that axis and its angular velocity is

$$\Omega_k = \left\{ \begin{array}{c} \dot{\phi}_k \\ \Omega_y \\ \Omega_z - z_{us}\Omega_x \end{array} \right\} . \tag{29.49}$$

The rotation and steering motion of each wheel are taken into account independently. Let the rotation angles of the wheels be χ_R and χ_L and the steering angles be δ_R and δ_L (R and L designate the right and left wheel of the axle).

If the wheel's rotation axis coincided with axis y_u of the unsprung mass, and both were parallel to the y axis of the axle, the angular velocity of the ith wheel in the reference frame of the kth unsprung mass would be

$$\Omega_{wi} = \Omega_k + \dot{\chi}_i \mathbf{e}_2 \qquad (i = L, R).$$ (29.50)

Generally speaking, the direction of the rotation axis of the wheel may be different (although usually not by much except for steering) from that of the y axis, making it possible to define the unit vector of the rotation axis $\mathbf{e}_{wi}$ in the reference frame of the unsprung mass. Such a unit vector does not depend upon the position of the suspension and thus is a function neither of ζ nor of ϕ_k. It follows that

$$\Omega_{wi} = \Omega_k + \dot{\chi}_i \mathbf{e}_{wi} \qquad (i = L, R).$$ (29.51)

The position of the rotation axis may be defined by introducing a rotation matrix $\mathbf{R}_{wi}$, allowing us to pass from the reference frame of the unsprung mass to a frame whose y axis coincides with the rotation axis of the wheel

$$\mathbf{e}_{wi} = \mathbf{R}_{wi} \mathbf{e}_2 .$$ (29.52)

If x_w, y_w are z_w the components of unit vector $\mathbf{e}_{wi}$[3], the value of the rotation matrix $\mathbf{R}_w$ is

$$\mathbf{R}_{wi} = \frac{1}{\sqrt{x_w^2 + y_w^2}} \begin{bmatrix} y_w & x_w\sqrt{x_w^2 + y_w^2} & -x_w z_w \\ -x_w & y_w\sqrt{x_w^2 + y_w^2} & -y_w z_w \\ 0 & z_w\sqrt{x_w^2 + y_w^2} & x_w^2 + y_w^2 \end{bmatrix} .$$ (29.53)

The rotation axis of the wheel is usually little inclined with respect to the horizontal direction. The trigonometric functions of the rotation axis included in matrix $\mathbf{R}_{wi}$ may be linearized. It follows thus

$$\mathbf{R}_{wi} = \begin{bmatrix} 1 & x_w & 0 \\ -x_w & 1 & -z_w \\ 0 & z_w & 1 \end{bmatrix} ,$$ (29.54)

where x_w coincides with the steering angle of the wheel (when the axle does not steer) with its sign changed (this angle is usually due to toe in and is very small), while z_w coincides with the camber angle of the wheel and is also small. For symmetry reasons, it follows that

$$x_{wR} = -x_{wL} , \qquad z_{wR} = -z_{wL} .$$ (29.55)

---

[3] Obviously $\sqrt{x_w^2 + y_w^2 + z_w^2} = 1$.

The angular velocity of the wheel in its reference frame, instead of the frame of the unsprung mass, is

$$\mathbf{\Omega}_{wi} = \mathbf{R}_{wi}{}^T \mathbf{\Omega}_k + \dot{\chi} \mathbf{e}_2 \ . \tag{29.56}$$

If the wheel steers, the reference frame of the ith wheel will no longer be parallel to the frame $x_u y_u z_u$ of the unsprung mass, but will be rotated by a steering angle δ_i. Assume that the kingpin axis of the wheel is parallel to axis z_u and define a further rotation matrix

$$\mathbf{R}_{4i} = \begin{bmatrix} \cos(\delta_i) & -\sin(\delta_i) & 0 \\ \sin(\delta_i) & \cos(\delta_i) & 0 \\ 0 & 0 & 1 \end{bmatrix} . \tag{29.57}$$

In this case, the kingpin axis is generally not parallel to the z axis of the axle. If $\mathbf{e}_k$ is the unit vector of the kingpin axis (its components will be indicated as x_k, y_k and $z_k{}^4$), which in solid axle suspensions may be considered as fixed, the rotation matrix $\mathbf{R}_{ki}$ allowing the reference frame of the unsprung mass to be rotated so that its z_u axis coincides with the kingpin axis of the ith wheel is

$$\mathbf{R}_{ki} = \frac{1}{\sqrt{x_w^2 + z_w^2}} \begin{bmatrix} z_w & -x_w y_w & x_w \sqrt{x_w^2 + z_w^2} \\ 0 & (x_w^2 + z_w^2) & y_w \sqrt{x_w^2 + z_w^2} \\ -x_w & -z_w y_w & z_w \sqrt{x_w^2 + z_w^2} \end{bmatrix} . \tag{29.58}$$

Usually the longitudinal inclination angle (the pitch angle of the kingpin axis) and the transversal inclination angle (the roll angle of the kingpin axis), are quite small, and the rotation matrix $\mathbf{R}_k$ reduces to

$$\mathbf{R}_{ki} \approx \begin{bmatrix} 1 & 0 & x_k \\ 0 & 1 & y_k \\ -x_k & -y_k & 1 \end{bmatrix} , \tag{29.59}$$

where x_k and y_k coincide respectively with the longitudinal inclination angle (not larger than about $1°$) and the transversal inclination angle changed in sign (usually not larger than about $10°$). For symmetry reasons, it follows that

$$x_{k_R} = x_{k_L} , \qquad y_{k_R} = -y_{k_L} . \tag{29.60}$$

The angular velocity of the wheel in the reference frame of the sprung mass is then

$$\mathbf{\Omega}_{wi} = \mathbf{\Omega}_k + \dot{\delta}_i \mathbf{R}_{ki} \mathbf{e}_3 + \dot{\chi}_i \mathbf{R}_{ki} \mathbf{R}_{4i} \mathbf{R}_{ki}^T \mathbf{R}_{wi} \mathbf{e}_2 \ . \tag{29.61}$$

To obtain the angular velocity of the wheel in its own reference frame, Eq. (29.61) must be premultiplied by $(\mathbf{R}_{ki} \mathbf{R}_{4i} \mathbf{R}_{ki}^T \mathbf{R}_{wi})^T$. Remembering that

$$\mathbf{R}_{4i} \mathbf{e}_3 = \mathbf{e}_3 \ ,$$

---

[4]Obviously $\sqrt{x_k^2 + y_k^2 + z_k^2} = 1$.

it follows that

$$\Omega_{wi} = \dot{\chi}\mathbf{e}_2 + \dot{\delta}\boldsymbol{\alpha}_1 + \boldsymbol{\alpha}_2\Omega_k. \tag{29.62}$$

where

$$\boldsymbol{\alpha}_1 = \mathbf{R}_{wi}^T\mathbf{R}_{ki}\mathbf{e}_3 \ , \qquad \boldsymbol{\alpha}_2 = \mathbf{R}_{wi}^T\mathbf{R}_{ki}\mathbf{R}_{4i}^T\mathbf{R}_{ki}^T \ . \tag{29.63}$$

It must be remembered that in a suspension there are two matrices $\mathbf{R}_{wi}$ and $\mathbf{R}_{ki}$ ($i = L$, R), one for each wheel.

Rotational kinetic energy

Because the wheel is a gyroscopic body (two of the principal moments of inertia are equal to each other), with a principal axis of inertia coinciding with the rotation axis, its inertia tensor has a peculiar form

$$\mathbf{J}_w = \mathrm{diag}\left(\begin{bmatrix} J_{tw} & J_{pw} & J_{tw} \end{bmatrix}\right) \ , \tag{29.64}$$

where J_{pw} and J_{tw} are respectively the polar and transversal moment of inertia of the wheel.

The rotational kinetic energy of the ith wheel is

$$\mathcal{T}_{wri} = \tfrac{1}{2}\Omega_k^T\boldsymbol{\alpha}_2^T\mathbf{J}_w\boldsymbol{\alpha}_2\Omega_k + \tfrac{1}{2}\dot{\chi}^2\mathbf{e}_2^T\mathbf{J}_w\mathbf{e}_2 + \tfrac{1}{2}\dot{\delta}^2\boldsymbol{\alpha}_1^T\mathbf{J}_w\boldsymbol{\alpha}_1 + \\ +\dot{\chi}\dot{\delta}\mathbf{e}_2^T\mathbf{J}_w\boldsymbol{\alpha}_1 + \dot{\chi}\mathbf{e}_2^T\mathbf{J}_w\boldsymbol{\alpha}_2\Omega_k + \dot{\delta}\boldsymbol{\alpha}_1^T\mathbf{J}_w\boldsymbol{\alpha}_2\Omega_k \ . \tag{29.65}$$

The first term is the rotational kinetic energy due to the angular velocity of the unsprung mass.

By neglecting the first term, which will later be included in the kinetic energy of the axle, and by linearizing and introducing the linearized expressions of the kinematic equations, the rotational kinetic energy of the ith wheel reduces to

$$\mathcal{T}_{wri} = \frac{1}{2}\dot{\chi}_i^2 J_{pw} + \frac{1}{2}\dot{\delta}_i^2 J_{tpw} + \dot{\chi}_i\dot{\delta}_i J_{pw}\left(y_k + z_w\right) + \tag{29.66}$$

$$+\dot{\chi}_i J_{pw}\left[\dot{\theta} + \phi\dot{\psi} + \left(x_w - \delta_i\right)\dot{\phi}_k + z_w\dot{\psi}\right] + \dot{\delta}J_{tw}\dot{\psi} \ . \tag{29.67}$$

The first term that was neglected above can be inserted into the rotational kinetic energy of the unsprung mass $\mathcal{T}_{ur}$:

$$\mathcal{T}_{ur} = \tfrac{1}{2}\Omega_k^T\mathbf{J}_u\Omega_k, \tag{29.68}$$

if the inertia tensor $\mathbf{J}_u$ also includes the inertia of the wheels, assumed to be non-rotating and non-steering.

Operating in this way, the variation of the inertia of the unsprung mass at the changing steering angle is neglected, but this approximation is acceptable. The inertia tensor of the unsprung mass has a structure similar to that of the sprung mass, because the suspension also has a symmetry plane coinciding with the $x_u z_u$ plane.

Performing the relevant computations, it follows that

$$\mathcal{T}_{ur} = \tfrac{1}{2}J_{x_u}\Omega_{xk}^2 + \tfrac{1}{2}J_{y_u}\Omega_{yk}^2 + \tfrac{1}{2}J_{z_u}\Omega_{zk}^2 - J_{xz_u}\Omega_{xk}\Omega_{zk} \tag{29.69}$$

and, by linearizing and including the linearized kinematic equations, the simple expression is obtained

$$\mathcal{T}_{ur} = \tfrac{1}{2}J^2_{x_u}\dot{\phi}_k + \tfrac{1}{2}J_{y_u}\dot{\theta}^2 + \tfrac{1}{2}J_{z_u}\dot{\psi}^2 + J_{xz_u}\dot{\phi}\dot{\psi} \ . \tag{29.70}$$

The total rotation kinetic energy of the axle is then

$$\mathcal{T}_{urt} = \mathcal{T}_{ur} + \mathcal{T}_{wrR} + \mathcal{T}_{wrL} \ . \tag{29.71}$$

Total kinetic energy

The kinetic energy of the axle is then

$$\mathcal{T}_{wri} = \dot{\chi}_i J_{pw}\left[\dot{\theta} + \phi\dot{\psi} + z_w\dot{\psi}\right] \ , \tag{29.72}$$

$$
\begin{aligned}
\mathcal{T}_u = {}& \tfrac{1}{2}m_u\left(v_x^2 + v_y^2 + v_k^2\right) + \tfrac{1}{2}\beta_{11}\dot{\theta}^2 + \tfrac{1}{2}\beta_{12}\dot{\psi}^2 + \tfrac{1}{2}\beta_{13}\dot{\phi}_k + \\
& + \beta_{14}\dot{\psi}\dot{\phi}_k + \tfrac{1}{2}\dot{\chi}_s^2 J_{pw} + \tfrac{1}{2}\dot{\delta}_s^2 J_{tw} + \tfrac{1}{2}\dot{\chi}_R^2 J_{pw} + \tfrac{1}{2}\dot{\delta}_R^2 J_{tw} + \\
& + \dot{\chi}_s\dot{\delta}_s J_{pw}\left(y_k + z_w\right) - \dot{\chi}_R\dot{\delta}_R J_{pw}\left(y_k + z_w\right) - \beta_{16}v_k\dot{\theta} \\
& + m_u v_x\left(\dot{\theta}\theta\beta_1 + \dot{\theta}Z_k - \dot{\theta}Z + \theta v_k - \theta v_z + \beta_3 v_k - \beta_3 v_z + \right. \\
& \left. + x_{Gu}\theta\dot{\theta} + \beta_5\dot{\psi}\phi_k\right) + m_u v_y\left(-\beta_5\dot{\phi}_k + x_{Gu}\dot{\psi}\right) + \\
& + \beta_{19}\dot{\delta}_L\dot{\psi} + \beta_{19}\dot{\delta}_R\dot{\psi} + \dot{\chi}_L\beta_{20}\left(x_w - \delta_s\right)\dot{\phi}_k - \dot{\chi}_R\beta_{20}\left(x_w + \delta_R\right)\dot{\phi}_k + \\
& + \dot{\chi}_L J_{pr}\left(\dot{\theta} + \phi\dot{\psi} + z_w\dot{\psi}\right) + \dot{\chi}_R J_{pr}\left(\dot{\theta} + \phi\dot{\psi} - z_w\dot{\psi}\right) \ ,
\end{aligned}
\tag{29.73}
$$

where

$$
\begin{aligned}
& \beta_5 = z_{Gu} \ , \quad \beta_{11} = m_u\left(z_{Gu}^2 + x_{Gu}^2\right) + J_{y_u} \ , \quad \beta_{12} = m_u x_{Gu}^2 + J_{z_u} , \\
& \beta_{13} = m_u z_{Gu}^2 + J_{x_u} \ , \qquad \beta_{14} = m_u z_{Gu} x_{Gu} - J_{xz_u} \ , \\
& \qquad\qquad \beta_{16} = m_u x_{Gu} \ , \quad \beta_{19} = J_{tw} \ , \quad \beta_{20} = J_{pw}.
\end{aligned}
$$

Potential energy

The height of the center of mass of the axle on the ground is

$$h_u = \mathbf{e}_3^T\overline{(\mathrm{G_N} - \mathrm{O'})} = \mathbf{e}_3^T\overline{(\mathrm{H} - \mathrm{O'})} + \mathbf{e}_3^T \mathbf{R}_1\mathbf{r} \ , \tag{29.74}$$

i.e.,

$$h_u = Z_0 + Z + r_z \ , \tag{29.75}$$

The expression of r_z, obtained by approximating vector $\mathbf{r}$ with its Taylor series truncated after the quadratic term in the small quantities and cancelling the constant terms that do not influence the equations of motion, is

$$r_z = Z_k - Z - \theta\beta_{22} - \frac{1}{2}\theta^2 z_{Gu} - \frac{1}{2}\phi_k^2 z_{Gu}, \tag{29.76}$$

where

$$\beta_{22} = z_{Gu}\left(\theta_0 - z_{us}\right) \ .$$

The gravitational potential energy

$$\mathcal{U}_g = m_u g h_u \tag{29.77}$$

is then

$$\mathcal{U}_g = m_u g \left(Z_k - \theta \beta_{22} - \frac{1}{2}\theta^2 z_{Gu} - \frac{1}{2}\phi_k^2 z_{Gu} \right) . \tag{29.78}$$

On each of the springs of the suspension it is possible to identify two points: one of these (A) is fixed to the body, while the other (B) is fixed to the axle. Considering $\mathbf{r}_A$ and $\mathbf{r}_B$ as the vectors defining their positions in the frame of the sprung mass ($\mathbf{r}_A$ is constant, while $\mathbf{r}_B$ depends on ζ and ϕ_k), it is possible to compute the shortening of the spring and then its elastic potential energy. In a similar way, it is possible to compute the potential energy of possible anti-roll bars applied to the axle.

In a linearized model of the suspension, the spring system linking the two rigid bodies can be reduced to a spring with stiffness K_ζ reacting to linear displacements, and a torsional spring with stiffness K_ϕ reacting to the relative rotation between sprung and unsprung masses $\phi - \phi_k$.

The general expression of the elastic potential energy of the whole suspension is

$$\mathcal{U}_m = \frac{1}{2}K_\zeta \left(\zeta + \zeta_0 \right)^2 + \frac{1}{2}K_\phi \left(\phi - \phi_k \right)^2 . \tag{29.79}$$

Note that for symmetry reasons the stiffness for heave and roll motions are fully independent from each other.

By introducing the expression for ζ as a function of the generalized coordinates, it follows that

$$\begin{aligned} \mathcal{U}_m = &\tfrac{1}{2}K_{11} \left(Z + Z_0 \right)^2 + \tfrac{1}{2}K_{22} \left(Z_k + Z_{k0} \right)^2 + \tfrac{1}{2}K_{33} \left(\theta + \theta_0 \right)^2 + \\ &-K_{12} \left(Z + Z_0 \right) \left(Z_k + Z_{k0} \right) - K_{13} \left(Z + Z_0 \right) \left(\theta + \theta_0 \right) + \\ &+K_{23} \left(Z_k + Z_{k0} \right) \left(\theta + \theta_0 \right) + \tfrac{1}{2}K_\phi \phi^2 + \tfrac{1}{2}K_\phi \phi_k^2 - K_\phi \phi \phi_k, \end{aligned} \tag{29.80}$$

where constants K_{ij} depend on both the elastic and geometric characteristics of the suspension.

This expression of the potential energy, along with the expression of the kinetic energy seen above, takes implicitly into account the actual trajectory of the suspension, or better, because the model is linearized, the tangent to the trajectory in the reference position. On the other hand, using a model of this kind makes it impossible to account for the deformation in longitudinal and lateral directions, and for the yaw and pitch compliance of the suspension.

In a similar way, the general expression of the elastic potential energy due to the deformation of the tires of the suspension is

$$\mathcal{U}_p = \frac{1}{2}K_{pz} \left(Z_k + Z_{k0} \right)^2 + \frac{1}{2}K_{p\phi} \phi_k^2 . \tag{29.81}$$

If K_p is the stiffness of a single tire and t is the track of the axle, for an axle with two wheels it follows that

$$K_{pz} = 2K_p , \qquad K_{p\phi} = \frac{1}{2}t^2 K_p . \tag{29.82}$$

Dissipation function

Operating with the same method used for the elastic potential energy of the suspension, the dissipation function due to the shock absorbers is

$$\mathcal{F}_a = \frac{1}{2}c_{11}v_z^2 + \frac{1}{2}c_{22}v_k^2 + \frac{1}{2}c_{33}\dot{\theta}^2 + \frac{1}{2}c_\phi\dot{\phi}^2 + \frac{1}{2}c_\phi\dot{\phi}_k^2 +$$
$$-c_{12}v_zv_k - c_{13}v_z\dot{\theta} + c_{23}v_k\dot{\theta} - c_\phi\dot{\phi}\dot{\phi}_k, \tag{29.83}$$

where constants c_{ij} depend both on the characteristics of the dampers and on the geometry of the suspension.

The dissipation function due to the damping of tires can be computed in the same way:

$$\mathcal{F}_p = \frac{1}{2}c_{pz}v_k^2 + \frac{1}{2}c_{p\phi}\Omega_k^2 . \tag{29.84}$$

29.2.4   General independent suspension

Geometry of the suspension

An axle with independent suspensions will be assumed to be made with two suspensions that are the mirror image of each other. Some parameters are identical for the two suspensions (for instance the mass m_i, some geometrical characteristics, some angles, etc.); others will be identical in modulus but with opposite sign (for instance, the product of inertia J_{xy}, some angles, etc.). In the latter case, reference will be made to the left suspension (subscript $i = L$), while the characteristics of the right suspension (subscript $i = R$) will have opposite sign.

The simplest case, although only an ideal one, is a suspension in which the unsprung masses can only move along a straight line in the direction of the z axis of the sprung mass. The sprung mass is the main body, while the single suspension (wheel, hub and all parts attached to it) is the secondary body (Fig. 29.5). The suspension is constrained to the main body by a prismatic guide whose axis is parallel to the z axis. The reference point C is on the axis of the guide and the distance ζ between the position C_1 of C belonging to the sprung mass and C_2 belonging to the unsprung mass is taken as a generalized coordinate. Obviously in the reference position with $\zeta = 0$, C_1 coincides with C_2. Moreover, assume that the directions of the axes of frame $x_uy_uz_u$ coincide with those of the axes of frame xyz.

Consider as reference point for translational coordinates the same point H already used to compute the kinetic energy of the sprung mass. The position of the center of mass of the suspension G_u in the inertial frame is

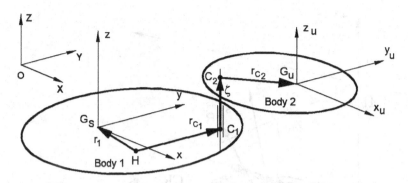

FIGURE 29.5. Sketch of an idealized independent suspension.

$$(\overline{G_u - O'}) = (\overline{H - O'}) + \mathbf{R}\,(\mathbf{r}_{C1} + \mathbf{r}_{Gu} + \zeta\mathbf{e}_3) = (\overline{H - O'}) + \mathbf{R}\mathbf{r}_2\,, \qquad (29.85)$$

where $\mathbf{R}$ is the rotation matrix defining the position of the xyz frame with respect to the inertial frame $X_i Y_i Z_i$, $\mathbf{e}_3$ is the unit vector of the z axis and

$$\mathbf{r}_2 = \mathbf{r}_{C1} + \mathbf{r}_{C2} + \zeta\mathbf{e}_3\,. \qquad (29.86)$$

If the steering of the wheel is accounted for, a part of the unsprung mass may rotate about the kingpin axis, which in the simplified model may be assumed to be parallel to z_u, and thus to the z, axis. The wheel also rotates about its own axis, which may be assumed to be parallel to the y axis when the steering angle is zero.

However, this model of independent suspension is too simple. No modern car has suspensions made by prismatic guides parallel to the z axis of the unsprung mass, nor is the kingpin axis parallel to the same axis, while the rotation of the wheels does not occur about an axis parallel to the y axis.

Each suspension has its own specific kinematics (as an example, an SLA suspension is shown in Fig. 29.6), or better, its own elasto-kinematics, because the various elements of the suspensions are rigid bodies only as a first approximation. However, while the exact elasto-kinematics is important in assessing the position of the wheel with respect to the ground and thus the forces they exchange, its effects on the inertia reactions of the various components of the suspension are usually very limited. It is then possible to neglect the deformation of the various links in the computation of the inertial part of the equations of motion and to introduce them later in the computation of the forces due to the tires.

If the deformation of the linkages is neglected, it is possible to define the trajectory of all points of the suspension in a reference frame fixed to the sprung mass. The trajectory of the center of mass, for instance, may be expressed by a function

$$\mathbf{r}_2 = \mathbf{r}_2(\zeta)\,, \qquad (29.87)$$

where ζ is a generalized coordinate that defines the position of the unsprung mass, with reference to a given position. In the extremely simplified case seen

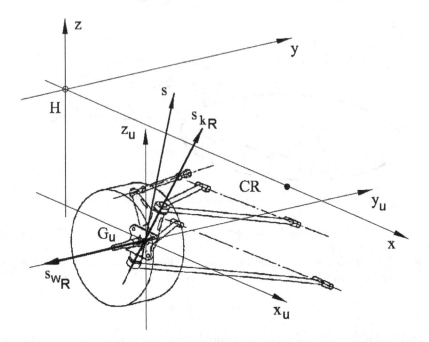

FIGURE 29.6. Sketch of an SLA suspension.

above, coordinate ζ is nothing other than the displacement of the unsprung mass in the z direction, and function $\mathbf{r}_2(\zeta)$ is the linear function expressed by Eq. (29.86).

The function expressed by Eq.(29.87) can be developed in McLaurin series about the reference position and the terms of an order higher than the first may be neglected. The position of the center of mass G_u of the suspension with respect to point H is

$$(\overline{G_u - H}) = \mathbf{r}_2 = \mathbf{r}_{2_0} + \left(\frac{d\mathbf{r}_2}{d\zeta}\right)_{\zeta=\zeta_0} \zeta . \tag{29.88}$$

Equation (29.88) coincides with Eq. (29.86) if vector

$$\mathbf{s}_0 = \left(\frac{d\mathbf{r}_2}{d\zeta}\right)_{\zeta=\zeta_0}$$

is substituted for $\mathbf{e}_3$, the unit vector of the z axis. The components of vector $\mathbf{r}_{2_0}$ will be indicated as x_{Gu}, y_{Gu} and z_{Gu}.

As an example, in the case of a trailing arms suspension hinged about an axis parallel to the y axis (Fig. 29.7) and with point C on the hinge axis in the oscillation plane of the center of mass, ζ_0 is the angle line CG_u makes with the x axis in the reference position, and coordinate ζ' is the angle the suspension rotates with respect to that position. The position of the center of mass may be thus defined by the function

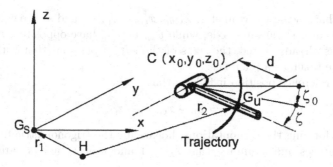

FIGURE 29.7. Sketch of a trailing arm suspension.

$$\mathbf{r}_2 = \left\{ \begin{array}{c} x_0 + d\cos\left(\zeta' + \zeta_0\right) \\ y_0 \\ z_0 - d\sin\left(\zeta' + \zeta_0\right) \end{array} \right\} . \tag{29.89}$$

If ζ' is small, the truncated series yielding the position of G_u is

$$\mathbf{r}_2 = \left\{ \begin{array}{c} x_0 + d\cos\left(\zeta_0\right) \\ y_0 \\ z_0 - d\sin\left(\zeta_0\right) \end{array} \right\} + \left\{ \begin{array}{c} -\sin\left(\zeta_0\right) \\ 0 \\ -\cos\left(\zeta_0\right) \end{array} \right\} d\zeta' . \tag{29.90}$$

Or, to give coordinate ζ the meaning of a displacement in the z direction of the unsprung mass, it is possible to state

$$\zeta = -\cos\left(\zeta_0\right) d\zeta' , \tag{29.91}$$

and then

$$\mathbf{r}_2 = \left\{ \begin{array}{c} x_0 + d\cos\left(\zeta_0\right) \\ y_0 \\ z_0 - d\sin\left(\zeta_0\right) \end{array} \right\} + \mathbf{s}_0\zeta , \tag{29.92}$$

where

$$\mathbf{s}_0 = \left[\begin{array}{ccc} \tan\left(\zeta_0\right) & 0 & 1 \end{array} \right]^T . \tag{29.93}$$

The generalized coordinate for the ith unsprung mass (right or left) may be the height on the ground of its center of mass instead of ζ. Such height is simply

$$c = \mathbf{e}_3^T \left[(\overline{H - O'}) + \mathbf{R}\mathbf{r}_2 \right] \qquad (i = L, R) . \tag{29.94}$$

Also, Z_{Gu_i} may be considered as the sum of a value taken at the reference position plus a displacement, one that is of the same order as the other small quantities (like Z):

$$Z_{Gu_i} = Z_{Gu_0} + Z_{u_i} = Z_0 + Z + \mathbf{e}_3^T \mathbf{R} \left\{ \begin{array}{c} x_{Gu} + \zeta s_x \\ y_{Gu} + \zeta s_y \\ z_{Gu} + \zeta s_z \end{array} \right\} . \tag{29.95}$$

Note that, owing to symmetry, Z_{Gu_0}, x_{Gu}, z_{Gu}, s_x, and s_z are equal for the two suspensions of the same axle, while y_{Gu} and s_y have opposite signs. In the following, as already stated, the signs of the left suspension (that with positive y_{Gu}) will be taken as a reference.

In the reference position it follows that

$$Z_{0_u} = Z_0 - x_{Gu}\theta_0 + z_{Gu} . \tag{29.96}$$

By performing the computations, linearizing the trigonometric functions of the small angles and assuming that $s_z = 1$ and that s_x and s_y are small, it follows that

$$Z_{u_i} = Z - x_{Gu}\theta + y_{Gu}\phi + \zeta \tag{29.97}$$

and thus the relationship linking ζ to Z_u is simply

$$\zeta = Z_{u_i} - Z + x_{Gu}\theta - y_{Gu}\phi . \tag{29.98}$$

Rotation of the wheels

Assume that the rotation axis of the wheels is fixed to the unsprung mass. Let χ be the angle of rotation of the wheel and $\dot{\chi}$ its angular velocity in a frame fixed to the suspension. The signs of angular velocities have been defined in such a way that, when $\dot{\chi}$ is positive, the wheel rotates in a direction that is consistent with a positive velocity v_x of the vehicle.

If the direction of the rotation axis of the wheel coincides with that of axis y_u of the unsprung mass, and then is parallel to the y axis of the vehicle (whose unit vector is $\mathbf{e}_2$), the absolute angular velocity of the wheel is, in the reference frame of the sprung mass,

$$\Omega_w = \Omega + \dot{\chi}\mathbf{e}_2 = \left\{ \begin{array}{c} \Omega_x \\ \Omega_y + \dot{\chi} \\ \Omega_z \end{array} \right\} . \tag{29.99}$$

However, the rotation axis of the wheel usually has a direction that may be different (only slightly, when the wheel is not steered) from the y axis, so that it is possible to define the unit vector of the rotation axis $\mathbf{e}_w$ in the reference frame of the sprung mass. Obviously such a vector depends on the position of the suspension and is then a function of ζ:

$$\mathbf{e}_w = \mathbf{e}_w (\zeta) . \tag{29.100}$$

As an alternative, the position of the rotation axis can be defined by stating a yaw angle ψ_w (steering due to the motion of the suspension) and a roll angle ϕ_w of the rotation axis of the wheel. A rotation matrix $\mathbf{R}_w$ is then written to pass from the reference frame of the sprung mass to a frame whose y axis coincides with the rotation axis of the wheel. Since the two ways of describing the position of the rotation axis must yield the same results, it follows that

$$\mathbf{e}_w = \mathbf{R}_w \mathbf{e}_2 . \tag{29.101}$$

With x_w, y_w and z_w being the components[5] of unit vector $\mathbf{e}_w$, the rotation matrix $\mathbf{R}_w$ is still expressed by Eq. (29.53). Because the rotation axis of the wheel is usually not far from horizontal, the trigonometric functions of ψ_w and ϕ_w can be linearized. Eq. (29.54), repeated here, still holds

$$\mathbf{R}_w = \begin{bmatrix} 1 & x_w & 0 \\ -x_w & 1 & -z_w \\ 0 & z_w & 1 \end{bmatrix} .$$

The angular velocity of the wheel in the reference frame of the sprung mass is

$$\mathbf{\Omega}_w = \mathbf{\Omega} + \dot{\chi}\mathbf{e}_w\left(\zeta\right) . \tag{29.102}$$

Actually, it is expedient to write the components of the angular velocity of the wheel in the frame fixed to the wheel instead of that fixed to the sprung mass.

The angular velocity of the wheel in its own reference frame is

$$\mathbf{\Omega}_w = \mathbf{R}_w{}^T\mathbf{\Omega} + \dot{\chi}\mathbf{e}_2 . \tag{29.103}$$

Steering

If the wheel steers, the reference frame of the wheel will no longer be parallel to frame $x_u y_u z_u$ of the unsprung mass, but will be rotated by a steering angle δ. Here two different approaches are possible: δ may be one of the variables of motion (free controls approach), or a constant or a known variable (locked controls approach).

Assume that the kingpin axis of the wheel is parallel to the z axis of the unsprung mass and define a further rotation matrix

$$\mathbf{R}_4 = \begin{bmatrix} \cos(\delta) & -\sin(\delta) & 0 \\ \sin(\delta) & \cos(\delta) & 0 \\ 0 & 0 & 1 \end{bmatrix} . \tag{29.104}$$

Assuming that the direction of the rotation axis does not change with the heave motion and is parallel to the z axis, the rotation velocity of the wheel, referred to its own reference frame, is

$$\mathbf{\Omega}_w = \dot{\chi}\mathbf{e}_2 + \dot{\delta}\mathbf{e}_3 + \mathbf{R}_4^T\mathbf{\Omega} . \tag{29.105}$$

Actually, the kingpin axis moves with changing ζ and in general is not parallel to the z axis of the unsprung mass, but its direction is defined by the unit vector $\mathbf{e}_k$, which is a function of ζ:

$$\mathbf{e}_k = \mathbf{e}_k\left(\zeta\right) . \tag{29.106}$$

---

[5] Obviously, they are functions of ζ and $\sqrt{x_w^2 + y_w^2 + z_w^2} = 1$.

The rotation matrix $\mathbf{R}_k$ to rotate the reference frame of the unsprung mass so that its z axis coincides with the kingpin axis

$$\mathbf{e}_k = \mathbf{R}_k \mathbf{e}_3 , \tag{29.107}$$

can also be written.

By defining a pitch angle θ_k of the kingpin axis (coinciding with the longitudinal inclination angle) and a roll angle ϕ_k (coinciding with the transversal inclination angle), and by indicating with x_k, y_k and z_k the components[6] of the unit vector $\mathbf{e}_k$, the rotation matrix $\mathbf{R}_k$ is still expressed by Eq. (31.67). If θ_k and ϕ_k are small angles, matrix $\mathbf{R}_k$ reduces to:

$$\mathbf{R}_k \approx \begin{bmatrix} 1 & 0 & x_k \\ 0 & 1 & y_k \\ -x_k & -y_k & 1 \end{bmatrix} . \tag{29.108}$$

The velocity of the wheel in the reference frame of the sprung mass is then

$$\mathbf{\Omega}_w = \mathbf{\Omega} + \dot{\delta}\mathbf{e}_k + \dot{\chi}\mathbf{R}_k\mathbf{R}_4\mathbf{R}_k^T\mathbf{e}_w . \tag{29.109}$$

To write it in the principal reference frame of the wheel, the expression of the angular velocity must be multiplied by $\left(\mathbf{R}_k\mathbf{R}_4\mathbf{R}_k^T\mathbf{R}_w\right)^T$:

$$\mathbf{\Omega}_{wi} = \dot{\chi}\mathbf{e}_2 + \dot{\delta}\boldsymbol{\alpha}_1 + \boldsymbol{\alpha}_2\mathbf{\Omega}, \tag{29.110}$$

where

$$\boldsymbol{\alpha}_1 = \mathbf{R}_{wi}^T\mathbf{R}_{ki}\mathbf{R}_4^T\mathbf{e}_3 , \qquad \boldsymbol{\alpha}_2 = \mathbf{R}_{wi}^T\mathbf{R}_{ki}\mathbf{R}_{4i}^T\mathbf{R}_{ki}^T . \tag{29.111}$$

The steering angle so defined does not coincide exactly with the steering angle defined in the preceding chapters, because it also has components along axes x and y.

Translational kinetic energy

The position of the center of mass of the unsprung mass in the inertial reference frame is

$$(\overline{G_u - O'}) = (\overline{H - O'}) + \mathbf{R}_1\mathbf{r} , \tag{29.112}$$

where

$$\mathbf{r} = \mathbf{R}_2\mathbf{R}_3 \left\{ \begin{array}{c} x_{Gu} + \zeta s_x \\ y_{Gu} + \zeta s_y \\ z_{Gu} + \zeta \end{array} \right\} . \tag{29.113}$$

By linearizing the expression for $\mathbf{r}$ and substituting its value for ζ, it follows that

$$\mathbf{r} = \left\{ \begin{array}{c} x_{Gu} + \theta_0 z_{Gu} + \theta z_{Gu} \\ y_{Gu} - \phi z_{Gu} \\ z_{Gu} - \theta_0 x_{Gu} + Z_u - Z \end{array} \right\} . \tag{29.114}$$

---

[6]Obviously, they are functions of ζ and $\sqrt{x_k^2 + y_k^2 + z_k^2} = 1$.

Its derivative, again approximated to the first-order term in the small quantities, is

$$\dot{\mathbf{r}} = \left\{ \begin{array}{c} \dot{\theta}z_{Gu} \\ -\dot{\phi}z_{Gu} \\ \dot{Z}_u - \dot{Z} \end{array} \right\} . \tag{29.115}$$

The speed of the center of mass of the unsprung mass, written in the inertial frame, is still expressed by Eq. (29.40) while the expression of the translational kinetic energy of the unsprung mass is identical to that seen for the rigid axle suspension (Eq. (29.44), where r_x, r_y, r_z are the components of vector $\mathbf{r}$). The equation is repeated here:

$$\mathcal{T}_t = \tfrac{1}{2}m_{ui} \left\{ v_x^2 + v_y^2 + v_z^2 + \dot{r}_x^2 + \dot{r}_y^2 + \dot{r}_z^2 + \dot{\psi}^2 \left(r_x^2 + r_y^2 \right) + \right.$$
$$\left. +2\left(v_x\dot{r}_x + v_y\dot{r}_y + v_z\dot{r}_z \right) + 2\dot{\psi}\left(-v_x r_y + v_y r_x - \dot{r}_x r_y + \dot{r}_y r_x \right) \right\}.$$

To obtain an expression containing all terms up to the quadratic in small quantities, the linearized expression of the components of $\mathbf{r}$ and their derivatives may be used, except for the term in $v_x\dot{r}_x$, where the quadratic terms must also be used

$$\dot{r}_x = \beta_1\dot{\theta} + x_{Gu}\dot{\theta}\theta + \beta_3\left(v_u - v_z \right) + \theta\left(v_u - v_z \right) + \dot{\theta}\left(Z_u - Z \right) - y_{Gu}s_x\dot{\phi} \tag{29.116}$$

where

$$\beta_1 = x_{Gu}s_x + z_{Gu} , \qquad \beta_3 = \theta_0 + s_x . \tag{29.117}$$

By linearizing the kinematic equations (A.111) and indicating with v_{ui} the derivative of Z_{ui}, the following expression of the translational kinetic energy of a single independent suspension is obtained:

$$\mathcal{T}_{ti} = \tfrac{1}{2}m_{ui} \left(v_x^2 + v_y^2 + v_{ui}^2 + \dot{\phi}^2 z_{Gu}^2 + \dot{\theta}^2 z_{Gu}^2 \right) +$$
$$+m_{ui}v_x\left[\beta_1\dot{\theta} + x_{Gu}\dot{\theta}\theta + \beta_3\left(v_{ui} - v_z \right) + \right.$$
$$\left. +\theta\left(v_{ui} - v_z \right) + \dot{\theta}\left(Z_{ui} - Z \right) - y_{Gu}s_x\dot{\phi} - \dot{\psi}y_{Gu} + z_{Gu}\phi\dot{\psi} \right] +$$
$$+m_{ui}v_y\left(\dot{\psi}x_{Gu} - \dot{\phi}z_{Gu} \right) - m_{ui}\dot{\psi}\dot{\theta}z_{Gu}y_{Gu} + m_{ui}\dot{\psi}\dot{\phi}z_{Gu}x_{Gu} . \tag{29.118}$$

Note that operating in this way the translational kinetic energy linked with steering has been neglected. This would be correct if the center of mass of the steering part of the suspension lies on the kingpin axis; however, the small error so introduced may be at least partially compensated for by introducing the moment of inertia of the steering parts about the kingpin axis instead of about a baricentric axis.

Rotational kinetic energy

The rotational kinetic energy of the wheel is

$$\mathcal{T}_{wr} = \tfrac{1}{2}\mathbf{\Omega}^T\mathbf{\alpha}_2^T\mathbf{J}_w\mathbf{\alpha}_2\mathbf{\Omega} + \tfrac{1}{2}\dot{\delta}^2\mathbf{\alpha}_1^T\mathbf{J}_w\mathbf{\alpha}_1 + \dot{\delta}\mathbf{\alpha}_1^T\mathbf{J}_w\mathbf{\alpha}_2\mathbf{\Omega} +$$
$$+\dot{\chi}\dot{\delta}\mathbf{e}_2^T\mathbf{J}_w\mathbf{\alpha}_1 + \tfrac{1}{2}\dot{\chi}^2\mathbf{e}_2^T\mathbf{J}_w\mathbf{e}_2 + \dot{\chi}\mathbf{e}_2^T\mathbf{J}_w\mathbf{\alpha}_2\mathbf{\Omega} . \tag{29.119}$$

Because the wheel is a gyroscopic body (two of its principal moments of inertia are equal to each other), with one of its principal axes of inertia coinciding with the rotation axis, its inertia matrix is diagonal and has a particular form

$$\mathbf{J}_w = \mathrm{diag}\left(\begin{bmatrix} J_{tw} & J_{pw} & J_{tw} \end{bmatrix}\right), \tag{29.120}$$

where J_{pw} and J_{tw} are the polar and transversal moments of inertia, respectively.

The rotational kinetic energy of the non-rotating parts of the suspensions is

$$\mathcal{T}_{nr} = \frac{1}{2}\dot{\delta}^2\boldsymbol{\alpha}_1^T\mathbf{J}_m\boldsymbol{\alpha}_1 + \frac{1}{2}\boldsymbol{\Omega}^T\boldsymbol{\alpha}_2^T\mathbf{J}_m\boldsymbol{\alpha}_2\boldsymbol{\Omega} + \dot{\delta}\boldsymbol{\alpha}_1^T\mathbf{J}_m\boldsymbol{\alpha}_2\boldsymbol{\Omega}.$$

where $\mathbf{J}_m$ is the inertia tensor of the non-rotating parts of the unsprung mass.

The first three terms of Eq. (29.119) may be directly included in the expression of $\mathcal{T}_{nr}$ if the inertia of the wheels is included in tensor $\mathbf{J}_m$.

Remembering that all angular velocities except for $\dot{\chi}$ are small quantities, the expression of the kinetic energy truncated to the second-order terms is fairly simplified.

Stating that $\mathbf{J}_m$ is the inertia tensor of the unsprung mass, which in general has no symmetry property, and remembering the peculiar structure of the inertia tensor of the wheel, the rotational kinetic energy of the unsprung mass (i.e., of one of the two unsprung masses of the axle) is

$$\begin{aligned}
\mathcal{T}_{ur} = &\tfrac{1}{2}J_{pw}\dot{\chi}^2 + \tfrac{1}{2}\dot{\delta}^2 J_{mz} + \tfrac{1}{2}\dot{\phi}^2 J_{mx} + \tfrac{1}{2}\dot{\theta}^2 J_{my} + \\
&+\tfrac{1}{2}\dot{\psi}^2 J_{mz} - \dot{\phi}\dot{\theta}J_{mxy} - \dot{\phi}\dot{\psi}J_{mxz} - \dot{\theta}\dot{\psi}J_{myz} + \\
&-\dot{\delta}\dot{\phi}J_{mxz} - \dot{\delta}\dot{\theta}J_{myz} + \dot{\delta}\dot{\psi}J_{mz} + \dot{\chi}\dot{\delta}J_{pw}\left(y_k + z_w\right) + \\
&+\dot{\chi}\dot{\phi}J_{pw}\left(x_w - \delta\right) + \dot{\chi}J_{pw}\left[\dot{\theta} + \dot{\psi}\left(z_w + \phi\right)\right].
\end{aligned} \tag{29.121}$$

As already stated, this expression is approximated for various reasons, and also because the parts of the suspension that do not steer have been neglected.

Total kinetic energy of the axle

Because different types of suspensions may be used on the same vehicle for front and rear axles, the equations of motion are best written with reference to coordinates that may be used for both rigid axle and independent suspensions. Consider the general kth axle made of two independent suspensions and assume

$$\begin{cases} Z_k = \dfrac{Z_{uL} + Z_{uR}}{2}, \\[2mm] \phi_k = \dfrac{Z_{uL} - Z_{uR}}{d_0}, \end{cases} \tag{29.122}$$

where as usual subscripts L and R designate the left and right suspension and d_0 is an arbitrary length, for instance the distance between the centers of mass of the two suspensions. Coordinate Z_k coincides with the vertical displacement of the center of mass of the system made by the two suspensions of the axle, and

ϕ_k is the roll rotation of a line passing through the two centers of mass (if d_0 is their distance). The coordinates are then the same used for rigid axles.

Taking into account the symmetry of the two suspensions, some terms in the kinetic energy are equal in modulus but have opposite signs (for instance $\beta_2 y_{Gu}\dot{\phi}$) and then cancel each other. Remembering that $m_u = 2m_{ui}$, substituting the coordinates Z_k and ϕ_k to Z_{ui}, the total kinetic energy of the system made by the two suspensions is

$$\mathcal{T}_u = \tfrac{1}{2}m_u\left(v_x^2 + v_y^2 + v_k^2\right) + \tfrac{1}{2}\beta_{10}\dot{\phi}^2 + \tfrac{1}{2}\beta_{11}\dot{\theta}^2 + \tfrac{1}{2}\beta_{12}\dot{\psi}^2 + $$
$$\tfrac{1}{2}\beta_{13}\dot{\phi}_k^2 + \beta_{15}\dot{\phi}\dot{\psi} + \tfrac{1}{2}J_{pw}\dot{\chi}_L^2 + \tfrac{1}{2}J_{pw}\dot{\chi}_R^2 + \tfrac{1}{2}\dot{\delta}_L^2 J_{mz} + \tfrac{1}{2}\dot{\delta}_R^2 J_{mz} + $$
$$+\dot{\chi}_L\dot{\delta}_L J_{pw}\left(y_k + z_w\right) - \dot{\chi}_R\dot{\delta}_R J_{pw}\left(y_k + z_w\right) + $$
$$+m_u v_x\left(\beta_1\dot{\theta} + \dot{\theta}Z_k - \dot{\theta}Z + \theta v_k - \theta v_z + \beta_3 v_k - \beta_3 v_z + \right.$$
$$\left. +\beta_4\dot{\psi}\phi + x_{Gu}\dot{\theta}\theta\right) + m_u v_y\left(x_{Gu}\dot{\psi} - \beta_4\dot{\phi}\right) - \beta_{17}\dot{\delta}_L\dot{\phi} - \beta_{17}\dot{\delta}_R\dot{\phi} + $$
$$-\beta_{18}\dot{\delta}_L\dot{\theta} + \beta_{18}\dot{\delta}_R\dot{\theta} + \beta_{19}\dot{\delta}_L\dot{\psi} + \beta_{19}\dot{\delta}_R\dot{\psi} + \dot{\chi}_L\dot{\phi}\beta_{21}\left(x_w - \delta_L\right) + $$
$$-\dot{\chi}_R\dot{\phi}\beta_{21}\left(x_w + \delta_R\right) + \dot{\chi}_L J_{pw}\left[\dot{\theta} + \dot{\psi}\left(z_w + \phi\right)\right] + \dot{\chi}_R J_{pw}\left[\dot{\theta} + \dot{\psi}\left(-z_w + \phi\right)\right],$$
$$(29.123)$$

where

$$\beta_4 = z_{Gu}, \quad \beta_{10} = m_u z_{Gu}^2 + 2J_{mx}, \quad \beta_{11} = m_u z_{Gu}^2 + 2J_{my},$$
$$\beta_{12} = 2J_{mz}, \quad \beta_{13} = \tfrac{1}{2}m_u d_0^2, \quad \beta_{15} = m_u z_{uG} x_{Gu} - 2J_{mxz},$$
$$\beta_{17} = J_{mxz}, \quad \beta_{18} = J_{myz}, \quad \beta_{19} = J_{mz}, \quad \beta_{21} = J_{pw}.$$

Potential energy

The height on the ground of the center of mass of one of the two suspensions is still expressed by Eq. (29.75):

$$h_u = Z_0 + Z + r_z.$$

The expression of r_z, obtained by approximating vector $\mathbf{r}$ with its series truncated after quadratic terms in the small quantities and eliminating the constant terms that do not affect the equations of motion, is

$$r_z = z_{Gu} - \frac{1}{2}\theta_0 z_{Gu} + Z_{ui} - Z - \theta\theta_0 z_{Gu} - \frac{1}{2}\theta^2 z_{Gu} - \frac{1}{2}\phi^2 z_{Gu}. \quad (29.124)$$

Neglecting constant terms, the gravitational potential energy of one of the two independent suspensions is

$$\mathcal{U}_{gi} = m_{ui}g\left(-\theta\theta_0 z_{Gu} + Z_{ui} - \frac{1}{2}\theta^2 z_{Gu} - \frac{1}{2}\phi^2 z_{Gu}\right). \quad (29.125)$$

The potential energy of the system made by the two suspensions is then

$$\mathcal{U}_g = m_u g\left(Z_k - \theta\beta_{22} - \frac{1}{2}\theta^2 z_{Gu} - \frac{1}{2}\phi^2 z_{Gu}\right) \quad (29.126)$$

where

$$\beta_{22} = \theta_0 z_{Gu} \ .$$

The expressions for the linearized elastic potential energy of the axle and the tires, as well as those of the dissipation functions, are those already seen for rigid axles. Obviously, the expressions of the coefficients describing the various stiffnesses and damping coefficients are different and must be computed in each case from the mechanical and geometrical characteristics of the suspensions, but in the linearized approach they are at any rate constant.

29.2.5 Comparison between independent and rigid axle suspensions

As already stated, the generalized coordinates here chosen may be used for both types of suspensions. The general expression of the kinetic energy of the axle is

$$
\begin{aligned}
\mathcal{T}_u &= \tfrac{1}{2} m_u \left(v_x^2 + v_y^2 + v_k^2 \right) + \tfrac{1}{2}\beta_{10}\dot{\phi}^2 + \tfrac{1}{2}\beta_{11}\dot{\theta}^2 + \tfrac{1}{2}\beta_{12}\dot{\psi}^2 + \tfrac{1}{2}\beta_{13}\dot{\phi}_k^2 + \\
&+ \beta_{14}\dot{\psi}\dot{\phi}_k + \beta_{15}\dot{\psi}\dot{\phi} + \tfrac{1}{2}\dot{\chi}_L^2 J_{pw} + \tfrac{1}{2}\dot{\delta}_L^2 J_{tw} + \tfrac{1}{2}\dot{\chi}_R^2 J_{pw} + \tfrac{1}{2}\dot{\delta}_R^2 J_{tw} + \\
&+ \dot{\chi}_L \dot{\delta}_L J_{pw} \left(y_k + z_w \right) - \dot{\chi}_R \dot{\delta}_R J_{pw} \left(y_k + z_w \right) - \beta_{16}\dot{Z}_k\dot{\theta} + \\
&+ m_u v_x \left(\dot{\theta}\beta_1 + \dot{\theta}Z_k - \dot{\theta}Z + \theta v_k - \theta v_z + \beta_3 v_k - \beta_3 v_z + \right. \\
&\left. + \beta_4 \dot{\phi}\dot{\psi} + x_{Gu}\dot{\theta}\theta + \beta_5\dot{\phi}_k\dot{\psi} \right) + m_u v_y \left(-\beta_5\dot{\phi}_k + x_{Gu}\dot{\psi} - \beta_4\dot{\phi} \right) + \\
&- \beta_{17}\dot{\delta}_L\dot{\phi} - \beta_{17}\dot{\delta}_R\dot{\phi} - \beta_{18}\dot{\delta}_L\dot{\theta} + \beta_{18}\dot{\delta}_R\dot{\theta} + \beta_{19}\dot{\delta}_L\dot{\psi} + \beta_{19}\dot{\delta}_R\dot{\psi} + \\
&+ \dot{\chi}_L\beta_{20} \left(x_w - \delta_L \right) \dot{\phi}_k - \dot{\chi}_R\beta_{20} \left(x_w + \delta_R \right) \dot{\phi}_k + \dot{\chi}_L\dot{\phi}\beta_{21} \left(x_w - \delta_L \right) + \\
&- \dot{\chi}_R\dot{\phi}\beta_{21} \left(x_w + \delta_R \right) + \dot{\chi}_L J_{pw} \left(\dot{\theta} + \dot{\phi}\dot{\psi} + z_w\dot{\psi} \right) + \dot{\chi}_R J_{pw} \left(\dot{\theta} + \dot{\phi}\dot{\psi} - z_w\dot{\psi} \right) \ .
\end{aligned}
$$

$$(29.127)$$

Some coefficients β_i vanish in the case of rigid axles (for instance β_{10} or β_{15}), while others vanish in independent suspensions (for instance β_{14} or β_{16}).

In a similar way, the gravitational potential energy can be written in the form

$$\mathcal{U}_g = m_u g \left(Z_u - \theta\beta_{22} - \frac{1}{2}\theta^2 z_{Gu} - \frac{1}{2}\phi^2\beta_{24} - \frac{1}{2}\phi_u^2\beta_{23} \right) \qquad (29.128)$$

where

$$\beta_{23} = z_{Gu} \ , \quad \beta_{24} = 0 \qquad (29.129)$$

in the case of rigid axles, and

$$\beta_{23} = 0 \ , \quad \beta_{24} = z_{Gu} \qquad (29.130)$$

for independent suspensions.

29.2.6   Lagrangian function of the whole vehicle

The Lagrangian function $\mathcal{L} = \mathcal{T} - \mathcal{U}$ of the whole vehicle can thus be computed without any difficulty:

$$
\begin{aligned}
\mathcal{L} = &\tfrac{1}{2}m\left(v_x^2 + v_y^2\right) + \tfrac{1}{2}m_s v_z^2 + \tfrac{1}{2}J_x\dot{\phi}^2 + \tfrac{1}{2}J_y\dot{\theta}^2 + \tfrac{1}{2}J_z\dot{\psi}^2 + \\
&-J_{xz}\dot{\psi}\dot{\phi} - m_s c v_z\dot{\theta} + v_x\dot{\theta}J_{s1} - J_{s3}v_y\dot{\phi} + J_{s3}v_x\dot{\phi}\dot{\psi} + \\
&+\textstyle\sum_{\forall k}\left[\tfrac{1}{2}m_u v_k^2 + \tfrac{1}{2}\beta_{13}\dot{\phi}_k^2 + \beta_{14}\dot{\psi}\dot{\phi}_k + \tfrac{1}{2}\dot{\chi}_L^2 J_{pw} + \tfrac{1}{2}\dot{\delta}_L^2 J_{tw} + \tfrac{1}{2}\dot{\chi}_R^2 J_{pw} + \right. \\
&+\tfrac{1}{2}\dot{\delta}_R^2 J_{tw} + \dot{\chi}_L\dot{\delta}_L J_{pw}\left(y_k + z_w\right) - \dot{\chi}_R\dot{\delta}_R J_{pw}\left(y_k + z_w\right) - \beta_{16}v_k\dot{\theta} + \\
&+m_u v_x\left(\dot{\theta}Z_k - \dot{\theta}Z + \theta v_k - \theta v_z + \beta_3 v_k - \beta_3 v_z + \beta_5\phi_k\dot{\psi}\right) + \\
&-m_u v_y\beta_5\dot{\phi}_k - \beta_{17}\dot{\delta}_L\dot{\phi} - \beta_{17}\dot{\delta}_R\dot{\phi} - \beta_{18}\dot{\delta}_L\dot{\theta} + \beta_{18}\dot{\delta}_R\dot{\theta} + \\
&+\beta_{19}\dot{\delta}_L\dot{\psi} + \beta_{19}\dot{\delta}_R\dot{\psi} + \dot{\chi}_L\beta_{20}\left(x_w - \delta_L\right)\dot{\phi}_k - \dot{\chi}_R\beta_{20}\left(x_w + \delta_R\right)\dot{\phi}_k + \\
&+\dot{\chi}_L\dot{\phi}\beta_{21}\left(x_w - \delta_L\right) - \dot{\chi}_R\dot{\phi}\beta_{21}\left(x_w + \delta_R\right) + \dot{\chi}_L J_{pw}\left(\dot{\theta} + \phi\dot{\psi} + z_w\dot{\psi}\right) + \\
&\left. +\dot{\chi}_R J_{pw}\left(\dot{\theta} + \phi\dot{\psi} - z_w\dot{\psi}\right)\right] - m_s g Z + M_{g1}\theta + \tfrac{1}{2}M_{g2}\theta^2 + \tfrac{1}{2}M_{g3}\phi^2 + \\
&-g\textstyle\sum_{\forall k}m_u\left(Z_u - \tfrac{1}{2}\beta_{23}\phi_k^2\right) - \sum_{\forall k}\left(\tfrac{1}{2}K_{11}\left(Z + Z_0\right)^2 + \right. \\
&+\tfrac{1}{2}\left(K_{22} + K_{pz}\right)\left(Z_k + Z_{k0}\right)^2 + \tfrac{1}{2}K_{33}\left(\theta + \theta_0\right)^2 + \\
&-K_{12}\left(Z + Z_0\right)\left(Z_k + Z_{k0}\right) - K_{13}\left(Z + Z_0\right)\left(\theta + \theta_0\right) + \\
&\left. +K_{23}\left(Z_k + Z_{k0}\right)\left(\theta + \theta_0\right) + \tfrac{1}{2}K_\phi\phi^2 + \tfrac{1}{2}\left(K_\phi + K_{p\phi}\right)\phi_k^2 - K_\phi\phi\phi_k\right) ,
\end{aligned}
$$
$$(29.131)$$

where

$$
\begin{aligned}
&m = m_s + \textstyle\sum_{\forall k}m_k , \qquad J_x = J_{x_L} + m_s h^2 + \sum_{\forall k}\beta_{10} \\
&J_y = J_{y_L} + m_s\left(h^2 + c^2\right) + \textstyle\sum_{\forall k}\beta_{11} , \qquad J_z = J_{z_L} + m_s c^2 + \sum_{\forall k}\beta_{12} \\
&J_{xz} = J_{xz_L} - m_s c h - \textstyle\sum_{\forall k}\beta_{15} , \qquad J_{s1} = -m_s\left(c\theta_0 - h\right) + \sum_{\forall k}m_k\beta_1 \\
&J_{s3} = m_s h + \textstyle\sum_{\forall k}m_k\beta_4 , \qquad M_{g1} = m_s g\left(c + h\theta_0\right) + g\sum_{\forall k}m_k\beta_{22} , \\
&M_{g2} = m_s g h + g\textstyle\sum_{\forall k}m_k z_{Gu} , \qquad M_{g3} = m_s g h + g\sum_{\forall k}m_k\beta_{24} .
\end{aligned}
$$
$$(29.132)$$

29.3   MODEL WITH 10 DEGREES OF FREEDOM WITH LOCKED CONTROLS

Consider a vehicle moving at a stated speed with a stated steering angle and neglect the longitudinal slip of the wheels. The forward speed V, which in the linearized approach (small value of the sideslip angle β) coincides with v_x, and its derivative $\dot{V}$ are imposed, and so are the steering angles of the wheels and their derivatives. The angular velocity of the wheels is simply

$$\dot{\chi}_i = \frac{V}{R_{e_i}} \tag{29.133}$$

where R_{e_i} is the effective rolling radius. This expression is approximated even if the rolling radius corresponding with the actual longitudinal slip was used, because the speed of the centers of the wheels does not coincide with the velocity V of the center of mass of the vehicle. Nonetheless, if motion takes place in conditions allowing the equations to be linearized, such assumptions can be accepted.

Moreover, assume that the derivatives $\dot{\delta}_i$ of the steering angles are vanishingly small, either because the steering angles are actually locked at a constant value or because the dynamic effects of their variation are negligible

29.3.1   Expression of the Lagrangian function and its derivatives

The expression of the Lagrangian function is much simplified and may be written as

$$\begin{aligned}
\mathcal{L} = &\tfrac{1}{2} m_e v_x^2 + \tfrac{1}{2} m v_y^2 + \tfrac{1}{2} m_s v_z^2 + \tfrac{1}{2} J_x \dot{\phi}^2 + \tfrac{1}{2} J_y \dot{\theta}^2 + \tfrac{1}{2} J_z \dot{\psi}^2 + \\
&- J_{xz} \dot{\psi}\dot{\phi} - m_s c v_z \dot{\theta} + v_x \dot{\theta} J_{s2} - J_{s3} v_y \dot{\phi} + J_{s3} v_x \dot{\phi}\dot{\psi} + \\
&+ \sum_{\forall k} \left\{ \tfrac{1}{2} m_k v_k^2 + \tfrac{1}{2} \beta_{13} \dot{\phi}_k^2 + \beta_{14} \dot{\psi}\dot{\phi}_k - \beta_{16} v_k \dot{\theta} - m_k v_y \beta_5 \dot{\phi}_k + \right. \\
&+ m_k v_x \left(\dot{\theta} Z_k - \dot{\theta} Z + \theta v_k - \theta v_z + \beta_3 v_k - \beta_3 v_z + \beta_5 \phi_k \dot{\psi} \right) + \\
&\left. + 2 \tfrac{v_x}{R_e} \left[-\beta_{20} \delta \dot{\phi}_k - \beta_{21} \delta \dot{\phi} + J_{pr} \dot{\phi}\dot{\psi} \right] \right\} - m_s g Z + M_{g1} \theta + \tfrac{1}{2} M_{g2} \theta^2 + \\
&+ \tfrac{1}{2} M_{g3} \phi^2 - g \sum_{\forall k} m_k \left(Z_k - \tfrac{1}{2} \beta_{23} \phi_k^2 \right) - \sum_{\forall k} \left(\tfrac{1}{2} K_{11} \left(Z + Z_0 \right)^2 + \right. \\
&+ \tfrac{1}{2} \left(K_{22} + K_{pz} \right) \left(Z_k + Z_{k0} \right)^2 + \tfrac{1}{2} K_{33} \left(\theta + \theta_0 \right)^2 + \\
&- K_{12} \left(Z + Z_0 \right) \left(Z_k + Z_{k0} \right) - K_{13} \left(Z + Z_0 \right) \left(\theta + \theta_0 \right) + \\
&\left. + K_{23} \left(Z_k + Z_{k0} \right) \left(\theta + \theta_0 \right) + \tfrac{1}{2} K_\phi \phi^2 + \tfrac{1}{2} \left(K_\phi + K_{p\phi} \right) \phi_k^2 - K_\phi \phi \phi_k \right),
\end{aligned} \tag{29.134}$$

where the equivalent mass and J_{s2} are

$$m_e = m + 2 \sum_{\forall k} J_{pw} \frac{1}{R_e^2} \ , \quad J_{s2} = J_{s1} + 2 \sum_{\forall k} \frac{J_{pw}}{R_e} \tag{29.135}$$

(coefficients 2 come from the assumption that each axle has two wheels) and δ is the average steering angle of the axle

$$\delta = \frac{\delta_L + \delta_R}{2} \ . \tag{29.136}$$

The derivatives of the Lagrangian function with respect to the generalized velocities and coordinates are

$$\frac{\partial \mathcal{L}}{\partial v_x} = m_e v_x + \dot{\theta} J_{s2} \ . \tag{29.137}$$

Note that the expression of this derivative has been further linearized by cancelling the terms of the same order of the squares of small quantities. For instance, the term in $\beta_3 v_k$ was cancelled because both β_3 and v_k are small quantities. Moreover, the use of the equivalent mass, which includes only the contribution to inertia due to the wheels, may be criticized because the transmission has not been modelled. Physically, this corresponds to considering the vehicle as pushed forward by an external force in the x direction, as in jet propelled record vehicles, instead of propelled by the driving torque applied to the wheels (or slowed by the braking torque).

$$\frac{\partial \mathcal{L}}{\partial v_y} = m v_y - J_{s3}\dot{\phi} - \sum_{\forall k} m_k \beta_5 \dot{\phi}_k, \tag{29.138}$$

$$\frac{\partial \mathcal{L}}{\partial v_z} = m_s \dot{Z} - m_s c\dot{\theta} - v_x \sum_{\forall k} m_k \left(\theta + \beta_3\right), \tag{29.139}$$

$$\frac{\partial \mathcal{L}}{\partial \dot{\phi}} = J_x \dot{\phi} - J_{xz}\dot{\psi} - J_{s3} v_y - 2\sum_{\forall k} \frac{v_x}{R_e}\beta_{21}\delta, \tag{29.140}$$

$$\frac{\partial \mathcal{L}}{\partial \dot{\theta}} = J_y \dot{\theta} - m_s c\dot{Z} + v_x J_{s2} + \sum_{\forall k}\left[-\beta_{16}\dot{Z}_k + m_k v_x \left(Z_k - Z\right)\right], \tag{29.141}$$

$$\frac{\partial \mathcal{L}}{\partial \dot{\psi}} = J_z \dot{\psi} - J_{xz}\dot{\phi} + J_{s3} v_x \phi + \sum_{\forall k}\left(\beta_{14}\dot{\phi}_k + m_k v_x \beta_5 \dot{\phi}_k + 2\frac{v_x}{R_e}J_{pw}\phi\right), \tag{29.142}$$

$$\frac{\partial \mathcal{L}}{\partial \dot{Z}_k} = m_k \dot{Z}_k - \beta_{16k}\dot{\theta} + m_k v_x \left(\theta + \beta_{3k}\right) \qquad \text{for } k = 1, 2, \tag{29.143}$$

$$\frac{\partial \mathcal{L}}{\partial \dot{\phi}_k} = \beta_{13k}\dot{\phi}_k + \beta_{14k}\dot{\psi} - m_k v_y \beta_{5k} - 2\frac{v_x}{R_{ek}}\beta_{20k}\delta_k \qquad \text{for } k = 1, 2, \tag{29.144}$$

$$\frac{\partial \mathcal{L}}{\partial X} = \frac{\partial \mathcal{L}}{\partial Y} = \frac{\partial \mathcal{L}}{\partial \psi} = 0, \tag{29.145}$$

$$\frac{\partial \mathcal{L}}{\partial Z} = -m_s g - \sum_{\forall k}\left[m_k v_x \dot{\theta} + K_{11}\left(Z + Z_0\right) + \right.$$
$$\left. -K_{12}\left(Z_k + Z_{0k}\right) - K_{13}\left(\theta + \theta_0\right)\right], \tag{29.146}$$

$$\frac{\partial \mathcal{L}}{\partial \theta} = M_{g1} + M_{g2}\theta + \sum_{\forall k}\left[m_u v_x \left(\dot{Z}_k - \dot{Z}\right) + \right.$$
$$\left. -K_{33}\left(\theta + \theta_0\right) + K_{13}\left(Z + Z_0\right) - K_{23}\left(Z_k + Z_{k0}\right)\right], \tag{29.147}$$

$$\frac{\partial \mathcal{L}}{\partial \phi} = J_{s3} v_x \dot{\psi} + M_{g3}\phi + \sum_{\forall k}\left(2\frac{v_x}{R_e}J_{pw}\dot{\psi} - K_\phi \phi + K_\phi \phi_k\right), \tag{29.148}$$

$$\frac{\partial \mathcal{L}}{\partial Z_k} = -gm_k + m_k v_x \dot{\theta} - (K_{22k} + K_{pzk})(Z_k + Z_{k0}) +$$

$$+K_{12k}(Z + Z_0) - K_{23k}(\theta + \theta_0), \qquad (29.149)$$

$$\frac{\partial \mathcal{L}}{\partial \phi_k} = +m_k v_x \beta_{5k} \dot{\psi} + gm_k \beta_{23k} \phi_k - (K_{\phi k} + K_{p\phi k})\phi_k + K_{\phi k}\phi . \qquad (29.150)$$

The last two derivatives must be computed for the various axles ($k = 1$, 2 for a two-axles vehicle).

29.3.2   Kinematic equations

Even if velocity V is stated, all 10 equations of motion must be written, because the generalized coordinates are 10. When all equations of motion have been obtained, it will be possible to state that the forward velocity is known and one of the equations can be eliminated.

The generalized coordinates for a two-axles vehicle are then

$$\mathbf{q} = \begin{bmatrix} X & Y & Z & \phi & \theta & \psi & z_1 & \phi_1 & z_2 & \phi_2 \end{bmatrix}^T . \qquad (29.151)$$

The vector containing the generalized velocities $\mathbf{w}$ is

$$\mathbf{w} = \begin{bmatrix} v_x & v_y & v_z & v_\phi & v_\theta & v_\psi & v_1 & v_{\phi 1} & v_2 & v_{\phi 2} \end{bmatrix}^T . \qquad (29.152)$$

Remark 29.1 *Velocities $\mathbf{w}$ are referred neither to a body-fixed frame nor to an inertial frame. Linear velocities are referred to the intermediate frame x^*y^*z, while the generalized velocities related to angular coordinates are the derivatives of Tait-Bryan angles. This may make the analysis more complicated, but only to a point.*

The relationship linking the velocities to the derivatives of the coordinates is, as usual

$$\mathbf{w} = \mathbf{A}^T \dot{\mathbf{q}} ,$$

where

$$\mathbf{A} = \begin{bmatrix} \cos(\psi) & -\sin(\psi) & \mathbf{0} \\ \sin(\psi) & \cos(\psi) & \\ \mathbf{0} & & \mathbf{I} \end{bmatrix} \qquad (29.153)$$

and $\mathbf{I}$ is an identity matrix of size 8×8.

The kinematic equations are the inverse transformation

$$\dot{\mathbf{q}} = \mathbf{A}^{-T}\mathbf{w} = \mathbf{B}\mathbf{w} . \qquad (29.154)$$

$\mathbf{A}$ is in this case a rotation matrix, so that

$$\mathbf{A}^{-1} = \mathbf{A}^T , \quad \mathbf{B} = \mathbf{A} . \qquad (29.155)$$

The equation of motion in the state space is made by the 10 equations of motion, plus the 10 kinematic equations and is then Eq. (A.101):

$$
\begin{cases}
\dfrac{\partial}{\partial t}\left(\left\{\dfrac{\partial \mathcal{L}}{\partial w}\right\}\right) + \mathbf{B}^T \mathbf{\Gamma}\left\{\dfrac{\partial \mathcal{L}}{\partial w}\right\} - \mathbf{B}^T\left\{\dfrac{\partial \mathcal{L}}{\partial q}\right\} + \left\{\dfrac{\partial \mathcal{F}}{\partial w}\right\} = \mathbf{B}^T \mathbf{Q}\ , \\[4mm]
\{\dot{q}_i\} = \mathbf{B}\,\{w_i\}\ .
\end{cases}
\tag{29.156}
$$

The column matrix $\mathbf{B}^T\mathbf{Q}$ containing the 10 components of the vector of the generalized forces will be computed later by writing the virtual work of the forces acting on the system. In the following equations its elements will be indicated with Q_x, Q_y, Q_z, Q_ϕ, Q_θ, Q_ψ, Q_{zk}, $Q_{\phi k}$.

The most complicated part of the computation is writing matrix $\mathbf{B}^T\mathbf{\Gamma}$. By performing fairly intricate computations, following the procedure described in Appendix A, it follows that

$$
\mathbf{B}^T\mathbf{\Gamma} =
\begin{bmatrix}
\begin{bmatrix}
0 & -\dot{\psi} \\
\dot{\psi} & 0 \\
0 & 0 \\
0 & 0 \\
0 & 0 \\
-v_y & v_x
\end{bmatrix} & \mathbf{0}_{6\times 8} \\[2mm]
\mathbf{0}_{4\times 2} & \mathbf{0}_{4\times 8}
\end{bmatrix}.
$$

By using the expressions of the derivatives with respect to the generalized velocities seen above, and differentiating again with respect to time, it follows that

$$
\frac{\partial}{\partial t}\left(\left\{\frac{\partial \mathcal{L}}{\partial w}\right\}\right) =
$$

$$
=
\begin{cases}
m_e \dot{v}_x + \ddot{\theta} J_{s2} \\
m\dot{v}_y - J_{s3}\ddot{\phi} - \sum_{\forall k} m\beta_{5k}\ddot{\phi}_k \\
m_s \ddot{Z} - m_s c\ddot{\theta} - \dot{v}_x \sum_{\forall k} m_k\left(\theta + \beta_{3k}\right) - v_x\dot{\theta}\sum_{\forall k} m_k \\
J_x\ddot{\phi} - J_{xz}\ddot{\psi} - J_{s3}\dot{v}_y - 2\dot{v}_x \sum_{\forall k} \frac{1}{R_{ek}}\beta_{21k}\delta_k \\
J_y\ddot{\theta} - m_s c\ddot{Z} + \dot{v}_x J_{s2} + \sum_{\forall k}\left[-\beta_{16k}\ddot{Z}_k + m_k\dot{v}_x\left(Z_k - Z\right) +\right. \\
\qquad\qquad \left. + m_k v_x \left(\dot{Z}_k - \dot{Z}\right)\right] \\
J_z\ddot{\psi} - J_{xz}\ddot{\phi} + J_{s3}\dot{v}_x\phi + J_{s3}v_x\dot{\phi} + \sum_{\forall k}\left[\beta_{14k}\ddot{\phi}_k + m_k\dot{v}_x\beta_{5k}\phi_k +\right. \\
\qquad\qquad \left. + m_k v_x\beta_{5k}\dot{\phi}_k + 2\dot{v}_x\frac{1}{R_{ek}}J_{pwk}\phi + 2v_x\frac{1}{R_{ek}}J_{pwk}\dot{\phi}\right] \\
m_k\ddot{Z}_k - \beta_{16k}\ddot{\theta} + m_k\dot{v}_x\left(\theta + \beta_{3k}\right) + m_k v_x\dot{\theta} \\
\beta_{13k}\ddot{\phi}_k + \beta_{14k}\ddot{\psi} - m_k\dot{v}_y\beta_{5k} - 2\dot{v}_x\frac{1}{R_{ek}}\beta_{20k}\delta_k
\end{cases}.
\tag{29.157}
$$

The last two equations refer to the coordinates of the axles, and then must be repeated for $k = 1,\ 2$.

$$\mathbf{B}^T\mathbf{\Gamma}\left\{\frac{\partial\mathcal{L}}{\partial w}\right\} = \left[\ -\dot{\psi}\frac{\partial\mathcal{L}}{\partial v_y}\quad \dot{\psi}\frac{\partial\mathcal{L}}{\partial v_x}\quad 0\quad 0\quad 0\quad -v_y\frac{\partial\mathcal{L}}{\partial v_x}+v_x\frac{\partial\mathcal{L}}{\partial v_y}\quad 0\quad 0\quad 0\quad 0\ \right]^T .$$

(29.158)

By introducing the values of the derivatives and linearizing, it follows that

$$\mathbf{B}^T\mathbf{\Gamma}\left\{\frac{\partial\mathcal{L}}{\partial w}\right\} = \left\{\begin{array}{c} \left\{\begin{array}{c} 0 \\ m_e v_x \dot{\psi} \end{array}\right\} \\ \mathbf{0}_{3\times1} \\ v_x\left[-J_{s3}\dot{\phi}-\sum_{\forall k}\left(2v_y J_{pwk}\frac{1}{R_e^2}+m_k\beta_5\dot{\phi}_k\right)\right] \\ \mathbf{0}_{4\times1} \end{array}\right\} .$$

(29.159)

Finally:

$$\mathbf{B}^T\left\{\frac{\partial\mathcal{L}}{\partial q}\right\} = \left\{\begin{array}{l} 0 \\ 0 \\ -m_s g - \sum_{\forall k}\left[m_k v_x\dot{\theta} + K_{11}\left(Z+Z_0\right)+\right. \\ \qquad\qquad \left. -K_{12}\left(Z_k+Z_{k0}\right)-K_{13}\left(\theta+\theta_0\right)\right] \\ J_{s3}v_x\dot{\psi}+M_{g3}\phi+\sum_{\forall k}\left(2v_x\frac{1}{R_e}J_{pwk}\dot{\psi}-K_\phi\phi+K_\phi\phi_k\right) \\ M_{g1}+M_{g2}\theta+\sum_{\forall k}\left[m_u v_x\left(\dot{Z}_k-\dot{Z}\right)-K_{33}\left(\theta+\theta_0\right)+\right. \\ \qquad\qquad \left. +K_{13}\left(Z+Z_0\right)-K_{23}\left(Z_k+Z_{k0}\right)\right] \\ 0 \\ -gm_k+m_k v_x\dot{\theta}-\left(K_{22k}+K_{pzk}\right)\left(Z_k+Z_{k0}\right)+ \\ \qquad\qquad +K_{12k}\left(Z+Z_0\right)-K_{23k}\left(Z+Z_0\right) \\ -m_k v_x\beta_{5k}\dot{\psi}-gm_{uk}\beta_{23k}\phi_k-K_{\phi k}\phi_k+K_{\phi k}\phi-K_{p\phi k}\phi_k \end{array}\right\} .$$

(29.160)

The last two equations refer to the axles and must be repeated for $k = 1,\ 2$.

The derivatives of the dissipation function are

$$\left\{\frac{\partial\mathcal{F}}{\partial w}\right\} = \left\{\begin{array}{l} 0 \\ 0 \\ \sum_{\forall k}\left(c_{11k}\dot{Z}-c_{12k}\dot{Z}_k-c_{13k}\dot{\theta}\right) \\ \sum_{\forall k}\left(c_{\phi k}\dot{\phi}-c_{\phi k}\dot{\phi}_k\right) \\ \sum_{\forall k}\left(c_{33k}\dot{\theta}-c_{13k}\dot{Z}+c_{23k}\dot{Z}_k\right) \\ 0 \\ \left(c_{22k}+c_{pzk}\right)\dot{Z}_k-c_{12k}\dot{Z}+c_{23k}\dot{\theta} \\ \left(c_{\phi k}+c_{p\phi k}\right)\dot{\phi}_k-c_{\phi k}\dot{\phi} \end{array}\right\} .$$

(29.161)

29.3.3 *Equations of motion*

First equation: longitudinal translation

By introducing the forward velocity of the vehicle V instead of v_x, the first equation becomes

$$m_e \dot{V} + \ddot{\theta} J_{s2} = Q_x . \qquad (29.162)$$

Second equation: lateral translation

$$m \dot{v}_y + m_e V \dot{\psi} - J_{s3} \ddot{\phi} - \sum_{\forall k} m_u \beta_5 \ddot{\phi}_k = Q_y . \qquad (29.163)$$

Third equation: vertical translation

$$m_s \ddot{Z} - m_s c \ddot{\theta} + \sum_{\forall k} \left[K_{11} \left(Z + Z_0 \right) - K_{12} \left(Z_k + Z_{k0} \right) + \right.$$
$$- K_{13} \left(\theta + \theta_0 \right) + c_{11k} \dot{Z} - c_{12k} \dot{Z}_k + \qquad (29.164)$$
$$\left. - c_{13k} \dot{\theta} - \dot{V} m_k \left(\theta + \beta_{3k} \right) \right] = -m_s g + Q_z .$$

Fourth equation: roll rotation

$$J_x \ddot{\phi} - J_{xz} \ddot{\psi} - J_{s3} \dot{v}_y - J_{s3} V \dot{\psi} - M_{g3} \phi + \sum_{\forall k} \left(-2\dot{V} \frac{1}{R_{ek}} \beta_{21k} \delta_k + \right.$$
$$\left. - 2 v_x \frac{1}{R_e} J_{pwk} \dot{\psi} + K_\phi \phi - K_\phi \phi_k + c_{\phi k} \dot{\phi} - c_{\phi k} \dot{\phi}_k \right) = Q_\phi . \qquad (29.165)$$

Fifth equation: pitch rotation

$$J_y \ddot{\theta} - m_s c \ddot{Z} + \dot{V} J_{s2} - M_{g2} \theta + \sum_{\forall k} \left[-\beta_{16} \ddot{Z}_k + \right.$$
$$+ m_k \dot{V} \left(Z_k - Z \right) + K_{33} \left(\theta + \theta_0 \right) - K_{13} \left(Z + Z_0 \right) + \qquad (29.166)$$
$$\left. + K_{23} \left(Z_k + Z_{k0} \right) + c_{33k} \dot{\theta} - c_{13k} \dot{Z} + c_{23k} \dot{Z}_k \right] = M_{g1} + Q_\theta .$$

Sixth equation: yaw rotation

$$J_z \ddot{\psi} - J_{xz} \ddot{\phi} + J_{s3} \dot{V} \phi + \sum_{\forall k} \left[\beta_{14k} \ddot{\phi}_k + m_k \dot{V} \beta_{5k} \phi_k + \right.$$
$$\left. + 2\dot{V} \frac{1}{R_e} J_{pwk} \phi + 2V \frac{1}{R_{ek}} J_{pwk} \dot{\phi} - 2V v_y J_{pwk} \frac{1}{R_{ek}^2} \right] = Q_\psi . \qquad (29.167)$$

Seventh and ninth equations: translation of axles

$$m_k \ddot{Z}_k - \beta_{16k} \ddot{\theta} + m_k \dot{V} \left(\theta + \beta_{3k} \right) +$$
$$+ \left(K_{22k} + K_{pzk} \right) \left(Z_k + Z_{k0} \right) - K_{12k} \left(Z + Z_0 \right) + \qquad (29.168)$$
$$+ K_{23k} \left(\theta + \theta_0 \right) + \left(c_{22k} + c_{pzk} \right) \dot{Z}_k - c_{12k} \dot{Z} + c_{23k} \dot{\theta} = -g m_k + Q_{zk} .$$

Eighth and tenth equations: rotation of axles

$$\beta_{13k}\ddot{\phi}_k + \beta_{14k}\ddot{\psi} - m_k\dot{v}_y\beta_{5k} - 2\dot{V}\frac{1}{R_{ek}}\beta_{20k}\delta_k - m_kv_x\beta_{5k}\dot{\psi} - gm_k\beta_{23k}\phi_k +$$
$$+ (K_{\phi k} + K_{p\phi k})\phi_k - K_{\phi k}\phi + (c_{\phi k} + c_{p\phi k})\dot{\phi}_k - c_{\phi k}\dot{\phi} = Q_{\phi k} \ .$$

$$(29.169)$$

29.3.4   Sideslip angles of the wheels.

The sideslip angles of the wheels can be computed directly from the components of the speed of the centers of the wheel-ground contact zone in the x^*y^*z frame. In the case of a solid axle suspension, the position of the center of the contact zone may be computed with the methods used for the center of mass of the axle, by substituting the coordinates of the center of the contact area x_{Cn}, y_{Cn}, z_{Cn} for those of the center of mass x_{Gu}, 0, z_{Gu} in Eq. (29.32):

$$(\overline{C_u - O'}) = (\overline{H - O'}) + \mathbf{R}_1\mathbf{r} \ ,$$

$$(29.170)$$

where

$$\mathbf{r} = \mathbf{R}_2\left[\mathbf{R}_3\left\{\begin{array}{c} x_u + \zeta s_x \\ 0 \\ \zeta \end{array}\right\} + \mathbf{R}_{us}\mathbf{R}_{3k}\left\{\begin{array}{c} x_{Cu} \\ y_{Cu} \\ z_{Cu} \end{array}\right\}\right] \ .$$

$$(29.171)$$

Obviously, vector x_{Cu}, y_{Cu}, z_{Cu} must be expressed in the same reference frame in which the coordinates of the center of mass x_{Gu}, 0, z_{Gu} were expressed. This procedure is approximate, because the deformations of the tire are neglected, but the approximation is not greater than those already introduced in the linearized model.

By introducing the linearized expression for ζ, it follows that

$$\mathbf{r} = \left\{\begin{array}{c} x_{Ct} + \theta z_{Cu} \\ y_{Cu} - \phi_k z_{Cu} \\ z_{Ct} + Z_k - Z - x_{Cu}\theta \end{array}\right\} \ ,$$

$$(29.172)$$

where

$$\begin{array}{rl} x_{Ct} = & x_u + x_{Cu} - z_{us}x_{Cu} + \theta_0 z_{Cu} \ , \\ z_{Ct} = & z_{Cu} + z_{us}x_{Cu} - \theta_0 x_{Cu}. \end{array}$$

$$(29.173)$$

In the case of independent suspensions, it follows (Eq. 29.113) that:

$$\mathbf{r} = \mathbf{R}_2\mathbf{R}_3\left\{\begin{array}{c} x_{Cu} + \zeta s_x \\ y_{Cu} + \zeta s_y \\ z_{Cu} + \zeta \end{array}\right\} \ .$$

$$(29.174)$$

By linearizing the expression for $\mathbf{r}$ and introducing the value of ζ, it follows that

$$\mathbf{r} = \left\{\begin{array}{c} x_{Ct} + \theta z_{Cu} \\ y_{Cu} - \phi z_{Cu} \\ z_{Ct} + Z_k - Z \end{array}\right\} \ ,$$

$$(29.175)$$

where

$$x_{Ct} = x_{Cu} + \theta_0 z_{Cu} , \quad z_{Ct} = z_{Cu} - \theta_0 x_{Cu}. \tag{29.176}$$

Note that the meaning of symbols x_{Cu}, y_{Cu}, z_{Cu} is different for the two suspension types.

To express $\mathbf{r}$ with a single equation, that holds in all cases, it is possible to write

$$\mathbf{r} = \left\{ \begin{array}{c} x_{Ct} + \theta z_{Cu} \\ y_{Cu} - \phi z_1 - \phi_k z_2 \\ z_{Ct} + Z_k - Z \end{array} \right\} , \tag{29.177}$$

where

$$z_1 = 0 , \quad z_2 = z_{Cu}, \tag{29.178}$$

in the case of solid axles, and

$$z_1 = z_{Cn} , \quad z_2 = 0, \tag{29.179}$$

in the case of independent suspensions

The velocity of the center of the contact area, expressed in the inertial frame, is

$$V_{C_N} = \left[\begin{array}{ccc} \dot{X} & \dot{Y} & \dot{Z} \end{array} \right]^T + \mathbf{R}_1 \dot{\mathbf{r}} + \dot{\mathbf{R}}_1 \mathbf{r} . \tag{29.180}$$

By premultiplying the velocity by $\mathbf{R}_1^T$ it is possible to obtain its value in the $x^* y^* z^*$ frame. Remembering Eq. (29.41), it follows that

$$V_{Cu} = \mathbf{V} + \dot{\mathbf{r}} + \dot{\psi} \mathbf{S} \mathbf{r} , \tag{29.181}$$

where the linearized expression for the derivative of $\mathbf{r}$ with respect to time is

$$\dot{\mathbf{r}} = \left\{ \begin{array}{c} \dot{\theta} z_{Cu} \\ -\dot{\phi} z_1 - \dot{\phi}_k z_2 \\ \dot{Z}_k - \dot{Z} - x_{Cu} \dot{\theta} \end{array} \right\} . \tag{29.182}$$

As a first approximation, it is possible to assume that the position of the center of the contact area of the ith tire P_i coincides with the projection on the ground of the center of the wheel. The velocity of the center of the contact area is then

$$V_{Pi} = V_{Cu} \left[\begin{array}{ccc} 1 & 0 & 0 \\ 0 & 1 & 0 \\ 0 & 0 & 0 \end{array} \right] = \left\{ \begin{array}{c} v_x + \dot{r}_x - \dot{\psi} r_y \\ v_y + \dot{r}_y + \dot{\psi} r_x \\ 0 \end{array} \right\} . \tag{29.183}$$

By performing the relevant computations and linearizing, it follows that

$$V_{Pi} = \left\{ \begin{array}{c} v_x + \dot{\theta} z_{Cu} - \dot{\psi} y_{Cu} \\ v_y - \dot{\phi} z_1 - \dot{\phi}_k z_2 + \dot{\psi} x_t \\ 0 \end{array} \right\} . \tag{29.184}$$

Because the mid-plane of the wheel is rotated by the steering angle δ_k (possibly increased by $(\delta_k)_{,\phi} (\phi - \phi_k)$ to account for roll steer) with respect to the x^*z plane, the usual linearizations allow writing

$$\alpha_k = \frac{v_y}{V} + \dot{\psi}\frac{x_{tk}}{V} - \dot{\phi}\frac{z_{1k}}{V} - \dot{\phi}_k\frac{z_{2k}}{V} - \delta_k - (\delta_k)_{,\phi}(\phi - \phi_k) , \qquad (29.185)$$

where subscript k refers to the axle.

The two wheels of the same axle are then at the same sideslip angle, as was the case for the rigid vehicle. Linearization again allows us to work in terms of axles instead of single wheels. The terms in $\dot{\phi}$ and $\dot{\phi}_k$ are usually small and will be neglected in the following equations.

The sideslip angles are then

$$\alpha_k = \frac{v_y}{V} + \dot{\psi}\frac{x_{tk}}{V} - \delta_k - (\delta_k)_{,\phi}(\phi - \phi_k) . \qquad (29.186)$$

This expression coincides with that obtained for the rigid vehicle, to which roll steer has been added.

29.3.5 Generalized forces

The generalized forces Q_k to be introduced into the equations of motion include only the forces due to tires, aerodynamic forces and possible forces that may be applied to the vehicle.

The virtual displacement of the left (right) wheel of the kth axle has an expression similar to Eq. (29.184):

$$\{\delta s_{P_{k_{L(R)}}}\}_{x^*y^*z} = \left\{ \begin{array}{c} \delta x^* + \delta\theta z_{Cu} - \delta\psi y_{Cu} \\ \delta y^* - \delta\phi z_1 - \delta\phi_k z_2 + \delta\psi x_t \\ 0 \end{array} \right\} . \qquad (29.187)$$

If coupling between vertical and horizontal displacements of the suspension must be accounted for, a term

$$\left(\frac{\partial x}{\partial z}\right)_k (\delta Z - x_t\delta\theta) = (x_k)_{,z}(\delta Z - x_t\delta\theta)$$

must be added to the x^* component of the virtual displacement.

If the forces exerted by the tire in the direction of the x^* and y^* axes are

$$F_x{}^* = F_{xp}\cos\left[\delta_i - (\delta_i)_{,\phi}\phi\right] - F_{yp}\sin\left[\delta_i - (\delta_i)_{,\phi}\phi\right] ,$$

$$F_y{}^* = F_{xp}\sin\left[\delta_i - (\delta_i)_{,\phi}\phi\right] + F_{yp}\cos\left[\delta_i - (\delta_i)_{,\phi}\phi\right] ,$$

and assuming that the longitudinal forces acting on the wheels of any axle are equal (if they are not, it is not difficult to add a yawing torque about the z axis), the expression of the virtual work is

$$\delta\mathcal{L}_k = \delta x^* F_x{}^* + \delta Z(x_k)_{,z}F_x{}^* + \delta\theta F_x{}^*(z_{Cu} - (x_k)_{,z}x_t) +$$
$$+\delta y^* F_y{}^* - \delta\phi F_y{}^* z_{1k} - \delta\phi_k F_y{}^* z_{2k} + \delta\psi\{F_y{}^* x_{tk} + M_z\} . \qquad (29.188)$$

The generalized forces may be obtained by differentiating the virtual work with respect to the virtual displacements δx^*, δy^*, $\delta \theta$, etc. The first two generalized forces are true forces directed along axes x^* and y^* and a suitable rotation matrix can be used to obtain the forces along the axes of the inertial frame.

Force $F_{y_{p_i}}$ on the ith tire may be expressed as a linear function of the sideslip and camber angles α_k and

$$\gamma_{0k} + (\gamma_k)_{,\phi}\phi + (\gamma_k)_{,\phi_k}\phi_k + (\gamma_k)_{,z}(Z - Z_k) .$$

Terms γ_{0k} and $(\gamma_k)_{,z}(Z - Z_k)$ for the two wheels of any axle cancel each other in the linearized model, because they produce equal and opposite forces. The side forces applied on each axle are then

$$F_{ypk} = -C_k\alpha_k + C_{\gamma k}\left[(\gamma_k)_{,\phi}\phi + (\gamma_k)_{,\phi k}\phi_k\right] , \qquad (29.189)$$

where both C_k and $C_{\gamma k}$ are referred to the whole axle.

Assume that aerodynamic forces are applied to the center of mass of the sprung mass. The virtual displacement of such a point in the x^*y^*z frame is

$$\{\delta s_{G_s}\}_{x^*y^*z} = \left\{ \begin{array}{c} \delta x^* + h\delta\theta + h\phi\delta\psi \\ \delta y^* - h\delta\phi + (c + h\theta_0 + h\theta)\,\delta\psi \\ \delta Z - c\delta\theta \end{array} \right\} . \qquad (29.190)$$

The virtual work of aerodynamic forces and moments is

$$\begin{aligned}
\delta\mathcal{L}_a = {}& F_{xa}\delta x^* + F_{ya}\delta y^* + F_{za}\delta Z + \\
& + \left(M'_{xa} - F_{ya}h\right)\delta\phi + \left(F_{xa}h - F_{za}c + M'_{ya} + M'_{za}\phi\right)\delta\theta + \\
& + \left[F_{xa}h\phi + F_{ya}\left(c + h\theta_0 + h\theta\right) - M'_{ya}(\theta_0 + \theta - \phi) + M'_{za}\right]\delta\psi .
\end{aligned} \qquad (29.191)$$

It is also possible to directly obtain the generalized forces by differentiating the virtual work with respect to the virtual displacements.

Because the aerodynamic forces are applied in the center of mass of the sprung mass instead of the center of mass of the vehicle, the aerodynamic moments referred to the former must be substituted for those defined in the usual way, that is, with reference to the center of mass of the vehicle:

$$\left\{ \begin{array}{l} M_{xa} = M'_{xa} - F_{ya}h , \\ M_{ya} = M'_{ya} + F_{xa}h , \\ M_{za} = M'_{za} - F_{ya}(c + h\theta_0) . \end{array} \right. \qquad (29.192)$$

Due to the linearization of the vehicle, force F_{xa} may be considered as a constant, while F_{ya}, M_{xa} and M_{za} may be considered as linear with angle β_a (if there is no side wind, with angle β), while F_{za} and M_{ya} may be considered as linear with angle θ. Neglecting small terms, it follows that

$$\delta\mathcal{L}_{aer} = F_{xa}\delta x^* + \frac{\partial F_{ya}}{\partial\beta}\beta\delta y^* + \frac{\partial F_{za}}{\partial\theta}\theta\delta Z + \qquad (29.193)$$

$$+ \left(\frac{\partial M_{xa}}{\partial\beta} - \frac{\partial F_{ya}}{\partial\beta}\beta h\right)\delta\phi + \frac{\partial M_{ya}}{\partial\theta}\delta\theta + \left(F_{xa}h\phi + \frac{\partial M_{za}}{\partial\beta}\beta\right)\delta\psi .$$

The vector of the generalized forces may be obtained by differentiating the virtual work with respect to the virtual displacements, and eliminating the terms containing generalized forces multiplied by variables of motion, which would lead to nonlinear terms once the generalized forces are expressed as functions of the same variables

$$
Q = \left\{
\begin{array}{c}
F_{x1} + F_{x2} + F_{xa} \\[2mm]
\sum_{\forall k} \{-C_k \alpha_k + C_{\gamma k} [(\gamma_k)_{,\phi} \phi + (\gamma_k)_{,\phi k} \phi_k]\} + \frac{\partial F_{ya}}{\partial \beta} \beta \\[2mm]
-(x_{i1})_{,z} F_{x1} - (x_{i2})_{,z} F_{x2} + \frac{\partial F_{za}}{\partial \theta} \theta + (F_{za})_{\theta=0} \\[2mm]
\sum_{\forall k} z_{1,k} \{-C_k \alpha_k + C_{\gamma k} [(\gamma_k)_{,\phi} \phi + (\gamma_k)_{,\phi k} \phi_k]\} - \frac{\partial M_{xa}}{\partial \beta} - \frac{\partial F_{ya}}{\partial \beta} \beta h \\[2mm]
F_{x1}(z_{Cn1} - (x_{i1})_{,z} x_{t1}) + F_{x2}(z_{Cn2} - (x_{i2})_{,z} x_{t2}) + \frac{\partial M_{ya}}{\partial \theta} \theta + \\ + (M_{ya})_{\theta=0} \\[2mm]
\sum_{\forall k} \left(\frac{\partial M_{x1}}{\partial \alpha} \alpha_1 + x_{wiA} \{-C_k \alpha_k + C_{\gamma k} [(\gamma_k)_{,\phi} \phi + (\gamma_k)_{,\phi k} \phi_k]\} \right) + \\ + F_{xa} h \phi + \frac{\partial M_{za}}{\partial \beta} \beta \\[2mm]
0 \\[2mm]
z_{2,1} \{-C_1 \alpha_1 + C_{\gamma 1} [(\gamma_1)_{,\phi} \phi + (\gamma_1)_{,\phi 1} \phi_1]\} \\[2mm]
0 \\[2mm]
z_{2,2} \{-C_2 \alpha_2 + C_{\gamma 2} [(\gamma_2)_{,\phi} \phi + (\gamma_k)_{,\phi 2} \phi_2]\}
\end{array}
\right\} .
$$

$$\tag{29.194}$$

The term

$$
F_{x1} [\delta_1 - (\delta_1)_{,\phi} (\phi - \phi_1)] + F_{x2} [\delta_2 - (\delta_2)_{,\phi \phi} (\phi - \phi_2)]
$$

should be included in the generalized force Q_y. It results from the component of the longitudinal force of the tire in the direction of the y axis of the vehicle due to the steering angle. It is a small term, owing to the small size of the longitudinal force F_x when compared to the cornering stiffness, and is usually neglected. A similar term should also be included in Q_ψ, but is usually neglected as well.

The vector of the generalized forces so obtained may be used directly in the equation of motion, because it is referred to the pseudo-coordinates x^*, y^* and to coordinates Z, ϕ, θ, ψ, etc.

29.3.6 Final form of the equations of motion

Remembering that the steering angles are small, it is easy to pass from the forces expressed in a frame fixed to the vehicle to one fixed to the tires. The equations of motion then take their form.

First equation: longitudinal translation

$$m_e \dot{V} + \dot{\theta} J_{s2} = F_{x1} + F_{x2} - \frac{1}{2}\rho V^2 SC_x \ . \tag{29.195}$$

Second equation: lateral translation

The sideslip angles, and then the cornering forces, may be easily expressed as functions of the variables of motion. Assuming that the steering angles of the axles are proportional to a reference value δ through constants K'_k:

$$\delta_k = K'_k \delta \tag{29.196}$$

and adding a side force F_{ye} applied to the vehicle, it follows that

$$Q_y = Y_v v_y + Y_w \dot{\psi} + Y_\phi \phi + Y_{\phi 1}\phi_1 + Y_{\phi 2}\phi_2 + Y_\delta \delta + F_{y_e} \ , \tag{29.197}$$

where

$$\begin{cases} Y_v = -\frac{1}{V}\sum_{\forall k} C_k + \frac{1}{2}\rho V_a S(C_y)_{,\beta} \ , \\ Y_w = -\frac{1}{V}\sum_{\forall k} x_{wk} C_k \ , \\ Y_\phi = \sum_{\forall k} C_i(\delta_k)_{,\phi} + \sum_{\forall k} C_{\gamma k}(\gamma_k)_{,\phi} \ , \\ Y_{\phi k} = C_{\gamma k}(\gamma_k)_{,\phi} \ , \\ Y_\delta = \sum_{\forall k} K'_k C_k \ . \end{cases} \tag{29.198}$$

The second equation then becomes

$$m \dot{v}_y - J_{s3}\ddot{\phi} - \sum_{\forall k} m_k \beta_{5k}\ddot{\phi}_k = Y_v v_y + Y_{\dot{\psi}}\dot{\psi} +$$

$$+ Y_\phi \phi + \sum_{\forall k} Y_{\phi_k}\phi_k + Y_\delta \delta + F_{y_e} \ , \tag{29.199}$$

where

$$Y_{\dot{\psi}} = Y_w - m_e V \ . \tag{29.200}$$

Third equation: vertical translation

By introducing the generalized forces into the third equation, it follows that

$$m_s \ddot{Z} - m_s c\ddot{\theta} + Z_{\dot{z}}\dot{Z} + Z_{\dot{\theta}}\dot{\theta} + Z_{\dot{z}1}\dot{Z}_1 + Z_{\dot{z}2}\dot{Z}_2 + Z_z Z + Z_\theta \theta +$$

$$+ Z_{z1}Z_1 + Z_{z2}Z_2 + \sum_{\forall k}(K_{11k}Z_0 - K_{12k}Z_{0k} - K_{13}\theta_0) = -m_s g + \tag{29.201}$$

$$+ \dot{V}\sum_{\forall k} m_k \beta_{3k} + \frac{1}{2}\rho V^2 S(C_Z)_{\theta=0} - (x_{i1})_{,z} F_{x1} - (x_{i2})_{,z} F_{x2} \ ,$$

where

$$
\begin{cases}
Z_{\dot{z}} = \sum_{\forall k} c_{11k} \ , & Z_{\dot{\theta}} = -\sum_{\forall k} c_{13k} \ , \\
Z_{\dot{z}k} = -c_{12k} \ , & Z_z = \sum_{\forall k} K_{11k} \ , \\
Z_\theta = -\sum_{\forall k} K_{13k} - \frac{1}{2}\rho V^2 S(C_z)_{,\theta} - \dot{V}\sum_{\forall k} m_k \ , \\
Z_{zk} = -K_{12k} \ .
\end{cases}
\tag{29.202}
$$

Fourth equation: roll rotation

Operating with the same methods used for the second equation, and linearizing the kinematic equations, the generalized forces may be written as functions of the variables of motion. The final form of the fourth equation can thus be obtained,

$$
J_x \ddot{\phi} - J_{xz}\ddot{\psi} - J_{s3}\dot{v}_y = L_v v_y + L_{\dot{\psi}}\dot{\psi} + L_{\dot{\phi}}\dot{\phi} + L_\phi \phi +
\tag{29.203}
$$

$$
+ \sum_{\forall k} L_{\dot{\phi}k}\dot{\phi}_k + \sum_{\forall k} L_{\phi k}\phi_k + L_\delta \delta \ ,
$$

where

$$
\begin{cases}
L_v = -\frac{1}{2}\rho V_a S\left[h(C_y)_{,\beta} + t(C_{M_x})_{,\beta}\right] - \frac{1}{V}\sum_{\forall k} C_k z_{1k} \ , \\
L_{\dot{\psi}} = -\frac{1}{V}\sum_{\forall k} x_{wk} z_{1k} C_k + \sum_{\forall k} 2V \frac{1}{R_{ek}} J_{pwk} + J_{s3} V \ , \\
L_{\dot{\phi}} = -\sum_{\forall k} c_{\phi k} \ , \\
L_\phi = M_{g3} - \sum_{\forall k}\left[K_{\phi k} + C_k(\delta_k)_{,\phi} z_{1k} + C_{\gamma k}(\gamma_k)_{,\phi} z_{1k}\right] \ , \\
L_{\dot{\phi}k} = c_{\phi k} \ , \\
L_{\phi k} = K_{\phi k} - C_{\gamma k}(\gamma_k)_{,\phi k} z_{1k} + C_k(\delta_k)_{,\phi} z_{1k} \ , \\
L_\delta = \sum_{\forall k}\left(K'_k C_k z_{1k} + 2\frac{\dot{V}}{R_e}\beta_{21}\right) \ .
\end{cases}
\tag{29.204}
$$

Fifth equation: pitch rotation

$$
J_y \ddot{\theta} - m_s c\ddot{Z} + \dot{V} J_{s2} - \sum_{\forall k}\beta_{16k}\ddot{Z}_k + M_{\dot{z}}\dot{Z} + M_{\dot{\theta}}\dot{\theta} + \sum_{\forall k} M_{\dot{z}k}\dot{Z}_k +
$$

$$
+ M_z Z + M_\theta \theta + \sum_{\forall k} M_{zk} Z_u + K_{33k}\theta_0 - K_{13k}Z_0 + K_{23k}Z_{k0} = M_{g1} +
$$

$$
+ \left(M_{y_{aer}}\right)_{\theta=0} - \sum_{\forall k} F_{x1}\left(z_{Cu1} - (x_{i1})_{,z} x_{t1}\right) + F_{x2}\left(z_{Cu2} - (x_{i2})_{,z} x_{t2}\right) \ ,
\tag{29.205}
$$

where

$$
\begin{cases}
M_{\dot{z}} = Z_{\dot{\theta}} = -\sum_{\forall k} c_{13k} \ , & M_{\dot{\theta}} = \sum_{\forall k} c_{33k} \ , \\
M_{\dot{z}k} = c_{23k} \ , & M_z = -\sum_{\forall k} K_{13k} - m_k \dot{V} \ , \\
M_\theta = -\frac{1}{2}\rho V_a^2 S(C'_{M_y})_{,\theta} - M_{g2} + \sum_{\forall k} K_{33k} \ , \\
M_{zk} = K_{23k} + m_k \dot{V} \ .
\end{cases}
\tag{29.206}
$$

Sixth equation: yaw rotation

$$
J_z \ddot{\psi} - J_{xz}\ddot{\phi} + \sum_{\forall k}\beta_{14k}\ddot{\phi}_k = N_v v_y + N_{\dot{\psi}}\dot{\psi} + N_{\dot{\phi}}\dot{\phi} +
\tag{29.207}
$$

$$
+ N_\phi \phi + \sum_{\forall k} N_{\phi k}\phi_k + N_\delta \delta + M_{ze} \ ,
$$

where

$$
\left\{
\begin{array}{l}
N_v = +\frac{1}{2}\rho V_a Sl(C'_{M_z}),_\beta + \frac{1}{V}\sum_{\forall k}\left[-x_{wk}C_k + (M_{zk}),_\alpha + 2J_{pwk}\left(\frac{V}{R_e}\right)^2\right] , \\
N_{\dot{\psi}} = \frac{1}{V}\sum_{\forall k}\left[x_{wk}^2 C_k + x_{wk}(M_{zk}),_\alpha\right] , \\
N_{\dot{\phi}} = -2V\sum_{\forall k}\frac{1}{R_e}J_{pwk} , \\
N_\phi = -J_{s3}\dot{V} - \frac{1}{2}\rho V^2 ShC_x + \sum_{\forall k}\left[x_{wk}C_k(\delta_k),_\phi + x_{wk}C_{\gamma k}(\gamma_k),_\phi \right. \\
\qquad\qquad \left. -(M_{zk}),_\alpha(\delta_k),_\phi - 2\dot{V}\frac{1}{R_e}J_{pwk}\right] , \\
N_{\phi k} = x_{wk}C_{\gamma k}(\gamma_k),_\phi - m_k\dot{V}\beta_{5k} - x_{wk}C_k(\delta_k),_\phi + (M_{zk}),_\alpha(\delta_k),_\phi , \\
N_\delta = \sum_{\forall k}\left[x_{wk}K'_k C_k - (M_{zk}),_\alpha\right] .
\end{array}
\right.
$$

$$(29.208)$$

Seventh and ninth equations: translations of axles

$$
m_k\ddot{Z}_k - \beta_{16k}\ddot{\theta} + Z_{kzk}Z_k + Z_{zk}Z + M_{zk}\theta +
$$

$$
+Z_{k\dot{z}k}\dot{Z}_k + Z_{\dot{z}k}\dot{Z} + M_{\dot{z}k}\dot{\theta} + (K_{22k} + K_{pzk})Z_{k0} + \qquad (29.209)
$$

$$
-K_{12k}Z_0 + K_{23k}\theta_0 = -m_k\dot{V}\beta_{3k} - gm_k ,
$$

where

$$
Z_{k\dot{z}k} = c_{22k} + c_{pzk} , \qquad\qquad Z_{kzk} = K_{22k} + K_{pzk} \qquad (29.210)
$$

and the other coefficients have already been defined.

Eighth and tenth equation: rotation of axles

$$
\beta_{13k}\ddot{\phi}_k + \beta_{14k}\ddot{\psi} - m_{uk}\dot{v}_y\beta_{5k} =
$$

$$
= L_{k\beta}\beta + L_{k\dot{\psi}}\dot{\psi} + L_{k\phi}\phi + L_{k\dot{\phi}}\dot{\phi} + L_{k\phi k}\phi_i + L_{k\dot{\phi}k}\dot{\phi}_i + L_{k\delta}\delta ,
$$

where

$$
\left\{
\begin{array}{l}
L_{kv} = -\frac{1}{V}z_{2k}C_k , \\
L_{k\dot{\psi}} = +m_k V\beta_{5k} - \frac{1}{V}x_{wk}z_{2k}C_k , \\
L_{k\dot{\phi}} = L_{\dot{\phi}k} , \\
L_{k\dot{\phi}k} = -(c_{\phi k} + c_{p\phi k}) , \\
L_{k\phi} = +K_{\phi k} - z_{2k}\left[C_k(\delta_k),_\phi + C_{\gamma k}(\gamma_k),_\phi\right] , \\
L_{k\phi k} = +gm_k\beta_{23k} - (K_{\phi k} + K_{p\phi k}) - z_{2k}\left[C_{\gamma k}(\gamma_k),_\phi - C_k(\delta_k),_\phi\right] , \\
L_{k\delta} = 2K'_k\dot{V}\frac{1}{R_{ek}}\beta_{20k} .
\end{array}
\right.
$$

$$(29.211)$$

29.3.7 Handling-comfort uncoupling

The 10 equations of motion ($6 + 2n$ equations in the generic case of a vehicle with n axles) obtained in the previous section constitute a set of linear second-order differential equations, even if the order of such a set is only 17 ($9 + 4n$) because three of the unknowns, namely x^*, y^* and ψ, are present only with their derivatives V, v_y and $\dot{\psi}$.

However, a detailed examination of such equations shows clearly that, if the speed V of the vehicle (which in the linearized model may be confused with its component v_x along the x^* axis) is a known function of time, the equations form two completely uncoupled sets of 5 ($3 + n$) equations each.

The first set contains only the generalized coordinates y^*, ψ, ϕ and ϕ_k (y^* is not a true but a pseudo-coordinate): as a consequence, it deals with the lateral behavior of the vehicle, or, as is usually said, its handling.

The second set contains the generalized coordinates x^*, Z, θ and Z_k, dealing with the "suspension motion" of the vehicle — its ride behavior. This set can be further uncoupled by separating the first equation, that regarding x^* coordinate (i.e. dealing with the longitudinal dynamics of the vehicle), and the following ($2 + n$) equations containing coordinates Z, θ and Z_k which allow ride comfort in a proper sense to be studied.

This uncoupling is an interesting result, even if it is strictly linked with a number of assumptions and, as a consequence, becomes inapplicable if one of them is dropped. The first assumption is the existence of a plane of symmetry, the xz plane. Usually the lack of inertial symmetry of the structure and the differences between the characteristics of the individual springs and shock absorbers located at opposite sides of the vehicle are small enough to be neglected. However, it can happen that the payload of the vehicle is placed asymmetrically, leading to a position of the centre of mass outside the symmetry plane and to non-vanishing moments of inertia J_{xy} and J_{yz}.

A second assumption is that of a perfect linearity of the behavior of the springs and shock absorbers. The linearity of the elastic behavior of springs and tires is an acceptable assumption in the motion about any equilibrium position, provided that its amplitude is small. The nonlinearity of the shock absorbers, on the other hand, cannot in principle be neglected even in the motion "in the small" if their force-velocity characteristic is unsymmetrical, because in the jounce and rebound movements they act with different damping coefficients even if the amplitude of the motion tends to zero. This issue has already been dealt with in detail in Part IV.

A third assumption regards angles β, α_k, θ_0, θ, ϕ and ϕ_k, which must be small enough to allow the linearization of their trigonometric functions. This assumption holds only for small displacements from the equilibrium position and also depends on the characteristics of the vehicle: The harder the suspensions, the more extended the range in which this assumption holds. In general, the mentioned angles are small enough in all normal driving conditions, except for vehicles with two wheels that may operate with large roll angles.

The linearization of the tire behavior in terms of the generation of longitudinal and cornering forces and aligning torques is not strictly required for uncoupling: Even if nonlinear laws $F_y(\alpha)$, $F_y(\gamma)$, $M_z(\alpha)$, etc. are introduced into the equations of motion, the two sets of equations for handling and ride would remain uncoupled, although nonlinear. This last statement is important, because the linear model for the behavior of the tires holds only for values of angles α and γ far smaller than those allowing the trigonometric functions to be linearized.

The kinetic energy linked with wheel rotation was taken into account in the model, with gyroscopic torques due to the wheels included in the equations. If their plane of rotation is close to the xz plane, this effect does not prevent uncoupling.

Some assumptions have been made on the modelling of the suspensions that are better suited for solid axles than for independent suspensions. While unavoidable kinematic errors cannot be accounted for in this way, it will be shown that this does not affect uncoupling.

The interaction between cornering forces and loads in the x and z direction on the tires should actually couple all equations. If the same approximated approach used for rigid vehicles is also adopted in the present case, however, it is possible to resort to uncoupled equations.

The uncoupled model, even if it represents only a first approximation, is important for two reasons. First, it sheds light on the actual behavior of road vehicles and gives a theoretical foundation to the practice of using separate approximate models for the study of handling and ride characteristics. Second, simple linearized models, allowing closed form solutions to be obtained, are well suited for optimization and parametric studies.

Clearly, there is no need to uncouple the equations. Comprehensive, detailed nonlinear models can be used if numerical simulations are performed. The limit in this case may well be the unavailability of good estimates of the numerical values of many parameters that must be entered into the equations.

29.3.8 Handling of a vehicle on elastic suspensions

The explicit formulation of the mathematical model for the handling of a vehicle with two axles is then

$$\mathbf{M}_1\ddot{\mathbf{q}}_1 + \mathbf{C}_1\dot{\mathbf{q}}_1 + \mathbf{K}_1\mathbf{q}_1 = \mathbf{F}_1 , \qquad (29.212)$$

where

$$\mathbf{q}_1 = \begin{bmatrix} y^* & \psi & \phi & \phi_1 & \phi_2 \end{bmatrix}^T ,$$

$$\mathbf{M}_1 = \begin{bmatrix} m & 0 & -J_{s3} & -m_1\beta_{5,1} & -m_2\beta_{5,2} \\ & J_z & -J_{xz} & \beta_{14,1} & \beta_{14,2} \\ & & J_x & 0 & 0 \\ & & & \beta_{13,1} & 0 \\ \text{symm.} & & & & \beta_{13,2} \end{bmatrix} ,$$

$$
\mathbf{C}_1 = \begin{bmatrix}
-Y_v & -Y_{\dot\psi} & 0 & 0 & 0 \\
-N_v & -N_{\dot\psi} & -N_{\dot\phi} & 0 & 0 \\
-L_v & -L_{\dot\psi} & -L_{\dot\phi} & -L_{\dot\phi_1} & -L_{\dot\phi_2} \\
-L_{1v} & -L_{1\dot\psi} & -L_{\dot\phi_1} & -L_{1\dot\phi 1} & 0 \\
-L_{2v} & -L_{2\dot\psi} & -L_{\dot\phi_2} & 0 & -L_{2\dot\phi 2}
\end{bmatrix} ,
$$

$$
\mathbf{K}_1 = \begin{bmatrix}
0 & 0 & -Y_\phi & -Y_{\phi_1} & -Y_{\phi_2} \\
0 & 0 & -N_\phi & -N_{\phi_1} & -N_{\phi_2} \\
0 & 0 & -L_\phi & -L_{\phi_1} & -L_{\phi_2} \\
0 & 0 & -L_{1\phi} & -L_{1\phi 1} & 0 \\
0 & 0 & -L_{2\phi} & 0 & -L_{1\phi 2}
\end{bmatrix} ,
$$

$$
\mathbf{F}_1 = \delta \begin{bmatrix} Y_\delta & N_\delta & L_\delta & L_{1\delta} & L_{2\delta} \end{bmatrix}^T + \begin{bmatrix} F_{y_e} & M_{z_e} & 0 & 0 & 0 \end{bmatrix}^T .
$$

As already stated, coordinates y^* and ψ are present only with their derivatives[7]: The order of the set of differential equations is then 8 instead of 10.

The mass matrix is symmetrical, as can be readily predicted. The other two matrices are not symmetrical; for instance, the damping matrix $\mathbf{C}_1$ contains symmetrical terms, like $L_{\dot\phi_1}$ and $L_{\dot\phi_2}$ that are linked with the roll damping of the axles, and skew symmetric terms, like

$$
2V \frac{1}{R_{ek}} J_{pwk} ,
$$

due to the gyroscopic moment of the wheels of each axle, contained in $L_{\dot\psi}$ and, with opposite sign, in $N_{\dot\phi}$. The other terms in J_{pwk} are not due to gyroscopic effects but wheel acceleration (terms in L_δ and $L_{k\delta}$) or the equivalent mass (term in N_v), and thus have no particular symmetry properties. Other terms due to generalized forces, such as the terms present in the stiffness matrix, are neither symmetrical nor skew symmetrical.

Even if it is possible to separate the symmetrical and the skew-symmetrical parts of the various matrices (so defining a gyroscopic and a circulatory matrix), the advantages so obtained do not justify the work.

If a state-space approach is used (Eq. (A.5)), by introducing the state variables $p = \dot\phi$, $p_1 = \dot\phi_1$, $p_2 = \dot\phi_2$ and $r = \dot\psi$, the relevant vectors and matrices are:

– State vector

$$
\mathbf{z} = \begin{bmatrix} v & r & p & p_1 & p_2 & \phi & \phi_1 & \phi_2 \end{bmatrix}^T , \tag{29.213}
$$

– Dynamic matrix

$$
\mathbf{A} = \begin{bmatrix}
-\mathbf{M}_1^{-1}\mathbf{C}_1 & -\mathbf{M}_1^{-1}\mathbf{K}_1^* \\
\begin{bmatrix} 0 & 0 & 1 & 0 & 0 \\ 0 & 0 & 0 & 1 & 0 \\ 0 & 0 & 0 & 0 & 1 \end{bmatrix} & \begin{bmatrix} 0 & 0 & 0 \\ 0 & 0 & 0 \\ 0 & 0 & 0 \end{bmatrix}
\end{bmatrix} , \tag{29.214}
$$

[7]They have no physical meaning.

where $\mathbf{K}_1^*$ is matrix $\mathbf{K}_1$ with the first two columns cancelled.
– input gain matrix

$$\mathbf{B} = \begin{bmatrix} -\mathbf{M}_1^{-1} \begin{bmatrix} Y_\delta & 1 & 0 \\ N_\delta & 0 & 1 \\ L_\delta & 0 & 0 \\ L_{1_\delta} & 0 & 0 \\ L_{2_\delta} & 0 & 0 \end{bmatrix} \\ [0]_{3\times 3} \end{bmatrix} , \tag{29.215}$$

– input vector

$$\mathbf{u} = \begin{bmatrix} \delta & F_{y_e} & M_{z_e} \end{bmatrix}^T . \tag{29.216}$$

This approach may be used for the study of the stability of the vehicle or for computing its response to the various inputs, as previously seen for rigid vehicle models. This model is only marginally more complex if numerical solutions are searched.

Even if the complexity of the model is not a factor, it is interesting to perform a simplification allowing one to reduce its size without sacrificing its applicability to actual problems. Because the stiffness of the tires in the z direction is much higher than that of the suspensions, their compliance becomes important only in high frequency motions, much higher than the frequencies involved in the handling of the vehicle. As a consequence, if the compliance of the tires is neglected, which amounts to stating that ϕ_1 and ϕ_2 and their derivatives are vanishingly small, the model reduces to a set of three equations (four first-order equations in the state-space approach) that retains most of the features of the complete, five equation set.

Because yawing moments due to load shift were not included in the present model, the three equations of motion may be obtained directly from those of the previous model by stating

$$\phi_1 = \phi_2 = 0 .$$

The sideslip angle β is often used in handling models instead of the lateral velocity v_y as a variable of motion, and the yaw velocity is indicated as r. Remembering that

$$v_y = V\beta , \quad \dot{\psi} = r ,$$

it follows that

$$\begin{cases} mV\dot{\beta} - J_{s3}\ddot{\phi} = \left(VY_v - m\dot{V} \right)\beta + Y_{\dot{\psi}}r + Y_\phi\phi + Y_\delta\delta + F_{y_e} , \\ J_z\dot{r} - J_{xz}\ddot{\phi} = N_vV\beta + N_{\dot{\psi}}r + N_{\dot{\phi}}\dot{\phi} + N_\phi\phi + N_\delta\delta + M_{z_e} , \\ J_x\ddot{\phi} - J_{xz}\dot{r} - J_{s3}V\dot{\beta} = \left(L_vV + J_{s3}\dot{V} \right)\beta + L_{\dot{\psi}}r + L_{\dot{\phi}}\dot{\phi} + L_\phi\phi + L_\delta\delta , \end{cases} \tag{29.217}$$

where terms Y_vV, N_vV and L_vV are often written as Y_β, N_β and L_β.

If the terms in $\dot{V}$ are dropped, the same set of equations frequently described in the literature[8] is obtained. There is, however, a difference: The model described here is obtained from the complete model of the vehicle with elastic suspensions through uncoupling and controlled simplifications, while that model is obtained through a number of more or less arbitrary assumptions. Moreover, this model accounts for the rotation of the wheels.

The study of either the stability or the response to a steering input or external force or moment is straightforward and follows the same lines seen for the rigid vehicle. Here the presence of an equation containing the first and second derivative of a generalized coordinate ϕ together with the coordinate itself may induce an oscillatory behavior. If roll oscillations are strongly coupled with those of the other variables of the motion (namely β and r), as may be caused by roll steer, the overall behavior may become strongly oscillatory and dynamic stability may be decreased.

The steady-state response of the vehicle is easily obtained from the following set of algebraic equations:

$$\begin{cases} m\dfrac{V^2}{R} = Y_\beta \beta + Y_{\dot{\psi}}\dfrac{V}{R} + Y_\phi \phi + Y_\delta \delta + F_{y_e} \,, \\[2ex] 0 = N_\beta \beta + N_{\dot{\psi}}\dfrac{V}{R} + N_\phi \phi + N_\delta \delta + M_{z_e} \,, \\[2ex] -J_{s3}\dfrac{V^2}{R} = L_\beta \beta + L_{\dot{\psi}}\dfrac{V}{R} + L_\phi \phi + L_\delta \delta \,, \end{cases} \qquad (29.218)$$

where the steady-state curvature of the trajectory

$$\frac{1}{R} = \frac{r}{V}$$

has been explicitly introduced.

By solving equations (29.218) in $1/R$ and neglecting external forces, the path curvature gain $1/R\delta$ is readily obtained,

$$\frac{1}{R\delta} = \frac{DC - AE}{V(BC - AF)} \,, \qquad (29.219)$$

where

$$A = N_\beta L_\phi - N_\phi L_\beta, \qquad B = J_{s3}VN_\phi - N_{\dot{\psi}}L_\phi + N_\phi L_{\dot{\psi}},$$
$$C = Y_\beta N_\phi - Y_\phi N_\beta, \qquad D = N_\delta L_\phi - N_\phi L_\delta,$$
$$E = Y_\delta N_\phi - Y_\phi N_\delta, \qquad F = mVN_\phi - Y_{\dot{\psi}}N_\phi + Y_\phi N_{\dot{\psi}} \,.$$

---

[8] See for example W. Steeds, *Mechanics of Road Vehicles*, ILIFFE & Sons, London, 1960.

29.3.9 Ride comfort

The explicit formulation of the mathematical model for ride comfort of a vehicle with two axles is

$$\mathbf{M}_2\ddot{\mathbf{q}}_2 + \mathbf{C}_2\dot{\mathbf{q}}_2 + \mathbf{K}_2\mathbf{q}_2 + \mathbf{K}_{2st}\mathbf{q}_{2st} = \mathbf{F}_2 + \mathbf{F}_{2st}\,, \qquad (29.220)$$

where

$$\mathbf{q}_2 = \begin{bmatrix} x^* & Z & \theta & Z_1 & Z_2 \end{bmatrix}^T,$$

$$\mathbf{q}_{2st} = \begin{bmatrix} 0 & Z_0 & \theta_0 & Z_{10} & Z_{20} \end{bmatrix}^T,$$

$$\mathbf{M}_2 = \begin{bmatrix} m_{at} & 0 & J_{s2} & 0 & 0 \\ & m_s & -m_s c & 0 & 0 \\ & & J_y & -\beta_{16,1} & -\beta_{16,2} \\ & & & m_1 & 0 \\ & \text{symm.} & & & m_2 \end{bmatrix},$$

$$\mathbf{C}_2 = \begin{bmatrix} 0 & 0 & 0 & 0 & 0 \\ 0 & Z_{\dot{z}} & Z_{\dot{\theta}} & Z_{\dot{z}1} & Z_{\dot{z}2} \\ 0 & & M_{\dot{\theta}} & M_{\dot{z}1} & M_{\dot{z}2} \\ 0 & & & Z_{1\dot{z}1} & 0 \\ 0 & & \text{symm.} & & Z_{2\dot{z}2} \end{bmatrix},$$

$$\mathbf{K}_2 = \begin{bmatrix} 0 & 0 & 0 & 0 & 0 \\ 0 & Z_z & Z_\theta & Z_{z1} & Z_{z2} \\ 0 & M_z & M_\theta & M_{z1} & M_{z2} \\ 0 & Z_{z1} & M_{z1} & Z_{1z1} & 0 \\ 0 & Z_{z2} & M_{z3} & 0 & Z_{2z2} \end{bmatrix},$$

$$\mathbf{K}_{st} = \begin{bmatrix} 0 & 0 & 0 & 0 & 0 \\ 0 & K_{11} & K_{13} & -K_{12,1} & -K_{12,2} \\ 0 & & K_{33} & K_{23,1} & K_{23,2} \\ 0 & & & K_{22,1} & 0 \\ 0 & & \text{symm.} & & K_{22,2} \end{bmatrix},$$

$$\mathbf{F}_2 = \left\{ \begin{array}{c} F_{x1} + F_{x2} - \frac{1}{2}\rho V^2 S C_x \\ \frac{1}{2}\rho V^2 S (C_Z)_{\theta=0} + \dot{V}\sum_{\forall k}[m_k \beta_{3k} - (x_{ik}),_z F_{xk}] \\ (M_{y\,aer})_{\theta=0} - \sum_{\forall k} F_{xk}(z_{Cuk} - (x_{ik}),_z x_{tk}) \\ -m_1 \dot{V}\beta_{3,1} \\ -m_2 \dot{V}\beta_{3,2} \end{array} \right\},$$

$$\mathbf{F}_{2st} = \begin{bmatrix} 0 & -m_s g & M_{g1} & -gm_1 & -gm_2 \end{bmatrix}^T.$$

In the reference condition, all variables of motion included in vector $\mathbf{q}_2$ vanish. It then follows that

$$\mathbf{K}_{2st}\mathbf{q}_{2st} = \mathbf{F}_{2st}\,. \qquad (29.221)$$

Equation (29.221) allows the values of Z_0, θ_0, etc., — the static equilibrium condition — to be computed. Although it is arbitrary to use the linearized equation for computing the static equilibrium condition, because Z_0 and Z_{k0} are not, generally, small quantities, this approximation influences the reference condition so obtained but is immaterial for the study of the small oscillations about that condition and thus does not detract from the dynamic study in the small here shown.

By introducing Eq. (29.221) into Eq. (29.220), it follows that

$$\mathbf{M}_2\ddot{\mathbf{q}}_2 + \mathbf{C}_2\dot{\mathbf{q}}_2 + \mathbf{K}_2\mathbf{q}_2 = \mathbf{F}_2 \ . \tag{29.222}$$

The mass and damping matrices are symmetrical. The stiffness matrix is symmetrical except for the terms in position 23 and 32: in Z_θ a term of aerodynamic origin is present, due to changes to aerodynamic lift caused by the pitch angle that is absent from M_z. A similar term in M_z would denote a change in the pitching moment due to vertical displacements that does not exist.

The first equation of the second set of five differential equations, that related to the longitudinal dynamics, is weakly coupled with the others and may be written in the form

$$\dot{V} = \frac{J_s\ddot{\theta} + \sum_{\forall i} F_{xi} + \frac{1}{2}\rho V_a^2 S C_x}{m} \ . \tag{29.223}$$

By introducing Eq. (29.223) into the other equations, the following set of four equations describing the suspension motions of a vehicle with two axles is obtained,

$$\mathbf{M}_3\ddot{\mathbf{q}}_3 + \mathbf{C}_3\dot{\mathbf{q}}_3 + \mathbf{K}_3\mathbf{q}_3 = \mathbf{F}_3 \ , \tag{29.224}$$

where

$$\mathbf{q}_3 = \begin{bmatrix} Z & \theta & Z_1 & Z_2 \end{bmatrix}^T ,$$

$$\mathbf{M}_3 = \begin{bmatrix} m_s & -m_s c & 0 & 0 \\ & J_y - \dfrac{J_{s2}^2}{m_{at}} & 0 & 0 \\ & & m_1 & 0 \\ & \text{symm.} & & m_2 \end{bmatrix} ,$$

matrices $\mathbf{C}_3$ and $\mathbf{K}_3$ coincide with matrices $\mathbf{C}_2$ and $\mathbf{K}_2$ without the first row and column, and

$$\mathbf{F}_3 = \left\{ \begin{array}{c} \frac{1}{2}\rho V^2 S (C_Z)_{\theta=0} + \dot{V}\sum_{\forall k}\left[m_k\beta_{3k} - (x_{ik}),_z F_{xk}\right] \\ \frac{1}{2}\rho V_a^2 S\left[-\frac{J_{s2}}{m_{at}}C_x + l(C_{M_y})_{\theta 0}\right] - \sum_{\forall k} F_{xk}\left(z_{Cuk} - (x_{ik}),_z x_{tk} + \frac{J_{s2}}{m_{at}}\right) \\ -m_1\dot{V}\beta_{3,1} \\ -m_2\dot{V}\beta_{3,2} \end{array} \right\} .$$

The expression for the generalized forces $\mathbf{F}_3$ was obtained assuming that the reference configuration corresponds to the static equilibrium position with the vehicle at standstill and with no force F_{xi}. In such a condition, all generalized coordinates are equal to zero, because they were defined as displacements from the same condition.

The equations were written with reference to coordinate Z, i.e., to the changes of the vertical displacement of point H in Fig. 29.3, which results in an inertial coupling. To study ride comfort it is better to refer to the vertical displacements of point H', so that the equations of motion have no inertial coupling, i.e. the mass matrix is diagonal. By introducing the coordinate

$$z_s = Z - c(\theta + \theta_0)$$

of point H', the mass matrix becomes

$$\mathbf{M}_3 = \begin{bmatrix} m_s & 0 & 0 & 0 \\ & J_y^* & 0 & 0 \\ & & m_1 & 0 \\ symm. & & & m_2 \end{bmatrix}, \qquad (29.225)$$

where

$$J_y^* = J_y - c^2 m_s - \frac{J_{s2}^2}{m_e} .$$

The damping and stiffness matrices are unchanged, provided that the distances x_i of wheels, springs and dampers are substituted by $x_i - c$. All matrices, except the stiffness matrix, remain symmetrical.

If the aerodynamic term causing the lack of symmetry of the stiffness matrix is neglected, which introduces only a small error because of the small size of the term, the system may be sketched as in Fig. 29.8a. The vehicle is modelled as a beam with elastic and damped supports, connected to the ground through the unsprung masses.

The quasi-static equilibrium attitude of the vehicle, which is different from the reference position because it takes into account both longitudinal forces on the tires and aerodynamic forces, can be immediately obtained from the steady-state solution of Eq. (29.224). Even if the acceleration of the vehicle does not appear explicitly in the equations, it is accounted for through forces F_{xi}.

The dynamic response of the vehicle to motion on uneven road is easily computed by assuming that points A and B move in a vertical direction with laws

$$h_A(t) = h(Vt) \quad \text{and} \quad h_B(t) = h(Vt + l) ,$$

where $h(x)$ is a function expressing the road profile. This amounts to exciting the two masses m_1 and m_2 with two forces equal to

$$K_{pz1} h_A(t) + c_{pz_1} \dot{h}_A(t) \quad \text{and} \quad K_{pz2} h_B(t + \tau) + c_{pz2} \dot{h}_B(t + \tau)$$

respectively, where $\tau = V/l$ is the delay due to the wheelbase.

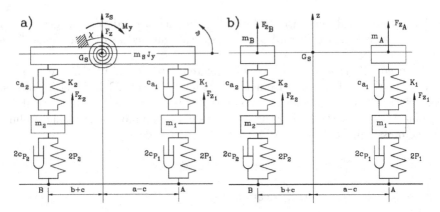

FIGURE 29.8. (a) Model with four degrees of freedom for the study of ride comfort; (b) model in which the sprung mass is simulated by two separate masses. Lengths a and b are the same as for the rigid vehicle, $a = x_1$ and $b = -x_2$. Note that the longitudinal positions of the springs and shock absorbers are assumed to be coincident $(x_i = x_{m_i} = x_{a_i})$..

29.3.10   Conclusions

The linearized model with 10 degrees of freedom for a vehicle with two axles, or more generally the model with $6 + 2n$ degrees of freedom for a vehicle with n axles, splits into three separate models, namely

- *Model for longitudinal behavior*, or performance model. The model includes a single degree of freedom, coordinate x^* (or better the forward speed V, because on a curved trajectory x^* is not a true coordinate), and allows the relationship between the longitudinal forces at the wheel-road contact and the vehicle speed to be computed, along with acceleration and braking performance. A detailed model of the tires may be introduced if their longitudinal slip is accounted for, as well as a model of the transmission and possibly of the engine. A model of the driver, intended as controller of the longitudinal motion through the accelerator and brake pedals, may also be introduced. Note that, owing to linearization, the longitudinal behavior on a curved trajectory coincides with that on straight road and thus with that studied in Chapter 23.

- *Model for lateral behavior*, or handling model. This model includes the degrees of freedom of lateral displacement (or better, lateral velocity, for the reasons cited above) and the yaw angle, which are the same degrees of freedom seen for the study of the handling of a rigid vehicle, plus the degrees of freedom related to the rolling of the vehicle body and of the axles. In such a model the input is the steering angle, but this can be easily modified to study the motion with free controls, possibly introducing a driver model as a steering controller as well. It has been assumed that the variations of the steering angle are slow enough to neglect its derivative $\dot{\delta}$. The presence of gyroscopic torques has no effect on the uncoupling between handling and comfort models, because a mass rotating about the y axis couples yaw and roll motions, both belonging to this model.

- *Model for suspension motions*, or ride comfort model. This model includes the degrees of freedom for vertical motion of the body (heave motion) and the axles, plus the pitch angle. Uncoupling between the longitudinal and comfort model is not complete, as shown by the pitching motions due to braking (dive) or driving (lift or squat). In the present chapter, the changes of longitudinal acceleration have been assumed to occur slowly. The changes of pitch angle may therefore be considered as a quasi static phenomenon, introduced into the equations by the longitudinal tire-road contact forces. In cases where longitudinal forces change quickly, ride comfort and longitudinal behavior (as well as transmission behavior) must be studied jointly.

29.4   MODELS OF DEFORMABLE VEHICLES

The assumption that the vehicle body can be considered as a rigid body is clearly an approximation that may be, in some cases, quite rough. This is particularly true of industrial vehicles and some passenger vehicles, such as open cars, whose stiffness is lower than usual.

If the body of the vehicle is not stiff, the position of any point P in the same inertial reference frame $OX_iY_iZ_i$ shown in Fig. 29.3 and already used to study the model based on rigid bodies, may be written in the form

$$(\overline{P-O'}) = (\overline{P_u-O'}) + \mathbf{s}_P \ , \qquad (29.226)$$

where $(\overline{P_u-O'})$ is the position of point P obtained by neglecting the deformation of the body and

$$\mathbf{s}_P = [u_x, u_y, u_z]_P^T \ , \qquad (29.227)$$

is the displacement function of time, due to compliance.

The position of P in undeflected conditions may be expressed by an equation similar to Eq. (31.55), and then

$$(\overline{P-O'}) = (\overline{H-O'}) + \mathbf{R}\,(\mathbf{r}_P + \mathbf{s}_P) \ , \qquad (29.228)$$

where vector

$$\mathbf{r}_P = [x, \ y, \ z]_P^T \ , \qquad (29.229)$$

is the vector, independent from time, leading from point H to point P, expressed in the reference frame of the sprung mass.

The velocity of point P is then

$$\mathbf{V}_P = \begin{bmatrix} \dot{X} & \dot{Y} & \dot{Z} \end{bmatrix}^T + \dot{\mathbf{R}}\,(\mathbf{r}_P + \mathbf{s}_P) + \mathbf{R}\dot{\mathbf{s}}_P \qquad (29.230)$$

or, by introducing the velocity $\mathbf{V}$ in the reference frame x^*y^*z:

$$\mathbf{V}_P = \mathbf{R}_1 \mathbf{V} + \dot{\mathbf{R}}\,(\mathbf{r}_P + \mathbf{s}_P) + \mathbf{R}\dot{\mathbf{s}}_P \ . \tag{29.231}$$

By remembering that pitch and roll angles are small, and introducing matrix

$$\mathbf{R}_{23} = \mathbf{R}_2 \mathbf{R}_3 \approx \begin{bmatrix} 1 & 0 & \theta_0 + \theta \\ 0 & 1 & -\phi \\ -\theta_0 - \theta & \phi & 1 \end{bmatrix} ,$$

it is possible to write

$$\mathbf{V}_P = \mathbf{R}_1 \left[\mathbf{V} + \left(\dot{\mathbf{R}}_{23} + \dot{\psi}\mathbf{S}\mathbf{R}_{23} \right) (\mathbf{r}_P + \mathbf{s}_P) + \mathbf{R}_{23}\dot{\mathbf{s}}_P \right] , \tag{29.232}$$

where

$$\mathbf{S} \approx \begin{bmatrix} 0 & -1 & 0 \\ 1 & 0 & 0 \\ 0 & 0 & 0 \end{bmatrix} .$$

The kinetic energy of the infinitesimal element of mass at coordinates x, y, z is

$$dT = \tfrac{1}{2}dm \left[\mathbf{V}^T\mathbf{V} + \mathbf{r}_P^T \left(\dot{\mathbf{R}}_{23} + \dot{\psi}\mathbf{S}\mathbf{R}_{23} \right)^T \left(\dot{\mathbf{R}}_{23} + \dot{\psi}\mathbf{S}\mathbf{R}_{23} \right) \mathbf{r}_P + \right.$$
$$+2\mathbf{V}^T \left(\dot{\mathbf{R}}_{23} + \dot{\psi}\mathbf{S}\mathbf{R}_{23} \right) \mathbf{r}_P + \mathbf{s}_P^T \left(\dot{\mathbf{R}}_{23} + \dot{\psi}\mathbf{S}\mathbf{R}_{23} \right)^T \left(\dot{\mathbf{R}}_{23} + \dot{\psi}\mathbf{S}\mathbf{R}_{23} \right) \mathbf{s}_P +$$
$$+2\mathbf{V}^T \left(\dot{\mathbf{R}}_{23} + \dot{\psi}\mathbf{S}\mathbf{R}_{23} \right) \mathbf{s}_P + \dot{\mathbf{s}}_P^T \mathbf{R}_{23}^T \mathbf{R}_{23}\dot{\mathbf{s}}_P + 2\mathbf{V}^T\mathbf{R}_{23}\dot{\mathbf{s}}_P +$$
$$\left. +2\dot{\mathbf{s}}_P^T \mathbf{R}_{23}^T \left(\dot{\mathbf{R}}_{23} + \dot{\psi}\mathbf{S}\mathbf{R}_{23} \right) (\mathbf{r}_P + \mathbf{s}_P) \right] .$$
$$\tag{29.233}$$

The sum of the first three terms (those not containing the deformation $\mathbf{s}_P$ or its derivatives) is the kinetic energy dT_R of the same mass element in a rigid motion. The other terms may be greatly simplified if the products of more than two small quantities are neglected. Because the deformations $\mathbf{s}_P$ and their derivatives are small quantities, it follows that

$$dT = dT_R + \tfrac{1}{2}dm \ (\dot{u}_x^2 + \dot{u}_y^2 + \dot{u}_z^2) + dm \ v_x \left[\dot{\theta}u_z - \dot{\psi}u_y + \right.$$
$$+\dot{u}_x + (\theta + \theta_o)\dot{u}_z] + dm \ \dot{u}_x \left(\dot{\theta}z - \dot{\psi}y \right) + \tag{29.234}$$
$$+dm \ \dot{u}_y \left(v_y - \dot{\phi}z + \dot{\psi}x \right) + dm \ \dot{u}_z \left(v_z - \dot{\theta}x + \dot{\phi}y \right) .$$

The deformation of the sprung mass can be expressed in terms of its modal coordinates, i.e., as a linear combination of the eigenfunctions of the undamped system. Note that this remains true even if the sprung mass is damped and even for nonlinear systems. Because the sprung mass has a plane of symmetry (in the present case the xz plane), its modes may be subdivided into symmetrical and skew-symmetrical modes, designated by subscripts s and a in the following equations.

The displacements $\mathbf{s_P} = [u_x, u_y, u_z]^T$ of point P(x, y, z) can thus be expressed as

$$\left\{ \begin{array}{c} u_x \\ u_y \\ u_z \end{array} \right\} = \mathbf{Q}_s(x, y, z)\boldsymbol{\eta}_s(t) + \mathbf{Q}_a(x, y, z)\boldsymbol{\eta}_a(t) , \qquad (29.235)$$

where $\mathbf{Q}_i(x, y, z)$ are matrices containing the eigenfunctions while $\boldsymbol{\eta}_i(t)$ are vectors containing the modal coordinates. Equation (29.235) is exact only if an infinity of eigenfunctions and modal coordinates are considered; however, a very good approximation is usually obtained by taking into account a small number of modes, particularly if the system is linear and lightly damped. The modes considered in the equation are those of the free structure; the rigid-body modes need not be considered because they have already been included in the rigid-body analysis already performed.

Instead of using the eigenfunctions, a set of arbitrary functions of the space coordinates may be used, as is common in the assumed modes methods for structural analysis; in this case, however, the number of coordinates needed to obtain a good approximation is higher and depends on the choice of the arbitrary functions: Moreover, the mass and stiffness matrices are not diagonal, as they are when using the eigenfunctions.

Let A and B be two points located in symmetrical positions with respect to the xz plane. It follows that

$$u_{xA} = u_{xB} , \quad u_{yA} = -u_{yB} , \quad u_{zA} = u_{zB}$$

for symmetrical modes and

$$u_{xA} = -u_{xB} , \quad u_{yA} = u_{yB} , \quad u_{zA} = -u_{zB}$$

for skew-symmetrical ones. As a consequence of symmetry, some relevant integrals extended to the whole unsprung mass may be written in a much simplified form

$$\int_m \left\{ \begin{array}{c} u_x \\ u_y \\ u_z \end{array} \right\} dm = \left\{ \begin{array}{c} \mathcal{A} \\ 0 \\ \mathcal{C} \end{array} \right\} \boldsymbol{\eta}_s + \left\{ \begin{array}{c} 0 \\ \mathcal{B} \\ 0 \end{array} \right\} \boldsymbol{\eta}_a ,$$

$$\int_m x \left\{ \begin{array}{c} u_x \\ u_y \\ u_z \end{array} \right\} dm = \left\{ \begin{array}{c} \mathcal{N} \\ 0 \\ \mathcal{F} \end{array} \right\} \boldsymbol{\eta}_s + \left\{ \begin{array}{c} 0 \\ \mathcal{D} \\ 0 \end{array} \right\} \boldsymbol{\eta}_a ,$$

$$\int_m y \left\{ \begin{array}{c} u_x \\ u_y \\ u_z \end{array} \right\} dm = \left\{ \begin{array}{c} 0 \\ \mathcal{H} \\ 0 \end{array} \right\} \boldsymbol{\eta}_s + \left\{ \begin{array}{c} \mathcal{E} \\ 0 \\ \mathcal{G} \end{array} \right\} \boldsymbol{\eta}_a , \qquad (29.236)$$

$$\int_m z \left\{ \begin{array}{c} u_x \\ u_y \\ u_z \end{array} \right\} dm = \left\{ \begin{array}{c} \mathcal{I} \\ 0 \\ \mathcal{L} \end{array} \right\} \boldsymbol{\eta}_s + \left\{ \begin{array}{c} 0 \\ \mathcal{M} \\ 0 \end{array} \right\} \boldsymbol{\eta}_a ,$$

$$\int_m (u_x^2 + u_y^2 + u_z^2) \, dm = \boldsymbol{\eta}_s^T \overline{\mathbf{M}}_s \boldsymbol{\eta}_s + \boldsymbol{\eta}_a^T \overline{\mathbf{M}}_a \boldsymbol{\eta}_a ,$$

where the diagonal matrices $\overline{\mathbf{M}}_s$ and $\overline{\mathbf{M}}_a$ are the modal mass matrices for symmetrical and skew-symmetrical modes, and where matrices from $\mathcal{A}$ to $\mathcal{N}$ are row matrices whose size is $1 \times n$, where n is the number of modal coordinates (either symmetrical or skew-symmetrical) that are considered.

By integrating Eq. (29.234), the kinetic energy of the sprung mass reduces to

$$\mathcal{T} = \mathcal{T}_R + \tfrac{1}{2}\dot{\boldsymbol{\eta}}_s^T \overline{\mathbf{M}}_s \dot{\boldsymbol{\eta}}_s + \tfrac{1}{2}\dot{\boldsymbol{\eta}}_a^T \overline{\mathbf{M}}_a \dot{\boldsymbol{\eta}}_a + \dot{\psi}\left(\mathcal{D} - \mathcal{E}\right)\dot{\boldsymbol{\eta}}_a + \dot{\theta}\left(\mathcal{I} - \mathcal{F}\right)\dot{\boldsymbol{\eta}}_s +$$

$$+\dot{\phi}\left(\mathcal{G} - \mathcal{M}\right)\dot{\boldsymbol{\eta}}_a + V\left[\mathcal{A}\dot{\boldsymbol{\eta}}_s + \dot{\theta}\mathcal{C}\boldsymbol{\eta}_s + \left(\theta + \theta_0\right)\mathcal{C}\dot{\boldsymbol{\eta}}_s - \dot{\psi}\mathcal{B}\boldsymbol{\eta}_a\right] + v_y \mathcal{B}\dot{\boldsymbol{\eta}}_a + v_z \mathcal{C}\dot{\boldsymbol{\eta}}_s \ . \tag{29.237}$$

The gravitational potential energy of the sprung mass may be expressed in the form

$$\mathcal{U}_{gS} = g \int_m Z_p dm = g \int_m \mathbf{e}_3^T \left[(\overline{\mathrm{H}\text{-}\mathrm{O'}}) + \mathbf{R}\left(\mathbf{r}_\mathrm{P} + \mathbf{s}_\mathrm{P}\right)\right] dm \ . \tag{29.238}$$

It then follows that

$$\mathcal{U}_{gS} = g\left(Z + Z_0\right) m_s + g \int_m \mathbf{e}_3^T \mathbf{R}_{23} \mathbf{r}_\mathrm{P} dm + g \int_m \mathbf{e}_3^T \mathbf{R}_{23} \mathbf{s}_\mathrm{P} dm \ . \tag{29.239}$$

The first two terms are the potential energy $\mathcal{U}_{gR}$ of the rigid body computed above. By introducing the linearized expression of matrix $\mathbf{R}_{23}$ it follows that

$$\mathcal{U}_{gS} = \mathcal{U}_{gR} + g \int_m \left[-\left(\theta + \theta_0\right)u_x + \phi u_y + u_z\right] dm \ , \tag{29.240}$$

or, by introducing the modal coordinates to express the deformation of the vehicle body,

$$\mathcal{U}_{gS} = \mathcal{U}_{gR} + g\left[-\left(\theta + \theta_0\right)\mathcal{A}\boldsymbol{\eta}_s + \phi\mathcal{B}\boldsymbol{\eta}_a + \mathcal{C}\boldsymbol{\eta}_s\right].$$

The deformation potential energy of the springs of the kth suspension is still expressed by Eq. (29.79), to which the terms linked with the deformation modes are added. Assuming that the points of attachment of the springs of the left (right) kth suspension are x_i, $\pm y_i$ and z_i, it follows that

$$\mathcal{U}_{mk} = \frac{1}{2} K_\zeta \left(\zeta + \zeta_0 - \mathbf{Q}_{zsk}\boldsymbol{\eta}_s\right)^2 + \frac{1}{2} K_\phi \left(\phi - \phi_k - \mathbf{Q}_{zak}\boldsymbol{\eta}_a\right)^2 \ , \tag{29.241}$$

where $\mathbf{Q}_{zsk}$ and $\mathbf{Q}_{zak}$ are the parts of the matrices of the eigenfunctions for symmetrical and skew-symmetrical modes linked to u_z computed at the point of coordinates x_i, y_i, z_i.

The potential energy of the kth suspension is then

$$\mathcal{U}_{mk} = \mathcal{U}_{mR} + \tfrac{1}{2}\boldsymbol{\eta}_s^T \mathbf{K}_{44}\boldsymbol{\eta}_s + \mathbf{K}_{14}\left(Z + Z_0\right)\boldsymbol{\eta}_s + \mathbf{K}_{24}\left(Z_k + Z_{k0}\right)\boldsymbol{\eta}_s +$$

$$+\mathbf{K}_{34}\left(\theta + \theta_0\right)\boldsymbol{\eta}_s + \tfrac{1}{2}K_\phi \boldsymbol{\eta}_a^T \mathbf{Q}_{zak}^T \mathbf{Q}_{zak}\boldsymbol{\eta}_a - K_\phi \phi \mathbf{Q}_{zak}\boldsymbol{\eta}_a + K_\phi \phi_k \mathbf{Q}_{zak}\boldsymbol{\eta}_a \ , \tag{29.242}$$

where $\mathbf{K}_{44}$ is a square matrix of size $n_s \times n_s$ (where n_s is the number of symmetrical modes considered), while $\mathbf{K}_{14}$, $\mathbf{K}_{24}$ and $\mathbf{K}_{34}$ are row matrices of size $1 \times n_s$.

The deformation potential energy of the tires of the kth suspension is expressed by Eq. (29.81) without any change. The potential energy due to the deformation of the sprung mass is obviously

$$\mathcal{U}_R = \frac{1}{2}\boldsymbol{\eta}_s^T \overline{\mathbf{K}}_s \boldsymbol{\eta}_s + \frac{1}{2}\boldsymbol{\eta}_a^T \overline{\mathbf{K}}_a \boldsymbol{\eta}_a , \qquad (29.243)$$

where $\overline{\mathbf{K}}_s$ and $\overline{\mathbf{K}}_a$ are the modal stiffness matrices.

In a similar way, the Raleigh dissipation function of the shock absorbers of the kth suspension is

$$\mathcal{F}_{ak} = \mathcal{F}_{akR} + \tfrac{1}{2}\dot{\boldsymbol{\eta}}_s^T \mathbf{c}_{44}\dot{\boldsymbol{\eta}}_s + \mathbf{c}_{14}\dot{Z}\dot{\boldsymbol{\eta}}_s + \mathbf{c}_{24}\dot{Z}_k\dot{\boldsymbol{\eta}}_s + \mathbf{c}_{34}\dot{\theta}\dot{\boldsymbol{\eta}}_s +$$
$$+ \tfrac{1}{2}c_\phi \dot{\boldsymbol{\eta}}_a^T \mathbf{Q}_{zaka}^T \mathbf{Q}_{zak}\dot{\boldsymbol{\eta}}_a - c_\phi \dot{\phi}\mathbf{Q}_{zaka}\dot{\boldsymbol{\eta}}_a + c_\phi \dot{\phi}_k \mathbf{Q}_{zaka}\dot{\boldsymbol{\eta}}_a , \qquad (29.244)$$

where $\mathcal{F}_{akR}$ is the function seen before for the rigid vehicle, matrices $\mathbf{c}_{ij}$ are similar to matrices $\mathbf{K}_{ij}$ seen above and $\mathbf{Q}_{zaka}$ is similar to $\mathbf{Q}_{zak}$, but referred to the points where the shock absorbers are attached.

The Raleigh dissipation function for the tires (Eq. 29.84) is unchanged, while that linked with deformation modes of the sprung mass is

$$\mathcal{F}_R = \frac{1}{2}\dot{\boldsymbol{\eta}}_s^T \overline{\mathbf{C}}_s \dot{\boldsymbol{\eta}}_s + \frac{1}{2}\dot{\boldsymbol{\eta}}_a^T \overline{\mathbf{C}}_a \dot{\boldsymbol{\eta}}_a , \qquad (29.245)$$

where $\overline{\mathbf{C}}_s$ and $\overline{\mathbf{C}}_a$ are the modal damping matrices. Note that the last expression is just an approximation, because modal damping matrices are not diagonal and may couple the various modes. Other approximations are linked to the way the presence of suspensions has been accounted for, but such approximations are similar to those already seen for the other linearized models.

If the virtual work of the external forces is computed neglecting displacements due to deformation modes, the expressions of the generalized forces are the same as those used for models based on rigid bodies. Such an assumption is well suited to the present linearized model, where the exact kinematics of suspensions has not been taken into account.

A detailed inspection of the expressions of the kinetic and potential energies and of the dissipation function shows that, if the forward velocity V is assumed to be a known function, the equations of motion divide into two separate sets, exactly as they do when the compliance of the sprung mass is neglected.

29.4.1 Handling model

A first set of equations contains generalized coordinates y^*, ψ, ϕ, ϕ_i and $\boldsymbol{\eta}_a$. If the vehicle has n axles and n_a skew-symmetrical modes are considered, their number is $3+n+n_a$. The differential equations modelling the lateral behavior are

$$\begin{bmatrix} \mathbf{M}_1 & \mathbf{M}_{a1}^T \\ \mathbf{M}_{a1} & \overline{\mathbf{M}}_a \end{bmatrix} \begin{Bmatrix} \ddot{\mathbf{q}}_1 \\ \ddot{\boldsymbol{\eta}}_a \end{Bmatrix} + \begin{bmatrix} \mathbf{C}_1 & \mathbf{C}_{a1}^T \\ \mathbf{C}_{a1} & \mathbf{C}_{aa} \end{bmatrix} \begin{Bmatrix} \dot{\mathbf{q}}_1 \\ \dot{\boldsymbol{\eta}}_a \end{Bmatrix} +$$

$$+ \begin{bmatrix} \mathbf{K}_1 & \mathbf{K}_{1a} \\ \mathbf{K}_{a1} & \mathbf{K}_{aa} \end{bmatrix} \begin{Bmatrix} \mathbf{q}_1 \\ \boldsymbol{\eta}_a \end{Bmatrix} = \begin{Bmatrix} \mathbf{F}_1 \\ \mathbf{0} \end{Bmatrix} , \qquad (29.246)$$

where $\mathbf{q}_1$, $\mathbf{M}_1$, $\mathbf{C}_1$, $\mathbf{K}_1$ and $\mathbf{F}_1$ are the same vectors and matrices seen in Eq. (31.124), and

$$\mathbf{M}_{a1} = \begin{bmatrix} \mathcal{B}^T & \mathcal{D}^T - \mathcal{E}^T & \mathcal{G}^T - \mathcal{M}^T & 0 & 0 \end{bmatrix} ,$$
$$\mathbf{C}_{a1} = \begin{bmatrix} 0 & -V\mathcal{B}^T & -\sum_{\forall k} c_{\phi k} \mathbf{Q}_{zaka}^T & c_{\phi 1} \mathbf{Q}_{za1a}^T & c_{\phi 2} \mathbf{Q}_{za2a}^T \end{bmatrix} ,$$
$$\mathbf{C}_{aa} = \overline{\mathbf{C}}_a + \sum_{\forall k} c_k \mathbf{Q}_{zaka}^T \mathbf{Q}_{zaka} ,$$
$$\mathbf{K}_{1a} = \begin{bmatrix} 0 & -\dot{V}\mathcal{B} & g\mathcal{B} - \sum_{\forall k} k_{\phi k} \mathbf{Q}_{zak} & k_{\phi 1} \mathbf{Q}_{za1} & k_{\phi 2} \mathbf{Q}_{za2} \end{bmatrix}^T ,$$
$$\mathbf{K}_{a1} = \begin{bmatrix} 0 & 0 & g\mathcal{B}^T - \sum_{\forall k} k_{\phi k} \mathbf{Q}_{zak}^T & k_{\phi 1} \mathbf{Q}_{za1}^T & k_{\phi 2} \mathbf{Q}_{za2}^T \end{bmatrix} ,$$
$$\mathbf{K}_{aa} = \overline{\mathbf{K}}_a + \sum_{\forall k} k_{\phi k} \mathbf{Q}_{zak}^T \mathbf{Q}_{zak} .$$

The expressions of the various matrices refer to a two-axle vehicle, but they may be easily generalized.

As in the previous models, coordinates y^* and ψ are present only with their derivatives: the order of the differential set of equations is then $4 + 2n + 2n_a$.

29.4.2   Ride comfort model

The second set of equations contains generalized coordinates x^*, Z, θ, Z_i and $\boldsymbol{\eta}_s$. If n_s symmetrical modes are included in the model, they are $3 + n + n_s$.

In this case the first equation, that describing longitudinal dynamics, is weakly coupled with the others, and may be studied separately. Its expression is still Eq. (29.223), with the term $\mathcal{A}\ddot{\boldsymbol{\eta}}_s$ added to the left side.

Neglecting the deformation corresponding to the static equilibrium condition (which may be computed using an equation of the type of Eq. (29.221), in which the modal coordinates corresponding to the static deformation and terms like $g\mathcal{C}$ are included), the set of $2 + n + n_s$ equations, remaining after separating the first equation, describes the suspension motions of the vehicle:

$$\begin{bmatrix} \mathbf{M}_3 & \mathbf{M}_{s3}^T \\ \mathbf{M}_{s3} & \overline{\mathbf{M}}_s \end{bmatrix} \begin{Bmatrix} \ddot{\mathbf{q}}_3 \\ \ddot{\boldsymbol{\eta}}_s \end{Bmatrix} + \begin{bmatrix} \mathbf{C}_3 & \mathbf{C}_{s3}^T \\ \mathbf{C}_{s3} & \mathbf{C}_{ss} \end{bmatrix} \begin{Bmatrix} \dot{\mathbf{q}}_3 \\ \dot{\boldsymbol{\eta}}_s \end{Bmatrix} +$$

$$+ \begin{bmatrix} \mathbf{K}_3 & \mathbf{K}_{s3}^T \\ \mathbf{K}_{s3} & \mathbf{K}_{ss} \end{bmatrix} \begin{Bmatrix} \mathbf{q}_3 \\ \boldsymbol{\eta}_s \end{Bmatrix} = \begin{Bmatrix} \mathbf{F}_3 \\ \mathbf{F}_s \end{Bmatrix} , \qquad (29.247)$$

where $\mathbf{q}_3$, $\mathbf{M}_3$, $\mathbf{C}_3$, $\mathbf{K}_3$ and $\mathbf{F}_3$ are the same matrices and vectors seen in Eq. (29.224), and

$$\mathbf{M}_{s3} = [\ \mathcal{C}^T \quad \mathcal{I}^T - \mathcal{F}^T \quad \mathbf{0} \quad \mathbf{0}\ ]\ ,$$

$$\mathbf{C}_{s3} = [\ \textstyle\sum_{\forall k} \mathbf{c}_{14k}^T \quad \textstyle\sum_{\forall k} \mathbf{c}_{34k}^T \quad \mathbf{c}_{24,1}^T \quad \mathbf{c}_{24,2}^T\ ]^T\ ,$$

$$\mathbf{C}_{ss} = \overline{\mathbf{C}}_s + \sum_{\forall i} \mathbf{c}_{44k}\ ,$$

$$\mathbf{K}_{s3} = [\ \textstyle\sum_{\forall k} \mathbf{K}_{14k}^T \quad \dot{V}\mathcal{C}^T - g\mathcal{A}^T + \textstyle\sum_{\forall i} \mathbf{K}_{34k}^T \quad \mathbf{K}_{24,1}^T \quad \mathbf{K}_{24,2}^T\ ]^T\ ,$$

$$\mathbf{K}_{ss} = \overline{\mathbf{K}}_s + \sum_{\forall i} \mathbf{K}_{44k}\ ,$$

$$\mathbf{F}_s = -\dot{V}\mathcal{C}\theta_0\ .$$

All matrices are symmetrical, except for the stiffness matrix resulting from the usual aerodynamic term included in $\mathbf{K}_3$. All aerodynamic forces have been assumed to be independent from the deformation modes: if this assumption were abandoned, the equations would change, but uncoupling would still hold.

29.4.3 Uncoupling of the equations of motion

Symmetrical and skew-symmetrical deformation modes thus play a very different role in the dynamic behavior of the vehicle. The first, like bending modes in the xz plane, affect riding comfort but have no importance in the study of handling. The most important skew-symmetrical modes are those related to torsional deformations. The significance of their influence on handling, particularly in sport cars and above all in Formula 1 racers, is well known. Transversal bending can have a similar effect.

Modal matrices $\overline{\mathbf{M}}_s$, $\overline{\mathbf{M}}_a$, $\overline{\mathbf{K}}_s$ and $\overline{\mathbf{K}}_a$ are diagonal, and describe the dynamic behavior of the vehicle body as a free compliant body. Usually the damping linked to the deformation modes of the body is not large, and it may be considered as an undamped, or at most a lightly damped, system. Neglecting damping, the natural frequencies of the symmetrical and skew-symmetrical modes are

$$\Omega_{sj} = \sqrt{\frac{\overline{\mathbf{K}}_{sj}}{\overline{\mathbf{M}}_{sj}}}\ , \quad \Omega_{aj} = \sqrt{\frac{\overline{\mathbf{K}}_{aj}}{\overline{\mathbf{M}}_{aj}}}\ . \tag{29.248}$$

If the coupling matrices $\mathbf{M}_{s3}$, $\mathbf{M}_{a1}$, $\mathbf{C}_{s3}$, $\mathbf{C}_{a1}$, etc. were negligible, the dynamic behavior of the vehicle could be studied by separating the dynamic behavior of the rigid vehicle on elastic suspensions (studied in the proceeding sections) from the dynamic behavior of the vehicle body, considered as a compliant body free in space.

In the case of passenger cars, the natural frequencies of deformation modes are much higher than those typical of a vehicle on elastic suspensions (as already stated, the typical frequencies of the sprung mass are slightly above 1 Hz while those linked to the unsprung masses are at most $8 - 10$ Hz). Coupling between the dynamic behavior of the vehicle as made of rigid bodies and of the compliant vehicle body is weak, and the vibration of the latter influences acoustic comfort more than handling or ride comfort.

Open cars and vans are often an exception: the stiffness of their bodies is lower, particularly in torsion and bending in the symmetry plane, and the related natural frequencies are little different than those linked with the behavior of the vehicle as a whole. The torsional deformation of the chassis and the body may have a strong effect on handling, usually making it worse. Bending of the body in its plane may have some effect on comfort, even if it is impossible to determine in general whether it improves or worsens it.

In the case of industrial vehicles, particularly trucks, some natural frequencies of deformation modes are usually low and the related modes may strongly interfere with handling modes (if skew-symmetrical) or with comfort modes (if symmetrical).

29.5   ARTICULATED VEHICLES

Consider an articulated vehicle, such as a tractor and a trailer or a semi-trailer. As already stated, it is possible to study its behavior using a multibody model, if its parts may be considered rigid. The number of degrees of freedom is six for each part constituting the body of the vehicle, plus a further degree of freedom for each independent suspension and two degrees of freedom for each solid axle suspension, minus the number of degrees of freedom constrained by the links connecting the various parts constituting the body.

As an example, the articulated truck in Fig. 29.1d has 21 degrees of freedom if the hitch connecting tractor and trailer may be modelled as a spherical hinge (24 degrees of freedom for the two rigid bodies and the 6 solid axles, minus 3 degrees of freedom constrained by the hinge). If the hitch were modelled as a cylindrical hinge with its axis in the vertical direction, the number of degrees of freedom would reduce to 19 (the hinge would constrain 5 of them instead of 3), but to constrain pitch and roll rotations of the trailer with respect to the tractor, the moments about the x and y axes the hinge would experience (and with negligible deformations, otherwise some deformation degrees of freedom would be needed) would be extremely high, creating a non-viable solution.

In normal operation the pitch, roll and yaw angles of the trailer with respect to the tractor are small and, as in the case of articulated vehicles it is possible to use linearized models. An articulated vehicle made by two rigid bodies plus compliant suspensions may be thought of as a single compliant system, whose deformation consists in the relative motion of the two bodies about the hinge. The rigid body modes of this compliant body may be considered similar to the deformation modes seen above, the only difference being that the relevant natural frequencies are zero, because the modal stiffness vanishes. This is obvious because there are no elastic systems applying restoring moments at the hitch.

The rigid body modes may also be subdivided into symmetrical and skew-symmetrical modes. In the case of an articulated truck the yaw rotation of the trailer (like angle θ in the model seen in Section 25.15) and the rolling motion of the trailer are skew-symmetric modes and thus couple with handling, while pitching rotations of the trailer couple with ride comfort. In the case of the truck and trailer system in Fig. 25.41a, it is possible to assume that the connection of the draw bar to the trailer and that between the dolly and the trailer are spherical hinges, even if the roll rotation between the trailer and the dolly may in some cases be considered locked. The pitch rotation of the dolly and the trailer thus enter the comfort model, while all yaw and roll rotations enter the handling model.

29.6   GYROSCOPIC MOMENTS AND OTHER SECOND ORDER EFFECTS

Gyroscopic moments due to wheel rotation were examined in the 10 degrees of freedom model for isolated vehicles with two axles. Within the frame of a linearized model, they have no effect on handling-comfort uncoupling and they enter only into the handling model. Gyroscopic moments are automatically present when the equations of motion are obtained through Lagrange equations, provided that the angular velocity of all rotating parts of the model is considered.

To evaluate the impact of gyroscopic moments caused by the wheels on handling, it is possible to assume, at least in the case of solid axle suspensions, that the angular velocity of the ith wheel $\dot{\chi}_i$ lies on axis y_k of the kth unsprung mass. Any angular velocity of the vehicle about the x_k and z_k axes will produce a gyroscopic moment due to the ith wheel that may be expressed in the x_k, y_k, z_k frame as

$$
M_g = \dot{\chi}_i J_{pi} \left\{ \begin{array}{c} \dot{\psi} \\ 0 \\ -\dot{\phi}_k \end{array} \right\} = J_{pi} \frac{V}{R_{ei}} \left\{ \begin{array}{c} \dot{\psi} \\ 0 \\ -\dot{\phi}_i \end{array} \right\} ,
\tag{29.249}
$$

where the angular velocity of the wheel has been assumed to be linked with the forward velocity by the usual relationship

$$
\dot{\chi}_i = \frac{V}{R_{ei}} .
\tag{29.250}
$$

As previously stated, the terms due to the gyroscopic moment are present in $L_{\dot{\psi}}$ and, with opposite sign, in $L_{\dot{\phi}k}$. In steady state motion, gyroscopic moments are usually quite small. Neglecting the camber angle, $\dot{\psi} = V/R$ (where R is the radius of the trajectory) and $\dot{\phi}_i = 0$. The component M_{g_z} of the gyroscopic moment vanishes, while

$$
M_{g_x} = \frac{V^2}{R} \sum_{\forall i} \frac{J_{pi}}{R_{ei}} ,
\tag{29.251}
$$

where the sum extends to all wheels. Gyroscopic moment is thus proportional to the centrifugal acceleration V^2/R, which is limited. If the sum of the polar moments of inertia of the wheels of an axle and R_e are equal, for instance, to 6 kg m^2 and 0.5 m respectively, values related to an industrial vehicle, and the centrifugal acceleration is 5 m/s^2, a gyroscopic moment of 60 Nm is obtained. If the track of the axle is 1.4 m, the load transfer due to the gyroscopic moment of the two wheels is roughly 43 N.

Gyroscopic wheel moments may, however, be more important in non-stationary conditions and, above all, can affect to a large extent the dynamics of the steering system: Their effect on free-control dynamics may thus be important. In solid axle suspensions of steering axles, strong reactions on the steering wheel due to gyroscopic wheel moments may be caused by travelling on uneven road. They can cause severe discomfort and make driving difficult.

Gyroscopic moments due to the engine or other rotating elements of the vehicle are usually less important, except in particular cases (usually not related with road vehicles), such as that of electric railway engines, in which they can cause an increase of wear of the wheel rims. As a last consideration, a mass rotating about an axis parallel to the z axis couples roll and pitch motions, making uncoupling between handling and comfort impossible even if the assumptions of small displacements, linearity and symmetry hold. This effect may even be exploited, as in the case of flywheel stabilizers in ships, where coupling allows the larger pitch moment of inertia to be used to limit roll oscillations. Also, a mass rotating about the x-axis has a coupling effect between pitch and yaw, while a mass rotating about the y-axis couples roll and yaw, already coupled in the handling behavior.

Other second-order dynamic effects may be of some importance. In non-stationary motion, for example, the angular acceleration of rotating masses may produce inertia torques of non-negligible size. Some of these effects have been included in the model seen above and are included in the terms containing $\dot{V}$ (because of the relationship assumed between the forward velocity V and the wheel velocity $\dot{\chi}$, the acceleration $\ddot{\chi}$ is proportional to $\dot{V}$).

In the previous models some second order effects have been neglected. For instance, the transmission of the driving torque to the wheels may cause a reaction torque that, being exerted between the parts constituting the vehicle, has no effect on its global dynamics, at least as a first approximation. This torque may, however, modify the configuration of the vehicle and affect the forces the vehicle exchanges with the ground or, although to a much lesser extent, with the air. In a vehicle with longitudinal engine and rear wheel drive with the differential on a solid axle, the driving torque causes a small roll angular displacement between the vehicle body and the solid axle. The small roll angle may induce roll steer that may affect handling. These effects are usually neglected, because they are small, but there is no difficulty in introducing them into the model.

A larger effect may be caused by the reaction torque exerted on suspensions when not directly transferred to the body by the suspension linkages. Instead it loads the suspension springs, causing lifting or sinking of the attachment points, as seen in *antidive, antilift* or *antisquat* configurations. For this effect to be present, the torque must be applied to the unsprung mass. For driving torques this occurs only in the case of live axles, while in braking torques the brakes are almost always located on the unsprung masses, the only exceptions being the little used layout in which they are placed close to the differential gear in De Dion axles, or when driving wheel suspension is independent. In such cases the suspension layout must allow a vertical movement when a torque is applied to it, i.e. the derivative $\partial z / \partial M_y$ must be other than zero, an example being that of trailing arm suspensions.

Similarly it is possible to take into account the deformation of unsprung masses without changing the general conclusion: This is important because configurations based on the compliance of the unsprung masses are increasingly common. With solid axles it is easy to evaluate which deformation modes of the unsprung masses are symmetrical, and thus are to be included in comfort dynamics, and which are skew-symmetrical and affect handling. In a suspension in which leaf springs are used as guiding elements, for instance, the lateral compliance of the springs gives way to skew symmetrical modes and influences handling, while their S deformation about the y axis is a symmetrical deformation and thus couples with comfort dynamics, or better longitudinal dynamics. The longitudinal compliance of the suspension may strongly affect comfort.

In the case of independent suspensions, the suspension of the whole axle, with its two rigid-body degrees of freedom, must be considered. The whole axle must be studied as well for deformation modes, as was done in Eq. (29.122). In this way it is again possible to distinguish between symmetrical and skew-symmetrical modes.

But uncoupling is a more general feature still. The above considerations may be applied to vehicles with two wheels, the only exceptions being that the roll angle can easily take values beyond the range in which linearization of trigonometric functions applies, while the lateral movements of the driver, aimed at displacing the centre of mass and producing unsymmetrical aerodynamic forces, can destroy the symmetry on which uncoupling is based.

No particular assumption about the nature of the forces supporting the vehicle has been made. The same uncoupling also holds for vehicles supported by hydrostatic, aerostatic or aerodynamic forces. In the first case, the assumption of the existence of a roll axis of the suspension is replaced by the assumption of a roll axis fixed to the hull in its undeflected configuration. The small roll oscillations are thus demonstrated to be uncoupled with pitch and bounce motions, which are coupled with each other. In the case of aircraft, roll and yaw oscillations are known to be coupled (dutch roll) while bounce and pitch oscillations are also coupled with each other. Even the presence of aerodynamic forces due to the deformations of the structure does not change the overall picture, provided that they can be assumed to depend linearly on the modal coordinates η_s and η_a.

30

TRANSMISSION MODELS

The take-off manoeuvre of a vehicle was studied in Section 23.9 using a simple model where the inertia of both engine and vehicle were modelled as two flywheels connected to each other by a rigid shaft and a friction clutch. This model can be made more realistic by adding the torsional compliance of the shaft, of the joints and possibly the gear wheels, as well as the rotational inertia of the various elements of the driveline. A model of the whole driveline is thus obtained, with the engine and vehicle modelled as two flywheels located at its ends.

However, the engine shaft is itself a compliant system. Moreover, its piston-connecting rod-crank systems should be modelled as systems with variable inertia in time. At the other end of the driveline, the dynamics of the transmission and the longitudinal dynamics of the vehicle are coupled by the tires, which are themselves compliant in torsion. The longitudinal compliance of the suspensions may affect the dynamics of the driveline and couples with the dynamics of the vehicle, which is in turn coupled with comfort dynamics.

Because many of the parts that may be included in the model of the driveline have a strongly nonlinear behavior, the model must include nonlinearities that prevent frequency domain solutions from being obtained if a high degree of detail is to be considered. In this case only time domain solutions can be obtained.

The mathematical models of the various parts of the transmission, from the engine to the vehicle, will be described in this chapter.

G. Genta, L. Morello, *The Automotive Chassis, Volume 2: System Design*,
Mechanical Engineering Series,
© Springer Science+Business Media B.V. 2009

30.1   COUPLING BETWEEN COMFORT AND DRIVELINE VIBRATION

As predictable in a system with many degrees of freedom, the driveline has many vibration modes and natural frequencies. The effects of the various modes are different, and a variety of models may be used for their study.

The most important mode for comfort is the first mode of the driveline, which usually has a natural frequency not much different from those typical of the comfort modes of the sprung mass related to heave and pitch. In this mode the transmission behaves as a massless torsional spring connecting two large inertias at its ends, those of the engine and the vehicle.

An extremely simple model may be used to study this mode, similar to those used earlier for the take-off manoeuvre, the difference being that the clutch may now be considered as a rigid joint. The natural frequencies of the crankshaft are much higher, and at these low frequencies the engine may be considered as a single moment of inertia.

In all reciprocating engines the driving torque changes in time, with a period depending on the duration of the thermodynamic cycle, lasting two revolutions of the crankshaft (in four-stroke cycle engines, one revolution in two-stroke cycle engines). These frequencies are higher, often much higher, than 10 Hz. The driving torque may be considered as constant at its average value computed over one cycle: The variability of the pressure of the working gases on the piston and the driving torque cannot excite vibration at such a low frequency.

Slower variations of the driving torque, however, such as those due to manipulation of the accelerator pedal, may have an important role in exciting low frequency vibration. A typical case is that of a manoeuvre usually called *tip-in, tip-out*: The driver pushes suddenly on the accelerator pedal while the vehicle is travelling at a constant speed, usually low, causing a driving torque step. The step increase of the driving torque may be followed by an equally sudden release of the accelerator pedal.

Some experimental results obtained during a tip-in, tip-out manoeuvre are shown in Fig. 30.1. The vehicle travels on a straight road at a given speed (in the figure, at a speed corresponding to an engine speed of about 1,500 rpm in top gear) and, when all parameters are constant, the accelerator pedal is pushed fully down. When the engine speed has increased by about 500 rpm the accelerator is fully released until the previous speed has again been attained. This manoeuvre is repeated several times, at different initial speeds and with different gears engaged. One of the cycles is shown in the figure; its duration is about 8s.

The results shown were filtered with a low-pass filter removing all frequencies higher than 25 Hz to make all phenomena occurring in the frequency range from 0 to 10 Hz more apparent. As can be seen, the vehicle velocity shows strong oscillations, causing longitudinal accelerations that were measured at two points important for comfort: The attachment points of the seat and its back.

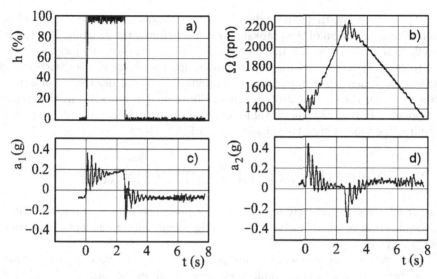

FIGURE 30.1. Experimental results obtained in a tip-in, tip-out manoeuver. a): throttle opening; b): engine speed; c): longitudinal acceleration measured at the attachment opoints of the seat; d): longitudinal acceleration at the back of the seat.

By analyzing the results it is possible to show that when the engine is accelerating the forced oscillations have a frequency of about 4 Hz, while when the vehicle slows their frequency is 3 Hz. This difference can be explained by the nonlinearity of some elements, such as the damper springs of the clutch disk, that perform more stiffly when heavily loaded in torsion.

As is common for step inputs, all frequencies of the system are in this way excited, particularly low frequencies, because the response of the engine is not immediate and smooths out what in theory should be a true step. The vehicle therefore does not accelerate (or decelerate) smoothly and torsional vibration of the driveline causes longitudinal oscillations of the whole vehicle, with vertical motions of the sprung and unsprung masses.

This manoeuvre may be performed at different speeds and in different ways, but the oscillations produced by it strongly reduce comfort, making it an important issue in vehicle testing. If problems appear, adequate correction must be introduced, usually by increasing the torsional natural frequencies of the driveline or its damping. A provision that was recently found to be quite effective is the use of a flywheel damper with two masses, as shown in Part II. The engine flywheel is divided into two parts, connected to each other by a low stiffness spring and adequate damping. Because the torsional oscillations of the transmission are triggered by manipulation of the accelerator, an effective solution is to modify the throttle control of the engine so that sudden increases of the engine torque are avoided. This is simple if a *by wire* engine control is used, because it is sufficient to introduce a smoother control algorithm into the system.

Apart from low frequency vibration, higher frequency vibration caused by the torsional vibration of the crankshaft of the engine or the gearbox, is possible. Its effect is to increase the noise produced by the gearbox (in jargon, *rattle*) and to cause fatigue problems in the crankshaft, the shafts of the gearbox and gearwheels. When this occurs, the useful provisions are, besides the use of a twin-mass flywheel, those typical of torsional vibration, i.e. inserting in the engine and possibly in the driveline suitable torsional dampers or compliant joints that uncouple the vibration of the various parts of the system.

30.2   DYNAMIC MODEL OF THE ENGINE

Almost all vehicles presently on the road are propelled by a reciprocating internal combustion engine. Machines containing reciprocating elements have some peculiar dynamic problems.

Most reciprocating machines, and practically all those used in the automotive industry, are based on a crank mechanism, often in the form of a crankshaft with several connecting rods and reciprocating elements. Such devices cannot, in general, be exactly balanced: The inertial forces they exert on the structure of the vehicle constitute a system of forces whose resultant is not insignificant and is variable in time. The geometric configuration of the system created by the crankshaft, the connecting rods, and the reciprocating elements can be quite complex. Crankshafts not only do not possess axial symmetry but often lack symmetry planes.

In these conditions, uncoupling among axial, torsional, and flexural behavior is not possible, in anything other than a rough approximation, and vibration modes become quite complicated. The external forces acting on the elements of reciprocating machines are usually variable in time, often following periodic laws, as the forces exerted by hot gases on the pistons of reciprocating internal-combustion engines demonstrate. Their period is equal to the rotation period in two-stroke cycle engines and is twice the rotation period in four-stroke cycle engines. Their periodic time histories are not harmonic but, once harmonic analysis has been performed, they may be considered as the sum of many harmonic components whose frequencies are usually multiples, by a whole number or a rational fraction, of the rotational speed of the machine. There may be many possibilities of resonance between these forcing functions and the natural frequencies of the system.

In general, the most dangerous vibrations are linked to modes that are essentially torsional. These couple with the modes of the driveline and the longitudinal dynamics of the vehicle

30.2.1   Equivalent system for a crank mechanism

The traditional approach to the study of torsional vibrations in reciprocating machines is based on the reduction of the actual system made of crankshafts,

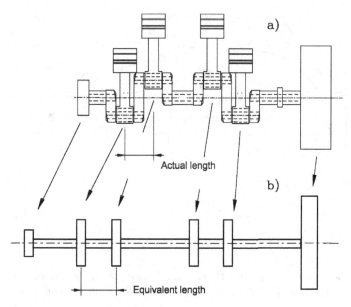

FIGURE 30.2. Sketch of the crankshaft: (a) actual system; (b) equivalent system, lumped-parameters model.

connecting rods, and reciprocating elements to an equivalent system. The latter is usually modeled as a lumped-parameters system whose torsional behavior can be studied separately[1] (Fig. 30.2).

Consider the crank mechanism sketched in Fig. 30.3. It is made of a disc, with a crankpin in B on which the connecting rod PB, whose center of mass is G, is articulated. The reciprocating parts of the machine are articulated to the connecting rod in P. The actual position of the center of mass of the reciprocating elements, which may include the piston as well as the crosshead and other parts, is not important in the analysis; in the following study this point will be assumed to be located directly in P. The axis of the cylinder, i.e., the line of motion of point P, does not necessarily pass through the axis of the shaft; the offset d will, however, be assumed to be small. Let J_d, J_b, m_b, and m_p be the moment of inertia of the disc that constitutes the crank, the moment of inertia of the connecting rod (about its center of gravity G) and the masses of the connecting rod and of the reciprocating parts, respectively.

---

[1]Torsional dynamics of reciprocating machinery is dealt with in many texts on vibration dynamics, like G. Genta, *Vibration of Structures and Machines*, Springer, New York, 1998. For a detailed study, specific texts on the subject can be found, such as E.J. Nestorides, *A handbook on torsional vibration*, Cambridge Univ. Press, 1958; K.E. Wilson, *Torsional vibration problems*, Chapman & Hall, 1963.

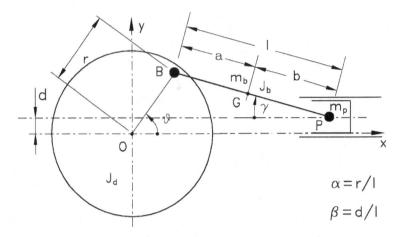

FIGURE 30.3. Sketch of the crank mechanism.

The coordinates of points B, G, and P can be expressed in the reference frame Oxy, shown in Fig. 30.3 as functions of the crank angle θ, as

$$(\overline{\text{B-O}}) = \left\{ \begin{array}{c} r\cos(\theta) \\ r\sin(\theta) \end{array} \right\}, \qquad (\overline{\text{G-O}}) = \left\{ \begin{array}{c} r\cos(\theta) + a\cos(\gamma) \\ r\sin(\theta) - a\sin(\gamma) \end{array} \right\}, \qquad (30.1)$$

$$(\overline{\text{P-O}}) = \left\{ \begin{array}{c} r\cos(\theta) + l\cos(\gamma) \\ d \end{array} \right\}.$$

Angle γ is linked to angle θ by the equation

$$r\sin(\theta) = d + l\sin(\gamma), \qquad (30.2)$$

i.e.

$$\sin(\gamma) = \alpha\sin(\theta) - \beta,$$

where

$$\alpha = \frac{r}{l}, \qquad \beta = \frac{d}{l}.$$

Ratios α and β are expressed by numbers smaller than 1, and in practice they are quite small; usually $\alpha \leq 0.3$ and $\beta = 0$.

Remark 30.1 *In the case of an ideal crank mechanism with an infinitely long connecting rod ($\alpha = 0$), with the axis of the cylinder passing through the axis of the crank ($\beta = 0$), the motion of the reciprocating masses is harmonic when the crank speed is constant.*

Because $\dot{\theta}$ is the angular velocity of the crank, its kinetic energy is simply

$$T_d = \frac{1}{2}J_d\dot{\theta}^2. \qquad (30.3)$$

The speed of the reciprocating masses can be easily obtained by differentiating the third equation (30.1) with respect to time and obtaining the expression for $\dot{\gamma}$ from equation (30.2):

$$V_p = -r\dot{\theta}\sin(\theta) - l\dot{\gamma}\sin(\gamma) = -r\dot{\theta}\left[\left(1 + \alpha\frac{\cos(\theta)}{\cos(\gamma)}\right)\sin(\theta) - \beta\frac{\cos(\theta)}{\cos(\gamma)}\right]. \quad (30.4)$$

The kinetic energy of the reciprocating masses is

$$T_p = \frac{1}{2}m_p r^2\dot{\theta}^2 f_1(\theta), \quad (30.5)$$

where

$$f_1(\theta) = \left[\sin(\theta) + \alpha\frac{\sin(2\theta)}{2\cos(\gamma)} - \beta\frac{\cos(\theta)}{\cos(\gamma)}\right]^2.$$

Instead of computing the kinetic energy of the connecting rod by writing the velocity of its center of gravity G, it is customary to replace the rod with a system made of two masses m_1 and m_2, located at the crankpin B and the wrist pin P, respectively, and a moment of inertia J_0. To simulate the connecting rod correctly, such a system must have the same total mass, moment of inertia, and center of mass position. These three conditions produce three equations yielding the following values for m_1, m_2, and J_0:

$$m_1 = m_b\frac{b}{l} \quad , \quad m_2 = m_b\frac{a}{l} \quad ,$$

$$J_0 = J_b - (m_1 a^2 + m_2 b^2) = J_b - m_b ab . \quad (30.6)$$

Generally speaking, the moment of inertia of masses m_1 and m_2 is greater than the actual moment of inertia of the connecting rod and, consequently, the term J_0 is negative. The kinetic energy of mass m_1 can be computed simply by adding a moment of inertia $m_1 r^2$ to that of the crank.

Remark 30.2 *The negative moment of inertia has no physical meaning in itself: The minus sign indicates that it is simply a term that must be subtracted in the expression of the kinetic energy.*

Similarly, the kinetic energy of mass m_2 can be accounted for by adding m_2 to the reciprocating masses. The effect of the moment of inertia J_0 can be easily computed

$$T_{J_0} = \frac{1}{2}J_0\dot{\gamma}^2 = \frac{1}{2}J_0\dot{\theta}^2 f_2(\theta), \quad (30.7)$$

where

$$f_2(\theta) = \alpha^2\left[\frac{\cos(\theta)}{\cos(\gamma)}\right]^2.$$

The total kinetic energy of the system shown in Fig. 30.3 is, consequently,

$$T = \frac{1}{2}\dot{\theta}^2\left[J_d + m_1 r^2 + (m_2 + m_p)r^2 f_1(\theta) + J_0 f_2(\theta)\right] = \frac{1}{2}J_{eq}(\theta)\dot{\theta}^2. \quad (30.8)$$

It is now clear that the whole system can be modeled, from the viewpoint of kinetic energy, by a single moment of inertia variable with the crank angle $J_{eq}(\theta)$, rotating at the angular velocity $\dot{\theta}$.

The equivalent moment of inertia is a periodic function of θ, with a period of 2π.

In the limiting case of $\alpha = \beta = 0$, corresponding to an infinitely long connecting rod (piston moving with harmonic time history), the expressions for $f_1(\theta)$ and $f_2(\theta)$ are particularly simple

$$f_1(\theta) = \sin^2(\theta) = \frac{1 - \cos(2\theta)}{2} \quad , \quad f_2(\theta) = 0 \ . \tag{30.9}$$

In practice, it is impossible to neglect the fact that the length of the connecting rod is finite, even if α is usually not greater than 0.3. At any rate there is no difficulty in expressing J_{eq} through a Fourier series

$$J_{eq} = J_0 + \sum_{i=1}^{n} J_{ci} \cos(i\theta) + \sum_{i=1}^{n} J_{si} \sin(i\theta) \ , \tag{30.10}$$

that is here truncated at the nth harmonics. Coefficients J_0, J_{ci} and J_{si} may be computed numerically without difficulty, by computing the values of functions $f_1(\theta)$ and $f_2(\theta)$ for a number of values of angle θ and then applying one of the standard FFT algorithms. The number of values of $J_{eq}(\theta)$ to be computed depends on the value of n and, if many harmonics are required, 2048 or 4096 values may be needed.

Traditionally, before the numerical computation of the coefficients of the Fourier series became straightforward, explicit expressions of the coefficients were used; these are discussed in several handbooks. The coefficients were expressed as power series in α and β; the number of terms needed depends on how many harmonics must be accounted for. To compute six harmonics, series with terms up to α^4 and β^4 were used.

If the axis of the cylinder passes through the center of the crank ($\beta = 0$), as is usually the case, $f_1(\theta)$ and $f_2(\theta)$ are even functions of θ for symmetry reasons. J_{eq} is then an even function and all coefficients J_{si} vanish. If $\alpha = 0$, the expression of the average equivalent moment of inertia reduces to

$$J_0 = J_d + r^2 \frac{2m_1 + m_2 + m_p}{2}. \tag{30.11}$$

30.2.2 Driving torque

A moment caused by the pressure of the gases contained in the cylinder $p(t)$ acts upon each crank, varying in time during the working cycle of the engine. Once the pressure $p(t)$ is known, the driving torque acting on the crankshaft can be computed from the virtual work $\delta\mathcal{L}$ performed by that pressure during a virtual displacement δs of the piston. A virtual displacement $\delta\theta$ of the crank corresponds to a displacement δs of the piston; the relationship between them is

$$\delta s = \frac{V_p}{\dot\theta}\delta\theta = r\sqrt{f_1(\theta)}\delta\theta \ . \tag{30.12}$$

The corresponding virtual work $\delta\mathcal{L}$ performed by this pressure can be expressed as

$$\delta\mathcal{L} = p(t)A\delta s = p(t)rA\sqrt{f_1(\theta)}\delta\theta, \tag{30.13}$$

where function $f_1(\theta)$ is given by equation (30.5) and A is the area of the piston. The generalized force M_m due to the pressure $p(t)$, i.e. the driving torque, is consequently

$$M_m = \frac{d(\delta\mathcal{L})}{d(\delta\theta)} = p(t)rA\sqrt{f_1(\theta)}. \tag{30.14}$$

In the case of two-stroke cycle engines working at constant speed, function $p(t)$ is periodic with a period equal to the time needed to perform one revolution of the crankshaft, i.e. its frequency is equal to the rotational speed Ω of the engine. In the case of four-stroke-cycle internal combustion engines, again assuming constant speed operation, the period of function $p(t)$ is doubled, i.e. its fundamental frequency is equal to $\Omega/2$. Because the generalized force (moment) $M_m(t)$ is periodic, with the same frequency of law $p(t)$, it can be expressed by a trigonometric polynomial, truncated after m harmonic terms

$$M_m(t) = M_0 + \sum_{k=1}^{m} M_{ck}\cos(k\omega't) + \sum_{k=1}^{m} M_{sk}\sin(k\omega't) \ , \tag{30.15}$$

where the frequency ω' of the fundamental harmonic is equal to Ω, except in the case of four-stroke-cycle internal-combustion engines, in which

$$\omega' = \frac{\Omega}{2} \ . \tag{30.16}$$

The coefficients of the polynomial may be computed starting from the theoretical or experimental law $p(t)$, and empirical expressions can be found in the literature. In any case, the driving torque depends upon working conditions. It is possible to assume that coefficients M_{ck} and M_{sk} are proportional to the average driving torque M_0 or to the product of half the capacity of the cylinder (the area of the piston times the crank radius) times the mean indicated pressure.

Angle θ may be used instead of time as an independent variable and the driving torque may be written as

$$M_m(\theta) = M_0 + \sum_{k=1}^{m} M_{ck}\cos(k\theta') + \sum_{k=1}^{m} M_{sk}\sin(k\theta') \ , \tag{30.17}$$

where θ' is equal to θ in two-stroke cycle engines and $\theta/2$ in four-stroke cycle engines.

Remark 30.3 *When the engine works at variable speed, it may be assumed that the speed variations are much slower than the phenomena occurring in the combustion chamber. Conditions at variable speed may be approximated by a sequence of constant speed operations at the various speeds.*

30.2.3 Forcing functions on the cranks of multicylinder machines

All motor vehicles other than motorcycles powered by single-cylinder engines are provided with reciprocating engines with a number of cylinders. The most common engine arrangement is in-line, but many engines have opposite cylinders or V arrangements.

In machines with a number of cranks, if the various cranks, reciprocating parts, and working cycles are all equal, the time histories of the moments acting on the various nodes of the equivalent system are all equal but are timed differently. Because each harmonic component of the moment acting on the cranks can be represented as the projection on the real axis of a vector rotating in the Argand plane with constant angular velocity, it is possible to draw, for each harmonic, a plot in which the various vectors acting on the different cranks of the machine are represented. Because, as already stated, the amplitudes of these vectors are equal, the diagram is useful only for comparing the phases of the vectors, which are traditionally plotted with unit amplitude. The phasing of the vectors depends on the geometric characteristics of the machine and, in the case of four-stroke-cycle engines, on the firing order. Such diagrams are usually referred to as *phase angle diagrams*.

Consider, for example, an in-line four-stroke-cycle internal-combustion engine. If the working cycles of the various cylinders are evenly spaced in time, the cranks that subsequently fire must be at an angle of $4\pi/n$ rad, where n is the number of cylinders. In a four-in-line engine, this angle is 180°, and the most common geometric configuration of the crankshaft is that shown in Fig. 30.4a, chosen because it allows the best balancing of inertia forces. In the same figure, the configuration of the crankshaft of a six-in-line engine is also shown.

In a four-cylinder engine, the possible firing orders are two: 1-2-3-4 and 1-3-4-2. In both cases, it is impossible to prevent two contiguous cylinders from immediately firing one after the other. The phase-angle diagrams for the first four harmonics are plotted in Fig. 30.4b for the second of the two firing orders.

If the order of the harmonic is a whole multiple of the number of cylinders, all rotating vectors are superimposed, i.e. the forcing functions acting on all cranks are all in phase. These harmonics are usually the most dangerous and are often referred to as major harmonics. The phase-angle diagrams for the $(n+i)$-th harmonic coincide with that related to the ith harmonic and, consequently, only the first n phase-angle diagrams are usually plotted.

Remark 30.4 *The phase-angle diagrams have been plotted in such a way that they supply the excitation phasing on the various cranks with respect to that acting on a crank chosen as reference, usually the first. Each harmonic then has*

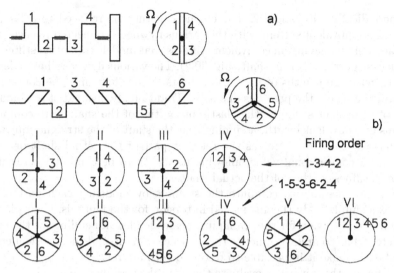

FIGURE 30.4. (a) Configuration of the crankshaft and crank angle diagrams for in-line four-stroke-cycle four- and six-cylinder internal combustion engines. In the latter case the configuration shown is just one of the possible choices; (b) phase-angle diagrams for the same engines.

a phasing with respect to the fundamental harmonic that must be considered when the effects of the various harmonics are added.

The forcing function acting on the jth crank may be approximated by the following series, truncated at the mth harmonic

$$M_{m_j} = \sum_{k=0}^{m} M_{m_k} e^{i\left(k\omega'i + \Phi_{m_k} + \delta_{j_k}\right)} ,$$ (30.18)

where

- M_{m_k} and Φ_{m_k} are the amplitude and phase of the kth harmonic of the driving torque, respectively. With reference to the series (30.15) approximating the driving torque, their values are

$$M_{m_k} = \sqrt{M_{ck}^2 + M_{sk}^2} \quad \text{and} \quad \Phi_{m_k} = \arctan(M_{ck}/M_{sk}) ,$$

respectively.

- δ_{j_k} is the phase of the kth harmonic acting on the jth crank, as obtained from the phase-angle diagram. If the diagram is referred to the first crank, $\delta_{j_k} = 0$ for $j = 1$.

30.2.4 Stiffness of the crankshaft

From the viewpoint of inertia forces, the cranks and reciprocating elements are equivalent to a number of concentrated flywheels, even if their moments of inertia vary periodically with angle θ. The engine can thus be reduced to a lumped-parameters equivalent system, with the various flywheels connected to each other by straight shafts having an equivalent stiffness that models the actual stiffness of the relevant portion of crankshaft (Fig. 30.2). The various flywheels have a length equal to zero: the lengths of the various parts of the shaft must be contiguous, each starting where the previous one ends. Traditionally, instead of reasoning in terms of equivalent stiffness, the elastic properties of the shaft were computed in terms of equivalent length, assuming that the shaft of the straight equivalent shaft has the same diameter as the relevant part of the actual shaft or, more often, has a conventional value, allowing its length to be computed so that its torsional stiffness is that of the actual shaft.

It is not possible to compute the stiffness by modeling each part of the crank as a simple body (beams loaded in torsion for the journals, beams loaded in bending and torsion for the crankpins, beams loaded in bending for the crank webs, etc.). The complex geometry, the presence of radii, and the low slenderness of the beams make it impossible to resort to such approach.

There are three ways to evaluate the equivalent stiffness:

1. experimental evaluation,

2. use of semi-empirical methods, and

3. numerical modeling, mainly using the FEM.

Experimental evaluation clearly gives the most reliable results, but it cannot be performed at the design stage without additional costs. Moreover, it increases in the time required for dynamic analysis because of the need to build models or prototypes.

Empirical and semi-empirical formulas, allowing at least approximate evaluations to be obtained, have been suggested by many authors and can be found in several handbooks[2].

Nowadays it is possible to build numerical models of a single crank and to evaluate their static stiffness by numerical methods, mainly the FEM. This is much simpler than the complete numerical simulation of the crankshaft using the same numerical approach. Only one crank (or half, for symmetry) needs to be modelled, assuming all cranks are equal, and the computation reduces to a static evaluation.

Nevertheless, the geometric complexity and uncertainties on how to constrain the mathematical model may make this computation more difficult than it appears.

Remark 30.5 *Strictly speaking, the lack of symmetry couples torsional and flexural deformations, and the stiffness of the crankcase and the presence of oil films in the bearings may affect the results.*

---

[2]See, for example, E.J. Nestorides, *A Handbook on Torsional Vibration*, Cambridge Univ. Press, 1958, or W. Ker Wilson, *Torsional Vibration Problems*, Chapman & Hall, 1963.

The equivalent stiffness and equivalent length, computed through any of the mentioned approaches, are linked through the obvious formula

$$k = G \frac{I_p}{l_{eq}}. \tag{30.19}$$

30.2.5 Damping of the system

If the damping present in the engine were mostly caused by the internal damping of the material constituting the crankshaft, there would be no difficulty introducing a proportional damping with modal damping ratio equal for all modes: $\zeta_j = \eta/2$, where η is the loss factor of the material of the crankshaft.

But damping is actually due to many causes, among which friction between moving parts (including that between the piston and the cylinder wall), electromagnetic forces (if an electric motor or generator is driven by the engine), and the presence of fluid in which some rotating parts move, can be important. Neglecting them would lead to a large underestimate of damping. It is usually necessary to resort to experimental results, obtained from machines similar to the one under study, and to empirical or semi-empirical formulas and numerical values reported in the literature.

The damping due to the crank mechanism is usually evaluated by introducing a damping force acting on the crankpin that is proportional to the area of the piston and the velocity of the crankpin. The damping moment acting on the jth crank is

$$M_d(t) = k'Ar^2 \,, \tag{30.20}$$

where k' is a coefficient whose dimension is a force multiplied by time and divided by the third power of a length. In S.I. units, it is expressed in Ns/m^3. Values of k' included in the range between 3,500 and 10,000 Ns/m^3 for in-line aircraft engines and between 15,000 and 1.5×10^6 Ns/m^3 for large internal-combustion engines can be found in the literature.

Remark 30.6 *The lower and upper values of these ranges are very different and must be regarded only as indicative values; only experimental results on machines similar to the one under study can be reliable.*

The use of equation (30.20) leads to the assumption that in each crank there is a viscous damper with damping coefficient equal to

$$c_j = k'Ar^2 \,. \tag{30.21}$$

In many cases, it is impossible to prevent the amplitude of torsional vibration from reaching values too large to insure safe operation of the machine or adequate vibrational and acoustic comfort solely by exploiting the damping properties of the system elements.

In such cases, torsional vibration dampers are applied at one end of the crankshaft. They are made of a flywheel (usually referred to as *seismic mass*)

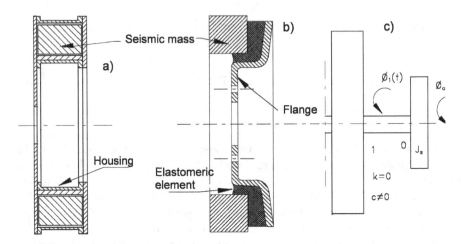

FIGURE 30.5. Dissipative torsional vibration damper; (a): viscous damper; (b): elastomeric damper; (c): sketch of the model for the dynamic study.

whose geometric configuration may draw on a wide variety of types, connected to the shaft by suitable elastic and damping elements.

Almost all torsional vibration dampers can be reduced to the concept of the damped vibration absorber. Without including all possible types, these can be subdivided into three categories: dissipative dampers, damped vibration absorbers, and rotating pendulum vibration absorbers. The latter are seldom used in automotive engines and will not be dealt with here.

A typical dissipative torsional damper used primarily in diesel engines for industrial vehicles is the viscous damper shown in Fig. 30.5a. It is applied to one end of the crankshaft and consists of a flywheel, generally shaped as a ring free to rotate within a casing filled with a high viscosity fluid, for example a silicon-based oil. Damping in this case is of the viscous type, i.e., the drag torque is proportional to the relative angular velocity between the ring and the housing, with the damping coefficient depending on the clearance between the two and on the characteristics of the fluid. The latter are greatly influenced by the fluid temperature. The model for the dynamic study of the system must be modified by adding the moment of inertia of the casing, in the node in which the damper is applied (node 1, Fig. 30.5c), and by adding a new node (node 0 in the same figure) in which the inertia of the ring is located. The two nodes are connected by a viscous damper and a spring with zero stiffness.

A viscous damper of this kind actually lacks stiffness only in static conditions and at very low frequency. The characteristics of the fluid are such that an elastic behavior of increasing strength occurs with increasing frequency, obviously adding to the damping behavior. This means that the torque the damper applies to the shaft depends not only on the relative velocity, but also on the angular displacement. A stiffness, which is a function of the frequency, must be added to the previous model, but this holds only in harmonic, or at least polyharmonic, vibration.

For a first-approximation evaluation of the optimum system damping, it may be assumed that the presence of the damper does not significantly affect the natural frequencies of the system. The above mentioned stiffness may be neglected. Under this assumption, the behavior of the damper may be studied separately, assuming the time history of the motion $\phi_1(t)$ of the node where it is applied. The equation of motion of node 0 is then

$$J_s \ddot{\phi}_0 + c\dot{\phi}_0 = c\dot{\phi}_{1_0} \ . \tag{30.22}$$

Assume that the time history at node 1, where the damper is applied, is harmonic

$$\phi_1(t) = \phi_{1_0} e^{i\omega t} \ .$$

The time history at node 0 is also harmonic

$$\phi_0(t) = \phi_{0_0} e^{i\omega t}$$

although not in phase with the excitation. The response can be computed using the frequency domain equation

$$\left(-J_s \omega^2 + i\omega c \right) \phi_{0_0} = i\omega c \phi_{1_0} \ . \tag{30.23}$$

By separating the real and imaginary parts of the response, it follows that

$$\begin{cases} \Re(\phi_{0_0}) = \phi_{1_0} \dfrac{c^2}{c^2 + J_s^2 \omega^2} \ , \\[3mm] \Im(\phi_{0_0}) = \phi_{1_0} \dfrac{-cJ_s\omega}{c^2 + J_s^2 \omega^2} \ . \end{cases} \tag{30.24}$$

If time $t = 0$ is chosen as the time when the angular displacement ϕ_1 reaches its maximum, i.e. if ϕ_{1_0} is real, the relative displacement can be expressed as

$$|\phi_0(t) - \phi_1(t)| = \sqrt{\left[\Re(\phi_{0_0}) - \phi_{1_0} \right]^2 + \left[\Im(\phi_{0_0}) \right]^2} \tag{30.25}$$

and then

$$|\phi_0(t) - \phi_1(t)| = \phi_{1_0} \frac{J_s^2 \omega^2}{\sqrt{c^2 + J_s^2 \omega^2}} \ . \tag{30.26}$$

The energy dissipated in a period by the damper is

$$E_d = \int_0^T c \left[\dot{\phi}_0(t) - \dot{\phi}_1(t) \right]^2 dt = c\phi_{1_0}^2 \pi \frac{J_s^2 \omega^4}{c^2 + J_s^2 \omega^2} . \tag{30.27}$$

It is easy to verify that both conditions $c = 0$ and $c \to \infty$ lead to a vanishingly small energy dissipation: In the first case because the seismic mass does not interact with the system, and in the second case because nodes 1 and 0 are rigidly connected. The value of the damping coefficient leading to a maximum energy dissipation can be obtained simply by differentiating equation (30.27) and equating the derivative to zero

$$J_s^2 \omega^2 - c^2 = 0 \ . \tag{30.28}$$

The value of the optimum damping so obtained is

$$c_{opt} = J_s \omega \ . \tag{30.29}$$

Even if the value of the optimum damping depends on the frequency, dampers of the type described here allow a substantial reduction of the amplitude of vibration in a large frequency range.

Because all dissipative dampers convert mechanical energy into heat, they are subject to potentially high temperatures. It is then necessary to verify that they can dissipate all the thermal energy they produce, which can be computed using formulas of the type of equation (30.27), at least in terms of average power over a given period of time.

A limit to the ratio between the thermal power and the external surface of the damper is usually assumed. For the type in Fig 30.5a, for example, it is suggested not to exceed 1.9×10^4 kW/m^2 in continuous operation and 5×10^4 kW/m^2 for short periods of time.

If the seismic mass is connected to the shaft with a torsional spring with non-vanishing stiffness, the device is a true dynamic vibration absorber. Such torsional vibration absorbers introduce a new natural frequency into the system and change the natural frequency on which they are tuned. The distance between the two resonance peaks increases with increasing moment of inertia of the seismic mass. Because an undamped vibration absorber is effective in a very narrow frequency range, outside of which it is not only ineffective but can cause new resonances, the seismic mass is connected to the shaft through a system that has a certain amount of damping. In such cases it is possible to obtain a response that is fairly flat in an ample range of frequencies.

From a practical viewpoint, all dampers shown in the previous section can be converted into damped vibration absorbers simply by adding an elastic element between the shaft and the seismic mass, which allows the damper to be tuned on the required frequency.

Elastomeric dampers are used on many automotive engines (Fig. 30.5b), particularly on small diesel engines. They may be considered as damped vibration absorbers. The elastomeric elements act as both springs and dampers, and can be designed so as to achieve the required dynamic characteristics. In this case the damper must be designed to take into account the heat generated within the damping element, particularly because the thermal conductivity and mechanical characteristics at high temperature of the rubber are both low. Overheating is particularly dangerous, because any increase of temperature leads to a decrease of the internal damping and then an increase of vibration amplitude. This leads to a further temperature increase until the damper is destroyed, something that could cause severe fatigue problems to the whole system.

30.2.6 Ancillary equipment

The crankshaft drives a number of ancillary devices that are increasingly common in modern cars. To the camshafts, the generator, the water pump and possibly the fan (which is, however, often driven directly by an electric motor) and other devices such as the power steering pump and the air conditioner compressor must be added. These devices are usually driven through a V or a timing belt, while camshafts are usually driven by a chain or a timing belt. Each shaft has its own speed, so that each transmission has its own transmission ratio. The transmission ratio of the cam shaft is always 1/2.

Ancillary devices may be included in the engine model as well as the driveline by adding other concentrated moments of inertia connected to the main system by secondary shafts that simulate the stiffness of the belt, chain or gearwheel transmission. The same procedure that will be shown when dealing with the driveline may be used to account for the different gear ratios. This approach may, however, be only a first approximation, because the belt usually moves more than one device and is kept tight by tensioners that have their own dynamics based on their mass and stiffness. The belt is then not equivalent to a number of shafts connecting the various devices to the crankshaft. Moreover, the behavior of the belt is often nonlinear.

Because of the presence of many ancillary devices, usually mounted on brackets with a limited stiffness, torsional vibration of the crankshaft causes vibration of the brackets and the masses mounted on them. These vibrations may adversely affect vibrational and acoustic comfort. Because these devices are usually located close to the cooling air intake, noise caused by their vibration propagates outside the vehicle and contributes to what is usually referred to as acoustic pollution.

Dynamic vibration absorbers or low stiffness joints are often located close to the pulleys driving the belts, to reduce both noise and dynamic stresses. The functions of elastic joint and damper are generally performed by a single device, containing two or more seismic masses, one of which is the outer rim of the pulley driving the belt.

Because the inertia of the ancillary device is not large and the stiffness of the belt or other transmission devices is high, their dynamics lies in a frequency range above that involving the sprung and unsprung mass modes, and primarily affects acoustic comfort.

30.2.7 Engine control

Variations of the engine torque in time due to the thermodynamic cycle may excite vibrations of the driveline at medium-high frequency. The fundamental harmonic has a frequency $\Omega/2$ in a four-stroke cycle engine, which at 1000 rpm already corresponds to 8 Hz. At the speeds at which the engine usually operates, frequencies are much higher. The other harmonics have a frequency that is a multiple of the fundamental frequency and is often quite high, because 20 or even 25 harmonics must usually be taken into account. At the first natural frequency of the driveline the engine torque can be considered constant in time, as far as the internal dynamics of the engine is concerned. The torque varies according to the commands given by the driver.

In traditional layouts, the engine is controlled by the driver through the accelerator pedal, with commands transmitted to the throttle or injection pump by a mechanical (cable) or hydraulic transmission. Even when the driver manoeuvres the accelerator quite quickly, the command is transferred directly to the engine. The engine torque increases rapidly following a sudden opening or closing of the throttle, with characteristic times typical of those of the thermodynamic cycle: A sort of step input is then occurring, exciting all frequencies up to values of some tens of Hz, or even more. The accelerator manoeuvre may then excite the first torsional frequency of the driveline.

In more modern layouts the transmission of the accelerator control is performed by a *by wire* device: The pedal is connected to a sensor supplying a position signal to the engine control system. In this case it is possible to prevent the transmission natural frequencies from being excited, avoiding high mechanical stresses to the involved elements.

The simplest strategy is to introduce a low-pass filter, cutting out frequencies above a value of about 1 or 2 Hz. This has the advantage of preventing the excitation of the driveline natural frequencies, but the engine becomes less responsive, both when accelerating and slowing down. More complex strategies based on filters that cut out specific frequencies, or provide laws of torque as a function of time to avoid excitation of resonant vibration, require a detailed knowledge of the dynamics of the driveline and accurate adjustments. Strategies of this type are called open loop or feedforward strategies, because they modify the input of the system (in this case the driving torque) without measuring its effects.

Such an open loop approach may be complemented by measuring the effects of the accelerator manoeuvre, (i.e. acceleration of the vehicle, torsional deformation of the driveline, etc.) and then modifying the command of the throttle or injection pump to limit vibration (closed loop or feedback control).

The model of the driveline must also contain in this case a mathematical model of the device controlling the engine, to simulate specific step input or tip-in, tip-out manoeuvres.

30.2.8 Engine suspension

The engine suspension system has two primary functions:

- supporting the static and dynamic loads due to the mass of the engine, its reciprocating and rotating elements and driving torque,

- isolating the structure of the vehicle from vibration and noise produced by the propulsion unit, which usually includes the engine, the gearbox and often the differential.

The suspension system should be stiff enough to perform the first task without allowing large displacements and rotations, the limiting case being mounting the engine stiffly on the vehicle body. On the other end, to effectively insulate the vehicle from vibration produced by the propulsion unit, its suspension should be as soft as possible. To strike a compromise between these contrasting requirements, it is of the utmost importance to locate the engine mounts suitably. These are usually elastomeric elements performing the tasks of spring and damper simultaneously.

If the mounts had low stiffness and damping, the engine would behave like a rigid body isolated in space and would transfer no vibration to the structure except that transmitted by the surrounding air. The motion of the engine could then be studied as that of an isolated rigid body on which the actions caused by the pressure of the working gases and the inertia forces caused primarily by imbalance of rotating elements as well as the linear inertia of reciprocating masses are exerted. In this analysis the forcing functions acting on the engine, which are periodic with a period equal to the duration of the thermodynamic cycle (fundamental frequency equal to half the rotation frequency in four-stroke cycle engines), may be developed in Fourier series. The static (zero-frequency) component due to weight and the constant component of the driving torque must be neglected in this computation, because it would lead to a non-periodic motion of the engine that is not supported in any way.

If the motion of the engine as an insulated rigid body under the action of dynamic forces were such that we could identify points where the amplitude vanishes, locating the supports in those points would allow the engine to be supported without transmitting dynamic loads. In actual conditions this is not possible, but analysis of the free motion allows us to identify points where the amplitude of that motion is relatively small. We can then identify configurations from which the optimization of the engine suspension geometry can begin, taking into account the compliance of the structure supporting the engine.

Among the possible layouts for engine suspensions, those based on three supports, and on two supports withstanding the weight of the engine plus a link, itself attached through elastomeric supports, to withstand the engine torque, must be mentioned. The latter solution allows a high rotational stiffness to be obtained, accompanied by a moderate translational stiffness.

The concept of complex stiffness may be used to identify an in-phase and an in-quadrature stiffness, both functions of frequency, and then to describe the elastic and damping behavior of the supports under a harmonic forcing function. The characteristics of one of the supports of an engine suspension are shown in Fig. 30.6 as an example. Supports with controllable characteristics and active supports have been built and applied on some vehicles.

Remark 30.7 *Engine suspension may have a strong influence not only on insulation from vibration and noise produced by the engine, but also on riding comfort, because the engine is quite a large mass suspended through an elastic*

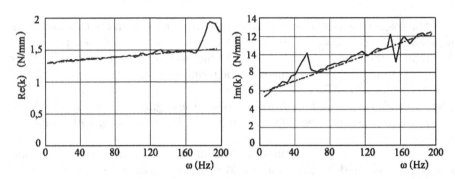

FIGURE 30.6. Real (in-phase) and imaginary (in quadrature) parts of the complex stiffness of a support for engine suspension versus the frequency.

and damping system that couples it to the heave and pitch dynamics of the vehicle and also, even if to a lesser extent, to the torsional dynamics of the driveline.

30.3 DRIVELINE

The driveline, including shafts, gear wheels, joints and other elements such as the clutch with related damper springs, may be modelled as a lumped parameters system (made by massless shafts where the elastic properties of the system are concentrated) with lumped masses modelling its inertial properties.

The damping of the system may be neglected altogether, or modelled by introducing suitable viscous dampers in parallel to the springs modelling the various parts of the shaft and joints.

If the clutch is assumed to be fully engaged and the gearbox is in a given gear, the configuration of the driveline is fixed. There is then no difficulty in building a simple mathematical model of the entire system.

Nor does the fact that the various elements of the driveline rotate at different speeds cause problems. Consider the system sketched in Fig. 30.7a, in which the two shafts are linked by a pair of gear wheels, with transmission ratio τ. For the study of the torsional vibrations of the system, it is possible to replace the system with a suitable equivalent , in which one of the two shafts is replaced by an expansion of the other (Fig. 30.7b).

Assuming as well that the deformation of gear wheels is negligible, the equivalent rotations ϕ_i^* may be obtained from the actual rotations ϕ_i simply by dividing the latter by the transmission ratio $\tau = \Omega_2/\Omega_1$,

$$\phi_i^* = \frac{\phi_i}{\tau}. \tag{30.30}$$

The kinetic energy of the ith flywheel, whose moment of inertia is J_i, and the elastic potential energy of the ith span of the shaft are, respectively,

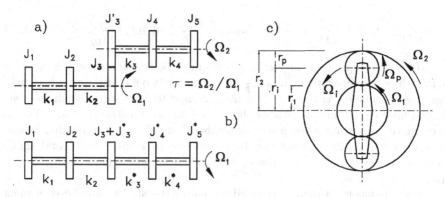

FIGURE 30.7. Geared system: Sketch of the (a) actual system and (b) equivalent system; (c) planetary gear train. Sketch of the system and notation.

$$\mathcal{T} = \tfrac{1}{2}J_i\dot{\phi}_i^2 = \tfrac{1}{2}J_i^*\dot{\phi}_i^{*2} \, ,$$
$$\mathcal{U} = \tfrac{1}{2}k_i\left(\phi_{i+1}^2 - \phi_i^2\right) = \tfrac{1}{2}k_i^*\left(\phi_{i+1}^{*2} - \phi_i^{*2}\right) ,$$

(30.31)

where the equivalent moment of inertia and stiffness are, respectively,

$$J_i^* = \tau^2 J_i \, , \qquad k_i^* = \tau^2 k_i \, .$$

(30.32)

The moments of inertia and the torsional stiffness of the various elements of the geared system can thus be reduced to the main system simply by multiplying them by the square of the gear ratio

$$c_i^* = \tau^2 c_i \, .$$

(30.33)

In the same way, if damping of the shafts is accounted for by introducing dampers in parallel to the springs, the damping coefficient must be multiplied by the square of the gear ratio.

If the system includes a planetary gear train, the computation can be performed without difficulties. The equivalent stiffness can be computed simply from the overall transmission ratio. The total kinetic energy of the rotating parts must be taken into account when computing the equivalent inertia. The angular velocities of the central gear Ω_1, of the ring gear Ω_2, of the revolving carrier Ω_i, and of the intermediate pinions Ω_p of the planetary gear shown in Figure 30.7c are linked by the equation

$$\frac{\Omega_1 - \Omega_i}{\Omega_2 - \Omega_i} = -\frac{r_2}{r_1} \quad , \qquad \Omega_p = (\Omega_1 - \Omega_i)\frac{r_1}{r_p} - \Omega_i.$$

(30.34)

The equivalent moment of inertia of the system made of the internal gear, with moment of inertia J_1, the ring gear, with moment of inertia J_2, the revolving carrier, with moment of inertia J_i, and n intermediate pinions, each with mass m_p and moment of inertia J_p, referred to the shaft of the internal gear is

$$J_{eq} = J_1 + J_2 \left(\frac{\Omega_2}{\Omega_1}\right)^2 + (J_i + n m_p r_i^2) \left(\frac{\Omega_i}{\Omega_1}\right)^2 + n J_p \left(\frac{\Omega_p}{\Omega_1}\right)^2. \qquad (30.35)$$

If the deformation of the meshing teeth must be accounted for, it is possible to introduce two separate degrees of freedom for the two meshing gear wheels into the model, modeled as two different inertias, and to introduce a shaft between them whose compliance simulates the compliance of the transmission. This is particularly important when a belt or flexible transmission of some kind is used instead of the stiffer gear wheels. In a driveline there may be several shafts connected to each other, in series or in parallel, by gear wheels with different transmission ratios.

The equivalent system is referred to one of the shafts and the equivalent inertias and stiffness of the elements of the others are all computed using the ratios between the speeds of the relevant element and the reference shaft. The equivalent system will then be made of a set of elements, in series or in parallel, following the scheme of the actual system, but with rotations that are all consistent.

If the compliance of the gears is to be accounted for in detail, the nonlinearities due to the contacts between the meshing teeth and backlash must be considered, as will be seen later.

The driveline may cause comfort problems not only in terms of its torsional compliance, but also its bending compliance. The propeller shaft and the wheel shafts have their own flexural natural frequencies and critical speeds. They may cause severe vibration when they operate close to a critical speed.

Without entering into details about the dynamic behavior of rotating elements[3], the following considerations can be advanced.

- The gyroscopic effects of transmission shafts are weak. Critical speeds are close to the natural frequencies when the system is not rotating.

- The balance conditions of the rotating elements have no effect on the natural frequency or the critical speed, but do determine the strength of the excitation at such speeds.

- The damping of the shaft (and the joints) has no effect in limiting the amplitude of vibration at the critical speeds, while the damping of the supports (non-rotating damping) is essential to this aim.

The critical speeds of the wheel shafts are generally beyond the working range and thus do not cause resonant vibration. However, the first critical speed of the propeller shaft in vehicles with front engine and rear wheel drive is in the working range. The critical speed of traditional propeller shafts, in two parts with a Hooke joint and elastic support in the middle, occurs when the vehicle

---

[3]See, for instance, G. Genta, *Dynamics of rotating systems*, Springer, New York, 2005.

travels at a low speed: The shaft then works normally in the supercritical regime, when self-centred.

The shaft and above all the joint must be accurately balanced, so as to go through the critical speed without strong vibration, while the central support must supply enough damping. The damping of the support is also needed to prevent the crossing of an instability threshold in high speed operation.

Cars with front engine and rear wheel drive are prone to vibrate strongly when passing through the critical speed of the propeller shaft if the balancing of the central joint deteriorates or the elastic and above all damping properties of the support become worse due to aging or wear.

Misalignment of the propeller shaft or wheel shafts may also cause the driveline to vibrate.

30.4   INERTIA OF THE VEHICLE

The tires may be considered as rigid bodies, allowing longitudinal slip to be neglected when performing a first approximation study. In this case the vehicle inertia and the resistance to motion may be accounted for as in Chapter 23 in the study of the take-off manoeuvre. The vehicle may then be modelled as a flywheel, connected after the wheel shafts that, in the equivalent system, rotate at the same speed as the engine.

Taking into account the inertia of all wheels, the moment of inertia of this flywheel is

$$J_v = \left(m + \sum_{\forall i} \frac{J_{ri}}{R_{ei}^2} \right) R_e^2 \tau^2 \, , \tag{30.36}$$

where J_{ri} is the moment of inertia of the ith wheel, which may have different equivalent rolling radii, R_e is the equivalent rolling radius of the driving wheels and τ is the overall gear ratio between engine and wheels.

The drag torque M_r applied to the flywheel simulating the vehicle is

$$M_r = F_r R_e \tau \, , \tag{30.37}$$

where the total resistance to motion (road load) F_r depends on the speed following Eq. (23.17):

$$F_r = A + BV^2 + CV^4 \, ,$$

and the expressions for constants A, B and C[4] are as reported in Chapter 23.

For a more detailed study, both the compliance of the tire and its longitudinal slip at the wheel-road contact must be accounted for. The simplest way to model the former is by simulating the tire as a rigid ring, with mass m_c and

---

[4]Parameters A, B and C used in the equation giving the road load must not be confused with the parameters with the same name included in the magic formula.

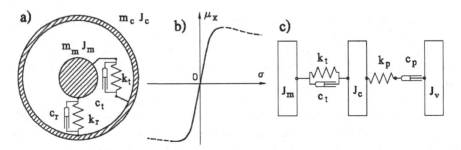

FIGURE 30.8. Model of the tire and the tire-road contact. a): Dynamic model of the tire; b): Force-longitudinal slip characteristic for the tire. c) Dynamic model of the tire-road contact (the moments of inertia and the torsional characteristics are drawn as masses and translational characteristics).

moment of inertia J_c, corresponding to the tread band and the belt beneath it. This is connected to the wheel hub, whose mass and moment of inertia are m_m and J_m through an elastic system having a radial and torsional stiffness equal to k_r and k_t respectively.

The rim and hub are also assumed to be rigid bodies. Viscous dampers with coefficients c_r and c_t (Fig. 30.8a) may be added in parallel to the springs. The masses and the radial stiffness and damping are included in the ride comfort models, as seen in the previous chapter, while the moments of inertia and the torsional stiffness and damping are included in the driveline and longitudinal models

The wheel-ground contact may be characterized by the plot of the longitudinal force coefficient versus the sideslip $\mu_x(\sigma)$ (Fig. 30.8b). Usually only the first part of the curve, approximated as a straight line, is used in the study of the driveline dynamics. The slope of the line may be easily obtained from the coefficients of the magic formula and is given by product BCD.

The longitudinal slip σ is linked to the ratio between the speed Ω_c of the wheel and the speed of the moment of inertia simulating the vehicle

$$\Omega_v = \frac{V}{R_e}$$

by the relationship

$$\sigma = \frac{\Omega_c}{\Omega_v} - 1 \ .$$

The longitudinal force the tire exerts is then

$$F_x = F_z \mu_x = b F_z \sigma = b F_z \left(\frac{\Omega_c - \Omega_v}{\Omega_v} \right) \ , \tag{30.38}$$

where

$$b = BCD \ .$$

The moment exerted on the wheel due to the longitudinal slip is

$$M = R_e F_x = \frac{b F_z R_e}{\Omega_v} (\Omega_c - \Omega_v) \ . \tag{30.39}$$

The wheel-ground contact may then be modelled as a viscous damper with damping coefficient

$$c_p = \frac{b F_z R_e^2}{V} \ . \tag{30.40}$$

Coefficient c_p depends first upon the vehicle speed and then upon the variable of motion Ω_v: the equation of motion is then nonlinear. For small velocity variations it is, however, possible to linearize the equations by using an average value of the speed in the expression of c_p. When the speed tends to zero, the damping coefficient tends to infinity: This linearized model cannot be used in the first instants of a take-off manoeuvre, when the vehicle is still stationary, because in these conditions the longitudinal slip is high, or better tends to infinity.

The tire model is usually complemented by adding a spring in series with the damper. Its stiffness is

$$k_p = \frac{b F_z R_e^2}{a} \ , \tag{30.41}$$

where a is a length equal to half the length of the contact zone.

Stiffness k_p can be suitably modified to take into account the longitudinal compliance of the suspension of the driving wheels.

The tire is then modelled with two moments of inertia connected to each other with a spring and a damper in parallel, and connected to the flywheel simulating the vehicle with a spring and a damper in series. The model is sketched in Fig. 30.8c, where the torsional springs and the moments of inertia are drawn as springs and masses.

30.5 LINEARIZED DRIVELINE MODEL

A driveline model from the engine to the vehicle can thus be assembled using the partial models seen above. As already stated, the elements to be taken into account depend upon the aim for which the model has been built. A relatively simple model is shown in Fig. 30.9a. If low frequency oscillations, such as those occurring in tip-in, tip-out manoeuvres, are to be studied, the engine can be modelled as a single moment of inertia. The two wheel shafts are modelled separately in this model, because in many cars with transversal front engine and front-wheel drive their stiffnesses are different. However, the two branches of the driveline can be joined if a first approximation study of the low frequency dynamics alone is required (Fig. 30.9b). This can be done by introducing inertias and stiffnesses equal to the sum of those of the single branches.

The model shown in Fig. 30.9a has 10 degrees of freedom. Because two of them have a vanishing associated mass, it has only 18 state variables. The

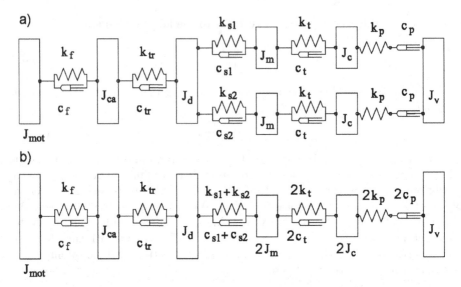

FIGURE 30.9. Model of the driveline for the study of low frequency dynamics a), and model in which the presence of two separate wheel shafts is neglected b).

generalized coordinates are the rotations of the various moments of inertia. It is possible to order them by separating the nodes where there is a mass from those that are massless. Thus

$$\mathbf{x} = \begin{bmatrix} \mathbf{x}_1^T & \mathbf{x}_2^T \end{bmatrix}^T , \tag{30.42}$$

where

$$\mathbf{x}_1 = \begin{bmatrix} \theta_{mot} & \theta_{ca} & \theta_d & \theta_{m1} & \theta_{m2} & \theta_{c1} & \theta_{c2} & \theta_v \end{bmatrix}^T ,$$

$$\mathbf{x}_2 = \begin{bmatrix} \theta_{i1} & \theta_{i2} \end{bmatrix}^T .$$

The mass matrix of the system may be partitioned in four parts as

$$\mathbf{M} = \begin{bmatrix} \mathbf{M}_{11} & \mathbf{M}_{12} \\ \mathbf{M}_{21} & \mathbf{M}_{22} \end{bmatrix} \tag{30.43}$$

where

$$\mathbf{M}_{11} = \mathrm{diag} \begin{bmatrix} J_{mot} & J_{ca}^* & J_d^* & J_m^* & J_m^* & J_c^* & J_c^* & J_v^* \end{bmatrix} \tag{30.44}$$

and all other sub-matrices are null.

The stiffness matrix may be partitioned in the same way

$$
\mathbf{K}_{11} =
\begin{bmatrix}
k_f & -k_f & 0 & 0 & 0 & 0 & 0 & 0 \\
 & k_f + k_{tr}^* & -k_{tr}^* & 0 & 0 & 0 & 0 & 0 \\
 & & k_1 & -k_{s1}^* & -k_{s2}^* & 0 & 0 & 0 \\
 & & & k_{s1}^* + k_t^* & 0 & -k_t^* & 0 & 0 \\
 & & & & k_{s2}^* + k_t^* & 0 & -k_t^* & 0 \\
 & & & & & k_t^* + k_p^* & 0 & 0 \\
 & & & & & & k_t^* + k_p^* & 0 \\
 & \text{symm.} & & & & & & 0
\end{bmatrix} ,
$$

(30.45)

where

$$
k_1 = k_{tr}^* + k_{s1}^* + k_{s2}^* ,
$$

$$
\mathbf{K}_{21} = \begin{bmatrix} \mathbf{0}_{2\times 5} & \begin{bmatrix} -k_p^* & 0 & 0 \\ 0 & -k_p^* & 0 \end{bmatrix} \end{bmatrix} ,
$$

(30.46)

$$
\mathbf{K}_{22} = \begin{bmatrix} k_p^* & 0 \\ 0 & k_p^* \end{bmatrix} , \quad \mathbf{K}_{12} = \mathbf{K}_{21}^T .
$$

(30.47)

In a similar way the submatrices of the damping matrix are

$$
\mathbf{C}_{11} =
\begin{bmatrix}
c_f & -c_f & 0 & 0 & 0 & 0 & 0 & 0 \\
 & c_f + c_{tr}^* & -c_{tr}^* & 0 & 0 & 0 & 0 & 0 \\
 & & c_1 & -c_{s1}^* & -c_{s2}^* & 0 & 0 & 0 \\
 & & & c_{s1}^* + c_t^* & 0 & -c_t^* & 0 & 0 \\
 & & & & c_{s2}^* + c_t^* & 0 & -c_t^* & 0 \\
 & & & & & c_t^* & 0 & 0 \\
 & & & & & & c_t^* & 0 \\
 & \text{symm.} & & & & & & 2c_p^*
\end{bmatrix} ,
$$

(30.48)

where

$$
c_1 = c_{tr}^* + c_{s1}^* + c_{s2}^* ,
$$

$$
\mathbf{C}_{21} = \begin{bmatrix} \mathbf{0}_{2\times 7} & \begin{bmatrix} -c_p^* \\ -c_p^* \end{bmatrix} \end{bmatrix} , \quad \mathbf{C}_{22} = \begin{bmatrix} c_p^* & 0 \\ 0 & c_p^* \end{bmatrix} , \quad \mathbf{C}_{12} = \mathbf{C}_{21}^T .
$$

(30.49)

The assumption that the damping of the various components of the driveline can be modelled as viscous is only approximate, but it cannot be modelled as hysteretic damping (which is not much better for elements like the clutch damper springs) because that would not allow the numerical simulation of manoeuvres such as the response to a step input. However, because the phenomenon here studied occurs at a well determined frequency, it is possible to approximate hysteretic damping with an equivalent viscous damping

$$
c_{eq} = \frac{\eta k}{\omega} ,
$$

(30.50)

where η and k are the loss factor and the stiffness of the relevant elements and ω is the frequency of the oscillations of the driveline. It is possible to perform a first computation with no damping (except that used to simulate tire slip) to compute a value for the frequency of the free oscillations, and then to proceed with calculations that include an equivalent damping.

The state vector can be written in the form

$$\mathbf{z} = \left[\begin{array}{ccc} \mathbf{v}_1^T & \mathbf{x}_1^T & \mathbf{x}_2^T \end{array} \right]^T , \tag{30.51}$$

where $\mathbf{v}_1$ contains the derivatives of coordinates $\mathbf{x}_1$.

The state equation is then

$$\left[\begin{array}{ccc} \mathbf{M}_{11} & \mathbf{0} & \mathbf{C}_{12} \\ \mathbf{0} & \mathbf{0} & \mathbf{C}_{22} \\ \mathbf{0} & \mathbf{I} & \mathbf{0} \end{array} \right] \dot{\mathbf{z}} = - \left[\begin{array}{ccc} \mathbf{C}_{11} & \mathbf{K}_{11} & \mathbf{K}_{12} \\ \mathbf{C}_{21} & \mathbf{K}_{21} & \mathbf{K}_{22} \\ -\mathbf{I} & \mathbf{0} & \mathbf{0} \end{array} \right] \mathbf{z} + \left\{ \begin{array}{c} \mathbf{F}_1 \\ \mathbf{0} \\ \mathbf{0} \end{array} \right\} , \tag{30.52}$$

where vector $\mathbf{F}_1$ contains the moments applied on the nodes whose coordinates are included in vector $\mathbf{x}_1$. Because only the driving and drag torques are present, it follows that

$$\mathbf{F}_1 = \left\{ \begin{array}{c} M_{mot} \\ \mathbf{0}_{6 \times 1} \\ M_r \end{array} \right\} . \tag{30.53}$$

The dynamic matrix of the system is then

$$\mathbf{A} = - \left[\begin{array}{ccc} \mathbf{M}_{11} & \mathbf{0} & \mathbf{C}_{12} \\ \mathbf{0} & \mathbf{0} & \mathbf{C}_{22} \\ \mathbf{0} & \mathbf{I} & \mathbf{0} \end{array} \right]^{-1} \left[\begin{array}{ccc} \mathbf{C}_{11} & \mathbf{K}_{11} & \mathbf{K}_{12} \\ \mathbf{C}_{21} & \mathbf{K}_{21} & \mathbf{K}_{22} \\ -\mathbf{I} & \mathbf{0} & \mathbf{0} \end{array} \right] . \tag{30.54}$$

Example 30.1 *Simulate a tip-in, tip-out manoeuvre in second gear with engine at 1,500 rpm. With the vehicle running at constant speed (driving torque equal to drag torque) increase suddenly the driving torque to its maximum value and keep such a value until 2,000 rpm. are reached. The accelerator is then fully released and the vehicle slows down.*

Data: maximum driving torque: $M_{\max} = 40$ Nm; braking torque of the engine: $M_f = -4$ Nm.

Vehicle: Mass $m = 950$ kg, $f_0 = 0.013$, $K = 6.5 \times 10^{-6}$ s^2/m^2, $C_x = 0.32$, $S = 1.7$ m^2, $R_e = 257$ mm, half-length of the contact area $a = 50$ mm. Neglect the efficiency of the transmission.

Moments of inertia: engine (including the flywheel) $J_{mot} = 0.125$ kg m^2, gear $J_{ca} = 0.0045$ kg m^2, differential $J_d = 0.065$ kg m^2, wheel hub $J_m = 0.3$ kg m^2, thread band $J_c = 0.32$ kg m^2.

Stiffnesses: Clutch disk (with damper springs, when driving), $k_f = 975$ Nm/rad, wheel shafts $k_{s1} = 7,500$ Nm/rad, $k_{s2} = 10,400$ Nm/rad, tire $k_t = 59,000$ Nm/rad.

Neglect the damping of the driveline elements.

Gear ratios: Final gear $\tau_p = 0.2884$; gearbox (second gear) $\tau_c = 23/63$.

By using the tire model previously used to compute the contact parameters, stiffness b is

$$b = A \left(\frac{K}{\alpha + d} \right)^{1/n} - D \,,$$

where $\alpha = 0$ (there is no sideslip angle), $A = 1.12$, $K = 46$, $n = 0.6$, $d = 5$, $D = 1$.

The compliance of the gearbox-differential connection is neglected and then a single moment of inertia $J_{ca} = 0.0695$ kg m^2 is assumed for gearbox and differential, located on the gearbox output shaft.

The values of the inertias and stiffnesses, reduced to the engine shaft (multiplied by the squares of the transmission ratios) are: $J_{ca}^* = 0.00926$ kg m^2, $J_m^* = 0.00333$ kg m^2, $J_c^* = 0.00355$ kg m^2, $k_{s1}^* = 83.14$ Nm/rad, $k_{s2}^* = 115.3$ Nm/rad, $k_t^* = 654.1$ Nm/rad.

The moment of inertia of the vehicle reduced to the engine shaft, including also the two free wheels as well, is $J_v^* = 0.696$ kg m^2.

The speed of the vehicle at 1,500 rpm is $V = 4.25$ m/s $= 15.30$ km/h.

Neglecting the term of the road load in V^4, the drag moment reduced to the engine shaft may be written as

$$M_r = RR_e \tau^2 = A_r + B_r \Omega_v^2 \,,$$

where $A_r = 3.28$ Nm, $B_r = 8.15 \times 10^{-6}$ Nms^2. The drag torque at 1,500 rpm is 3.48 Nm: Predictably, the quadratic term as a small effect at such a low speed.

Coefficients c_p and k_p are $c_p = 1,913$ Nms/rad and $k_p = 163,000$ Nm/rad at a speed of 1,500 rpm. It then follows that $c_p^* = 21.21$ Nms/rad and $k_p^* = 1,803$ Nm/rad.

Natural frequencies are immediately obtained from the eigenvalues of the dynamic matrix. Three eigenvalues are equal to zero, because rigid body modes are possible. The eigenvalue with the smallest imaginary part , i.e. that corresponding to motion with the lowest natural frequency, is

$$s = -1.69 \pm 35.58 \; i \; 1/s \,,$$

and the corresponding frequency of the damped free oscillations is 5.66 Hz. It is the low frequency mode typical of the phenomenon under study.

The following eigenvalue is

$$s = -4.32 \pm 347 \; i \; 1/s \,,$$

corresponding to a damped oscillation with a frequency of 55.23 Hz. This is almost ten times the frequency of the lowest mode. To study oscillations at this frequency it is advisable to use a more detailed model.

The results of the tip-in, tip-out manoeuvre are shown in Fig. 30.10. In a) the time histories of the velocity of engine and vehicle are reported, while the longitudinal acceleration of the vehicle is plotted in b). By comparing the results of the simulation with those shown in Fig. 30.1, a qualitative similarity is found, although the results refer to different vehicles. The longitudinal acceleration computed in the simulation is that of the wheel hub, while the experimental results were measured in the passenger compartment and then filtered by the structure of the vehicle. Moreover, the mathematical model does not take into account the damping of the driveline, but only that caused by tire slip.

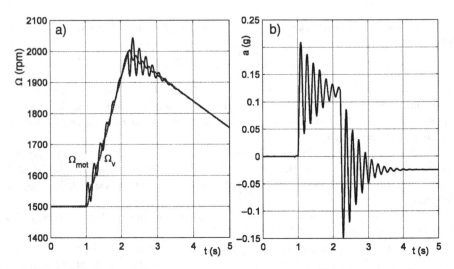

FIGURE 30.10. Results of a tip-in, tip-out manoeuvre. a): Engine and wheel speeds (reduced to the engine shaft); b) longitudinal acceleration of the vehicle.

Example 30.2 *Repeat the simulation of the previous example, taking into account n hysteretic damping with a loss factor $\eta = 0.2$ for the clutch damper springs and $\eta = 0.05$ for all other elements. To convert hysteretic into viscous damping, the equivalent damping at a frequency of 35.58 rad/s is computed.*

The eigenvalue with the lowest imaginary part, i.e. that corresponding to the oscillations with the lowest frequency, is now

$$s = -2.88 \pm 35.43 \; i \; 1/s \; ,$$

corresponding to oscillations at a frequency of 5.64 Hz. By comparing this result with that seen above, it is clear that the decay rate has increased considerably (it almost doubled), while the frequency is essentially the same.

The results of the simulation are reported in Fig. 30.11. The effect of damping is fairly limited: the oscillations damp out in a shorter time, but the maximum values of the acceleration are little changed.

30.6   NON-TIME-INVARIANT MODELS

If the torsional vibration of the driveline must be studied with the reciprocating masses in the engine in mind, a model with inertial properties that are variable in time must be used.

Three different approaches are possible. Listing them in order of increasing complexity, they are

1. Traditional approach. The effect of the variable component of the moments of inertia of the crank systems is modelled as a torque with known time

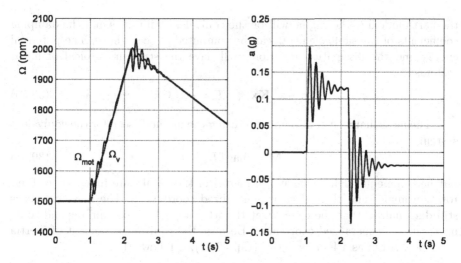

FIGURE 30.11. Results of a tip-in, tip-out manoeuvre, computed while taking into account the damping of the various elements. a): Engine and wheel speeds (reduced to the engine shaft); b) longitudinal acceleration of the vehicle.

history applied to the node where the crank is located. Because the model is linearized, it is possible to write the frequency domain equations and to solve them analytically.

2. Approach in which the torsional deformations are considered small and the corresponding angles can be neglected in the computation of the inertia of the cranks. The equations of motion, nonlinear and non-time-invariant, have coefficients with a known time history. No closed form of the equations of motion is possible, and numerical integration is needed.

3. Approach with no particular simplifying assumption. The equations of motion must be solved numerically, but the solution is much more difficult than in case (2).

30.6.1 Equations of motion

The equation of motion of the engine-driveline system can be obtained in the usual way, using Lagrange equations. The system may be modelled starting from a lumped parameters approach, obtaining a model of the type shown in Fig. 30.9a, with the difference that now the engine is not lumped in a single inertia, but is modelled as another lumped parameter system, with the various moments of inertia modelling cranks, flywheel, ancillary devices, dampers, etc. connected to each other by torsional springs and dampers. The system may be in-line, like that in Fig. 30.9b, or multiply connected, like that in Fig. 30.9a.

There is no difficulty in linearizing the elastic and dissipative part of the system by writing stiffness and damping matrices $\mathbf{K}$ and $\mathbf{C}$. The stiffnesses of the part modelling the engine will be the equivalent stiffnesses and the damping coefficients of the same elements can be computed as seen above. The potential energy and the dissipation functions will have the structure typical of linear systems

$$\mathcal{U} = \frac{1}{2}\mathbf{x}^T\mathbf{K}\mathbf{x} \ , \quad \mathcal{F} = \frac{1}{2}\dot{\mathbf{x}}^T\mathbf{C}\dot{\mathbf{x}} \ . \tag{30.55}$$

The mass matrix can be computed using the methods seen for the equivalent system

$$\mathbf{M} = \text{diag}\,[J_{eq_i}] \ , \tag{30.56}$$

but here the equivalent moments of inertia of the cranks are functions of their rotation angles and then of their generalized coordinates. The mass matrix is still diagonal, so that the element at the ith row and ith column depends only upon the ith generalized coordinate. Because the equivalent moment of inertia must be written as a Fourier series (Eq. 30.12), it follows that

$$\mathbf{M} = \mathbf{M}_0 + \mathbf{M}_1\,(\mathbf{x}) = \text{diag}\,[J_{0i}] + \text{diag}\,[J_{1i}\,(\theta_i)] \ , \tag{30.57}$$

where:

- $\mathbf{M}_0$ is a diagonal matrix containing the average values of the moments of inertia of the cranks and the moments of inertia at all other nodes. Except for the nodes with which no inertia is associated, the elements are all non-zero and their values are J_{0i}.

-
$$J_{1i}\,(\theta_i) = \sum_{k=1}^{m} [J_{cik}\cos(k\theta_i + \delta_{ik}) + J_{sik}\sin(k\theta_i + \delta_{ik})]\,; \tag{30.58}$$

  where

  i is the subscript referring to the node ($i = 1, ..., n$);

- k is the subscript referring to the relevant harmonics and thus spans from 1 to m, the number of the harmonics considered in the series (in theory $m = \infty$);

- θ_i is the rotation angle of the ith node, and thus is the ith element of vector $\mathbf{x}$;

- J_{cik} and J_{sik} are the coefficients of the terms in sine and cosine of the Fourier series for the equivalent moments of inertia. These vanish for all the nodes where no crank is located and are equal for all cranks (they do not depend on subscript i) if the crank systems are all equal;

- δ_{ik} are the phases of the various harmonics in the various cranks, as given by the phase angle diagrams.

The kinetic energy is then

$$\mathcal{T} = \frac{1}{2}\dot{\mathbf{x}}^T \mathbf{M}\dot{\mathbf{x}} . \tag{30.59}$$

Assuming that the rotation angle θ_i of the ith node is given by the sum of an average angle θ_0 of the driveline and a torsion angle ϕ_i :

$$x_i = \theta_0 + \phi_i , \tag{30.60}$$

for the first $n-1$ nodes and

$$x_n = \theta_{0v} + \phi_n , \tag{30.61}$$

for the last node where the moment of inertia simulating the vehicle is located.

Remark 30.8 *The latter node must be kept separate, because, owing to the longitudinal slip of the tire, the average rotations of the driveline and the vehicle θ_0 and θ_{0v} diverge in time.*

Let $\mathbf{S}$ be a vector of order $n-1$, whose components are all equal to 1. It is then possible to write

$$\mathbf{x} = \left\{ \begin{array}{c} \theta_0 \mathbf{S} \\ \theta_{0v} \end{array} \right\} + \boldsymbol{\phi} . \tag{30.62}$$

Remembering that the last row and the last column of matrix $\mathbf{K}$ vanish (Eq. (30.45), where all nodes are present), the potential energy is

$$\mathcal{U} = \frac{1}{2}\theta_0^2 \mathbf{S}^T \mathbf{K}^* \mathbf{S} + \frac{1}{2}\boldsymbol{\phi}^{*T}\mathbf{K}^*\boldsymbol{\phi}^* + \theta_0 \mathbf{S}^T \mathbf{K}^*\boldsymbol{\phi}^* , \tag{30.63}$$

where $\mathbf{K}^*$ and $\boldsymbol{\phi}^*$ are the stiffness matrix and the vector of the generalized coordinates without the the last row and column and without the last row respectively.

Rotation expressed by vector $\theta_0 \mathbf{S}$ is a rigid rotation. Because the driveline is free to rotate, product $\mathbf{K}^*\mathbf{S}$ is null and thus it follows that

$$\mathcal{U} = \frac{1}{2}\boldsymbol{\phi}^{*T}\mathbf{K}^*\boldsymbol{\phi}^* , \tag{30.64}$$

i.e.

$$\mathcal{U} = \frac{1}{2}\boldsymbol{\phi}^T \mathbf{K}\boldsymbol{\phi} . \tag{30.65}$$

In a similar way,

$$\dot{\mathbf{x}} = \left\{ \begin{array}{c} \Omega_t \mathbf{S} \\ \Omega_v \end{array} \right\} + \dot{\boldsymbol{\phi}} , \tag{30.66}$$

where Ω_t and Ω_v are the average velocities of the driveline and the vehicle.

The damping matrix may be subdivided into three parts

$$\mathbf{C} = \mathbf{C}_1 + \mathbf{C}_2 + \mathbf{C}_3 \tag{30.67}$$

where

- C_1 is a diagonal matrix, where all dampings toward the ground are listed, that is, all terms expressed by Eq. (30.21) to simulate the energy losses due to the absolute rotation of the nodes,

- C_2 is a matrix with the structure shown in Eq. (30.48) where all nodes are present and the damping c_p simulating the tire slip is not included,

- C_3 is a matrix of type shown in Eq. (30.48) where all nodes are present; it contains only the damping c_p simulating the tire slip.

The dissipation function is then

$$\mathcal{F} = \frac{1}{2} \left(\left\{ \begin{array}{c} \Omega_t \mathbf{S} \\ \Omega_v \end{array} \right\} + \dot{\phi} \right)^T (\mathbf{C}_1 + \mathbf{C}_2 + \mathbf{C}_3) \left(\left\{ \begin{array}{c} \Omega_t \mathbf{S} \\ \Omega_v \end{array} \right\} + \dot{\phi} \right), \qquad (30.68)$$

that is, performing the products and remembering the properties of the involved matrices

$$\mathcal{F} = \frac{1}{2} \left\{ \begin{array}{c} \Omega_t \\ \Omega_v \end{array} \right\} \left[\begin{array}{cc} \sum_{i=1}^{n-1} c_{1i} + 2c_p & -2c_p \\ -2c_p & 2c_p \end{array} \right] \left\{ \begin{array}{c} \Omega_t \\ \Omega_v \end{array} \right\} + $$
$$+ \left\{ \begin{array}{c} \Omega_t \mathbf{S} \\ \Omega_v \end{array} \right\}^T (\mathbf{C}_1 + \mathbf{C}_3) \dot{\phi} + \frac{1}{2} \dot{\phi}^T \mathbf{C} \dot{\phi}. \qquad (30.69)$$

In a similar way, remembering that the mass matrix is diagonal, it is possible to write

$$\mathcal{T} = \frac{1}{2} \sum_{i=1}^{n-1} [J_{0i} + J_{1i}(\theta_i)] \left(\Omega_t + \dot{\phi}_i \right)^2 + \frac{1}{2} J_{0n} \left(\Omega_v + \dot{\phi}_n \right)^2. \qquad (30.70)$$

30.6.2   Rigid-body motion of the driveline

The generalized coordinates are the average rotations of the driveline and the flywheel simulating the vehicle (or better, their derivatives Ω_t and Ω_v) and the torsional rotations ϕ_i of the various elements. It is possible to assume that the low frequency dynamics (actually a non-periodic dynamics) may be studied separately from the torsional dynamics of the system.

When studying the first, the equations simplify: Not only the terms containing ϕ vanish, but if the cranks are all equal and (angularly) uniformly spaced, it follows that

$$\sum_{i=1}^{n-1} J_{1i}(\theta_i) = 0 \quad , \quad J_{tot} = \sum_{i=1}^{n-1} J_{0i}. \qquad (30.71)$$

Neglecting torsional rotations, it follows that

$$\mathcal{U} = 0 \, , \qquad (30.72)$$

$$\mathcal{F} = \frac{1}{2} \left\{ \begin{array}{c} \Omega_t \\ \Omega_v \end{array} \right\} \left[\begin{array}{cc} \sum_{i=1}^{n-1} c_{1i} + 2c_p & -2c_p \\ -2c_p & 2c_p \end{array} \right] \left\{ \begin{array}{c} \Omega_t \\ \Omega_v \end{array} \right\}, \tag{30.73}$$

$$\mathcal{T} = \frac{1}{2}\Omega_t^2 J_{tot} + \frac{1}{2}\Omega_v^2 J_v \tag{30.74}$$

and the equation of motion is simply

$$\left[\begin{array}{cc} J_{tot} & 0 \\ 0 & J_v \end{array} \right] \left\{ \begin{array}{c} \dot{\Omega}_t \\ \dot{\Omega}_v \end{array} \right\} + \left[\begin{array}{cc} \sum_{i=1}^{n-1} c_{1i} + 2c_p & -2c_p \\ -2c_p & 2c_p \end{array} \right] \left\{ \begin{array}{c} \Omega_t \\ \Omega_v \end{array} \right\} = \left\{ \begin{array}{c} M_m \\ M_v \end{array} \right\}.$$
$$\tag{30.75}$$

M_m is the total driving torque, or better the constant component of the total driving torque.

Remark 30.9 *This result is trivial, and could be obtained from the previous, simpler, models.*

30.6.3 Torsional dynamics of the engine and driveline

The following assumptions can be made in the study of torsional dynamics:

- The average speed of the engine and the vehicle are known functions of time;

- The changes in the average speeds are slow enough to neglect the derivatives of the average speeds with respect to time.

The derivative of the kinetic energy with respect to the ith generalized velocity is

$$\frac{\partial \mathcal{T}}{\partial \dot{\phi}_i} = [J_{0i} + J_{1i}(\theta_i)]\left(\Omega_t + \dot{\phi}_i\right), \tag{30.76}$$

for $i = 1, ..., n-1$ and

$$\frac{\partial \mathcal{T}}{\partial \dot{\phi}_n} = J_{0n}\left(\Omega_v + \dot{\phi}_n\right). \tag{30.77}$$

Differentiating with respect to time, it follows that

$$\frac{d}{dt}\left(\frac{\partial \mathcal{T}}{\partial \dot{\phi}_i}\right) = [J_{0i} + J_{1i}(\theta_i)]\ddot{\phi}_i + \frac{\partial J_{1i}(\theta_i)}{\partial t}\left(\Omega_t + \dot{\phi}_i\right), \tag{30.78}$$

i.e.

$$\frac{d}{dt}\left(\frac{\partial \mathcal{T}}{\partial \dot{\phi}_i}\right) = [J_{0i} + J_{1i}(\theta_i)]\ddot{\phi}_i + \frac{\partial J_{1i}(\theta_i)}{\partial \theta_i}\left(\Omega_t + \dot{\phi}_i\right)^2, \tag{30.79}$$

for $i = 1, ..., n-1$ and

$$\frac{d}{dt}\left(\frac{\partial \mathcal{T}}{\partial \dot{\phi}_n}\right) = J_{0n}\ddot{\phi}_n. \tag{30.80}$$

Finally, the derivatives of the kinetic energy with respect to the generalized coordinates are

$$\frac{\partial \mathcal{T}}{\partial \phi_i} = \frac{1}{2} \frac{\partial J_{1i}(\theta_i)}{\partial \theta_i} \left(\Omega_t + \dot{\phi}_i\right)^2 , \qquad (30.81)$$

for $i = 1, ..., n - 1$ and

$$\frac{\partial \mathcal{T}}{\partial \phi_n} = 0 . \qquad (30.82)$$

Remembering that $J_{1i}(\theta_i)$ vanishes in the last equation of motion (it actually vanishes in all equations regarding nodes where no crank is located), the inertial part of all equations of motion is

$$\frac{d}{dt}\left(\frac{\partial \mathcal{T}}{\partial \dot{\phi}_i}\right) - \frac{\partial \mathcal{T}}{\partial \phi_i} = \left[J_{0i} + J_{1i}(\theta_i)\right]\ddot{\phi}_i + \frac{1}{2}\frac{\partial J_{1i}(\theta_i)}{\partial \theta_i}\left(\Omega_t + \dot{\phi}_i\right)^2 , \qquad (30.83)$$

while in the last equation only the term $J_{0n}\ddot{\phi}_n$ is present.

The other terms of the equations of motion are

$$\left\{\frac{\partial \mathcal{U}}{\partial \phi_i}\right\} = \mathbf{K}\phi , \qquad (30.84)$$

$$\frac{\partial \mathcal{F}}{\partial \dot{\phi}_i} = \mathbf{C}\dot{\phi} + (\mathbf{C}_1 + \mathbf{C}_3)\left\{\begin{array}{c}\Omega_t \mathbf{S} \\ \Omega_v\end{array}\right\} . \qquad (30.85)$$

The final form of the equation of motion is then

$$\left[\mathbf{M}_0 + \mathbf{M}_1(\mathbf{x})\right]\ddot{\phi} + \frac{1}{2}\frac{\partial \mathbf{M}_1(\mathbf{x})}{\partial \mathbf{x}}\left\{\left(\Omega_t + \dot{\phi}_i\right)^2\right\} +$$

$$+\mathbf{C}\dot{\phi} + \mathbf{K}\phi = \mathbf{F} - (\mathbf{C}_1 + \mathbf{C}_3)\left\{\begin{array}{c}\Omega_t \mathbf{S} \\ \Omega_v\end{array}\right\} , \qquad (30.86)$$

where vector $\mathbf{F}$ contains the driving torques applied to the various cranks and the drag torque applied to the flywheel simulating the vehicle.

30.6.4   Traditional approach

Not only does Eq. (30.86) contain coefficients that are varying in time, (through term θ_0 included in the total rotations $\mathbf{x}$ and then in $\mathbf{M}_1(\mathbf{x})$), but it is also nonlinear, both because ϕ_i are present in the total rotations $\mathbf{x}$ and then in $\mathbf{M}_1(\mathbf{x})$), and because it includes the squares of the generalized displacements ϕ_i.

The traditional approach is based on the following simplifications:

- $\mathbf{M}_1(\mathbf{x})$ is neglected with respect to $\mathbf{M}_0$ in the term in $\ddot{\phi}$,

- $\dot{\phi}_i$ is neglected with respect to Ω_t in the term in $\left(\Omega_t + \dot{\phi}_i\right)^2$,

- $M_1(x)$ is considered as a function of θ_0 but not of ϕ_i: The inertia of the crank system is considered as a function of the average rotation of the crankshaft, but not of the torsional rotation of the various cranks,

- usually, even if it is not strictly needed, the angular velocity Ω_t is assumed to be constant, and then $\theta_0 = \Omega_t t$.

In these conditions, the term

$$\frac{\partial M_1(x)}{\partial x}\left\{ \left(\Omega_t + \dot{\phi}_i\right)^2 \right\}$$

becomes a known function of time and then is brought to the right-hand side of the equation, together with the forcing functions.

The equation of motion reduces to

$$M_0\ddot{\phi} + C\dot{\phi} + K\phi = F + F_{in} + (C_1 + C_3)\left\{ \begin{array}{c} \Omega_t S \\ \Omega_v \end{array} \right\} , \tag{30.87}$$

where the terms

$$F_{in} = -\frac{1}{2}\Omega_t^2 \frac{\partial M_1(\theta_0)}{\partial \theta_0} S \tag{30.88}$$

are usually defined as *inertia torques* of the cranks. By introducing the value of M_1 in the equation, the various inertia torques are

$$F_{in_i} = -\frac{1}{2}\Omega_t^2 \sum_{k=1}^{m} k\left[-J_{cik}\sin(k\theta_0 + \delta_{ik}) + J_{sik}\cos(k\theta_0 + \delta_{ik}) \right] . \tag{30.89}$$

Vector F contains the driving torque, which is periodic with period equal to the time needed to perform two revolutions of the crankshaft in a four-stroke cycle engine (fundamental frequency $\Omega_t/2$). F_{in} is also periodic, but its fundamental frequency is Ω_t. Usually, F_{in} is written as if its fundamental frequency were $\Omega_t/2$, with the amplitudes of all odd harmonics, including the fundamental one, set to zero. In this way the sum $F + F_{in}$ has a simpler structure.

The homogeneous equation associated with Eq. (30.87) does not take into account the variability in time of the equivalent moments of inertia of the crank systems. The natural frequencies are thus those of a system with constant inertia.

30.6.5 Numerical approach

Equation (30.86) may be solved directly by numerical integration. If nonlinearity, and above all the dependence of the various parameters upon time, may make it difficult to proceed with the integration, the number of degrees of freedom of the system is nonetheless low (usually no more than 20), so the computation is not difficult. The time histories of the rotations of the various nodes can thus be obtained along with those of the stresses in the various parts of the crankshaft and driveline. The time histories can then be developed in Fourier series and the various harmonic contents extracted. However, if the time histories of the motion of the various elements of the driveline must be obtained, it now seems more expedient to resort directly to multibody computer codes instead of writing the equations of motion of the driveline and building *ad hoc* programs.

30.7 MULTIBODY DRIVELINE MODELS

The linearized models seen in the previous section have the advantage of yielding closed form, frequency domain solutions and of correctly simulating manoeuvres like tip-in, tip-out in a simple way. However, the assumption that the stiffness and damping characteristics of elements such as the clutch damper springs or elastomeric dampers may be simulated as linear springs and viscous dampers leads to a poor approximation.

The torsional characteristic of a clutch plate with damper springs is shown in Fig. 30.12a: Not only is it possible to identify three fields where the stiffness takes different values, but a well-defined hysteresis that has the characteristics of a dry friction is also present. The fact that the stiffness is variable in the three ranges does not prevent the elastic behavior about a particular working condition from being linearized, while the nature of the damping, which may be dealt with as dry friction, makes it impossible to linearize the behavior of this element if damping is accounted for. As an example, the hysteresis cycle for displacements about a condition in which the driving torque is 50 Nm is shown in Fig. 30.12b.

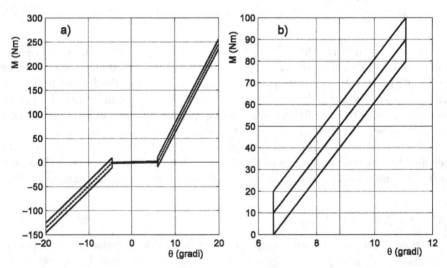

FIGURE 30.12. Angular displacement-torque characteristics of a clutch with damper springs. a) Characteristics with engine braking (left) and driving (right); b): zoom on the zone of the plot for a condition where the torque cycles about a value of the driving torque of 50 Nm.

If a torsional displacement-torque characteristic like that shown in Fig. 30.12 prevents the use of linearized models, or better, introduces large errors if a linearization is attempted, it may be included without difficulty in a multibody model. Multibody codes operate by numerically integrating in time the nonlinear equations of motion and thus are ideal for cases like this. It is then possible to include the detailed mathematical models of nonlinear elements and proceed to simulate the behavior of the driveline in detail. The engine may be modelled directly with the pistons, connecting rods and cranks, while its elastic supports may also be included. On the other end of the driveline, the suspension connecting the wheel to the vehicle body may also be included in the model with its compliance in the longitudinal direction, which often introduces a coupling between longitudinal motion of the vehicle and ride comfort. Multibody models may also include the engine control to correctly simulate various manoeuvres.

In the case of models restricted to the engine, multibody models that take the actual geometry of the reciprocating and rotating elements into account are increasingly replacing the traditional approach based on the modelling of the inertia torques applied to the cranks as external forcing functions. Obviously, their basic drawback of not allowing closed form solutions to be reached remains.

31

MODELS FOR TILTING BODY VEHICLES

The models seen in the previous chapters dealt with vehicles that maintain their symmetry plane more or less perpendicular to the ground; i.e. they move with a roll angle that is usually small. Moreover, the pitch angle was also assumed to be small, with the z axis remaining close to perpendicular to the ground. Since pitch and roll angles are small, stability in the small can be studied by linearizing the equations of motion in a position where $\theta = \phi = 0$.

Two-wheeled vehicles are an important exception. Their roll angle is defined by equilibrium considerations and, particularly at high speed, may be very large. To study the stability in the small, it is still possible to resort to linearization of the equations of motion, but now about a position with $\theta = 0$, $\phi = \phi_0$, where ϕ_0 is the roll angle in the equilibrium condition. An example of this method is shown in Appendix B, where the equation of motion of motorcycles is discussed.

Two-wheeled vehicles aside, this condition also occurs when the body of the vehicle is inclined with respect to the perpendicular to the road; this may be accomplished manually, as in motorcycles, or by devices (usually an active control system) that hold the roll angle to a value determined by a well-defined strategy. Vehicles of this type are usually defined as *tilting body vehicles*.

The most common application of tilting body vehicles today is in rail transportation, but road vehicles following the same strategy, particularly those with three wheels, have been built.

Rolling may be controlled according to two distinct strategies: by keeping the z-axis in the direction of the local vertical or by insuring that the load shift between wheels of the same axle vanishes. In the case of two-wheeled vehicles, the latter strategy results in maintaining roll equilibrium The two strategies coincide only if the roll axis is located on the ground and no rolling moments act on the vehicle, so that the wheels in particular produce no gyroscopic moment.

Tilting body vehicles arouse much interest because they allow us to build tall vehicles that, although having a limited width (or better having a large height/width ratio), have good dynamic performance, particularly in terms of high speed handling. It is thus possible to build vehicles that combine the typical advantages of motorcycles (good handling in heavy traffic conditions, low road occupation, ease of parking) with those of cars (ease of driving, active and passive safety, shelter from bad weather, no equilibrium problem when operating with frequent stops, etc.).

As always occurs when new concepts are experimented with, many configurations are considered both for geometry and mechanical solutions as well as hardware and software for the tilt control. No mutually agreed upon solution has yet arisen.

Most such vehicles are three-wheeled, both for legal and fiscal reasons (in many countries vehicles with three wheels have particular fiscal advantages). They are also much simpler and potentially lower in cost. If a two-wheel axle is needed to control tilting (solutions using a gyroscope to control tilting and thus do away with the need for an axle with two wheels, were proposed but seldom tested), having a single wheel on the other axle simplifies the mechanical layout, reducing weight, cost and size. Body tilting eliminates the stability problems typical of three-wheeled vehicles by reducing or eliminating load shift. In some solutions the single wheel is at the front, while in others it is at the back.

There are solutions where the roll axis is physically identified by a true cylindrical hinge located between a rigid axle and the vehicle body. The two-wheeled axle may be a solid axle or made by two independent suspensions with limited excursion, particularly for roll motions, connected to a frame that in turn carries the cylindrical hinge connected to the body (Fig. 31.1a). If the vehicle has four wheels, the roll centers of the two axles, materialized by two cylindrical hinges, identify the roll axis. If the vehicle has three wheels, the roll axis is

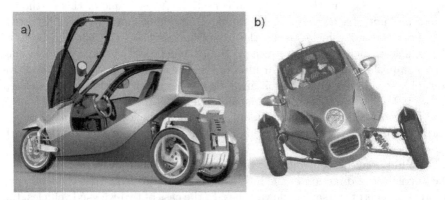

FIGURE 31.1. Prototypes of tilting vehicles. a): BMW C.L.E.V.E.R; b) Mercedes F 300. http://it.cars.yahoo.com/06062006/254/t/bmw-c-l-v-r-concept.html; http://www.3wheelers.com/mercedes.html.

identified by the center of the tire-road contact zone of the single wheel and the center of the cylindrical hinge on the two-wheeled axle. In this way the roll axis remains in a more or less fixed position in roll motion.

Usually, however, a different solution is found: The axle with two wheels has an independent suspension that allows large roll rotations of the body and behaves like a roll hinge (Fig. 31.1b). The roll center of the suspension is virtual, because it is not physically identified by a hinge; its position changes during roll motion. The roll center is then a fixed point only for small angles about the symmetric position (vanishing roll angle). In the case of large roll angles the roll center, and the roll axis as well, lies outside the symmetry plane of the body.

31.1 SUSPENSIONS FOR HIGH ROLL ANGLES

The wheels remain more or less perpendicular to the ground (the inclination angle of the wheels, here confused with the camber angle, is small) in those cases where the roll axis is defined by a physical hinge located between the frame carrying the suspension and the vehicle body. When independent suspensions directly attached to the vehicle body are used, on the other hand, it is possible to maintain the midplane of the wheels parallel to the symmetry plane of the body, i.e. $\phi = \gamma$, or $\partial\gamma/\partial\phi = 1$ or, at least, to obtain a large camber angle.

In such cases the possibility of setting the wheels at a large camber angle is interesting: Since the vehicle tilts towards the inside of the turn, camber forces add to sideslip forces, as in two-wheeled vehicles. Moreover, it is possible to exploit the difference in camber angles of the wheels of the two axles to modify the handling characteristics of the vehicle.

In the following sections two layouts will be considered: Trailing arms and transversal quadrilateral suspensions[1].

31.1.1 Trailing arms suspensions

Suspensions of this kind are characterized by

$$\frac{\partial t}{\partial z} = \frac{\partial \gamma}{\partial z} = \frac{\partial t}{\partial \phi} = 0 \ , \quad \frac{\partial \gamma}{\partial \phi} = 1$$

for small angles about the symmetrical conditions.

The track, defined as the distance between the centers of the contact areas of the two wheels of an axle, and the camber angle remain constant even at large vertical displacements. The camber angle also remains equal to the roll angle for large values of the latter. Indeed, the track is no longer constant at large roll angles, but becomes

---

[1]The term SLA suspension does not apply here, since the upper and lower arms have roughly the same length.

$$t = \frac{t_0}{\cos{(\phi)}}.$$

The changes in track, which are negligible for small values of the roll angle, increase with ϕ. When $\phi = 45°$ (a value still reasonable in motorcycles), the track increases by 40%. The roll center remains on the ground, so that a suspension of this type behaves like a single wheel in the symmetry plane, except for the changes of track. However, the wheels move in a longitudinal direction, both for vertical and roll displacements, and changes in the direction of the kingpin axis also occur, if the suspension is used for steering wheels. Such displacements depend on the length of the arms and their position in the reference conditions.

31.1.2   Transversal quadrilateral suspensions

If the wheels must be maintained parallel to the symmetry plane, the transversal quadrilaterals must actually be parallelograms: the upper and lower arms must have the same length and be parallel to each other. In this case it follows that

$$\frac{\partial \gamma}{\partial z} = 0 \ , \quad \frac{\partial \gamma}{\partial \phi} = 1,$$

in any condition. If the links connecting the body with the wheel hub are horizontal (Fig. 31.2a), the roll center of the suspension lies on the ground for $\phi = 0$.

As usual, the suspension has two degrees of freedom, designated as ϕ_1 and ϕ_2 in Fig. 31.2b.

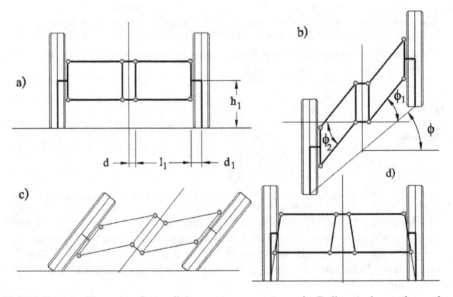

FIGURE 31.2. Transversal parallelograms suspension. a): Roll axis located on the ground and geometrical definitions; b) skew-symmetric deformation corresponding to roll; c): suspension in high roll conditions; d) configuration equivalent to a).

If angles ϕ_i are positive when the wheel moves in the up direction (with respect to the body), the roll angle and the displacement in the direction of the z axis of the body is easily computed:

$$\phi = \text{artg} \left(\frac{l_1 \left[\sin\left(\phi_1\right) - \sin\left(\phi_2\right) \right]}{2\left(d + d_1\right) + l_1 \left[\cos\left(\phi_1\right) + \cos\left(\phi_2\right) \right]} \right) ,$$

$$\Delta z = -l_1 \frac{\left(d + d_1\right) \left[\sin\left(\phi_1\right) + \sin\left(\phi_2\right) \right] + l_1 \sin\left(\phi_1 + \phi_2\right)}{2\left(d + d_1\right) + l_1 \left[\cos\left(\phi_1\right) + \cos\left(\phi_2\right) \right]} .$$

(31.1)

It is also possible to identify a symmetrical mode, linked with vertical displacement, and a skew-symmetrical mode, linked with roll. The former is characterized by $\phi_2 = \phi_1$, the latter by $\phi_2 = -\phi_1$. The skew symmetrical mode causes no vertical displacements of the body and the symmetrical one causes no roll, even for angle values that go beyond linearity.

Remark 31.1 *The possibility of expressing a generic motion as the sum of a symmetric and a skew-symmetrical mode is limited to conditions where the superimposition principle holds, that is, to conditions where it is possible to linearize the trigonometric functions of the angles.*

Let

$$t_0 = 2\left(d + d_1 + l_1\right)$$

be the reference value for the track; in a symmetrical mode the track depends on ϕ_1 through the relationship

$$t = 2 \left[d + d_1 + l_1 \cos(\phi_1) \right] = t_0 - 2l_1 \left[1 - \cos(\phi_1) \right] . \tag{31.2}$$

Only when $\phi_1 = 0$ do the track variations vanish, i.e.,

$$\frac{\partial t}{\partial z} = 0 .$$

Because the vertical displacement is

$$z = -l_1 \sin(\phi_1) \tag{31.3}$$

it follows that

$$t = t_0 - 2l_1 \left[1 - \sqrt{1 - \left(\frac{z}{l_1}\right)^2} \right] . \tag{31.4}$$

In the skew-symmetrical roll mode, the relationship between ϕ and ϕ_1 is

$$\tan\left(\phi\right) = \frac{l_1 \sin\left(\phi_1\right)}{d + d_1 + l_1 \cos(\phi_1)} \tag{31.5}$$

and the track is

$$t = 2 \frac{\left[d + d_1 + l_1 \cos(\phi_1)\right]}{\cos\left(\phi\right)} . \tag{31.6}$$

Equation (31.5) may be inverted, producing an equation allowing ϕ_1 to be computed as a function of ϕ,

$$\tan^2\left(\frac{\phi_1}{2}\right) - 2\frac{l_1}{(d + d_1 - l_1)\tan(\phi)}\tan\left(\frac{\phi_1}{2}\right) + \frac{d + d_1 + l_1}{d + d_1 - l_1} = 0. \qquad (31.7)$$

In the ideal case where $d + d_1 = 0$, it follows that

$$\phi_1 = \phi , \qquad (31.8)$$

and the track remains constant even for large values of the roll angle

$$\frac{\partial t}{\partial \phi} = 0 ;$$

otherwise the track remains constant only for small deviations from the symmetrical condition.

As already stated, the roll center remains on the ground only if in the reference condition the upper and lower links are horizontal, that is, if angle ϕ_1 and ϕ_2 have equal moduli and opposite signs. If, on the contrary, the symmetrical reference condition is characterized by positive values of ϕ_1 and ϕ_2 (the body is in a lower position with respect to the situation mentioned above), the roll center is below the road surface and vice-versa. These considerations are based on the assumption that the tire can be considered as a rigid disk; if, on the contrary, the compliance of the tire is accounted for, the position of the roll center is lower. If the transversal profile of the tires is curved, so that in roll motion they roll sideways on the ground, the roll center remains on the ground but is displaced sideways, outside the symmetry plane of the tire.

If the vehicle is controlled so that the local vertical remains in the symmetry plane, the load on the suspension changes with the roll angle (if, for instance, $\phi = 45°$, the centrifugal force is equal to the weight. The load is then equal to the static load multiplied by $\sqrt{2} \approx 1,4$). The suspension is compressed with increasing ϕ and the roll center goes deeper in the ground. To prevent this from occurring, devices able to control the compression of the suspensions must be used.

If the direction of the upper and lower links of the suspension is important in the kinematics of the suspension, the direction of the links modelling the vehicle body and the wheel hub is immaterial. The suspensions of Figs. 31.2a and 31.2d behave in the same way.

31.1.3   Tilting control

Consider a vehicle equipped with a tilting control system. Assume that such a device is integrated with the suspension springs, as shown in Fig. 31.3a: A rotary actuator with axis at point C rotates the arm CB to which the suspension springs AB and A'B are connected. Consider the rotation ϕ_c of the actuator arm as the control variable.

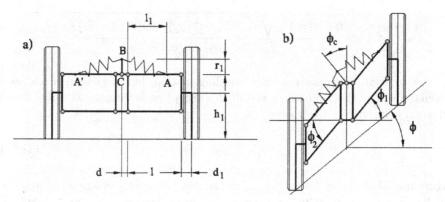

FIGURE 31.3. Sketch of the control of the transversal parallelograms suspension.

Assuming angles ϕ_i as positive when the suspensions move upwards with respect to the body, the coordinates of points A, A' and B in a system with origin in C and whose axes are parallel to the y and z axes are

$$(A - C) = \left\{ \begin{array}{c} d + l_2 \cos(\phi_1) \\ l_2 \sin(\phi_1) \end{array} \right\} \ , \quad (A' - C) = \left\{ \begin{array}{c} -d - l_2 \cos(\phi_2) \\ l_2 \sin(\phi_2) \end{array} \right\} \ , \quad (31.9)$$

$$(B - C) = \left\{ \begin{array}{c} -r_1 \sin(\phi_c) \\ r_1 \cos(\phi_c) \end{array} \right\} \ . \quad (31.10)$$

The length of the springs is then

$$\overline{A - B} = l_R = \sqrt{\beta_1 + \beta_2 \cos(\phi_1) + \beta_3 \sin(\phi_c) - \beta_4 \sin(\phi_1 - \phi_c)} \ ,$$

$$(31.11)$$

$$\overline{A' - B} = l_L = \sqrt{\beta_1 + \beta_2 \cos(\phi_2) - \beta_3 \sin(\phi_c) - \beta_4 \sin(\phi_2 + \phi_c)} \ ,$$

where subscripts L and R designate the left and right suspensions and

$$\begin{array}{ll} \beta_1 = d^2 + r_1^2 + l_2^2 \ , & \beta_3 = 2dr_1 \ , \\ \beta_2 = 2dl_2 \ , & \beta_4 = 2l_2r_1 \ . \end{array} \quad (31.12)$$

The length of the springs in the reference condition ($\phi_1 = \phi_2 = \phi_c = 0$) is

$$l_0^2 = l_{0L}^2 = l_{0R}^2 = \beta_1 + \beta_2 \ . \quad (31.13)$$

First consider the springs as rigid bodies. The relationships yielding angles ϕ_1 and ϕ_2 as functions of ϕ_c may be obtained equating l_R and l_L to l_0 :

$$-\beta_2 + \beta_2 \cos(\phi_1) + \beta_3 \sin(\phi_c) - \beta_4 \sin(\phi_1 - \phi_c) = 0 \ ,$$

$$(31.14)$$

$$-\beta_2 + \beta_2 \cos(\phi_2) - \beta_3 \sin(\phi_c) - \beta_4 \sin(\phi_2 + \phi_c) = 0 \ .$$

Equations (31.14) may be solved in ϕ_1 and ϕ_2 obtaining

$$\tan\left(\frac{\phi_1}{2}\right) = \frac{\beta_4 \cos(\phi_c) - \sqrt{\beta_4^2 - \beta_3^2 \sin^2(\phi_c) + 2\beta_2(\beta_3 + \beta_4) \sin(\phi_c)}}{(\beta_3 - \beta_4) \sin(\phi_c) - 2\beta_2} \ ,$$

$$(31.15)$$

$$\tan\left(\frac{\phi_2}{2}\right) = \frac{\beta_4 \cos\left(\phi_c\right) - \sqrt{\beta_4^2 - \beta_3^2 \sin^2\left(\phi_c\right) - 2\beta_2\left(\beta_3 + \beta_4\right)\sin\left(\phi_c\right)}}{\left(\beta_4 - \beta_3\right)\sin\left(\phi_c\right) - 2\beta_2}.$$

$$(31.16)$$

A rotation ϕ_c causes not only a rolling motion, but in general produces a displacement in the z direction as well. An exception is the case with $d = 0$ and thus $\beta_2 = \beta_3 = 0$. In this case

$$\phi_1 = -\phi_2 = \phi_c .$$

$$(31.17)$$

Remark 31.2 *If $d = 0$ a rotation of the control actuator produces a roll rotation of the vehicle (skew-symmetrical mode) but no displacement in the z direction. This statement amounts to saying that the roll center remains on the ground for all roll angles. The center of mass obviously lowers, because the roll center is on the ground, but the suspension behaves like a motorcycle wheel.*

Example 31.1 *Consider a transversal parallelogram suspension with the following data: $d_1 = 81.5$ mm, $r_1 = 138$ mm, $l_1 = 414$ mm, $l_2 = 388$ mm.*

Compute angles ϕ_1 and ϕ_2 as functions of ϕ_c and the displacements of the roll center along the z axis for three values of d, namely 0, 25 and 50 mm.

The results, computed using the above mentioned equations, are shown in Fig. 31.4.

As expected, if $d = 0$ rotation ϕ_c causes rolling of the vehicle body about the roll center that remains on the ground. If, on the contrary, $d \neq 0$, ϕ_1 is not equal to ϕ_2 and a displacement along the z direction (positive, in the sense that the body moves in the direction of the positive z axis) occurs. This displacement may reach 100 mm for $d = 50$ mm and $\phi_c = 50°$.

The center of mass obviously moves downwards when the vehicle rolls, but less than when d is zero.

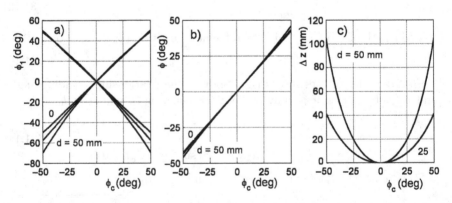

FIGURE 31.4. Transversal parallelograms suspension. a) Angles ϕ_1 and ϕ_2; b) roll angle ϕ and c) displacement in z direction of the roll center as a function of ϕ_c for three values of d: $d = 0$; $d = 25$ mm and $d = 50$ mm.

31.1.4 Suspension stiffness

The elastic potential energy of the springs, referred to the condition with $\phi_1 = \phi_2 = \phi_c = 0$, is

$$\mathcal{U}_m = \frac{1}{2} K \left[(l_R - l_0)^2 + (l_L - l_0)^2 \right] , \qquad (31.18)$$

where K is the stiffness of the springs.

First consider a suspension with $d = 0$. In this case $\phi_1 = -\phi_2$ and $\Delta z = 0$, when the springs are in the reference condition.

Let angles ϕ_1 and ϕ_2 vary about this condition by the small quantities $d\phi_1$ and $d\phi_2$. The roll angle and the displacement in the z direction may be obtained from Eq. (31.1):

$$\text{tg}\,(\phi + d\phi) = \frac{l_1 \left[\sin\,(\phi_1 + d\phi_1) - \sin\,(\phi_2 + d\phi_2) \right]}{2d_1 + l_1 \left[\cos\,(\phi_1 + d\phi_1) + \cos\,(\phi_2 + d\phi_2) \right]} , \qquad (31.19)$$

$$\Delta z + d\Delta z = l_1 \frac{d_1 \left[\sin\,(\phi_1 + d\phi_1) + \sin\,(\phi_2 + d\phi_2) \right] + l_1 \sin\,(\phi_1 + d\phi_1 + \phi_2 + d\phi_2)}{d_1 + l_1 \left[\cos\,(\phi_1 + d\phi_1) + \cos\,(\phi_2 + d\phi_2) \right]} .$$
$$(31.20)$$

Rolling motion

Assume that

$$d\phi_1 = -d\phi_2 . \qquad (31.21)$$

Because angle $d\phi_1$ and $d\phi_2$ are small and $\Delta z = 0$, it follows that

$$\text{tg}\,(\phi + d\phi) = \frac{l_1 \sin\,(\phi_1) + l_1 d\phi_1 \cos\,(\phi_1)}{d_1 + l_1 \cos\,(\phi_1) - l_1 d\phi_1 \sin\,(\phi_1)}, \qquad (31.22)$$

$$d\Delta z = 0 . \qquad (31.23)$$

The motion of the suspension is then rolling. Some computations are needed to obtain a relationship linking $d\phi$ to $d\phi_1$. They yield

$$\frac{d\phi_1}{d\phi} = \frac{d_1^2 + l_1^2 + 2d_1 l_1 \cos\,(\phi_1)}{l_1^2 + d_1 l_1 \cos\,(\phi_1)}. \qquad (31.24)$$

The derivative $d\mathcal{U}_m/d\phi$, i.e. the restoring moment due to the spring system, is

$$\frac{d\mathcal{U}_m}{d\phi} = K \left[(l_R - l_0) \frac{dl_R}{d\phi_1} + (l_L - l_0) \frac{dl_L}{d\phi_2} \frac{d\phi_2}{d\phi_1} \right] \frac{d\phi_1}{d\phi} \qquad (31.25)$$

where

$$\frac{\partial l_R}{\partial \phi_1} = \frac{1}{2l_R} \left[-\beta_4 \cos\,(\phi_1 - \phi_c) \right] ,$$
$$(31.26)$$
$$\frac{dl_L}{d\phi_2} \frac{d\phi_2}{d\phi_1} = \frac{1}{2l_L} \left[\beta_4 \cos\,(\phi_1 - \phi_c) \right] .$$

Because it has been assumed that $d = 0$, the above mentioned equations may be simplified, obtaining

$$\frac{\partial \mathcal{U}_m}{\partial \phi} = K l_2 r_1 l_0 \cos (\phi_1 - \phi_c) \frac{\partial \phi_1}{\partial \phi} \times$$

$$\times \frac{\sqrt{\beta_1 + \beta_4 \sin (\phi_1 - \phi_c)} - \sqrt{\beta_1 - \beta_4 \sin (\phi_1 - \phi_c)}}{\sqrt{\beta_1^2 - \beta_4^2 \sin^2 (\phi_1 - \phi_c)}} . \qquad (31.27)$$

As expected, if $\phi_1 = \phi_c$ the moment due to the springs vanishes, i.e.,

$$\frac{\partial \mathcal{U}_m}{\partial \phi} = 0 .$$

If the configuration is changed by a small angle about this equilibrium position, i.e. if

$$\phi_1 = \phi_c + \Delta \phi_1 ,$$

the rolling moment is

$$\frac{\partial \mathcal{U}_m}{\partial \phi} = K l_2 r_1 l_0 \frac{\partial \phi_1}{\partial \phi} \frac{\sqrt{\beta_1 + \beta_4 \Delta \phi_1} - \sqrt{\beta_1 - \beta_4 \Delta \phi_1}}{\beta_1} \qquad (31.28)$$

and then

$$\frac{\partial \mathcal{U}_m}{\partial \phi} = 2K \frac{l_2^2 r_1^2}{l_2^2 + r_1^2} \frac{d_1^2 + l_1^2 + 2d_1 l_1 \cos (\phi_1)}{l_1^2 + d_1 l_1 \cos (\phi_1)} \Delta \phi_1. \qquad (31.29)$$

The rolling moment is proportional to angle $\Delta \phi_1$ and thus to the roll angle ϕ about the reference position. The rolling stiffness of the suspension is then

$$K_\phi = \frac{1}{\phi} \frac{\partial \mathcal{U}_m}{\partial \phi} = \frac{1}{\Delta \phi_1} \frac{\partial \phi_1}{\partial \phi} \frac{\partial \mathcal{U}_m}{\partial \phi}, \qquad (31.30)$$

i.e.,

$$K_\phi = 2K \frac{l_2^2 r_1^2}{l_2^2 + r_1^2} \left(\frac{d_1^2 + l_1^2 + 2d_1 l_1 \cos (\phi_1)}{l_1^2 + d_1 l_1 \cos (\phi_1)} \right)^2 . \qquad (31.31)$$

If d_1 is also equal to zero,

$$\frac{\partial \phi_1}{\partial \phi} = 1$$

and the vehicle tilts, when there is no rolling moment, until an angle equal to ϕ_c has been reached.

Motion in the z direction

If the deformation is symmetrical, i.e. if

$$d\phi_1 = d\phi_2, \qquad (31.32)$$

it is possible to write

$$tg\,(\phi + \Delta\phi) = tg\,(\phi)\,, \tag{31.33}$$

$$d\Delta z = l_1 d\phi_1 \frac{d_1 \cos{(\phi_1)} + l_1}{d_1 + l_1 \cos{(\phi_1)}}\,. \tag{31.34}$$

The derivative $d\mathcal{U}_m/d\Delta z$, i.e. the force in the z direction due to the suspension springs, is

$$\frac{d\mathcal{U}_m}{d\Delta z} = K\left[(l_R - l_0)\frac{dl_R}{d\phi_1} + (l_L - l_0)\frac{dl_L}{d\phi_2}\right]\frac{d\phi_1}{d\Delta z}. \tag{31.35}$$

Remembering that $\phi_1 = -\phi_2$, it follows that

$$\frac{dl_L}{d\phi_2} = \frac{1}{2l_L}\left[\beta_4 \cos{(\phi_1 - \phi_c)}\right]\,,$$

$$\frac{d\phi_1}{d\Delta z} = \frac{d_1 + l_1 \cos{(\phi_1)}}{l_1 d_1 \cos{(\phi_1)} + l_1^2}\,. \tag{31.36}$$

This result may also be simplified, obtaining

$$\frac{\partial\mathcal{U}_m}{\partial\Delta z} = K l_2 r_1 l_0 \cos{(\phi_1 - \phi_c)}\frac{\partial\phi_1}{\partial\Delta z}\times$$

$$\times \frac{\sqrt{\beta_1 + \beta_4 \sin{(\phi_1 - \phi_c)}} - \sqrt{\beta_1 - \beta_4 \sin{(\phi_1 - \phi_c)}}}{\sqrt{\beta_1^2 - \beta_4^2 \sin^2{(\phi_1 - \phi_c)}}}\,. \tag{31.37}$$

Because condition $\phi_1 = \phi_c$ was assumed to be an equilibrium condition, the force in the z direction vanishes if $\phi_1 = \phi_c$. Operating in the same way as a rolling condition, assuming that

$$\phi_1 = \phi_c + \Delta\phi_1\,,$$

the value of the force in the z direction is obtained:

$$\frac{\partial\mathcal{U}_m}{\partial\Delta z} = 2K\frac{l_2^2 r_1^2}{l_2^2 + r_1^2}\frac{d_1 + l_1 \cos{(\phi_1)}}{l_1 d_1 \cos{(\phi_1)} + l_1^2}\Delta\phi_1. \tag{31.38}$$

The force in the z direction is then proportional to angle $\Delta\phi_1$ and thus to the displacement Δz. The stiffness of the suspension in the z direction is then

$$K_z = \frac{1}{\Delta z}\frac{\partial\mathcal{U}_m}{\partial\Delta z} = \frac{1}{\Delta\phi_1}\frac{\partial\phi_1}{\partial\Delta z}\frac{\partial\mathcal{U}_m}{\partial\Delta z}\,, \tag{31.39}$$

i.e.,

$$K_z = 2K\frac{l_2^2 r_1^2}{l_2^2 + r_1^2}\left(\frac{d_1 + l_1 \cos{(\phi_1)}}{l_1 d_1 \cos{(\phi_1)} + l_1^2}\right)^2. \tag{31.40}$$

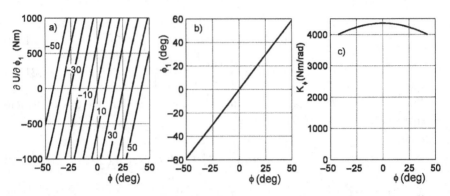

FIGURE 31.5. Transversal parallelograms suspension. a): Restoring moment due to the suspension springs versus the roll angle ϕ for various values of the control variable ϕ_c. b): Relationship between ϕ and ϕ_c. c): Stiffness for small roll oscillations about the static equilibrium condition.

Example 31.2 *Consider a transversal parallelogram suspension with the following data: $d = 0$, $d_1 = 81.5$ mm, $r_1 = 138$ mm, $l_1 = 414$ mm, $l_2 = 388$ mm.*

Compute the relationship linking ϕ to ϕ_1 and plot the restoring moment due to the suspension springs $\partial \mathcal{U}_m / \partial \phi$ versus ϕ, for various values of ϕ_c and the stiffness of the suspension K_ϕ versus ϕ_c.

The results are reported in Fig. 31.5.

From Fig. 31.5a it is clear that the restoring moment $\partial \mathcal{U}_m / \partial \phi$ is linear with the roll angle ϕ, while the stiffness depends only slightly on the position about which the motion occurs (Fig. 31.5c). Also the dependence of ϕ_1 from ϕ is almost linear, as shown by Fig. 31.5b. Because $d = 0$, it follows that in the equilibrium condition $\phi_1 = \phi_c$.

31.1.5 Roll damping of the suspension

Consider a damper system made by two shock absorbers located in parallel to the springs between points A and B and points A' and B.

The dissipation function of the suspension is then

$$\mathcal{F} = \frac{1}{2}c\left\{\left[\frac{d\left(\overline{A - B}\right)}{dt}\right]^2 + \left[\frac{d\left(\overline{A' - B}\right)}{dt}\right]^2\right\} . \qquad (31.41)$$

Remembering that lengths $l_D = \left(\overline{A - B}\right)$ and $l_L = \left(\overline{A' - B}\right)$ are functions of ϕ_c and ϕ_1, the dissipation function can be computed as

$$\mathcal{F} = \frac{1}{2}c\left\{\left[\left(\frac{\partial l_R}{\partial \phi_1}\frac{\partial \phi_1}{\partial \phi}\dot{\phi} + \frac{\partial l_R}{\partial \phi_c}\dot{\phi}_c\right)\right]^2 + \left[\left(\frac{\partial l_L}{\partial \phi_1}\frac{\partial \phi_1}{\partial \phi}\dot{\phi} + \frac{\partial l_L}{\partial \phi_c}\dot{\phi}_c\right)\right]^2\right\} . \qquad (31.42)$$

The previous equation may be written in the form

$$\mathcal{F} = \frac{1}{2}\left(c_{11}\dot{\phi}^2 + c_{22}\dot{\phi}_c^2 + 2c_{12}\dot{\phi}\dot{\phi}_c\right) , \qquad (31.43)$$

where

$$c_{11} = c \left[\left(\frac{\partial l_R}{\partial \phi_1} \right)^2 + \left(\frac{\partial l_L}{\partial \phi_1} \right)^2 \right] \left(\frac{\partial \phi_1}{\partial \phi} \right)^2 ,$$

$$c_{12} = c \left(\frac{\partial l_R}{\partial \phi_1} \frac{\partial \phi_1}{\partial \phi} \frac{\partial l_R}{\partial \phi_c} + \frac{\partial l_L}{\partial \phi_1} \frac{\partial \phi_1}{\partial \phi} \frac{\partial l_L}{\partial \phi_c} \right) , \tag{31.44}$$

$$c_{22} = c \left[\left(\frac{\partial l_R}{\partial \phi_c} \right)^2 + \left(\frac{\partial l_2}{\partial \phi_c} \right)^2 \right] .$$

Some of the derivatives are reported in Eq. 31.26; the others are

$$\frac{\partial l_R}{\partial \phi_c} = -\frac{\partial l_L}{\partial \phi_c} = \frac{1}{2 l_L} \left[\beta_3 \cos (\phi_c) + \beta_4 \cos (\phi_1 - \phi_c) \right] . \tag{31.45}$$

With the control locked, i.e. with $\dot{\phi}_c = 0$, the damping coefficient of the suspension coincides with c_{11}.

If

$$d = 0 ,$$

it can immediately be derived that

$$c_{11} = c_{22} = -c_{12} = k \frac{c}{K} , \tag{31.46}$$

where k is the roll stiffness of the suspension, while c and K are the characteristics of the damper and the spring.

Example 31.3 *Compute the rolling damping coefficient of the suspension of the previous example, with locked controls, as a function of the static equilibrium position.*

The result is shown in Fig. 31.6. The linearized characteristics of the suspension depend little on the position, in terms of damping.

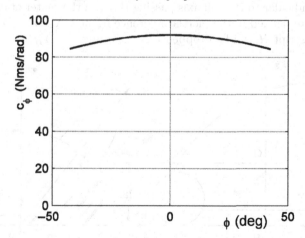

FIGURE 31.6. Damping cefficient of the suspension of the previous example for small movements about the equilibrium position.

31.2   LINEARIZED RIGID BODY MODEL

The simplest model for a tilting body vehicle is one with four degrees of freedom. It may be obtained from the model with 10 degrees of freedom of Fig. 29.3 (Section 29.2.2), locking the degrees of freedom θ and Z of the sprung mass and the symmetrical motions of the suspensions.

In the case of a two-wheeled vehicle, the kinematics is much simplified because:

- the mid-plane of the wheels remains parallel to the symmetry plane of the vehicle (actually coinciding with it);

- the roll axis is on the ground and in a fixed position, at least as a first approximation, if the effect of the transversal profile of the tires is neglected.

These considerations do not hold in the case of tilting body vehicles with more than two wheels. The roll axis is determined by the characteristics of the suspensions or by the position of a true cylindrical hinge: In the first case the very concept of a roll is inappropriate because of the large roll angles vehicles of this type can manage. The roll axis is an axis of instantaneous rotation, one that has no meaning in case of large rotations.

Assume that the suspensions are designed so that the mid-plane of the wheels remains parallel to the symmetry plane of the vehicle and the roll axis remains on the ground, at the intersection of the symmetry plane and the ground plane, as in simplified motorcycle models (See Appendix B).

The roll axis now coincides with the x^*-axis of the $x^*y^*z^*$ reference frame, seen in the previous section (Fig. 31.7). In this case the generalized coordinates for translations are the coordinates X_H, Y_H (coordinate Z_H vanishes) of point H, instead of the coordinates of the center of mass. Point H is on the ground, on the perpendicular to the roll axis passing through the center of mass G. Such coordinates are defined in the inertial reference frame $OX_iY_iZ_i$. To simplify the notation, subscript H will be dropped ($X = X_H$ and $Y = Y_H$).

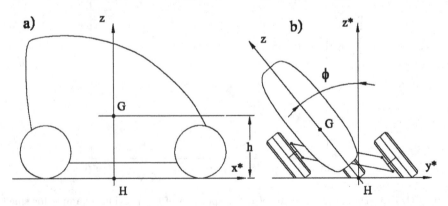

FIGURE 31.7. Reference frames for the sprung mass and definition of point H.

The generalized coordinates for rotations are the yaw angle ψ and the roll angle ϕ. As usual, the assumption of small angles (particularly for the sideslip angle β) allows the component of the velocity v_{x^*} to be confused with the forward velocity V. Angular velocities $\dot{\psi}$ and $\dot{\phi}$ will be considered small quantities as well.

31.2.1 Kinetic and potential energy

Because the pitch rotation is not included in the model, the roll axis is horizontal. The rotation matrix allowing us to change from the body-fixed frame $Gxyz$ to the inertial frame $X_iY_iZ_i$ is

$$\mathbf{R} = \mathbf{R}_1\mathbf{R}_2 \,, \tag{31.47}$$

where

$$\mathbf{R}_1 = \begin{bmatrix} \cos(\psi) & -\sin(\psi) & 0 \\ \sin(\psi) & \cos(\psi) & 0 \\ 0 & 0 & 1 \end{bmatrix} \,, \quad \mathbf{R}_2 = \begin{bmatrix} 1 & 0 & 0 \\ 0 & \cos(\phi) & -\sin(\phi) \\ 0 & \sin(\phi) & \cos(\phi) \end{bmatrix} \,.$$

The derivative of the rotation matrix is

$$\dot{\mathbf{R}} = \dot{\mathbf{R}}_1\mathbf{R}_2 + \mathbf{R}_1\dot{\mathbf{R}}_2 \,. \tag{31.48}$$

The components of the angular velocity in the direction of the body-fixed axes are linked with the derivatives of the coordinates by the equation

$$\left\{ \begin{array}{c} \Omega_x \\ \Omega_y \\ \Omega_z \end{array} \right\} = \begin{bmatrix} 1 & 0 \\ 0 & \sin(\phi) \\ 0 & \cos(\phi) \end{bmatrix} \left\{ \begin{array}{c} \dot{\phi} \\ \dot{\psi} \end{array} \right\} \,. \tag{31.49}$$

The vector of the generalized coordinates is

$$\mathbf{q} = \begin{bmatrix} X & Y & \phi & \psi \end{bmatrix}^T \,. \tag{31.50}$$

The generalized velocities for translational degrees of freedom are the components of the velocity in the $x^*y^*z^*$ frame. The derivatives of coordinates ϕ and ψ, that will be referred to as v_ϕ and v_ψ, will be used for the rotational degrees of freedom. The generalized velocities are then

$$\mathbf{w} = \begin{bmatrix} v_x & v_y & v_\phi & v_\psi \end{bmatrix}^T \,. \tag{31.51}$$

The relationship between generalized velocities and derivatives of the generalized coordinates may be written in the usual form

$$\mathbf{w} = \mathbf{A}^T\dot{\mathbf{q}} \,, \tag{31.52}$$

where matrix $\mathbf{A}^2$ is

$$\mathbf{A} = \begin{bmatrix} \cos(\psi) & -\sin(\psi) & 0 & 0 \\ \sin(\psi) & \cos(\psi) & 0 & 0 \\ 0 & 0 & 1 & 0 \\ 0 & 0 & 0 & 1 \end{bmatrix}. \tag{31.53}$$

Because in this case $\mathbf{A}$ is a rotation matrix, the inverse transformation is

$$\dot{\mathbf{q}} = \mathbf{Bw} = \mathbf{Aw} .$$

The vector defining the position of the center of the sprung mass G_S with respect to point H is, in the body-fixed frame,

$$\mathbf{r}_1 = h \begin{bmatrix} 0 & 0 & 1 \end{bmatrix}^T . \tag{31.54}$$

In the inertial frame the position of the same point is

$$(\overline{G_S - O'}) = (\overline{H - O'}) + \mathbf{R}\mathbf{r}_1. \tag{31.55}$$

Because $\mathbf{r}_1$ is constant, the velocity of point G_S is

$$\mathbf{V}_{G_S} = \begin{bmatrix} \dot{X} & \dot{Y} & 0 \end{bmatrix}^T + \dot{\mathbf{R}}\mathbf{r}_1 , \tag{31.56}$$

i.e.

$$\mathbf{V}_{G_S} = \mathbf{R}_1\mathbf{V} + \dot{\mathbf{R}}\mathbf{r}_1 , \tag{31.57}$$

and then the translational kinetic energy of the sprung mass is

$$\mathcal{T}_t = \frac{1}{2}m \left(\mathbf{V}^T\mathbf{V} + \mathbf{r}_1{}^T\dot{\mathbf{R}}^T\dot{\mathbf{R}}\mathbf{r}_1 + 2\mathbf{V}^T\mathbf{R}_1^T\dot{\mathbf{R}}\mathbf{r}_1 \right) . \tag{31.58}$$

Because plane xz is a symmetry plane for the sprung mass, its inertia tensor is

$$\mathbf{J} = \begin{bmatrix} J_x & 0 & -J_{xz} \\ 0 & J_y & 0 \\ -J_{xz} & 0 & J_z \end{bmatrix} . \tag{31.59}$$

The rotational kinetic energy of the sprung mass is then

$$\mathcal{T}_r = \frac{1}{2}\boldsymbol{\Omega}^T\mathbf{J}\boldsymbol{\Omega} . \tag{31.60}$$

By performing the relevant computations, expressing the components of the angular velocity as functions of the derivatives of the coordinates and neglecting the terms containing powers of small quantities higher than the second, it follows that

---

[2]Matrix $\mathbf{A}$ here defined must not be confused with the dynamic matrix in the state space, which is also usually referred to as $\mathbf{A}$.

$$T = \tfrac{1}{2}m\left(v_x^2 + v_y^2\right) + \tfrac{1}{2}J_x^*\dot{\phi}^2 + \tfrac{1}{2}\left[J_y^* \sin^2(\phi) + J_z \cos^2(\phi)\right]\dot{\psi}^2$$
$$-J_{xz}\cos(\phi)\,\dot{\psi}\,\dot{\phi} + mv_x\,h\dot{\psi}\sin(\phi) - mv_y h\dot{\phi}\cos(\phi)\ , \tag{31.61}$$

where

$$J_x^* = mh^2 + J_x\ ,\ \ J_y^* = mh^2 + J_y\ .$$

The height of the center of mass of the sprung mass on the ground is

$$Z_G = h\cos(\phi)\ , \tag{31.62}$$

and then the gravitational potential energy of the vehicle is

$$\mathcal{U}_g = mgh\cos(\phi)\ . \tag{31.63}$$

The potential energy reduces to its gravitational components in the case of a two-wheeled vehicle. In vehicles with three or more wheels with suspensions, the elastic potential energy due to the springs must also be accounted for. In the following study the elastic potential energy will be assumed to depend only on the roll angle; however, it is not a simple quadratic function as in the case of linearized models, because the roll angle may be large. In general, it is possible to state that

$$\mathcal{U}_s = \mathcal{U}_s(\phi)\ . \tag{31.64}$$

If the vehicle has suspensions for the roll motion and the latter are provided with dampers, a dissipative function may be defined,

$$\mathcal{F} = \mathcal{F}\left(\phi, \dot{\phi}\right)\ . \tag{31.65}$$

It must be expressly stated that the equations above were obtained without resorting to the assumption that all variables of motion, with the exception of the roll angle ϕ, are small quantities. Moreover, these equations are more general and hold even if the roll axis does not lie on the ground or is exactly horizontal, provided that the angle between the roll axis and the ground plane (referred to as θ_0 in the previous chapters) is a small angle and that h is the distance between the center of mass and the roll axis instead of its height on the ground.

31.2.2 Rotation of the wheels

Because it has been assumed that, as in the case of vehicles with two wheels (see Appendix B), the rotation axis of the wheels is perpendicular to the symmetry plane, the absolute angular velocity of the ith wheel expressed in the reference frame of the sprung mass is

$$\Omega_i = \left\{ \begin{array}{c} \Omega_x \\ \Omega_y + \dot{\chi}_i \\ \Omega_z \end{array} \right\}\ , \tag{31.66}$$

where χ_i is the rotation angle of the wheel.

If the wheel steers, the reference frame of the ith wheel will be rotated by a steering angle δ_i about an axis, the kingpin axis, that in general is not perpendicular to the ground. If $\mathbf{e}_k$ is the unit vector of the kingpin axis (its components will be indicated as x_k, y_k and z_k)[3], the rotation matrix $\mathbf{R}_{ki}$ to rotate the reference frame fixed to the sprung mass in such a way that its z axis coincides with the kingpin axis of the ith wheel is

$$
\mathbf{R}_{ki} = \frac{1}{\sqrt{x_k^2 + z_k^2}} \begin{bmatrix} z_k & -x_k y_k & x_k \sqrt{x_k^2 + z_k^2} \\ 0 & (x_k^2 + z_k^2) & y_k \sqrt{x_k^2 + z_k^2} \\ -x_k & -z_k y_k & z_k \sqrt{x_k^2 + z_k^2} \end{bmatrix} . \tag{31.67}
$$

The caster and the inclination angles of the kingpin are usually small in suspensions for two-wheeled axles and, as seen in the previous sections, rotation matrix $\mathbf{R}_{ki}$ reduces to

$$
\mathbf{R}_{ki} \approx \begin{bmatrix} 1 & 0 & x_k \\ 0 & 1 & y_k \\ -x_k & -y_k & 1 \end{bmatrix} , \tag{31.68}
$$

where x_k and y_k are the caster and the inclination angles (the latter changed in sign) of the kingpin axis. For symmetry reasons

$$
x_{k_D} = x_{k_S} , \quad y_{k_D} = -y_{k_S} . \tag{31.69}
$$

In motorcycles y_k is zero, while the caster angle x_k may be large. In the following parts of this section this possibility will not be considered.

A further rotation matrix

$$
\mathbf{R}_{4i} = \begin{bmatrix} \cos(\delta_i) & -\sin(\delta_i) & 0 \\ \sin(\delta_i) & \cos(\delta_i) & 0 \\ 0 & 0 & 1 \end{bmatrix} \tag{31.70}
$$

can be defined for the rotation of the wheel about the kingpin axis.

The angular velocity of the wheel in the reference frame of the sprung mass is then

$$
\mathbf{\Omega}_{wi} = \mathbf{\Omega} + \dot{\delta}_i \mathbf{R}_{ki} \mathbf{e}_3 + \dot{\chi}_i \mathbf{R}_{ki} \mathbf{R}_{4i} \mathbf{R}_{ki}^T \mathbf{e}_2 . \tag{31.71}
$$

Eq. (31.71) must be premultiplied by $(\mathbf{R}_{ki} \mathbf{R}_{4i} \mathbf{R}_{ki}^T)^T$ to obtain the angular velocity of the wheel in its own reference frame. Remembering that $\mathbf{R}_{4i} \mathbf{e}_3 = \mathbf{e}_3$, it follows that

$$
\mathbf{\Omega}_{wi} = \dot{\chi}_i \mathbf{e}_2 + \dot{\delta}_i \boldsymbol{\alpha}_1 + \boldsymbol{\alpha}_2 \mathbf{\Omega} , \tag{31.72}
$$

where

$$
\boldsymbol{\alpha}_1 = \mathbf{R}_{ki} \mathbf{e}_3 , \quad \boldsymbol{\alpha}_2 = \mathbf{R}_{ki} \mathbf{R}_{4i}^T \mathbf{R}_{ki}^T . \tag{31.73}
$$

---

[3]Obviously $\sqrt{x_k^2 + y_k^2 + z_k^2} = 1$.

Because the wheel is a gyroscopic body (two of its principal moments of inertia are equal) with a principal axis of inertia coinciding with its rotation axis, its inertia matrix is diagonal and has the form

$$\mathbf{J}_{wi} = \operatorname{diag}\left(\begin{bmatrix} J_{ti} & J_{pi} & J_{ti} \end{bmatrix}\right) , \tag{31.74}$$

where J_{pi} is the polar moment of inertia and J_{ti} is the transversal moment of inertia of the ith wheel.

The rotational kinetic energy of the ith wheel is

$$\begin{aligned} \mathcal{T}_{wri} = \tfrac{1}{2}\boldsymbol{\Omega}^T\boldsymbol{\alpha}_2^T\mathbf{J}_{wi}\boldsymbol{\alpha}_2\boldsymbol{\Omega} + \tfrac{1}{2}\dot{\chi}_i^2\mathbf{e}_2^T\mathbf{J}_{wi}\mathbf{e}_2 + \tfrac{1}{2}\dot{\delta}_i^2\boldsymbol{\alpha}_1^T\mathbf{J}_{wi}\boldsymbol{\alpha}_1 + \\ +\dot{\chi}_i\dot{\delta}_i\mathbf{e}_2^T\mathbf{J}_{wi}\boldsymbol{\alpha}_1 + \dot{\chi}_i\mathbf{e}_2^T\mathbf{J}_{wi}\boldsymbol{\alpha}_2\boldsymbol{\Omega} + \dot{\delta}_i\boldsymbol{\alpha}_1^T\mathbf{J}_{wi}\boldsymbol{\alpha}_2\boldsymbol{\Omega} . \end{aligned} \tag{31.75}$$

By performing the relevant computations and assuming that all variables of motion, except for ϕ and χ_i, are small, it follows that

$$\begin{aligned} \mathcal{T}_{wri} = \tfrac{1}{2}J_{ti}\dot{\phi}^2 + \tfrac{1}{2}\left[J_{pi}\sin^2\left(\phi\right) + J_{ti}\cos^2\left(\phi\right)\right]\dot{\psi}^2 + \tfrac{1}{2}J_{pi}\dot{\chi}_i^2 + \\ +\tfrac{1}{2}\dot{\delta}_i^2 J_{ti} - J_{pi}\delta_i\dot{\phi}\dot{\chi}_i + J_{pi}\,y_{ki}\dot{\chi}_i\dot{\delta}_i + J_{pi}\sin\left(\phi\right)\dot{\psi}\dot{\chi}_i + J_{ti}\cos\left(\phi\right)\dot{\psi}\dot{\delta}_i . \end{aligned} \tag{31.76}$$

The first two terms express the rotational kinetic energy of the wheel due to angular velocity of the vehicle and thus have already been included in the expression of the kinetic energy of the vehicle, if the moments of inertia of the wheels have been taken into account when computing the total inertia.

31.2.3 Lagrangian function

The Lagrangian function of the vehicle is then

$$\begin{aligned} \mathcal{L} = \tfrac{1}{2}m\left(v_x^2 + v_y^2\right) + \tfrac{1}{2}J_x^*\dot{\phi}^2 + \tfrac{1}{2}\left[J_y^*\sin^2\left(\phi\right) + J_z\cos^2\left(\phi\right)\right]\dot{\psi}^2 + \\ -J_{xz}\cos\left(\phi\right)\dot{\psi}\,\dot{\phi} + mv_x\,h\dot{\psi}\sin\left(\phi\right) - mv_y h\dot{\phi}\cos\left(\phi\right) + \\ +\sum_{\forall i}\left[\tfrac{1}{2}J_{pi}\dot{\chi}_i^2 + \tfrac{1}{2}\dot{\delta}_i^2 J_{ti} - J_{pi}\delta_i\dot{\phi}\dot{\chi}_i + J_{pi}\,y_{ki}\dot{\chi}_i\dot{\delta}_i + \right. \\ \left. +J_{pi}\sin\left(\phi\right)\dot{\psi}\dot{\chi}_i + J_{ti}\cos\left(\phi\right)\dot{\psi}\dot{\delta}_i\right] - mgh\cos\left(\phi\right) - \mathcal{U}_s\left(\phi\right) . \end{aligned} \tag{31.77}$$

If the longitudinal slip of the wheels is neglected, their angular velocity is

$$\dot{\chi}_i = \frac{V}{R_{e_i}} . \tag{31.78}$$

In a way similar to our treatment of the four-wheeled vehicle, the kinetic energy linked with the steering velocity $\dot{\delta}$ may be neglected in the locked control motion. The Lagrangian reduces to

$$\begin{aligned} \mathcal{L} = \tfrac{1}{2}m_{at}V^2 + \tfrac{1}{2}mv_y^2 + \tfrac{1}{2}J_x^*\dot{\phi}^2 + \tfrac{1}{2}\left[J_y^*\sin^2\left(\phi\right) + J_z\cos^2\left(\phi\right)\right]\dot{\psi}^2 + \\ -J_{xz}\cos\left(\phi\right)\dot{\psi}\,\dot{\phi} + VJ_s\dot{\psi}\sin\left(\phi\right) - mv_y h\dot{\phi}\cos\left(\phi\right) + \\ -V\sum_{\forall i}\tfrac{J_{pi}}{R_{e_i}}\delta_i\dot{\phi} - mgh\cos\left(\phi\right) - \mathcal{U}_s\left(\phi\right) , \end{aligned} \tag{31.79}$$

where

$$m_{at} = m + \sum_{\forall i} \frac{J_{pi}}{R_{e_i}^2} , \quad J_s = m\,h + \sum_{\forall i} \frac{J_{pi}}{R_{e_i}} ,$$

$$J_x^* = mh^2 + J_x , \quad J_y^* = mh^2 + J_y .$$

The derivatives of the Lagrangian function are then

$$\frac{\partial \mathcal{L}}{\partial V} = m_{at}V + J_s\dot{\psi}\sin(\phi) , \tag{31.80}$$

$$\frac{\partial \mathcal{L}}{\partial v_y} = mv_y - mh\dot{\phi}\cos(\phi) , \tag{31.81}$$

$$\frac{\partial \mathcal{L}}{\partial \dot{\phi}} = J_x^*\dot{\phi} - J_{xz}\cos(\phi)\,\dot{\psi} - mv_yh\cos(\phi) - V\sum_{\forall i} \frac{J_{pi}}{R_{e_i}}\delta_i , \tag{31.82}$$

$$\frac{\partial \mathcal{L}}{\partial \dot{\psi}} = \left[J_y^*\sin^2(\phi) + J_z\cos^2(\phi) \right]\dot{\psi} - J_{xz}\cos(\phi)\,\dot{\phi} + VJ_sh\sin(\phi) . \tag{31.83}$$

The derivative with respect to time of the derivatives with respect to the generalized velocities contains products that are themselves the products of two or more small quantities, and thus must be neglected in the linearization process. Also $\dot{V}$ may be considered as a small quantity, and then terms containing, for instance, product $\dot{V}\delta$ may be neglected. It then follows that

$$\frac{d}{dt}\left(\frac{\partial \mathcal{L}}{\partial V} \right) = m_{at}\dot{V} + J_s\ddot{\psi}\sin(\phi) , \tag{31.84}$$

$$\frac{d}{dt}\left(\frac{\partial \mathcal{L}}{\partial v_y} \right) = m\dot{v}_y - mh\ddot{\phi}\cos(\phi) , \tag{31.85}$$

$$\frac{d}{dt}\left(\frac{\partial \mathcal{L}}{\partial \dot{\phi}} \right) = J_x^*\ddot{\phi} - J_{xz}\cos(\phi)\,\ddot{\psi} - m\dot{v}_yh\cos(\phi) , \tag{31.86}$$

$$\frac{d}{dt}\left(\frac{\partial \mathcal{L}}{\partial \dot{\psi}} \right) = \left[J_y^*\sin^2(\phi) + J_z\cos^2(\phi) \right]\ddot{\psi} - J_{xz}\cos(\phi)\ddot{\phi}+$$

$$+ J_s\dot{V}\sin(\phi) + J_sV\,\cos(\phi)\,\dot{\phi} , \tag{31.87}$$

$$\frac{\partial \mathcal{L}}{\partial x^*} = \frac{\partial \mathcal{L}}{\partial y^*} = \frac{\partial \mathcal{L}}{\partial \psi} = 0 , \tag{31.88}$$

$$\frac{\partial \mathcal{L}}{\partial \phi} = J_sV\dot{\psi}\cos(\phi) + mgh\sin(\phi) - \frac{\partial \mathcal{U}_s(\phi)}{\partial \phi} . \tag{31.89}$$

31.2.4   Kinematic equations

Matrix $\mathbf{A}$ is what we have already seen for the model with 10 degrees of freedom, except that the last six rows and columns are not present here.

The equation of motion in the configuration space is

$$\frac{\partial}{\partial t}\left(\left\{\frac{\partial\mathcal{L}}{\partial w}\right\}\right) + \mathbf{B^T}\mathbf{\Gamma}\left\{\frac{\partial\mathcal{L}}{\partial w}\right\} - \mathbf{B}^T\left\{\frac{\partial\mathcal{L}}{\partial q}\right\} + \left\{\frac{\partial\mathcal{F}}{\partial w}\right\} = \mathbf{B}^T\mathbf{Q} . \qquad (31.90)$$

The column matrix $\mathbf{B}^T\mathbf{Q}$ containing the four components of the generalized forces vector will be computed later, when the virtual work of the forces acting on the system is described. In the following its elements will be written as Q_x, Q_y, Q_ϕ, Q_ψ.

As usual, the most difficult part is writing matrix $\mathbf{B^T}\mathbf{\Gamma}$. By performing somewhat complex computations, following the procedure outlined in Appendix A, it follows that

$$\mathbf{B^T}\mathbf{\Gamma} = \left[\begin{array}{cc} \begin{bmatrix} 0 & -\dot{\psi} \\ \dot{\psi} & 0 \\ 0 & 0 \\ -v_y & v_x \end{bmatrix} & \mathbf{0}_{4\times 2} \end{array}\right] .$$

By introducing the values of the derivatives and linearizing, it follows that

$$\mathbf{B^T}\mathbf{\Gamma}\left\{\frac{\partial\mathcal{L}}{\partial w}\right\} = \left\{\begin{array}{c} 0 \\ m_{at}V\dot{\psi} \\ 0 \\ V\left[-mh\dot{\phi}\cos(\phi) - v_y\sum_{\forall k}\left(J_{pr}\frac{1}{R_e^2}\right)\right] \end{array}\right\} . \qquad (31.91)$$

Finally

$$\mathbf{B}^T\left\{\frac{\partial\mathcal{L}}{\partial q}\right\} = \left\{\frac{\partial\mathcal{L}}{\partial q}\right\} . \qquad (31.92)$$

31.2.5   Equations of motion

First equation: longitudinal translation

$$m_{at}\dot{V} + J_s\ddot{\psi}\sin(\phi) = Q_x . \qquad (31.93)$$

Second equation: lateral translation

$$m\dot{v}_y + m_{at}V\dot{\psi} - mh\ddot{\phi}\cos(\phi) = Q_y . \qquad (31.94)$$

Third equation: roll rotation

$$J_x^* \ddot{\phi} - J_{xz} \cos(\phi) \ddot{\psi} - m\dot{v}_y h \cos(\phi) - J_s V \dot{\psi} \cos(\phi) +$$

$$-mgh \sin(\phi) + \frac{\partial U_s(\phi)}{\partial \phi} + \frac{\partial \mathcal{F}(\phi, \dot{\phi})}{\partial \phi} = Q_\phi . \tag{31.95}$$

Fourth equation: yaw rotation

$$\left[J_y^* \sin^2(\phi) + J_z \cos^2(\phi) \right] \ddot{\psi} - J_{xz} \cos(\phi) \ddot{\phi} +$$

$$+ J_s \dot{V} \sin(\phi) + V \cos(\phi) \dot{\phi} \sum_{\forall i} \frac{J_{pi}}{R_{e_i}} - V v_y \sum_{\forall k} \frac{J_{pi}}{R_{e_i}^2} = Q_\psi . \tag{31.96}$$

31.2.6 Sideslip angles of the wheels

The sideslip angles of the wheels may be computed from the components of the velocities of the centers of the contact areas of the wheels in the $x^* y^* z$ frame. If the roll axis lies on the ground, some simplifications may be introduced: The roll angle and the roll velocity do not appear in the expression of the velocity of the wheel-ground contact points, if the track variations due to roll are neglected. The expression of the sideslip angle coincides with that seen for the rigid vehicle, except for the term containing the steering angle. Assuming that the sideslip angle is small, it follows that

$$\alpha_k = \frac{v_y}{V} + \dot{\psi} \frac{x_{Pk}}{V} - \delta_k \cos(\phi) - \delta_k(\phi) \cos(\phi) , \tag{31.97}$$

where subscript k refers to the axle, because the two wheels of the same axle have the same sideslip angle.

The term $\cos(\phi)$ multiplying the steering angle is linked to the circumstance that the steering loses its effectiveness with increasing roll angle, and was computed assuming that the kingpin axis is, when the roll angle vanishes, essentially perpendicular to the ground. If it is not, the caster and inclination angles had to be taken into account, together with their variation with the roll angle. The term $\delta_k(\phi)$ is roll steer that, in case of large roll angles, may be too large to be linearized.

31.2.7 Generalized forces

The generalized forces Q_k to be introduced into the equations of motion include the forces due to the tires, the aerodynamic forces and possible forces applied on the vehicle by external agents.

The virtual displacement of the center of the contact area of the left (right) wheel of the kth axle is

$$\{\delta s_{Pk_{L(R)}}\}_{x^*y^*z} = \left\{\begin{array}{c} \delta x^* - \delta\psi y_{Pk} \\ \delta y^* + \delta\psi x_{Pk} \\ 0 \end{array}\right\}, \tag{31.98}$$

where x_{Pk} and y_{Pk} are the coordinates of the center of the contact area in the reference frame $x^*y^*z^*$.

By writing as F_x^* and F_y^* the forces exerted by the tire in the direction of the x^* and y^* axes, assuming that the longitudinal forces acting on the wheels of the same axle are equal, the expression of the virtual work is

$$\delta\mathcal{L}_k = \delta x^* F_x^* + \delta y^* F_y^* + \delta\psi \left[F_y^* x_{Pk} + M_z\right]. \tag{31.99}$$

Because of the small steering angle, forces F_x^* and F_y^* will be confused in the following sections with the forces expressed in the reference frame of the wheel.

In a similar way, the virtual displacement of the center of mass for the computation of the aerodynamic forces is, in the $x^*y^*z^*$ frame,

$$\{\delta s_{G_S}\}_{x^*y^*z^*} = \left\{\begin{array}{c} \delta x^* + h\sin\left(\phi\right)\delta\psi \\ \delta y^* - h\cos\left(\phi\right)\delta\phi \\ -h\sin\left(\phi\right)\delta\phi \end{array}\right\}. \tag{31.100}$$

The aerodynamic forces and moments are referred to the xyz frame and not to the $x^*y^*z^*$ frame. Force F_{za}, for example, lies in the symmetry plane of the vehicle and is not perpendicular to the road. In this way it may be assumed that aerodynamic forces do not depend on the roll angle ϕ. A rotation of the reference frame is then needed:

$$\left\{\begin{array}{c} F_{xa}^* \\ F_{ya}^* \\ F_{za}^* \end{array}\right\} = \left\{\begin{array}{c} F_{xa} \\ F_{ya}\cos\left(\phi\right) - F_{za}\sin\left(\phi\right) \\ F_{ya}\sin\left(\phi\right) + F_{za}\cos\left(\phi\right) \end{array}\right\}, \tag{31.101}$$

$$\left\{\begin{array}{c} M_{xa}^* \\ M_{ya}^* \\ M_{za}^* \end{array}\right\} = \left\{\begin{array}{c} M_{xa} \\ M_{ya}\cos\left(\phi\right) - M_{za}\sin\left(\phi\right) \\ M_{ya}\sin\left(\phi\right) + M_{za}\cos\left(\phi\right) \end{array}\right\}. \tag{31.102}$$

The virtual work of the aerodynamic forces and moments is then

$$\delta\mathcal{L}_a = F_{xa}\delta x^* + \left[F_{ya}\cos\left(\phi\right) - F_{za}\sin\left(\phi\right)\right]\delta y^* +$$

$$+ \left(M_{xa}' - F_{ya}h\right)\delta\phi + \left[\left(F_{xa}h + M_{ya}\right)\sin\left(\phi\right) + M_{za}\cos\left(\phi\right)\right]\delta\psi. \tag{31.103}$$

It then follows that

$$\mathbf{Q} = \left\{\begin{array}{c} \sum_{\forall k} F_{xk} + F_{xa} \\ \sum_{\forall k} F_{yk} + F_{ya}\cos\left(\phi\right) - F_{za}\sin\left(\phi\right) \\ M_{xa}' - F_{ya}h \\ \sum_{\forall k}\left(F_y^* x_{Pk} + M_z\right) + \left(F_{xa}h + M_{ya}\right)\sin\left(\phi\right) + M_{za}\cos\left(\phi\right) \end{array}\right\}. \tag{31.104}$$

Because of the linearization of the model, forces F_{xa} and F_{za} may be considered as constant, while F_{y_a}, M_{xa} and M_{za} may be considered as linear with angle β_a, or if there is no side wind, angle β.

The force F_{yk} on the kth axle may be considered as a linear function of the sideslip angle and a more complex function of the camber angle, because the latter was assumed to coincide with the roll angle ϕ and is therefore not small. It then follows that

$$F_{ypk} = -C_k \alpha_k + F_{y\gamma k}(\phi) , \qquad (31.105)$$

where both C_k and $F_{y\gamma k}(\phi)$ are referred to the whole axle.

In the following the camber thrust will be assumed to be linear with the camber angle, even for large values of the latter, and the side force will be written as

$$F_{ypk} = -C_k \alpha_k + C_{\gamma k} \phi . \qquad (31.106)$$

This is doubtless an approximated expression, but it must be made if searching for closed form results. Roll steer will also be neglected.

31.2.8   Final form of the equations of motion

First equation: longitudinal translation

$$m_{at} \dot{V} + J_s \ddot{\psi} \sin(\phi) = F_{x1} + F_{x2} - \frac{1}{2} \rho V^2 S C_x . \qquad (31.107)$$

Second equation: lateral translation

$$m \dot{v}_y + m_{at} V \dot{\psi} - m h \ddot{\phi} \cos(\phi) = [Y_v + \cos(\phi) Y_{v1}] v_y + Y_{\dot{\psi}} \dot{\psi} +$$

$$+ Y_\phi \phi + \cos(\phi) Y_\delta \delta - \frac{1}{2} \rho V^2 S C_z \sin(\phi) + F_{y_e} , \qquad (31.108)$$

where

$$\begin{cases} Y_v = -\frac{1}{V} \sum_{\forall k} C_k , \\[2mm] Y_{v1} = \frac{1}{2} \rho V_a S (C_y)_{,\beta} , \\[2mm] Y_{\dot{\psi}} = -\frac{1}{V} \sum_{\forall k} x_{Pk} C_k , \\[2mm] Y_\phi = \sum_{\forall k} C_{\gamma k} , \\[2mm] Y_\delta = \sum_{\forall k} K_k' C_k . \end{cases} \qquad (31.109)$$

Third equation: roll rotation

$$J_x^* \ddot{\phi} - J_{xz} \cos(\phi) \ddot{\psi} - m \dot{v}_y h \cos(\phi) - J_s V \dot{\psi} \cos(\phi) +$$

$$- m g h \sin(\phi) + \frac{\partial \mathcal{U}_s(\phi)}{\partial \phi} + \frac{\partial \mathcal{F}(\phi, \dot{\phi})}{\partial \phi} = L_v v_y , \qquad (31.110)$$

where

$$L_v = \frac{1}{2}\rho V S \left[-h(C_y)_{,\beta} + t(C_{M_x})_{,\beta}\right] . \tag{31.111}$$

Fourth equation: yaw rotation

$$\left[J_y^* \sin^2(\phi) + J_z \cos^2(\phi)\right] \ddot{\psi} - J_{xz} \cos(\phi) \ddot{\phi}+$$

$$+J_s \dot{V} \sin(\phi) + V \cos(\phi) \dot{\phi} \sum_{\forall i} \frac{J_{pi}}{R_{e_i}} =$$

$$= [N_v + \cos(\phi) Y_{v1}] v_y + N_{\dot\psi}\dot\psi + N_\phi \phi + \cos(\phi) N_\delta \delta+$$

$$+\tfrac{1}{2}\rho V^2 S(-hC_x + lC_{M_y})\sin(\phi) + M_{ze} ,$$

where

$$\begin{cases} N_v = \frac{1}{V}\sum_{\forall k}\left[-x_{Pk}C_k + (M_{zk})_{,\alpha} + 2J_{pr}\left(\frac{V}{R_e}\right)^2\right] , \\[2mm] N_{v1} = \frac{1}{2}\rho V_a Sl(C'_{M_z})_{,\beta} , \\[2mm] N_{\dot\psi} = \frac{1}{V}\sum_{\forall k}\left[-x_{Pk}^2 C_k + x_{r_k}(M_{z_k})_{,\alpha}\right] , \\[2mm] N_\phi = \sum_{\forall k} x_{rk}C_{\gamma k} , \\[2mm] N_\delta = \sum_{\forall k}\left[x_{Pk}K'_k C_k - (M_{zk})_{,\alpha}\right] . \end{cases} \tag{31.113}$$

31.2.9 Steady-state equilibrium conditions

Consider a vehicle in which control of the roll angle is performed in such a way that the transversal load vanishes. The condition that must be stated is that the equilibrium to roll rotations is granted without the suspension exerting any roll torque.

In steady-state conditions accelerations $\dot{V}$, $\dot{v}_y$, $\ddot\phi$ and $\ddot\psi$ and velocity $\dot\phi$ vanish, and the condition in which the suspension exerts no roll torques is

$$\frac{\partial \mathcal{U}_s(\phi)}{\partial \phi} = \frac{\partial \mathcal{F}\left(\phi, \dot\phi\right)}{\partial \dot\phi} = 0 .$$

The equilibrium equation to roll becomes

$$-J_s V\dot\psi \cos(\phi) - mgh\sin(\phi) = L_v v_y . \tag{31.114}$$

In steady-state, the yaw velocity $\dot\psi$ is linked to the forward velocity V and to the radius of the path (which is circular) R by the usual relationship

$$\dot\psi = \frac{V}{R} , \tag{31.115}$$

and then the equilibrium equation reduces to

$$-J_s \frac{V^2}{R}\cos(\phi) - mgh\sin(\phi) = L_v v_y . \qquad (31.116)$$

By introducing the value of J_s into the last equation, it follows that

$$\left(mh + \sum_{\forall i} \frac{J_{pi}}{R_{e_i}}\right) \frac{V^2}{R}\cos(\phi) + mgh\sin(\phi) + L_v v_y = 0 . \qquad (31.117)$$

The third term, due to aerodynamic actions, is small when compared with the others and may, at least initially, be neglected. Eq. (31.117) then allows the steady-state roll angle to be computed:

$$\phi_0 = -\text{artg}\left[\frac{V^2}{Rg}\left(1 + \frac{1}{mh}\sum_{\forall i}\frac{J_{pi}}{R_{e_i}}\right)\right] , \qquad (31.118)$$

which coincides with the expression obtained from the simplified ideal steering model.

31.2.10   Motion about the steady-state equilibrium position

Consider a vehicle working in a condition close to the above computed equilibrium condition. The roll angle may be expressed as

$$\phi = \phi_0 + \phi_1 ,$$

where ϕ_1 is a small angle. The trigonometric functions of the roll angle may then be approximated as

$$\sin(\phi_0 + \phi_1) \approx \sin(\phi_0) + \phi_1\cos(\phi_0) ,$$
$$\cos(\phi_0 + \phi_1) \approx \cos(\phi_0) - \phi_1\sin(\phi_0) .$$

The elastic and damping behavior of the suspension may be linearized about the equilibrium position, stating

$$\frac{\partial \mathcal{U}_s(\phi)}{\partial \phi} = k(\phi_0)\phi_1 , \quad \frac{\partial \mathcal{F}\left(\phi,\dot\phi\right)}{\partial \dot\phi} = c(\phi_0)\dot\phi_1 . \qquad (31.119)$$

Neglecting the term in L_v, the equations of motion become

$$m_{at}\dot{V} + J_s\ddot{\psi}\sin(\phi_0) = F_{x1} + F_{x2} - \frac{1}{2}\rho V^2 SC_x , \qquad (31.120)$$

$$m\dot{v}_y + m_{at}V\dot\psi - mh\ddot\phi_1\cos(\phi_0) = Y_v v_y + Y_{\dot\psi}\dot\psi +$$

$$+Y_\phi\phi_0 + Y_\phi\phi_1 + \cos(\phi_0) Y_{v1}v_y + \cos(\phi_0) Y_\delta\delta + \qquad (31.121)$$

$$-\tfrac{1}{2}\rho V^2 SC_z \sin(\phi_0) - \tfrac{1}{2}\rho V^2 SC_z\phi_1\cos(\phi_0) + F_{y_e} ,$$

$$J_x^* \ddot{\phi}_1 - J_{xz} \cos(\phi_0) \ddot{\psi} - m \dot{v}_y h \cos(\phi_0) +$$

$$-mgh\phi_1 \cos(\phi_0) + k(\phi_0)\phi_1 + C(\phi_0)\dot{\phi}_1 = 0 , \tag{31.122}$$

$$\left[J_y^* \sin^2(\phi_0) + J_z \cos^2(\phi_0) \right] \ddot{\psi} - J_{xz} \cos(\phi_0) \ddot{\phi}_1 + J_s \dot{V} \sin(\phi_0) +$$

$$+ V \cos(\phi_0) \dot{\phi}_1 \sum_{\forall i} \frac{J_{pi}}{R_{e_i}} = N_v v_y + N_{\dot{\psi}} \dot{\psi} + N_\phi \phi_0 + N_\phi \phi_1 +$$

$$+ \cos(\phi_0) N_\delta \delta + N_{v1} v_y \cos(\phi_0) + \tfrac{1}{2} \rho V^2 S(-hC_x + lC_{M_y}) \sin(\phi_0) + \tag{31.123}$$

$$+ \tfrac{1}{2} \rho V^2 S(-hC_x + lC_{M_y})\phi_1 \cos(\phi_0) + M_{ze} .$$

In a more synthetic way, it is possible to write

$$\mathbf{M}\ddot{\mathbf{q}} + \mathbf{C}\dot{\mathbf{q}} + \mathbf{K}\mathbf{q} = \mathbf{F} + \mathbf{F}_1 , \tag{31.124}$$

where

$$\mathbf{q}_1 = \begin{bmatrix} x^* & y^* & \phi_1 & \psi \end{bmatrix}^T ,$$

$$\mathbf{M} = \begin{bmatrix} m_{at} & 0 & 0 & J_s \sin(\phi_0) \\ & m & -mh\cos(\phi_0) & 0 \\ & & J_x^* & -J_{xz}\cos(\phi_0) \\ & \text{symm.} & & J_y^*\sin^2(\phi_0) + J_z\cos^2(\phi_0) \end{bmatrix} ,$$

$$\mathbf{C} = \begin{bmatrix} 0 & 0 & 0 & 0 \\ 0 & -Y_v - \cos(\phi_0)Y_{v1} & 0 & m_{at}V - Y_{\dot{\psi}} \\ 0 & 0 & c(\phi_0) & 0 \\ 0 & -N_v - N_{v1}\cos(\phi_0) & V\cos(\phi_0)\sum_{\forall i}\frac{J_{pi}}{R_{e_i}} & -N_{\dot{\psi}} \end{bmatrix} ,$$

$$\mathbf{K} = \begin{bmatrix} 0 & 0 & 0 & 0 \\ 0 & 0 & -Y_\phi + \tfrac{1}{2}\rho V SC_z \cos(\phi_0) & 0 \\ 0 & 0 & -mgh\cos(\phi_0) + k(\phi_0) & 0 \\ 0 & 0 & \tfrac{1}{2}\rho V_a S(-hC_x + lC_{M_y})\cos(\phi_0) - N_\phi & 0 \end{bmatrix} ,$$

$$\mathbf{F} = \begin{Bmatrix} F_{x1} + F_{x2} - \tfrac{1}{2}\rho V^2 SC_x \\ Y_\phi \phi_0 - \tfrac{1}{2}\rho V^2 SC_z \sin(\phi_0) \\ 0 \\ +N_\phi \phi_0 + \tfrac{1}{2}\rho V^2 S(-hC_x + lC_{M_y})\sin(\phi_0) \end{Bmatrix} ,$$

$$\mathbf{F}_1 = \delta \begin{Bmatrix} 0 \\ \cos(\phi_0)Y_\delta \\ 0 \\ \cos(\phi_0)N_\delta \end{Bmatrix} + \begin{Bmatrix} 0 \\ F_{y_e} \\ 0 \\ M_{ze} \end{Bmatrix} .$$

As already stated, coordinates x^*, y^* and ψ are present only in the form of their derivatives: The order of the differential set of equations is then 5 rather than 8.

The mass matrix is symmetrical, as could be easily predicted, while the two other matrices are not.

31.2.11 Steady-state handling

In steady-state conditions, the first equation reduces to

$$F_{x1} + F_{x2} - \frac{1}{2}\rho V^2 S C_x = 0 \,,$$

which coincides with the equation seen for the motor vehicle working with small roll angles.

As expected, the third equation yields simply

$$\phi_1 = 0 \,.$$

The other two equations reduce to

$$\left[\begin{array}{cc} -Y_v - \cos{(\phi_0)}\, Y_{v1} & m_{at}V - Y_{\dot\psi} \\ -N_v - N_{v1}\cos{(\phi_0)} & -N_{\dot\psi} \end{array} \right] \left\{ \begin{array}{c} v_y \\ \dot\psi \end{array} \right\} = \delta\cos{(\phi_0)} \left\{ \begin{array}{c} Y_\delta \\ N_\delta \end{array} \right\} + \quad (31.125)$$

$$+ \left\{ \begin{array}{c} Y_\phi \phi_0 - \frac{1}{2}\rho V S^2 C_z \sin{(\phi_0)} \\ +N_\phi \phi_0 + \frac{1}{2}\rho V^2 S(-h C_x + l C_{M_y})\sin{(\phi_0)} \end{array} \right\} + \left\{ \begin{array}{c} F_{y_e} \\ M_{z_e} \end{array} \right\} \,.$$

Because in steady-state

$$v_y = V\beta \,, \quad \dot\psi = \frac{V}{R} \,,$$

the radius of the trajectory and the sideslip angle may be computed at any given steering angle. As an alternative, the steering and sideslip angles may be computed as functions of the radius of the trajectory. In the latter case, it follows that

$$\left[\begin{array}{cc} -Y_v - \cos{(\phi_0)}\, Y_{v1} & -Y_\delta \cos{(\phi_0)} \\ -N_v - N_{v1}\cos{(\phi_0)} & -N_\delta \cos{(\phi_0)} \end{array} \right] \left\{ \begin{array}{c} v_y \\ \delta \end{array} \right\} = -\frac{V}{R} \left\{ \begin{array}{c} m_{at}V - Y_{\dot\psi} \\ -N_{\dot\psi} \end{array} \right\} +$$

$$\hspace{10cm} (31.126)$$

$$+ \left\{ \begin{array}{c} Y_\phi \phi_0 - \frac{1}{2}\rho V^2 S C_z \sin{(\phi_0)} \\ +N_\phi \phi_0 + \frac{1}{2}\rho V^2 S(-h C_x + l C_{M_y})\sin{(\phi_0)} \end{array} \right\} + \left\{ \begin{array}{c} F_{y_e} \\ M_{z_e} \end{array} \right\} \,.$$

The model is nonlinear at ϕ_0, making it impossible to compute gains independent from the conditions of motion.

It is, at any rate, interesting to write Eq. (31.126) assuming that angle ϕ_0 is small enough to linearize its trigonometric functions and that the gyroscopic effect of the wheels is negligible. In this case

$$\phi_0 = -\mathrm{artg}\left(\frac{V^2}{Rg}\right) \approx -\frac{V^2}{Rg}$$

and, if no external forces and moments act on the vehicle, Eq. (31.126) becomes

$$\left[\begin{array}{cc} -Y_v^* & -Y_\delta \\ -N_v^* & -N_\delta \end{array} \right] \left\{ \begin{array}{c} v_y \\ \delta \end{array} \right\} = \frac{V}{R} \left\{ \begin{array}{c} Y_{\dot\psi} - m_{at}V \\ N_{\dot\psi} \end{array} \right\} - \frac{V^2}{Rg} \left\{ \begin{array}{c} Y_\phi^* \\ N_\phi^* \end{array} \right\} \,, \quad (31.127)$$

where

$$Y_v^* = Y_v + Y_{v1} , \qquad Y_\phi^* = Y_\phi - \tfrac{1}{2}\rho V S^2 C_z ,$$
$$N_v^* = N_v + N_{v1} , \qquad N_\phi^* = N_\phi + \tfrac{1}{2}\rho V S^2(-hC_x + lC_{M_y}) . \tag{31.128}$$

The path curvature gain is then

$$\frac{1}{R\delta} = \frac{1}{V} \frac{N_v^* Y_\delta - N_\delta Y_v^*}{\left[Y_v^* N_{\dot\psi} + N_v^* \left(m_{at}V - Y_{\dot\psi} \right) \right] + \frac{V}{g}\left[N_v^* Y_\phi^* - N_\phi^* Y_v^* \right]} . \tag{31.129}$$

The result is identical to that seen for the non-tilting vehicle (Y_v^* and N_v^* also coincide with the values computed in Chapter 25) except for the term in braces at the denominator, containing terms Y_ϕ^* and N_ϕ^* due to the camber stiffness of the tires, plus some aerodynamic terms. It is interesting to note that the tilt of the vehicle and thus the camber thrust (because it has been assumed that $\gamma = \phi$) affects its behavior even if the roll angle tends to zero.

Remark 31.3 *This outcome should be obvious: If the vehicle does not tilt, the side force is due only to the sideslip of the wheels, while if $\gamma = \phi$, roll produces a camber thrust that adds to the sideslip force. If $R \to \infty$, both the components of the side force tend to zero, but their ratio remains constant.*

Example 31.4 *Consider a three-wheeled vehicle with two wheels at the front axle, with the following characteristics:*
Geometrical data: $l = 1.720$ m, $a = 0.77$ m, $h = 576$ mm, $R_{e1} = R_{e2} = 310$ mm.
Inertial data: $m = 358$ kg, $J_x = 31$ kg m^2, $J_y = 125$ kg m^2, $J_z = 111$ kg m^2, $J_{xz} = 0$, $J_{p1} = J_{p2} = 0.18$ kg m^2.
Aerodynamic data: $\rho = 1.29$ kg/m^3, $S = 1$ m^2, $C_x = 0.35$, $C_{My} = (C_{Mx})_{,\beta} = (C_{Mz})_{,\beta} = C_z = 0$, $(C_y)_{,\beta} = 0.026$.
Tire data: $f_0 = 0.01$, $K = 4 \times 10^{-6}$ s^2/m^2, $C_1/F_z = C_2/F_z = 17.9$ 1/rad, $(M_{z1})_{,\alpha}/F_z = (M_{z2})_{,\alpha}/F_z = 0.21$ m/rad, $C_{\gamma 1}/F_z = C_{\gamma 2}/F_z = -1.1$ 1/rad.
Compute the steady state roll angle as a function of the ratio between centrifugal and gravitational accelerations, and the path curvature gain for different values of the radius of the trajectory.
Steady-state roll angle. The result, computed both by taking gyroscopic moments into account and neglecting them, is reported in Fig. 31.8.
From the plot it is clear that the gyroscopic effect of the wheels has little influence in determining the steady-state roll angle and that the conditions of no load shift and local vertical aligned with the z axis coincide.
Trajectory curvature gain. The results, computed on a trajectory with a radius tending to infinity, and equal to 1,000, 500, 200, 100 and 50 m are reported in Fig. 31.9, together with the roll angle on the same radii. The dashed line, labelled $\phi_0 = 0$, refers to a non-tilting vehicle.
The non-tilting vehicle is strongly understeer (traction has not been accounted for). Tilting allows the vehicle to travel on the curve with smaller sideslip angles of the wheels. At large radii, the vehicle even becomes oversteer.

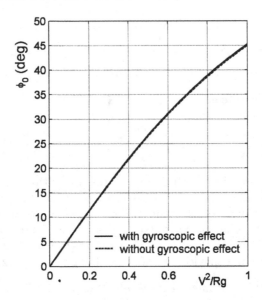

FIGURE 31.8. Steady state roll angle as a function of centrifugal acceleration, computed both by considering the gyrosoping moments of the wheels and neglecting them.

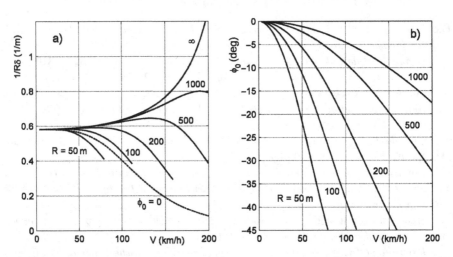

FIGURE 31.9. Path curvature gain $1/R\delta$ and steady-state roll angle ϕ_0 versus the speed V on trajectories with different radii.

With decreasing path radius (and then at equal speed with increasing centrifugal acceleration and roll angle) the vehicle first becomes less oversteer and then increasingly understeer, the result of the term in $\cos(\phi_0)$ multiplying the steering angle δ. In the figure the tilt is limited to 45°, with the curve stopping at a given speed in the case of a path with small radius.

31.2.12   Stability about the steady-state condition

Assume that the vehicle is travelling at a constant speed V on a circular trajectory in steady-state conditions characterized by the values v_{y0},

$$\dot{\psi}_0 = \frac{V}{R}$$

and ϕ_0 of the variables of motion and by the corresponding value δ_0 of the steering angle. Assume also that the external forces F_{y_e} and M_{z_e} vanish. The small perturbations v_{y1}, $\dot{\psi}_1$ and ϕ_1 add to the above mentioned values of the parameters.

Uncoupling, at least as a first approximation, the first equation dealing with longitudinal motion, the remaining three equations of motion (31.124) become

$$m\dot{v}_{y1} - mh\cos\left(\phi_0\right)\ddot{\phi}_1 - \left[Y_v + \cos\left(\phi_0\right)Y_{v1}\right]\left(v_{y1} + v_{y0}\right) +$$

$$+ \left(m_{at}V - Y_{\dot{\psi}}\right)\left(\dot{\psi}_1 + \dot{\psi}_0\right) + \left[Y_\phi + Y_{\phi1}\cos\left(\phi_0\right)\right]\phi_1 = \qquad (31.130)$$

$$= Y_\phi\phi_0 - \tfrac{1}{2}\rho V S^2 C_z \sin\left(\phi_0\right) + \cos\left(\phi_0\right)Y_\delta\delta_0 \ ,$$

$$-mh\cos\left(\phi_0\right)\dot{v}_{y1} + J_x^*\ddot{\phi}_1 - J_{xz}\cos\left(\phi_0\right)\ddot{\psi}_1 +$$

$$+c(\phi_0)\dot{\phi}_1 + \left[-mgh\cos\left(\phi_0\right) + k(\phi_0)\right]\phi_1 = 0 \ , \qquad (31.131)$$

$$-J_{xz}\cos\left(\phi_0\right)\ddot{\phi}_1 + \left[J_y^*\sin^2\left(\phi_0\right) + J_z\cos^2\left(\phi_0\right)\right]\ddot{\psi}_1 +$$

$$+ \left[-N_v - N_{v1}\cos\left(\phi_0\right)\right]\left(v_{y1} + v_{y0}\right) + V\cos\left(\phi_0\right)\sum_{\forall i}\frac{J_{pi}}{R_{e_i}}\dot{\phi}_1 +$$

$$-N_{\dot{\psi}}\left(\dot{\psi}_1 + \dot{\psi}_0\right) - \left[N_{\phi1}\cos\left(\phi_0\right) + N_\phi\right]\phi_1 = N_\phi\phi_0 + \qquad (31.132)$$

$$+\tfrac{1}{2}\rho V^2 S(-hC_x + lC_{M_y})\sin\left(\phi_0\right) + \cos\left(\phi_0\right)N_\delta\delta_0 \ ,$$

where

$$N_{\phi1} = \frac{1}{2}\rho V^2 S(-hC_x + lC_{M_y}) \ ,$$

$$Y_{\phi1} = -\frac{1}{2}\rho V^2 SC_z \ .$$

Because motion takes place about the static equilibrium condition, it is possible to eliminate the parameters related to the latter by using Equations (31.125) and (31.124), obtaining

$$m\dot{v}_{y1} - mh\cos\left(\phi_0\right)\ddot{\phi}_1 - \left[Y_v + \cos\left(\phi_0\right)Y_{v1}\right]v_{y1} +$$

$$+ \left(m_{at}V - Y_{\dot{\psi}}\right)\dot{\psi}_1 - \left[Y_\phi + Y_{\phi1}\cos\left(\phi_0\right)\right]\phi_1 = 0 \ , \qquad (31.133)$$

$$-mh\cos(\phi_0)\,\dot{v}_{y1} + J_x^*\ddot{\phi}_1 - J_{xz}\cos(\phi_0)\,\ddot{\psi}_1+$$

$$+c(\phi_0)\dot{\phi}_1 + [-mgh\cos(\phi_0) + k(\phi_0)]\,\phi_1 = 0 , \tag{31.134}$$

$$-J_{xz}\cos(\phi_0)\,\ddot{\phi}_1 + \left[J_y^*\sin^2(\phi_0) + J_z\cos^2(\phi_0)\right]\ddot{\psi}_1+$$

$$- [N_v + N_{v1}\cos(\phi_0)]\,v_{y1} + V\cos(\phi_0)\sum_{\forall i}\frac{J_{pi}}{R_{e_i}}\dot{\phi}_1+ \tag{31.135}$$

$$-N_{\dot{\psi}}\dot{\psi}_1 - [N_{\phi1}\cos(\phi_0) + N_\phi]\,\phi_1 = 0 .$$

The equations may then be written in the state space in the form

$$\mathbf{A}_2\dot{\mathbf{z}} = \mathbf{A}_1\mathbf{z} , \tag{31.136}$$

where

$$\mathbf{z} = \begin{bmatrix} v_y & v_\phi & v_\psi & \phi \end{bmatrix}^T ,$$

$$v_\phi = \dot{\phi} , \quad v_\psi = \dot{\psi} ,$$

and

$$\mathbf{A}_2 = \begin{bmatrix} m & -mh\cos(\phi_0) & 0 & 0 \\ -mh\cos(\phi_0) & J_x^* & -J_{xz}\cos(\phi_0) & 0 \\ 0 & -J_{xz}\cos(\phi_0) & J_y^*\sin^2(\phi_0) + J_z\cos^2(\phi_0) & 0 \\ 0 & 0 & 0 & 1 \end{bmatrix} ,$$

$$\mathbf{A}_1 = \begin{bmatrix} Y_v^* & 0 & -m_{at}V + Y_{\dot{\psi}} & Y_\phi + Y_{\phi1}\cos(\phi_0) \\ 0 & -c(\phi_0) & J_sV\cos(\phi_0) & mgh\cos(\phi_0) - k(\phi_0) \\ N_v^* & N_\phi^* & N_{\dot{\psi}} & N_{\phi1}\cos(\phi_0) + N_\phi \\ 0 & 1 & 0 & 0 \end{bmatrix} ,$$

$$Y_v^* = Y_v + Y_{v1}\cos(\phi_0) , \quad N_v^* = N_v + N_{v1}\cos(\phi_0) ,$$

$$N_\phi^* = -V\cos(\phi_0)\sum_{\forall i}\frac{J_{pi}}{R_{e_i}} .$$

The dynamic matrix, whose eigenvalues allow the stability to be studied, is then

$$\mathbf{A} = \mathbf{A}_2^{-1}\mathbf{A}_1 . \tag{31.137}$$

Example 31.5 *Study the stability of the vehicle of the previous example, assuming that the stiffness and the damping of the suspension are constant with varying roll angle. Use the values $k = 4.000$ Nm/rad and $c = 90$ Nms/rad.*

The real and imaginary parts of the eigenvalues are plotted versus the speed together with the roots locus for various path curvature radii in Fig. 31.10. As can be seen, the vehicle is stable in all conditions.

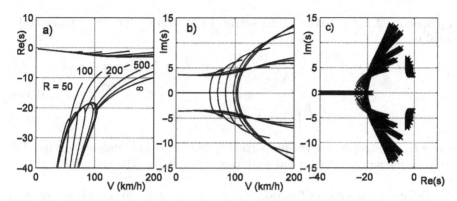

FIGURE 31.10. Real (a) and imaginary (b) parts of the eigenvalues of the dynamic matrix versus the speed and (c) roots locus for various path curvature radii ($R = 50$, 100, 200, 500, 1000 m and $R \to \infty$).

31.3 DYNAMIC TILTING CONTROL

Assume that the vehicle is provided with a tilt control device able to maintain load shift at a zero value or to keep the local vertical in the symmetry plane. In the previous section it was shown that in steady state conditions these two goals almost coincide, at least with the usual values of the gyroscopic moments of the wheels and of aerodynamic actions (the two curves in Fig. 31.8 are practically superimposed upon each other).

If it is easy to define the roll angle to satisfy this requirement in steady state conditions, it is much more difficult to identify a control strategy to do the same in non-steady state conditions.

Assume that the actuator dynamics may be expressed by the equation

$$J_a \ddot{\phi}_c + c_{22} \dot{\phi}_c - c_{21} \dot{\phi}_s + k_{22} \phi_c - k_{21} \phi_s = M_c , \qquad (31.138)$$

where ϕ_s is the rotation angle of the actuator corresponding to roll angle ϕ when the spring exerts no force, J_a is the moment of inertia of the actuator, M_c is the torque it exerts, both reduced to its output shaft, and c_{ij} and k_{ij} are the suspension damping coefficients and stiffnesses, which obviously are functions of ϕ and ϕ_c.

If the error is defined as

$$e = \phi + \mathrm{artg}\left(\dot{\psi}\frac{V J_s}{gmh}\right) , \qquad (31.139)$$

a proportional, integrative and derivative (PID) strategy leads to a moment M_c equal to

$$M_c = -K_p \left[\phi + \mathrm{artg}\left(\dot{\psi}\frac{VJ_s}{gmh} \right) \right] - K_d \left(\dot{\phi} + \ddot{\psi}\frac{VJ_s}{gmh} \right) +$$
$$-K_i \int \left[\phi + \mathrm{artg}\left(\dot{\psi}\frac{VJ_s}{gmh} \right) \right] dt \ . \tag{31.140}$$

where K_p, K_d and K_i are the proportional, derivative and integrative gains. The error for the derivative gain was simplified by conflating the arctangent with its argument.

Because ϕ_s is a known function of ϕ, it is possible to add the control equation to those of the vehicle, thus studying the dynamics of the controlled system.

In the following pages it will be assumed for simplicity that $d = d_1 = 0$, and then $\phi_s = \phi$. In this case $c_{22} = c_{21} = c_\phi$ and $k_{22} = k_{21} = k_\phi$ and the equation of motion of the controlled actuator becomes

$$J_a \ddot{\phi}_c + K_d \ddot{\psi}\frac{VJ_s}{gmh} + c_\phi \dot{\phi}_c - (c_\phi - K_d)\dot{\phi} + K_p \mathrm{artg}\left(\dot{\psi}\frac{VJ_s}{gmh} \right) +$$
$$+k_\phi \phi_c - (k_\phi - K_p)\phi + K_i \int \left[\phi + \mathrm{artg}\left(\dot{\psi}\frac{VJ_s}{gmh} \right) \right] dt = 0 \ . \tag{31.141}$$

The equation of motion of the controlled system in the state space may be written in the form

$$\mathbf{A}_2 \dot{\mathbf{z}} = \mathbf{A}_1 \mathbf{z} + \mathbf{f} \ , \tag{31.142}$$

where to the states of the vehicle

$$V \ , \ v_y \ , \ v_\phi = \dot{\phi} \ , \ v_\psi = \dot{\psi} \ , \ \phi \ ,$$

other states must be added, namely ϕ_c and its derivative $v_{\phi c} = \dot{\phi}_c$ plus a state linked with the error of the derivative branch of the control

$$e_i = \int \left[\phi + \mathrm{artg}\left(\dot{\psi}\frac{VJ_s}{gmh} \right) \right] dt \ .$$

The state vector is then

$$\mathbf{z} = \begin{bmatrix} V & v_y & v_\phi & v_\psi & v_{\phi c} & \phi & \phi_c & e_i \end{bmatrix}^T \ .$$

The other terms included in the state space equation are

$$\mathbf{A}_2 = \begin{bmatrix} m_{at} & 0 & 0 & J_s \sin(\phi) & 0 & 0 & 0 & 0 \\ & m & -mh\cos(\phi) & 0 & 0 & 0 & 0 & 0 \\ & & J_x^* & -J_{xz}\cos(\phi) & 0 & 0 & 0 & 0 \\ & & & J_z^* & 0 & 0 & 0 & 0 \\ & & & K_d\frac{VJ_s}{gmh} & J_a & 0 & 0 & 0 \\ & & & & & 1 & 0 & 0 \\ & & & & & & 1 & 0 \\ & \text{symm.} & & & & & & 1 \end{bmatrix} ,$$

where

$$J_z^* = J_y^* \sin^2(\phi) + J_z \cos^2(\phi) \ ,$$

$$
\mathbf{A}_1 =
\begin{bmatrix}
0 & 0 & 0 & 0 & 0 & 0 & 0 & 0 \\
0 & Y_v^* & 0 & -m_{at}V + Y_{\dot\psi} & 0 & Y_\phi & 0 & 0 \\
0 & L_v & -c_\phi & J_s V \cos(\phi) & c_\phi & -k_\phi & k_\phi & 0 \\
0 & N_v^* & N_{\dot\phi}^* & N_{\dot\psi} & 0 & N_\phi & 0 & 0 \\
0 & 0 & (c_\phi - K_d) & K_p \dfrac{V}{g} & -c_\phi & k_\phi - K_p & -k_\phi & -K_i \\
0 & 0 & 1 & 0 & 0 & 0 & 0 & 0 \\
0 & 0 & 0 & 0 & 1 & 0 & 0 & 0 \\
0 & 0 & 0 & 0 & 0 & 1 & 0 & 0
\end{bmatrix} \ ,
$$

$$Y_v^* = Y_v + Y_{v1}\cos(\phi) \ , \quad N_v^* = N_v + N_{v1}\cos(\phi) \ ,$$

$$N_{\dot\phi}^* = -V\cos(\phi)\sum_{\forall i}\frac{J_{pi}}{R_{e_i}} \ ,$$

and

$$
\mathbf{f} =
\left\{
\begin{array}{c}
F_{x1} + F_{x2} - \frac{1}{2}\rho V^2 S C_x \\
\cos(\phi) Y_\delta \delta - \frac{1}{2}\rho V^2 S C_z \sin(\phi) + F_{y_e} \\
mgh\sin(\phi) \\
\cos(\phi) N_\delta \delta + \frac{1}{2}\rho V^2 S(-hC_x + lC_{M_y})\sin(\phi) + M_{ze} \\
-K_p \mathrm{atan}\left(\dot\psi \frac{V J_s}{gmh}\right) \\
0 \\
0 \\
\mathrm{atan}\left(\dot\psi \frac{V J_s}{gmh}\right)
\end{array}
\right\} \ .
$$

Note that matrix $\mathbf{A}_2$ is not fully symmetrical owing to the term K_d in position 5,4.

Example 31.6 *Using the vehicle of the previous example, study the response to a steering step, assuming that the actuator's moment of inertia, reduced to the output shaft, is $J_a = 0.001$ kg/m^2. Assume control gains $K_p = 60,000$ Nm/rad, $K_d = 6,000$ Nms/rad, $K_i = 10,000$ Nm/(s rad). The manoeuvre is performed at a speed of 120 km/h and the steering angle $\delta = 1°$ is given at $t = 0$.*

Because the manoeuvre is performed at constant speed, the first equation may be considered uncoupled from the others and is therefore not considered.

The results are reported in Fig. 31.11. From the plot it is clear that the vehicle reaches steady-state conditions in about 1 s. After 2 s the values of ϕ and ϕ_c are respectively 39.85° and 39.98°, while the steady state value on the same path ($R = 136.24$ m) is 39.98° for both. The values of β (0.175°) and $\dot\psi$ (0.2447 rad/s) at the end of the manoeuvre coincide with those computed for steady-state operation.

Because the input is a step, the sideslip angle becomes strongly negative at the beginning and the center of mass moves to the outside of the curve, because the vehicle

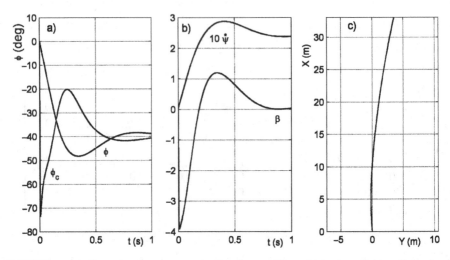

FIGURE 31.11. Response to a step steering input. Time histories of the roll angle ϕ and rotation angle of the actuator ϕ_c (a) and of the sideslip angle β and yaw velocity $\dot{\psi}$ (b). (c): Path.

starts overturning. The controller immediately reacts with a high value of ϕ_c and starts a correction that prevents the vehicle from rolling over: After several much damped oscillations, equilibrium is restored.

31.4 HANDLING-COMFORT COUPLING

The dynamics of tilting body vehicles was studied in the previous sections in terms of handling using a model with four degrees of freedom. However, this approach can only be considered a rough approximation, because uncoupling between handling and comfort is no longer applicable when the assumption of small angles does not hold.

The present section will be devoted to developing a model similar to the previous, but with two added degrees of freedom linked with comfort: heave and pitch. It is thus a model with sixdegrees of freedom, still based on the assumption of rigid tires, that could be extended to nine or 10 degrees of freedom (for vehicles with three or four wheels respectively) by including the compliance of the tires.

The assumptions that the roll axis remains on the ground during heave motion and that it remains in the same position shown for the vehicle without suspension will be made. The displacement of the center of mass of the vehicle, which will at any rate be considered a small quantity, will occur in the direction of the z axis of the body-fixed reference frame. Pitch rotation will occur about the baricentric y axis, which is perpendicular to the symmetry plane in its undeformed position.

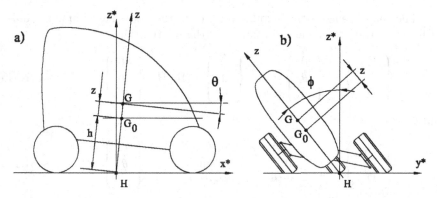

FIGURE 31.12. Reference frames for the sprung mass and definition of points G, G_0 and H.

The roll axis will then display no pitch rotation. The generalized coordinates for translations of the sprung mass will again be coordinates X_H, Y_H of point H located on the ground, on the perpendicular to the roll axis passing through the centre of mass G. The z coordinate (Fig. 31.12), and the yaw ψ, roll ϕ and then pitch θ, will be added as generalized coordinates. The three angles will be taken in this order, with the latter considered as a small angle. Note that although the order is different from the usual, these are still Tait–Bryan angles.

As usual, the assumption of small angles (particularly for the sideslip angle β) allows the component v_{x^*} of the velocity to be conflated with the forward velocity V. Linear velocities v_y and $\dot{z}$ and and angular velocities $\dot{\psi}$, $\dot{\phi}$ and $\dot{\theta}$ will be considered as small quantities too. The small size of displacements z and θ make the order in which these two displacements (linear and angular) are performed immaterial.

31.4.1 Kinetic and potential energies

Because pitch rotation was not considered in the definition of the roll axis and the latter is horizontal, the order of the rotations is now yaw, roll and pitch. The rotation matrix allowing us to pass from the body-fixed frame $Gxyz$ to the inertial frame $X_iY_iZ_i$ is:

$$\mathbf{R} = \mathbf{R}_1\mathbf{R}_2\mathbf{R}_3 , \qquad (31.143)$$

where to matrices $\mathbf{R}_1$ and $\mathbf{R}_2$ seen in the previous section, a pitch matrix must be added

$$\mathbf{R}_3 = \begin{bmatrix} \cos(\theta) & 0 & \sin(\theta) \\ 0 & 1 & 0 \\ -\sin(\theta) & 0 & \cos(\theta) \end{bmatrix} .$$

The time derivative of the rotation matrix is

$$\dot{\mathbf{R}} = \dot{\mathbf{R}}_1\mathbf{R}_2\mathbf{R}_3 + \mathbf{R}_1\dot{\mathbf{R}}_2\mathbf{R}_3 + \mathbf{R}_1\mathbf{R}_2\dot{\mathbf{R}}_3 . \qquad (31.144)$$

The components of the angular velocity in the body-fixed frame are linked with the derivatives of the coordinates by the relationship

$$
\left\{ \begin{array}{c} \Omega_x \\ \Omega_y \\ \Omega_z \end{array} \right\} = \left\{ \begin{array}{c} 0 \\ \dot{\theta} \\ 0 \end{array} \right\} + \mathbf{R}_3^T \left\{ \begin{array}{c} \dot{\phi} \\ 0 \\ 0 \end{array} \right\} + \mathbf{R}_3^T \mathbf{R}_2^T \left\{ \begin{array}{c} 0 \\ 0 \\ \dot{\psi} \end{array} \right\},
\tag{31.145}
$$

and then

$$
\left\{ \begin{array}{c} \Omega_x \\ \Omega_y \\ \Omega_z \end{array} \right\} = \left[\begin{array}{ccc} \cos(\theta) & 0 & \sin(\theta)\cos(\phi) \\ 0 & 1 & \sin(\phi) \\ -\sin(\theta) & 0 & \cos(\theta)\cos(\phi) \end{array} \right] \left\{ \begin{array}{c} \dot{\phi} \\ \dot{\theta} \\ \dot{\psi} \end{array} \right\}.
\tag{31.146}
$$

The vector of the generalized coordinates is

$$
\mathbf{q} = \left[\begin{array}{cccccc} X & Y & z & \phi & \theta & \psi \end{array} \right]^T.
\tag{31.147}
$$

Let the generalized velocities for translational degrees of freedom be the components of the velocity v_x and v_y, referred to frame $x^*y^*z^*$, plus component v_z in the direction of axis z. The velocities for the rotational degrees of freedom are SIMPLY the derivatives of the coordinates ϕ, θ and ψ. They will be designated as v_ϕ, v_θ and v_ψ respectively.

The vector of the generalized velocities is then

$$
\mathbf{w} = \left[\begin{array}{cccccc} v_x & v_y & v_z & v_\phi & v_\theta & v_\psi \end{array} \right]^T.
\tag{31.148}
$$

The relationship between generalized velocities and derivatives of coordinates is the usual one

$$
\mathbf{w} = \mathbf{A}^T \dot{\mathbf{q}},
\tag{31.149}
$$

where matrix $\mathbf{A}$[4] is:

$$
\mathbf{A} = \left[\begin{array}{cc} \mathbf{R}_1 & \mathbf{0}_{3\times 3} \\ \mathbf{0}_{3\times 3} & \mathbf{I}_{3\times 3} \end{array} \right].
\tag{31.150}
$$

Because $\mathbf{A}$ is a rotation matrix, the inverse transformation is

$$
\dot{\mathbf{q}} = \mathbf{Bw} = \mathbf{Aw}.
$$

The vector defining the position of the center of mass of the sprung mass G_S with respect to point H is

$$
\mathbf{r}_1 = (h + z) \left[\begin{array}{ccc} 0 & 0 & 1 \end{array} \right]^T,
\tag{31.151}
$$

and then the absolute position of G_S is

$$
(\overline{G_S - O'}) = (\overline{H - O'}) + \mathbf{R}\mathbf{r}_1.
\tag{31.152}
$$

---

[4]Again, matrix $\mathbf{A}$ has nothing to do with the dynamic matrix of the system in the state space, usually referred to as $\mathbf{A}$.

The velocity of G_S may be written as

$$\mathbf{V}_{G_S} = \begin{bmatrix} \dot{X} & \dot{Y} & 0 \end{bmatrix}^T + \dot{\mathbf{R}}\mathbf{r}_1 + \mathbf{R}\dot{\mathbf{r}}_1 , \qquad (31.153)$$

i.e.

$$\mathbf{V}_{G_S} = \mathbf{R}_1\mathbf{V} + \dot{\mathbf{R}}\mathbf{r} + \mathbf{R}\dot{\mathbf{r}}_1 . \qquad (31.154)$$

The translational kinetic energy of the sprung mass is then

$$\mathcal{T}_t = \tfrac{1}{2}m \left(\mathbf{V}^T\mathbf{V} + \mathbf{r}_1{}^T\dot{\mathbf{R}}^T\dot{\mathbf{R}}\mathbf{r}_1 + \dot{\mathbf{r}}_1{}^T\mathbf{R}^T\mathbf{R}\dot{\mathbf{r}}_1 \right) +$$

$$+ m \left(\mathbf{V}^T\mathbf{R}_1^T\dot{\mathbf{R}}\mathbf{r}_1 + \mathbf{V}^T\mathbf{R}_1^T\mathbf{R}\dot{\mathbf{r}}_1 + \mathbf{r}_1{}^T\dot{\mathbf{R}}^T\mathbf{R}\dot{\mathbf{r}}_1 \right) . \qquad (31.155)$$

Because plane xz coincides with the symmetry plane of the sprung mass, the inertia tensor of the latter is

$$\mathbf{J} = \begin{bmatrix} J_x & 0 & -J_{xz} \\ 0 & J_y & 0 \\ -J_{xz} & 0 & J_z \end{bmatrix} . \qquad (31.156)$$

The rotational kinetic energy of the sprung mass is

$$\mathcal{T}_r = \frac{1}{2}\mathbf{\Omega}^T\mathbf{J}\mathbf{\Omega} . \qquad (31.157)$$

By performing the relevant computations, expressing the angular velocity as functions of the variables of motion and neglecting all terms containing powers of small quantities higher than the second, it follows that

$$\mathcal{T} = \tfrac{1}{2}m \left(v_x^2 + v_y^2 + v_z^2 \right) + \tfrac{1}{2}J_x^*\dot{\phi}^2 + \tfrac{1}{2}\left[J_y^* \sin^2(\phi) + J_z \cos^2(\phi) \right] \dot{\psi}^2 +$$

$$- J_{xz} \cos(\phi)\,\dot{\psi}\,\dot{\phi} + J_y \sin(\phi)\,\dot{\psi}\dot{\theta} + mv_x \left\{ \theta\dot{z} + (h+z) \left[\dot{\theta} + \dot{\psi}\sin(\phi) \right] \right\} +$$

$$+ \tfrac{1}{2}J_y^*\dot{\theta}^2 - mv_y \left[\dot{z}\sin(\phi) + h\dot{\phi}\cos(\phi) \right] , \qquad (31.158)$$

where

$$J_x^* = mh^2 + J_x , \quad J_y^* = mh^2 + J_y .$$

Note that in the present model the unsprung mass is neglected, making m both the total mass of the vehicle and the mass of the body.

It can easily be seen that the expression of the kinetic energy coincides with the expression obtained for the model with four degrees of freedom (Eq. 31.61), plus the term

$$\Delta\mathcal{T} = \tfrac{1}{2}mv_z^2 + \tfrac{1}{2}J_y^*\dot{\theta}^2 + J_y \sin(\phi)\,\dot{\psi}\dot{\theta} +$$

$$+ mv_x \left[\theta v_z + \dot{\theta}(h+z) + z\dot{\psi}\sin(\phi) \right] - mv_y\dot{z}\sin(\phi) . \qquad (31.159)$$

The height of the center of mass on the ground is

$$Z_G = (h + z) \cos(\phi) \cos(\theta) \, , \tag{31.160}$$

and then the gravitational potential energy of the vehicle is, with the usual approximations due to the smallness of θ,

$$\mathcal{U}_g = mg(h + z) \cos(\phi) \left(1 - \frac{\theta^2}{2} \right) . \tag{31.161}$$

While in the previous model the potential energy due to suspensions was a function of the roll angle only, here it depends also on the pitch angle and the vertical displacement. However, it can be assumed that the suspensions are such that it is possible to keep the two contributions separate:

$$\mathcal{U}_s = \mathcal{U}_{s1}(\phi) + \mathcal{U}_{s2}(z, \theta) . \tag{31.162}$$

The potential energy is then what was seen in the previous model, plus a contribution due to the two additional degrees of freedom

$$\Delta \mathcal{U} = mgz \cos(\phi) - mg \cos(\phi) \frac{\theta^2}{2} + \mathcal{U}_{s2}(z, \theta) . \tag{31.163}$$

In a similar way, also the dissipation function may be modified by simply adding the term

$$\Delta \mathcal{F} = \mathcal{F}_2\left(\dot{z}, \dot{\theta}\right) . \tag{31.164}$$

Because generalized coordinates z and θ are small quantities, functions $\mathcal{U}_{s2}$ and $\mathcal{F}_2$ are those of a linear system. $\mathcal{F}_2$ in particular does not depend on z and θ, but only on their derivatives.

It is possible to assume, at least as a first approximation, that the two added degrees of freedom have no effect on the kinetic energy of the wheels. In that case the total Lagrangian function of the system is that of the previous model, to which the term

$$\Delta \mathcal{L} = \Delta \mathcal{T} - \Delta \mathcal{U} \tag{31.165}$$

is added.

The derivatives of the added terms in the Lagrangian function are

$$\frac{\partial \Delta \mathcal{L}}{\partial V} = m\dot{\theta}h \, , \quad \frac{\partial \Delta \mathcal{L}}{\partial v_y} = -mv_z \sin(\phi) \, , \tag{31.166}$$

$$\frac{\partial \Delta \mathcal{L}}{\partial v_z} = mv_z - mv_y \sin(\phi) \, , \quad \frac{\partial \Delta \mathcal{L}}{\partial \dot{\phi}} = 0 \, , \tag{31.167}$$

$$\frac{\partial \Delta \mathcal{L}}{\partial \dot{\theta}} = J_y^* \dot{\theta} + J_y \sin(\phi) \dot{\psi} + mv_x(h + z) \, , \tag{31.168}$$

$$\frac{\partial \Delta \mathcal{L}}{\partial \dot{\psi}} = J_y \sin(\phi) \dot{\theta} + mv_x z \sin(\phi) . \tag{31.169}$$

Always remembering that no term containing the products of two or more small quantities may be present in the equations of motion, it follows that

$$\frac{d}{dt}\left(\frac{\partial \Delta \mathcal{L}}{\partial V}\right) = m\ddot{\theta}h \ , \quad \frac{d}{dt}\left(\frac{\partial \Delta \mathcal{L}}{\partial v_y}\right) = -m\dot{v}_z \sin(\phi) \ , \tag{31.170}$$

$$\frac{d}{dt}\left(\frac{\partial \Delta \mathcal{L}}{\partial v_z}\right) = m\dot{v}_z - m\dot{v}_y \sin(\phi) \ , \quad \frac{d}{dt}\left(\frac{\partial \Delta \mathcal{L}}{\partial \dot{\phi}}\right) = 0 \ , \tag{31.171}$$

$$\frac{d}{dt}\left(\frac{\partial \Delta \mathcal{L}}{\partial \dot{\theta}}\right) = J_y^*\ddot{\theta} + J_y \sin(\phi)\,\ddot{\psi} + m\dot{V}(h+z) + mVv_z \ , \tag{31.172}$$

$$\frac{d}{dt}\left(\frac{\partial \Delta \mathcal{L}}{\partial \dot{\psi}}\right) = J_y \sin(\phi)\,\ddot{\theta} + m\dot{V}z\sin(\phi) + mV\dot{z}\sin(\phi) \ , \tag{31.173}$$

$$\frac{\partial \Delta \mathcal{L}}{\partial x^*} = \frac{\partial \Delta \mathcal{L}}{\partial y^*} = \frac{\partial \Delta \mathcal{L}}{\partial \psi} = 0 \ , \tag{31.174}$$

$$\frac{\partial \Delta \mathcal{L}}{\partial z} = -mg\cos(\phi) - \frac{\partial \mathcal{U}_{s2}(z,\theta)}{\partial z} \ , \tag{31.175}$$

$$\frac{\partial \Delta \mathcal{L}}{\partial \theta} = +mVv_z + mgh\cos(\phi)\,\theta - \frac{\partial \mathcal{U}_{s2}(z,\theta)}{\partial \theta} \ , \tag{31.176}$$

$$\frac{\partial \Delta \mathcal{L}}{\partial \phi} = mgz\sin(\phi) \ . \tag{31.177}$$

31.4.2 Equations of motion

Matrix $\mathbf{B}^{\mathsf{T}}\boldsymbol{\Gamma}$ is identical to that of the previous model, apart from the different number of rows and columns

$$\mathbf{B}^{\mathsf{T}}\boldsymbol{\Gamma} = \begin{bmatrix} \begin{bmatrix} 0 & -\dot{\psi} \\ \dot{\psi} & 0 \end{bmatrix} & \mathbf{0}_{2\times 4} \\ \mathbf{0}_{3\times 2} & \mathbf{0}_{3\times 4} \\ \begin{bmatrix} -v_y & v_x \end{bmatrix} & \mathbf{0}_{1\times 4} \end{bmatrix} \ .$$

Matrix $\mathbf{B}^{\mathsf{T}}\boldsymbol{\Gamma}\left\{\frac{\partial \mathcal{L}}{\partial w}\right\}$ is the same too, except for a term that must be introduced in the last equation that may be written as

$$\mathbf{B}^{\mathsf{T}}\boldsymbol{\Gamma}\left\{\frac{\partial \Delta \mathcal{L}}{\partial w}\right\} = \left\{\begin{array}{c} \mathbf{0}_{5\times 1} \\ -mVv_z\sin(\phi) \end{array}\right\} \ . \tag{31.178}$$

By adding the relevant terms, the following equations may be obtained:

First equation: Longitudinal translation

$$m_{at}\dot{V} + m\ddot{\theta}h + J_s\ddot{\psi}\sin(\phi) = Q_x \ . \tag{31.179}$$

Second equation: Lateral translation

$$m\dot{v}_y + m_{at}V\dot{\psi} - m\dot{v}_z \sin{(\phi)} + mv_z\dot{\phi}\cos{(\phi)} - mh\ddot{\phi}\cos{(\phi)} = Q_y . \quad (31.180)$$

Third equation: Translation in the z direction

$$m\dot{v}_z - m\dot{v}_y \sin{(\phi)} + mg\cos{(\phi)} + \frac{\partial \mathcal{F}_2\left(\dot{z}, \dot{\theta}\right)}{\partial \dot{z}} + \frac{\partial \mathcal{U}_{s2}\left(z, \theta\right)}{\partial z} = Q_z . \quad (31.181)$$

Fourth equation: Roll rotation

$$J_x^* \ddot{\phi} - J_{xz}\cos{(\phi)}\ddot{\psi} - m\dot{v}_y h\cos{(\phi)} - J_s V\dot{\psi}\cos{(\phi)} +$$

$$-mgh\sin{(\phi)} - mgz\sin{(\phi)} + \frac{\partial \mathcal{U}_s\left(\phi\right)}{\partial \phi} + \frac{\partial \mathcal{F}\left(\phi, \dot{\phi}\right)}{\partial \dot{\phi}} = Q_\phi . \quad (31.182)$$

Fifth equation: Pitch rotation

$$J_y^* \ddot{\theta} + J_y \sin{(\phi)}\ddot{\psi} + m\dot{V}\left(h + z\right) - mgh\cos{(\phi)}\theta +$$

$$+\frac{\partial \mathcal{F}_2\left(\dot{z}, \dot{\theta}\right)}{\partial \dot{\theta}} + \frac{\partial \mathcal{U}_{s2}\left(z, \theta\right)}{\partial \theta} = Q_\theta . \quad (31.183)$$

Sixth equation: Yaw rotation

$$\left[J_y^* \sin^2{(\phi)} + J_z \cos^2{(\phi)}\right]\ddot{\psi} - J_{xz}\cos{(\phi)}\ddot{\phi} + J_y \sin{(\phi)}\ddot{\theta} + m\dot{V}z\sin{(\phi)} +$$

$$+J_s\dot{V}\sin{(\phi)} + V\cos{(\phi)}\dot{\phi}\sum_{\forall i}\frac{J_{pi}}{R_{ei}} - Vv_y\sum_{\forall i}\frac{J_{pi}}{R_{ei}^2} = Q_\psi . \quad (31.184)$$

31.4.3   Final form of the equations of motion

The sideslip angles of the wheels and the generalized forces due to tires are identical to those seen in the previous model.

The aerodynamic forces and moments are referred to the xyz frame: Because the two added degrees of freedom cause a virtual displacement of the center of mass in the z direction equal to δz, a virtual rotation $\delta \theta$ about the y axis and an additional displacement proportional to $\delta \theta$ in the x direction, the virtual work of aerodynamic forces and moments is

$$\delta L_a = F_{xa}\delta x^* + [F_{ya}\cos(\phi) - F_{za}\sin(\phi)]\,\delta y^* + F_{za}\delta z + (F_{xa}h + M_{ya})\,\delta\theta^* +$$

$$+ (M'_{xa} - F_{ya}h)\,\delta\phi + [(F_{xa}h + M_{ya})\sin(\phi) + M_{za}\cos(\phi)]\,\delta\psi\ . \tag{31.185}$$

In the following equations the generalized aerodynamic forces included in Q_z and Q_θ will be assumed to be constant.

First equation: Longitudinal translation

$$m_{at}\dot{V} + + m\ddot{\theta}h + J_s\ddot{\psi}\sin(\phi) = F_{x1} + F_{x2} - \frac{1}{2}\rho V^2 SC_x\ . \tag{31.186}$$

Second equation: Lateral translation

$$m\dot{v}_y + m_{at}V\dot{\psi} - m\dot{v}_z\sin(\phi) + mv_z\dot{\phi}\cos(\phi) - mh\ddot{\phi}\cos(\phi) =$$

$$= [Y_v + \cos(\phi)\,Y_{v1}]\,v_y + Y_{\dot{\psi}}\dot{\psi} + Y_\phi\phi + \cos(\phi)\,Y_\delta\delta - \frac{1}{2}\rho V^2 SC_z\sin(\phi) + F_{y_e}\ , \tag{31.187}$$

where

$$\begin{cases} Y_v = -\frac{1}{V}\sum_{\forall k}C_k\ , \\[2mm] Y_{v1} = \frac{1}{2}\rho V_a S(C_y)_{,\beta}\ , \\[2mm] Y_{\dot{\psi}} = -\frac{1}{V}\sum_{\forall k}x_{Pk}C_k\ , \\[2mm] Y_\phi = \sum_{\forall k}C_{\gamma k}\ , \\[2mm] Y_\delta = \sum_{\forall k}K'_k C_k\ . \end{cases} \tag{31.188}$$

Third equation: Translation in the z direction

$$m\dot{v}_z - m\dot{v}_y\sin(\phi) + mg\cos(\phi) + \frac{\partial \mathcal{F}_2\left(\dot{z},\dot{\theta}\right)}{\partial \dot{z}} + \frac{\partial \mathcal{U}_{s2}\left(z,\theta\right)}{\partial z} = \frac{1}{2}\rho V^2 SC_z\ . \tag{31.189}$$

Fourth equation: Roll rotation

$$J_x^*\ddot{\phi} - J_{xz}\cos(\phi)\,\ddot{\psi} - m\dot{v}_y h\cos(\phi) - J_s V\dot{\psi}\cos(\phi) +$$

$$-mgh\sin(\phi) - mgz\sin(\phi) + \frac{\partial \mathcal{U}_s(\phi)}{\partial \phi} + \frac{\partial \mathcal{F}\left(\phi,\dot{\phi}\right)}{\partial \dot{\phi}} = L_v v_y\ , \tag{31.190}$$

where

$$L_v = \frac{1}{2}\rho V S\left[-h(C_y)_{,\beta} + t(C_{M_x})_{,\beta}\right]\ . \tag{31.191}$$

Fifth equation: Pitch rotation

$$J_y^* \ddot{\theta} + J_y \sin{(\phi)}\, \ddot{\psi} + m\dot{V}\,(h+z) - mgh\cos{(\phi)}\,\theta +$$

$$+ \frac{\partial \mathcal{F}_2\left(\dot{z}, \dot{\theta}\right)}{\partial \dot{\theta}} + \frac{\partial \mathcal{U}_{s2}\left(z, \theta\right)}{\partial \theta} = \frac{1}{2}\rho V^2 S\left(hC_z + lC_{My}\right)\ . \tag{31.192}$$

Sixth equation: Yaw rotation

$$\left[J_y^* \sin^2{(\phi)} + J_z \cos^2{(\phi)}\right]\ddot{\psi} - J_{xz}\cos{(\phi)}\,\ddot{\phi} + J_y \sin{(\phi)}\,\ddot{\theta} +$$

$$+ m\dot{V}z\sin{(\phi)} + J_s\dot{V}\,\sin{(\phi)} + V\cos{(\phi)}\,\dot{\phi}\sum_{\forall i}\frac{J_{pi}}{R_{e_i}} = \tag{31.193}$$

$$= \left[N_v + \cos{(\phi)}\,Y_{v1}\right]v_y + N_{\dot{\psi}}\dot{\psi} + N_{\phi}\phi + \cos{(\phi)}\,N_{\delta}\delta +$$

$$+ \tfrac{1}{2}\rho V^2 S(-hC_x + lC_{My})\sin{(\phi)} + M_{ze}\ ,$$

where

$$\begin{cases} N_v = \frac{1}{V}\sum_{\forall k}\left[-x_{Pk}C_k + (M_{zk}),_{\alpha} + 2J_{pr}\left(\frac{V}{R_e}\right)^2\right]\ , \\[2mm] N_{v1} = \frac{1}{2}\rho V_a Sl(C'_{M_z}),_{\beta}\ , \\[2mm] N_{\dot{\psi}} = \frac{1}{V}\sum_{\forall k}\left[-x_{Pk}^2 C_k + x_{r_k}(M_{zk}),_{\alpha}\right]\ , \\[2mm] N_{\phi} = \sum_{\forall k} x_{rk}C_{\gamma k}\ , \\[2mm] N_{\delta} = \sum_{\forall k}\left[x_{Pk}K'_k C_k - (M_{zk}),_{\alpha}\right]\ . \end{cases} \tag{31.194}$$

31.4.4 *Motion about the steady-state equilibrium configuration*

Proceeding as in the previous model, a value for the roll angle in steady-state conditions that coincides with that already computed is obtained. If the expressions so obtained are directly compared, they appear different, because in the present case there is a term

$$mgz\sin{(\phi)}$$

that was not present in the earlier model. However, this term has the same order of magnitude of the term

$$mzV\dot{\psi}\cos{(\phi)}\ ,$$

which was neglected, because it contained the product of two small quantities (z and $\dot{\psi}$) (actually, if the roll angle is less than 45° this product is even smaller). The problem lies in the fact that once angle ϕ is no longer considered as a small quantity, to consider other variables as such is no longer correct, leading to problems that cannot be solved within the frame of models of this kind. The only solution is to neglect the term $mgz \sin{(\phi)}$ as well.

Assuming that the coordinates are expressed as the sum of a steady state contribution (subscript 0) plus a contribution that varies in time (subscript 1), the equations of motion may be written as

$$m_{at}\dot{V} + + m\ddot{\theta}_1 h + J_s\ddot{\psi}_1 \sin{(\phi_0)} = F_{x1} + F_{x2} - \frac{1}{2}\rho V^2 SC_x \ , \qquad (31.195)$$

$$m\dot{v}_{y1} + m_{at}V\dot{\psi}_1 + m_{at}V\dot{\psi}_0 - m\dot{v}_{z1}\sin{(\phi_0)} - mh\ddot{\phi}_1\cos{(\phi_0)} =$$

$$= [Y_v + \cos{(\phi_0)}\, Y_{v1}]\,(v_{y0} + v_{y1}) + Y_{\dot{\psi}}\left(\dot{\psi}_0 + \dot{\psi}_1\right) + Y_\phi\,(\phi_0 + \phi_1) +$$

$$+ \cos{(\phi_0)}\, Y_\delta\,(\delta_0 + \delta_1) - \tfrac{1}{2}\rho V^2 SC_z\sin{(\phi_0)} - \tfrac{1}{2}\rho V^2 SC_z\phi_1\cos{(\phi_0)} + F_{y_e} \ ,$$
$$\qquad (31.196)$$

$$m\dot{v}_{z1} - m\dot{v}_{y1}\sin{(\phi_0)} + mg\cos{(\phi_0)} - mg\phi_1\sin{(\phi_0)} + \frac{\partial \mathcal{F}_2\left(\dot{z}_1, \dot{\theta}_1\right)}{\partial \dot{z}} +$$

$$+ \frac{\partial \mathcal{U}_{s2}\,(z_0 + z_1, \theta_0 + \theta_1)}{\partial z} = \frac{1}{2}\rho V^2 SC_z \ ,$$
$$\qquad (31.197)$$

$$J_x^*\ddot{\phi}_1 - J_{xz}\cos{(\phi_0)}\,\ddot{\psi}_1 - m\dot{v}_{y1}h\cos{(\phi_0)} - J_sV\dot{\psi}_1\cos{(\phi_0)} +$$

$$- J_sV\dot{\psi}_0\cos{(\phi_0)} - mgh\sin{(\phi_0)} - mgh\phi_1\cos{(\phi_0)} - mgz_1\sin{(\phi_0)} +$$

$$+ \frac{\partial \mathcal{U}_s\,(\phi)}{\partial \phi} + \frac{\partial \mathcal{F}\left(\phi, \dot{\phi}\right)}{\partial \dot{\phi}} = L_v\,(v_{y0} + v_{y1}) \ ,$$
$$\qquad (31.198)$$

$$J_y^*\ddot{\theta}_1 + J_y\sin{(\phi_0)}\,\ddot{\psi}_1 + m\dot{V}\,(h + z_0 + z_1) - mgh\cos{(\phi_0)}\,(\theta_0 + \theta_1) +$$

$$\qquad (31.199)$$

$$+ \frac{\partial \mathcal{F}_2\left(\dot{z}, \dot{\theta}\right)}{\partial \dot{\theta}} + \frac{\partial \mathcal{U}_{s2}\,(z_0 + z_1, \theta_0 + \theta_1)}{\partial \theta} = \frac{1}{2}\rho V^2 S\,(hC_z + lC_{My}) \ ,$$

$$\left[J_y^* \sin^2{(\phi_0)} + J_z \cos^2{(\phi_0)}\right] \ddot{\psi}_1 - J_{xz} \cos{(\phi_0)} \, \dot{\phi}_1 + J_y \sin{(\phi_0)} \, \ddot{\theta}_1 +$$

$$+ m\dot{V} \left(z_0 + z_1\right) \sin{(\phi_0)} + J_s \dot{V} \sin{(\phi_0)} + J_s \dot{V} \phi_1 \cos{(\phi_0)} + V \dot{\cos}{(\phi_0)} \, \dot{\phi}_1 \sum_{\forall i} \frac{J_{pi}}{R_{e_i}} =$$

$$= \left[N_v + \cos{(\phi)} \, Y_{v1}\right] \left(v_{y0} + v_{y1}\right) + N_{\dot\psi} \left(\dot{\psi}_0 + \dot{\psi}_1\right) + N_\phi \left(\phi_0 + \phi_1\right) +$$

$$+ \cos{(\phi_0)} \, N_\delta \left(\delta_0 + \delta_1\right) + \tfrac{1}{2}\rho V^2 S (-hC_x + lC_{M_y}) \sin{(\phi_0)} +$$

$$+ \tfrac{1}{2}\rho V^2 S (-hC_x + lC_{M_y}) \phi_1 \sin{(\phi_0)} + M_{ze} \ . \tag{31.200}$$

Steady-state conditions

$$F_{x1} + F_{x2} - \frac{1}{2}\rho V^2 S C_x = 0 \ , \tag{31.201}$$

$$m_{at} V \dot{\psi}_0 = \left[Y_v + \cos{(\phi)} \, Y_{v1}\right] v_{y0} + Y_{\dot\psi} \dot{\psi}_0 + Y_\phi \phi_0 +$$
$$+ \cos{(\phi_0)} \, Y_\delta \delta - \tfrac{1}{2}\rho V^2 S C_z \sin{(\phi_0)} + F_{y_e} \ , \tag{31.202}$$

$$mg \cos{(\phi_0)} + \frac{\partial \mathcal{U}_{s2}(z, \theta)}{\partial z} = \frac{1}{2}\rho V^2 S C_z \ , \tag{31.203}$$

$$-J_s V \dot{\psi} \cos{(\phi_0)} - mgh \sin{(\phi_0)} = L_v v_{y0} \ , \tag{31.204}$$

$$-mgh \cos{(\phi_0)} \, \theta_0 + \frac{\partial \mathcal{U}_{s2}(z, \theta)}{\partial \theta} = \frac{1}{2}\rho V^2 S \left(hC_z + lC_{My}\right) \ , \tag{31.205}$$

$$\left[N_v + \cos{(\phi)} \, Y_{v1}\right] v_{y0} + N_{\dot\psi} \dot{\psi}_0 + N_\phi \phi_0 + \cos{(\phi_0)} \, N_\delta \delta +$$
$$+ \tfrac{1}{2}\rho V^2 S (-hC_x + lC_{M_y}) \sin{(\phi_0)} + M_{ze} = 0 \ . \tag{31.206}$$

The first, second, fourth and sixth equations coincide with those previously seen, and may be used to compute first the driving forces needed to travel at speed V, then the roll angle ϕ_0 and v_{y0} (or, better, the sideslip angle β) and the yaw velocity $\dot{\psi}_0$ (or better the radius of the path).

Finally, the third and fifth equations allow z_0 and θ_0 to be computed.

Remark 31.4 *The steady-state condition is not influenced by the presence of suspensions, even if the uncoupling between handling and comfort cannot be managed because the roll angle is not small.*

Motion about the steady-state condition

The equation of motion in the state space is

$$\mathbf{A}_2 \dot{\mathbf{z}} = \mathbf{A}_1 \mathbf{z} \ , \tag{31.207}$$

where:

$$\mathbf{z} = \begin{bmatrix} V & v_y & v_z & v_\phi & v_\theta & v_\psi & z & \phi & \theta \end{bmatrix}^T ,$$

$$v_z = \dot{z} , \quad v_\phi = \dot{\phi} , \quad v_\theta = \dot{\theta} , \quad v_\psi = \dot{\psi}$$

and

$$\mathbf{A}_2 = \begin{bmatrix} \begin{bmatrix} m_{at} & 0 & 0 & 0 & mh & J_s s \\ 0 & m & -ms & -mhc & 0 & 0 \\ 0 & -ms & m & 0 & 0 & 0 \\ 0 & -mhc & 0 & J_x^* & 0 & -J_{xz}c \\ m(h+z_0) & 0 & 0 & 0 & J_y^* & J_y s \\ J_s s & 0 & 0 & -J_{xz}c & J_y s & J_y^* s^2 + J_z c^2 \end{bmatrix} & \mathbf{0}_{6 \times 3} \\ \mathbf{I}_{3 \times 6} & \mathbf{I}_{3 \times 3} \end{bmatrix} ,$$

$$\mathbf{A}_1 = \begin{bmatrix} 0 & 0 & 0 & 0 & 0 & 0 & 0 & 0 & 0 \\ 0 & Y_v^* & 0 & 0 & 0 & m' & 0 & Y_\phi + Y_{\phi 1}c & 0 \\ 0 & 0 & -c_{11} & 0 & -c_{12} & 0 & -k_{11} & -mg\phi_1 s & -k_{12} \\ 0 & L_v & 0 & -c_\phi & 0 & J_s Vc & +mgs & m'' & 0 \\ 0 & 0 & -c_{12} & 0 & -c_{22} & 0 & m\dot{V} - k_{12} & 0 & -k_{22}^* \\ 0 & N_v^* & 0 & N_\phi^* & 0 & N_{\dot\psi} & m\dot{V}s & N_\phi^* & 0 \\ 0 & 0 & 1 & 0 & 0 & 0 & 0 & 0 & 0 \\ 0 & 0 & 0 & 1 & 0 & 0 & 0 & 0 & 0 \\ 0 & 0 & 0 & 0 & 1 & 0 & 0 & 0 & 0 \end{bmatrix} ,$$

$$c = \cos(\phi_0) , \quad s = \sin(\phi_0) , \quad Y_v^* = Y_v + Y_{v1}c , \quad k_{22}^* = k_{22} - mghc ,$$

$$m' = -m_{at} + Y_{\dot\psi} , \quad m'' = mghc - k_\phi ,$$

$$N_v^* = N_v + N_{v1}c , \quad N_\phi^* = -Vc\sum_{\forall i} \frac{J_{pi}}{R_{ei}} , \quad N_\phi^* = N_{\phi 1}c + N_\phi - J_s\dot{V}c .$$

Remark 31.5 *As could be predicted, handling and comfort are not uncoupled, but all coupling terms contain the sine of angle ϕ_0, and thus vanish when the roll angle is small.*

The coupled handling and comfort model can also be used for the study of the controlled system by adding the equations describing the behavior of the roll controller.

Appendix A
EQUATIONS OF MOTION
IN THE STATE
AND CONFIGURATION SPACES

A.1 EQUATIONS OF MOTION OF DISCRETE LINEAR SYSTEMS

A.1.1 Configuration space

Consider a system with a single degree of freedom and assume that the equation expressing its dynamic equilibrium is a second order ordinary differential equation (ODE) in the generalized coordinate x. Assume as well that the forces entering the dynamic equilibrium equation are

- a force depending on acceleration (inertial force),

- a force depending on velocity (damping force),

- a force depending on displacement (restoring force),

- a force, usually applied from outside the system, that depends neither on coordinate x nor on its derivatives, but is a generic function of time (external forcing function).

If the dependence of the first three forces on acceleration, velocity and displacement respectively is linear, the system is linear. Moreover, if the constants of such a linear combination, usually referred to as mass m, damping coefficient c and stiffness k do not depend on time, the system is time-invariant. The dynamic equilibrium equation is then

$$m\ddot{x} + c\dot{x} + kx = f(t) \, . \tag{A.1}$$

If the system has a number n of degrees of freedom, the most general form for a linear, time invariant set of second order ordinary differential equations is

$$\mathbf{A}_1\ddot{\mathbf{x}} + \mathbf{A}_2\dot{\mathbf{x}} + \mathbf{A}_3\mathbf{x} = \mathbf{f}(t) \,, \tag{A.2}$$

where:

- $\mathbf{x}$ is a vector of order n (n is the number of degrees of freedom of the system) where the generalized coordinates are listed;

- $\mathbf{A}_1$, $\mathbf{A}_2$ and $\mathbf{A}_3$ are matrices, whose order is $n \times n$; they contain the characteristics (independent of time) of the system;

- $\mathbf{f}$ is a vector function of time containing the forcing functions acting on the system.

Matrix $\mathbf{A}_1$ is usually symmetrical. The other two matrices in general are not. They can be written as the sum of a symmetrical and a skew-symmetrical matrices

$$\mathbf{M}\ddot{\mathbf{x}} + (\mathbf{C} + \mathbf{G})\,\dot{\mathbf{x}} + (\mathbf{K} + \mathbf{H})\,\mathbf{x} = \mathbf{f}(t) \,, \tag{A.3}$$

where:

- $\mathbf{M}$, the *mass matrix* of the system, is a symmetrical matrix of order $n \times n$ (coincides with $\mathbf{A}_1$). Usually it is not singular.

- $\mathbf{C}$ is the real symmetric *viscous damping matrix* (the symmetric part of $\mathbf{A}_2$).

- $\mathbf{K}$ is the real symmetric *stiffness matrix* (the symmetric part of $\mathbf{A}_3$).

- $\mathbf{G}$ is the real skew-symmetric *gyroscopic matrix* (the skew-symmetric part of $\mathbf{A}_2$).

- $\mathbf{H}$ is the real skew-symmetric *circulatory matrix* (the skew-symmetric part of $\mathbf{A}_3$).

Remark A.1 *Actually it is possible to write the set of linear differential Equations (A.2) in such a way that no matrix is either symmetric or skew symmetric (it is enough to multiply one of the equations by a constant other than 1). A better way to say this is that* $\mathbf{M}$, $\mathbf{C}$, *and* $\mathbf{K}$ *can be reduced to symmetric matrices by the same linear transformation that reduces* $\mathbf{G}$ *and* $\mathbf{H}$ *into skew-symmetric matrices.*

Remark A.2 *The same form of Equation (A.2) may result from mathematical modeling of physical systems whose equations of motion are obtained by means of space discretization techniques, such as the well-known finite elements method.*

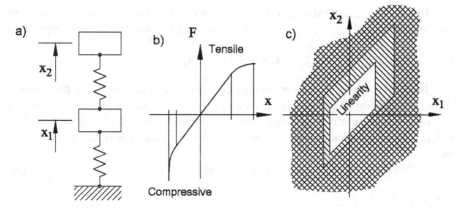

FIGURE A.1. Sketch of a system with two degrees of freedom (a) made by two masses and two springs, whose characteristics (b) are linear only in a zone about the equilibrium position. Three zones can be identified in the configuration space (c): in one the system behaves linearily, in another the system is nonlinear. The latter zone is surrounded by a 'forbidden' zone.

x is a vector in the sense it is a column matrix. Indeed, any set of n numbers may be interpreted as a vector in an n-dimensional space. This space containing vector **x** is usually referred to as *configuration space*, because any point in this space may be associated with a configuration of the system. Actually, not all points of the configuration space, intended to be an infinite n-dimensional space, correspond to configurations that are physically possible for the system: It is then possible to define a subset of possible configurations. Moreover, even systems that are dealt with using linear equations of motion are linear only for configurations little displaced from a reference configuration (usually the equilibrium configuration) and thus the linear equation (A.2) applies in an even smaller subset of the configuration space.

A simple system with two degrees of freedom is shown in Fig. A.1a; it consists of two masses and two springs whose behavior is linear in a zone around the equilibrium configuration with $x_1 = x_2 = 0$, but behave in a nonlinear way to fail at a certain elongation. In the configuration space, which in the case of a system with two degrees of freedom has two dimensions and thus is a plane, there is a linearity zone, surrounded by a zone where the system behaves in nonlinear way. Around the latter is another zone where the system loses its structural integrity.

A.1.2   State space

A set of n second order differential equations is a set of order $2n$ that can be expressed in the form of a set of $2n$ first order equations.

In a way similar to above, a generic linear differential equation with constant coefficients can be written in the form of a set of first order differential equations

$$\mathbf{A}_1\dot{\mathbf{x}} + \mathbf{A}_2\mathbf{x} = \mathbf{f}(t) \ . \tag{A.4}$$

In system dynamics this set of equations is usually solved in the first derivatives (monic form) and the forcing function is written as the linear combination of the minimum number of functions expressing the *inputs* of the system. The independent variables are said to be *state variables* and the equation is written as

$$\dot{\mathbf{z}} = \mathbf{A}\mathbf{z} + \mathbf{B}\mathbf{u} \ , \tag{A.5}$$

where

- $\mathbf{z}$ is a vector of order m, in which the state variables are listed (m is the number of the state variables);

- $\mathbf{A}$ is a matrix of order $m \times m$, independent of time, called the *dynamic matrix*;

- $\mathbf{u}$ is a vector function of time, where the inputs acting on the system are listed (if r is the number of inputs, its size is $r \times 1$);

- $\mathbf{B}$ is a matrix independent of time that states how the various inputs act in the various equations. It is called the input gain matrix and its size is $m \times r$.

As was seen for vector $\mathbf{x}$, $\mathbf{z}$ is also a column matrix that may be considered as a vector in an m-dimensional space. This space is usually referred to as the *state space*, because each point of this space corresponds to a given state of the system.

Remark A.3 *The configuration space is a subspace of the space state.*

If Eq. (A.5) derives from Eq. (A.2), a set of n auxiliary variables must be introduced to transform the system from the configuration to the state space. Although other choices are possible, the simplest choice is to use the derivatives of the generalized coordinates (generalized velocities) as auxiliary variables. Half of the state variables are then the generalized coordinates $\mathbf{x}$, while and the other half are the generalized velocities $\dot{\mathbf{x}}$.

If the state variables are ordered with velocities first and then coordinates, it follows that

$$\mathbf{z} = \left\{ \begin{array}{c} \dot{\mathbf{x}} \\ \mathbf{x} \end{array} \right\} \ .$$

A number n of equations expressing the link between coordinates and velocities must be added to the n equations (A.2). By using symbol $\mathbf{v}$ for the generalized velocities $\dot{\mathbf{x}}$, and solving the equations in the derivatives of the state variables, the set of $2n$ equations corresponding to Eq. (A.3) is then

$$\begin{cases} \dot{\mathbf{v}} = -\mathbf{M}^{-1}\left(\mathbf{C} + \mathbf{G}\right)\mathbf{v} - \mathbf{M}^{-1}\left(\mathbf{K} + \mathbf{H}\right)\mathbf{x} + \mathbf{M}^{-1}\mathbf{f}(t) \\ \dot{\mathbf{x}} = \mathbf{v} \, . \end{cases} \tag{A.6}$$

Assuming that inputs $\mathbf{u}$ coincide with the forcing functions $\mathbf{f}$, matrices $\mathbf{A}$ and $\mathbf{B}$ are then linked to $\mathbf{M}, \mathbf{C}, \mathbf{K}, \mathbf{G}$ and $\mathbf{H}$ by the following relationships

$$\mathbf{A} = \begin{bmatrix} -\mathbf{M}^{-1}\left(\mathbf{C} + \mathbf{G}\right) & -\mathbf{M}^{-1}\left(\mathbf{K} + \mathbf{H}\right) \\ \mathbf{I} & \mathbf{0} \end{bmatrix}, \tag{A.7}$$

$$\mathbf{B} = \begin{bmatrix} \mathbf{M}^{-1} \\ \mathbf{0} \end{bmatrix}. \tag{A.8}$$

The first n out of the $m = 2n$ equations constituting the state equation (A.5) are the dynamic equilibrium equations. These are usually referred to as dynamic equations. The other n express the relationship between the position and the velocity variables. These are usually referred to as kinematic equations.

Often what is more interesting than the state vector $\mathbf{z}$ is a given linear combination of states $\mathbf{z}$ and inputs $\mathbf{u}$, usually referred to as the *output vector*. The state equation (A.5) is then associated with an *output equation*

$$\mathbf{y} = \mathbf{Cz} + \mathbf{Du} \, , \tag{A.9}$$

where

- $\mathbf{y}$ is a vector where the output variables of the system are listed (if the number of outputs is s, its size is $s \times 1$);

- $\mathbf{C}$ is a matrix of order $s \times m$, independent of time, called the *output gain matrix*;

- $\mathbf{D}$ is a matrix independent of time that states how the inputs enter the linear combination yielding the output of the system. It is called the *direct link matrix* and its size is $s \times r$. In many cases the inputs do not enter the linear combination yielding the outputs, and $\mathbf{D}$ is nil.

The four matrices $\mathbf{A}, \mathbf{B}, \mathbf{C}$ and $\mathbf{D}$ are usually referred to as the quadruple of the dynamic system.

Summarizing, the equations that define the dynamic behavior of the system, from input to output, are

$$\begin{cases} \dot{\mathbf{z}} = \mathbf{Az} + \mathbf{Bu} \\ \mathbf{y} = \mathbf{Cz} + \mathbf{Du}. \end{cases} \tag{A.10}$$

Remark A.4 *While the state equations are differential equations, the output equations are algebraic. The dynamics of the system is then concentrated in the former.*

The input-output relationship described by Eq. (A.10) may be described by the block diagram shown in Fig. A.2.

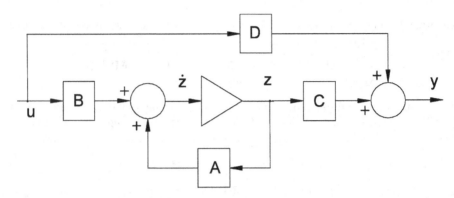

FIGURE A.2. Block diagram corresponding to Eq. (A.10).

A.2   STABILITY OF LINEAR DYNAMIC SYSTEMS

The linearity of a set of equations allows one to state that a solution exists and is unique. The general solution of the equation of motion is the sum of the general solution of the homogeneous equation associated with it and a particular solution of the complete equation. This is true for any differential linear set of equations, even if it is not time-invariant.

The former is the free response of the system, the latter the response to the forcing function.

Consider the equation of motion written in the configuration space (A.2). As already stated, matrix $\mathbf{A}_1$ is symmetrical, while the other two may not be.

The homogeneous equation

$$\mathbf{A}_1 \ddot{\mathbf{x}}(t) + \mathbf{A}_2 \dot{\mathbf{x}}(t) + \mathbf{A}_3 \mathbf{x}(t) = 0 \qquad (A.11)$$

describes the free motion of the system and allows its stability to be studied.

The solution of Eq. (A.11) may be written as

$$\mathbf{x}(t) = \mathbf{x}_0\, e^{st} \ , \qquad (A.12)$$

where $\mathbf{x}_0$ and s are a vector and a scalar, respectively, both complex and constant. To state the time history of the solution allows the differential equation to be transformed into an algebraic equation

$$\left(\mathbf{A}_1 s^2 + \mathbf{A}_2 s + \mathbf{A}_3\right) \mathbf{x}_0 = \mathbf{0} \ . \qquad (A.13)$$

This is a set of linear algebraic homogeneous equations, whose coefficients matrix is a second order *lambda matrix*[1]; it is square and, because the mass matrix $\mathbf{A}_1 = \mathbf{M}$ is not singular, the lambda matrix is said to be *regular*.

---

[1]The term *lambda matrix* comes from the habit of using the symbol λ for the coefficient appearing in the solution $\mathbf{q}(t) = \mathbf{q}_0\, e^{\lambda t}$. Here symbol s has been used instead of λ, following a more modern habit.

The equation of motion (A.11) has solutions different from the trivial

$$\mathbf{x}_0 = \mathbf{0} \tag{A.14}$$

if and only if the determinant of the matrix of the coefficients vanishes:

$$\det\left(\mathbf{A}_1 s^2 + \mathbf{A}_2 s + \mathbf{A}_3\right) = 0 . \tag{A.15}$$

Equation (A.15) is the characteristic equation of a generalized eigenproblem. Its solutions s_i are the eigenvalues of the system and the corresponding vectors $\mathbf{x}_{0_i}$ are its eigenvectors. The rank of the matrix of the coefficients obtained in correspondence of each eigenvalue s_i defines its multiplicity: If the rank is $n - \alpha_i$, the multiplicity is α_i. The eigenvalues are $2n$ and, correspondingly, there are $2n$ eigenvectors.

A.2.1 Conservative natural systems

If the gyroscopic matrix $\mathbf{G}$ is not present the system is said to be *natural*. If the damping and circulatory matrices $\mathbf{C}$ and $\mathbf{H}$ also vanish the system is *conservative*. A system with $\mathbf{G} = \mathbf{C} = \mathbf{H} = \mathbf{0}$ (or, as is usually referred to, an MK system) is then both natural and conservative. The characteristic equation reduces to the algebraic equation

$$\det\left(\mathbf{M} s_i^2 + \mathbf{K}\right) = 0 . \tag{A.16}$$

The eigenproblem can be reduced in canonical form

$$\mathbf{D}\mathbf{x}_i = \mu_i \mathbf{x}_i, \tag{A.17}$$

where the dynamic matrix in the configuration space $\mathbf{D}$ (not to be confused with the dynamic matrix in the state space $\mathbf{A}$) is

$$\mathbf{D} = \mathbf{M}^{-1}\mathbf{K} , \tag{A.18}$$

and the parameter in which the eigenproblem is written is

$$\mu_i = -s_i^2 . \tag{A.19}$$

Because matrices $\mathbf{M}$ and $\mathbf{K}$ are positive defined (or, at least, semi-defined), the n eigenvalues μ_i are all real and positive (or zero) and then the eigenvalues in terms of s_i are $2n$ imaginary numbers in pairs with opposite sign

$$(s_i, \overline{s}_i) = \pm i\sqrt{\mu_i} . \tag{A.20}$$

The n eigenvectors $\mathbf{x}_i$ of size n are real vectors.

When the eigenvalue s_i is imaginary, the solution (A.12) reduces to an undamped harmonic oscillation

$$\mathbf{x}(t) = \mathbf{x}_0 \, e^{i\omega t} , \tag{A.21}$$

where
$$\omega = is = \sqrt{\mu} \tag{A.22}$$

is the (circular) frequency.

The n values of ω_i, computed from the eigenvalues μ_i, are the natural frequencies or eigenfrequencies of the system, usually referred to as ω_{n_i}.

If $\mathbf{M}$ or $\mathbf{K}$ are not positive defined or semidefined, at least one of the eigenvalues μ_i is negative, making one of the pair of solutions in s real, being made of a positive and a negative value. As will be seen below, the real negative solution corresponds to a time history that decays in time in a non-oscillatory way, the positive solution to a time history that increases in time in an unbounded way. The system is then unstable.

A.2.2 Natural nonconservative systems

If matrix $\mathbf{C}$ does not vanish while $\mathbf{G} = \mathbf{H} = \mathbf{0}$, the system is still natural and non-circulatory, but is no longer conservative.

The characteristic equation (A.15) cannot be reduced to an eigenproblem in canonical form in the configuration space and the state space formulation must be used.

The general solution of the homogeneous equation associated with Eq. (A.5) is of the type
$$\mathbf{z} = \mathbf{z}_0 e^{st} , \tag{A.23}$$

where s is generally a complex number. Its real and imaginary parts are usually indicated with symbols ω and σ
$$\begin{aligned} \omega &= \Im(s) \\ \sigma &= \Re(s) \end{aligned} \tag{A.24}$$

and represent the frequency of the free oscillations and the decay rate. Solution (A.23) can in fact be written in the form
$$\mathbf{z} = \mathbf{z}_0 e^{\sigma t} e^{i\omega t} , \tag{A.25}$$

or, because both σ and ω are real numbers,
$$\mathbf{z} = \mathbf{z}_0 e^{\sigma t} [\cos(\omega t) + i\sin(\omega t)] . \tag{A.26}$$

By introducing solution (A.23) into the homogeneous equation associated with Eq. (A.5), the latter transforms from a set of differential equations to a (homogeneous) set of algebraic equations
$$s\mathbf{z}_0 = \mathbf{A}\mathbf{z}_0 , \tag{A.27}$$

i.e.
$$(\mathbf{A} - s\mathbf{I})\mathbf{z}_0 = 0 . \tag{A.28}$$

As seen for the equation of motion in the configuration space, the homogeneous equations will have solutions other than the trivial solution $z_0 = 0$ only if the determinant of the coefficients matrix vanishes

$$\det\left(\mathbf{A} - s\mathbf{I}\right) = 0 . \tag{A.29}$$

Equation (A.29) can be interpreted as an algebraic equation in s, i.e. the characteristic equation of the dynamic systems. It is an equation of power $2n$, yielding the $2n$ values of s. The $2n$ values of s are the eigenvalues of the system and the corresponding $2n$ values of z_0 are the eigenvectors. In general, both eigenvalues and eigenvectors are complex.

If matrix $\mathbf{A}$ is real, as is usually the case, the solutions are either real or complex conjugate. The corresponding time histories are (Fig. A.3):

- Real solutions ($\omega = 0$, $\sigma \neq 0$): Either exponential time histories, with monotonic decay of the amplitude if the solution is negative (stable, non-oscillatory behavior), or exponential time histories, with monotonic increase of the amplitude if the solution is positive (unstable, non-oscillatory behavior).

- Complex conjugate solutions ($\omega \neq 0$, $\sigma \neq 0$): Oscillating time histories, expressed by Eq. (A.26) with amplitude decay if the real part of the solution

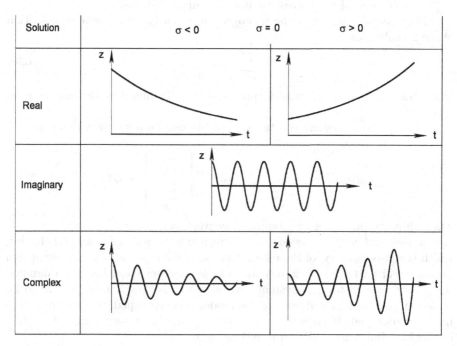

FIGURE A.3. Time history of the free motion for the various types of the eigenvalues of the system

is negative (stable, oscillatory behavior) or amplitude increase in time if the real part of the solution is positive (unstable, oscillatory behavior). If the system is stable, stability is asymptotic.

To these two cases, that previously seen for conservative systems may be added:

- Imaginary solutions ($\omega \neq 0$, $\sigma = 0$): Harmonic time histories (sine or cosine waves, undamped oscillatory behavior). In this case stability is non-asymptotic.

The necessary and sufficient condition for stable behavior is thus that the real part of all eigenvalues is negative.

If any one of the real parts of the eigenvalues is zero, the behavior is still stable (because the amplitude does not grow uncontrolled in time) but not asymptotically stable.

If at least one of the real parts of the eigenvalues is positive, the system is unstable.

If the system is little damped, i.e. the eigenvalues are conjugate and the decay rates σ are small, the values of the natural frequencies ω are close to those of the corresponding undamped system, i.e. to those of the MK system obtained by simply neglecting the damping matrix $\mathbf{C}$. In this case the natural frequencies ω_{n_i} are still those of the corresponding undamped systems.

The general solution of the homogeneous equation is a linear combination of the $2n$ solutions

$$\mathbf{z} = \sum_{i=1}^{2n} C_i \mathbf{z}_{0i} e^{s_i t} , \tag{A.30}$$

where the $2n$ constants C_i must be obtained from the initial conditions, i.e. from vector $\mathbf{z}(0)$.

The equation allowing constants C_i to be computed can be written as

$$\mathbf{z}(0) = \begin{bmatrix} \mathbf{z}_{01} & \mathbf{z}_{02} & \cdots & \mathbf{z}_{02n} \end{bmatrix} \begin{Bmatrix} C_1 \\ C_2 \\ \cdots \\ C_{2n} \end{Bmatrix} = \mathbf{\Phi C} , \tag{A.31}$$

where $\mathbf{\Phi}$ is the matrix of the complex eigenvectors.

A real and negative eigenvalue corresponds to an *overdamped* behavior, which is non-oscillatory, of the relevant mode. If the eigenvalue is complex (with negative real part) the mode has an *underdamped* behavior, i.e. has a damped oscillatory time history. A system with all underdamped modes is said to be underdamped, while if only one of the modes is overdamped, the system is said to be overdamped. If all modes are overdamped, the system cannot have free oscillations, but can oscillate if forced to do so.

It must be noted that if all matrices $\mathbf{M}$, $\mathbf{K}$ and $\mathbf{C}$ are positive defined (or at least semidefined), as in the case of a structure with viscous damping with positive stiffness and damping, there is no eigenvalue with positive real part and hence the system is stable. If all matrices are strictly positive defined, there is no eigenvalue with vanishing real part and the system is asymptotically stable.

A.2.3 Systems with singular mass matrix

If matrix $\mathbf{M}$ is singular, it is impossible to write the dynamic matrix in the usual way. This usually occurs because a vanishingly small inertia is associated with some degrees of freedom, as for instance in the case of the driveline models shown in Fig. 30.9, where the tire is modelled as a spring and a damper in series, with no mass between them. Clearly the problem may be circumvented by associating a very small mass with the relevant degrees of freedom: A new very high natural frequency that has no physical meaning is thus introduced and, if this is done carefully, no numerical instability problem results. However, it makes little sense to resort to tricks of this kind when it is possible to overcome the problem in a more correct and essentially simple way.

The degrees of freedom can be subdivided into two sets: A vector $\mathbf{x}_1$ containing those with which a non-vanishing inertia is associated, and a vector $\mathbf{x}_2$, containing all others. All matrices and forcing functions may be similarly split . The mass matrix $\mathbf{M}_{22}$ vanishes, and if the mass matrix is diagonal, $\mathbf{M}_{12}$ and $\mathbf{M}_{21}$ also vanish.

Assuming that $\mathbf{M}_{12}$ and $\mathbf{M}_{21}$ are zero, the equations of motion become

$$\begin{cases} \mathbf{M}_{11}\ddot{\mathbf{x}}_1 + \mathbf{C}_{11}\dot{\mathbf{x}}_1 + \mathbf{C}_{12}\dot{\mathbf{x}}_2 + \mathbf{K}_{11}\mathbf{x}_1 + \mathbf{K}_{12}\mathbf{x}_2 = \mathbf{f}_1(t) \\ \mathbf{C}_{21}\dot{\mathbf{x}}_1 + \mathbf{C}_{22}\dot{\mathbf{x}}_2 + \mathbf{K}_{21}\mathbf{x}_1 + \mathbf{K}_{22}\mathbf{x}_2 = \mathbf{f}_2(t) \ . \end{cases} \tag{A.32}$$

To simplify the equations of motion neither the gyroscopic nor the circulator matrices are explicitly written, but in what follows no assumption on the symmetry of the stiffness and damping matrices will be made. The equations also hold for gyroscopic and circulatory systems.

By introducing the velocities $\mathbf{v}_1$ together with generalized coordinates $\mathbf{x}_1$ and $\mathbf{x}_2$ as state variables, the state equation is

$$\mathbf{M}^* \begin{Bmatrix} \mathbf{v}_1 \\ \mathbf{x}_1 \\ \mathbf{x}_2 \end{Bmatrix} = \mathbf{A}^* \begin{Bmatrix} \mathbf{v}_1 \\ \mathbf{x}_1 \\ \mathbf{x}_2 \end{Bmatrix} + \begin{bmatrix} \mathbf{I} & \mathbf{0} \\ \mathbf{0} & \mathbf{I} \\ \mathbf{0} & \mathbf{0} \end{bmatrix} \begin{Bmatrix} \mathbf{f}_1(t) \\ \mathbf{f}_2(t) \end{Bmatrix} , \tag{A.33}$$

where

$$\mathbf{M}^* = \begin{bmatrix} \mathbf{M}_{11} & \mathbf{0} & \mathbf{C}_{12} \\ \mathbf{0} & \mathbf{0} & \mathbf{C}_{22} \\ \mathbf{0} & \mathbf{I} & \mathbf{0} \end{bmatrix} , \quad \mathbf{A}^* = - \begin{bmatrix} \mathbf{C}_{11} & \mathbf{K}_{11} & \mathbf{K}_{12} \\ \mathbf{C}_{21} & \mathbf{K}_{21} & \mathbf{K}_{22} \\ -\mathbf{I} & \mathbf{0} & \mathbf{0} \end{bmatrix} . \tag{A.34}$$

The dynamic matrix and the input gain matrix are

$$\mathbf{A} = \mathbf{M}^{*-1}\mathbf{A}^* , \quad \mathbf{B} = \mathbf{M}^{*-1} \begin{bmatrix} \mathbf{I} & \mathbf{0} \\ \mathbf{0} & \mathbf{I} \\ \mathbf{0} & \mathbf{0} \end{bmatrix} . \tag{A.35}$$

Alternatively, the expressions of $\mathbf{M}^*$ and $\mathbf{A}^*$ can be

$$\mathbf{M}^* = \begin{bmatrix} \mathbf{M}_{11} & \mathbf{C}_{11} & \mathbf{C}_{12} \\ \mathbf{0} & \mathbf{C}_{21} & \mathbf{C}_{22} \\ \mathbf{0} & \mathbf{I} & \mathbf{0} \end{bmatrix} , \quad \mathbf{A}^* = - \begin{bmatrix} \mathbf{0} & \mathbf{K}_{11} & \mathbf{K}_{12} \\ \mathbf{0} & \mathbf{K}_{21} & \mathbf{K}_{22} \\ -\mathbf{I} & \mathbf{0} & \mathbf{0} \end{bmatrix} . \quad \text{(A.36)}$$

If vector $\mathbf{x}_1$ contains n_1 elements and $\mathbf{x}_2$ contains n_2 elements, the size of the dynamic matrix $\mathbf{A}$ is $2n_1 + n_2$.

A.2.4   Conservative gyroscopic systems

If matrix $\mathbf{G}$ is not zero, while both $\mathbf{C}$ and $\mathbf{H}$ vanish, the dynamic matrix reduces to

$$\mathbf{A} = \begin{bmatrix} -\mathbf{M}^{-1}\mathbf{G} & -\mathbf{M}^{-1}\mathbf{K} \\ \mathbf{I} & \mathbf{0} \end{bmatrix} . \quad \text{(A.37)}$$

By premultiplying the first n equations by $\mathbf{M}$ and the other n by $\mathbf{K}$, it follows that

$$\mathbf{M}^*\dot{\mathbf{z}} + \mathbf{G}^*\mathbf{z} = \mathbf{0} , \quad \text{(A.38)}$$

where

$$\mathbf{M}^* = \begin{bmatrix} \mathbf{M} & \mathbf{0} \\ \mathbf{0} & \mathbf{K} \end{bmatrix} , \quad \mathbf{G}^* = \begin{bmatrix} \mathbf{G} & \mathbf{K} \\ -\mathbf{K} & \mathbf{0} \end{bmatrix} . \quad \text{(A.39)}$$

The first matrix is symmetrical, while the second is skew symmetrical.

By introducing solutions (A.23) into the equation of motion, the following homogeneous equation

$$s\mathbf{M}^*\mathbf{z}_0 + \mathbf{G}^*\mathbf{z}_0 = \mathbf{0} \quad \text{(A.40)}$$

is obtained.

The corresponding eigenproblem has imaginary solutions like those of an MK system, even if the structure of the eigenvectors is different. In any case the time history of the free oscillations is harmonic and undamped, because the decay rate $\sigma = \Re(s)$ is zero.

A.2.5   General dynamic systems

The situation is similar to that seen for natural non-conservative systems, in the sense that the time histories of the free oscillations are those seen in Fig. A.3 and stability is dominated by the sign of the real part of s.

Remark A.5 *In general, the presence of a gyroscopic matrix does not reduce the stability of the system, while the presence of a circulatory matrix has a destabilizing effect.*

Consider, for instance, a two degrees of freedom system made by two independent MK system; each with a single degree of freedom, and assume that the two masses are equal. The equations for free motion are

$$\begin{cases} m\ddot{x}_1 + k_1 x_1 = 0 \\ m\ddot{x}_2 + k_2 x_2 = 0 \ . \end{cases} \tag{A.41}$$

Introduce now a coupling term in both equations, introducing for instance a spring with stiffness k_{12} between the two masses. The equations of motion become

$$\begin{cases} m\ddot{x}_1 + (k_1 + k_{12})\, x_1 - k_{12} x_2 = 0 \ , \\ m\ddot{x}_2 - k_{12} x_1 + (k_1 + k_{12})\, x_2 = 0 \ . \end{cases} \tag{A.42}$$

By introducing parameters

$$\omega_0^2 = \frac{k_1 + k_2 + 2k_{12}}{2m} \ , \quad \alpha = \frac{k_2 - k_1}{2m\Omega_0^2} \ , \quad \epsilon = \frac{k_{12}}{m\Omega_0^2} \ , \tag{A.43}$$

the equation of motion can be written as

$$\left\{ \begin{matrix} \ddot{x} \\ \ddot{y} \end{matrix} \right\} + \omega_0^2 \begin{bmatrix} 1 - \alpha & \epsilon \\ \epsilon & 1 + \alpha \end{bmatrix} \left\{ \begin{matrix} x \\ y \end{matrix} \right\} = 0 \ . \tag{A.44}$$

Note that

$$-1 \leq \alpha \leq 1 \ . \tag{A.45}$$

The matrix that multiplies the generalized coordinates is symmetrical and is thus a true stiffness matrix. The coupling is said in this case to be *non-circolatory* or *conservative*. Because there is no damping matrix and the stiffness matrix is positive defined ($-1 \leq \alpha \leq 1$), the eigenvalues are imaginary and the system is stable, even if it is not asymptotically stable as it would be if a positive defined damping matrix were present.

The natural frequencies of the system, made nondimensional by dividing them by ω_0, depend upon two parameters, α and ϵ. They are shown in Fig. A.4(a) as functions of α for some values of ϵ. The distance between the two curves (one for $\omega > \omega_0$ and the other for $\omega < \omega_0$) increases if the coupling term ϵ increases. For this reason this type of coupling is said to be *repulsive*.

Consider now the case with coupling term ϵ in the form

$$\left\{ \begin{matrix} \ddot{x} \\ \ddot{y} \end{matrix} \right\} + \Omega_0^2 \begin{bmatrix} 1 - \alpha & \epsilon \\ -\epsilon & 1 + \alpha \end{bmatrix} \left\{ \begin{matrix} x \\ y \end{matrix} \right\} = 0 \ . \tag{A.46}$$

The terms outside the main diagonal of the stiffness matrix now have the same modulus but opposite sign. The matrix multiplying the displacements is made up of a symmetrical part (the stiffness matrix) and a skew-symmetrical part (the circulatory matrix). A coupling of this type is said to be *circulatory* or *non-conservative*.

While in the previous case the effect could be caused by the presence of a spring between the two masses, it cannot be due to springs or similar elements here. There are situations of practical interest where circulatory coupling occurs.

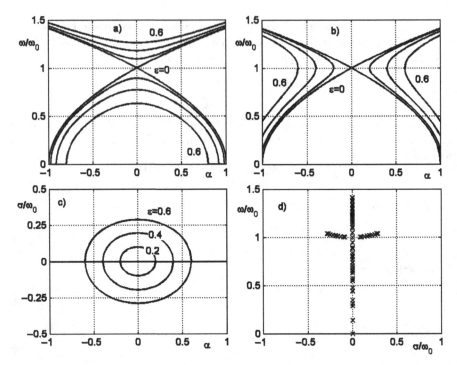

FIGURE A.4. Nondimensional natural frequencies as functions of parameters α and ϵ for a system with two degrees of freedom with non-circulatory (a) and circulatory (b) coupling. Decay rate (c) and roots locus (d) for the system with circulatory coupling.

The natural frequencies of the system in this case also depend on the two parameters α and ϵ. These are plotted in nondimensional form, by dividing them by ω_0, in Fig. A.4(b) as functions of α for some values of ϵ. The two curves now close on each other. Starting from the condition with $\alpha = -1$, the two curves meet for a certain value of α in the interval $(-1,\ 0)$. There is a range, centered in the point with $\alpha = 0$, where the solutions of the eigenproblem are complex. Beyond this range the two curves separate again.

Because the two curves approach each other and finally meet, this type of coupling is said to be *attractive*.

In the range where the values of s are complex, one of the two solutions has a positive real part: It follows that an unstable solution exists, as can be seen from the decay rate plot in Fig. A.4(c) and from the roots locus in Fig. A.4(d).

Remark A.6 *Instability is linked with the skew-symmetric matrix due to coupling, i.e. because of the fact that a circulatory matrix exists.*

A.3 CLOSED FORM SOLUTION OF THE FORCED RESPONSE

The particular solution of the complete equation depends on the time history of the forcing function (input) $u(t)$. In case of harmonic input

$$\mathbf{u} = \mathbf{u}_0 e^{i\omega t} , \tag{A.47}$$

the response is harmonic as well

$$\mathbf{z} = \mathbf{z}_0 e^{i\omega t} , \tag{A.48}$$

and has the same frequency as the forcing function ω. As usual, by introducing the time history of the forcing function and the response into the equation of motion, it transforms into an algebraic equation

$$(\mathbf{A} - i\omega \mathbf{I}) \, \mathbf{z}_0 + \mathbf{B} \mathbf{u}_0 = 0, \tag{A.49}$$

that allows the amplitude of the response to be computed

$$\mathbf{z}_0 = - \, (\mathbf{A} - i\omega \mathbf{I})^{-1} \mathbf{B} \mathbf{u}_0 . \tag{A.50}$$

If the input is periodic, it may be decomposed in Fourier series and the response to each of its harmonic components computed. The results are then added. This is possible only because the system is linear.

If the input is not harmonic or at least periodic, it is possible to resort to Laplace transforms or the Duhamel integral. These techniques apply only to linear systems.

Remark A.7 *Linear models allow closed form solutions to be obtained and stability, in particular, to be studied. In linear systems, moreover, stability is a property of the system and not of its peculiar working conditions.*

A.4 NONLINEAR DYNAMIC SYSTEMS

The state equations of dynamic systems are often nonlinear. The reasons for the presence of nonlinearities may differ, owing to the presence of elements behaving in an intrinsically nonlinear way (e.g. springs producing a force dependent in a nonlinear way on the displacement), or the presence of trigonometric functions of some of the generalized coordinates in the dynamic or kinematic equations. If inertial forces are linear in the accelerations, the equations of motion can be written in the form

$$\mathbf{M}\ddot{\mathbf{x}} + \mathbf{f}_1(\mathbf{x}, \dot{\mathbf{x}}) = \mathbf{f}(t) . \tag{A.51}$$

Function $\mathbf{f}_1$ may often be considered as the sum of a linear and nonlinear part. The equation of motion can then be written as

$$\mathbf{M}\ddot{\mathbf{x}} + (\mathbf{C} + \mathbf{G})\,\dot{\mathbf{x}} + (\mathbf{K} + \mathbf{H})\,\mathbf{x} + \mathbf{f}_2(\mathbf{x},\,\dot{\mathbf{x}}) = \mathbf{f}(t)\,, \tag{A.52}$$

where function $\mathbf{f}_2$ contains only the nonlinear part of the dynamic system.

The state equations corresponding to Eq. (A.51) and Eq. (A.52) are

$$\dot{\mathbf{z}} = \mathbf{f}_1(\mathbf{z}) + \mathbf{B}\mathbf{u}\,, \tag{A.53}$$

or, by separating the linear from the nonlinear part,

$$\dot{\mathbf{z}} = \mathbf{A}\mathbf{z} + \mathbf{f}_2(\mathbf{z}) + \mathbf{B}\mathbf{u}\,. \tag{A.54}$$

Another way to express the equation of motion or the state equation of a nonlinear system is by writing equations (A.3) or (A.10), where the various matrices are functions of the generalized coordinates and their derivatives, or of the state variables. In the state space it follows that

$$\begin{cases} \dot{\mathbf{z}} = \mathbf{A}(\mathbf{z})\mathbf{z} + \mathbf{B}(\mathbf{z})\mathbf{u} \\ \mathbf{y} = \mathbf{C}(\mathbf{z})\mathbf{z} + \mathbf{D}(\mathbf{z})\mathbf{u}\,. \end{cases} \tag{A.55}$$

If the system is not time-invariant, the various matrices may also be explicit functions of time

$$\begin{cases} \dot{\mathbf{z}} = \mathbf{A}(\mathbf{z},\,t)\mathbf{z} + \mathbf{B}(\mathbf{z},\,t)\mathbf{u} \\ \mathbf{y} = \mathbf{C}(\mathbf{z},\,t)\mathbf{z} + \mathbf{D}(\mathbf{z},\,t)\mathbf{u}\,. \end{cases} \tag{A.56}$$

Remark A.8 *It is not possible to obtain a closed form solution of nonlinear systems, and concepts like natural frequency or decay rate lose their meaning. It is not even possible to distinguish between free and forced behavior, in the sense that the free oscillations depend upon the zone of the state space where the system operates.*

In some zones of the state space the behavior of the system may be stable, while in others it may be unstable.

In any case it is often possible to linearize the equations of motion about any given working conditions, i.e. any given point of the state space, and to use the linearized model so obtained in that area of the space state to study the motion of the system and above all its stability. In this case the motion and stability are studied *in the small*. It is, however, clear that no general result may be obtained in this way.

If the state equation is written in the form (A.53), its linearization about a point of coordinates $\mathbf{z}_0$ in the state space is

$$\dot{\mathbf{z}} = \left(\frac{\partial \mathbf{f}_1}{\partial \mathbf{z}}\right)_{\mathbf{z}=\mathbf{z}_0} \mathbf{z} + \mathbf{B}\mathbf{u}\,, \tag{A.57}$$

where $\left(\frac{\partial \mathbf{f}_1}{\partial \mathbf{z}}\right)_{\mathbf{z}=\mathbf{z}_0}$ is the Jacobian matrix of function $\mathbf{f}_1$ computed in $\mathbf{z}_0$.

If the formulation (A.55) is used, the linearized dynamics of the system about point $\mathbf{z}_0$ may be studied through the linear equation

$$\begin{cases} \dot{\mathbf{z}} = \mathbf{A}(\mathbf{z}_0)\mathbf{z} + \mathbf{B}(\mathbf{z}_0)\mathbf{u} \\ \mathbf{y} = \mathbf{C}(\mathbf{z}_0)\mathbf{z} + \mathbf{D}(\mathbf{z}_0)\mathbf{u} \,. \end{cases} \tag{A.58}$$

Remark A.9 *While the motion and stability in the small can be studied in closed form, studying the motion* in the large *requires resorting to the numerical integration of the equations of motion, that is, resorting to numerical simulation.*

A.5 LAGRANGE EQUATIONS IN THE CONFIGURATION AND STATE SPACE

In relatively simple systems it is possible to write the equations of motion directly in the form of Eq. (A.3), by writing all forces, internal and external to the system, acting on its various parts. However, if the system is complex, and in particular if the number of degrees of freedom is large, it is expedient to resort to the methods of analytical mechanics.

One of the simplest approaches to writing the equations of motion of multi-degrees of freedom systems is by resorting to Lagrange equations. Consider a generic mechanical system with n degrees of freedom, i.e. one whose configuration may be expressed using n generalized coordinates x_i. Its equations of motion can in general be written in the form

$$\frac{d}{dt}\left(\frac{\partial \mathcal{T}}{\partial \dot{x}_i}\right) - \frac{\partial \mathcal{T}}{\partial x_i} + \frac{\partial \mathcal{U}}{\partial x_i} + \frac{\partial \mathcal{F}}{\partial \dot{x}_i} = Q_i \qquad (i = 1,...., n), \tag{A.59}$$

where:

- $\mathcal{T}$ is the kinetic energy of the system. This allows inertial forces to be written in a synthetic way. In general,

$$\mathcal{T} = \mathcal{T}(\dot{x}_i, x_i, t) \,.$$

The kinetic energy is basically a quadratic function of the generalized velocities

$$\mathcal{T} = \mathcal{T}_0 + \mathcal{T}_1 + \mathcal{T}_2 \,, \tag{A.60}$$

where $\mathcal{T}_0$ does not depend on the velocities, $\mathcal{T}_1$ is linear and $\mathcal{T}_2$ is quadratic.

In linear systems, the kinetic energy must contain terms of the velocities and coordinates having no powers higher than 2 or products of more than two of them. As a consequence, $\mathcal{T}_2$ cannot contain displacements

$$\mathcal{T}_2 = \frac{1}{2}\sum_{i=1}^{n}\sum_{j=1}^{n} m_{ij}x_i x_j = \frac{1}{2}\dot{\mathbf{x}}^T \mathbf{M}\dot{\mathbf{x}} \,, \tag{A.61}$$

where the terms m_{ij} don't depend on either $\mathbf{x}$ or $\dot{\mathbf{x}}$. If the system is time-invariant, $\mathbf{M}$ is constant.

$\mathcal{T}_1$ is linear in the velocities, and thus, if the system is linear, cannot contain terms other than constant or linear in the displacements

$$\mathcal{T}_1 = \frac{1}{2}\dot{\mathbf{x}}^T \left(\mathbf{M}_1\mathbf{x} + \mathbf{f}_1\right) , \tag{A.62}$$

where matrix $\mathbf{M}_1$ and vector $\mathbf{f}_1$ do not contain the generalized coordinates, even if $\mathbf{f}_1$ may be a function of time even in time-invariant systems.

$\mathcal{T}_0$ does not contain generalized velocities but, in the case of linear systems, only contains terms with power not greater than 2 in the generalized coordinates:

$$\mathcal{T}_o = \frac{1}{2}\mathbf{x}^T\mathbf{M}_g\mathbf{x} + \mathbf{x}^T\mathbf{f}_2 + e , \tag{A.63}$$

where matrix $\mathbf{M}_g$, vector $\mathbf{f}_2$ and scalar e are constant. Constant e does not enter the equations of motion. As will be seen later, the structure of $\mathcal{T}_o$ is similar to that of the potential energy. The term

$$\mathcal{U} - \mathcal{T}_0$$

is often referred to as *dynamic potential*.

- $\mathcal{U}$ is the potential energy. It allows conservative forces to be expressed in a synthetic form. In general,
$$\mathcal{U} = \mathcal{U}(x_i) .$$

In linear systems, the potential energy is a quadratic form in the generalized coordinates and, apart from a constant term that does not enter the equations of motion and thus has no importance, can be written as

$$\mathcal{U} = \frac{1}{2}\mathbf{x}^T\mathbf{K}\mathbf{x} + \mathbf{x}^T\mathbf{f}_0 , \tag{A.64}$$

By definition the potential energy does not depend on the generalized velocities and its derivatives with respect to the generalized velocities $\dot{x}_i$ vanish. Equation (A.59) is often written with reference to the *Lagrangian function*

$$\mathcal{L} = \mathcal{T} - \mathcal{U}$$

and becomes

$$\frac{d}{dt}\left[\frac{\partial\mathcal{L}}{\partial\dot{x}_i}\right] - \frac{\partial\mathcal{L}}{\partial x_i} + \frac{\partial\mathcal{F}}{\partial\dot{x}_i} = Q_i . \tag{A.65}$$

- $\mathcal{F}$ is the Raleigh dissipation function. It allows some types of damping forces to be expressed in a synthetic form. In many cases $\mathcal{F} = \mathcal{F}(\dot{x}_i)$, but it may also depend upon the generalized coordinates. In linear systems, the dissipation function is a quadratic form in the generalized velocities and, apart from terms not depending upon $\dot{x}_i$ that do not enter the equation of motion and thus have no importance, may be written as

$$\mathcal{F} = \frac{1}{2}\dot{\mathbf{x}}^T\mathbf{C}\dot{\mathbf{x}} + \frac{1}{2}\dot{\mathbf{x}}^T \left(\mathbf{C}_1\mathbf{x} + \mathbf{f}_3\right) . \tag{A.66}$$

- Q_i are generalized forces that cannot be expressed using the above mentioned functions. In general, $Q_i = Q_i(\dot{q}_i, q_i, t)$. In the case of linear systems, these forces do not depend on the generalized coordinates and velocities, and then

$$Q_i = Q_i(t) \ . \tag{A.67}$$

In linear systems, by performing the relevant derivatives

$$\frac{\partial(\mathcal{T} - \mathcal{U})}{\partial \dot{x}_i} = \mathbf{M}\dot{x} + \frac{1}{2}\left(\mathbf{M}_1 x + \mathbf{f}_1\right) \ , \tag{A.68}$$

$$\frac{d}{dt}\left[\frac{\partial(\mathcal{T} - \mathcal{U})}{\partial \dot{x}_i}\right] = \mathbf{M}\ddot{x} + \frac{1}{2}\mathbf{M}_1\dot{x} + \dot{\mathbf{f}}_1 \ , \tag{A.69}$$

$$\frac{\partial(\mathcal{T} - \mathcal{U})}{\partial x_i} = \frac{1}{2}\mathbf{M}_1^T x + \mathbf{M}_g x - \mathbf{K}x + \mathbf{f}_2 - \mathbf{f}_0 \ , \tag{A.70}$$

$$\frac{\partial \mathcal{F}}{\partial \dot{x}_i} = \mathbf{C}\dot{x} + \mathbf{C}_1 x + \mathbf{f}_3 \ , \tag{A.71}$$

the equation of motion becomes

$$\mathbf{M}\ddot{x} + \frac{1}{2}\left(\mathbf{M}_1 - \mathbf{M}_1^T\right)\dot{x} + \mathbf{C}\dot{x} + (\mathbf{K} - \mathbf{M}_g + \mathbf{C}_1)x = -\dot{\mathbf{f}}_1 + \mathbf{f}_2 - \mathbf{f}_3 - \mathbf{f}_0 + \mathbf{Q} \ . \tag{A.72}$$

Matrix $\mathbf{M}_1$ is normally skew-symmetric. However, even if it is not, it may be written as the sum of a symmetrical and a skew-symmetrical part

$$\mathbf{M}_1 = \mathbf{M}_{1symm} + \mathbf{M}_{1skew} \ . \tag{A.73}$$

By introducing this form into Eq. (A.72), the term

$$\mathbf{M}_1 - \mathbf{M}_1^T$$

becomes

$$\mathbf{M}_{1symm} + \mathbf{M}_{1skew} - \mathbf{M}_{1symm} + \mathbf{M}_{1skew} = 2\mathbf{M}_{1skew} \ .$$

Only the skew-symmetric part of $\mathbf{M}_1$ is included in the equation of motion. $\mathbf{C}_1$ is usually skew-symmetrical.

Writing $\mathbf{M}_{1skew}$ as $\mathbf{G}$ and $\mathbf{C}_1$ (or at least its skew-symmetric part; if a symmetric part existed, it could be included into matrix $\mathbf{K}$) as $\mathbf{H}$, and including vectors $\mathbf{f}_0$, $\mathbf{f}_1$, $\mathbf{f}_2$ and $\mathbf{f}_3$ into forcing functions $\mathbf{Q}$, the equation of motion becomes

$$\mathbf{M}\ddot{x} + (\mathbf{C} + \mathbf{G})\dot{x} + (\mathbf{K} - \mathbf{M}_g + \mathbf{H})x = \mathbf{Q} \ . \tag{A.74}$$

The mass, stiffness, gyroscopic and circulatory matrices $\mathbf{M}$, $\mathbf{K}$, $\mathbf{G}$ and $\mathbf{H}$ have already been defined. The symmetric matrix $\mathbf{M}_g$ is often defined as a *geometric matrix*[2].

---

[2]Here the symbol $\mathbf{M}_g$ is used instead of the more common $\mathbf{K}_g$ to emphasize that it comes from the kinetic energy.

As already stated, a system in which T_1 is not present is said to be *natural*. Its equation of motion does not contain a gyroscopic matrix. In many cases T_0 also is absent and the kinetic energy is expressed by Eq. (A.61).

The linearized equation of motion of a nonlinear system can be written in two possible ways. The first is by writing the complete expression of the energies, performing the derivatives obtaining the complete equations of motion and then cancelling nonlinear terms.

The second is by reducing the expression of the energies to quadratic forms, developing their expressions in power series and then truncating them after the quadratic terms. The linearized equations of motion are then directly obtained.

Remark A.10 *These two approaches yield the same result, but the first is usually more computationally intensive. At any rate, a set of n second order equations are obtained: These are either linear or nonlinear depending on the system under study.*

To write the state equations, a number n of kinematic equations must be written

$$\dot{x}_i = v_i \qquad (i = 1,...., n). \tag{A.75}$$

If the state vector is defined in the usual way

$$\mathbf{z} = \left\{ \begin{array}{c} \mathbf{v} \\ \mathbf{x} \end{array} \right\},$$

this procedure is straightforward.

A.6   HAMILTON EQUATIONS AND PHASE SPACE

If the generalized momenta are used as auxiliary variables instead of the generalized velocities, the equations are written with reference to the *phase space* and *phase vector* instead of the state space and vector.

The generalized momenta are defined, starting from the Lagrangian $\mathcal{L}$, as

$$\mathbf{p} = \frac{\partial \mathcal{L}}{\partial \dot{x}_i}. \tag{A.76}$$

If the system is a natural linear system, this definition reduces to the usual one

$$\mathbf{p} = \mathbf{M}\dot{\mathbf{x}}. \tag{A.77}$$

By including the forces coming from the dissipation function in the generalized forces Q_i, the Lagrange equation simplifies as

$$\dot{p}_i = \frac{\partial \mathcal{L}}{\partial x_i} + Q_i. \tag{A.78}$$

A function $\mathcal{H}(\dot{x}_i, x_i, t)$, called the *Hamiltonian function*, is defined as

$$\mathcal{H} = \mathbf{p}^T \dot{\mathbf{x}} - \mathcal{L} \, . \tag{A.79}$$

Because $\mathcal{H}$ is a function of p_i, x_i and t ($\mathcal{H}(p_i, x_i, t)$), the differential $\delta\mathcal{H}$ is

$$\delta\mathcal{H} = \sum_{i=1}^{n} \left(\frac{\partial\mathcal{H}}{\partial p_i} \delta p_i + \frac{\partial\mathcal{H}}{\partial x_i} \delta x_i \right) \, . \tag{A.80}$$

On the other hand, Eq. (A.79) yields

$$
\begin{aligned}
\delta\mathcal{H} &= \sum_{i=1}^{n} \left(p_i \delta \dot{x}_i + \dot{x}_i \delta p_i - \frac{\partial\mathcal{L}}{\partial x_i} \delta x_i - \frac{\partial\mathcal{L}}{\partial \dot{x}_i} \delta \dot{x}_i \right) = \\
&= \sum_{i=1}^{n} \left(\dot{x}_i \delta p_i - \frac{\partial\mathcal{L}}{\partial x_i} \delta x_i \right) \, ,
\end{aligned} \tag{A.81}
$$

and then

$$\frac{\partial\mathcal{H}}{\partial p_i} = \dot{x}_i \, , \qquad \frac{\partial\mathcal{H}}{\partial x_i} = -\frac{\partial\mathcal{L}}{\partial x_i} \, . \tag{A.82}$$

The $2n$ phase space equations are then

$$
\begin{cases}
\dot{x}_i = \dfrac{\partial\mathcal{H}}{\partial p_i} \\[4mm]
\dot{p}_i = -\dfrac{\partial\mathcal{H}}{\partial x_i} + Q_i \, .
\end{cases} \tag{A.83}
$$

A.7   LAGRANGE EQUATIONS IN TERMS OF PSEUDO COORDINATES

While in vehicle dynamics Hamilton equations are seldom used, the state equations are often written with reference to generalized velocities that are not simply the derivatives of the generalized coordinates. In particular, it is often expedient to use suitable combinations of the derivatives of the coordinates $v_i = \dot{x}_i$ as generalized velocities

$$\{w_i\} = \mathbf{A}^T \{\dot{x}_i\} \, , \tag{A.84}$$

where the coefficients of the linear combinations included into matrix $\mathbf{A}^T$ may be constant, but in general are functions of the generalized coordinates.

Equation (A.84) may be inverted, obtaining

$$\{\dot{x}_i\} = \mathbf{B} \{w_i\} \, , \tag{A.85}$$

where

$$\mathbf{B} = \mathbf{A}^{-T} \tag{A.86}$$

and symbol $\mathbf{A}^{-T}$ indicates the inverse of the transpose of matrix $\mathbf{A}$.

In some cases matrix $\mathbf{A}^{T}$ is a rotation matrix whose inverse coincides with its transpose. In such cases

$$\mathbf{B} = \mathbf{A}^{-T} = \mathbf{A} \ .$$

However, this generally does not occur and

$$\mathbf{B} \neq \mathbf{A} \ .$$

While v_i are the derivatives of the coordinates x_i, it is usually not possible to express w_i as the derivatives of suitable coordinates. Eq. (A.84) can be written in the infinitesimal displacements dx_i

$$\{d\theta_i\} = \mathbf{A}^{T} \{dx_i\} \ , \tag{A.87}$$

obtaining a set of infinitesimal displacements $d\theta_i$, corresponding to velocities w_i. Equations (A.87) can be integrated, yielding displacements θ_i corresponding to the velocities w_i, only if

$$\frac{\partial a_{js}}{\partial x_k} = \frac{\partial a_{ks}}{\partial x_j} \ .$$

Otherwise equations (A.87) cannot be integrated and velocities w_i cannot be considered as the derivatives of true coordinates. In such cases they are said to be the derivatives of pseudo-coordinates.

As a first consequence of the non-existence of coordinates corresponding to velocities w_i, Lagrange equation (A.59) cannot be written directly using velocities w_i (which cannot be considered as derivatives of the new coordinates), but must be modified to allow the use of velocities and coordinates that are not direct derivatives of each other.

The use of pseudo-coordinates is fairly common, particularly in vehicle dynamics. If, for instance, the generalized velocities in a reference frame following the body in its motion are used in the dynamics of a rigid body, while the coordinates x_i are the displacements in an inertial frame, matrix $\mathbf{A}^{T}$ is simply the rotation matrix allowing passage from one reference frame to the other. Matrix $\mathbf{B}$ then coincides with $\mathbf{A}$, but neither is symmetrical. The velocities in the body-fixed frame cannot therefore be considered as the derivatives of the displacements in that frame.

Remark A.11 *The body-fixed frame rotates continuously so that it is not possible to integrate the velocities along the body-fixed axes to obtain the displacements along the same axes. This fact notwithstanding, it is possible to use the components of the velocity along the body-fixed axes to write the equations of motion.*

The kinetic energy can be written in general in the form

$$\mathcal{T} = \mathcal{T}(w_i, x_i, t) \ .$$

The derivatives $\frac{\partial T}{\partial \dot{x}_i}$ included into the equations of motion are

$$\frac{\partial T}{\partial \dot{x}_k} = \sum_{i=1}^{n} \frac{\partial T}{\partial w_i} \frac{\partial w_i}{\partial \dot{x}_k} , \tag{A.88}$$

i.e., in matrix form,

$$\left\{ \frac{\partial T}{\partial \dot{x}} \right\} = \mathbf{A} \left\{ \frac{\partial T}{\partial w} \right\} , \tag{A.89}$$

where

$$\left\{ \frac{\partial T}{\partial \dot{x}} \right\} = \left[\begin{array}{ccc} \frac{\partial T}{\partial \dot{x}_1} & \frac{\partial T}{\partial \dot{x}_2} & \cdots \end{array} \right]^T ,$$

$$\left\{ \frac{\partial T}{\partial w} \right\} = \left[\begin{array}{ccc} \frac{\partial T}{\partial w_1} & \frac{\partial T}{\partial w_2} & \cdots \end{array} \right]^T .$$

By differentiating with respect to time, it follows that

$$\frac{\partial}{\partial t} \left(\left\{ \frac{\partial T}{\partial \dot{x}} \right\} \right) = \mathbf{A} \frac{\partial}{\partial t} \left(\left\{ \frac{\partial T}{\partial w} \right\} \right) + \dot{\mathbf{A}} \left\{ \frac{\partial T}{\partial w} \right\} , \tag{A.90}$$

The generic element $\dot{a}_{jk}$ of matrix $\dot{\mathbf{A}}$ is

$$\dot{a}_{jk} = \sum_{i=1}^{n} \frac{\partial a_{jk}}{\partial x_i} \dot{x}_i = \dot{\mathbf{x}}^T \left\{ \frac{\partial a_{jk}}{\partial x} \right\} , \tag{A.91}$$

and then

$$\dot{a}_{jk} = \mathbf{w}^T \mathbf{B}^T \left\{ \frac{\partial a_{jk}}{\partial x} \right\} . \tag{A.92}$$

The various $\dot{a}_{jk}$ so computed can be written in matrix form

$$\dot{\mathbf{A}} = \left[\mathbf{w}^T \mathbf{B}^T \left\{ \frac{\partial a_{jk}}{\partial x} \right\} \right] . \tag{A.93}$$

The computation of the derivatives of the generalized coordinates $\left\{ \frac{\partial T}{\partial x} \right\}$ is usually less straightforward. The generic derivative $\frac{\partial T}{\partial x_k}$ is

$$\frac{\partial T^*}{\partial x_k} = \frac{\partial T}{\partial x_k} + \sum_{i=1}^{n} \frac{\partial T}{\partial w_i} \frac{\partial w_i}{\partial x_k} = \frac{\partial T}{\partial x_k} + \sum_{i=1}^{n} \frac{\partial T}{\partial w_i} \sum_{j=1}^{n} \frac{\partial a_{ij}}{\partial x_k} \dot{x}_j , \tag{A.94}$$

where T^* is the kinetic energy expressed as a function of the generalized coordinates and their derivatives (the expression to be introduced into the Lagrange equation in its usual form), while T is expressed as a function of the generalized coordinates and of the velocities in the body-fixed frame. Equation (A.94) can be written as

$$\frac{\partial T^*}{\partial x_k} = \frac{\partial T}{\partial x_k} + \mathbf{w}^T \mathbf{B}^T \frac{\partial \mathbf{A}}{\partial x_k} \left\{ \frac{\partial T}{\partial w} \right\} , \tag{A.95}$$

where product $\mathbf{w}^T \mathbf{B}^T \frac{\partial \mathbf{A}}{\partial x_k}$ yields a row matrix with n elements, which multiplied by the column matrix $\left\{ \frac{\partial T}{\partial w} \right\}$ yields the required number.

By combining these row matrices, a square matrix is obtained

$$\left[\mathbf{w}^T \mathbf{B}^T \frac{\partial \mathbf{A}}{\partial x_k} \right] , \tag{A.96}$$

and then the column containing the derivatives with respect to the generalized coordinates is

$$\left\{ \frac{\partial T^*}{\partial x} \right\} = \left\{ \frac{\partial T}{\partial x} \right\} + \left[\mathbf{w}^T \mathbf{B}^T \frac{\partial \mathbf{A}}{\partial x} \right] \left\{ \frac{\partial T}{\partial w} \right\} . \tag{A.97}$$

By definition, the potential energy does not depend on the generalized velocities. Thus the term $\frac{\partial \mathcal{U}}{\partial x_i}$ is not influenced by the way the generalized velocities are written. Finally, the derivatives of the dissipation function are

$$\left\{ \frac{\partial \mathcal{F}}{\partial \dot{x}} \right\} = \mathbf{A} \left\{ \frac{\partial \mathcal{F}}{\partial w} \right\} \tag{A.98}$$

The equation of motion (A.59) is then

$$\mathbf{A} \frac{\partial}{\partial t} \left(\left\{ \frac{\partial T}{\partial w} \right\} \right) + \mathbf{\Gamma} \left\{ \frac{\partial T}{\partial w} \right\} - \left\{ \frac{\partial T}{\partial x} \right\} + \left\{ \frac{\partial \mathcal{U}}{\partial x} \right\} + \mathbf{A} \left\{ \frac{\partial \mathcal{F}}{\partial w} \right\} = \mathbf{Q} , \tag{A.99}$$

where

$$\mathbf{\Gamma} = \left[\mathbf{w}^T \mathbf{B}^T \left\{ \frac{\partial a_{jk}}{\partial x} \right\} \right] - \left[\mathbf{w}^T \mathbf{B}^T \frac{\partial \mathbf{A}}{\partial x_k} \right] \tag{A.100}$$

and $\mathbf{Q}$ is a vector containing the n generalized forces Q_i.

By premultiplying all terms by matrix $\mathbf{B}^T = \mathbf{A}^{-1}$ and attaching the kinematic equations to the dynamic equations, the final form of the state space equations is obtained

$$\begin{cases} \dfrac{\partial}{\partial t} \left(\left\{ \dfrac{\partial T}{\partial w} \right\} \right) + \mathbf{B}^T \mathbf{\Gamma} \left\{ \dfrac{\partial T}{\partial w} \right\} - \mathbf{B}^T \left\{ \dfrac{\partial T}{\partial x} \right\} + \mathbf{B}^T \left\{ \dfrac{\partial \mathcal{U}}{\partial x} \right\} + \left\{ \dfrac{\partial \mathcal{F}}{\partial w} \right\} = \mathbf{B}^T \mathbf{Q} \\ \\ \{\dot{q}_i\} = \mathbf{B} \{w_i\} . \end{cases}$$

$$\tag{A.101}$$

A.8   MOTION OF A RIGID BODY

A.8.1   Generalized coordinates

Consider a rigid body free in tri-dimensional space. Define an inertial reference frame $OXYZ$ and a frame $Gxyz$ fixed to the body and centred in its center of mass. The position of the rigid body is defined once the position of frame $Gxyz$ is defined with respect to $OXYZ$, that is, once the transformation leading $OXYZ$ to coincide with $Gxyz$ is defined. It is well known that the motion of the second frame can be considered as the sum of a displacement plus a rotation. The parameters to be defined are therefore 6: 3 components of the displacement, two of the components of the unit vector defining the rotation axis (the third component need not be defined and may be computed from the condition that the unit vector has unit length) and the rotation angle. A rigid body thus has six degrees of freedom in tri-dimensional space.

There is no problem in defining the generalized coordinates for the translational degrees of freedom, because the coordinates of the center of mass G in any inertial reference frame (in particular, in frame $OXYZ$) are usually the simplest, and the most obvious, choice. For the other generalized coordinates the choice is much more complicated. It is possible to resort, for instance, to two coordinates of a second point and to one of the coordinates of a third point (not on a straight line through the other two), but this choice is far from being the most expedient.

An obvious way to define the rotation of frame $Gxyz$ with respect to $OXYZ$ is to directly express the rotation matrix linking the two reference frames. It is a square matrix of size 3×3 (in tri-dimensional space) and thus has 9 elements. Three of these are independent, while the other 6 may be obtained from the first 3 using suitable equations.

Alternatively, the position of the body-fixed frame can be defined with a sequence of three rotations about the axes. Because rotations are not vectors, the order in which they are performed must be specified.

Start rotating, for instance, the inertial frame about the X-axis. The second rotation may be performed about axes Y or Z (obviously in the position they take after the first rotation), but not about X-axis, because in the latter case the two rotations would simply add to each other and would amount to a single rotation. Assume, for instance, that the frame is rotated about the Y-axis. The third rotation may occur about either the X-axis or the Z-axis (in the new position, taken after the second rotation), but not about the Y-axis.

The possible rotation sequences are 12, but may be subdivided into two types: Those like $X \rightarrow Y \rightarrow X$ or $X \rightarrow Z \rightarrow X$, where the third rotation occurs about the same axis as the first, and those like $X \rightarrow Y \rightarrow Z$ or $X \rightarrow Z \rightarrow Y$, where the third rotation is performed about a different axis.

In the first cases the angles are said to be *Euler angles*, because they are of the same type as the angles Euler proposed to study the motion of gyroscopes (precession ϕ about the Z-axis, nutation θ about the X-axis and rotation ψ, again about the Z-axis). In the second case they are said to be *Tait-Bryan angles*[3].

The possible rotation sequences are reported in the following table

First	X				Y				Z			
Second	Y		Z		X		Z		X		Y	
Third	X	Z	X	Y	Y	Z	Y	X	Z	Y	Z	X
Type	E	TB	E	TB	E	TB	E	TB	E	TB	E	TB

In the case of vehicle dynamics Euler angles have the drawback of being indeterminate when plane xy of the rigid body is parallel to THE XY-plane of the inertial frame. They also yield indications that are less intuitively clear.

In the dynamics of vehicles the most common approach is to use Tait-Bryan angles of the type $Z \to Y \to X$ so defined (Fig. A.5):

- Rotate frame XYZ (whose XY plane is parallel to the ground) about the Z-axis until axis X coincides with the projection of the x-axis on plane XY (Fig. A.5a). Such a position of the X-axis can be indicated as x^*; the rotation angle between axes X and x^* is the yaw angle ψ. The rotation matrix allowing passage from the x^*y^*Z frame, which will be defined as the *intermediate frame*, to the inertial frame XYZ is

$$\mathbf{R}_1 = \begin{bmatrix} \cos(\psi) & -\sin(\psi) & 0 \\ \sin(\psi) & \cos(\psi) & 0 \\ 0 & 0 & 1 \end{bmatrix}. \tag{A.102}$$

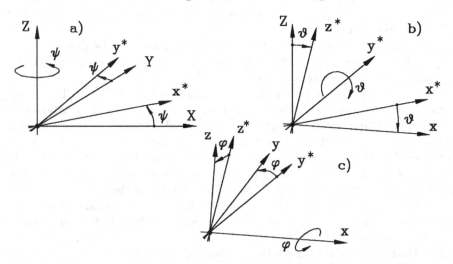

FIGURE A.5. Definition of angles: yaw ψ (a), pitch θ (b) and roll ϕ (c).

---

[3]Sometimes all sets of three ordered angles are said to be Euler angles. With this wider definition Tait-Brian angles are also considered as Euler angles.

- The second rotation is the pitch rotation θ about the y^*-axis, so that axis x^* reaches the position of the x-axis (Fig. A.5b). The rotation matrix is

$$\mathbf{R}_2 = \begin{bmatrix} \cos(\theta) & 0 & \sin(\theta) \\ 0 & 1 & 0 \\ -\sin(\theta) & 0 & \cos(\theta) \end{bmatrix} . \tag{A.103}$$

- The third rotation is the roll rotation ϕ about the x-axis, so that axes y^* and z^* coincide with axes y and z (Fig. A.5c). The rotation matrix is

$$\mathbf{R}_3 = \begin{bmatrix} 1 & 0 & 0 \\ 0 & \cos(\phi) & -\sin(\phi) \\ 0 & \sin(\phi) & \cos(\phi) \end{bmatrix} . \tag{A.104}$$

The rotation matrix allowing any vector in the body-fixed frame xyz to be rotated to the inertial frame XYZ is clearly the product of the three matrices

$$\mathbf{R} = \mathbf{R}_1 \mathbf{R}_2 \mathbf{R}_3 . \tag{A.105}$$

Deriving the product of the rotation matrices, it follows that

$$\mathbf{R} = \begin{bmatrix} c(\psi)c(\theta) & c(\psi)s(\theta)s(\phi) - s(\psi)c(\phi) & c(\psi)s(\theta)c(\phi) + s(\psi)s(\phi) \\ s(\psi)c(\theta) & s(\psi)s(\theta)s(\phi) + c(\psi)c(\phi) & s(\psi)s(\theta)c(\phi) - c(\psi)s(\phi) \\ -s(\theta) & c(\theta)s(\phi) & c(\theta)c(\phi) \end{bmatrix} , \tag{A.106}$$

where symbols cos and sin have been replaced by c and s.

Roll and pitch angles are sometimes small. In this case it is expedient to keep the last two rotations separate from the first ones, which cannot usually be linearized.

The product of the rotation matrices related to the last two rotations is

$$\mathbf{R}_2\mathbf{R}_3. = \begin{bmatrix} \cos(\theta) & \sin(\theta)\sin(\phi) & \sin(\theta)\cos(\phi) \\ 0 & \cos(\phi) & -\sin(\phi) \\ -\sin(\theta) & \cos(\theta)\sin(\phi) & \cos(\theta)\cos(\phi) \end{bmatrix} , \tag{A.107}$$

which becomes, in the case of small angles

$$\mathbf{R}_2\mathbf{R}_3 \approx \begin{bmatrix} 1 & 0 & \theta \\ 0 & 1 & -\phi \\ -\theta & \phi & 1 \end{bmatrix} . \tag{A.108}$$

The angular velocities $\dot{\psi}$, $\dot{\theta}$ and $\dot{\phi}$ are not applied along the x, y and z axes, and thus are not the components Ω_x, Ω_y and Ω_z of the angular velocity in the body-fixed reference frame[4]. Their directions are those of axes Z, y^* and x, and then the angular velocity in the body-fixed frame is

---

[4] Symbols p, q and r are often used for the components of the angular velocity in the body-fixed frame.

$$\left\{ \begin{array}{c} \Omega_x \\ \Omega_y \\ \Omega_z \end{array} \right\} = \dot{\phi} \mathbf{e}_x + \dot{\theta} \mathbf{R}_3^T \mathbf{e}_y + \dot{\psi} \left[\mathbf{R}_2 \mathbf{R}_3 \right]^T \mathbf{e}_z , \qquad (A.109)$$

where the unit vectors are obviously

$$\mathbf{e}_x = \left\{ \begin{array}{c} 1 \\ 0 \\ 0 \end{array} \right\} , \quad \mathbf{e}_y = \left\{ \begin{array}{c} 0 \\ 1 \\ 0 \end{array} \right\} , \quad \mathbf{e}_z = \left\{ \begin{array}{c} 0 \\ 0 \\ 1 \end{array} \right\} . \qquad (A.110)$$

By deriving the products, it follows that

$$\left\{ \begin{array}{l} \Omega_x = \dot{\phi} - \dot{\psi} \sin(\theta) \\ \Omega_y = \dot{\theta} \cos(\phi) + \dot{\psi} \sin(\phi) \cos(\theta) \\ \Omega_z = \dot{\psi} \cos(\theta) \cos(\phi) - \dot{\theta} \sin(\phi) , \end{array} \right. \qquad (A.111)$$

or, in matrix form

$$\left\{ \begin{array}{c} \Omega_x \\ \Omega_y \\ \Omega_z \end{array} \right\} = \left[\begin{array}{ccc} 1 & 0 & -\sin(\theta) \\ 0 & \cos(\phi) & \sin(\phi)\cos(\theta) \\ 0 & -\sin(\phi) & \cos(\phi)\cos(\theta) \end{array} \right] \left\{ \begin{array}{c} \dot{\phi} \\ \dot{\theta} \\ \dot{\psi} \end{array} \right\} . \qquad (A.112)$$

If the pitch and roll angles are small enough to linearize the relevant trigonometric functions, the components of the angular velocity may be approximated as

$$\left\{ \begin{array}{l} \Omega_x = \dot{\phi} - \theta\dot{\psi} \\ \Omega_y = \dot{\theta} + \phi\dot{\psi} \\ \Omega_z = \dot{\psi} - \phi\dot{\theta} . \end{array} \right. \qquad (A.113)$$

Other alternatives sometimes used in vehicle modelling, such as quaternions, will not be dealt with here.

A.8.2   Equations of motion - Lagrangian approach

Consider a rigid body in tri-dimensional space and chose as generalized coordinates the displacements X, Y and Z of its center of mass and angles ψ, θ and ϕ. Assuming that the body axes xyz are principal axes of inertia, the kinetic energy of the rigid body is

$$\begin{aligned} T = &\tfrac{1}{2}m \left(\dot{X}^2 + \dot{Y}^2 + \dot{Z}^2 \right) + \tfrac{1}{2}J_x \left[\dot{\phi} - \dot{\psi} \sin(\theta) \right]^2 + \\ &+ \tfrac{1}{2}J_y \left[\dot{\theta} \cos(\phi) + \dot{\psi} \sin(\phi) \cos(\theta) \right]^2 + \\ &+ \tfrac{1}{2}J_z \left[\dot{\psi} \cos(\theta) \cos(\phi) - \dot{\theta} \sin(\phi) \right]^2 . \end{aligned} \qquad (A.114)$$

Introducing the kinetic energy into the Lagrange equations

$$\frac{d}{dt} \left(\frac{\partial T}{\partial \dot{q}_i} \right) - \frac{\partial T}{\partial q_i} = Q_i ,$$

and performing the relevant derivatives, the six equations of motion are directly obtained. The three equations for translational motion are

$$
\begin{cases}
m\ddot{X} = Q_X \\
m\ddot{Y} = Q_Y \\
m\ddot{Z} = Q_Z \ .
\end{cases}
\tag{A.115}
$$

The equations for rotational motion are much more complicated

$$
\begin{aligned}
&\ddot{\psi}\left[J_x \sin^2(\theta) + J_y \sin^2(\phi)\cos^2(\theta) + J_z \cos^2(\phi)\cos^2(\theta)\right] + \\
&-\dot{\phi} J_x \sin(\theta) + \ddot{\theta}\,(J_y - J_z)\sin(\phi)\cos(\phi)\cos(\theta) + \\
&+\dot{\phi}\dot{\theta}\cos(\theta)\left\{\left[1 - 2\sin^2(\phi)\right](J_y - J_z) - J_x\right\} + \\
&+2\dot{\phi}\dot{\psi}\,(J_y - J_z)\cos(\phi)\cos^2(\theta)\sin(\phi) + \\
&+2\dot{\theta}\dot{\psi}\sin(\theta)\cos(\theta)\left[J_x - \sin^2(\phi)J_y - \cos^2(\phi)J_z\right] + \\
&+\dot{\theta}^2\,(-J_y + J_z)\sin(\phi)\cos(\phi)\sin(\theta) = Q_\psi,
\end{aligned}
$$

$$
\begin{aligned}
&\ddot{\psi}\,(J_y - J_z)\sin(\phi)\cos(\theta)\cos(\phi) + \ddot{\theta}\left[J_y \cos^2(\phi) + J_z \sin^2(\phi)\right] + \\
&+2\dot{\phi}\dot{\theta}\,(J_z - J_y)\sin(\phi)\cos(\phi) + \dot{\phi}\dot{\psi}\,(J_y - J_z)\cos(\theta)\left[1 - 2\sin^2(\phi)\right] + \\
&+\dot{\psi}\dot{\phi}J_x \cos(\theta) - \dot{\psi}^2 \sin(\theta)\cos(\theta)\left[J_x - J_y \sin^2(\phi) - J_z \cos^2(\phi)\right] = Q_\theta,
\end{aligned}
\tag{A.116}
$$

$$
\begin{aligned}
&J_x \ddot{\phi} - \sin(\theta)J_x \ddot{\psi} - \dot{\theta}\dot{\psi}J_z \sin^2(\phi)\cos(\theta) + \\
&-\dot{\psi}\dot{\theta}\cos(\theta)\left\{J_x + J_y\left[1 - 2\sin^2(\phi)\right] - J_z \cos^2(\phi)\right\} + \\
&+\dot{\theta}^2\,(J_y - J_z)\sin(\phi)\cos(\phi) - \dot{\psi}^2\,(J_y - J_z)\cos(\phi)\cos^2(\theta)\sin(\phi) = Q_\phi \ .
\end{aligned}
$$

Angle ψ does not appear explicitly in the equations of motion. If the roll and pitch angles are small all trigonometric functions can be linearized. If the angular velocities are also small, the equations of motion for rotations reduce to

$$
\begin{cases}
J_z \ddot{\psi} = Q_\psi \\
J_y \ddot{\theta} = Q_\theta \\
J_x \ddot{\phi} = Q_\phi \ .
\end{cases}
\tag{A.117}
$$

In this case, the kinetic energy may be directly simplified, by developing the trigonometric functions in Taylor series and neglecting all terms containing products of three or more small quantities. For instance, the term

$$
\left[\dot{\phi} - \dot{\psi}\sin(\theta)\right]^2
$$

reduces to

$$
\left[\dot{\phi} - \dot{\psi}\theta + \dot{\psi}\theta^3/6 + ...\right]^2
$$

and then to $\dot{\phi}^2$, because all other terms contain products of at least three small quantities. The kinetic energy then reduces to

$$
T \approx \frac{1}{2}m\left(\dot{X}^2 + \dot{Y}^2 + \dot{Z}^2\right) + \frac{1}{2}\left(J_x \dot{\phi}^2 + J_y \dot{\theta}^2 + J_z \dot{\psi}^2\right) \ .
\tag{A.118}
$$

Remark A.12 *This approach is simple only if the roll and pitch angles are small. If they are not, the equations of motion obtained in this way in terms of angular velocities $\dot{\phi}$, $\dot{\theta}$ and $\dot{\psi}$ are quite complicated and another approach is more expedient.*

A.8.3 Equations of motion using pseudo-coodinates

Because the forces and moments applied to the rigid body are often written with reference to the body-fixed frame, the equations of motion are best written with reference to the same frame. The kinetic energy can then be written in terms of the components v_x, v_y and v_z (often referred to as u, v and w) of the velocity and Ω_x, Ω_x e Ω_x (often referred to as p, q and r) of the angular velocity.

If the body fixed frame is a principal frame of inertia, the expression of the kinetic energy is

$$T = \frac{1}{2} m \left(v_x^2 + v_y^2 + v_z{}^2 \right) + \frac{1}{2} \left(J_x \Omega_x^2 + J_y \Omega_y^2 + J_z \Omega_z^2 \right) \ .$$

The components of the velocity and the angular velocity in the body fixed frame are not the derivatives of coordinates, but are linked to the coordinates by the six kinematic equations

$$\left\{ \begin{array}{c} v_x \\ v_y \\ v_z \end{array} \right\} = \mathbf{R}^T \left\{ \begin{array}{c} \dot{X} \\ \dot{Y} \\ \dot{Z} \end{array} \right\} , \tag{A.119}$$

$$\left\{ \begin{array}{c} \Omega_x \\ \Omega_y \\ \Omega_z \end{array} \right\} = \left[\begin{array}{ccc} 1 & 0 & -\sin(\theta) \\ 0 & \cos(\phi) & \sin(\phi)\cos(\theta) \\ 0 & -\sin(\phi) & \cos(\theta)\cos(\phi) \end{array} \right] \left\{ \begin{array}{c} \dot{\phi} \\ \dot{\theta} \\ \dot{\psi} \end{array} \right\} , \tag{A.120}$$

that is, in more compact form,

$$\mathbf{w} = \mathbf{A}^T \dot{\mathbf{q}} , \tag{A.121}$$

where the vectors of the generalized velocities and of the derivatives of the generalized coordinates are

$$\mathbf{w} = \left[\begin{array}{cccccc} v_x & v_y & v_z & \Omega_x & \Omega_y & \Omega_z \end{array} \right]^T , \tag{A.122}$$

$$\dot{\mathbf{q}} = \left[\begin{array}{cccccc} \dot{X} & \dot{Y} & \dot{Z} & \dot{\phi} & \dot{\theta} & \dot{\psi} \end{array} \right]^T \tag{A.123}$$

and matrix $\mathbf{A}$ is

$$\mathbf{A} = \left[\begin{array}{cc} \mathbf{R} & \mathbf{0} \\ \mathbf{0} & \left[\begin{array}{ccc} 1 & 0 & -\sin(\theta) \\ 0 & \cos(\phi) & \sin(\phi)\cos(\theta) \\ 0 & -\sin(\phi) & \cos(\theta)\cos(\phi) \end{array} \right]^T \end{array} \right] . \tag{A.124}$$

Note that the second submatrix is not a rotation matrix (the first submatrix is) and then

$$\mathbf{A}^{-1} \neq \mathbf{A}^T \; ; \qquad \mathbf{B} \neq \mathbf{A} \; . \tag{A.125}$$

The inverse transformation is Eq. (A.85)

$$\dot{\mathbf{q}} = \mathbf{B}\mathbf{w} \; ,$$

where $\mathbf{B} = \mathbf{A}^{-\mathbf{T}}$.

None of the velocities included in vector $\mathbf{w}$ can be integrated to obtain a set of generalized coordinates, and must all be considered as derivatives of pseudo-coordinates.

The state space equation, made up of the six dynamic and the six kinematic equations, is then equation (A.101), simplified because in the present case neither the potential energy nor the dissipation function are present

$$\begin{cases} \dfrac{\partial}{\partial t}\left(\left\{\dfrac{\partial \mathcal{T}}{\partial w}\right\}\right) + \mathbf{B}^T\mathbf{\Gamma}\left\{\dfrac{\partial \mathcal{T}}{\partial w}\right\} - \mathbf{B}^T\left\{\dfrac{\partial \mathcal{T}}{\partial q}\right\} = \mathbf{B}^T\mathbf{Q} \\[2mm] \{\dot{q}_i\} = \mathbf{B}\{w_i\} \; . \end{cases} \tag{A.126}$$

Here $\mathbf{B}^T\mathbf{Q}$ is simply a column matrix containing the three components of the force and the three components of the moment applied to the body along the body-fixed axes x, y, z.

The most difficult part of the computation is writing matrix $\mathbf{B^T\Gamma}$. Performing rather difficult computations it follows that

$$\mathbf{B^T\Gamma} = \begin{bmatrix} \widetilde{\mathbf{\Omega}} & \mathbf{0} \\ \widetilde{\mathbf{V}} & \widetilde{\mathbf{\Omega}} \end{bmatrix} \; . \tag{A.127}$$

where $\widetilde{\mathbf{\Omega}}$ and $\widetilde{\mathbf{V}}$ are skew-symmetric matrices containing the components of the angular and linear velocities

$$\widetilde{\mathbf{\Omega}} = \begin{bmatrix} 0 & -\Omega_z & \Omega_y \\ \Omega_z & 0 & -\Omega_x \\ -\Omega_y & \Omega_x & 0 \end{bmatrix} \; , \quad \widetilde{\mathbf{V}} = \begin{bmatrix} 0 & -v_z & v_y \\ v_z & 0 & -v_x \\ -v_y & v_x & 0 \end{bmatrix} \; . \tag{A.128}$$

If the body-fixed axes are principal axes of inertia, the dynamic equations are simply

$$\begin{cases} m\dot{v}_x = m\Omega_z v_y - m\Omega_y v_z + F_x \\ m\dot{v}_y = m\Omega_x v_z - m\Omega_z v_x + F_y \\ m\dot{v}_z = m\Omega_y v_x - m\Omega_x v_y + F_z \\ J_x\dot{\Omega}_x = \Omega_y\Omega_z\left(J_y - J_z\right) + M_x \\ J_y\dot{\Omega}_y = \Omega_x\Omega_z\left(J_z - J_x\right) + M_y \\ J_z\dot{\Omega}_z = \Omega_x\Omega_y\left(J_x - J_y\right) + M_x \end{cases} \tag{A.129}$$

Remark A.13 *The equations so obtained are much simpler than equations (A.116). The last three equations are nothing other than Euler equations.*

Appendix B
DYNAMICS OF MOTOR CYCLES

When studying the handling behavior of a two-wheeled vehicle, rolling motions and, to a lesser extent, gyroscopic moments must not be neglected. A linearized model similar in many respects to that seen in Part IV for single-track vehicles may be built.

Linearization obviously requires that the roll angle be small, severely limiting the applicability of such a model to the study of stability on straight roads and operating conditions where the lateral acceleration is small compared to gravitational acceleration.

The mass of the driver, who controls the vehicle not only by acting on the steering but also displacing his body, can be a substantial fraction of the total mass. Moreover, a two-wheeled vehicle is intrinsically unstable. The driver thus has to perform as a stabilizer for the capsize mode.

Finally, the body of the driver, acting as an aerodynamic brake or control surface, contributes in a substantial way to aerodynamic forces. To model a two-wheeled vehicle without modelling the driver is merely a first approximation approach, useful for conditions in which only low performance is required.

In such cases the vehicle can be modelled as a rigid body that also includes the driver. A sketch of the vehicle model is shown in Fig. B.1. The reference frame Hxyz is fixed to such a rigid body, with origin at point H defined in the same way as for the model of the vehicle on elastic suspensions. Its position is defined by the yaw and roll angles ψ and ϕ; the first is defined as for a vehicle with four wheels. The roll angle is defined as the angle between the z axis and the perpendicular to the ground. The roll axis is assumed to pass through the centers of the contact areas of the tires, a rough approximation only because motor cycle

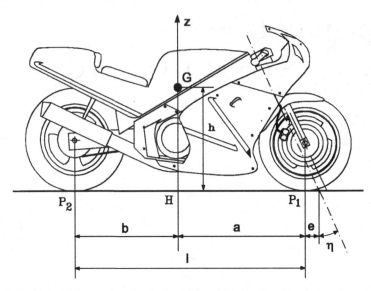

FIGURE B.1. Model for a two wheeled vehicle; reference frames and main geometrical definitions

tires usually have a considerably rounded transversal profile. In locked control dynamics the steering angle δ is an input, while in free control dynamics it is one of the variables of motion.

The main difficulty is linked to the high values, even larger than 45°, that the roll angle may take: In these conditions, assumption of small angles does not hold. The kinematic of the steering system is further complicated by the large values that the caster angle (η in Fig. B.1) may take. The caster offset, shown in the figure with symbol e, may be relatively large and is an important parameter in the study of the behavior of motor cycles.

Since angle η may be not small, the steering angle δ_s measured on the ground does not coincide with the steering angle δ at the handlebar. If the roll angle is small, it follows that

$$\delta_s \approx \delta \cos(\eta) \ . \tag{B.1}$$

The trajectory curvature gain in kinematic conditions is then

$$\left(\frac{1}{R\delta}\right)_c \approx \frac{1}{l}\delta \cos(\eta) \ . \tag{B.2}$$

It follows that the more the steering axis is inclined with respect to the vertical, the lower the trajectory curvature gain and the less manoeuvrable the vehicle.

B.1   BASIC DEFINITIONS

The generalized coordinates are the coordinates X and Y of point H in the inertial frame XYZ and the yaw ψ and roll ϕ angles. The steering angle δ may be considered as a variable of the motion (free controls) or an input (locked controls).

The components of the velocity in the body-fixed frame u and v are linked to the derivatives of the generalized coordinates $\dot{X}$, $\dot{Y}$ in the inertial frame by the usual relationship

$$\left\{ \begin{array}{c} u \\ v \\ 0 \end{array} \right\} = \mathbf{R}_1^T \left\{ \begin{array}{c} \dot{X} \\ \dot{Y} \\ 0 \end{array} \right\} = \left[\begin{array}{ccc} \cos(\psi) & \sin(\psi) & 0 \\ -\sin(\psi) & \cos(\psi) & 0 \\ 0 & 0 & 1 \end{array} \right] \left\{ \begin{array}{c} \dot{X} \\ \dot{Y} \\ 0 \end{array} \right\}, \tag{B.3}$$

where $\mathbf{R}_1$ is the yaw rotation matrix.

The angular velocities Ω_x, Ω_y and Ω_z about the body axes are linked to the roll velocity $p = \dot{\phi}$ and yaw velocity $r = \dot{\psi}$ by the relationship

$$\mathbf{\Omega} = \left\{ \begin{array}{c} \Omega_x \\ \Omega_y \\ \Omega_z \end{array} \right\} = \left\{ \begin{array}{c} \dot{\phi} \\ 0 \\ 0 \end{array} \right\} + \mathbf{R}_2^T \left\{ \begin{array}{c} 0 \\ 0 \\ \dot{\psi} \end{array} \right\}, \tag{B.4}$$

where $\mathbf{R}_2$ is the roll rotation matrix

$$\mathbf{R}_2 = \left[\begin{array}{ccc} 1 & 0 & 0 \\ 0 & \cos(\phi) & -\sin(\phi) \\ 0 & \sin(\phi) & \cos(\phi) \end{array} \right].$$

Performing the relevant computations, the relationship linking the angular velocities about the axes of the body-fixed frame and the derivatives of the generalized coordinates is

$$\mathbf{\Omega} = \left\{ \begin{array}{c} \Omega_x \\ \Omega_y \\ \Omega_z \end{array} \right\} = \mathbf{A}^T \left\{ \begin{array}{c} \dot{\phi} \\ \dot{\psi} \end{array} \right\} = \left\{ \begin{array}{c} \dot{\phi} \\ \dot{\psi}\sin(\phi) \\ \dot{\psi}\cos(\phi) \end{array} \right\}, \tag{B.5}$$

where

$$\mathbf{A}^T = \left[\begin{array}{cc} 1 & 0 \\ 0 & \sin(\phi) \\ 0 & \cos(\phi) \end{array} \right]. \tag{B.6}$$

The angular velocity of the steering system $\mathbf{\Omega}_1$ must be explicitly computed when studying motion in free controls conditions

$$\mathbf{\Omega}_1 = \left\{ \begin{array}{c} 0 \\ 0 \\ \dot{\delta} \end{array} \right\} + \mathbf{R}_3^T \mathbf{R}_\eta^T \left\{ \begin{array}{c} \dot{\phi} \\ 0 \\ 0 \end{array} \right\} + \mathbf{R}_3^T \mathbf{R}_\eta^T \mathbf{R}_2^T \left\{ \begin{array}{c} 0 \\ 0 \\ \dot{\psi} \end{array} \right\}, \tag{B.7}$$

i.e.

$$\mathbf{\Omega_1} = \left\{ \begin{array}{c} 0 \\ 0 \\ \dot\delta \end{array} \right\} + \mathbf{R}_3^T \mathbf{R}_\eta^T \mathbf{\Omega} \, , \tag{B.8}$$

where matrices $\mathbf{R}_3$ (steering rotation) and $\mathbf{R}_\eta^T$ (matrix defining the direction of the steering axis) are

$$\mathbf{R}_3 = \begin{bmatrix} \cos(\delta) & -\sin(\delta) & 0 \\ \sin(\delta) & \cos(\delta) & 0 \\ 0 & 0 & 1 \end{bmatrix}, \quad \mathbf{R}_\eta = \begin{bmatrix} \cos(\eta) & 0 & -\sin(\eta) \\ 0 & 1 & 0 \\ \sin(\eta) & 0 & \cos(\eta) \end{bmatrix}.$$

The final expression of the angular velocity of the steering system is

$$\mathbf{\Omega_1} = \left\{ \begin{array}{c} \dot\phi \cos(\eta)\cos(\delta) + \dot\psi \left[\cos(\phi)\sin(\eta)\cos(\delta) + \sin(\phi)\sin(\delta)\right] \\ -\dot\phi \cos(\eta)\sin(\delta) + \dot\psi \left[-\cos(\phi)\sin(\eta)\sin(\delta) + \sin(\phi)\cos(\delta)\right] \\ \dot\delta - \dot\phi \sin(\eta) + \dot\psi \cos(\phi)\cos(\eta) \end{array} \right\} . \tag{B.9}$$

The position of the mass center G is

$$(\overline{G - O}) = \left\{ \begin{array}{c} X \\ Y \\ 0 \end{array} \right\} + \mathbf{R}_1 \mathbf{R}_2 \left\{ \begin{array}{c} 0 \\ 0 \\ h \end{array} \right\} . \tag{B.10}$$

The velocity of the same point is then

$$V_G = \left\{ \begin{array}{c} \dot X \\ \dot Y \\ 0 \end{array} \right\} + \left(\dot{\mathbf{R}}_1 \mathbf{R}_2 + \mathbf{R}_1 \dot{\mathbf{R}}_2 \right) \left\{ \begin{array}{c} 0 \\ 0 \\ h \end{array} \right\} , \tag{B.11}$$

that is

$$V_G = \left\{ \begin{array}{c} \dot X + h\dot\psi \cos(\psi)\sin(\phi) + h\dot\phi \sin(\psi)\cos(\phi) \\ \dot Y + h\dot\psi \sin(\psi)\sin(\phi) - h\dot\phi \cos(\psi)\cos(\phi) \\ -h\dot\phi \sin(\phi) \end{array} \right\} . \tag{B.12}$$

In the study of free controls dynamics, the position and velocity of the center of mass will be assumed to be unaffected by the steering angle δ.

The translational and rotational kinetic energies are respectively:

$$\mathcal{T}_t = \frac{1}{2} m V_G^2 \, , \tag{B.13}$$

$$\mathcal{T}_r = \frac{1}{2} \left\{ \begin{array}{c} \Omega_x \\ \Omega_y \\ \Omega_z \end{array} \right\}^T \begin{bmatrix} J_x & 0 & J_{xz} \\ 0 & J_y & 0 \\ J_{xz} & 0 & J_z \end{bmatrix} \left\{ \begin{array}{c} \Omega_x \\ \Omega_y \\ \Omega_z \end{array} \right\} ,$$

i.e.

$$\mathcal{T}_r = \frac{1}{2} \left\{ \begin{array}{c} \dot\phi \\ \dot\psi \end{array} \right\}^T \mathbf{A} \begin{bmatrix} J_x & 0 & J_{xz} \\ 0 & J_y & 0 \\ J_{xz} & 0 & J_z \end{bmatrix} \mathbf{A}^T \left\{ \begin{array}{c} \dot\phi \\ \dot\psi \end{array} \right\} .$$

The expression of the kinetic energy of the system is

$$T = \frac{1}{2}m\left(\dot{X}^2 + \dot{Y}^2\right) + \frac{1}{2}\dot{\phi}^2 J_x^* + \frac{1}{2}\dot{\psi}^2\left[J_z\cos^2(\phi) + J_y^*\sin^2(\phi)\right] +$$

$$+ \dot{\psi}\dot{\phi}J_{xz}\cos(\phi) + mh\left[\dot{X}\dot{\psi}\sin(\phi) - \dot{Y}\dot{\phi}\cos(\phi)\right]\cos(\psi) + \qquad \text{(B.14)}$$

$$+ mh\left[\dot{X}\dot{\phi}\cos(\phi) + \dot{Y}\dot{\psi}\sin(\phi)\right]\sin(\psi) ,$$

where

$$J_x^* = J_x + mh^2, \quad J_y^* = J_y + mh^2$$

are the roll and pitch moments of inertia with respect to a reference frame set on the ground.

The kinetic energy of the steering system due to steering motion, needed to study the free controls behavior, is

$$\mathcal{T}_{r_1} = \frac{1}{2}\boldsymbol{\Omega}_1^T \mathbf{J}_1 \boldsymbol{\Omega}_1 , \qquad \text{(B.15)}$$

where $\mathbf{J}_1$ is the inertia tensor of the steering system. The kinetic energy of the steering system and the front wheel were already partly taken into account in the expression of the kinetic energy of the vehicle.

Since the steering angle is small in normal vehicle use, the trigonometric functions of δ will be linearized when computing the kinetic energy $\mathcal{T}_{r_1}$. It then follows that

$$\mathcal{T}_{r_1} = \mathcal{T}_{0_1} + \frac{1}{2}J_{z1}\dot{\delta}^2 + \dot{\delta}\dot{\psi}\left[J_{z1}\cos(\eta) + J_{xz1}\sin(\eta)\cos(\phi)\right] + \qquad \text{(B.16)}$$

$$+ \dot{\delta}\dot{\psi}\left[-J_{z1}\sin(\eta) + J_{xz1}\cos(\eta)\right] + A_1\dot{\delta}\dot{\psi}^2 + A_2\dot{\delta}\dot{\phi}^2 + A_3\dot{\delta}\dot{\psi}\dot{\phi} ,$$

where the terms that do not depend on δ, and thus have already been accounted for in the expression used for locked controls motion, are included in $\mathcal{T}_{0_1}$. Terms A_i are:

$$A_1 = \left(J_{x1} - J_{y1}\right)\sin(\phi)\cos(\phi)\sin(\eta) + J_{xz1}\left[\sin(\phi)\cos(\phi)\cos(\eta)\right],$$
$$A_2 = J_{xz1}\sin(\phi), \qquad \text{(B.17)}$$
$$A_3 = \left(J_{x1} - J_{y1}\right)\cos(\eta)\sin(\phi) - J_{xz1}\sin^2(\eta) .$$

These will be neglected in the following equations.

The kinetic energy of the wheels due to rotation about their axis must be computed to take into account their gyroscopic moments as well.

If χ_i is the rotation angle of the ith wheel, the angular velocity of the rear wheel is

$$\boldsymbol{\Omega}_{w2} = \left\{\begin{array}{c}\dot{\phi}\\\dot{\chi}_2\\0\end{array}\right\} + \mathbf{R}_2^T\left\{\begin{array}{c}0\\0\\\dot{\psi}\end{array}\right\} = \boldsymbol{\Omega} + \left\{\begin{array}{c}0\\\dot{\chi}_2\\0\end{array}\right\} , \qquad \text{(B.18)}$$

i.e.,

$$\mathbf{\Omega}_{w2} = \left\{ \begin{array}{c} \dot{\phi} \\ \dot{\psi}\sin(\phi) + \dot{\chi}_2 \\ \dot{\psi}\cos(\phi) \end{array} \right\} . \tag{B.19}$$

Things are more complicated for the front wheel, because it can steer:

$$\mathbf{\Omega}_{w1} = \mathbf{\Omega}_1 + \left\{ \begin{array}{c} 0 \\ \dot{\chi}_1 \\ 0 \end{array} \right\} . \tag{B.20}$$

In locked controls motion the kinetic energy of the ith wheel is

$$\mathcal{T}_{r_i} = \frac{1}{2}\mathbf{\Omega}_{wi}^T \mathbf{J}_{wi}\mathbf{\Omega}_{wi} , \tag{B.21}$$

where, because the wheels are gyroscopic solids (two of their moments of inertia are equal to each other) the inertia matrix $\mathbf{J}_{r_i}$ reduces to

$$\mathbf{J}_{w_i} = \left[\begin{array}{ccc} J_{t_i} & 0 & 0 \\ 0 & J_{p_i} & 0 \\ 0 & 0 & J_{t_i} \end{array} \right] .$$

Stating $\mathbf{\Omega}_1 = \mathbf{\Omega}$, and remembering that, at least as a first approximation, the angular velocity of the wheel is

$$\dot{\chi}_i = \frac{V}{R_{e_i}} , \tag{B.22}$$

the kinetic energy is

$$\mathcal{T}_{r_i} = \frac{1}{2}J_{t_i}\dot{\phi}^2 + \frac{1}{2}J_{t_i}\dot{\psi}^2\cos^2(\phi) + \frac{1}{2}J_{p_i}\dot{\psi}^2\sin^2(\phi) + \frac{1}{2}V^2\frac{J_{pi}^2}{R_{ei}^2} + V\dot{\psi}\sin(\phi)\frac{J_{p_i}}{R_{e_i}} . \tag{B.23}$$

The first three terms were already included in the rotational kinetic energy of the vehicle. It then becomes possible to account for the energy due to wheel rotation simply by adding the term

$$\Delta\mathcal{T} = \frac{1}{2}V^2\left(\frac{J_{p_1}^2}{R_{e_1}^2} + \frac{J_{p_2}^2}{R_{e_2}^2}\right) + V\dot{\psi}\sin(\phi)\left(\frac{J_{p_1}}{R_{e_1}} + \frac{J_{p_2}}{R_{e_2}}\right) , \tag{B.24}$$

to the already computed value of the kinetic energy.

If the steering control is free, the expression of the kinetic energy is much more complicated. With somewhat complex computations, assuming that angle δ is small, a further increase of the kinetic energy is obtained

$$\Delta\mathcal{T}_1 = -V\frac{J_{p_1}}{R_{e_1}}\delta\left[\dot{\psi}\cos(\phi)\sin(\eta) + \dot{\phi}\cos(\eta)\right] . \tag{B.25}$$

The gravitational potential energy is:

$$\mathcal{U} = mgh\cos(\phi) . \tag{B.26}$$

B.2 LOCKED CONTROLS MODEL

B.2.1 Equations of motion

The locked controls Lagrangian function is

$$\mathcal{L} = \frac{1}{2}m\left(\dot{X}^2 + \dot{Y}^2\right) + \frac{1}{2}\dot{\phi}^2 J_x^* + \frac{1}{2}\dot{\psi}^2\left[J_z\cos^2(\phi) + J_y^*\sin^2(\phi)\right] +$$

$$+ \dot{\psi}\dot{\phi}J_{xz}\cos(\phi) + mh\left[\dot{X}\dot{\psi}\sin(\phi) - \dot{Y}\dot{\phi}\cos(\phi)\right]\cos(\psi) + \qquad \text{(B.27)}$$

$$+ mh\left[\dot{X}\dot{\phi}\cos(\phi) + \dot{Y}\dot{\psi}\sin(\phi)\right]\sin(\psi) + \frac{1}{2}V^2\left(\frac{J_{p_1}^2}{R_{e_1}^2} + \frac{J_{p_2}^2}{R_{e_2}^2}\right) +$$

$$+ V\dot{\psi}\sin(\phi)\left(\frac{J_{p_1}}{R_{e_1}} + \frac{J_{p_2}}{R_{e_2}}\right) - mgh\cos(\phi) \ .$$

First two equations of motion

The derivatives entering the first two equations are

$$\frac{\partial\mathcal{L}}{\partial\dot{X}} = m\left[\dot{X} + h\dot{\psi}\sin(\phi)\cos(\psi) + h\dot{\phi}\cos(\phi)\sin(\psi)\right] \ ,$$

$$\frac{\partial\mathcal{L}}{\partial\dot{Y}} = m\left[\dot{Y} + h\dot{\psi}\sin(\phi)\sin(\psi) - h\dot{\phi}\cos(\phi)\cos(\psi)\right] \ , \qquad \text{(B.28)}$$

$$\frac{\partial\mathcal{L}}{\partial X} = \frac{\partial\mathcal{L}}{\partial Y} = 0 \ .$$

Remembering that

$$\left\{ \begin{array}{c} \dot{X} \\ \dot{Y} \end{array} \right\} = \left[\begin{array}{cc} \cos(\psi) & -\sin(\psi) \\ \sin(\psi) & \cos(\psi) \end{array} \right] \left\{ \begin{array}{c} u \\ v \end{array} \right\} = \mathbf{R}_1\left\{ \begin{array}{c} u \\ v \end{array} \right\} \ , \qquad \text{(B.29)}$$

it follows that

$$\frac{\partial\mathcal{T}}{\partial\dot{X}} = m\left[u + h\dot{\psi}\sin(\phi)\right]\cos(\psi) - m\left[v - h\dot{\phi}\cos(\phi)\right]\sin(\psi) \ ,$$

$$\frac{\partial\mathcal{T}}{\partial\dot{Y}} = m\left[u + h\dot{\psi}\sin(\phi)\right]\sin(\psi) + m\left[v - h\dot{\phi}\cos(\phi)\right]\cos(\psi) \ . \qquad \text{(B.30)}$$

By performing the derivatives with respect to time, and collecting the terms in $\cos(\psi)$ and $\sin(\psi)$, the following equations of motion can be obtained

$$m\left[\begin{array}{cc} \cos(\psi) & -\sin(\psi) \\ \sin(\psi) & \cos(\psi) \end{array} \right] \times \qquad \text{(B.31)}$$

$$\times \left\{ \begin{array}{l} \dot{u} - h\dot{\psi}v + h\ddot{\psi}\sin(\phi) + 2h\dot{\psi}\dot{\phi}\cos(\phi) \\ \dot{v} + u\dot{\psi} - h\ddot{\phi}\cos(\phi) + h\dot{\phi}^2\sin(\phi) + h\dot{\psi}^2\sin(\phi) \,. \end{array} \right\} = \left\{ \begin{array}{l} Q_X \\ Q_Y \end{array} \right\}.$$

Remembering that

$$\left\{ \begin{array}{l} Q_X \\ Q_Y \end{array} \right\} = \left[\begin{array}{cc} \cos(\psi) & -\sin(\psi) \\ \sin(\psi) & \cos(\psi) \end{array} \right] \left\{ \begin{array}{l} Q_x \\ Q_y \end{array} \right\} = \mathbf{R}_1 \left\{ \begin{array}{l} Q_x \\ Q_y \end{array} \right\}, \qquad (B.32)$$

the first two equations reduce to

$$\left\{ \begin{array}{l} m\left[\dot{u} - h\dot{\psi}v + h\ddot{\psi}\sin(\phi) + 2h\dot{\psi}\dot{\phi}\cos(\phi)\right] = Q_x\,, \\ m\left[\dot{v} + u\dot{\psi} - h\ddot{\phi}\cos(\phi) + h\dot{\phi}^2\sin(\phi) + h\dot{\psi}^2\sin(\phi)\right] = Q_y\,. \end{array} \right. \qquad (B.33)$$

Third equation of motion

The third equation, describing the yaw angle ψ, can be obtained in the same way. The derivatives are

$$\frac{\partial \mathcal{L}}{\partial \dot{\psi}} = \dot{\psi}\left[J_z\cos^2(\phi) + J_y^*\sin^2(\phi)\right] + \dot{\phi}J_{xz}\cos(\phi) +$$
$$+ mh\sin(\phi)\left[\dot{X}\cos(\psi) + \dot{Y}\sin(\psi)\right] + V\sin(\phi)\left(\frac{J_{p1}}{R_{e_1}} + \frac{J_{p2}}{R_{e_2}}\right),$$
$$\frac{\partial \mathcal{L}}{\partial \psi} = -mh\left[\dot{X}\dot{\psi}\sin(\phi) - \dot{Y}\dot{\phi}\cos(\phi)\right]\sin(\psi) +$$
$$+ mh\left[\dot{X}\dot{\phi}\cos(\phi) + \dot{Y}\dot{\psi}\sin(\phi)\right]\cos(\psi)\,, \qquad (B.34)$$

and then

$$\frac{d}{dt}\left(\frac{\partial \mathcal{L}}{\partial \dot{\psi}}\right) = \ddot{\psi}\left[J_z\cos^2(\phi) + J_y^*\sin^2(\phi)\right] + 2\dot{\psi}\dot{\phi}\left(-J_z + J_y^*\right)\sin(\phi)\cos(\phi)$$
$$+ \ddot{\phi}J_{xz}\cos(\phi) - \dot{\phi}^2 J_{xz}\sin(\phi) + mh\dot{\phi}\cos(\phi)\left[\dot{X}\cos(\psi) + \dot{Y}\sin(\psi)\right] +$$
$$+ mh\sin(\phi)\left[\ddot{X}\cos(\psi) + \ddot{Y}\sin(\psi)\right] + + V\dot{\phi}\cos(\phi)\left(\frac{J_{p1}}{R_{e_1}} + \frac{J_{p2}}{R_{e_2}}\right) +$$
$$+ mh\dot{\psi}\sin(\phi)\left[-\dot{X}\sin(\psi) + \dot{Y}\cos(\psi)\right] + \dot{V}\sin(\phi)\left(\frac{J_{p1}}{R_{e_1}} + \frac{J_{p2}}{R_{e_2}}\right). \qquad (B.35)$$

The third equation is then

$$\ddot{\psi}\left[J_z\cos^2(\phi) + J_y^*\sin^2(\phi)\right] + 2\dot{\psi}\dot{\phi}\left(-J_z + J_y^*\right)\sin(\phi)\cos(\phi) +$$
$$+ \ddot{\phi}J_{xz}\cos(\phi) - \dot{\phi}^2 J_{xz}\sin(\phi) + + mh\sin(\phi)\left[\ddot{X}\cos(\psi) + \ddot{Y}\sin(\psi)\right] + \qquad (B.36)$$
$$+ \dot{V}\sin(\phi)\left(\frac{J_{p1}}{R_{e_1}} + \frac{J_{p2}}{R_{e_2}}\right) + V\dot{\phi}\cos(\phi)\left(\frac{J_{p1}}{R_{e_1}} + \frac{J_{p2}}{R_{e_2}}\right) = Q_\psi\,.$$

Because

$$\ddot{X}\cos(\psi) + \ddot{Y}\sin(\psi) = \dot{u} - v\dot{\psi}, \qquad (B.37)$$

its final form is

$$
\begin{aligned}
&\ddot{\psi}\left[J_z\cos^2(\phi)+J_y^*\sin^2(\phi)\right]+2\dot{\psi}\dot{\phi}\left(-J_z+J_y^*\right)\sin(\phi)\cos(\phi)+\\
&\quad+\ddot{\phi}J_{xz}\cos(\phi)-\dot{\phi}^2J_{xz}\sin(\phi)+mh\sin(\phi)\left(\dot{u}-v\dot{\psi}\right)+\\
&\quad+\dot{V}\sin(\phi)\left(\frac{J_{p1}}{R_{e_1}}+\frac{J_{p2}}{R_{e_2}}\right)+V\dot{\phi}\cos(\phi)\left(\frac{J_{p1}}{R_{e_1}}+\frac{J_{p2}}{R_{e_2}}\right)=Q_\psi .
\end{aligned}
\tag{B.38}
$$

Fourth equation of motion

The fourth equation, describing the roll angle ϕ, may be obtained in the same way. The derivatives are

$$
\begin{aligned}
\frac{\partial\mathcal{L}}{\partial\dot{\phi}}&=\dot{\phi}J_x^*+\dot{\psi}J_{xz}\cos(\phi)+mh\cos(\phi)\left[-\dot{Y}\cos(\psi)+\dot{X}\sin(\psi)\right],\\
\frac{\partial\mathcal{L}}{\partial\phi}&=\dot{\psi}^2\left[-J_z+J_y^*\right]\cos(\phi)\sin(\phi)-\dot{\psi}\dot{\phi}J_{xz}\sin(\phi)+\\
&\quad+mh\left[\dot{X}\dot{\psi}\cos(\phi)+\dot{Y}\dot{\phi}\sin(\phi)\right]\cos(\psi)+mgh\sin(\phi)+\\
&\quad+mh\left[-\dot{X}\dot{\phi}\sin(\phi)+\dot{Y}\dot{\psi}\cos(\phi)\right]\sin(\psi)+V\dot{\psi}\cos(\phi)\left(\frac{J_{p1}}{R_{e_1}}+\frac{J_{p2}}{R_{e_2}}\right),
\end{aligned}
\tag{B.39}
$$

and then

$$
\begin{aligned}
\frac{d}{dt}\left(\frac{\partial\mathcal{L}}{\partial\dot{\phi}}\right)&=\ddot{\phi}J_x^*+\ddot{\psi}J_{xz}\cos(\phi)-\dot{\psi}\dot{\phi}J_{xz}\sin(\phi)+\\
&\quad-mh\dot{\phi}\sin(\phi)\left[-\dot{Y}\cos(\psi)+\dot{X}\sin(\psi)\right]+mh\cos(\phi)\left[-\ddot{Y}\cos(\psi)+\right.\\
&\quad\left.+\ddot{X}\sin(\psi)\right]++mh\dot{\psi}\cos(\phi)\left[\dot{Y}\sin(\psi)+\dot{X}\cos(\psi)\right].
\end{aligned}
\tag{B.40}
$$

The fourth equation is

$$
\begin{aligned}
&\ddot{\phi}J_x^*+\ddot{\psi}J_{xz}\cos(\phi)-\dot{\psi}^2\left[-J_z+J_y^*\right]\cos(\phi)\sin(\phi)+\\
&\quad+mh\cos(\phi)\left[-\ddot{Y}\cos(\psi)+\ddot{X}\sin(\psi)\right]+\\
&\quad-V\dot{\psi}\cos(\phi)\left(\frac{J_{p1}}{R_{e_1}}+\frac{J_{p2}}{R_{e_2}}\right)-mgh\sin(\phi)=Q_\phi .
\end{aligned}
\tag{B.41}
$$

Because

$$
\ddot{Y}\cos(\psi)+\ddot{X}\sin(\psi)=\dot{v}+u\dot{\psi},
\tag{B.42}
$$

it may be written in the form

$$
\begin{aligned}
&\ddot{\phi}J_x^*+\ddot{\psi}J_{xz}\cos(\phi)-\dot{\psi}^2\left[-J_z+J_y^*\right]\cos(\phi)\sin(\phi)+\\
&\quad-mh\cos(\phi)\left(\dot{v}+u\dot{\psi}\right)-V\dot{\psi}\cos(\phi)\left(\frac{J_{p1}}{R_{e_1}}+\frac{J_{p2}}{R_{e_2}}\right)-mgh\sin(\phi)=Q_\phi .
\end{aligned}
\tag{B.43}
$$

B.2.2   Linearization of the equations of motion

If the values of ϕ and v are small, it is possible to linearize the equations of motion. As usual in linearized models, V and u are interchangeable and the terms containing the products of small quantities may be neglected. It then follows that

$$
\begin{cases}
m\dot{V} = Q_x \ , \\
m\dot{v} + mV\dot{\psi} - mh\ddot{\phi} = Q_y \ , \\
J_z\ddot{\psi} + J_{xz}\ddot{\phi} + m\phi h\dot{V} + V\dot{\phi}\left(\frac{J_{p_1}}{R_{e_1}} + \frac{J_{p_2}}{R_{e_2}}\right) + \dot{V}\phi\left(\frac{J_{p_1}}{R_{e_1}} + \frac{J_{p_2}}{R_{e_2}}\right) = Q_\psi \ , \\
J_x\ddot{\phi} + J_{xz}\ddot{\psi} - mh\dot{v} - mhV\dot{\psi} - V\dot{\psi}\left(\frac{J_{p_1}}{R_{e_1}} + \frac{J_{p_2}}{R_{e_2}}\right) - mgh\phi = Q_\phi \ .
\end{cases}
\tag{B.44}
$$

B.2.3   Generalized forces

The forces at the wheel-ground contact may be computed in a way similar to that for the vehicle with two axles, with the monotrack model being no longer a simplification but a realistic model.

The sideslip angle of the ith wheel is

$$
\alpha_i = \arctan\left(\frac{v + \dot{\psi}x_i}{u}\right) - \delta_i \ .
\tag{B.45}
$$

The rear wheel usually does not steer, while the front wheel steers about an axis that is inclined with respect to the vertical. If the inclination angle is γ, it follows that

$$
\delta_1 = \delta \cos(\gamma) \ , \qquad \delta_2 = 0 \ .
\tag{B.46}
$$

The linearized expression of the sideslip angle is then identical to that of two-axles vehicles

$$
\begin{cases}
\alpha_1 = \beta + \dfrac{a}{V}r - \delta\cos(\eta) \ , \\
\alpha_2 = \beta - \dfrac{b}{V}r \ .
\end{cases}
\tag{B.47}
$$

The virtual displacement of the center of the contact zone of the ith wheel in the reference frame of the vehicle is

$$
\delta u_i = \left\{ \begin{array}{c} \delta x \\ \delta y + \delta\psi x_i \end{array} \right\} \ .
\tag{B.48}
$$

The virtual work of forces F_{x_i} and F_{y_i} and of the moment M_{z_i} exchanged between wheel and ground is

$$
\delta\mathcal{L} = F_{x_i}\delta x + F_{y_i}\left(\delta y + \delta\psi x_i\right) + M_{z_i}\delta\psi \ .
\tag{B.49}
$$

The aerodynamic forces are applied at the center of mass of the vehicle. Assuming that the force components F_{xa}, F_{ya} and F_{za} and the components of the moment M_{za} and M_{xa} are referred to the x and y axes laying on the ground and to a vertical z axis, the virtual displacement of the center of mass of the vehicle is

$$\delta u_G = \left\{ \begin{array}{c} \delta x + h\delta\psi \sin(\phi) \\ \delta y - h\delta\phi \cos(\phi) \\ -h\delta\phi \sin(\phi) \end{array} \right\} . \tag{B.50}$$

The virtual work of the aerodynamic forces and moments is then

$$\begin{aligned} \delta\mathcal{L} = \ & F_{xa}\left[\delta x + h\delta\psi \sin(\phi)\right] + F_{ya}\left[\delta y - h\delta\phi \cos(\phi)\right] + \\ & + F_{za}\left[-h\delta\phi \sin(\phi)\right] + M_{za}\delta\psi + M_{xa}\delta\phi . \end{aligned} \tag{B.51}$$

The generalized forces that must be introduced into the equations are then

$$\left\{ \begin{array}{l} Q_x = F_{x_1} + F_{x_2} + F_{xa} , \\ Q_y = F_{y_1} + F_{y_2} + F_{ya} , \\ Q_\psi = F_{y_1}a - F_{y_2}b + M_{z_1} + M_{z_2} - F_{xa}h \sin(\phi) + M_{za} , \\ Q_\phi = -hF_{ya}\cos(\phi) - F_{za}h \sin(\phi) + M_{xa} . \end{array} \right. \tag{B.52}$$

Side forces depend in this case not only on the slip angle but also on the camber angle, which is here equal to the roll angle.

B.2.4 Linearized expression of the generalized forces

As in the case of four-wheeled vehicles, the generalized forces may be linearized. Remembering that for small angles

$$\beta = \frac{v}{V} , \tag{B.53}$$

and proceeding as seen in the previous models, it follows that

$$Q_y = Y_v v + Y_r \dot{\psi} + Y_\phi \phi + Y_\delta \delta + F_{y_e} , \tag{B.54}$$

where

$$\left\{ \begin{array}{l} Y_v = \frac{1}{V}\left[-C_1 - C_2 + \frac{1}{2}\rho V_r^2 S(C_y)_{,\beta}\right] , \\ \\ Y_r = -\frac{1}{V}\left(aC_1 - bC_2\right) , \\ \\ Y_\phi = (F_{y_1})_{,\gamma} + (F_{y_2})_{,\gamma} , \\ \\ Y_\delta = C_1 \cos(\eta) , \end{array} \right. \tag{B.55}$$

and where $(F_{y_i})_{,\gamma}$ is the camber stiffness. Moreover,

$$Q_\psi = N_v v + N_r r + N_\phi \phi + N_\delta \delta + M_{z_e} , \tag{B.56}$$

where

$$\begin{cases} N_v = -aC_1 + bC_2 + (M_{z_1}),_\alpha + (M_{z_1}),_\alpha + \frac{1}{2}\rho V_r^2 S(C_{M_z}),_\beta \ , \\ N_r = \frac{1}{V}\left[-a^2 C_1 - b^2 C_2 + (M_{z_1}),_\alpha a - (M_{z_2}),_\alpha b\right] \ , \\ N_\phi = a(F_{y_1}),_\gamma - b(F_{y_2}),_\gamma - \frac{1}{2}\rho V_r^2 ShC_x \ , \\ N_\delta = [C_1 a - (M_{z_1}),_\alpha] \cos(\gamma) \ . \end{cases} \tag{B.57}$$

At last

$$Q_\phi = L_v v + L_\phi \phi \ , \tag{B.58}$$

where

$$\begin{cases} L_v = \frac{1}{2}\rho V_r S\left[l(C_{M_x}),_\beta - h(C_y),_\beta\right] \ , \\ L_\phi = -\frac{1}{2}\rho V_r^2 ShC_z \ . \end{cases} \tag{B.59}$$

B.2.5 Final expression of the linearized equations of motion

The linearized equations may be uncoupled, as in the case of four wheeled vehicles, by assuming that the forward velocity V is a known function of time instead of being a variable of motion. The first equation is the usual one

$$m\dot{V} = F_{x_1} + F_{x_2} + F_{xa} \ . \tag{B.60}$$

It allows the driving (or braking) force needed to follow a given law $V(t)$ to be computed.

The other three equations may be written as

$$\begin{bmatrix} m & 0 & -mh \\ 0 & J_z & J_{xz} \\ -mh & J_{xz} & J_x \end{bmatrix}\begin{Bmatrix} \dot{v} \\ \ddot{\psi} \\ \ddot{\phi} \end{Bmatrix} + \begin{bmatrix} -Y_v & mV - Y_r & 0 \\ -N_v & -N_r & N_g \\ -L_v & -mhV - VN_g & 0 \end{bmatrix}\begin{Bmatrix} v \\ \dot{\psi} \\ \dot{\phi} \end{Bmatrix} +$$

$$+ \begin{bmatrix} 0 & 0 & -Y_\phi \\ 0 & 0 & mh\dot{V} + \dot{V}\left(\frac{J_{p_1}}{R_{e_1}} + \frac{J_{p_2}}{R_{e_2}}\right) - N_\phi \\ 0 & 0 & -mgh - L_\phi \end{bmatrix}\begin{Bmatrix} y \\ \psi \\ \phi \end{Bmatrix} = \begin{Bmatrix} Y_\delta \delta + F_{y_e} \\ N_\delta \delta + M_{z_e} \\ 0 \end{Bmatrix} \ , \tag{B.61}$$

where

$$N_g = V\left(\frac{J_{p_1}}{R_{e_1}} + \frac{J_{p_2}}{R_{e_2}}\right) \ . \tag{B.62}$$

The three matrices included in the equation of motion in the configuration space are normally defined as mass, stiffness and damping matrices ($\mathbf{M}$, $\mathbf{C}$ and $\mathbf{K}$). The first is symmetrical, while the other two are not. As usual in vehicle dynamics, the first two columns of the stiffness matrix vanish, because coordinates

y^1 and ψ do not appear directly in the equations. The order of the linearized set of equations is then 4 and not 6, and the state space model is made by four first-order differential equations.

The state space open loop model is the usual one

$$\dot{\mathbf{z}} = \mathbf{A}\mathbf{z} + \mathbf{B}_c\mathbf{u}_c + \mathbf{B}_e\mathbf{u}_e \ , \tag{B.63}$$

where

$$\mathbf{z} = \begin{bmatrix} v & r & p & v_\delta \end{bmatrix}^T \ ,$$

r and p are the derivatives of ψ and ϕ with respect to time;

$$\mathbf{A} = \begin{bmatrix} -\mathbf{M}^{-1}\mathbf{C} & -\mathbf{M}^{-1}\mathbf{K}^* \\ 0 \quad 0 \quad 1 & 0 \end{bmatrix} ,$$

where $\mathbf{K}^*$ has been obtained by cancelling the first two columns of $\mathbf{K}$, and

$$\mathbf{B}_c = \begin{bmatrix} \mathbf{M}^{-1} \begin{bmatrix} Y_\delta & N_\delta & 0 \end{bmatrix}^T \\ 0 \end{bmatrix} \ , \quad \mathbf{B}_e = \begin{bmatrix} \mathbf{M}^{-1} \\ 0 \end{bmatrix} \ ,$$

$$\mathbf{u}_c = \delta \ , \qquad \mathbf{u}_e = \left\{ \begin{array}{c} F_{y_e} \\ M_{z_e} \\ 0 \end{array} \right\} \ .$$

B.3   LOCKED CONTROLS STABILITY

Stability with locked controls may be studied simply by searching the eigenvalues of the dynamic matrix $\mathbf{A}$. The eigenproblem usually yields two real eigenvalues and one complex conjugated pair. Of the real solutions, one is negative and has little importance in the behavior of the system, while the other is positive and hence unstable. The latter corresponds to the capsize mode and must be stabilized by the driver or by some control device. This eigenvalue decreases with increasing speed, as gyroscopic moments of the wheels reduce the velocity at which the motorcycle leans to the side.

The two complex conjugate pairs are related to the so-called weave mode; this mode is primarily a yaw oscillation of the whole vehicle but it also involves the roll and steering degrees of freedom. Weave oscillation is usually damped, at least in locked control motion. At low speed it may not involve a true oscillation (the imaginary part of the eigenvalue may be equal to zero, but in any case the real part is negative).

Remark B.1 *Weave motion usually becomes less stable with increasing speed, i.e. the modulus of its real part decreases while the frequency increases. At high speed it may be difficult for the driver to control the motion, because its frequency is high enough to produce instability.*

---

[1]v is the derivative of a pseudo-coordinate here and thus y has no physical meaning.

B.3.1 *Capsize motion*

An extremely simplified model for capsize motion is an inverted pendulum (Fig. B.2a). The linearized equation of a pendulum with length h and baricentric moment of inertia J_x is the usual one

$$\left(J_x + mh^2\right)\ddot{\phi} - mgh\phi = 0 , \qquad (\text{B.64})$$

where a (-) sign has been introduced in the gravitational term to take into account the fact that the pendulum is inverted (the suspension point is below the center of mass).

The characteristic equation may be obtained by introducing expression

$$\phi = \phi_0 e^{st} ,$$

into the equation of motion. It is

$$s^2 \left(J_x + mh^2\right) - mgh = 0 , \qquad (\text{B.65})$$

which yields

$$s = \pm\sqrt{\frac{mgh}{J_x + mh^2}} . \qquad (\text{B.66})$$

The positive solution shows that the capsize motion is unstable. Its time constant is

$$\tau = \frac{1}{s} = \pm\sqrt{\frac{J_x + mh^2}{mgh}} . \qquad (\text{B.67})$$

Actually the motorcycle does not behave as an inverted pendulum and the time constant increases (that is, s decreases) with increasing speed. This is due both to gyroscopic effect and the camber thrust of the tires caused by roll.

A way to factor in the latter effect is to build a simple model made by an inverted pendulum whose supporting point is free to move horizontally (Fig. B.2b). The position of point P is

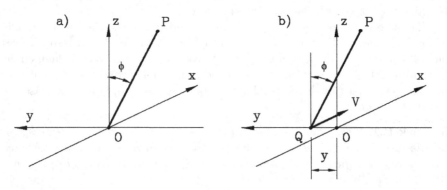

FIGURE B.2. Simplified models for capsize motion. a) Inverted pendulum; b) inverted pendulum with moving supporting point

$$(\overline{P - O}) = \left\{ \begin{array}{c} 0 \\ h\cos(\phi) \\ y - h\sin(\phi) \end{array} \right\} \tag{B.68}$$

and its velocity is

$$V_P = \left\{ \begin{array}{c} 0 \\ -h\dot{\phi}\sin(\phi) \\ \dot{y} - h\dot{\phi}\cos(\phi) \end{array} \right\} . \tag{B.69}$$

The kinetic energy of the system is then

$$\mathcal{T} = \frac{1}{2}m\left[\dot{y}^2 - 2h\dot{y}\dot{\phi}\cos(\phi) + h^2\dot{\phi}^2\right] + \frac{1}{2}J_x\dot{\phi}^2 . \tag{B.70}$$

The gravitational potential energy is

$$\mathcal{U} = mgh\cos(\phi) . \tag{B.71}$$

The equations of motion obtained through Lagrange equations by performing the relevant derivatives of the Lagrangian function $\mathcal{T} - \mathcal{U}$ and linearizing are

$$\left\{ \begin{array}{l} m\ddot{y} - mh\ddot{\phi} = Q_y , \\ -mh\ddot{y} + \left(mh^2 + J_x\right)\ddot{\phi} - mgh\phi = Q_\phi . \end{array} \right. \tag{B.72}$$

The side force acting on point Q is due to the tires, which work with both a sideslip and camber angle. If the virtual displacement of point Q is δy and the usual linearized expression is used for the side force F_y

$$F_y = -C\alpha + (F_y)_{,\gamma}\,\phi , \tag{B.73}$$

where the cornering stiffness C and the camber stiffness $(F_y)_{,\gamma}$ are those of the whole vehicle (the sum of the stiffness referred to the two tires), the virtual work is

$$\delta\mathcal{L} = \delta y\left[-C\alpha + (F_y)_{,\gamma}\,\phi\right] . \tag{B.74}$$

Assuming that the system moves along the x axis at a speed V, the velocity of point Q is

$$V_Q = \left[\begin{array}{ccc} V & \dot{y} & 0 \end{array}\right]^T . \tag{B.75}$$

The sideslip angle of the wheels is then

$$\alpha = \mathrm{artg}\left(\frac{\dot{y}}{V}\right) \approx \frac{\dot{y}}{V} . \tag{B.76}$$

By differentiating the virtual work with respect to the virtual displacement the generalized forces are immediately obtained,

$$Q_y = -C\frac{\dot{y}}{V} + (F_y)_{,\gamma}\,\phi , \quad Q_\phi = 0 . \tag{B.77}$$

The equation of motion is then

$$
\begin{bmatrix} m & -mh \\ -mh & mh^2 + J_x \end{bmatrix} \begin{Bmatrix} \ddot{y} \\ \ddot{\phi} \end{Bmatrix} + \frac{1}{V} \begin{bmatrix} C & 0 \\ 0 & 0 \end{bmatrix} \begin{Bmatrix} \dot{y} \\ \dot{\phi} \end{Bmatrix} +
$$

$$
+ \begin{bmatrix} 0 & -(F_y)_{,\gamma} \\ 0 & -mgh \end{bmatrix} \begin{Bmatrix} y \\ \phi \end{Bmatrix} = \mathbf{0} \ . \tag{B.78}
$$

This equation coincides with Eq. (B.61), where the second row and column have been cancelled in all matrices and aerodynamic terms have been neglected.

The characteristic equation allowing the natural frequencies to be computed is

$$
\det \begin{bmatrix} ms^2 + \frac{C}{V}s & -mhs^2 - (F_y)_{,\gamma} \\ -mhs^2 & (mh^2 + J_x)\,s^2 - mgh \end{bmatrix} = 0 \ , \tag{B.79}
$$

and then

$$
s \left\{ mJ_x V s^3 + C\left(mh^2 + J_x\right)s^2 + mhV\left[-\left(F_y\right)_{,\gamma} - mg\right]s - Cmgh \right\} = 0 \ . \tag{B.80}
$$

One solution is obviously $s = 0$; out of the other three, one is real and positive (capsize motion) while the other two are complex with a negative real part. The latter represent a kind of weave motion, but because the model does not take yaw rotation into account, an important factor in weave motion, these solutions have no physical meaning. In this way the time constant of capsize motion becomes a function of speed.

B.3.2   Weave motion

It is possible to build a much simplified model for weave motion as well. Because weave motion primarily involves the vehicle body and not the steering system, a model based on an almost horizontal pendulum hinged on the steering axis (axis H_1H_2 in Fig. B.3) can be built. The length of such a pendulum is GH_1,

$$
\overline{GH_1} = l_1 = [a + e - h \tan{(\eta)}] \cos{(\eta)} \ . \tag{B.81}
$$

The distance of the point where the side force of the rear tire is applied from the hinge axis is P_2H_2:

$$
\overline{P_2H_2} = l_2 = (l + e) \cos{(\eta)} \ . \tag{B.82}
$$

If the rotation angle of the pendulum about the steering axis is θ, the steering angle with respect to the xz plane of the rear wheel is

$$
\theta \cos{(\eta)}
$$

and its lateral velocity is $\dot{\theta} l_2$. The sideslip angle of the rear wheel is then

$$
\alpha_2 = \theta \cos{(\eta)} + \dot{\theta}\frac{l_2}{V} \ . \tag{B.83}
$$

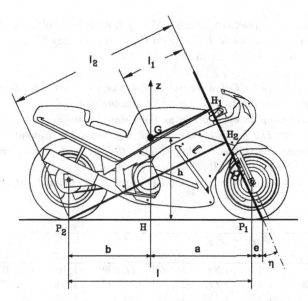

FIGURE B.3. Simplified model for weave motion

Because the roll angle has been assumed to be zero, the side force on the tire is

$$F_{y_2} = -C_2\alpha_2 = -C_2 \left[\theta \cos{(\eta)} + \dot{\theta}\frac{l_2}{V} \right] .$$

(B.84)

The equation of motion of the vehicle body about the steering axis is

$$J\ddot{\theta} + \frac{C_2 l_2^2}{V}\dot{\theta} + l_2 C_2 \theta \cos{(\eta)} = 0 ,$$

(B.85)

where J is its moment of inertia about the same axis.

A rough approximation is

$$J = J_z + m l_1^2$$

where J_z is the moment of inertia of the whole vehicle about the z axis and m refers to the whole vehicle.

The pole for weave motion is

$$s = \frac{-C_2 l_2^2 \pm \sqrt{C_2^2 l_2^4 - 4V^2 J l_2 C_2 \cos{(\eta)}}}{2VJ} .$$

(B.86)

At low speed the two solutions are both real and negative, so that weave is stable and not oscillatory (weave is a misnomer in this case). Starting from a speed

$$V = \frac{l_2}{2}\sqrt{\frac{C_2 l_2}{J \cos{(\eta)}}}$$

(B.87)

the roots become complex conjugate and the motion is oscillatory. At low speed the motion is damped, but at high speed the real part may become positive and weave motion may become unstable.

Example B.1 *Study the locked controls stability of the motorcycle of Appendix E.10. Plot the eigenvalues as functions of the speed and the roots locus.*

To correctly compute the roots locus the cornering stiffness of the tires must be computed correctly. The forces on the ground F_{z_i} were then computed, factoring in aerodynamic forces and rolling resistance. The effect of driving forces can also be accounted for, at least using the elliptical approximation.

The values of the cornering and camber stiffness at standstill, at 100 and at 200 km/h are

V km/h	F_{z1} N	F_{z2} N	C_1 N/rad	C_2 N/rad	$(F_{y1})_\gamma$ N/rad	$(F_{y2})_\gamma$ N/rad
0	1,451	1,394	39,600	41,400	$-1,708$	$-1,905$
100	1,384	1,414	37,700	38,600	$-1,630$	$-1,933$
200	1,183	1,473	32,300	32,300	$-1,392$	$-2,013$

The roots locus and the plot of the eigenvalues versus the speed are reported in Fig. B.4.

Weave is oscillatory from about 100 km/h and becomes unstable slightly above 240 km/h, while capsize is, as is normal for motorcycles, unstable.

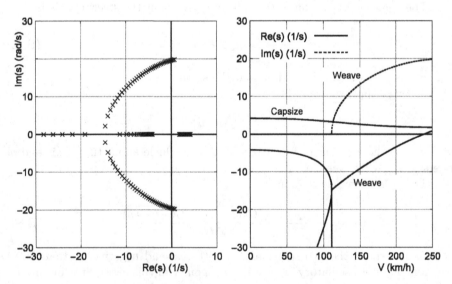

FIGURE B.4. Locked controls stability of a motor cycle. Roots locus and plot of the eigenvalues versus the forward speed

The values of the natural frequencies and of the time constant of capsize motion at three values of speed are

V	s_{cap}	τ_{cap}	s_{weave}	τ^*_{cap}	τ^{**}_{cap}	s^*_{weave}
km/h	1/s	s	1/s	s	s	1/s
0	4.20	0.24	–	0.33	0.33	–
100	3.44	0.29	-8.76	0.33	0.37	$-10.8\pm17.6i$
200	2.00	0.50	$-3.45\pm18.0\,i$	0.33	0.41	$-4.26\pm17.8i$

*The values of the capsize time constants and of the eigenvalue for weave computed using the simplified models (values with * and **), are also reported. The latter model largely underestimates the speed at which weave becomes oscillatory (52 km/h against about 100 in the model with three degrees of freedom), but once motion is oscillatory, allows its frequency to be computed in a way that is surprisingly consistent with the value obtained from the more complex model.*

B.4   STEADY-STATE MOTION

Equation (B.61) may be used to compute the steady-state response by simply assuming that v, r, ϕ, δ, and the speed V are constant. A simple way to compute the steady-state response is to fix a certain value of either r or δ and then solve the equations for v, ϕ, δ or r. Perhaps the most immediate approach is the first, because in steady-state conditions, to state r means to state a value of the radius of the path R. The problem is then stated in this form: Given a certain circular path, a value of the speed and perhaps also of external forces, compute sideslip, roll and steering angles ($\beta = v/V$, ϕ and δ). The relevant equation is

$$\begin{bmatrix} C_{11} & K_{13} & -Y_\delta \\ C_{21} & K_{23} & -N_\delta \\ C_{31} & K_{33} & 0 \end{bmatrix} \begin{Bmatrix} V\beta \\ \phi \\ \delta \end{Bmatrix} = r \begin{Bmatrix} -C_{12} \\ -C_{22} \\ -C_{32} \end{Bmatrix} + \begin{Bmatrix} F_{y_e} \\ M_{z_e} \\ 0 \end{Bmatrix} . \qquad (B.88)$$

By solving Eq. (B.88) it is possible to compute

- the path curvature gain:

$$\frac{1}{R\delta} = \frac{r}{V\delta} \; ;$$

- the sideslip angle gain

$$\frac{\beta}{\delta} = \frac{v}{V\delta} \; ;$$

- the roll angle gain

$$\frac{\phi}{\delta} \; .$$

Owing to the linearity of the model, if the external forces are assumed to be equal to zero, the mentioned gains are a function of the velocity only; i.e., they do not depend on the radius of the trajectory. Equation (B.88) can thus be assumed to hold for very high speeds as well, if a large enough radius R is considered.

Once the trajectory curvature gain has been obtained, the usual definitions of understeer, neutral steer and oversteer can be applied.

The present model is linearized and hence holds only if the relevant angles are small enough to allow the linearization of their trigonometric functions. The limit in this case is primarily due to the roll angle ϕ which can easily exceed the value of about 20° that can be considered a limit for linearization.

The computation of the steady-state response assumes that the motion is possible, implying that the driver stabilizes the capsize mode, which is unstable in open-loop operation.

Example B.2 *Compute the steady-state steering response of the motorcycle of Appendix E.10, assuming that no external force is present. Compute the values of β, δ, and ϕ on a curve with a radius of 200 m at a speed of 80 km/h.*

The various gains are plotted versus the speed in Fig. B.5. The nondimensional curves hold for any value of the radius R: They are plotted for speeds up to 240 km/h, although on a curve with a radius of 200 m the limit for linearization occurs at speeds slightly in excess of 80 km/h.

At 80 km/h their values are $1/R\delta = 0.56 \ 1/m$; $\beta/\delta = 0.77$, and $\phi/\delta = -28.5$. On a curve of 200 m radius the steering angle is $\delta = 0.0090 \ rad = 0.52$ °. Because the

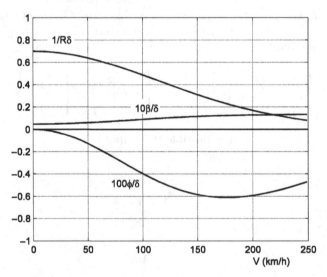

FIGURE B.5. Values of the path curvature gain $1/R\delta = r/V\delta$, the sideslip angle gain $\beta/\delta = v/V\delta$ and the roll angle gain ϕ/δ for the motorcycle of Appendix E.10 versus the speed V

wheelbase is 1.316 m, the kinematic value of the trajectory curvature gain is 0.70 and the steering angle is then $\delta_c = 0.0071$ rad $= 0.41°$. The vehicle is then understeer.

The other values are $\beta = 0.0069$ rad $= 0.40°$, and $\phi = -0.257$ rad $= -14.7°$. The value of the roll angle is quite close to the limit for linearization.

B.5   FREE CONTROLS MODEL

The Lagrangian function for a system with free controls is that already seen for the locked controls model, with some additional terms

$$
\Delta\mathcal{L} = \frac{1}{2}J_{z1}\dot{\delta}^2 + \dot{\delta}\dot{\psi}\left[J_{z1}\cos(\eta) + J_{xz1}\sin(\eta)\cos(\phi)\right] +
$$
$$
+ \dot{\delta}\dot{\phi}\left[-J_{z1}\sin(\eta) + J_{xz1}\cos(\eta)\right] + A_1\delta\dot{\psi}^2 + A_2\delta\dot{\phi}^2 +
$$
$$
+ A_3\delta\dot{\psi}\dot{\phi} - V\frac{J_{p1}}{R_{e_1}}\delta\left[\dot{\psi}\cos(\phi)\sin(\eta) + \dot{\phi}\cos(\eta)\right] .
$$

First two equations of motion
$\Delta\mathcal{L}$ Does not contain either X or Y or their derivatives. The first two equations are not changed by the free controls assumptions.
Third equation of motion
When linearizing the equation, terms in A_i disappear. The derivatives included in the third equation are

$$
\frac{\partial\Delta\mathcal{L}}{\partial\dot{\psi}} = \dot{\delta}\left[J_{z1}\cos(\eta) + J_{xz1}\sin(\eta)\right] - V\frac{J_{p1}}{R_{e_1}}\delta\sin(\eta),
$$
$$
\frac{\partial\Delta\mathcal{L}}{\partial\psi} = 0
$$

(B.89)

and then

$$
\frac{d}{dt}\left(\frac{\partial\Delta\mathcal{L}}{\partial\dot{\psi}}\right) = \ddot{\delta}\left[J_{z1}\cos(\eta) + J_{xz1}\sin(\eta)\right] - \dot{V}\frac{J_{p1}}{R_{e_1}}\delta\sin(\eta) - V\frac{J_{p1}}{R_{e_1}}\dot{\delta}\sin(\eta) .
$$

(B.90)

The third equation is then

$$
J_z\ddot{\psi} + J_{xz}\ddot{\phi} + \ddot{\delta}\left[J_{z1}\cos(\eta) + J_{xz1}\sin(\eta)\right] + m\phi h\dot{V} + V\dot{\phi}\left(\frac{J_{p1}}{R_{e_1}} + \frac{J_{p2}}{R_{e_2}}\right) +
$$
$$
-V\frac{J_{p1}}{R_{e_1}}\dot{\delta}\sin(\eta) + \dot{V}\phi\left(\frac{J_{p1}}{R_{e_1}} + \frac{J_{p2}}{R_{e_2}}\right) - \dot{V}\frac{J_{p1}}{R_{e_1}}\delta\sin(\eta) = Q_\psi .
$$

(B.91)

Fourth equation of motion

The fourth equation, dealing with the roll angle ϕ, may be obtained in the same way. The derivatives are

$$\frac{\partial \Delta \mathcal{L}}{\partial \dot{\phi}} = \dot{\delta} \left[-J_{z1} \sin{(\eta)} + J_{xz1} \cos{(\eta)} \right] - V \frac{J_{p1}}{R_{e1}} \delta \cos{(\eta)} ,$$

$$\frac{\partial \Delta \mathcal{L}}{\partial \phi} = 0$$

$$(B.92)$$

and then

$$\frac{d}{dt} \left(\frac{\partial \Delta \mathcal{L}}{\partial \dot{\phi}} \right) = \ddot{\delta} \left[-J_{z1} \sin{(\eta)} + J_{xz1} \cos{(\eta)} \right] - V \frac{J_{p1}}{R_{e1}} \dot{\delta} \cos{(\eta)} - \dot{V} \frac{J_{p1}}{R_{e1}} \delta \cos{(\eta)} \ .$$

$$(B.93)$$

The fourth equation is then

$$J_x \ddot{\phi} + J_{xz} \ddot{\psi} + \ddot{\delta} \left[-J_{z1} \sin{(\eta)} + J_{xz1} \cos{(\eta)} \right] - mh\dot{v} - mhV\dot{\psi} +$$

$$-V\dot{\psi} \left(\frac{J_{p1}}{R_{e1}} + \frac{J_{p2}}{R_{e2}} \right) - V \frac{J_{p1}}{R_{e1}} \dot{\delta} \cos{(\eta)} - \dot{V} \frac{J_{p1}}{R_{e1}} \delta \cos{(\eta)} - mgh\phi = Q_\phi \ .$$

$$(B.94)$$

Fifth equation of motion

A further equation describing the motion of the steering system must be added to the first four equations. The relevant derivatives are

$$\frac{\partial \Delta \mathcal{L}}{\partial \dot{\delta}} = J_{z1} \dot{\delta} + \dot{\psi} \left[J_{z1} \cos{(\eta)} + J_{xz1} \sin{(\eta)} \right] + \dot{\phi} \left[-J_{z1} \sin{(\eta)} + J_{xz1} \cos{(\eta)} \right] \ ,$$

$$\frac{\partial \Delta \mathcal{L}}{\partial \delta} = -V \frac{J_{p1}}{R_{e1}} \left[\dot{\psi} \sin{(\eta)} + \dot{\phi} \cos{(\eta)} \right] \ ,$$

$$(B.95)$$

and then

$$\frac{d}{dt} \left(\frac{\partial \Delta \mathcal{L}}{\partial \dot{\delta}} \right) = J_{z1} \ddot{\delta} + \ddot{\psi} \left[J_{z1} \cos{(\eta)} + J_{xz1} \sin{(\eta)} \right] +$$

$$+ \ddot{\phi} \left[-J_{z1} \sin{(\eta)} + J_{xz1} \cos{(\eta)} \right] \ . \qquad (B.96)$$

The fifth equation is

$$J_{z1} \ddot{\delta} + \ddot{\psi} \left[J_{z1} \cos{(\eta)} + J_{xz1} \sin{(\eta)} \right] + \ddot{\phi} \left[-J_{z1} \sin{(\eta)} + J_{xz1} \cos{(\eta)} \right] +$$

$$+ V \frac{J_{p1}}{R_{e1}} \left[\dot{\psi} \sin{(\eta)} + \dot{\phi} \cos{(\eta)} \right] = Q_\delta \ . \qquad (B.97)$$

B.5.1   Generalized forces

The virtual displacement of the center of the contact zone of the rear wheel is the same as seen for the case of locked controls motion. The sideslip angle of the front wheel, assuming that the steering angle is small, is

$$\alpha_1 = \arctan\left(\frac{v + \dot{\psi}a - \dot{\delta}e}{u}\right) - \delta\cos\left(\eta\right) . \tag{B.98}$$

Its linearized expression is then

$$\alpha_1 = \beta + \frac{a}{V}r - \frac{e}{V}\dot{\delta} - \delta\cos\left(\eta\right) . \tag{B.99}$$

The virtual displacement of the center of the contact area of the front wheel is

$$\delta u_1 = \left\{ \begin{array}{c} \delta x \\ \delta y + \delta\psi a - e\delta\delta \end{array} \right\} . \tag{B.100}$$

The virtual work of forces F_{x_i} and F_{y_i} and of the moment M_{z_i} acting between wheel and ground is

$$\delta\mathcal{L}_1 = F_{x_1}\delta x + F_{y_1}\left[\delta y + \delta\psi x_i - e\delta\delta\right] + M_{z_i}\left[\delta\psi + \delta\delta\cos\left(\eta\right)\right] . \tag{B.101}$$

The generalized forces to be introduced into the first equations of motion are those already seen in the previous model. The force included in the fifth equation is

$$Q_\delta = -eF_{y_1} + M_{z_i}\cos\left(\eta\right) . \tag{B.102}$$

Proceeding as in the previous models, it follows that

$$Q_y = Y_v v + Y_r \dot{\psi} + Y_{\dot{\delta}} \dot{\delta} + Y_\phi \phi + Y_\delta \delta + F_{y_e} , \tag{B.103}$$

$$Q_\psi = N_v v + N_r \dot{\psi} + N_{\dot{\delta}} \dot{\delta} + N_\phi \phi + N_\delta \delta + M_{z_e} , \tag{B.104}$$

where the already defined derivatives of stability are not changed,

$$Y_{\dot{\delta}} = C_1 \frac{e}{V} , \tag{B.105}$$

$$N_{\dot{\delta}} = \left[C_1 a - (M_{z_1})_{,\alpha}\right] \frac{e}{V} . \tag{B.106}$$

The expression of Q_ϕ is the same as seen for the locked controls model

$$Q_\phi = M_v v + M_r \dot{\psi} + M_{\dot{\delta}} \dot{\delta} + M_\phi \phi + M_\delta \delta , \tag{B.107}$$

where

$$\left\{ \begin{array}{l} M_v = \frac{1}{V}\left[C_1 e + (M_{z_1})_{,\alpha}\cos\left(\eta\right)\right] , \\[2mm] M_r = \frac{a}{V}C_1\left[C_1 e + (M_{z_1})_{,\alpha}\cos\left(\eta\right)\right] , \\[2mm] M_{\dot{\delta}} = -\frac{e}{V}\left[C_1 e + (M_{z_1})_{,\alpha}\cos\left(\eta\right)\right] , \\[2mm] M_\phi = -e(F_{y_1})_{,\gamma} , \\[2mm] M_\delta = -\left[C_1 e + (M_{z_1})_{,\alpha}\cos\left(\eta\right)\right]\cos\left(\eta\right) . \end{array} \right. \tag{B.108}$$

B.5.2   Final expression of the linearized, free controls equations

The four equations describing the lateral free controls motions are

$$
\begin{bmatrix}
m & 0 & -mh & 0 \\
 & J_z & J_{xz} & J_{z1}\cos(\eta) + J_{xz1}\sin(\eta) \\
 & & J_x & -J_{z1}\sin(\eta) + J_{xz1}\cos(\eta) \\
 & \text{symm.} & & J_{z1}
\end{bmatrix}
\begin{Bmatrix}
\dot{v} \\ \ddot{\psi} \\ \ddot{\phi} \\ \ddot{\delta}
\end{Bmatrix} +
$$

$$
+
\begin{bmatrix}
-Y_v & mV - Y_r & 0 & -Y_{\dot\delta} \\
-N_v & -N_r & N_g & -N_{\dot\delta} - S^* \\
-L_v & -mhV - VN_g & 0 & -VC^* \\
-M_v & -M_r + VS^* & +VC^* & -M_{\dot\delta} + c_\delta
\end{bmatrix}
\begin{Bmatrix}
v \\ \dot{\psi} \\ \dot{\phi} \\ \dot{\delta}
\end{Bmatrix} +
\qquad \text{(B.109)}
$$

$$
+
\begin{bmatrix}
0 & 0 & -Y_\phi & -Y_\delta \\
0 & 0 & mh\dot V + \dot V\left(\frac{J_{p1}}{R_{e_1}} + \frac{J_{p2}}{R_{e_2}}\right) - N_\phi & -N_\delta - \dot V S^* \\
0 & 0 & -mgh - L_\phi & -\dot V C^* \\
0 & 0 & -M_\phi & -M_\delta
\end{bmatrix}
\begin{Bmatrix}
y \\ \psi \\ \phi \\ \delta
\end{Bmatrix} =
$$

$$
=
\begin{Bmatrix}
Y_\delta \delta + F_{y_e} \\
N_\delta \delta + M_{z_e} \\
0 \\
M_g
\end{Bmatrix},
$$

where

$$
S^* = \frac{J_{p1}}{R_{e_1}}\sin(\eta) , \quad C^* = \frac{J_{p1}}{R_{e_1}}\cos(\eta) .
$$

From matrices $\mathbf{M}$, $\mathbf{C}$ and $\mathbf{K}$, defined in the configurations space, it is immediately possible to obtain the dynamic matrix in the state space. In this case the first two equations are first- order differential equations, while the others are second-order equations. The order of the set of equations is then 6 and the dynamic matrix has 6 rows and columns.

B.5.3   Free controls stability

The stability with free controls may be studied simply by searching the eigenvalues of the' dynamic matrix $\mathbf{A}$. The eigenproblem usually yields two real eigenvalues and two complex conjugated pairs.

Out of the real solutions, one is negative and has practically no importance in the behavior of the system, while the other is positive and hence unstable. As in the locked control model, the latter corresponds to the capsize mode and must be stabilized by the driver or by some control device. This eigenvalue decreases with increasing speed, as gyroscopic moments of the wheels reduce the velocity at which the motorcycle falls on its side.

The two complex conjugate pairs are related to the so-called weave and wobble modes: The first, already seen in the locked control study, is primarily a yaw oscillation of the whole vehicle but involves the roll and steering degrees of freedom as well, while the other is primarily an oscillation of the steering system about its axis. The weave frequency is lower than the wobble frequency, and is usually more dependent on speed.

While the first mode is usually damped, the latter can become unstable, particularly at high speed. To stabilize wobble motion it is possible to introduce a steering damper, which has been included in the present mathematical model. The damper has the effect of reducing wobble instability, but may affect weave negatively. Too large a damping can trigger weave instability: The value of coefficient c_δ must be chosen with care and the present mathematical model can supply useful guidelines.

Wobble motion

A simplified model may also be built for wobble motion. If the steering system is considered separately from the vehicle using a model similar to that seen in Fig. B.3 for the vehicle body, the following equation of motion is obtained,

$$J\ddot{\delta} + \frac{C_1 e^2 \cos^2(\eta)}{V}\dot{\delta} + eC_1\delta \cos^2(\eta) = 0, \tag{B.110}$$

where J is the moment of inertia of the steering system about the steering axis. If a steering damper is present, its damping coefficient must be added to the term

$$\frac{C_1 e^2 \cos^2(\eta)}{V}$$

multiplying $\dot{\delta}$. Note that the damping of the steering damper must also be added in the equation related to weave.

The pole for wobble motion is

$$s = \cos(\eta) \frac{-C_1 e^2 \cos(\eta) \pm \sqrt{C_1^2 e^4 \cos^2(\eta) - 4V^2 J e C_1}}{2VJ}. \tag{B.111}$$

At low speed there are two negative real poles, so that wobble motion is stable. Starting from the speed

$$V = \frac{e\cos(\eta)}{2}\sqrt{\frac{C_1 e}{J}}, \tag{B.112}$$

the roots become complex and the motion is oscillatory. It is initially damped, but with increasing speed the real part may become positive, producing an unstable motion.

Example B.3 *Study the free controls stability of the motorcycle studied in the previous example and plot the eigenvalues as functions of the speed and the roots locus.*

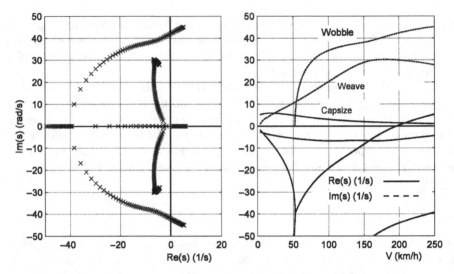

FIGURE B.6. Free controls stability study for the motorcycle of Appendix E.10. Roots locus and plot of the eigenvalues versus the speed

Operating as seen in the previous examples, the plots shown in Fig. B.6 are obtained.

Wobble motion becomes unstable starting from about 200 km/h.

The values of the natural frequencies and the time constant for capsize at three values of the speed are

V km/h	s_{cap} 1/s	τ_{cap} s	s_{weave} 1/s	s_{wob} 1/s
0	4.37	0.228	–	–
100	3.07	0.326	$-6.74\pm20.29\,i$	$-19.04\pm34.46\,i$
200	1.40	0.711	$-6.11\pm29.95\,i$	$+0.43\pm42.35\,i$

V km/h	τ_{cap}^{*} s	τ_{cap}^{**} s	s_{weave}^{*} 1/s	s_{wob}^{*} 1/s
0	0.33	0.33	–	–
100	0.33	0.37	$-10.8\pm17.6i$	$-2.60\pm38.89\,i$
200	0.33	0.41	$-4.26\pm17.8\,i$	$-1.11\pm36.03\,i$

*Wobble motion becomes unstable at about 200 km/h. The values obtained using the simplified models (values with *) are reported together with those obtained using the present model.*

The computation was repeated assuming that a steering damper with $c_\delta = 25$ Nms/rad is present. The relevant results are plotted in Fig. B.7. The damper stabilizes wobble motion up to more than 250 km/h, but reduces weave stability.

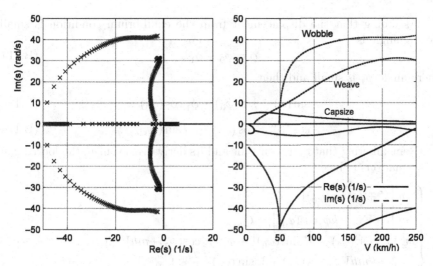

FIGURE B.7. Free controls stability study for the motor cycle of Appendix E.10. Roots locus and plot of the eigenvalus versus the speed. Steering damper with $c_\delta = 25$ Nms/rad

B.5.4 Steady-state response

The steady-state response may be obtained directly from Eq. (B.109) by introducing the steering moment M_s as an unknown. The equation is

$$
\begin{bmatrix}
C_{11} & K_{13} & K_{14} & 0 \\
C_{21} & K_{23} & K_{24} & 0 \\
C_{31} & K_{33} & K_{34} & 0 \\
C_{41} & K_{43} & K_{44} & -1
\end{bmatrix}
\left\{
\begin{array}{c}
V\beta \\
\phi \\
\delta \\
M_s
\end{array}
\right\}
= r
\left\{
\begin{array}{c}
-C_{12} \\
-C_{22} \\
-C_{32} \\
-C_{42}
\end{array}
\right\}
+
\left\{
\begin{array}{c}
F_{y_e} \\
M_{z_e} \\
-hF_{y_e} \\
eF_{y_{e_1}}
\end{array}
\right\}. \quad \text{(B.113)}
$$

It is then possible to obtain, together with the path curvature gain, the sideslip angle gain and the roll angle gain, as well as the steering moment gain M_s/δ.

B.6   STABILITY AT LARGE ROLL ANGLES

B.6.1 Locked controls motion

In fast driving the roll angle of a motorcycle may become even larger than $45°$. In these conditions the linearized model seen in the previous sections cannot be used. However, even when managing a curve at high speed, the sideslip angles and the steering angle are not large, and the only source of nonlinearity is the roll angle. In these conditions stability may be studied by linearizing the equations about a static equilibrium condition, characterized by a value ϕ_0 of the roll angle, that may be computed for example using Eq. (25.45).

Assuming that the displacement from the equilibrium condition is small, the roll angle is

$$\phi = \phi_0 + \phi_1 , \tag{B.114}$$

where angle ϕ_1 is small and then

$$\sin(\phi) = \sin(\phi_0) + \phi_1 \cos(\phi_0) , \tag{B.115}$$

$$\cos(\phi) = \cos(\phi_0) - \phi_1 \sin(\phi_0) . \tag{B.116}$$

Remembering that ϕ_0 is constant and its derivatives vanish, the locked controls equations reduce to

$$
\begin{cases}
m\left[\dot{V} + h\ddot{\psi}\sin(\phi_0)\right] = Q_x \\
m\left[\dot{v} + V\dot{\psi} - h\ddot{\phi}_1\cos(\phi_0)\right] = Q_y \\
\ddot{\psi}\left[J_z\cos^2(\phi_0) + J_y^*\sin^2(\phi_0)\right] + \ddot{\phi}_1 J_{xz}\cos(\phi_0) + mh\dot{V}\sin(\phi_0) + \\
\quad + mh\dot{V}\phi_1\cos(\phi_0) + \dot{V}\sin(\phi_0)\left(\frac{J_{p_1}}{R_{e_1}} + \frac{J_{p_2}}{R_{e_2}}\right) + \\
\quad + \dot{V}\phi_1\cos(\phi_0)\left(\frac{J_{p_1}}{R_{e_1}} + \frac{J_{p_2}}{R_{e_2}}\right) + V\dot{\phi}_1\cos(\phi_0)\left(\frac{J_{p_1}}{R_{e_1}} + \frac{J_{p_2}}{R_{e_2}}\right) = Q_\psi \\
\ddot{\phi}J_x^* + \ddot{\psi}J_{xz}\cos(\phi_0) - mh\cos(\phi_0)\dot{v} - V\dot{\psi}\cos(\phi_0)\left(mh + \frac{J_{p_1}}{R_{e_1}} + \frac{J_{p_2}}{R_{e_2}}\right) + \\
\quad - mgh\sin(\phi_0) - mgh\phi_1\cos(\phi_0) = Q_\phi .
\end{cases}
\tag{B.117}
$$

The generalized forces are

$$
\begin{cases}
Q_x = F_{x_1} + F_{x_2} + F_{xa} , \\
Q_y = F_{y_1} + F_{y_2} + F_{ya} , \\
Q_\psi = F_{y_1}a - F_{y_2}b + M_{z_1} + M_{z_2} - F_{xa}h\sin(\phi_0) - F_{xa}h\phi_1\cos(\phi_0) + M_{za} , \\
Q_\phi = -hF_{ya}\cos(\phi_0) - F_{za}h\sin(\phi_0) - F_{za}h\phi_1\cos(\phi_0) + M_{xa} .
\end{cases}
\tag{B.118}
$$

Assuming that the expression for the sideslip angles of the wheels is that seen for small roll angles(an assumption that may be justly criticized at least for the front wheel), the expression for Q_y and for the derivatives of stability Y_v, Y_r, Y_ϕ and Y_δ are those already seen.

Q_ψ and the derivatives of stability N_v, N_r and N_δ are not changed, but derivative N_ϕ becomes

$$N_\phi = a(F_{y_1})_{,\gamma} - b(F_{y_2})_{,\gamma} - \frac{1}{2}\rho V_r^2 ShC_x\cos(\phi_0) \tag{B.119}$$

and a moment

$$N_0 = -\frac{1}{2}\rho V_r^2 SC_x h\sin(\phi_0), \tag{B.120}$$

due to aerodynamic drag, must be added to the external moment M_{z_e}.

As far as Q_ϕ is concerned, derivatives L_v and L_ϕ become

$$
\begin{cases}
L_v = \frac{1}{2}\rho V_r S\left[l(C_{M_x})_{,\beta} - h(C_y)_{,\beta}\cos(\phi_0)\right] , \\
\\
L_\phi = -\frac{1}{2}\rho V_r^2 ShC_z\cos(\phi_0) ,
\end{cases}
\tag{B.121}
$$

and a term

$$L_0 = -\frac{1}{2}\rho V_r^2 S C_z h \sin(\phi_0)$$
(B.122)

must be added.

B.6.2   Equilibrium condition

The computation of the steady-state equilibrium condition at constant speed reduces to the computation of either v, $\dot\psi$ and ϕ_0 (in this condition ϕ_1 is equal to zero) at a given value of δ or v, δ and ϕ_0 for a given value of $\dot\psi$. As usual the linearized equations may be uncoupled and longitudinal dynamics does not affect handling.

The other three equations may be written as

$$\begin{cases} Y_v v + (Y_r - mV)\,\dot\psi + Y_\phi \phi_0 + Y_\delta \delta + F_{y_e} = 0\,, \\ N_v v + N_r r + N_\phi \phi_0 + N_\delta \delta + N_0 + M_{z_e} = 0\,, \\ V\dot\psi \cos(\phi_0)\left(mh + \frac{J_{p_1}}{R_{e_1}} + \frac{J_{p_2}}{R_{e_2}}\right) + L_\phi \phi_0 + \\ \qquad + L_v v + L_0 + mgh \sin(\phi_0) = 0\,. \end{cases}$$
(B.123)

Because unknown ϕ_0 is present in an explicit way in the trigonometric functions and also in an implicit way in N_ϕ, N_0, L_v, L_ϕ and L_0, it may be expedient to choose a value of ϕ_0 and then solve the equations in v, $\dot\psi$ and δ, i.e., choose a value of the roll angle and then compute the parameters of the resulting path, obviously after also choosing a value for the speed. If the steady-state values are v_0 and $\dot\psi_0$, the equation becomes

$$\begin{bmatrix} -Y_v & mV - Y_r & -Y_\delta \\ -N_v & -N_r & -N_\delta \\ -L_v & -V\cos(\phi_0)\left(mh + \frac{J_{p_1}}{R_{e_1}} + \frac{J_{p_2}}{R_{e_2}}\right) & 0 \end{bmatrix} \begin{Bmatrix} v_0 \\ \dot\psi_0 \\ \delta_0 \end{Bmatrix} =$$
(B.124)

$$= \begin{Bmatrix} Y_\phi \phi_1 + F_{y_e} \\ N_\phi \phi_1 + N_\delta \delta + N_0 + M_{z_e} \\ L_\phi \phi_0 + L_0 + mgh \sin(\phi_0) \end{Bmatrix}\,.$$

It is immediately clear that, if the aerodynamic terms are neglected, the last equation uncouples and becomes

$$\tan(\phi_0) = -\frac{V\dot\psi}{mgh}\left(mh + \frac{J_{p_1}}{R_{e_1}} + \frac{J_{p_2}}{R_{e_2}}\right)\,,$$
(B.125)

which coincides with Eq. (25.45)[2].

---

[2]The $(-)$ sign comes from the fact that if the curve is toward the left (positive $\dot\psi$) the vehicle tilts to the left (negative ϕ).

B.6.3 Stability at constant speed

The term $\ddot{\psi}$ present in the first equation of motion weekly couples this equation to the others. If this term is neglected, and stating that

$$v = v_0 + v_1 , \quad \dot{\psi} = \dot{\psi}_0 + \dot{\psi}_1$$

(because motion occurs with locked controls $\delta = \delta_0$), it follows that

$$\begin{bmatrix} m & 0 & -mh\cos(\phi_0) \\ 0 & J_z\cos^2(\phi_0) + J_y^*\sin^2(\phi_0) & J_{xz}\cos(\phi_0) \\ -mh\cos(\phi_0) & \dot{\psi}J_{xz}\cos(\phi_0) & J_x^* \end{bmatrix} \begin{Bmatrix} \dot{v}_1 \\ \ddot{\psi}_1 \\ \ddot{\phi}_1 \end{Bmatrix} + \tag{B.126}$$

$$+ \begin{bmatrix} -Y_v & mV - Y_r & 0 \\ -N_v & -N_r & N_g\cos(\phi_0) \\ -L_v & -(mhV+N_g)\cos(\phi_0) & 0 \end{bmatrix} \begin{Bmatrix} v_1 \\ \dot{\psi}_1 \\ \dot{\phi}_1 \end{Bmatrix} + \tag{B.127}$$

$$+ \begin{bmatrix} 0 & 0 & -Y_\phi \\ 0 & 0 & -N_\phi \\ 0 & 0 & -mgh\cos(\phi_0) - L_\phi \end{bmatrix} \begin{Bmatrix} y \\ \psi \\ \phi_1 \end{Bmatrix} = \mathbf{0}.$$

Locked controls stability in the small for a motorcycle managing a curve in steady-state condition with an arbitrarily large roll angle can thus be studied.

B.6.4 Free controls model

Operating as seen in previous models, the following equations may be written

$$\begin{cases} m\left[\dot{V} + h\ddot{\psi}\sin(\phi_0)\right] = Q_x , \\[2mm] m\left[\dot{v}+V\dot{\psi} - h\ddot{\phi}_1\cos(\phi_0)\right] = Q_y , \\[2mm] \ddot{\psi}\left[J_z\cos^2(\phi_0) + J_y^*\sin^2(\phi_0)\right] + \ddot{\delta}\left[J_{z1}\cos(\eta) + J_{xz1}\sin(\eta)\cos(\phi_0)\right] + \\[2mm] + \ddot{\phi}_1 J_{xz}\cos(\phi_0) - V\frac{J_{p1}}{R_{e_1}}\dot{\delta}\sin(\eta)\cos(\phi_0) - \dot{V}\frac{J_{p1}}{R_{e_1}}\delta\sin(\eta)\cos(\phi_0) + \\[2mm] + mh\dot{V}\phi_1\cos(\phi_0) + \dot{V}\sin(\phi_0)\left(\frac{J_{p1}}{R_{e_1}} + \frac{J_{p2}}{R_{e_2}}\right) + mh\dot{V}\sin(\phi_0) + \\[2mm] + \dot{V}\phi_1\cos(\phi_0)\left(\frac{J_{p1}}{R_{e_1}} + \frac{J_{p2}}{R_{e_2}}\right) + V\dot{\phi}_1\cos(\phi_0)\left(\frac{J_{p1}}{R_{e_1}} + \frac{J_{p2}}{R_{e_2}}\right) = Q_\psi , \end{cases} \tag{B.128}$$

$$\begin{cases} \ddot{\phi}J_x^* + \ddot{\psi}J_{xz}\cos(\phi_0) + \ddot{\delta}\left[-J_{z1}\sin(\eta) + J_{xz1}\cos(\eta)\right] - mh\dot{v}\cos(\phi_0)+ \\[2mm] \qquad -V\dot{\psi}\cos(\phi_0)\left(mh + \frac{J_{p1}}{R_{e_1}} + \frac{J_{p2}}{R_{e_2}}\right) - V\frac{J_{p1}}{R_{e_1}}\dot{\delta}\cos(\eta) + \\[2mm] \qquad - \dot{V}\frac{J_{p1}}{R_{e_1}}\delta\cos(\eta) - mgh\sin(\phi_0) - mgh\phi_1\cos(\phi_0) = Q_\phi\ , \\[3mm] J_{z1}\ddot{\delta} + \ddot{\psi}\left[J_{z1}\cos(\eta) + J_{xz1}\sin(\eta)\cos(\phi_0)\right] + \ddot{\phi}\left[-J_{z1}\sin(\eta) + \right. \\[2mm] \qquad \left. +J_{xz1}\cos(\eta)\right] + V\frac{J_{p1}}{R_{e_1}}\left[\dot{\psi}\sin(\eta)\cos(\phi_0) + \dot{\phi}\cos(\eta)\right] = Q_\delta\ . \end{cases}$$

The four equations describing the lateral free controls motion about a position defined by the values v_0, $\dot{\psi}_0$, ϕ_0 and δ_0 (in this case δ is a variable and may be written as $\delta_0 + \delta_1$) and thus allowing free controls stability to be studied, are

$$\begin{bmatrix} m & 0 & -mh\cos(\phi_0) & 0 \\ & J_z^{**} & J_{xz}\cos(\phi_0) & J_{z1}\cos(\eta) + J_{xz1}\sin(\eta)\cos(\phi_0) \\ & & J_x^* & -J_{z1}\sin(\eta) + J_{xz1}\cos(\eta) \\ \text{symm.} & & & J_{z1} \end{bmatrix} \begin{Bmatrix} \dot{v}_1 \\ \ddot{\psi}_1 \\ \ddot{\phi}_1 \\ \ddot{\delta}_1 \end{Bmatrix} +$$

$$+ \begin{bmatrix} -Y_v & mV - Y_r & 0 & -Y_{\dot{\delta}} \\ -N_v & -N_r & N_g\cos(\phi_0) & -N^* \\ -L_v & -(mhV + N_g)\cos(\phi_0) & 0 & -VC^* \\ -M_v & -M_r + VS^* & VC^* & -M_{\dot{\delta}} + c_\delta \end{bmatrix} \begin{Bmatrix} v_1 \\ \dot{\psi}_1 \\ \dot{\phi}_1 \\ \dot{\delta}_1 \end{Bmatrix} +$$

(B.129)

$$+ \begin{bmatrix} 0 & 0 & -Y_\phi & -Y_\delta \\ 0 & 0 & -N_\phi & -N_\delta \\ 0 & 0 & -mgh\cos(\phi_0) - L_\phi & 0 \\ 0 & 0 & -M_\phi & -M_\delta \end{bmatrix} \begin{Bmatrix} y \\ \psi \\ \phi_1 \\ \delta_1 \end{Bmatrix} = \mathbf{0}\ ,$$

where

$$J_z^{**} = J_z\cos^2(\phi_0) + J_y^*\sin^2(\phi_0)\ , \quad S^* = \frac{J_{p1}}{R_{e_1}}\sin(\eta)\ , \qquad \text{(B.130)}$$

$$C^* = \frac{J_{p1}}{R_{e_1}}\cos(\eta)\ , \quad N^* = N_{\dot{\delta}} + VS^*\cos(\phi_0)\ . \qquad \text{(B.131)}$$

Appendix C
WHEELED VEHICLES FOR EXTRATERRESTRIAL ENVIRONMENTS

Humankind has been exploring space surrounding its home planet, the Earth, and the nearest celestial bodies since the late 1950s. Such exploration is not an occasional enterprise, but the beginning of a trend that will see our species build a true spacefaring civilization, one based on many worlds, initially in our solar system and then beyond it.

There is no doubt that the great difficulties now being encountered and those that will emerge in the future will make this expansion a slow one, requiring decades or more likely centuries to yield important results. Owing to the size of the Universe, it is an expansion that will never end. Certainly it will require technological advancements that today cannot even be imagined. But it is also true that such difficulties are only partially technological. Economical, social and political factors will play an even more significant role.

Beginning with the first experiences of robotic and human exploration of other celestial bodies, a need was felt for vehicles able to carry instruments, objects of various kinds and finally the explorers themselves.

The vehicles for planetary exploration used up to now have all been robotic vehicles operating on wheels, designed and built using technologies that have little in common with standard automotive technologies. This is due in part to their small size and above all to their low speed, which ranges from a maximum of 1,5 km/h, for the Lunokhod, to some tens of m/h, for the automatic vehicles used for Mars exploration.

At these speeds suspensions based on spring and damper systems are not necessary, although articulated systems to suitably distribute the load on the ground were needed, because all these vehicles have more than 3 wheels.

Kinematic steering is more than adequate to control the trajectory.

The only exception is the vehicle used in the last *Apollo* missions to grant the needed mobility to the astronauts. Even if more than 30 years have passed since this vehicle was used, it remains an interesting case.

C.1 THE LUNAR ROVING VEHICLE (LRV) OF THE APOLLO MISSIONS

The LRV is an electric four wheel drive and steering (4WDS) vehicle able to carry two people in space suits at a speed of 18 km/h for a maximum distance of 120 km (Fig. C.1).

The main characteristics of the LRV were[1]

- Mass: 210 kg,

- Payload: 450 kg,

- Length (overall): 3,099 mm,

- Wheelbase: 2,286 mm,

- Track: 1,829 mm,

- Maximum speed: 18 km/h,

- Maximum manageable slope: 25°,

FIGURE C.1. The *Lunar Roving Vehicle* (LRV) of the last three *Apollo* missions on the Lunar surface (NASA image)

---

[1]A. Ellery, *An Introduction to Space Robotics*, Springer Praxis, Chichester, 2000, – – –, *Lunar Roving Vehicle Operations Handbook*, ttp://www.hq.nasa.gov/office /pao/History/alsj/lrvhand.html.

- Maximum manageable obstacle: 300 mm height,

- Maximum manageable crevasse: 700 mm length,

- Range: 120 km in four traverses,

- Operating life: 78 h.

At the time it was designed the LRV was a concentrate of high automotive technology adapted to the peculiar operating conditions that could be found on the Lunar surface. It marked the first time a vehicle with four steering and driving wheels was built, with each wheel having its own motor in the hub. Steering was by wire.

The project was deeply conditioned by the mass and size constraints driven by the need to be carried to the Moon along with the astronauts by the *Lunar Excursion Module* (LEM). Apart from the need to minimize its structural mass, these constraints forced designers to use a foldable architecture; it is likely that these considerations were determinant in the use of a by-wire configuration, in that era a completely immature technology.

A short analysis of the various subsystems and some considerations on what is still viable today and what, on the contrary, has been out-paced by technological advances, is reported in the following sections.

C.1.1 Wheels and tires

Pneumatic or solid rubber tires were discarded primarily to reduce the vehicle mass. In their place tires made with an open steel wire mesh, with a number of titanium alloy plates acting as treads in the ground contact zone, were built. Inside the tire a smaller, more rigid frame acted as a stop to avoid excessive deformation under high impact loads. The outer diameter of the tire was 818 mm.

Although the possibility of using standard pneumatic tires on a planet without an atmosphere may be raised, the lack of air is not the point. The short duration of use predicted (a single mission and few working hours) allowed such an innovative technology to be chosen without the need to perform long duration tests. In future missions aimed at building permanent outposts, things will be different and more traditional solutions may be needed. It must be noted that the version of the vehicle used to train the astronauts on Earth had standard pneumatic tires.

C.1.2 Drive and brake system

The LRV had 4 independent, series wound DC brush electric motors mounted in the wheel hubs together with harmonic drive reduction gears[2]. Each motor

---

[2] *Harmonic drives* are reduction gears based on a compliant gear wheel that meshes inside an internal gear having one tooth more than the former wheel. An eccentric (wave generator)

was rated 180 W, with a maximum speed of 17,000 rpm. The gear ratio of the harmonic drives was 80:1. The nominal input voltage was 36 V, controlled by PWM from the Electronic Control Unit. The drive unit was sealed, maintaining an internal pressure of about 0.5 bar for proper lubrication and brush operation.

The energy needed for motion was supplied by two silver-zinc primary batteries with a nominal voltage of 36 V and a capacity of 115 Ah (4,14 kWh) each. The use of primary batteries derives from the fact that each vehicle had to be used for a single mission, and for a limited time.

The LRV had four cable actuated drum brakes directly mounted on the wheels. As will be seen later, braking torques are low in vehicles operating in low gravity conditions and drum brakes are a reasonable solution. They are also preferable because of dust, which may create problems for disc brakes because they are more exposed to external contamination. Even if the braking power is much smaller than that typical of vehicles used on Earth, the lack of air may cause overheating problems.

The driver interface for longitudinal control was the same T handle that actuated the steering. Advancing the joystick actuated the motors forward. Pulling it back actuated reverse, but only if the reverse switch were engaged. Actuating the brakes required the handle to be pivoted backward about the brake pivot point.

The wheels could be disengaged from the drive-brake system into a free-wheeling condition.

C.1.3 Suspensions

Suspensions were fairly standard SLA suspensions, with the upper and lower arms almost parallel. No anti-dive or anti-squat provisions seem to have been included. Springs were torsion bars applied to the two arms, while a conventional shock absorber was located as a diagonal of the quadrilateral. The ground clearance varied from full load and unloaded conditions by 76 mm (between 356 and 432 mm). These values yield a vertical stiffness of the suspension-tire assembly of 2.40 kN/m, a very low value. Assuming that the tire was much harder than the suspension, the natural frequency in bounce was just 0.6 Hz for the fully loaded vehicle or 1.1 Hz in empty conditions.

C.1.4 Steering

Steering controlled all wheels and was electrically actuated (steer by wire). The geometry was designed with kinematic steering in mind: Ackermann steering on each axle and opposite steering of the rear axle with equal angles at front

---

deforms the first wheel so that for each revolution of the eccentric the flexible wheel moves by one pitch. Very high gear ratios can thus be obtained while maintaing good efficiency and little backlash, but the high cost of such devices allows their use only in selected applications, primarily in robotics.

and rear wheels. The kinematic wall-to-wall steering radius was 3.1 m. Each steering mechanism was actuated by an electric motor through a reduction gear and a spur gear sector; in case of malfunction of one of the two steering devices, the relevant steering could be centered and blocked and the vehicle driven with steering on one axle only.

The same T handle controlling the motors and brakes operated the steering by lateral displacement. A feedback loop insured that the wheels were steered by an angle proportional to the lateral displacement of the handle, but there was no force feedback except for a restoring force increasing linearly with the steering angle up to a 9° handle angle, then increasing with a step and finally increasing again linearly with greater stiffness.

Steering control may be the most outdated part of the LRV, although in a way it was a forerunner of the 4WS and steer by wire systems currently available. The 4WS logic was based strictly on kinematic steering. The low top speed justifies this choice, but only to a point: The low gravitational acceleration of the Moon implies that sideslip angles are much higher, for a given trajectory and a given speed, than on Earth. The very concept of kinematic steering is thus applicable only at speeds much lower than on Earth. Because centrifugal acceleration is proportional to the square of the speed, the top speed $V_{max} = 18$ km/h is equivalent, from this viewpoint, to the speed at which dynamic effect starts to appear. No modern 4WS vehicle has such prominent rear axle steering. Today a much more sophisticated rear steering strategy is present in even the simplest vehicle with all-wheel steering.

A second difference is that today steer by wire systems are reversible and the driver interface is haptic, i.e. there is an actuator supplying a feedback that allows the driver to feel the wheel reaction, as in conventional mechanical steering systems. The fact that it is possible to drive without a feedback is proven by the fact that vehicles in videogames and radio controlled model cars can be operated (the control of the LRV has surprising similarities with that of R/C model cars) but it is considered unsafe and difficult to operate a full size car in this way. Because *Apollo* astronauts needed much training to operate the LRV, a purposely designed trainer simulating its performance had to be built, it being impossible to operate the lunar vehicle on Earth.

C.2   TYPES OF MISSIONS

The types of missions requiring the use of vehicles on celestial bodies can be tentatively subdivided into the following classes[3].

1. Robotic exploration missions

2. Robotic exploitation missions

---

[3]G. Genta, M. A. Perino, *Teleoperation Support for Early Human Planetary Missions*, New Trends in Astrodynamics and Applications II, Princeton, June 2005.

3. Human exploration missions with robotic rovers to help humans in exploration duties

4. Human exploration missions with vehicles to enhance human mobility

5. Human exploration-exploitation missions requiring construction-excavation devices.

Vehicle for unmanned missions (1 and 2) are usually very small and slow, more similar to moving robots than to vehicles, while those for missions 2 and 5 are more similar to earth moving and construction machinery. Even if a good knowledge of vehicle dynamics is needed to design them properly, these are machines having little in common with automotive technology. Missions of type 3 require small machines, more similar to robots than to vehicles.

Missions of type 4 require true vehicles that may differ from those used on our planet due to the different environmental conditions in which they must operate. But because the goal is essentially the same, namely to transport personnel and goods from one place to the other, with the required velocity and safety and in an economical and energetically efficient way, there will be many points in common.

The first essential difference is that of operating in an environment in which human life is impossible without suitable protection. The designer has to chose between two alternatives: A simple mobility device, without any life-support capability, whose task is to carry humans protected by their own space suit, and a vehicle providing a true shirt-sleeve environment.

The first solution, like the LRV, allows the us of small and very light vehicles, the minimum size being that of a city car. Vehicles of the second type, which are true mobile habitats, are larger and more complex machines. A modular approach may be followed, with a chassis containing all devices offering the required mobility, on which a habitat may be mounted. Current vehicles similar to this solution are military vehicles in which the inner space is completely insulated from the outside to provide some protection from chemical or bacteriological attacks. The smallest size for vehicles of this type is that of a small van.

C.3 ENVIRONMENTAL CONDITIONS

The first difference between Earth and other celestial bodies is gravitational acceleration. All celestial bodies to be explored (with the exception of Venus, where $g = 8,87$ m/s^2, but whose environment is so difficult in terms of temperature, atmospheric density and pressure that no operation on its surface is planned in the foreseeable future) are characterized by a gravitational acceleration much lower than that of Earth. (Table C.1).

All asteroids except the largest have a very low gravitational acceleration, and their irregular form causes the gravitational acceleration to be variable from

TABLE C.1. Values of the gravitational acceleration, in m/s^2, on the surface of the Moon, some rocky planets, satellites of giant planets and asteroids.

Planets	Mars	3,77	Mercury	3,59
Satellites	Moon	1,62		
	Io	1,80	Europa	1,32
	Ganimedes	1,43	Callisto	1,24
	Titan	1,35	Triton	0,78
Asteroids	Pallas	0,31	Vesta	0,28
	Ceres	0,26	Juno	0,09

place to place and not perpendicular to the ground. For instance, the gravitational acceleration on Eros varies between 0.0023 and 0.0055 m/s^2.

A low gravitational acceleration g has some advantages, from the possibility of human explorers carrying very bulky life support equipment to reduced structural stresses and to reduced power requirements for motion. Another important advantage is that the weight the vehicle-ground contact must carry is low, making floating possible. But performance of vehicles whose mobility is produced by friction forces is drastically reduced.

Although this does not limit mobility in a strict sense, i.e. the possibility for the vehicle to move, it does lower the longitudinal and lateral acceleration of the vehicle and the top speed it can travel. The maximum longitudinal and lateral accelerations of the vehicle can be assumed to be proportional to the gravitational acceleration, if the ideal braking and steering approach is used.

The second point is the thin atmosphere. On the Moon there is no atmosphere at all, as is true of all asteroids, comets and most satellites of the outer planets. The atmosphere on Mars is so thin that it can be neglected when dealing with ground locomotion, although thick enough to support aerodynamic vehicles or balloons. The only body among those mentioned in Table C.1 having an atmosphere comparable in density to Earth is Titan.

The lack of an atmosphere with a non-negligible density means no aerodynamic drag and thus no constraint to the outer shape of the vehicle. This may, however, be a negligible advantage if the speed of the vehicle is low due to poor traction. Moreover, the lack of an atmosphere makes it impossible to use aerodynamic forces to increase the forces on the ground and thus traction.

A third point is the absence of humidity. The Moon, Mars and asteroid surfaces are completely of from water. Because the presence of mud produces the most difficult mobility for wheeled vehicles, this is without any doubt a positive feature. The surfaces of comets and outer planets' satellites are rich in ice but conditions are such that the ice is far from its melting point and the pressure of the vehicle on the ground is low. The surface of the ice does not melt under the wheels, offering good traction in these conditions. The soil of Titan may contain liquid hydrocarbons, but it is unknown whether these produce conditions similar to muddy ground on Earth.

A final point is the absence of products of biological origin, important in giving the peculiar characteristics to the wide variety of soils that can be found on Earth. Celestial bodies likely to be explored first are covered with regolith, pulverized rock and gravel, with characteristics that are less variable than those of soils encountered on Earth.

The Moon, for instance, is covered with a layer of regolith about 2 or 3 m thick on the so-called maria and 3 to 16 m thick on the highlands. The mean size of grains varies between 40 and 270 μm, with a preponderance of the smallest size. The porosity of the soil is high at the surface (40 $\div$ 43%), leading to a density as low as about 1000 kg/m^3. Porosity reduces with depth and only 200 mm below the surface the density doubles to about 2000 kg/m^3.

The most important parameter for locomotion is the cohesive bearing strength, which is about 300 Pa on the surface, increasing about ten fold at 200 mm depth[4].

Maria are fairly flat, offering few obstacles and a good mobility, as demonstrated by the last *Apollo* missions. On the highlands, on the other hand, there are many craters and mountains with high slopes and obstacles. To travel through them expeditiously it will be necessary to prepare tracks as soon as the first outposts are built.

The surface of Mars is also covered with regolith, with even thinner grains owing to a more complex geological history, with erosion due to water (present in very ancient times) and wind. The types of obstacles are much more varied than on the Moon, and mobility may be quite difficult above all in the zones that are more scientifically interesting, owing to ravines, canyons, cliffs and boulders of all sizes.

C.4   MOBILITY

The low value of the pressure on the ground and the lack of water, and thus of mud, make the use of tracks less convenient than on Earth. The choice of wheeled vehicles is also suggested by their lower mass (until it is possible to build vehicles there, vehicles will be carried from Earth), complexity and energy consumption, along with their greater reliability. Other solutions, such as legged vehicles, were often considered for robots, but never for transportation vehicles.

The experience gained with the LRV confirms these considerations: That vehicle never experienced mobility problems and showed no tendency to sink into the regolith. All the pictures taken on the Moon show that the wheels of the vehicle, even when fully loaded, ride lightly on the surface (Fig. C.2a).

The sinking of a wheel with a contact area 100 mm in width is plotted versus the pressure on the ground for 3 different values of the friction angle in Fig. C.2b. The ratio between tractive and normal force F_t/F_n, in conditions of

---

[4]A. Ellery, *An Introduction to Space Robotics*, Springer Praxis, Chichester, 2000.

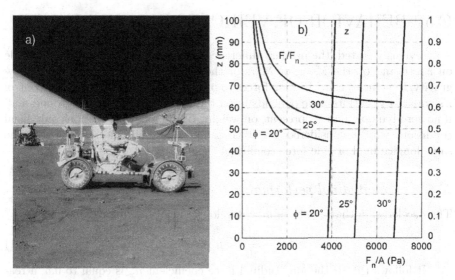

FIGURE C.2. a) Image of the Lunar Rover (LRV) with astronaut Eugene Cernan on board taken during the third extra-vehicular activity (EVA) in the *Apollo* 17 mission (NASA image). b): Sinking z and ratio between tractive and normal force F_t/F_n as functions of the normal pressure on the ground for 3 different values of the friction angle for a wheel 100 mm wide

no sinking[5] is also reported versus pressure in the same figure. The values of the friction angle here considered are reasonable, owing to the lack of humidity.

The carrying capacity of the ground seems to be low, but it is more than enough to support quite a massive vehicle owing to low gravity. Moreover, the density and the cohesive strength of regolith increase with depth. Thus the carrying capacity increases quickly with small values of sinking, much more than that shown in the figure. The figure was plotted assuming that the properties of the ground are constant, equal to those characterizing the surface layer.

Finally, rolling resistance is expected to be quite low, owing to the limited sinking. Its order of magnitude should be similar to that found on roads with a natural surface on Earth and thus, for a pneumatic wheel, f_0 should be about 0,05.

Similar considerations should also hold for Mars, where the soil is also composed of dry regolith.

In terms of obstacle management, wheels can manage at low speed obstacles whose height is on the order of magnitude of their radius or more if particular types of suspensions are used, and can traverse ditches whose width is no more than 70% of their diameter, unless the vehicle layout allows operation with some wheels off the ground.

---

[5]G. Genta, *Design of Planetary Exploration Vehicles*, ESDA 2006, Torino, Luglio 2006.

C.5   BEHAVIOR OF VEHICLES IN LOW GRAVITY

As previously stated, the primary difference between operating a ground vehicle on Earth and on the Moon or Mars or the other bodies to be explored is low gravity. In vehicles using friction forces to propel and brake the vehicle or control its trajectory, performance depends upon the forces exerted on the ground and, if no aerodynamic force is present, on weight. In the following sections only ideal conditions will be considered; i.e. it is assumed that all wheels work with the same longitudinal or side force coefficient.

C.5.1   Longitudinal performance

The deceleration that a vehicle may develop in ideal braking conditions is

$$\frac{dV}{dt} = \mu_x g \ . \tag{C.1}$$

It follows that if the longitudinal force coefficient μ_x is equal to 0.5, a reasonable value on non-prepared ground, the maximum deceleration that may be obtained on the Moon ($g = 1,62$ m/s^2) is 0,8 m/s^2. Obviously the deceleration the vehicle may actually reach is lower, and may be computed in the usual way.

Load transfer does not depend upon the gravitational acceleration, but only upon the longitudinal force coefficient: If a vehicle brakes with a certain value of μ_x, it has the same load shift (obviously in a relative sense) on Earth and on the Moon. The curve $\eta_f(\mu_x)$ is then the same, for a given braking system. Note that the variability of μ_x may be lower on the Moon than on Earth, because it is influenced by the presence of water or ice on the ground (possible only on Earth). A good knowledge of the lunar ground thus makes it possible to design braking system operating at higher efficiency than usual.

The distance needed to stop the vehicle, computed assuming uniformly decelerated motion for values of the deceleration included between 0,2 and 1 m/s^2, is shown in Fig. C.3.

The same holds for acceleration: On Earth the limitations in acceleration are set primarily by the available power, except for very powerful vehicles or extremely poor road conditions; on the Moon it is likely that the limitations come from the power that can be transferred to the ground. A layout with all wheel drive is advisable to ensure reasonable performance in acceleration.

The low traction available and the simultaneous presence of longitudinal load transfer makes the use of ABS and traction control devices necessary, even more so than on Earth, although the lessened variability of ground characteristics make these systems less useful.

Reduced gravitational acceleration has no effect on the ability to manage slopes, but does reduce the required power. The steepest slope that can be managed by a 4WD vehicle, assuming ideal driving conditions (i.e. that all wheels work with the same longitudinal force coefficient) is given by the usual formula

$$\tan(\alpha_{max}) = i_{max} = \mu_{x_p} \ . \tag{C.2}$$

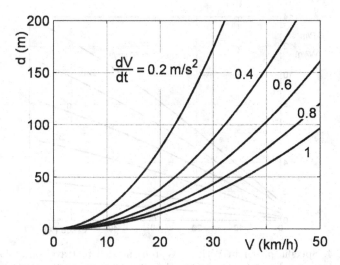

FIGURE C.3. Distance needed to stop the vehicle at constant deceleration versus the speed for values of the deceleration included between 0,2 and 1 m/s².

Assuming a value of the force coefficients $\mu_x = 0.5$, a maximum slope $\alpha_{max} = 27°$, i.e. a 46% grade, is obtained. This can, however, be obtained only in ideal conditions, and only when using accurate slip control of the wheels.

Taking the difference of the force coefficient at the front and rear wheels into account, if the height of the center of mass on the ground is $h_G = a/2$, the values $\alpha_{max} = 22°$, corresponding to $i_{max} = 38\%$, are obtained for a 4WD vehicle. These values are fairly good, particularly if the low value of the force coefficient used is considered.

The specific power in W/kg (kW/ton) needed to travel on a slope on the Moon is reported in Fig. C.4 as a function of the slope. A power of just 2 kW allows a 1 ton vehicle to travel on level road at more than 50 km/h and to overcome a 40 % grade at 10 km/h.

C.5.2   Handling

Referring to the concept of ideal steering, if there are no aerodynamic forces and the road is level, the maximum centrifugal acceleration is

$$\frac{V^2}{R} = g\mu_y , \qquad (C.3)$$

where the value of μ_y is that of the whole vehicle. The relationship between the minimum radius of the trajectory and the speed in the lunar environment for various values of μ_y is shown in Fig. C.5. It has been assumed in plotting the figure that the value of μ_y for the vehicle is 70% of that referred to the tires.

Again, in spite of the poor performance, the transversal load transfer is not less than on Earth, with higher centrifugal accelerations but the same value of

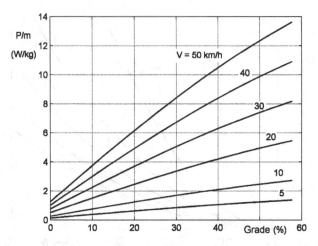

FIGURE C.4. Specific power in W/kg (kW/ton) needed to travel on a slope on the Moon at various speeds, with $f_0 = 0,05$ and $\eta_t = 0,9$.

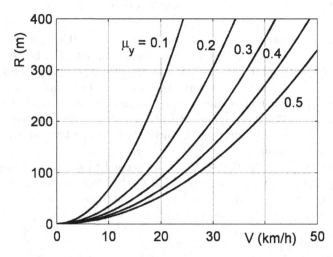

FIGURE C.5. Relationship between the radius of the trajectory and the maximum speed for a vehicle on the Moon with different values of the cornering force coefficient μ_y. The sliding factor is assumed to be 70% of the cornering force coefficient of the wheel

the force coefficient μ_y. The only point is that the tires are likely to be oversized owing to the low load (see below), and may work in the part of the $C\left(F_z\right)$ (cornering stiffness-load) curve where load transfer is less important, at least using linearized models.

In spite of the low speed, the very low cornering forces available may make the use of modern stability enhancement (ESP, VDC, etc.) systems worthwhile.

C.5.3 Comfort

Comfort is little influenced by the gravitational acceleration of the planet on which the vehicle moves: The criteria for bounce and pitch motions, suspension damping and other suspension characteristics developed for vehicles on Earth hold on other worlds. There is even an advantage: One limitation to spring softness in vehicular suspensions comes from the need to limit suspension travel with changing load. In low gravity, if the springs are designed with dynamic considerations in mind, the static deflection under load is small presenting no constraints upon suspension softness. The only limit in this area is the need to avoid bounce and pitch frequencies that are too low. These may cause motion sickness.

The human side is another matter. We know little about how a human body accustomed to no gravity after a few days of space travel will react to vibration under low gravity conditions. The usual guidelines may not apply as they do on Earth. The LRV, for example, had a bounce frequency in full loaded conditions that was too low for comfort, but it is not known that any astronaut suffered from motion sickness while driving on the Moon. Further studies are needed, but they must be conducted on site, because low gravity cannot be properly simulated on Earth.

However, low gravity does cause an unwanted effect on bounce and pitch motion: The wheels tend to lift from the ground, as is apparent in the movies taken in the *Apollo* missions[6]. It is obvious that at reduced gravitational acceleration, inertia forces become more important with respect to weight, causing the difficulty in maintaining a good wheel-ground contact when travelling on uneven ground to increase. This effect can be evidenced by plotting the ratio between the minimum vertical force (static value minus amplitude of the dynamic force) and the static force as a function of the frequency when driving on a road with a harmonic profile of a given amplitude.

A plot of this type for a quarter car with two degrees of freedom with a sprung mass of 240 kg, unsprung mass of 38 kg, tire stiffness 135 kN/m, suspension stiffness 15,7 kN/m and optimum damping (1,52 kNs/m) is shown in Fig. C.6. The results for the lunar environment are compared with those obtained for Earth.

From the figure it is clear that while on Earth the suspension maintains contact with the ground even at an amplitude of 10 mm, on the Moon an amplitude as small as 2 mm causes the wheel to bounce at a frequency of about 9 Hz. For larger amplitudes, the contact of the wheel on the ground is quite uncertain, with ensuing reduction of the already poor traction and cornering forces. This problem may be lessened by increasing the damping of the shock absorber or decreasing the stiffness of the spring, but this would in turn lessen comfort.

---

[6]If a further clue were needed, those moves demonstrate that the astronauts were actually on the Moon. At the speed reached by the LRV, the dynamic behavior of the vehicle is inconsistent with conditions on Earth.

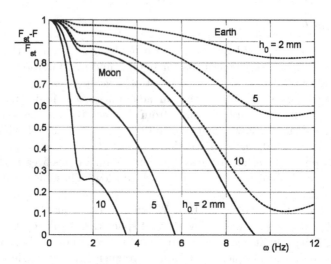

FIGURE C.6. Ratio between the minimum vertical force on the ground and the static force for a quarter car model as a function of the frequency when driving on Earth (dashed lines) and on the Moon (full lines) on a road with harmonic profile at a given amplitude h_0.

The ground contact problem could suggest the use of at least semi-active suspensions, or even fully active ones. An interesting possibility is the use of electromagnetic damping because of the difficulties of cooling standard shock absorbers in the vacuum of space (Moon) or in a very thin atmosphere (Mars).

C.6   POWER SYSTEM

Travelling in low gravity conditions requires much less power than on our planet: Rolling resistance is much lower, aerodynamic drag is absent (and at low speeds like those that can be reached in these conditions it would be negligible even in an Earth-like atmosphere). Low traction prevents accelerations requiring high power from being obtained.

The low top speed, linked to the difficulties in acceleration, braking and cornering, reduces the power requirements in unstationary conditions. In these conditions electric drive powered by batteries seems to be a good solution, especially because the cost problems that prevent the use of high performance batteries on normal vehicles are much less important. If the vehicle must be used for prolonged periods of time, rechargeable batteries will be used instead of the primary batteries used on the LRV. They will be charged by the solar (or better nuclear) power system of the outpost.

An alternative may be fuel cells, using locally produced oxidizer and fuel.

Other more conventional solutions, such as internal combustion engines of a more or less traditional type fed by locally produced fuel are also viable. On

Mars, for instance, production of oxygen and methane from ice, surely present in many places on the planet, and from the abundant atmospheric carbon dioxide seems a good choice.

Radioisotope thermoelectric generators (RTG) may be a good choice for small vehicles, while larger ones may use, at least in principle, a small nuclear reactor directly, even if the use of batteries recharged from a stationary reactor seems more convenient.

Some power is also needed for thermal control. On the Moon, for instance, thermal excursions are extreme and the vehicle, even if not operating, must be protected from the cold during a night that lasts 14 days. The thermal control systems are not different from those used on space vehicles, and will not be dealt with here. The same is true of life support systems, although these may not be a part of the vehicle, but may be a part of the habitat carried on board.

C.7 CONCLUSIONS

The examples above dealt mostly with the Moon, the body where the most likely need for ground transportation will first arise. On Mars the effects will be less severe, but the situation will be more or less similar.

When vehicles for the exploration and the exploitation of asteroids are designed, significantly different problems will be encountered, because of the low gravitational acceleration and the small size of these bodies. The first feature makes it necessary to proceed with extreme caution to ensure that inertial forces do not make the vehicle lift off and become lost in space. The second makes it useless to move at speeds above a low value. It is likely that vehicles used on asteroids and comets will have little in common with automotive technology and will be small space vehicles.

On the other hand, the vehicles that will carry explorers and then colonists on the surface of the Moon and then on Mars will be essentially similar to motor vehicles and their designers will need a good knowledge of automotive technology. Unlike of what was feared in 1960, lunar and martian soil is not at all difficult and, at least in the flatter zones, offers good carrying capacity and fair traction because of the lack of humidity. There is no problem in moving at moderate speed, even without building permanent infrastructures.

It will be harder to reach places in more difficult areas, which on Mars are the most interesting from a scientific viewpoint, and it is likely that it will be necessary to build, at the least, dirt roads or tracks.

Worse problems will be encountered when mobility at higher speed is necessary. Here the worst problems will come from the low gravitational acceleration. Roads with a good surface and large radii may be a partial solution, but it is likely that other approaches will be followed, such as hopping vehicles (performing parabolic flights is not energetically infeasible, owing to low gravity), and above all guided vehicles on rails or, more likely, maglev vehicles. The latter may be particularly suitable to low gravity conditions, but these are solutions for a more distant future, when traffic density will justify building costly infrastructures.

In the short and medium term robotic and manned vehicles will be based on more or less standard automotive technology, benefiting from the recent advances of the latter. Drive and brake by wire may allow performance satisfying the needs of planetary exploration with the required safety, while semi-active and active suspensions will increase comfort to acceptable levels.

Appendix D
PROBLEMS RELATED TO ROAD ACCIDENTS

As discussed in Part III, the cost of accidents due to the use of motor vehicles, both in terms of human lives and economic losses, is quite high. the goal of increasing the safety of motor vehicles is generally considered a technical and social priority.

The actions taken to reduce both the number and severity of accidents are both technical and legal and involve many disciplines. Automotive engineers are involved both in the design of the vehicle and, as a consultant of courts or law makers, in the reconstruction of accidents. Their aim is to ascertain responsibilities and introduce into the standards and rules those provisions necessary to prevent accidents or reduce their consequences.

It must be noted that the reconstruction of an accident is often a difficult task. The expert, who usually acts as a consultant of the court or of one of the persons involved, may have only a partial knowledge of the situation. The only known data are in many cases the positions of the vehicles after the accident plus any marks that may remain on the road, and sometimes even these are uncertain and affected by large errors. A road accident, occurring in a very short time and usually in an unexpected way, can have a large psychological impact on witnesses and protagonists. This, together with the limited technical knowledge of the persons involved and the economic interests at stake, may make the reports of witnesses difficult to interpret and weigh correctly, particularly when they lead to conflicting reconstructions.

The traditional methods used in the reconstruction of accidents are based on rough approximations, which are, however, justified by the uncertainties and sometimes the quick variations of the parameters of the problem. Only the use of more elaborated numerical models, implemented on computers, can produce higher accuracy.

D.1   VEHICLE COLLISION: IMPULSIVE MODEL

A frequent scenario in the reconstruction of accidents is a collision between vehicles or between a vehicle and an obstacle. The simplest model deals with the collision as an impulsive phenomenon, i.e., it assumes that the time during which the vehicles remain in contact (typically on the order of 0.1 s) is vanishingly short and that the forces they exchange are infinitely large. Mathematically the forces may be represented by a Dirac impulse function. Their study is based on the momentum theorem, stating that the variation of the momenta of the vehicles is equal to the impulse of the forces they exchange.

Because the forces the vehicles exchange during the collision are larger by orders of magnitude than the other forces acting on them, this approach is quite correct, even if it has the disadvantage of not allowing us to study what happens during the impact but only how the motion changes between instant t_1 preceding the collision and instant t_2 following it.

Remark D.1 *It is meaningless to ask what happens during a phenomenon whose duration is zero by definition.*

D.1.1   Central head-on collision

The simplest case is that of a head-on collision in which the velocities of the centres of mass of the vehicles lie along the same straight line (Fig. D.1a). Actually, if the velocities $\dot{x}_A$ and $\dot{x}_B$ have the same sign a rear collision occurs while a head-on collision is characterized by opposite signs of the velocities.

The relative velocity

$$V_R = \dot{x}_B - \dot{x}_A \tag{D.1}$$

must be negative; otherwise the vehicles do not approach each other.

During the collision the absolute value of the relative velocity decreases between times t_1 and time t_i, when the centres of mass are at their minimum distance. In the rebound phase, between times t_i and t_2, the relative velocity

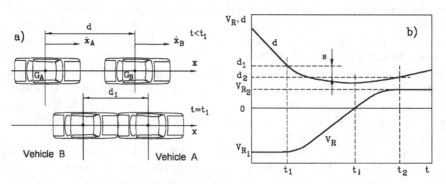

FIGURE D.1. Central head-on collision. (a) Situations before the collision and at time t_1; (b) relative velocity and distance between the centres of gravity as functions of time.

becomes positive (Fig. D.1b). The distance between the centres of mass is plotted in the same figure. The final crush s of the vehicles is the difference between d_1 and d_2.

As already said, however, the impulsive model does not show what happens between t_1 and t_2. The conservation of momentum allows the relationship to be written as

$$m_A \dot{x}_{A_1} + m_B \dot{x}_{B_1} = m_A \dot{x}_{A_2} + m_B \dot{x}_{B_2} . \tag{D.2}$$

Because the collision is generally inelastic, the kinetic energy is not conserved

$$\frac{1}{2} m_A \dot{x}_{A_1}^2 + \frac{1}{2} m_B \dot{x}_{B_1}^2 \geq \frac{1}{2} m_A \dot{x}_{A_2}^2 + \frac{1}{2} m_B \dot{x}_{B_2}^2 ; \tag{D.3}$$

the two sides being equal in the case of a perfect elastic collision.

The position and the velocity of the centre of mass of the system made by the two vehicles are simply

$$x_G = \frac{m_A x_A + m_B x_B}{m_A + m_B} , \tag{D.4}$$

$$\dot{x}_G = \frac{m_A \dot{x}_A + m_B \dot{x}_B}{m_A + m_B} . \tag{D.5}$$

A consequence of the conservation of the momentum is the conservation of the velocity of the centre of mass of the system:

$$\dot{x}_{G_1} = \dot{x}_{G_2} .$$

The velocities of the vehicles can thus be expressed as functions of the velocity of the centre of mass, which remains constant, and of the relative velocity, a function of time

$$\dot{x}_A = \dot{x}_G - V_R \frac{m_B}{m_A + m_B} , \qquad \dot{x}_B = \dot{x}_G + V_R \frac{m_A}{m_A + m_B} . \tag{D.6}$$

The kinetic energy of the system can thus be written in the form

$$T = \frac{1}{2} \left[\dot{x}_G^2 (m_A + m_B) + V_R^2 \frac{m_A m_B}{m_A + m_B} \right] . \tag{D.7}$$

As the velocity of the centre of mass is constant, the maximum energy dissipation occurs when the relative velocity vanishes. It is therefore possible to state a lower and an upper bound to the kinetic energy after the collision

$$\frac{1}{2} \dot{x}_G^2 (m_A + m_B) \leq T_2 \leq \frac{1}{2} \left[\dot{x}_G^2 (m_A + m_B) + V_{R_1}^2 \frac{m_A m_B}{m_A + m_B} \right] . \tag{D.8}$$

The minimum kinetic energy (expression on the left) occurs when the collision is perfectly inelastic and the two vehicles remain attached to each other.

In the case of a perfectly elastic collision, the absolute value of the relative velocity after the impact is equal to that at time t_1:

$$V_{R_2} = -V_{R_1} .$$

A restitution coefficient e^* is usually defined as[1]

$$e^* = -\frac{V_{R_2}}{V_{R_1}} . \tag{D.9}$$

In case of a perfectly elastic collision $e^* = 1$, while $e^* = 0$ for inelastic impacts. In all actual impacts $0 < e^* < 1$.

The energy dissipated during the collision is then

$$T_2 - T_1 = \frac{1}{2}\frac{m_A m_B}{m_A + m_B}\left(V_{R_1}^2 - V_{R_2}^2\right) = \frac{1}{2}\frac{m_A m_B}{m_A + m_B}V_{R_1}^2\left(1 - e^{*2}\right) . \tag{D.10}$$

In the first part of the collision, from time t_1 to time t_i, the relative velocity decreases to zero and the kinetic energy reduces to a minimum, given by the expression on the left side of Eq. (D.8). The energy related to velocity V_{R_1} transforms partly into elastic deformation energy, with part of it dissipated as heat. In the second part of the collision, from time t_i to time t_2, a fraction of this energy is transformed back to kinetic energy. The ratio between the energy restituted in the second phase and that subtracted from the kinetic energy in the first is e^{*2}.

In motor vehicle collisions the value of e^* is low, typically in the range of $0.05 \div 0.2$ for impacts with large permanent deformations. It depends, however, on the relative velocity and may be higher in low speed collisions, tending to unity when no permanent deformations are left.

The velocities of the vehicles after the impact are

$$\begin{cases} \dot{x}_{A_2} = \dot{x}_{A_1} + m_B V_{R_1}\dfrac{1 + e^*}{m_A + m_B} \\[4mm] \dot{x}_{B_2} = \dot{x}_{B_1} - m_A V_{R_1}\dfrac{1 + e^*}{m_A + m_B} . \end{cases} \tag{D.11}$$

D.1.2   Oblique collision

Seldom is the situation as described in Fig. D.1; usually the velocity vectors of each of the two vehicles do not pass through the centre of mass of the other as in Fig. D.2. If the collision is considered as an impulsive phenomenon, the conservation of momentum still holds

$$\begin{cases} m_A \dot{x}_{A_1} + m_B \dot{x}_{B_1} = m_A \dot{x}_{A_2} + m_B \dot{x}_{B_2} \\[3mm] m_A \dot{y}_{A_1} + m_B \dot{y}_{B_1} = m_A \dot{y}_{A_2} + m_B \dot{y}_{B_2} , \end{cases} \tag{D.12}$$

---

[1]Symbol e^* will be used for the restitution coefficient instead of e to avoid confusion with the base of natural logarithms.

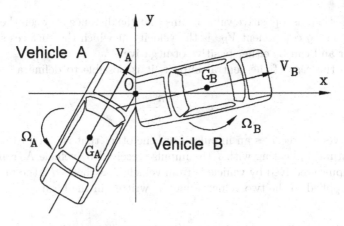

FIGURE D.2. Oblique collision.

but does not allow us to solve the problem of projecting the situation after
the collision from that preceding it. Nor does the assessment of a coefficient of
restitution allow us to obtain a solution, because the conditions to be stated are
two and not one (Eq. (D.12) contains two conditions and 4 unknowns).

Assume that the contact surface at the moment of maximum deformation
is flat and state a reference frame Oxy centred at point O in which the resultant
of the contact forces is applied, with the x-axis perpendicular to the contact
area. Note that the components of velocities V_i of the vehicles are plotted in the
figure in the direction of the positive axes: Practically speaking, some of them
are negative. The vehicles may also have an angular velocity Ω; if not at time
t_1, then surely at time t_2.

The impulsive model implies that the duration of the impact is zero: It may
thus be thought of as the collision between two rigid bodies, predeformed as
shown in the figure.

The velocity of point O, considered as belonging alternatively to vehicle A
and vehicle B, is

$$\vec{V}_{A_O} = \left\{ \begin{array}{c} \dot{x}_{G_A} - \Omega_A y_A \\[2mm] \dot{y}_{G_A} + \Omega_A x_A \end{array} \right\}, \qquad \vec{V}_{B_O} = \left\{ \begin{array}{c} \dot{x}_{G_B} - \Omega_B y_B \\[2mm] \dot{y}_{G_B} + \Omega_B x_B \end{array} \right\}, \qquad (D.13)$$

where x_A, y_A, etc. are the coordinates of the centres of mass in the Oxy frame.
Note that both x_A and y_A are negative in the figure.

The relative velocity of vehicle B with respect to vehicle A at point O is

$$\vec{V}_R = \vec{V}_{B_O} - \vec{V}_{A_O} = \left\{ \begin{array}{c} \dot{x}_{G_B} - \dot{x}_{G_A} - \Omega_B y_B + \Omega_A y_A \\[2mm] \dot{y}_{G_B} - \dot{y}_{G_A} + \Omega_B x_B - \Omega_A x_A \end{array} \right\}. \qquad (D.14)$$

The x component of the relative velocity, hereafter written as $V_{R_\perp}$ is the
velocity at which the two surfaces approach each other, or better, move away

from each other, as a positive value means that the distance between the surfaces increases. The y component $V_{R_\parallel}$ is the velocity at which the surfaces slide over each other and can be either positive or negative.

As in the case of the central impact, it is possible to define a restitution coefficient

$$e^* = -\frac{V_{R_{\perp 2}}}{V_{R_{\perp 1}}} .$$

Each vehicle receives an impulse from the other that is equal to its change of momentum. Indicating with $\vec{I}$ the impulse received by vehicle A from vehicle B (the impulse received by vehicle B from vehicle A is then $-\vec{I}$), the momentum theorem applied to the two vehicles may be written in the form

$$\begin{cases} I_x = m_A \left(\dot{x}_{A_2} - \dot{x}_{A_1} \right) \\ I_y = m_A \left(\dot{y}_{A_2} - \dot{y}_{A_1} \right) \end{cases} \qquad \begin{cases} -I_x = m_B \left(\dot{x}_{B_2} - \dot{x}_{B_1} \right) \\ -I_y = m_B \left(\dot{y}_{B_2} - \dot{y}_{B_1} \right) . \end{cases} \qquad (D.15)$$

Similarly the conservation of the angular momentum can be written in the form

$$\begin{cases} I_x y_A - I_y x_A = J_A \left(\Omega_{A_2} - \Omega_{A_1} \right) \\ \\ I_x y_B - I_y x_B = -J_B \left(\Omega_{B_2} - \Omega_{B_1} \right) . \end{cases} \qquad (D.16)$$

A single relationship is still needed. It is possible to relate the components of the impulse in a direction perpendicular to the impact surface I_x to the tangential component I_y using a simple relationship

$$I_y = \lambda I_x , \qquad (D.17)$$

where λ is a kind of friction coefficient in the zone where the two vehicles are in contact. Because the vehicles usually interlock with each other, its value may be far higher than that of an actual coefficient of friction. Moreover, its value changes in time and only an average value is required. Its sign is the same as that of ratio $V_{R_\perp}/V_{R_\parallel}$.

By introducing equations (D.15) and (D.16) into Eq. (D.14), it follows that

$$V_{R_{\perp 2}} = V_{R_{\perp 1}} - I_x \left(a - \lambda b \right) , \qquad (D.18)$$

where

$$\begin{cases} a = \dfrac{1}{m_A} + \dfrac{1}{m_B} + \dfrac{y_A^2}{J_A} + \dfrac{y_B^2}{J_B} \\ \\ b = \dfrac{x_A y_A}{J_A} + \dfrac{x_B y_B}{J_B} . \end{cases}$$

The value of the x component of the impulse is then

$$I_x = V_{R_{\perp 1}} \frac{1 + e^*}{a - \lambda b} . \qquad (D.19)$$

The direct problem, i.e. that of finding the conditions after the collision (time t_2) from those at time t_1 is now easily solved. Once the components $V_{R_{\perp_1}}$ and $V_{R_{\parallel_1}}$ are computed, the sign of λ is known and the impulse can be computed from equations (D.19) and (D.17). The velocities after the collision are thus computed through equations (D.15) and (D.16).

For the inverse problem, i.e. that of finding the conditions before the collision (time t_1) from those at time t_2, Eq. (D.19) can be modified as

$$I_x = -V_{R_{\perp_2}} \frac{1 + e^*}{e^* (a - \lambda b)} \tag{D.20}$$

and a procedure similar to the previous one can be followed.

The problem is the assessment of the values of coefficients λ and e^* and of the positions of the centres of mass of the vehicles with respect to point O, because the position of the latter cannot usually be evaluated with precision. It is possible to repeat the computation with different values of the uncertain parameters in order to define zones in the parameter space where the solutions must lie and to rule out possible reconstructions of the accident. Some solutions can be ruled out by remembering that $a - \lambda b$ must be positive, as both I_x and V_{i_1} are negative, if vehicle A is in the half plane with negative x.

D.1.3   Collision against a fixed obstacle

A particular case is that of the collision against a fixed obstacle (Fig. D.3). The equations seen above still apply, provided that

$$\frac{1}{m_B} = \frac{1}{J_B} = \dot{x}_B = \dot{y}_B = 0 \,,$$

i.e. the fixed obstacle is considered as a vehicle with infinite mass travelling at zero speed.

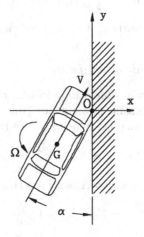

FIGURE D.3. Collision against a fixed obstacle.

It then follows that

$$\vec{V}_R = \left\{ \begin{array}{c} -V\cos(\alpha) - \Omega y_G \\ \\ -V\sin(\alpha) + \Omega x_G \end{array} \right\} , \tag{D.21}$$

$$I = \frac{mr^2(1+e^*)[-V_1\cos(\alpha) - \Omega_1 y_G]}{r^2 + y_G(y_G - \lambda x_G)} , \tag{D.22}$$

where

$$r = \sqrt{\frac{J}{m}}$$

is the radius of gyration of the vehicle.

If the angular velocity Ω_1 is negligible, the expressions for the velocity after the impact are

$$\left\{ \begin{array}{l} \dot{x}_2 = \dot{x}_1 \dfrac{y_G(y_G - \lambda x_G) - r^2 e^*}{y_G(y_G - \lambda x_G) + r^2} \\ \\ \dot{y}_2 = \dot{y}_1 - \dot{x}_1 \dfrac{\lambda r^2(1+e^*)}{y_G(y_G - \lambda x_G) + r^2} \\ \\ \Omega_2 = \dot{x}_1 \dfrac{(y_G - \lambda x_G)(1+e^*)}{y_G(y_G - \lambda x_G) + r^2} . \end{array} \right. \tag{D.23}$$

If $\dot{x}_2 > 0$ or $\Omega_2 > 0$ the vehicle has a second collision with the obstacle with the rear part of the body or, in practice, it undergoes a deformation that displaces the point of contact rearward, with a change in the values of x_G and y_G: It is impossible that $\dot{x}_2 > 0$ after a collision. In case of a perfectly inelastic impact the vehicle finally slides along the obstacle. The deformations are such that

$$y_G - \lambda x_G \approx 0 , \quad \dot{x}_2 = \Omega_2 = 0 , \quad \dot{y}_2 = \dot{y}_1 - \lambda \dot{x}_1.$$

If the obstacle is a flat surface, λ is the coefficient of friction.

D.1.4   Non-central head-on collision

Consider a head-on collision in which the velocities of the centres of mass of the vehicles do not lie along the same straight line (Fig. D.4).

If the vehicles have no angular velocity before the collision, it follows that

$$\dot{y}_{A_1} = \dot{y}_{B_1} = \Omega_A = \Omega_B = 0 , \quad \dot{x}_{A_1} = V_A , \quad \dot{x}_{B_1} = V_B.$$

The components of the relative velocity are then

$$V_{R_{\perp_1}} = V_B - V_A , \quad V_{R_{\parallel_1}} = 0$$

and, because the slip velocity is nil, λ can be assumed to vanish.

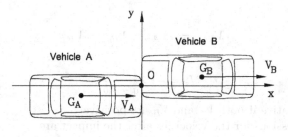

FIGURE D.4. Non-central head-on collision.

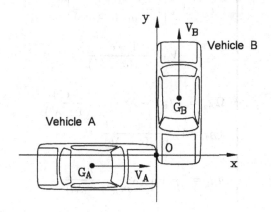

FIGURE D.5. Lateral collision.

The expressions for the velocities after the impact are

$$
\begin{cases}
\dot{x}_{A_2} = V_A + (V_B - V_A)\dfrac{1+e^*}{am_A} \\[2ex]
\dot{x}_{B_2} = V_B - (V_B - V_A)\dfrac{1+e^*}{am_B} \\[2ex]
\Omega_{A_2} = (V_B - V_A)\dfrac{y_A(1+e^*)}{aJ_A} \\[2ex]
\Omega_{B_2} = -(V_B - V_A)\dfrac{y_B(1+e^*)}{aJ_B} \;.
\end{cases}
\tag{D.24}
$$

D.1.5   Lateral collision

Consider a lateral collision in which the velocities of the centres of mass of the vehicles are perpendicular to each other (Fig. D.5).

  If the angular velocities of the vehicles are vanishingly small before the collision

$$
\dot{y}_{A_1} = \dot{x}_{B_1} = \Omega_A = \Omega_B = 0 \;, \quad \dot{x}_{A_1} = V_A \;, \quad \dot{y}_{B_1} = V_B \;,
$$

it follows that

$$V_{R_{\perp_1}} = -V_A , \quad V_{R_{\parallel_1}} = V_B$$

and then

$$I_x = -V_A \frac{1 + e^*}{a - \lambda b} , \tag{D.25}$$

where λ is negative if both V_A and V_B are positive.

The expressions for the velocities after the impact are

$$
\begin{cases}
\dot{x}_{A_2} = V_A \left[1 - \dfrac{1 + e^*}{m_A(a - \lambda b)} \right] \\[2ex]
\dot{y}_{A_2} = -V_A \lambda \dfrac{1 + e^*}{m_A(a - \lambda b)} \\[2ex]
\Omega_{A_2} = -(y_A - \lambda x_A) \dfrac{(1 + e^*)}{J_A(a - \lambda b)} \\[2ex]
\dot{x}_{B_2} = V_A \dfrac{1 + e^*}{m_B(a - \lambda b)} \\[2ex]
\dot{y}_{B_2} = V_B + V_A \lambda \dfrac{1 + e^*}{m_A(a - \lambda b)} \\[2ex]
\Omega_{B_2} = (y_B - \lambda x_B) \dfrac{(1 + e^*)}{J_A(a - \lambda b)} .
\end{cases}
\tag{D.26}
$$

D.1.6   Simplified approach

The problem is often simpler. If the directions of the velocity vectors of the vehicles before and after the impact can be estimated independently, there is no need to assume the values of e^* and λ. With reference to Fig. D.6, Eq. (D.12) can be written as

$$
\begin{cases}
m_A V_{A_1} \cos(\theta_{A_1}) + m_B V_{B_1} \cos(\theta_{B_1}) = m_A V_{A_2} \cos(\theta_{A_2}) + m_B V_{B_2} \cos(\theta_{B_2}) \\[2ex]
m_A V_{A_1} \sin(\theta_{A_1}) + m_B V_{B_1} \sin(\theta_{B_1}) = m_A V_{A_2} \sin(\theta_{A_2}) + m_B V_{B_2} \sin(\theta_{B_2}) .
\end{cases}
\tag{D.27}
$$

If all angles θ_i at times t_1 and t_2 are known or can be assumed, both the direct or the inverse problem are easily solved. Once the velocities have been computed it is possible to obtain the angular velocities, the components of the impulse, e^* and λ.

In the inverse problem, the velocity after the impact is usually obtained from the distance travelled by the two vehicles between time t_2 and the instant t_3 in which all motion ceases. The wheels are often assumed to be blocked, which is correct only if the deformations due to the collision are sufficiently large, or if the sideslip angles are sizeable enough to cause the wheels to slide on the ground.

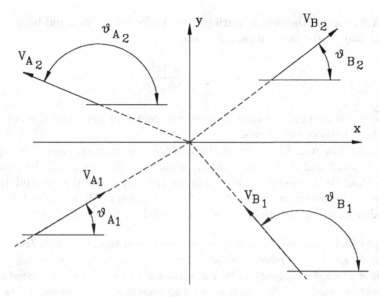

FIGURE D.6. Simplified approach: velocities at times t_1 and t_2.

If the vehicle slides for a distance d before coming to a stop without hitting any other obstacle, the velocity V_2 can be computed by equating the kinetic energy of the vehicle after the collision with the energy dissipated by friction, that is the product of the friction force mgf by the distance travelled d:

$$V_2 = \sqrt{2gdf}\,, \qquad (D.28)$$

where f is the coefficient of friction between the tires and the road.

If at time t_3 the vehicle hits an obstacle with a speed V_3 after sliding for a distance d, it follows that

$$V_2 = \sqrt{2gdf + V_3^2}\,. \qquad (D.29)$$

Velocity V_3 may be assessed only from the damage suffered by the vehicle during the secondary collision, a difficult and process that is affected by large uncertainties and errors.

Remark D.2 *The larger the value of the friction coefficient f or the distance d, the less significant are the errors in the estimate of V_3.*

Another source of uncertainties is the evaluation of f, which is affected not only by road conditions and the possibility that some of the wheels are rolling instead of slipping but also by the actual motion of the vehicle.

The motion of the vehicle on the road after the collision is actually not a simple translational motion but a combination of translation and rotation. As already seen, the angular velocity of the vehicle after the collision may be computed even if only approximately for the position of the point of application of the shock load can only be guessed. At first glance it would seem that the rotation $\psi_3 - \psi_2$ of the vehicle during the time from t_2 to t_3 could be computed the same way as the translational motion

$$\psi_3 - \psi_2 = \frac{J_z \Omega_2^2}{2mg\overline{d}f} , \tag{D.30}$$

where $\overline{d}$ is the average distance between the centre of mass and the centres of the contact areas of the wheels.

This is, however, incorrect as the coefficient of friction cannot be applied to translational and rotational motions separately. But as it will be shown in Section D.3.1, it is possible to approximate the correct results by studying the two motions separately with two "equivalent" friction coefficients, both of which are smaller in magnitude than the actual coefficient f.

Example D.1 *Two cars collided at the intersection of two urban three-lanes streets. Evaluate the speed of the vehicles at the instant of the collision, knowing only the final positions at which they stopped and their directions at time t_1. Vehicle A stopped without any secondary collision while vehicle B ended up against a wall. From the deformations caused by the secondary impact, a value $V_3 \approx 11 \ m/s \approx 40 \ km/h$ is assumed.*

With reference to the xy frame shown in Fig. D.7, the coordinates of the centres of mass of the vehicles at time t_3 are $x_{A_3} = 9.2 \ m$, $y_{A_3} = 12.8 \ m$, $x_{B_3} = 13.2 \ m$ and $y_{B_3} = 12.0 \ m$. The masses of the vehicles are $m_A = 850 \ kg$ and $m_B = 870 \ kg$.

Because the exact position of the vehicles at time t_1 is unknown, each of them will be assumed to have traveled in the centre of the right, centre and left lane. As a result, nine possible positions of the impact point will be considered. With simple geometrical computations, the values of y_{A_1} and y_{B_1} as functions of the position of vehicle A and those of x_{A_1} and x_{B_1} as functions of the position of vehicle B are

| | Vehicle A || Vehicle B ||
Lane	y_{A_1}	y_{B_1}	x_{A_1}	x_{B_1}
Right	-9.8	10.2	6.0	8.6
centre	-6.3	-6.7	2.6	5.2
Left	-2.8	-3.2	-0.8	1.8

It is straightforward to compute the distances d_A and d_B for the nine cases under study. Vehicle A stops without hitting any obstacle. As its velocity was almost perpendicular to its longitudinal axis and some rotation occurred, owing to the fact that the front wheels were blocked as a consequence of the impact, a fairly high value of the friction coefficient can be assumed, namely $f_A = 0,45$.

Vehicle B had only one wheel locked and moved in a direction less inclined with respect to its longitudinal axis; an average value of the friction coefficient, $f_B = 0,30$ is then assumed. In all nine cases the values of V_{A_2} and V_{B_2} can be easily obtained

$$V_{A_2} = \sqrt{2gd_A f_A} , \qquad V_{B_2} = \sqrt{2gd_B f_B + V_{B_3}^2} .$$

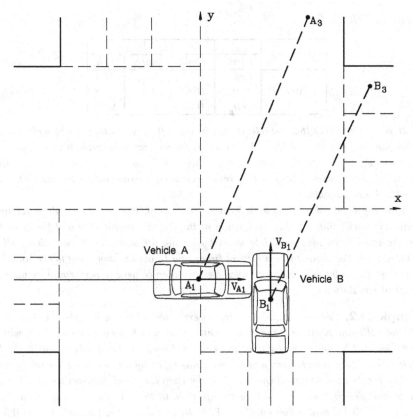

FIGURE D.7. Example D.1; positions of the vehicles.

Angles θ_{A_1} and θ_{B_1} are respectively $\theta_{A_1} = 0$ and $\theta_{B_1} = 90°$. Angles θ_{A_2} and θ_{B_2} can be computed as

$$\theta_{A_2} = \arcsin\left(\frac{y_{A_3} - y_{A_1}}{d_A}\right) \,, \qquad \theta_{B_2} = \arcsin\left(\frac{y_{B_3} - y_{B_1}}{d_B}\right) \,.$$

By solving Eq. (D.27) in V_{A_1} and V_{B_1}, the following values of the velocities are obtained:
Velocity V_{A_1} (in km/h)

		Vehicle A		
	Lane	Right	centre	Left
	Left	49.3	53.3	59.0
Vehicle B	centre	34.9	38.5	42.8
	Left	19.1	21.2	24.1

Velocity V_{B_1} (in km/h)

		Vehicle A		
	Lane	Right	centre	Left
	Left	99.7	91.4	81.7
Vehicle B	centre	103.3	95.7	86.8
	Left	105.5	98.6	90.7

It is very unlikely that vehicle B was in the left lane, as this would yield too low a value for the velocity of vehicle A. However, its position has little effect on the value of the velocity of vehicle B, which, in this case, is the most important parameter to be determined. Some uncertainty in this result remains, because values between 81.7 and 105.5 km/h are possible.

This approach, which may be defined as traditional, still carries a wide uncertainty margin, primarily linked to the evaluation of the friction coefficients and the velocities after the impact, in case a vehicle stops by hitting an obstacle. Its simplicity allows one to perform the computations several times and to obtain upper and lower bounds of the results. The values of the initial velocities can then be used to perform more accurate numerical simulations.

Example D.2 *Two cars collided at the intersection of two urban streets (Fig. D.8). Both the collision point and the final positions at which they stopped (both vehicles stopped without hitting any obstacle) are known. Compute the speeds at which the two vehicles reached the intersection and the positions taken by the vehicles after the impact.*

The inertial properties of the vehicles, the coordinates of their centre of mass and their yaw angles at times t_2 and t_3 with reference to the frame xy shown in the figure are $m_A = 1130$ kg, $m_B = 890$ kg, $J_A = 1780$ kg m^2, $J_B = 1400$ kg m^2, $x_{A_1} = 0.3$ m, $y_{A_1} = 0.6$ m, $x_{B_1} = 0.7$ m, $y_{B_1} = 1.4$ m, $x_{A_3} = -4.4$ m, $y_{A_3} = 8.5$ m, $x_{B_3} = 1.7$ m, $y_{B_3} = 4.9$ m, $\psi_{A_1} = 90°$, $\psi_{B_1} = 180°$, $\psi_{A_3} = 120°$ and $\psi_{B_3} = 20°$.

The positions of the centres of mass of the vehicle with reference to frame $x'y'$ centred on the impact zone are $x'_{G_A} = -0.825$ m, $y'_{G_A} = -0.80$ m, $x'_{G_B} = 1.575$ m and $y'_{G_B} = 0$.

For the computation of the velocities after the impact a value $f_A = 0.15$ was assumed for the first vehicle (after the impact it travelled with little sideslip) and $f_B = 0.50$ was assumed for the second. As the distances travelled after the collision are $d_A = 9.123$ m and $d_B = 3.640$ m, the velocities at time t_2, computed through Eq. (D.28), are $V_{A_2} = 18.7$ m/s $= 67.3$ km/h and $V_{B_2} = 5.98$ m/s $= 21.5$ km/h.

The velocity vectors are then

$$\vec{V}_{A_2} = \left\{ \begin{array}{c} -9.65 \\ 16.22 \end{array} \right\} m/s, \qquad \vec{V}_{B_2} = \left\{ \begin{array}{c} -1.64 \\ 5.75 \end{array} \right\} m/s.$$

The velocities before the collision are easily obtained from Eq. (D.27)

$$\vec{V}_{A_1} = \left\{ \begin{array}{c} 0 \\ 20.75 \end{array} \right\} m/s, \qquad \vec{V}_{B_1} = \left\{ \begin{array}{c} -13.89 \\ 0 \end{array} \right\} m/s.$$

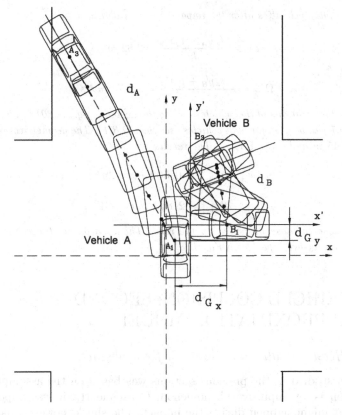

FIGURE D.8. Example D.2; positions of the vehicles at times t_1 and t_3 and reconstruction of the accident; computed positions at times 0.05, 0.1, 0.15, 0.2, 0.3 and 0.5 s from time t_2.

The two cars were then travelling at 74.7 km/h and 50.0 km/h.

This result was obtained using a much simplified model. A more detailed approach can be used to confirm the results, using equations (D.25) and following, the only difference being that in this case vehicle B hits vehicle A. Because $V_{R_{\perp_1}} = V_{B_1} = 13.89$ m/s and $V_{R_{\parallel_1}} = -V_{A_1} = -20.75$ m/s, it follows that

$$I_x' = V_B \frac{1+e^*}{a - \lambda b} = -\frac{13,890(1+e^*)}{2.368 - 0.371\lambda}.$$

The components of the impulse can be easily computed from the velocities before and after the impact: $I_x = -10,905$ Ns and $I_y = -5,119$ Ns. The value $\lambda = 0.469$ is easily obtained, a high but realistic value owing to the possibility of the vehicles becoming interlocked.

By equating the two values of I_x the value $e^* = 0.723$ for the restitution coefficient is obtained. It is also quite high, but is again justified by the small permanent deformations found on the vehicles.

The angular velocities after the impact are then obtained

$$\Omega_{A_2} = \frac{I_x y_A' - I_y x_A'}{J_A} = 2.52 \ rad/s \ ,$$

$$\Omega_{B_2} = \frac{-I_x y_B' + I_y x_B'}{J_B} = -5.75 \ rad/s \ .$$

To compute the rotations of the vehicles during the accident Eq. (D.30) can be used. The value $f = 0.7$ can be used for both vehicles (see Sec. D.3.1). The geometrical parameters are $\overline{d}_A = 1.45 \ m$ and $\overline{d}_B = 1.35 \ m$. The results are

$$|\psi_{A_3} - \psi_{A_2}| = 0,502 \ rad = 29° \ ,$$

$$|\psi_{B_3} - \psi_{B_2}| = 2,813 \ rad = 161° \ .$$

which are close to those measured on the road. The positions of the vehicle at times between t_2 and t_3 are reported in Fig. D.8.

D.2   VEHICLE COLLISION: SECOND APPROXIMATION MODEL

D.2.1   Head on collision against a fixed obstacle

The model studied in the previous sections was based on the assumption that the collision is an impulsive phenomenon. Consequently, it was impossible to assess what might happen during the impact. The shock, however, has a short but finite duration, making it possible to study how the displacement, velocity and acceleration change between t_1 and t_2.

Consider first a head-on collision against a fixed obstacle, like that occurring during a crash test (Fig. D.9). The force that the vehicle receives from the obstacle has a time history of the type shown in Fig. D.10a. The curve is not smooth because the compliance of the front of the vehicle changes strongly as it is crushed, owing to geometric nonlinearities, buckling and other phenomena. The experimental law may, however, be approximated by a smooth curve $F(t)$ while retaining the most important features of the actual behavior of the vehicle (dashed curve in the figure).

The linkage of force and acceleration is complex, as the vehicle's configuration changes in time, with each point of the vehicle having its own acceleration. However, with the exception of the front part, which is crushed, the vehicle may be considered as a rigid body.

The position of the centre of mass may be considered as fixed to the undeformed part of the vehicle. The acceleration can thus be obtained directly from the force $F(t)$; in an actual crash test the acceleration is usually measured by accelerometers located on the vehicle and the force is obtained from these readings.

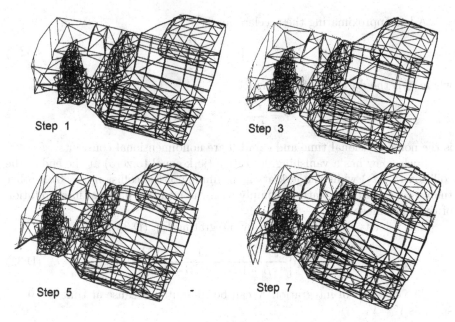

FIGURE D.9. Numerical simulation of a crash test against a rigid barrier: Deformation of the vehicle at various instants (fuinite element method).

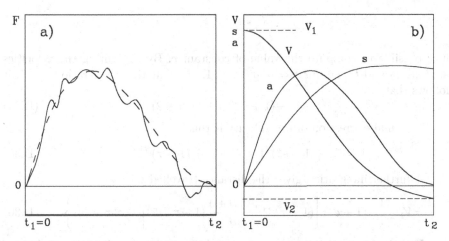

FIGURE D.10. (a) Force the vehicle receives from the obstacle during a crash test as a function of time. Experimental curve and mathematical empirical law. (b) Time histories $V(t)$, $a(t)$ and $s(t)$ obtained from the empirical law $F(t)$.

From the acceleration it is straightforward to compute the velocity and the deformation of the vehicle.

A law approximating the acceleration is[2]

$$a = \tau(1-\tau)^\beta \frac{cV_1}{t_2} \,, \tag{D.31}$$

where t_2 is the duration of the impact,

$$\tau = \frac{t}{t_2} \qquad (0 \le \tau \le 1)$$

is the nondimensional time and c and β are nondimensional constants.

Such a law has a vanishing derivative (jerk equal to zero) at the end of the collision $(t = t_2)$ while the jerk at the beginning of the collision $(t = 0)$ is other than zero; both these features comply with the intuitive physical interpretation of the phenomenon.

The velocity may be obtained by integrating Eq. (D.31):

$$V = -cV_1 \left[\frac{(1-\tau)^{\beta+1}}{\beta+1} - \frac{(1-\tau)^{\beta+2}}{\beta+2} \right] + K \,. \tag{D.32}$$

The constant of integration K can be computed because at time $t = 0$ the speed is $V = V_1$

$$K = V_1 \left[1 + \frac{c}{(\beta+1)(\beta+2)} \right] \,. \tag{D.33}$$

By remembering the definition of the restitution coefficient

$$e^* = -\frac{V_2}{V_1} \,,$$

it is possible to compute the value of constant c. By computing the velocities at times t_2 and t_1, i.e. for $\tau = 0$ and $\tau = 1$, and equating their ratio to $-e^*$, it follows that

$$c = -(1+e^*)(\beta+1)(\beta+2) \,. \tag{D.34}$$

The final expression of the velocity is thus

$$V = V_1 \left\{ (1+e^*) \left[1 + \tau(\beta+1) \right] (1-\tau)^{\beta+1} - e^* \right\} \,. \tag{D.35}$$

A further integration gives the distance travelled s

$$s = V_1 t_2 \left\{ -(1+e^*) \left[(1-\tau)^{\beta+2} - \frac{\beta+1}{\beta+3}(1-\tau)^{\beta+3} \right] - e^*\tau + K_1 \right\} \,. \tag{D.36}$$

If at time $t = 0$ the distance is assumed to be nil, s describes the crushing of the front part of the vehicle. This statement allows the value of the integration constant K_1 to be computed

$$K_1 = \frac{2(1+e^*)}{\beta+3} \,. \tag{D.37}$$

---

[2]R.H. Macmillan, *Dynamic of Vehicle Collisions*, Inderscience Enterprises, Jersey 1983.

The final expression of the displacement is thus

$$s = V_1 t_2 \left(\frac{1 + e^*}{\beta + 3} \left\{ 2 - (1 - \tau)^{\beta+2} \left[2 + (\beta + 1)\tau \right] \right\} - e^* \tau \right) . \qquad (D.38)$$

At time $\tau = 1$, the displacement directly yields the residual crushing of the vehicle s_2

$$s_2 = V_1 t_2 \left[\frac{2(1 + e^*)}{\beta + 3} - e^* \right] . \qquad (D.39)$$

In case of an elastic collision, the residual crushing must vanish: If $e^* = 1$ then $s_2 = 0$. From this statement a first relationship between β and e^* can be stated: $\beta = 1$ if $e^* = 1$.

Parameters β and e^* characterizing the impact depend on many factors, beginning with the structural characteristics of the vehicle and including the type of impact and, in the case of head-on collision against a fixed obstacle, the impact velocity V_1 and time t_2.

A primary characteristic of the vehicle is its stiffness at the instant it enters into contact with the obstacle, namely its *crushing modulus* (Fig. D.11). With simple kinematic considerations it follows that

$$K = m \left(\frac{da}{ds} \right)_{\tau=0} = m \left(\frac{da}{dt} \right)_{\tau=0} \left(\frac{ds}{dt} \right)_{\tau=0}^{-1} = \frac{m}{V_1} \left(\frac{da}{dt} \right)_{\tau=0} . \qquad (D.40)$$

By introducing the expression (D.31) of the acceleration into Eq. (D.40), it follows that

$$K = m \frac{(1 + e^*)(\beta + 1)(\beta + 2)}{t_2^2} . \qquad (D.41)$$

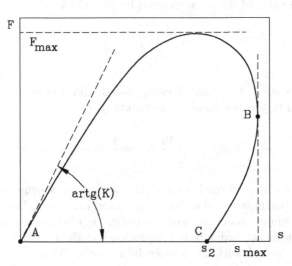

FIGURE D.11. Force received by the vehicle during a head-on collision against the obstacle as a function of the crushing s.

The value of K can be obtained from crash tests. Values between 1 and 2 MN/m are found in the literature.

The crushing process is far from linear. A qualitative plot of force F against the displacement s is reported in Fig. D.11: The inelastic behavior of the vehicle and the hysteresis cycle are clearly shown. The area below the line from point A to point B is the energy absorbed by the vehicle from time t_1 to the instant t_i at which the maximum displacement is reached. In the case of collision against a fixed rigid obstacle it is equal to the kinetic energy of the vehicle.

The area below the line from point B to point C is the energy that is transformed back into kinetic energy during the rebound phase from time t_i to t_2. If $e^* = 0$ such an area vanishes and $s_2 = s_{max}$. If, on the contrary $e^* = 1$ line BC is superimposed to line AB and the area of the hysteresis cycle vanishes.

The plot is usually obtained from a crash test, either performed on an actual vehicle or simulated by computer. The numerical simulation is a complex task, owing to the complex geometry of the front part of the vehicle and the nonlinearities of all types involved. A large and fast computer is thus required. Some results obtained through the finite element method, which at present is the only method allowing problems of this complexity to be tackled, are reported in Fig. D.9. The computation has been performed through step by step integration in time of the equations of motion; the deformed mesh at four different time values has been reported in the figure.

The law $F(s)$, obtained through the empirical law $F(t)$ defined above, is often assumed to be independent of the deformation rate ds/dt. Such an assumption has no theoretical background but is justified by the fact that it is substantiated by experimental evidence, at least when the deformation rate is low.

The mean value of the force received by the vehicle

$$\overline{F} = \frac{1}{t_2} \int_0^{t_2} F \, dt \tag{D.42}$$

can be easily computed by remembering that the total impulse received by the vehicle is equal to the change of the momentum

$$\overline{F} = m\frac{V_2 - V_1}{t_2} = -m\frac{V_1}{t_2}(1 + e^*) . \tag{D.43}$$

If the mean value of force F is small, the permanent deformations are usually negligible and the impact is close to being elastic, i.e. $e^* \approx 1$. With increasing $\overline{F}$ the collision becomes more and more inelastic, i.e. e^* decreases with increasing $\overline{F}$. It is possible to approximate the dependence of the restitution coefficient e^* on the average force $\overline{F}$ with an exponential law (Fig. D.12)

$$e^* = e^{-\overline{F}/K_r} , \tag{D.44}$$

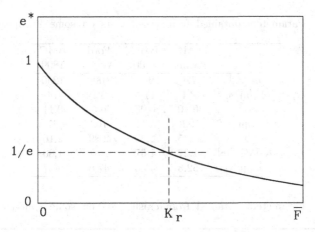

FIGURE D.12. Law $e^*(\overline{F})$ approximated by Eq. (D.44).

where K_r, usually referred to as the *impact resistance modulus*, is the value of $-\overline{F}$ at which the coefficient of restitution takes the value

$$e^* = \frac{1}{e} = 0.368 \ .$$

K_r can also be obtained from a crash test. From Equations. (D.41) and (D.44), it follows that

$$K_r = m\frac{V_1(1+e^*)}{t_2 \ln(1/e^*)} \ , \tag{D.45}$$

which allows K_r to be calculated from quantities that can be measured or computed. Values between 40 and 100 kN have been reported for K_r in the literature.

As seen above, constant β depends on e^*: It takes a unit value when $e^* = 1$ and increases when e^* decreases. Assume that, at least in the case of strongly inelastic impacts, i.e. near the condition $e^* = 0$, the law $\beta(e^*)$ may be approximated as

$$\beta = \beta_0 - (\beta_0 - 1)e^* \ , \tag{D.46}$$

which, although assumed for $e^* \approx 0$, gives the correct result also for $e^* = 1$.

Because $\beta > 1$, β_0 is always larger than unity. Also β_0, which is nondimensional, may be considered as a characteristic of the vehicle, and will be referred to as the *structural index*. A large value of β_0 characterizes vehicles with a very stiff front section, while a compliant front section is typical of vehicles with low β_0. Its values are usually close to 2.

Parameters K, K_r and β_0 completely characterize a vehicle in terms of head-on collision against a rigid obstacle. Once they have been measured, it is possible to compute the laws $F(t)$, $a(t)$ and $s(t)$ from the collision conditions, namely the velocity V_1 and the mass of the vehicle m.

TABLE D.1. Parameters obtained from crash tests on some European passenger vehicles.

	BMC Mini	BMC 1100	Ford Anglia	BMC 1800
m [kg]	720	950	1000	1250
V_1 [m/s]	14	11,5	14	14
e^*	0,10	0,08	0,5	0,11
t_2 [ms]	92	102	103	97
β_0	2,35	2,79	2,89	2,16
K [MN/m]	1,27	1,67	1,81	1,80
K_r [kN]	52,3	45,8	47,6	90,7

TABLE D.2. Parameters obtained from crash tests on some American passenger vehicles.

	Sub compact	Compact	Interm.	Standard 71/72	Standard 73/74
m [kg]	1135	1545	1820	2045	2045
V_1 [m/s]	12	12	10,5	10	10,3
e^*	0,01	0,01	0,20	0,20	0,20
t_2 [ms]	102	102	151	147	143
β_0	2,50	2,50	2,00	2,00	2,00
K [MN/m]	1,60	2,17	1,03	1,09	1,16
K_r [kN]	28,6	38,8	96,7	93,6	108,8

The values reported in Tables D.1 and D.2 have been obtained from crash tests published by manufacturers[3]. The values in the first table are not exactly comparable with the others because they have been computed from well documented tests, while not all the parameters were known for the others and the values of β_0 and s_2 had to be assumed.

To solve the direct problem, i.e. to obtain the conditions after the collision, mainly s_2 and V_2, from V_1, it is possible to compute t_2 from Eq. (D.45) and to introduce it into Eq. (D.41), obtaining

$$KmV_1^2 = K_r^2 \left[\ln \left(\frac{1}{e^*} \right) \right] \frac{(\beta + 1)(\beta + 2)}{1 + e^*} . \tag{D.47}$$

Substituting Eq. (D.46) into Eq. (D.47) the latter yields

$$\frac{KmV_1^2}{K_r^2} = \left[\ln \left(\frac{1}{e^*} \right) \right] \frac{(\beta_0 + 1)(\beta_0 + 2) - e^*(\beta_0 - 1)(2\beta_0 + 3) + e^{*^2}(\beta_0 - 1)^2}{1 + e^*} ,$$

$$\tag{D.48}$$

which can be solved numerically in e^* and then allows the direct problem to be solved.

---

[3]R.H. Macmillan, *Dynamic of Vehicle Collisions*, Inderscience Enterprises, Jersey 1983.

The inverse problem, i.e. obtaining the conditions before the collision, primarily V_1, from the crushing s_2, is also easily solved. From Equations. (D.37), (D.41) and (D.45) it follows that

$$s_2 \frac{K}{K_r} = \ln\left(\frac{1}{e^*}\right) \frac{(\beta+1)(\beta+2)\left[2(1+e^*) - (\beta+3)e^*\right]}{\beta+3}. \tag{D.49}$$

Substituting Eq. (D.46) into Eq. (D.49) the latter yields

$$s_2 \frac{K}{K_r} = \ln\left(\frac{1}{e^*}\right)\left[e^{*^2}(\beta_0 - 1)^2 - e^*(2\beta_0^2 + \beta_0 - 3) + 3\beta_0 + \beta_0 + 2\right]$$

$$\times \frac{e^*(e^* - 1)(\beta_0 - 1) + 2}{-e^*(\beta_0 - 1) + \beta_0 + 3}, \tag{D.50}$$

which may be solved numerically in e^* and allowing the inverse problem to be solved.

Example D.3 *Consider a car with $\beta_0 = 2$, $K = 1.2$ MN/m, $K_r = 65$ kN and $m = 1000$ kg impacting an obstacle at 20 m/s. Compute the parameters of the impact and laws $F(t)$, $a(t)$ $V(t)$ and $s(t)$.*

The numerical solution of Eq. (D.48) yields $e^* = 0.0416$ and then $\beta = 1.958$, $t_2 = 0.1$ s. The laws $F(t)$, $a(t)$ $V(t)$ and $s(t)$ for this case are plotted in Fig. D.13; the residual crush is $s_2 = 757$ mm.

D.2.2   Head-on collision between vehicles

The head-on collision between vehicles may be studied using the same model seen in the previous section, provided the contact surface is assumed to remain planar

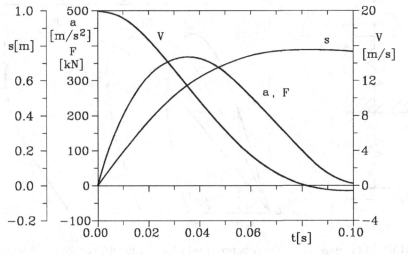

FIGURE D.13. Laws $F(t)$, $a(t)$, $V(t)$ and $s(t)$ for Example D.3.

and the characteristics of the impact, and particularly law $F(s)$, are independent of the deformation rate. Each vehicle can thus be assumed to impact against a moving massless obstacle (Fig. D.13a). Each vehicle is modelled as a point mass, provided with a nonlinear spring, whose characteristics (K, K_r and β_0) are those seen for the collision against a fixed obstacle.

The deformation rates

$$
\begin{cases}
\dot{s}_A = \dot{x}_b - \dot{x}_A \\
\\
\dot{s}_B = -\dot{x}_b + \dot{x}_B
\end{cases}
\tag{D.51}
$$

are positive when the springs are compressed.

The relative velocity V_R is then

$$
V_R = \dot{x}_B - \dot{x}_A = -(\dot{s}_A + \dot{s}_B) . \tag{D.52}
$$

If the curves $F(s)$ for the two vehicles are known, a plot of the type shown in Fig. D.14b can be drawn. As the total force acting on the virtual obstacle must vanish,

$$
|F_A| = |F_B|
$$

for each value of time. The intersections of the curves $|F_A(s_A)|$ and $|F_B(s_B)|$ with any line $F = $ constant thus yield the deformations s_A and s_B in the same instant. A third curve in which the force is plotted as a function of the total deformation $s = s_A + s_B$, i.e. of the change of distance between the centres of mass, can be plotted.

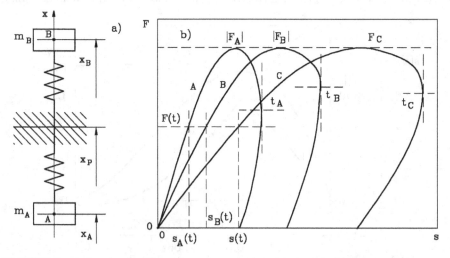

FIGURE D.14. Head-on collision between vehicles. (a) model; (b) forces as functions of the crush of the vehicles.

The maximum values of the deformations need not be reached at the same time: In the figure the maximum deformation is reached at time t_A and t_B for the two vehicles, while the distance between the centres of mass is at its minimum at time t_C.

Because the forces acting on the vehicles are simply

$$F_A = m_A \ddot{x}_A , \quad F_B = m_B \ddot{x}_B,$$

it follows that

$$F_A = -F_B = \dot{V}_R \frac{m_A m_B}{m_A + m_B} . \tag{D.53}$$

If the derivative of the relative velocity has a time history which is of the same type as that assumed for the acceleration in Eq. (D.31),

$$\dot{V}_R = -\frac{cV_{R_1}}{t_2} \tau (1 - \tau)^\beta , \tag{D.54}$$

where, as usual,

$$\tau = \frac{t}{t_2} ,$$

the same model used for the collision against an obstacle is sufficient for the present case as well.

Instead of m, K, V and s, the relevant equations now contain

$$m_C = \frac{m_A m_B}{m_A + m_B} , \qquad K_C = \frac{K_A K_B}{K_A + K_B} , \qquad V_R , \qquad s_C = s_A + s_B .$$

Assume that the characteristics of the vehicles and the residual deformation of one of them, say s_{A_2}, are known. Because curve $F(s)$ has been assumed to be the same as that characterizing the impact against the barrier, from s_{A_2} it is possible to compute e_A^* directly from Eq. (D.50) and then velocity V'_{A_1}[4] which causes the same residual deformation s_{A_2} in an impact with a rigid obstacle.

The maximum value F_{max_A} of the force received by the first vehicle is

$$F_{max_A} = K_{r_A} \beta_A^{\beta_A} \ln \left(\frac{1}{e_A^*} \right) \frac{(\beta_A + 1)(\beta_A + 2)}{(\beta_A + 1)^{\beta_A + 1}} . \tag{D.55}$$

The absolute value of such force is equal to the maximum force acting on vehicle B. It is thus possible to write an equation identical to Eq. (D.55) with the characteristics of the second vehicle. Because

$$\beta_B = \beta_{0_B} - (\beta_{0_B} - 1)e_B^* ,$$

it is possible to obtain e_B^* and then t_{B_2}, V_{B_1} and s_{B_2}.

---

[4]V' is the velocity in the equivalent collision against a rigid obstacle.

The two equivalent collisions against a rigid obstacle are thus completely characterized. As the curves $F(s)$ are the same in the actual collision and its equivalents, the energy dissipated in the former is equal to that dissipated in the latter

$$\Delta E = \Delta E_A + \Delta E_B = \frac{1}{2} m_A V_{A_1}^2 \left(1 - e_A^{*2}\right) + \frac{1}{2} m_B V_{B_1}^2 \left(1 - e_B^{*2}\right). \quad (\text{D.56})$$

It is thus possible to consider the actual collision, described by curve C in Fig. D.14b, by using the same equations seen in the previous chapter with the mentioned substitutions. The maximum force, already expressed by Eq. (D.55), and the energy dissipated take the values

$$F_{max} = \frac{s_{C_2} K_C \beta_C^{\beta_C} (\beta_C + 3)}{(\beta_C + 1)^{\beta_C + 1} [2 - e_C^*(\beta_C + 1)]}, \quad (\text{D.57})$$

$$\Delta E = \frac{1}{2} m_C V_{C_1}^2 \left(1 - e_C^{*2}\right). \quad (\text{D.58})$$

Equations (D.57) and (D.58) contain 3 unknowns β_C, e_C^* and V_R (or V_{C_1}, which is the same). A third equation may be obtained considering that force F_{max} can also be expressed as

$$F_{max} = K_C V_R t_2 \frac{\beta_C^{\beta_C}}{(\beta_C + 1)^{\beta_C + 1}}. \quad (\text{D.59})$$

Because

$$t_2^2 = \frac{m_C}{K_C} (1 + e_C^*)(\beta_C + 1)(\beta_C + 2), \quad (\text{D.60})$$

computing V_R from Eq. (D.58) and substituting the value so obtained into Equations. (D.59) and (D.60), it follows that

$$F_{max}^2 = K_C \frac{2\Delta E \beta^{2\beta} (\beta + 2)}{(1 - e^*)(\beta + 1)^{2\beta + 1}}. \quad (\text{D.61})$$

The last equation can be solved in e^*, obtaining

$$e^* = 1 - K_C \frac{2\Delta E \beta^{2\beta} (\beta + 2)}{F_{max}^2 (\beta + 1)^{2\beta + 1}}. \quad (\text{D.62})$$

Finally, introducing Eq. (D.62) into Eq. (D.57), the equation

$$F_{max} = s_C K_C \frac{\beta^\beta (\beta + 3)}{(\beta + 1)^{\beta + 1}} \left[1 - \beta + \frac{2\Delta E K_C}{F_{max}^2} (\beta + 2) \left(\frac{\beta}{\beta + 1}\right)^{2\beta}\right]^{-1}, \quad (\text{D.63})$$

is obtained, which may be easily solved numerically in β. It is thus possible to obtain the values of e^* and V_R, solving the problem.

The inverse problem, consisting in obtaining the parameters characterizing the collision once the relative velocity V_R is known, is more difficult as it must be solved in an iterative way. A value of the final crushing s_2^* of one of the two vehicles is assumed and from it the relative velocity V_R^* can be computed as seen above. A new value of the residual crushing, for example obtained as

$$s_2^{**} = s_2^* \frac{V_R}{V_R^*} ,$$

can then be computed and a new relative velocity, which is closer to the correct one, is obtained. The procedure should converge quickly to the required result. The velocities of the vehicle after the collisions can then be obtained without further problems.

Example D.4 *Consider the head-on collision between two cars whose characteristics are known from crash tests, e.g. the vehicles of the first and fourth columns of Table D.1.*

The characteristics are then $m_A = 720$ kg, $m_B = 1250$ kg, $\beta_{0_A} = 2.35$, $\beta_{0_B} = 2.16$, $K_A = 1.27$ MN/m, $K_B = 1.80$ MN/m, $K_{r_A} = 52.3$ kN and $K_{r_B} = 90.7$ kN. The residual crush of the first vehicle is $s_{A_2} = 400$ mm. Compute the parameters of the impact and the relative velocity at time t_1.

The parameters characterizing the collision are $m_C = 456.85$ kg and $K_C = 0.745$ MN/m. The coefficient of restitution for the first vehicle is easily computed from the residual crush by numerically solving Eq. (D.50): $e_A^* = 0.105$. The values of β_A, t_{A_2} and V_{A_1} are then $\beta_A = 2.208$, $t_{A_2} = 0.092$ s and $V_{A_1} = 13.63$ m/s.

Equations (D.55), (D.46), (D.41) and (D.45) yield $F_{max} = 218$ kN, $e_B^* = 0.252$, $\beta_B = 1.87$, $t_{B_2} = 0.098$ s, $V_{B_1} = 7.85$ m/s and $s_{B_2} = 202$ mm.

The energy dissipated and the total crush are then $\Delta E = 102$ kJ and $s_C = 602$ mm.

Solving Eq. (D.63) in β and using Equations (D.58) and (D.62), the final results are obtained:

$$\beta = 2,037 , \quad e^* = 0,159 , \quad V_R = 21,43 \ m/s .$$

Example D.5 *Consider the vehicles of the previous examples colliding head-on with velocities $V_{A_1} = 26.7$ m/s $= 96$ km/h and $V_{B_1} = -13.1$ m/s $= -47$ km/h. Compute the velocities after the impact.*

The relative velocity is $V_R = 39.8$ m/s. By assuming a residual crush of the first vehicle $s_{A_2} = 400$ mm in the previous example a relative velocity $V_R = 21.43$ m/s has been obtained.

Correcting the assumed residual crush linearly, a new trial value $s_{A_2} = 743$ mm is obtained, which yields a relative velocity $V_R = 36.16$ m/s that is already close to the correct values.

With two further iterations a crush $s_{A_2} = 823$ mm is obtained, together with $e^* = 0.048$, $F_{max} = 378$ kN and $\Delta E = 360$ kJ.

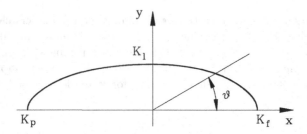

FIGURE D.15. Polar diagram $K(\theta)$ approximated by two arcs of ellipse.

As final results, the velocities after the collision are obtained:

$$V_{A_2} = 0,234 \ m/s = 0,84 \ km/h \ ,$$
$$V_{B_2} = 2,144 \ m/s = 7,72 \ km/h \ .$$

D.2.3   Oblique collision between vehicles

The first issue in the study of oblique collisions is the evaluation of the characteristics of the vehicle: If it is already difficult to obtain the values of K, K_r and β_0 for head-on collisions, it is almost impossible to obtain them for a generic impact direction. If it is possible to find the relevant values for side or rear impacts, the dependence of the characteristics upon angle θ (Fig. D.15) may be approximated by two arcs of ellipse, one for the front and one for the rear.

Following the notation of the figure, function $K(\theta)$ may be expressed as

$$\frac{K_f K_l}{\sqrt{K_f^2 \sin^2(\theta) + K_l^2 \cos^2(\theta)}} \qquad \text{for} \quad 0 \leq \theta \leq 90°,$$

$$\frac{K_p K_l}{\sqrt{K_p^2 \sin^2(\theta) + K_l^2 \cos^2(\theta)}} \qquad \text{for} \quad 90° \leq \theta \leq 180° \ .$$

$$(D.64)$$

Similar relationships may be written to approximate the other characteristics of the vehicle.

If the collision, albeit oblique, were central, i.e. the two velocities were aligned, and the two angular velocities were equal to zero (Fig. D.16a), the procedure seen above for the head-on collision would still hold, provided that the correct characteristics of the vehicles were used.

In case of non-central oblique collisions the vehicles are subject also to angular accelerations about the yaw axis z. An approximated but simple way to take this into account is to substitute the mass of the vehicle against which each vehicle collides with an effective mass, lower than the actual one. It is questionable whether it is worthwhile to attempt to refine this model further, because the assumptions on which the present models are based and the uncertainties in the data do not allow high precision to be obtained.

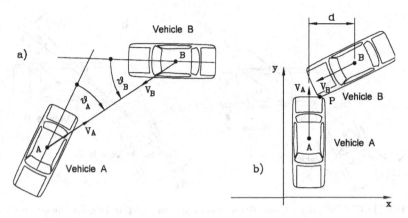

FIGURE D.16. (a) Central oblique collision. (b) Oblique collision in which vehicle A hits the front part of vehicle B.

Consider the situation of Fig. D.16b: Vehicle A hits the front part of vehicle B. Neglecting the friction in the contact area ($\lambda = 0$), the x component of the momentum of the second vehicle does not change and the impulse reduces to its y component. Vehicle B undergoes an acceleration in the y direction and an angular acceleration

$$\ddot{y}_{G_B} = \frac{F}{m_B} , \qquad \ddot{\theta}_B = \frac{Fd}{J_{z_B}} . \tag{D.65}$$

The acceleration of point P, seen as belonging to vehicle B, is then

$$\ddot{y}_{P_B} = \ddot{y}_{G_B} + d\ddot{\theta}_B = \frac{F}{m_B}\left(1 - \frac{d^2}{r_B^2}\right) , \tag{D.66}$$

where r_B is the radius of gyration. The acceleration of the contact point is thus equal to the acceleration of the centre of mass of a vehicle having a *reduced mass*

$$m_{r_B} = m_B \frac{r_B^2}{r_B^2 + d^2} .$$

If neither vehicle collides head-on with the other, reference must be made to the surface in collision, as seen in Section D.1.2. If no allowance is taken for the friction between the vehicles, the reduced masses of the two vehicles can be computed with reference to the perpendicular to these surfaces (Fig. D.17a). If friction is taken into account, reference must be made to a direction inclined at an angle $\arctan(\lambda)$ with respect to the perpendicular to the collision surface (Fig. D.17b).

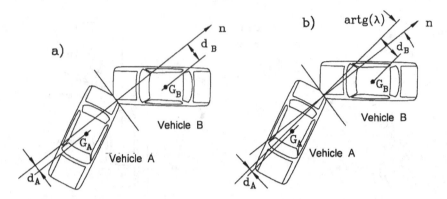

FIGURE D.17. Oblique collision; definition of distances d_A and d_B for the computation of the reduced mass. (a) No friction ($\lambda = 0$) and (b) $\lambda \neq 0$.

D.3   MOTION AFTER THE COLLISION

D.3.1   Vehicle with locked wheels

If, after the collision, the motion of the vehicle is simply translational, the distance travelled may be easily computed as seen in Section D.1.6. This, however, is seldom the case: After the impact a certain yaw velocity $\Omega = \dot{\psi}$ is usually present. It is thus incorrect to take translational and rotational motion into account independently.

Consider a vehicle that, after the collision, moves with the wheels completely locked, as if the brakes were fully applied or the deformations of the body were sufficient to prevent the wheels from rotating. With reference to Fig. D.18a, the components of the velocity of the centre of contact P_i of the ith wheel u_i and v_i in the directions of axes x and y fixed to the vehicle are

$$\begin{cases} u_i = V \cos(\beta) - \Omega r_i \sin(\chi_i) \\ \\ v_i = V \sin(\beta) + \Omega r_i \cos(\chi_i) \, , \end{cases} \tag{D.67}$$

where r_i and χ_i are linked to the coordinates x_i and y_i of point P_i by the obvious relationships

$$r_i = \sqrt{x_i^2 + y_i^2} \, , \qquad \chi_i = \arctan\left(\frac{y_i}{x_i}\right) \, .$$

The absolute value of force F_i exchanged by the ith locked wheel is simply

$$|F_i| = f Z_i$$

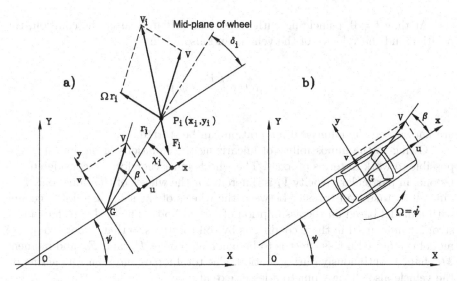

FIGURE D.18. (a) Velocity of the centre of contact of the i-th wheel; (b) inertial reference frame and variables of motion.

and its direction is equal to that of the velocity V_i, with opposite sign. The components of force F_i are then

$$
\begin{cases}
F_{x_i} = -fZ_i\dfrac{u_i}{|V_i|} = fZ_i\dfrac{-V\cos(\beta) + \Omega r_i\sin(\chi_i)}{|V_i|} \\[4mm]
F_{y_i} = -fZ_i\dfrac{v_i}{|V_i|} = -fZ_i\dfrac{V\sin(\beta) + \Omega r_i\cos(\chi_i)}{|V_i|} \ ,
\end{cases}
\tag{D.68}
$$

where

$$
|V_i| = \sqrt{V^2 + \Omega^2 r_i^2 + 2V\Omega r_i\sin(\beta - \chi_i)} \ .
\tag{D.69}
$$

The moment of force F_i about the yaw axis z is

$$
M_i = F_{y_i} r_i\cos(\chi_i) - F_{X_i} r_i\sin(\chi_i) = fZ_i r_i\dfrac{\Omega r_i + V\sin(\beta - \chi_i)}{|V_i|} \ .
\tag{D.70}
$$

The trajectory can thus be easily computed by numerical integration of the equations of motion. No linearization is possible in this case, because the slip angle of the vehicle β may be quite large. The equations of motion are Eq. (25.65), where the forces acting on the wheels are those expressed by Eq. (D.68) rotated in the inertial reference frame by multiplying them by a suitable rotation matrix. The model here used is essentially a three-degrees of freedom, rigid body model in which all forces except those due to tire-road interaction have been neglected. It would in any case be difficult to take aerodynamic forces into account when angle β_a is large and rapidly varying.

At time $t = 0$, coinciding with time t_2 immediately after the collision, the position and the velocity of the vehicle and also angle

$$\beta = \arctan\left(\frac{\dot{x}'}{\dot{y}'} - \psi\right)$$

are known and the numerical integration can be started.

Owing to the impossibility of linearizing the equations of motion, it is impossible to work in terms of axles. The wheels must therefore be considered one by one; in particular velocity V_i is different for the wheels of the same axle. It is difficult to take load transfer between the wheels of the same axle into account with a model based on the assumption of a rigid body. The load transfer can be strongly influenced in these conditions by roll rotations so that a more complete model is required if this effect is to be included. Forces F_{x_i} and F_{y_i} and moment M_i change continuously during motion. The total forces and moments acting on the vehicle also change, but to a lesser extent.

A simple approach that can be used to study the motion of the vehicle without having to perform a numerical simulation is to substitute the actual contact area between the vehicle and the ground with a circle having a radius r equal to the average distance of the centres of the contact areas of the wheels and the centre of mass (Fig. D.19a).

The force and the moment exerted on an arc of amplitude $d\theta$ of such a circumference may be expressed by equations of the same type as equations (D.68) and (D.70). Assuming that the vertical load mg exerted on the contact area is evenly distributed on the circumference, the components of the force and the yawing momentare

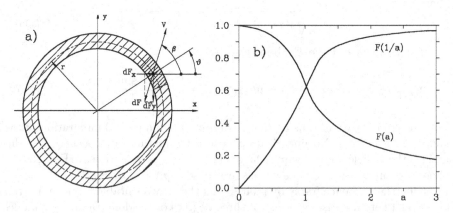

FIGURE D.19. (a) Simplified model for the study of the trajectory of a vehicle moving with locked wheels. Circular contact area that is substituted for the actual contact area. (b) Functions $F(a)$ and $F(1/a)$ for a spanning between 0 and 3.

$$\begin{cases} dF_x = f\dfrac{mg}{2\pi}\dfrac{-V\cos(\beta)+\Omega r\sin(\theta)}{\sqrt{V^2+\Omega^2 r^2+2V\Omega r\sin(\beta-\theta)}}d\theta \\[4mm] dF_y = -f\dfrac{mg}{2\pi}\dfrac{V\sin(\beta)+\Omega r\cos(\theta)}{\sqrt{V^2+\Omega^2 r^2+2V\Omega r\sin(\beta-\theta)}}d\theta \\[4mm] dM = -f\dfrac{mgr}{2\pi}\dfrac{V\sin(\beta-\theta)+\Omega r}{\sqrt{V^2+\Omega^2 r^2+2V\Omega r\sin(\beta-\theta)}}d\theta \, . \end{cases} \qquad (\text{D.71})$$

Force dF may be decomposed along directions parallel and perpendicular to the velocity V. The first component, which is tangential to the trajectory, is

$$dF_\parallel = dF_x\cos(\beta)+dF_y\sin(\beta) = -f\dfrac{mg}{2\pi}\dfrac{V+\Omega r\sin(\beta-\theta)}{\sqrt{V^2+\Omega^2 r^2+2V\Omega r\sin(\beta-\theta)}}d\theta \, .$$
$$(\text{D.72})$$

Its effect is to reduce the speed of the vehicle. The second component, acting in a direction perpendicular to the trajectory and thus bending the path of the vehicle, is

$$dF_\perp = -dF_x\sin(\beta)+dF_y\cos(\beta) = -f\dfrac{mg}{2\pi}\dfrac{\Omega r\cos(\beta-\theta)}{\sqrt{V^2+\Omega^2 r^2+2V\Omega r\sin(\beta-\theta)}}d\theta \, .$$
$$(\text{D.73})$$

By integrating the expression of the forces and moments, it follows that

$$\begin{cases} F_\parallel = -f\dfrac{mg}{2\pi}\displaystyle\int_0^{2\pi}\dfrac{1+a\sin(\zeta)}{\sqrt{1+a^2+2a\sin(\zeta)}}d\zeta \\[4mm] F_\perp = -f\dfrac{mg}{2\pi}\displaystyle\int_0^{2\pi}\dfrac{a\cos(\zeta)}{\sqrt{1+a^2+2a\sin(\zeta)}}d\zeta = 0 \\[4mm] M = -f\dfrac{mgr}{2\pi}\displaystyle\int_0^{2\pi}\dfrac{1+\frac{1}{a}\sin(\zeta)}{\sqrt{1+\left(\frac{1}{a}\right)^2+\frac{2}{a}\sin(\zeta)}}d\zeta \, , \end{cases} \qquad (\text{D.74})$$

where

$$\zeta = \beta - \theta$$

(because β does not depend on θ, $d\zeta = d\theta$) and the nondimensional parameter a is the ratio between the component of the velocity due to rotation and velocity V

$$a = \dfrac{\Omega r}{V} \, .$$

The component of force F perpendicular to the trajectory is equal to zero: This means that the trajectory is straight, at least within the assumptions used in the present model. If a more accurate model were used the trajectory would bend, although only slightly.

TABLE D.3. Values of functions $F(a)$ and $F(1/a)$ for some values of a.

a	$F(a)$	$F(1/a)$	a	$F(a)$	$F(1/a)$
0	1	0	1,4	0,3860	0,8566
0,2	0,9899	0,1005	1,6	0,3306	0,8936
0,4	0,9587	0,2043	1,8	0,2900	0,9177
0,6	0,9028	0,3158	2,0	0,2587	0,9342
0,8	0,8125	0,4441	2,5	0,2043	0,9587
1,0	0,6366	0,6366	3,0	0,1691	0,9716
1,2	0,4685	0,7926	∞	0	1

The integrals which appear in the expressions of the forces and the moments are functions of parameter a only. By introducing the function $F(a)$ defined as

$$F(a) = \frac{1}{2\pi} \int_0^{2\pi} \frac{1 + a\sin(\zeta)}{\sqrt{1 + a^2 + 2a\sin(\zeta)}} d\zeta , \qquad (D.75)$$

the expressions of $F_\parallel$ and M are simply

$$\begin{cases} F_\parallel = -fmgF(a) \\ \\ M = -fmgrF\left(\frac{1}{a}\right) . \end{cases} \qquad (D.76)$$

Function $F(a)$ must be obtained numerically. Its plot is reported in Fig. D.19b and some values are reported in Table D.3.

The equations governing the motion of the vehicle are two differential equations

$$\begin{cases} \dfrac{dV}{dt} = -fgF\left(\dfrac{r\Omega}{V}\right) \\ \\ \dfrac{d\Omega}{dt} = -fgr\dfrac{m}{J_z}F\left(\dfrac{V}{r\Omega}\right) . \end{cases} \qquad (D.77)$$

These must be integrated numerically, because parameter a changes continuously during motion and function $F(V/r\Omega)$ cannot be expressed in closed form. Once laws $V(t)$ and $\Omega(t)$ are known, the yaw angle and the position on the trajectory can be obtained by further integration

$$\psi = \int_0^t \Omega(u)du , \qquad s = \int_0^t V(u)du . \qquad (D.78)$$

Example D.6 *Consider a vehicle with $m = 1000$ kg, $J_z = 2000$ kg m^2, $a = 1.4$ m, $b = 1.4$ m and $t = 1.08$ m. After the collision, the vehicle moves with a speed $V = 7.36$ m/s, angular velocity $\Omega = -5.19$ rad/s and angle β equal to $53°$. Compute the trajectory and the positions taken by the vehicle until it stops.*

The results obtained by numerically integrating the equations of motion are reported in Fig. D.20a, curves A. A value of 0.7 for the friction coefficient has been

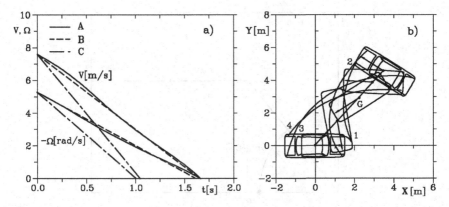

FIGURE D.20. Motion of a vehicle with locked wheels after a collision. (a) Time histories $V(t)$ and $\Omega(t)$ computed through numerical integration of the equations of motion (curves A), numerical integration of Eq. (D.77) (B) and by considering translations and rotations (C) separately. (b) Trajectory corresponding to curve (A).

assumed. The integration was performed using a time step $\Delta t = 0.01$ s. The trajectory is reported in Fig. D.20b. The vehicle stops in a time of 1.70 s, after a displacement of 6.33 m and a rotation of 4.19 rad ($240°$).

Similar results are obtained by integrating Eq. (D.77) (curves B). The values of function $F(a)$ used for the integration were taken from Table D.3 and interpolated linearly. The time needed to extinguish the motion, the displacement and the rotation are respectively 1.62 s, 6.11 m and 4.06 rad ($233°$).

Incorrect results would have been obtained by considering translational and rotational motion (curves C) separately. The time needed to extinguish the motion, the displacement and the rotation would have been respectively 1.01 s, 3.94 m and 2.62 rad ($150°$).

Laws $V(t)$ and $\Omega(t)$ are almost linear. If they were exactly linear the value of a would remain constant during the motion and no numerical integration would be required: A constant rate deceleration with the values of dV/dt and $d\Omega/dt$ given by Eq. (D.77) would occur. Note that this is equivalent to studying the motion as a translation and a rotation occurring separately, with "reduced" coefficients of friction equal to $fF(a)$ and $fF(1/a)$ respectively. In the example, immediately after the collision the value of a is 1.05, $F(a) = 0.62$ and $F(1/a) = 0.66$. By multiplying the friction coefficient by these values and considering the two motions separately, a time of 1.7 s for coming to a standstill is obtained. This value is close to that obtained through more complex models.

D.3.2   Vehicle with free wheels

In most cases the wheels of the vehicle, or at least some of them, remain free to rotate after the collision. There is little difficulty in numerically integrating the equations of motion, obviously written without any linearization. Equation (25.65) can be used, together with Eq. (25.94) yielding the sideslip angles of the

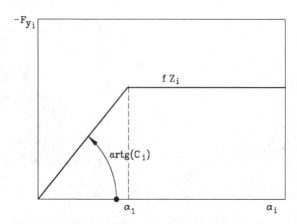

FIGURE D.21. Cornering force of the tire as a function of the sideslip angle. Simplified model in which the curve $F_y(\alpha)$ is approximated by two straight lines.

wheels. A model of the tire which can be used for all values of α from 0 to 360° is needed. The "magic formula" is at present probably the best and most precise choice.

The only difference from the nonlinear model is that some of the wheels may be locked. However, if the angular velocity of the vehicle is high, the sideslip angles remain at high values for a long time and the importance of accurately modelling the behavior of the tires at low sideslip angle is not great. In this situation a simple model for the cornering force of the tire as the one shown in Fig. D.21 may be used. With reference to the figure, the force the tire receives from the road is

$$
\begin{cases}
F_{x_i} = F_{z_i} \dfrac{\mu_i}{|\mu_i|} \left[-f_r \cos(\delta_i) - f \sin(\delta_i) \right] \\
\qquad\qquad\qquad\qquad\qquad\qquad\qquad \text{if } \alpha > \alpha_1 \ , \\
F_{y_i} = F_{z_i} \dfrac{\mu_i}{|\mu_i|} \left[-f_r \sin(\delta_i) + f \cos(\delta_i) \right] \\
F_{x_i} = F_{z_i} \dfrac{\mu_i}{|\mu_i|} \left[-f_r \cos(\delta_i) - \dfrac{\alpha_i}{C_i} f \sin(\delta_i) \right] \\
\qquad\qquad\qquad\qquad\qquad\qquad\qquad \text{if } \alpha \le \alpha_1 \ , \\
F_{y_i} = F_{z_i} \dfrac{\mu_i}{|\mu_i|} \left[-f_r \sin(\delta_i) + \dfrac{\alpha_i}{C_i} f \cos(\delta_i) \right]
\end{cases}
\tag{D.79}
$$

where f_r is the rolling coefficient.

The moment about the z-axis is expressed by the first part of Eq. (D.70). The rolling drag is usually small compared to the other forces and might be neglected but, because the equation must in any case be integrated numerically, there is no need to do so.

Example D.7 *Repeat the study of Example D.6 assuming that the wheels are free and all steer angles δ_i are equal to zero.*

By assuming a law $F_y(\alpha)$ of the type shown in Fig. D.21 with $\alpha_1 = 8°$ and $f_r = 0.02$ the results shown in Fig. D.22 are obtained through numerical integration. The

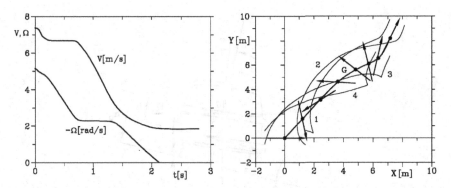

FIGURE D.22. Motion of a vehicle with free wheels after a collision. (a) Time histories $V(t)$ and $\Omega(t)$ computed through numerical integration of the equations of motion and (b) trajectories of the centre of mass and of the points of contact of the wheels.

computations were repeated with different values of angle α_1, obtaining practically the same results in all cases. With $\alpha_1 = 0$ the laws $V(t)$ and $\Omega(t)$ do not change, except for some oscillations due to numerical problems.

The vehicle at the end of the simulation is aligned with its velocity and rolls forward freely. In other cases the simulation may end with the vehicle rolling away in reverse.

D.4   ROLLOVER

D.4.1   Quasi-static rollover

As discussed in Part IV, rollover of a rigid vehicle in static or quasi-static conditions is usually impossible, except in the case of vehicles with particularly narrow track or high centre of mass: The conditions for slipping are reached well before those needed for rollover.

Rollover on a flat surface is controlled by parameter

$$\frac{t}{2h_G} \, ,$$

which is sometimes referred to as the rollover threshold: This constitutes the limit to the ratio between the lateral acceleration and the gravitational acceleration a_y/g. If the road has a transversal slope

$$i_t = \tan(\alpha_t)$$

the threshold becomes

$$\frac{t}{2h_G} - i_t \, .$$

The threshold increases linearly with the slope if the external part of the curve is raised and decreases otherwise.

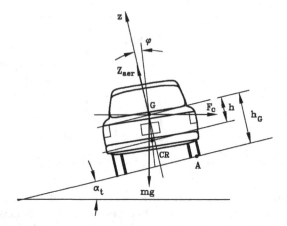

FIGURE D.23. Quasi-static rollover of a vehicle on elastic suspensions.

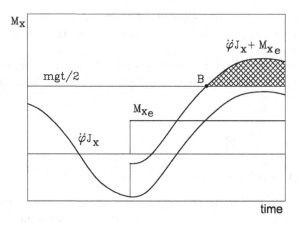

FIGURE D.24. Roll inertia torques causing rollover: In B the conditions for rollover are reached.

If

$$i_t = \frac{t}{2h_G}$$

the vehicle overturns naturally without the application of external forces, but this can occur only in case of off-road motion, because the adverse slope must be quite large.

To release the assumption of a rigid vehicle, the roll of the vehicle body must be accounted for. With reference to Fig. D.24, the forces acting in the direction parallel and perpendicular to the road surface are

$$\begin{cases} F_{\parallel} = \dfrac{mV^2}{R} \cos(\alpha_t) - mg\sin(\alpha_t) \\[2ex] F_{\perp} = \dfrac{mV^2}{R} \sin(\alpha_t) + mg\cos(\alpha_t) - Z_{aer} \, . \end{cases} \qquad \text{(D.80)}$$

If the inertia of the unsprung masses and the compliance of the tires are neglected and the roll angle is small enough, its value is

$$\phi \approx \frac{F_{\parallel} h}{K_t} , \tag{D.81}$$

where h is the height of the centre of mass over the roll axis and K_T is the torsional stiffness of the vehicle.

The vehicle rolls over if

$$\frac{F_{\parallel}}{F_{\perp}} > \frac{t - 2h\phi}{2h_G} . \tag{D.82}$$

Neglecting aerodynamic lift and assuming that the road surface is flat, the above equations may be solved in the lateral acceleration, yielding

$$\frac{V^2}{Rg} > \frac{t}{2h_G} \frac{1}{1 + \dfrac{mgh^2}{K_t h_G}} . \tag{D.83}$$

The fraction on the right hand side is a factor smaller than one expressing the reduction of resistance to rollover due to the presence of suspensions. The rollover threshold decreases with increasing roll compliance $1/K_t$ of the suspensions and distance h between the centre of mass and the roll centre. A stiff suspension with high roll centre can substantially reduce rollover danger and bring the rollover conditions close to those characterizing the rigid vehicle.

The rollover threshold can be reduced, owing to roll compliance, by a few percent (typically 5%) in passenger vehicles, with lower reductions in sports cars and greater in large luxury saloon cars.

The compliance of the tires increases this effect slightly, and some effects may be due to the exact geometry of the suspensions, particularly in terms of their lateral deflection, the lateral deflection of the tires and the inclination of the roll axis due to the differences between front and rear suspensions. To account for all these factors, a detailed mathematical model of the vehicle must be built and analyzed on a computer.

Alternatively, the whole vehicle can be put on a "tilt-table", i.e. on a platform that can be inclined laterally, and the lateral slope can be increased until the load on the less loaded wheels reduces almost to zero, denoting that the threshold of rollover has been reached. Note that the tilt-table arrangement exactly simulates rollover due to the lateral slope of the road, as may occur in off-road driving, but does not exactly reproduce conditions on flat road. The difference may be small if the vehicle rolls over on a lateral slope that is not too steep, i.e. about $20° \div 25°$, while the errors build up when the lateral slope reaches values as high as $45°$, where errors of about 30% may be present. To avoid this problem, a "cable-pull" test can be devised, in which the lateral forces are applied by cables directly to the vehicle at the location of the centre of mass.

In spite of these effects, static rollover remains a rare occurrence for all vehicles but those with very high centre of mass and narrow track, because the available lateral forces on the tires are not high enough to prevent lateral slipping before rollover.

D.4.2 Dynamic rollover

Dynamic conditions linked to roll oscillations, on the other hand, can lead to rollover if the external rolling moment adds to an inertia torque due to roll accelerations.

If a step roll input is applied to the vehicle, the response is a damped roll oscillation about the static equilibrium position that would have been reached if the same input had been applied slowly. Note that an input is applied slowly if its characteristic times are larger than the period of the lowest natural frequency involved; because the roll period is usually on the order of one second, most roll inputs, being faster than that, have the characteristics of a dynamic load.

Remark D.3 *If the vehicle were critically damped in roll no oscillations would take place. The roll angle would increase monotonically, reaching the static value asymptotically.*

Because all vehicles are underdamped in roll, the roll angle overshoots and then oscillates; the lower the damping ratio of the system the higher the overshoot and thus the danger of rollover. The presence of anti-roll bars makes things worse from this viewpoint. When anti-roll bars are added to an existing suspension the roll stiffness is increased but roll damping is usually unchanged, because the damping of shock absorbers is optimized for bounce and pitch behavior. The effect is an increase of the natural frequency and the underdamped behavior of the system. While the static roll angle is decreased by the added stiffness, both the overshoot and the natural frequency of roll oscillations are increased, the latter effect increasing roll inertia torques.

Consider a vehicle undergoing roll oscillations (Fig. D.24). The inertia torque $J_x\ddot{\phi}$ due to roll oscillations is sketched as a function of time in the figure. It is always lower than $mgt/2$, the restoring moment due to weight on level road, and hence no danger of rollover is present. However, if an external force causing the rolling moment M_{x_e} is applied, the conditions for incipient rollover occur in B.

From this point on the wheels at one side lose contact with the ground. The weight stabilizes, until a large enough roll rotation to bring the weight force outside point A in Fig. D.24 is completed, but the vehicle continues the roll over motion. Note that this simple model is inadequate to predict what happens when some of the wheels are no longer in contact with the ground and the roll angle is magnified.

A fast-reacting driver may prevent rollover even in these conditions by giving appropriate steering inputs, as testified by stunt drivers who can proceed with the car on two wheels, without either rolling over or rolling back onto four wheels, but it is most unlikely that a normal driver in normal conditions would perform stunts of this type. In most cases, once rollover starts it proceeds to its inevitable consequences.

Roll dynamics is closely coupled with yaw dynamics and rollover is no exception. During a sinusoidal steering input the lateral tire forces on the rear wheels are delayed with respect to those on the front wheels; this delay is larger in longer vehicles and the effect is even more pronounced in articulated vehicles. The combined roll-yaw dynamics with its resonances may facilitate rollover, particularly when selected frequencies are present in the excitation.

Tractor-trailer combinations are particularly subject to rollover due to sudden steering inputs: The roll of the trailer is delayed with respect to that of the tractor and can be large enough to cause the former to rollover. A hitch providing roll coupling between trailer and tractor helps in this case, because the latter collaborates in resisting the tendency of the former to overturn.

D.4.3 Lateral collision with the curb

Consider a vehicle modelled as a rigid body and assume that its velocity V is not contained in the plane of symmetry (Fig. D.25). Both wheels on one side enter into contact at time t_1 with the curb. At time t_1 the components of the velocity are[5]

$$u = u_1 , \quad v = v_1 , \quad w = p = q = r = 0 ,$$

i.e. the velocity of the vehicle is contained in a plane parallel to the road and the angular velocity is nil.

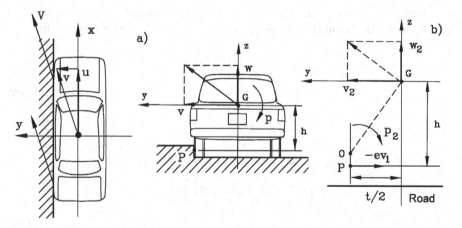

FIGURE D.25. Side impact against the curb. (a) Sketch; (b) determination of the centre of rotation at time t_2.

---

[5]The components Ω_x, Ω_y and Ω_z of the angular velocity are here indicated as p, q and r, and the component V_z of the velocity is indicated with w.

If friction between the vehicle and the curb is neglected, the component u of the velocity along the x-axis remains constant and the components q and r of the angular velocity along the y and z axes remain vanishingly small. The other components of the velocity after the impact, i.e. at time t_2, are

$$v = v_2 , \quad w = w_2 , \quad p = p_2 .$$

The component of the velocity in a direction perpendicular to the impact surface in P at time t_1 is

$$V_{\perp_1} = v_1 .$$

Assuming a partially inelastic impact with coefficient of restitution e^*, at time t_2 the same velocity is

$$V_{\perp_2} = -e^* v_1 .$$

If the condition that the motion of point P between time t_1 and t_2 occurs in a plane parallel to the ground is added, the position of the centre of rotation at time t_2 can be easily found and a relationship linking w_2 and p_2 to v_1 can be obtained.

With reference to Fig. D.25b it follows that

$$\begin{cases} p_2 = -\dfrac{v_2 + e^* v_1}{h} \\[3mm] w_2 = -p_2 \dfrac{t}{2} = (v_2 + e^* v_1)\dfrac{t}{2h} , \end{cases} \tag{D.84}$$

The components of the impulse the vehicle receives during the impact can be related to the variations of the momentum and the angular momentum

$$\begin{cases} I_y = m(v_1 - v_2) \\[3mm] I_z = -m w_2 \\[3mm] I_y h - I_z \dfrac{t}{2} = -J_x p_2 . \end{cases} \tag{D.85}$$

By introducing Eq. (D.84) into Eq. (D.85) it follows that

$$\begin{cases} v_2 = -v_1 \dfrac{A^2 - e^*(1 + B^2)}{1 + A^2 + B^2} \\[3mm] p_2 = -v_1 \dfrac{A^2(1 + e^*)}{h(1 + A^2 + B^2)} , \end{cases} \tag{D.86}$$

where

$$A = \frac{2h}{t} , \qquad B = \frac{2}{t}\sqrt{\frac{J_x}{m}} .$$

Rollover motion starts at time t_2. If the component of the velocity of point P in the y direction vanishes rapidly owing to the friction of the road, rollover actually occurs if the centre of mass crosses a line perpendicular to the road in P. This means that the vehicle rolls over only if its centre of mass moves above a distance Δh equal to

$$\Delta h = \sqrt{\frac{t^2}{4} + h^2} - h = h\left(\sqrt{1 + \frac{1}{A^2}} - 1\right) . \tag{D.87}$$

This can occur only if the kinetic energy associated with velocities v_2, w_2 and p_2 is at least equal to the potential energy $mg\Delta h$. With simple computations it can be shown that the condition for completing rollover is

$$v_2^2 + w_2^2 + \frac{J_x}{m}p_2^2 \geq 2hg\left(\sqrt{1 + \frac{1}{A^2}} - 1\right) . \tag{D.88}$$

Because

$$v_2^2 + w_2^2 = p_2^2\left(\frac{t^2}{4} + h^2\right) ,$$

the condition for rollover becomes

$$v_1^2 \frac{A^2(1 + e^*)^2}{1 + A^2 + B^2} \geq 2hg\left(\sqrt{1 + \frac{1}{A^2}} - 1\right) . \tag{D.89}$$

Consider for instance a vehicle with $A = 0.6$ and $B = 0.7$ hitting a curb in a perfectly elastic way ($e^* = 1$), which is the most dangerous condition. From Eq. (D.89) the rollover condition is

$$v_1^2 \geq 2.42gh .$$

Equation (D.89) may be written explicitly in terms of forward velocity V and impact angle θ

$$V^2 \sin^2(\theta) \geq 2hg\frac{1 + A^2 + B^2}{A^2(1 + e^*)^2}\left(\sqrt{1 + \frac{1}{A^2}} - 1\right) . \tag{D.90}$$

If $h = 0.36$ m and the impact angle is $\theta = 15°$, the vehicle in question will roll over for forward velocities greater than 11 m/s (40 km/h). Note that this result depends strongly on the value of e^*: The ratio between the velocities V needed for rollover when $e^* = 0$ and $e^* = 1$ is equal to $\sqrt{2}$. It makes sense that in this case the restitution coefficient e^* is greater than in the case of impacts between vehicles, but in general it depends upon the impact conditions.

It must also be noted that if the height of the curb is low, and consequently the value of h is large, the vehicle tends not to engage against the curb but to drive over it, making the whole study inapplicable. The assumption of a rigid body is also questionable: The impact usually occurs between the curb and the unsprung mass and the latter can undergo plastic and elastic deformations, accompanied by deformations of the tires.

D.4.4 Effect of the transversal slope and the curvature of the road

The above model is based upon the assumptions that the road is flat and that the curb is straight. To account for the transversal slope of the road it is sufficient to assume that the y-axis is inclined. The only difference is that of changing the expression of the potential energy due to the vertical displacement of the centre of mass of the vehicle, because a vertical line passing through point P is no longer perpendicular to the road.

If the curb is not straight the motion is more complex and it is not possible to study the motion in the yz plane independently of that in the x direction. However, if the curb follows a circular path with a radius far greater than the length of the vehicle and of the displacements in the y direction involved in the rollover motions, it is possible to simplify the problem.

Consider the situation shown in Fig. D.26a. The motion in the yz plane can be studied with reference to the non-inertial xyz frame, rotating about line CC' with angular velocity

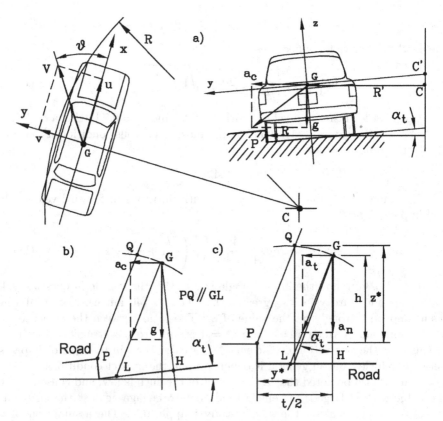

FIGURE D.26. Side impact against a curb following a circular path. (a) Sketch at time t_2; (b) accelerations broken into vertical and horizontal directions and (c) in directions parallel and perpendicular to the road.

$$\frac{V\cos(\theta)}{R} .$$

Both centrifugal and Coriolis acceleration must be introduced in the study of the motion starting from time t_2.

Centrifugal acceleration is directed radially while Coriolis acceleration acts in the x direction. Their values are, respectively,

$$\begin{cases} a_c = \dfrac{u^2}{R'} = \dfrac{V^2}{R'}\cos^2(\theta) \\[3mm] a_{cor} = -2\dfrac{V}{R'}\cos(\theta)\left[v_2\cos(\alpha_t) + w_2\sin(\alpha_t)\right] . \end{cases} \tag{D.91}$$

Only centrifugal acceleration needs to be considered in the study of the motion in the yz plane. It can be considered constant, neglecting the fact that R' and u change during the rollover motion. The method used in the previous case may be repeated by simply introducing the resultant of centrifugal and gravitational accelerations (here broken into directions parallel and perpendicular to the road surface)

$$\begin{cases} a_\perp = g\cos(\alpha_t) - a_c\sin(\alpha_t) \\[2mm] a_\parallel = g\sin(\alpha_t) + a_c\cos(\alpha_t) . \end{cases} \tag{D.92}$$

for the gravitational acceleration (Fig. D.26a,b). The vehicle will roll over if its centre of mass goes beyond line PQ, whose direction is that of the resultant of the two accelerations. The coordinates of point Q are then linked to each other by the relationship

$$\frac{y^*}{z^*} = \frac{a_\parallel}{a_\perp} .$$

Point Q lies on a circle centred in P passing through point G. It then follows that

$$y^{*^2} + z^{*^2} = h^2 + \frac{t^2}{4} . \tag{D.93}$$

By intersecting the circle with line PQ it follows that

$$y^* = h\frac{\sqrt{1+\frac{1}{A^2}}}{\sqrt{1+\left(\frac{a_\perp}{a_\parallel}\right)^2}} , \qquad z^* = h\frac{\sqrt{1+\frac{1}{A^2}}}{\sqrt{1+\left(\frac{a_\parallel}{a_\perp}\right)^2}} , \tag{D.94}$$

where

$$A = \frac{2h}{t} .$$

The total increase of potential energy in the motion from G to Q to be substituted for $mg\Delta h$ in the rollover condition is

$$\Delta \mathcal{U} = mha_\perp\left\{\sqrt{\left(1+\frac{1}{A^2}\right)\left[1+\left(\frac{a_\perp}{a_\parallel}\right)^2\right]} - 1 - \frac{a_\parallel}{Aa_\perp}\right\} , \tag{D.95}$$

The final form of the rollover condition is thus

$$v_1^2 \frac{A^2(1+e^*)^2}{1+A^2+B^2} \geq 2ha_\perp \left\{ \sqrt{\left(1+\frac{1}{A^2}\right)\left[1+\left(\frac{a_\perp}{a_\parallel}\right)^2\right]} - 1 - \frac{a_\parallel}{Aa_\perp} \right\}. \quad \text{(D.96)}$$

Clearly if

$$\frac{a_\perp}{a_\parallel} = A,$$

the vehicle rolls over even for $v_1 = 0$: Points G and Q coincide and the system is statically on the verge of instability. If the curb is along a straight line ratio $a_\parallel/a_\perp$ is equal to $\tan(\alpha_t) = i_t$, transversal slope of the road.

If on the contrary the road is flat but curved,

$$\frac{a_\parallel}{a_\perp} = \frac{a_c}{g} = V^2 \frac{\cos^2(\theta)}{Rg}.$$

The nondimensional velocity $v_1/\sqrt{a_\perp h}$ needed for rollover is plotted against ratio $a_\parallel/a_\perp$ for various values of e^* in Fig. D.27. The plot has been obtained for $A = 0.6$ and $B = 0.7$.

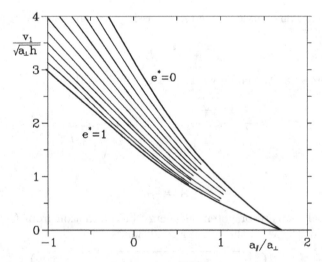

FIGURE D.27. Nondimensional velocity $v_1/\sqrt{a_\perp h}$ needed for rollover as a function of $a_\parallel/a_\perp$ for various values of e^*. Vehicle with $A = 0.6$ and $B = 0.7$.

D.5 MOTION OF TRANSPORTED OBJECTS DURING THE IMPACT

D.5.1   Free objects

The study of the motion of objects onboard vehicles during a collision is important both for understanding their effects on passengers and for devising suitable restraint systems for people and packaging for objects.

If the object is simply supported on a surface parallel to the road, when the vehicle stops abruptly because of a collision (which in the following will be considered as a central impact against a fixed object) it will continue its motion until it collides with the front wall of the load compartment. This first part of the collision process is easily studied under the assumption that the mass of the object is negligible with respect to that of the vehicle. The motion of the former does not influence that of the latter.

Consider the situation sketched in Fig. D.28, in which the elastic and damping properties of the transported object are modelled by a spring-damper system. The motion of the vehicle is assumed to follow the model described in Section D.2.1 and the motion of the object will be studied with reference to the xz frame centred in the position of point P, belonging to the front part of the object, at time t_1, when the vehicle hits the obstacle.

Neglecting friction, the motion of point P immediately after the collision is described by the law

$$x_P = V_1 t = V_1 t_2 \tau \,, \tag{D.97}$$

where V_1 is the velocity of the vehicle before the collision and the nondimensional time

$$\tau = \frac{t}{t_2}$$

is defined with respect to the duration of the collision t_2.

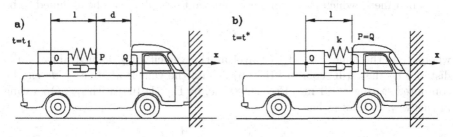

FIGURE D.28. Motion of the object O free to move on a vehicle hitting a rigid barrier. The elastic and damping properties of the object are modelled by a spring-damper system. (a) Situation at time t_1. (b) Situation at time t^*, when the object hits the front wall of the load compartment.

Assuming the model of Section D.2.1 to describe the motion of the vehicle, the x coordinate of point Q is

$$x_Q = d + V_1 t_2 \left(\frac{1 + e^*}{\beta + 3} \left\{ 2 - [2 + (\beta + 1)\tau] (1 - \tau)^{\beta+2} \right\} - e^* \tau \right). \qquad (D.98)$$

The object hits the front wall of the load compartment at time t^*, when the distance between points P and Q vanishes, i.e. when

$$\tau^* - \frac{2 - [2 + (\beta + 1)\tau^*] (1 - \tau^*)^{\beta+2}}{\beta + 3} = \frac{d}{V_1 t_2 (1 + e^*)}. \qquad (D.99)$$

Equation (D.99) can be easily solved in τ^*; however this solution holds only if $\tau^* \leq 1$, i.e. if the object collides before time t_2. If this condition does not hold the secondary collision inside the vehicle occurs when the latter has started its motion backward after rebounding from the obstacle and Eq. (D.98) does not apply. The condition for which Eq. (D.99) holds, i.e. for which $\tau^* \leq 1$, is simply

$$d \leq V_1 t_2 (1 + e^*) \frac{\beta + 1}{\beta + 3}. \qquad (D.100)$$

If condition (D.100) is satisfied, the relative velocity is easily computed by considering that at time τ^* the velocities of the object and the vehicle are both known

$$V_R = V_1 (1 + e^*) \left\{ 1 - [1 + (\beta + 1)\tau^*] (1 - \tau^*)^{\beta+1} \right\}. \qquad (D.101)$$

If, on the contrary, condition (D.100) is not satisfied, the computation is even simpler. If the rebound velocity is considered constant,

$$V_2 = -e^* V_1 ,$$

after time t_2, the relative velocity is simply

$$V_R = V_1 (1 + e^*) .$$

The time at which the secondary collision takes place can be assumed to be

$$t^* = t_2 + t_3 ,$$

where t_3 is the time needed to travel at the velocity V_R defined above for a distance d_2 that still separates the object and the vehicle at time t_2. d_2 can be computed stating $\tau = 1$ into Eq. (D.98), i.e. by Eq. (D.100) with '='. Operating in this way, it follows that

$$t_3 = \frac{d}{V_1 (1 + e^*)} - t_2 \frac{\beta + 1}{\beta + 3}. \qquad (D.102)$$

The motion of the object after it hits the wall of the load compartment may be analyzed in different ways.

It is possible to resort to a semi-empirical time history, as was done for the vehicle colliding against an obstacle, or to model the mechanical properties of the object and to use that model to obtain the equation of motion.

By assuming that the object and the wall are rigid bodies with an interposed linear spring-damper system, as sketched in Fig. D.28, the equation of motion of point O for $t > t^*$ is

$$m\ddot{x}_O + c(\dot{x}_O - \dot{x}_Q) + k(x_O + l - x_Q) = 0 , \qquad (D.103)$$

with the initial conditions

$$x_O = x_Q - l \quad \text{and} \quad \dot{x}_O = V_1 \quad \text{for} \quad t = t^* . \qquad (D.104)$$

By introducing the nondimensional coordinate

$$\chi = \frac{x_O + l - x_Q}{V_1 t_2}$$

and the nondimensional time τ, the equation of motion reduces to

$$\frac{d^2\chi}{d\tau^2} + \frac{ct_2}{m}\frac{d\chi}{d\tau} + \frac{kt_2^2}{m}\chi = \frac{d^2\chi_Q}{d\tau^2} , \qquad (D.105)$$

with the initial conditions

$$\chi = 0 \quad \text{and} \quad \frac{d\chi}{d\tau} = 1 - \left(\frac{d\chi_Q}{d\tau}\right)_{\tau=\tau^*} \quad \text{for} \quad t = t^*.$$

The expressions of $d^2\chi_Q/d\tau^2$ and of $d\chi_Q/d\tau$ can be easily computed from t^* to t_2 $(0 \leq \tau \leq 1)$ from Equations. (D.31) and (D.35) while for $t > t_2$ $(\tau > 1)$ they are simply

$$\frac{d^2\chi_Q}{d\tau^2} = 0 , \quad \frac{d\chi_Q}{d\tau} = -e^* .$$

If the secondary collision occurs before time t_2 the equation of motion of the object is

$$\frac{d^2\chi}{d\tau^2} + \frac{ct_2}{m}\frac{d\chi}{d\tau} + \frac{kt_2^2}{m}\chi = (\beta+1)(\beta+2)(1+e^*)\tau(1-\tau)^\beta \qquad \text{for} \quad \tau^* < \tau < 1,$$

$$\frac{d^2\chi}{d\tau^2} + \frac{ct_2}{m}\frac{d\chi}{d\tau} + \frac{kt_2^2}{m}\chi = 0 \qquad \text{for} \quad \tau > 1,$$

$$\qquad (D.106)$$

with the initial conditions $\chi = 0$ and

$$\frac{d\chi}{d\tau} = (1+e^*)\left\{1 - [1 + (\beta+1)\tau^*](1-\tau)^{\beta+1}\right\}$$

for $t = t^*$.

If the secondary shock occurs during the rebound phase, after the vehicle has lost contact with the obstacle, the equation of motion is the second of Equations. (D.106) and the initial condition on $d\chi/d\tau$ reduces to

$$\frac{d\chi}{d\tau} = (1 + e^*) .$$

These equations of motion hold only up to the time in which the object rebounds, losing contact with the wall of the load compartment. This instant can be easily computed by looking for the time at which the acceleration of the object vanishes, because the force acting between points P and Q reduces to zero.

The acceleration may can be computed as

$$\frac{d^2\chi_0}{d\tau^2} = \frac{V_1}{t_2} \left[\frac{d^2\chi}{d\tau^2} - (\beta+1)(\beta+2)(1+e^*)\tau(1-\tau)^\beta \right] \qquad \text{if } \tau^* < \tau < 1,$$

$$\frac{d^2\chi_0}{d\tau^2} = \frac{V_1}{t_2}\frac{d^2\chi}{d\tau^2} \qquad\qquad\qquad \text{if } \tau > 1.$$

$$\text{(D.107)}$$

This expression of the acceleration is one of the most interesting results of this study, because the aim of the elastic system with stiffness k and damping c is to allow the object to survive the shock of the collision, which means reducing its acceleration within allowable limits.

Another important result is the value of the maximum displacement χ, i.e. the maximum compression of the spring. This value cannot be higher than a given allowable limit, beyond which the elastic system is crushed or, at least, shows nonlinear characteristics with increasing stiffness. In practice, to limit the acceleration the spring must be soft but the decrease of the value of k is limited by the available space because it causes the maximum travel to increase.

The whole process is governed by five nondimensional parameters: β and e^*, linked to the way the vehicle collides with the obstacle, ratio

$$\frac{d}{V_1 t_2} ,$$

linked to the clearance between the object and the wall, and

$$\frac{k t_2^2}{m} \quad \text{and} \quad \frac{c t_2}{m} ,$$

related to the elastic and damping characteristics of the object.

Parameter $k t_2^2/m$ can be written in the form

$$\frac{k t_2^2}{m} = (\omega_n t_2)^2 = 4\pi^2 \left(\frac{t_2}{T_n} \right)^2 , \qquad\qquad \text{(D.108)}$$

where ω_n and T_n are the circular frequency and the period of the undamped free oscillations of a spring-mass system with mass m and stiffness k. If this parameter has a value equal to π^2 the half-period of the undamped oscillations is equal to t_2.

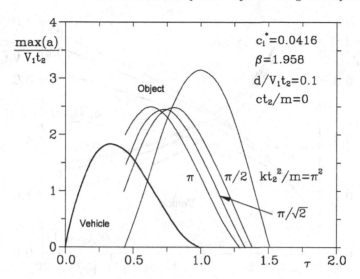

FIGURE D.29. Time history of the acceleration for the vehicle studied in Fig. D.13 hitting a fixed obstacle at 20 m/s and for an object carried by it.

Parameter ct_2/m is linked to kt_2^2/m and to the damping ratio ζ by the relationship

$$\frac{ct_2}{m} = \zeta\sqrt{2}\sqrt{\frac{kt_2^2}{m}} . \tag{D.109}$$

The absolute value of the acceleration of an object onboard the vehicle studied in Fig. D.13 hitting a fixed obstacle at 20 m/s is plotted against time in Fig. D.29. The value of kt_2^2/m has been assumed to be equal to π^2; the various curves correspond to different values of ct_2/m. In the case studied, the lowest maximum acceleration is obtained for

$$\frac{ct_2}{m} = \pi/\sqrt{2} ,$$

i.e. for $\zeta = 0.5$.

This value of the damping ratio coincides with the "optimum value" defined in Section 6.8.1 for the quarter car model with a single degree of freedom. As the two models are different, such instances cannot be generalized and the values of ζ minimizing the acceleration must be obtained for each case, as can be verified by plotting the same figure with different values of the parameters.

The maximum values of the acceleration and the displacement are plotted versus parameter kt_2^2/m in Fig. D.30. The plot has been obtained with the same values of β and e^* used for Fig. D.29, with the added assumption of a damping ratio equal to 0.5. From the figure it is clear that the distance d must be kept to a minimum to provide low values for the acceleration. From the maximum allowable value of the displacement χ it is possible to obtain the minimum value of k and then the value of the peak acceleration occurring during the impact.

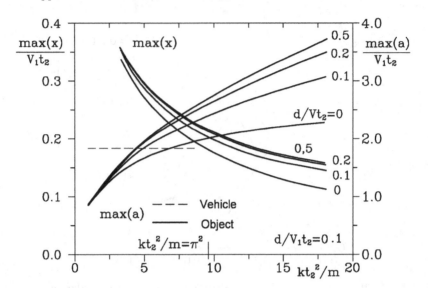

FIGURE D.30. Peak values of the acceleration and of the displacement as a function of the stiffness of the spring for various values of the clearance d. Same vehicle as in Fig.D.29; damping equal to half of the critical damping ($\zeta = 0.5$).

The results shown in Figs. D.29 and D.30 were obtained by numerical integration of the equation of motion in time.

The simple models here shown allow basic and straightforward qualitative assessments, with a limited number of parameters involved. There is, however, no difficulty in building more complete models that result in good quantitative predictions of the actual behavior of the system.

D.5.2 Constrained objects

The behavior of an object constrained onboard a vehicle is not qualitatively different from that seen for free objects. In the case of constrained objects the distance d, i.e. the clearance of the restraining device, is smaller and may even be equal to zero.

The motion of a constrained object can be studied in four distinct phases (Fig. D.31).

1. The first phase begins in the instant the vehicle encounters the obstacle starting its deceleration and ends at time t^* when the object contacts the restraining system. Usually, if the clearance is not too large, time t^* occurs before time t_2.

2. The second phase extends between time t^* and time t_2. The collision of the vehicle against the obstacle is completed and the object is retained by the constraining device. At the end the vehicle rebounds freely.

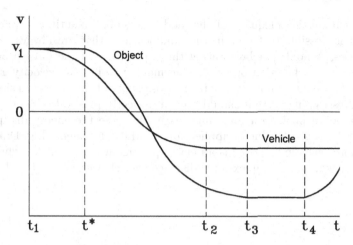

FIGURE D.31. Time history of velocity of the motion of the vehicle and of an object constrained on board.

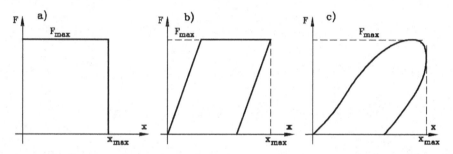

FIGURE D.32. Force exerted by the restraining system as a function of the deformation x; the energy dissipated is the area under the curve. (a) Perfectly plastic, (b) elasto-plastic and (c) actual behaviour.

3. The third phase is between time t_2 and time t_3. The object rebounds backward and, at the end, loses contact with the restraining system. The vehicle moves backward, with a speed that is low if the collision is strongly inelastic.

4. The fourth phase goes from time t_3 to time t_4 when the object, moving backward at a speed higher than that of the vehicle, hits the back wall of the load compartment.

The study may then continue, because at this point there may be a further rebound, which can be particularly dangerous.

A case of particular importance that can be studied, at least as a first approximation, with the present technique, is that of the motion of a person wearing a seat belt. The collision with the seat at time t_4 can be the most dangerous event under these circumstances.

To increase the adequacy of the model to supply quantitative information as well it is possible to factor in the nonlinearities that are always present in actual cases. A nonlinear law stating the dependence of the force provided by the restraint as a function of the displacement x and of the velocity $\dot{x}$ may be introduced without adding greatly to the complexity of the numerical simulation.

The restraining system should have a perfectly plastic behavior to reduce the acceleration peak to a minimum, as in this case the energy dissipation is maximum for a given value of applied force and displacement (Fig. D.32a).

A more realistic behavior is an elasto-plastic law (Fig. D.32b) while actual systems show a more complex force-displacement characteristic (Fig. D.32c).

Appendix E
DATA ON VARIOUS VEHICLES

This appendix contains fairly complete data on different vehicles: The small cars of the A, B, C and D sections, two sports cars, a van, an articulated truck and a racing motorcycle

Some of the data here reported were used throughout the text in the examples and may be used by the reader to repeat the computations shown for different kinds of motor vehicles.

Although not an exact description of any actual vehicle, the characteristics shown here are typical.

E.1 Small car (a)

The vehicle is a typical late model small car with five seats and a 1.2 liter spark ignition engine.

The primarygeometrical data and inertial properties of the vehicle are:

length = 3,540 mm   width = 1,580 mm   height = 1,540 mm

$a = 923$ mm $b = 1,376$ mm $l = 2,299$ mm

$m = 860 \div 1,020^1$ kg $t_1 = 1,370$ mm $t_2 = 1,360$ mm

$J_x = 500$ kg m^2 $J_y = 1,000$ kg m^2 $J_z = 1,225$ kg m^2

$h_G = 600$ mm

J_r(each) = 0.59 kg m^2 $J_m = 0.088$ kg m^2 $J_t = 0.05$ kg m^2

---

[1]The first value is the mass of the empty vehicle; the second includes two passengers and is consistent with the values of the moments of inertia the height of the center of mass.

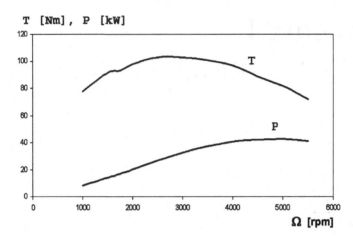

FIGURE E.1. Maximum power and torque curves of the engine.

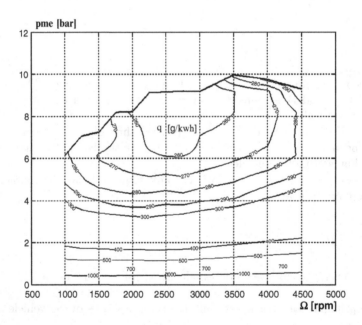

FIGURE E.2. Map of the specific fuel consumption of the engine; the consumption at idle (700 rpm) is 340 g/h.

The maximum engine power and torque are

$P_{max} = 43$ kW at 5000 rpm; $T_{max} = 103$ Nm at 3000 rpm.

The performance curves of the engine are reported in Fig. E.1; the map of specific fuel consumption is shown in Fig. E.2.

The data for the computation of rolling resistance and for the computation of longitudinal and lateral forces (equations (2.34, 2.35, 2.36, 2.37)) for the 155/70 R 13 tires used on this car are reported in the following table:

$$R_e = 279 \text{ mm}$$

$f_0 = 0.011$

$p_{Cx1} = 1.62$

$p_{Dx1} = 0.941$

$p_{Dx2} = 5.84 \cdot 10^{-2}$

$p_{Dx3} = -2.45 \cdot 10^{-5}$

$p_{Ex1} = 0.804$

$p_{Ex2} = -0.312$

$p_{Ex3} = -0.416$

$p_{Ex4} = -3.79 \cdot 10^{-6}$

$p_{Kx1} = 41.3$

$p_{Kx2} = 15.5$

$p_{Kx3} = -5.44 \cdot 10^{-3}$

$p_{Hx1} = -5.27 \cdot 10^{-8}$

$p_{Hx2} = 5.60 \cdot 10^{-8}$

$p_{Vx1} = -8.54 \cdot 10^{-8}$

$p_{Vx2} = 3.48 \cdot 10^{-7}$

$K = 2.6 \cdot 10^{-8} \text{ s}^2/\text{m}^2$

$p_{Cy1} = 1.711$

$p_{Dy1} = 0.900$

$p_{Dy2} = -0.273$

$p_{Dy3} = 5.45 \cdot 10^{-4}$

$p_{Ey1} = 5.84 \cdot 10^{-2}$

$p_{Ey2} = -0.616$

$p_{Ey3} = -1.49 \cdot 10^{-5}$

$p_{Ey4} = 2.59 \cdot 10^{-5}$

$p_{Ky1} = -12.4$

$p_{Ky2} = 1.08$

$p_{Ky3} = 0.767$

$p_{Hy1} = -2.26 \cdot 10^{-8}$

$p_{Hy2} = 2.07 \cdot 10^{-7}$

$p_{Hx3} = -5.36 \cdot 10^{-7}$

$p_{Vy1} = 6.37 \cdot 10^{-8}$

$p_{Vy2} = 8.13 \cdot 10^{-7}$

$p_{Vy3} = -1.40 \cdot 10^{-6}$

$p_{Vy4} = 7.80 \cdot 10^{-7}$

Aerodynamic data:

$$S = 2.04 \text{ m}^2 \qquad C_x = 0.33$$

Gear ratios (front wheel drive):

$\tau_I = 3.909 \qquad \tau_{II} = 2.157 \qquad \tau_{III} = 1.48 \qquad \eta_t = 0.93$

$\tau_{IV} = 1.121 \qquad \tau_V = 0.897 \qquad \tau_f = 3.438$

E.2 Small car (b)

The vehicle is a typical 1990s small car with five seats and a 1 liter spark ignition engine of the 1990s. The performance curves of the engine and map of specific fuel consumption are shown in Fig. E.3.

The primary geometrical data and the inertial properties of the vehicle, wheels and driveline are:

length = 3,640 mm   width = 1,560 mm   height = 1,410 mm

a = 870 mm b = 1,290 mm l = 2,160 mm

m = 830 kg t_1 = 1,284 mm t_2 = 1,277 mm

$J_x = 290$ kg m^2 $J_y = 1,094$ kg m^2 $J_z = 1,210$ kg m^2

$J_{xz} = -84$ kg m^2

J_r(each) = 0.4 kg m^2 $J_m = 0.085$ kg m^2 $J_t = 0.05$ kg m^2

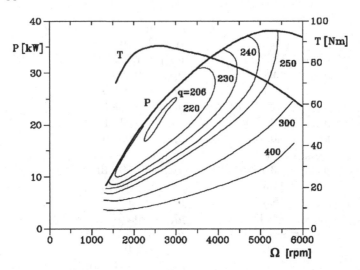

FIGURE E.3. Maximum torque and power curves and map of the specific fuel consumption of the engine; the consumption at idle (700 rpm) is 27 g/h.

The maximum engine power and torque are

$P_{max} = 38.3$ kW at 5200 rpm; $T_{max} = 87$ Nm at 3,000 rpm.

The data for the computation of rolling resistance and longitudinal and lateral forces (equations (2.23, 2.28, 2.29)) for the 145 R 13 tires used on this car are reported in the following table:

$f_0 = 0.013$	$K = 6.5 \cdot 10^{-6}$ s^2/m^2	$R_e = 257$ mm
$a_0 = 1.30$	$a_1 = -53.3$	$a_2 = 1,190$
$a_3 = 588$	$a_4 = 2.52$	$a_5 = 0$
$a_6 = -0.519$	$a_7 = 1.00$	$a_8 = 0$
$a_9 = 0$	$a_{10} = 0$	$a_{111} = 0$
$a_{112} = 0$	$a_{12} = 0$	$a_{13} = 0$
$c_0 = 2.40$	$c_1 = -4.40$	$c_2 = -1.36$
$c_3 = -4.10$	$c_4 = -3.28$	$c_5 = 0.245$
$c_6 = 0$	$c_7 = -0.0792$	$c_8 = 0$
$c_9 = 1.00$	$c_{10} = 0$	$c_{11} = 0$
$c_{12} = 0$	$c_{13} = 0$	$c_{14} = 0$
$c_{15} = 0$	$c_{16} = 0$	$c_{17} = 0$
$A = 1.12$	$C = 0.625$	$D = 1$
$n = 0.6$	$k = 46$	$d = 5$

Aerodynamic data:

$S = 1.7$ m^2	$C_x = 0.32$	$C_z = -0.21$
$C_{M_y} = -0.09$	$(C_y)_{,\beta} = -2.2$ 1/rad	$(C_{M_z})_{,\beta} = -0.6$ 1/rad

Driveline (front wheels drive):

$$\tau_I = 4.64 \qquad \tau_{II} = 2.74 \qquad \tau_{III} = 1.62$$
$$\tau_{IV} = 1.05 \qquad \tau_f = 3.47 \qquad \eta_t = 0.91$$

E.3   Small car (c)

The vehicle is a typical late-model small car with five seats and a 1.3 liter direct injection diesel engine.

The main geometrical data and inertial properties of the vehicle, wheels and engine are:

length = 4,030 mm	width = 1,690 mm	height = 1,490 mm
a = 958 mm	b = 1,552 mm	l = 2,510 mm
m = 1,090÷1,250^2 kg	t_1 = 1,440 mm	t_2 = 1,420 mm
J_x = 960 kg m^2	J_y = 2,058 kg m^2	J_z = 2,663 kg m^2
h_G = 550 mm		
J_r(each) = 0.77 kg m^2	J_m = 0.16 kg m^2	J_t = 0.06 kg m^2

The maximum engine power and torque are
P_{max} = 51 kW at 4,000 rpm; T_{max} = 180 Nm at 1,800 rpm.

The performance curves of the engine are reported in Fig. E.4; the map of specific fuel consumption is shown in Fig. E.5.

The data for the computation of rolling resistance and longitudinal and lateral forces (equations (2.34, 2.35, 2.36, 2.37)) for the 175/65 R 15 tires used on this car are reported in the following table:

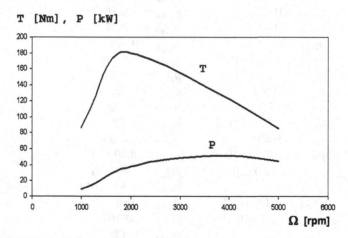

FIGURE E.4. Maximum power and torque curves of the engine.

---

[2]The first value is the mass of the empty vehicle; the second includes two passengers and is consistent with the values of the moments of inertia and height of the center of mass.

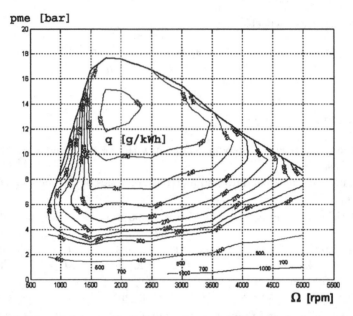

FIGURE E.5. Map of the specific fuel consumption of the engine; the consumption at idle (800 rpm) is 300 g/h.

$$R_e = 296 \text{ mm}$$

$$f_0 = 0.0096 \qquad\qquad K = 2.7 \cdot 10^{-8} \text{ s}^2/\text{m}^2$$

$$p_{Cx1} = 1.45 \qquad\qquad p_{Cy1} = 1.11$$

$$p_{Dx1} = 1.00 \qquad\qquad p_{Dy1} = 0.982$$

$$p_{Dx2} = 0.0 \qquad\qquad p_{Dy2} = -0.16$$

$$p_{Dx3} = 0.0 \qquad\qquad p_{Dy3} = -2.47$$

$$p_{Ex1} = 8.58 \cdot 10^{-7} \qquad\qquad p_{Ey1} = 0.571$$

$$p_{Ex2} = -7.60 \cdot 10^{-6} \qquad\qquad p_{Ey2} = -0.261$$

$$p_{Ex3} = -2.24 \cdot 10^{-5} \qquad\qquad p_{Ey3} = -0.323$$

$$p_{Ex4} = -202 \qquad\qquad p_{Ey4} = -22.66$$

$$p_{Kx1} = 25.1 \qquad\qquad p_{Ky1} = -20.22$$

$$p_{Kx2} = -5.11 \cdot 10^{-6} \qquad\qquad p_{Ky2} = 2.72$$

$$p_{Kx3} = 0.399 \qquad\qquad p_{Ky3} = 0.0811$$

$$p_{Hx1} = -8.71 \cdot 10^{-6} \qquad\qquad p_{Hy1} = 3.39 \cdot 10^{-3}$$

$$p_{Hx2} = -1.0 \cdot 10^{-5} \qquad\qquad p_{Hy2} = 1.58 \cdot 10^{-3}$$

$$p_{Hy3} = 9.83 \cdot 10^{-2}$$

$$p_{Vx1} = 1.42 \cdot 10^{-5} \qquad\qquad p_{Vy1} = 4.51 \cdot 10^{-4}$$

$$p_{Vx2} = 8.62 \cdot 10^{-5} \qquad\qquad p_{Vy2} = 0.0161$$

$$p_{Vy3} = 0.438$$

$$p_{Vy4} = -0.378$$

Aerodynamic data:

$$S = 2.14 \text{ m}^2 \qquad\qquad C_x = 0.35$$

Driveline (front wheels drive):

$$\tau_I = 3.909 \qquad \tau_{II} = 2.238 \qquad \tau_{III} = 1.444 \qquad \eta_t = 0.93$$
$$\tau_{IV} = 1.029 \qquad \tau_V = 0.767 \qquad \tau_f = 3.563$$

E.4 Medium size saloon car (a)

The vehicle is a typical late-model saloon car with five seats and a 1.9 liter direct injection diesel engine.

The main geometrical data and inertial properties of the vehicle, wheels and engine are:

length = 4,250 mm   width = 1,760 mm   height = 1,530 mm
$a = 1,108$ mm $b = 1,492$ mm $l = 2,600$ mm
$m = 1,320 \div 1,400^3$ kg $t_1 = 1,506$ mm $t_2 = 1,498$ mm
$J_x = 545$ kg m^2 $J_y = 1,936$ kg m^2 $J_z = 2,038$ kg m^2
$h_G = 565$ mm
J_r(each) = 0.86 kg m^2 $J_m = 0.27$ kg m^2 $J_t = 0.08$ kg m^2

The maximum engine power and torque are
$P_{max} = 85$ kW at 4,000 rpm; $T_{max} = 280$ Nm at 2,000 rpm.

The performance curves of the engine are reported in Fig. E.6; the map of specific fuel consumption is shown in Fig. E.7.

The data for the computation of rolling resistance and longitudinal and lateral forces (equations (2.34, 2.35, 2.36, 2.37)) for the 195/65 R 15 tires used on this car are reported in the following table:

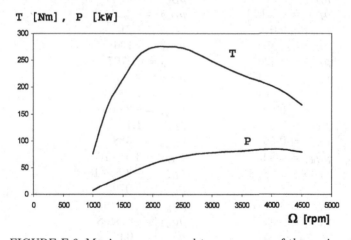

FIGURE E.6. Maximum power and torque curves of the engine.

---

[3]The first value is the mass of the empty vehicle; the second includes two passengers and is consistent with the values of the moments of inertia and height of the center of mass.

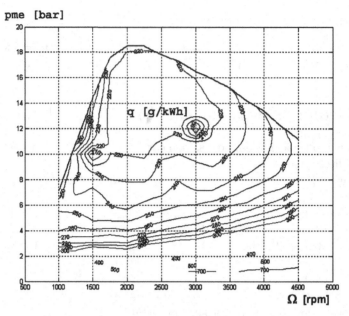

FIGURE E.7. Map of the specific fuel consumption of the engine; the consumption at idle (850 rpm) is 520 g/h

$$R_e = 307 \text{ mm}$$

$f_0 = 0.0124$ $K = 0.23 \cdot 10^{-8} \text{ s}^2/\text{m}^2$

$p_{Cx1} = 1.53$ $p_{Cy1} = 0.0486$

$p_{Dx1} = 1.20$ $p_{Dy1} = 18.1$

$p_{Dx2} = -0.129$ $p_{Dy2} = -2.21$

$p_{Dx3} = 0.0$ $p_{Dy3} = 1.71$

$p_{Ex1} = 0.456$ $p_{Ey1} = 0.106$

$p_{Ex2} = 0.420$ $p_{Ey2} = 0.0767$

$p_{Ex3} = 0.024$ $p_{Ey3} = -1.57$

$p_{Ex4} = 0.174$ $p_{Ey4} = 28.9$

$p_{Kx1} = 60.9$ $p_{Ky1} = -24.03$

$p_{Kx2} = 0.0807$ $p_{Ky2} = 1.70$

$p_{Kx3} = 0.567$ $p_{Ky3} = 0.388$

$p_{Hx1} = -5.99 \cdot 10^{-5}$ $p_{Hy1} = 3.07 \cdot 10^{-3}$

$p_{Hx2} = 5.76 \cdot 10^{-4}$ $p_{Hy2} = -3.55 \cdot 10^{-3}$

     $p_{Hx3} = 0.0796$

$p_{Vx1} = 0.0347$ $p_{Vy1} = -0.0153$

$p_{Vx2} = 0.0221$ $p_{Vy2} = -0.0468$

     $p_{Vy3} = -0.0522$

     $p_{Vy4} = 0.217$

Aerodynamic data:

$$S = 2.15 \text{ m}^2 \qquad C_x = 0.34$$

Driveline (front wheels drive):

$$\tau_I = 3.58 \qquad \tau_{II} = 1.89 \qquad \tau_{III} = 1.19 \qquad \eta_t = 0.93$$
$$\tau_{IV} = 0.85 \qquad \tau_V = 0.69 \qquad \tau_f = 3.63$$

E.5 Medium size saloon car (b)

The vehicle is a typical saloon car with a 2 liter spark ignition supercharged engine from the late 1980s.

The performance curves of the engine are reported in Fig. E.8.

The main geometrical data and inertial properties of the vehicle, wheels and driveline are:

length = 4,690 mm   width = 1,830 mm   height = 1,450 mm
$a = 1,064$ mm $b = 1,596$ mm $l = 2,660$ mm
$m = 1,150$ kg $t_1 = 1,490$ mm $t_2 = 1,482$ mm
$J_x = 530$ kg m^2 $J_y = 1,630$ kg m^2 $J_z = 1,850$ kg m^2
$h_G = 570$ mm
$J_{xz} = -120$ kg m^2
J_r(each) $= 0.6$ kg m^2 $J_m = 0.19$ kg m^2 $J_t = 0.07$ kg m^2

The maximum engine power and torque are

$P_{max} = 122$ kW at 5,500 rpm; $T_{max} = 255$ Nm at 2,500 rpm, with the possibility of short surges up to 284 Nm at 2,750 rpm (*overboost*).

The data for the computation of rolling resistance and longitudinal and lateral forces (equations (2.23, 2.28, 2.29)) for the 195 R 14 tires used on this car are reported in the following table:

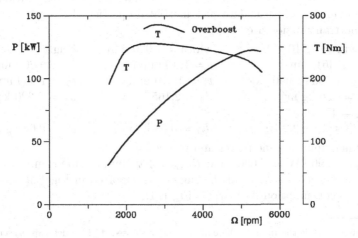

FIGURE E.8. Maximum power and torque curves of the engine

$f_0 = 0.013$	$K = 6.5 \times 10^{-6}$ s^2/m^2	$R_e = 287$ mm
$a_0 = 1.69$	$a_1 = -55.2$	$a_2 = 1,270$
$a_3 = 1.600$	$a_4 = 6.49$	$a_5 = 4.80 \cdot 10^{-3}$
$a_6 = -0.388$	$a_7 = 1.00$	$a_8 = -4.54 \cdot 10^{-2}$
$a_9 = 4.28 \cdot 10^{-3}$	$a_{10} = 8.65 \cdot 10^{-2}$	$a_{111} = -7.97$
$a_{112} = -0.223$	$a_{12} = 7.61$	$a_{13} = 45.9$
$b_0 = 1.65$	$b_1 = -7.61$	$b_2 = 1,122$
$b_3 = -7.36 \cdot 10^{-3}$	$b_4 = 145$	$b_5 = -7.66 \cdot 10^{-2}$
$b_6 = -3.86 \cdot 10^{-3}$	$b_7 = 8.50 \cdot 10^{-2}$	$b_8 = 7.57 \cdot 10^{-2}$
$b_9 = 2.36 \cdot 10^{-2}$	$b_{10} = 2.36 \cdot 10^{-2}$	
$c_0 = 2.22$	$c_1 = -3.04$	$c_2 = -9.23$
$c_3 = 0.500$	$c_4 = -5.57$	$c_5 = -0.260$
$c_6 = -1.30 \cdot 10^{-3}$	$c_7 = -0.358$	$c_8 = 3.74$
$c_9 = -15.2$	$c_{10} = 2.11 \cdot 10^{-3}$	$c_{11} = 3.46 \cdot 10^{-4}$
$c_{12} = 9.14 \cdot 10^{-3}$	$c_{13} = -0.244$	$c_{14} = 0.101$
$c_{15} = -1.40$	$c_{16} = 0.444$	$c_{17} = -0.999$

Aerodynamic data:

$S = 2.06$ m^2	$C_x = 0.36$	$C_z = -0.12$
$C_{M_y} = -0.05$	$(C_y)_{,\beta} = -1.8$ 1/rad	$(C_{M_z})_{,\beta} = -0.5$ 1/rad

Driveline (front wheel drive)

$\tau_I = 3.750$	$\tau_{II} = 2.235$	$\tau_{III} = 1.518$	$\eta_t = 0.91$
$\tau_{IV} = 1.132$	$\tau_V = 0.82$	$\tau_f = 2.95$	

E.6   Sports car (a)

The vehicle is a recent two-seater sports car with a 4.2 liter spark ignition engine.

The main geometrical data and inertial properties of the vehicle, wheels, engine and transmission are

length = 4,510 mm	width = 1,920 mm	height = 1,210 mm
$a = 1,461$ mm	$b = 1,199$ mm	$l = 2,660$ mm
$m = 1,590 \div 1,690^4$ kg	$t_1 = 1,670$ mm	$t_2 = 1,600$ mm
$J_x = 626$ kg m^2	$J_y = 2,165$ kg m^2	$J_z = 2,220$ kg m^2
$h_G = 470$ mm		
J_r (total) = 2.2 kg m^2	$J_m = 0.28$ kg m^2	$J_t = 0.07$ kg m^2

The maximum engine power and torque are

$P_{max} = 290$ kW at 7,000 rpm; $T_{max} = 460$ Nm at 4,400 rpm.

The performance curves of the engine are reported in Fig. E.9; the map of specific fuel consumption is shown in Fig. E.10.

---

[4]The first value is the mass of the empty vehicle; the second includes two passengers and is consistent with the values of the moments of inertia and height of the center of mass.

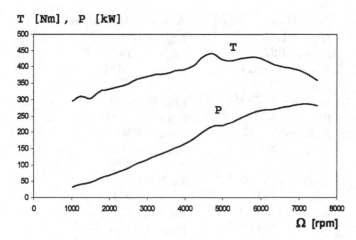

FIGURE E.9. Maximum power and torque curves of the engine.

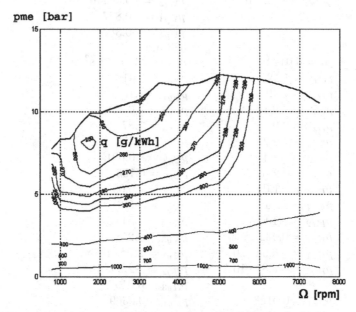

FIGURE E.10. Map of the specific fuel consumption of the engine; the consumption at idle (850 rpm) is 520 g/h

The data for the computation of rolling resistance and longitudinal and lateral forces (equations (2.34, 2.35, 2.36, 2.37)) for the front 245/40 R 18 and rear 285/40 R 18 tires used on this car are reported in the following table:

$$f_0 = 0.014 \qquad\qquad K = 0.5 \cdot 10^{-7}~\mathrm{s^2/m^2}$$

$$
\begin{aligned}
p_{Cx1} &= 1.44 & p_{Cy1} &= 1.84 \\
p_{Dx1} &= 1.27 & p_{Dy1} &= 1.19 \\
p_{Dx2} &= -0.0824 & p_{Dy2} &= -0.181 \\
p_{Dx3} &= 0.0 & p_{Dy3} &= 3.85 \\
p_{Ex1} &= -2.37 \cdot 10^{-5} & p_{Ey1} &= -1.89 \cdot 10^{-4} \\
p_{Ex2} &= -3.93 \cdot 10^{-5} & p_{Ey2} &= 5.17 \cdot 10^{-5} \\
p_{Ex3} &= 8.57 \cdot 10^{-6} & p_{Ey3} &= 1060 \\
p_{Ex4} &= -201 & p_{Ey4} &= 1490 \\
p_{Kx1} &= 35.8 & p_{Ky1} &= -3050 \\
p_{Kx2} &= 4.73 \cdot 10^{-4} & p_{Ky2} &= 299 \\
p_{Kx3} &= 0.143 & p_{Ky3} &= 0.224 \\
p_{Hx1} &= -7.69 \cdot 10^{-5} & p_{Hy1} &= -4.70 \cdot 10^{-4} \\
p_{Hx2} &= 1.35 \cdot 10^{-4} & p_{Hy2} &= -1.24 \cdot 10^{-4} \\
 & & p_{Hy3} &= -8.04 \cdot 10^{-3} \\
p_{Vx1} &= 3.32 \cdot 10^{-4} & p_{Vy1} &= -0.0153 \\
p_{Vx2} &= 5.72 \cdot 10^{-4} & p_{Vy2} &= -0.00210 \\
 & & p_{Vy3} &= -0.857 \\
 & & p_{Vy4} &= -0.380
\end{aligned}
$$

$$R_e = 333~\mathrm{mm}$$

$$f_0 = 0.0124 \qquad\qquad K = 0.23 \cdot 10^{-8}~\mathrm{s^2/m^2}$$

$$
\begin{aligned}
p_{Cx1} &= 1.52 & p_{Cy1} &= 1.41 \\
p_{Dx1} &= 1.49 & p_{Dy1} &= 1.18 \\
p_{Dx2} &= -0.102 & p_{Dy2} &= -0.158 \\
p_{Dx3} &= 0.0 & p_{Dy3} &= 1.87 \\
p_{Ex1} &= -0.277 & p_{Ey1} &= 0.358 \\
p_{Ex2} &= -0.422 & p_{Ey2} &= -0.691 \\
p_{Ex3} &= 0.251 & p_{Ey3} &= 0.196 \\
p_{Ex4} &= -1.04 & p_{Ey4} &= -6.03 \\
p_{Kx1} &= 27.6 & p_{Ky1} &= -60.1 \\
p_{Kx2} &= -10.1 & p_{Ky2} &= 3.51 \\
p_{Kx3} &= 0.432 & p_{Ky3} &= 0.271 \\
p_{Hx1} &= -6.34 \cdot 10^{-4} & p_{Hy1} &= 2.74 \cdot 10^{-3} \\
p_{Hx2} &= -3.87 \cdot 10^{-4} & p_{Hy2} &= -2.29 \cdot 10^{-4} \\
 & & p_{Hy3} &= 0.0197 \\
p_{Vx1} &= 0.0127 & p_{Vy1} &= 0.0369 \\
p_{Vx2} &= 0.0214 & p_{Vy2} &= -0.0262 \\
 & & p_{Vy3} &= -0.47 \\
 & & p_{Vy4} &= -0.565
\end{aligned}
$$

Aerodynamic data:

$$S = 2.21~\mathrm{m^2} \qquad C_x = 0.35$$

Driveline (rear wheel drive):

$$\tau_I = 3.29 \qquad \tau_{II} = 2.16 \qquad \tau_{III} = 1.61 \qquad \tau_{IV} = 1.27 \qquad \eta_t = 0.91$$
$$\tau_V = 1.04 \qquad \tau_{VI} = 0.88 \qquad \tau_f = 4.19$$

E.7   Sports car (b)

The vehicle is a mid-engine two-seater sports car with a 3.5 liter spark ignition supercharged engine.

The performance curves of the engine and the map of specific fuel consumption are shown in Fig. E.11.

The main geometrical data and inertial properties of the vehicle, wheels, engine and transmission are

length = 4,250 mm	width = 1,900 mm	height = 1,160 mm
m = 1,480 kg	t_1 = 1,502 mm	t_2 = 1,578 mm
J_x = 590 kg m^2	J_y = 1,730 kg m^2	J_z = 1,950 kg m^2
J_{xz} = −50 kg m^2	J_{r_1} (each) = 7 kg m^2	J_{r_2} (each) = 7 kg m^2
J_m = 0.7 kg m^2	J_t = 0.08 kg m^2	

P_{max} = 235 kW at 7,200 rpm; T_{max} = 324 Nm at 5,000 rpm.

Data for rolling coefficient and for the "magic formula" for lateral forces, longitudinal forces and aligning torque of tires:

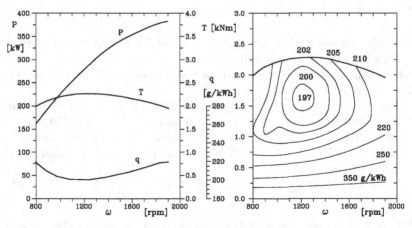

FIGURE E.11. Maximum torque and power curves and map of the specific fuel consumption of the engine.

$f_0 = 0.013$ $K = 6.5 \times 10^{-6}\ \mathrm{s^2/m^2}$ $R_{e_1} = 310\ \mathrm{mm}$

$R_{e_2} = 315\ \mathrm{mm}$

$a_0 = 1.7990$	$a_1 = 0$	$a_2 = 1688.0000$
$a_3 = 4140.0000$	$a_4 = 6.0260$	$a_5 = 0$
$a_6 = -0.3589$	$a_7 = 1.0000$	$a_8 = 0$
$a_9 = -6.1110 \times 10^{-3}$	$a_{10} = -3.2240 \times 10^{-2}$	$a_{111} = 0$
$a_{112} = 0$	$a_{12} = 0$	$a_13 = 0$
$b_0 = 1.65$	$b_1 = 0$	$b_2 = 1688$
$b_3 = 0$	$b_4 = 229$	$b_5 = 0$
$b_6 = 0$	$b_7 = 0$	$b_8 = -10$
$b_9 = 0$	$b_{10} = 0$	
$c_0 = 2.0680$	$c_1 = -6.4900$	$c_2 = -21.850$
$c_3 = 0.4160$	$c_4 = -21.3100$	$c_5 = 2.9420 \times 10^{-2}$
$c_6 = 0$	$c_7 = -1.1970$	$c_8 = 5.2280$
$c_9 = -14.8400$	$c_{10} = 0$	$c_{11} = 0$
$c_{12} = -3.7360 \times 10^{-3}$	$c_{13} = 3.8910 \times 10^{-2}$	$c_{14} = 0$
$c_{15} = 0$	$c_{16} = 0.6390$	$c_{17} = 1.6930$

Aerodynamic data:

$S = 1.824\ \mathrm{m^2}$ $C_x = 0.335$ $C_z = -0.34$

$C_{M_y} = 0$ $(C_y)_{,\beta} = -2.3\ \mathrm{1/rad}$ $(C_N)_{,\beta} = -0.3\ \mathrm{1/rad}$

Transmission (rear wheel drive) (the value of τ_f is inclusive of the reduction gears located between engine and gearbox):

$\tau_I = 1/3.214$ $\tau_{II} = 1/2.105$ $\tau_{III} = 1/1.458$

$\tau_{IV} = 1/1.094$ $\tau_V = 1/0.861$ $\tau_f = 1/4.051$

$\eta_t = 0.87$

E.8   Van

The vehicle is a van with a carrying capacity of 1.4 t (fully loaded mass 3,500 kg) with a 2.3 l direct injection diesel engine.

The main geometrical data and inertial properties of the vehicle, wheels, engine and transmission are

length = 5,400 mm   width = 2,050 mm   height = 2,500 mm

$a = 1,204\ \mathrm{mm}$ $b = 2,246\ \mathrm{mm}$ $l = 3,450\ \mathrm{mm}$

$m = 2,020 \div 2,100^5\ \mathrm{kg}$ $t_1 = 1,810\ \mathrm{mm}$ $t_2 = 1,790\ \mathrm{mm}$

$J_x = 1,400\ \mathrm{kg\ m^2}$ $J_y = 6,000\ \mathrm{kg\ m^2}$ $J_z = 6,230\ \mathrm{kg\ m^2}$

$h_G = 679\ \mathrm{mm}$

$J_r(\text{each}) = 1.07\ \mathrm{kg\ m^2}$ $J_m = 0.335\ \mathrm{kg\ m^2}$ $J_t = 0.32\ \mathrm{kg\ m^2}$

---

[5]The first value is the mass of the empty vehicle; the second includes one passenger and luggage and is consistent with the values of the moments of inertia and height of the center of mass. The effect of the payload must then be added.

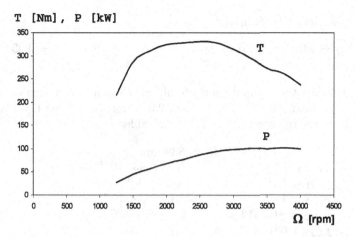

FIGURE E.12. Maximum power and torque curves of the engine.

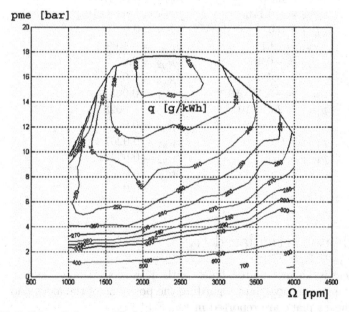

FIGURE E.13. Map of the specific fuel consumption of the engine; the consumption at idle (800 rpm) is 500 g/h.

The maximum engine power and torque are

$P_{max} = 103$ kW at 3,750 rpm; $T_{max} = 330$ Nm at 2,500 rpm.

The performance curves of the engine are reported in Fig. E.12; the map of specific fuel consumption is shown in Fig. E.13.

Aerodynamic data:

$$S = 4.4 \text{ m}^2 \qquad C_x = 0.37$$

Driveline (rear wheels drive):

$$\tau_I = 4.99 \qquad \tau_{II} = 2.6 \qquad \tau_{III} = 1.52 \qquad \eta_t = 0.93$$
$$\tau_{IV} = 1.00 \qquad \tau_V = 0.777 \qquad \tau_f = 3.72$$

The data for the computation of rolling resistance and longitudinal and lateral forces (equations (2.34, 2.35, 2.36, 2.37)) for the 225/65 R 16 tires used on this vehicle are reported in the following table:

$$R_e = 339 \text{ mm}$$

$f_0 = 0.0094$	$K = 0.43 \cdot 10^{-8} \text{ s}^2/\text{m}^2$
$p_{Cx1} = 1.76$	$p_{Cy1} = 1.3$
$p_{Dx1} = 1.09$	$p_{Dy1} = -0.857$
$p_{Dx2} = -0.0312$	$p_{Dy2} = -0.0702$
$p_{Dx3} = 0.0$	$p_{Dy3} = 0.0$
$p_{Ex1} = 0.552$	$p_{Ey1} = 0.00117$
$p_{Ex2} = 0.370$	$p_{Ey2} = -0.0106$
$p_{Ex3} = -0.170$	$p_{Ey3} = -8.68 \cdot 10^{-5}$
$p_{Ex4} = 0.0$	$p_{Ey4} = 0.0$
$p_{Kx1} = 19.1$	$p_{Ky1} = -15.3$
$p_{Kx2} = -0.466$	$p_{Ky2} = 2.93$
$p_{Kx3} = 0.483$	$p_{Ky3} = 0.0$
$p_{Hx1} = -5.45 \cdot 10^{-4}$	$p_{Hy1} = 0.00571$
$p_{Hx2} = 2.09 \cdot 10^{-4}$	$p_{Hy2} = 0.00283$
	$p_{Hy3} = 0.0$
$p_{Vx1} = 0.0$	$p_{Vy1} = 0.0207$
$p_{Vx2} = 0.0$	$p_{Vy2} = 0.00421$
	$p_{Vy3} = 0.0$
	$p_{Vy4} = 0.0$

E.9   Heavy articulated truck

The vehicle is an articulated truck with a two-axle tractor and a three-axle trailer. The geometrical data regarding the positions of the axles and centers of mass in the xz plane are reported in Fig. E.14.

E.9.1   Tractor

The main geometrical data and inertial properties of the vehicle, wheels, engine and transmission are:

$m = 7{,}150$ kg	$t_1 = 2{,}100$ mm	$t_2 = 1{,}835$ mm
$J_x = 4{,}500$ kg m^2	$J_y = 25{,}800$ kg m^2	$J_z = 27{,}000$ kg m^2
$J_{xz} = -3{,}800$ kg m^2	$J_r = 2.5$ kg m^2 (each)	
$J_m = 2.55$ kg m^2	$J_t = 1.1$ kg m^2	

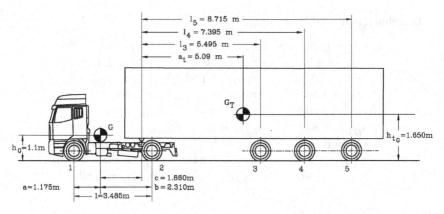

FIGURE E.14. Sketch of an articulated truck with 5 axles.

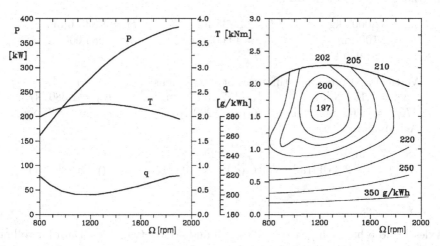

FIGURE E.15. Maximum torque and power curves and map of the specific fuel consumption of the engine.

The vertical and torsional stiffnesses of the suspensions are:

$$K_1 = 2{,}100 \text{ N/m} \qquad K_2 = 2{,}800 \text{ N/m} \qquad K_{t_1} = 3{,}790 \text{ Nm/rad}$$
$$K_{t_2} = 3{,}790 \text{ Nm/rad}$$

Aerodynamic data:

$$S = 5.14 \text{ m}^2 \qquad C_x = 0.45 \text{ (tractor)} \qquad C_x = 0.65 \text{ (whole vehicle)}$$
$$C_z = 0 \qquad (C_y)_{,\beta} = -2.2 \text{ 1/rad} \qquad (C_{M_z})_{,\beta} = -1.5 \text{ 1/rad}$$

The curves of the maximum power and torque of the engine are reported in Fig. E.15 together with the map of the specific fuel consumption

Driveline (rear wheels drive):

$$\tau_I = 12.5 \qquad \tau_{II} = 8.35 \qquad \tau_{III} = 6.12$$
$$\tau_{IV} = 4.56 \qquad \tau_V = 3.38 \qquad \tau_{VI} = 2.47$$
$$\tau_{VII} = 2.14 \qquad \tau_{VIII} = 1.81 \qquad \tau_{IX} = 1.57$$
$$\tau_X = 1.35 \qquad \tau_{XI} = 1.17 \qquad \tau_{XII} = 1.00$$
$$\tau_{XIII} = 0.87 \qquad \tau_f = 4.263 \qquad \eta_{tI} = 0.81$$
$$\eta_{tII} = 0.84 \qquad \eta_{tIII} = 0.84 \qquad \eta_{tIV} = 0.84$$
$$\eta_{tV} = 0.89 \qquad \eta_{tVI} = 0.87 \qquad \eta_{tVII} = 0.84$$
$$\eta_{tVIII} = 0.87 \qquad \eta_{tIX} = 0.84 \qquad \eta_{tX} = 0.87$$
$$\eta_{tXI} = 0.84 \qquad \eta_{tXII} = 0.93 \qquad \eta_{tXIII} = 0.89$$

E.9.2 Trailer

Inertial properties of the trailer:

$$m = 32,000^6 \text{ kg} \qquad t_3 = 1,835 \text{ mm} \qquad t_4 = 1,835 \text{ mm}$$
$$t_5 = 2,100 \text{ mm} \qquad J_x = 30,000 \text{ kg m}^2 \qquad J_y = 285,000 \text{ kg m}^2$$
$$J_z = 285,000 \text{ kg m}^2$$

Vertical and torsional stiffnesses of the suspensions:
$$K_3 = 2,150 \text{ N/m} \qquad K_4 = 2,150 \text{ N/m} \qquad K_5 = 1,380 \text{ N/m}$$
$$K_{t3} = 3,200 \text{ Nm/rad} \qquad K_{t4} = 3,200 \text{ Nm/rad} \qquad K_{t5} = 2,800 \text{ Nm/rad}$$
Aerodynamic data:

$$S = 7.5 \text{ m}^2 \qquad C_x = 0.25 \qquad C_z = 0$$
$$(C_y)_{,\beta} = -2.35 \text{ 1/rad} \qquad (C_{M_z})_{,\beta} = -0.6 \text{ 1/rad} \qquad \text{(referred a } l_5)$$

E.9.3 Tires

Axles 1 and 5 have single tires, axles 2, 3 and 4 have twin tires. Data for rolling coefficient and basic data for cornering forces and aligning torque using a simplified magic formula (equations (2.23, 2.28, 2.29)):

$$f_0 = 0.008 \qquad K = 0 \qquad R_e = 460 \text{ mm}$$
$$a_3 = 5019.3 \qquad a_4 = 65.515$$
$$c_3 = -0.6100 \qquad c_4 = -2.3400 \qquad c_5 = 0.02727$$

E.10   Racing motorcycle

The vehicle is a racing motorcycle. The geometrical sketch of the vehicle and the torque and power curves of the engine are reported in Fig. E.16.

---

[6]Fully loaded mass.

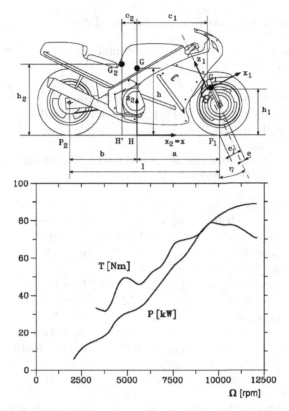

FIGURE E.16. Geometrical sketch of the vehicle and torque and power curves of the engine.

The main geometrical data and inertial properties of the vehicle (moment of inertia J_x is referred to an axis lying on the ground), wheels, engine, transmission are:

$m_1 = 20$ kg	$m_2 = 200$ kg	$m_d = 70$ kg (driver)
$J_x = 80$ kg m^2	$J_y = 110$ kg m^2	$J_z = 40$ kg m^2
$J_{xz} = 0$	$J_{1_z} = 2$ kg m^2	$J_{1_{xz}} = 1$ kg m^2
$J_{p_1} = 0.4$	$J_{p_2} = 0.4$ kg m^2	$J_e = 0.08$ kg m^2
$l = 1320$ mm	$a = 642$ mm	$b = 678$ mm
$c_1 = 561.5$ mm	$c_2 = 54.2$ mm	$e = 50$ mm
$e_1 = 95$ mm	$h = 495.6$ mm	$h_1 = 432$ mm
$h_2 = 500$ mm	$\eta = 23°$	$c_\delta = 8$ Nms/rad

$P_{max} = 88$ kW a 11.900 rpm; $T_{max} = 76$ Nm a 9.750 rpm.

Wheels and tires, geometrical and linearized data:

$R_{e1} = 300$ mm $\qquad R_{e2} = 300$ mm $\qquad r_{t_1} = 70$ mm

$r_{t_2} = 110$ mm $\qquad r_t$ (average) $= 90$ mm

$f_0 = 0.01$ $\qquad K = 4 \cdot 10^{-6}$ s^2/m^2 $\qquad \mu_{x_{max}} = 1.1$

$(C_1)_{,Z} = 27.27$ 1/rad $\qquad (C_2)_{,Z} = 30.0$ 1/rad

$(M_{z_2})_{,\alpha,Z} = 0.228$ m/rad $\qquad (F_{y_1})_{,\gamma,Z} = -1.177$ 1/rad

$(M_{z_1})_{,\alpha,Z} = 0.210$ m/rad $\qquad (F_{y_2})_{,\gamma,Z} = -1.367$ 1/rad

Data for the "magic formula" for lateral forces, longitudinal forces and aligning torque, front tire (Equations (2.23, 2.28, 2.29)):

$a_0 = 1.50$	$a_1 = -27.9$	$a_2 = 1,280$
$a_3 = -1000$	$a_4 = 4.00$	$a_5 = 0.015$
$a_6 = -0.35$	$a_7 = -1.99$	$a_8 = 0.058$
$a_9 = 0$	$a_{10} = 0$	$a_{11} = 5$
$a_{12} = 0$	$a_{13} = 0$	
$b_3 = 49.6$	$b_4 = 226$	$b_5 = 0.069$
$c_0 = 2.40$	$c_1 = -2.72$	$c_2 = -2.28$
$c_3 = 1.86$	$c_4 = 2.73$	$c_5 = 0.11$
$c_6 = 0.030$	$c_7 = -0.07$	$c_8 = 0.643$
$c_9 = -4.04$	$c_{10} = -0.07$	$c_{11} = -0.015$
$c_{12} = 0$	$c_{13} = 0$	$c_{14} = -0.066$
$c_{15} = 0.945$	$c_{16} = 0$	$c_{17} = 0$

Data for the "magic formula" for lateral forces, longitudinal forces and aligning torque, rear tire (Equations (2.23, 2.28, 2.29)):

$a_0 = 1.50$	$a_1 = -27.9$	$a_2 = 1,275$
$a_3 = -1.100$	$a_4 = 4.00$	$a_5 = 0.010$
$a_6 = -0.35$	$a_7 = -1.99$	$a_8 = 0.058$
$a_9 = 0$	$a_{10} = 0$	$a_{11} = 5$
$a_{12} = 0$	$a_{13} = 0$	
$b_3 = 49.6$	$b_4 = 226$	$b_5 = 0.069$
$c_0 = 2.40$	$c_1 = -2.72$	$c_2 = -2.28$
$c_3 = 1.86$	$c_4 = 2.73$	$c_5 = 0.11$
$c_6 = 0.03$	$c_7 = -0.070$	$c_8 = 0.643$
$c_9 = -4.04$	$c_{10} = -0.07$	$c_{11} = -0.015$
$c_{12} = 0$	$c_{13} = 0$	$c_{14} = -0.066$
$c_{15} = 0.945$	$c_{16} = 0$	$c_{17} = 0$

Aerodynamic data:

$S = 1$ m^2 $\qquad C_x = 0.23$ $\qquad C_z = 0.10$

$C_{M_y} = 0$ $\qquad (C_y)_{,\beta} = 0.026$ 1/rad $\qquad (C_{M_z})_{,d} = 0.065$ 1/rad

Transmission (the values of τ_i are inclusive of the reduction gear located between engine and gearbox):

$$\tau_I = 4.91 \qquad \tau_{II} = 3.84 \qquad \tau_{III} = 3.22$$
$$\tau_{IV} = 2.81 \qquad \tau_V = 2.5 \qquad \tau_{VI} = 2.29$$
$$\tau_f = 3.00 \qquad \eta_t = 0.88$$

BIBLIOGRAPHY OF VOLUME 2

Part III

1. Data S., Ugo A., *Objective Evaluation of Steering System Quality*, FISITA International Congress, Prague, 1996.

2. The Auto - Oil II Program, *A report from the services of the European Commission*, Bruxelles, 2000.

3. Lenz H. P. *et alii*, *Transport emissions in EU - 15*, ACEA, Bruxelles, 2002.

4. Caviasso G., Data S., Pascali L., Tamburro A., *Customer Orientation in Advanced Vehicle Design*, SAE Paper 2002-01-1576, 2002.

5. Data S., Frigerio F., *Objective Evaluation of Handling Quality*, Journal of Automotive Engineering, 3, 2002.

6. - , *Road Accidents 1980 - 2000*, ACEA, Bruxelles, 2002.

7. - , *Panorama of Transport*, Eurostat, Luxenbourg, 2003.

8. - , *Autoincifre 2004*, ANFIA, Torino, 2004.

9. Bargero R., Brizio P., Celiberti L., Falasca V., *Objective Evaluation of Vibroacustical Quality*, Congresso SAE, Detroit, 2004.

10. - , *Car Park 1995 - 2002*, ACEA, Bruxelles, 2004.

11. - , *Energy, Transport and Environment Indicators*, Eurostat, Luxenbourg, 2005.

12. Putignano C., *Statistiche dei trasporti 2002 - 2003*, ISTAT, Rome, 2005.

13. - , *European Automotive Industry Report 2005*, ACEA, Bruxelles, 2005.

14. - , *Segments and Bodies 1990 - 2004*, ACEA, Bruxelles, 2005.

Part IV

1. E. Koenig, R. Fachsenfield, *Aerodynamik des Kraftfahrzeuge*, Verlag der Motor, Rundshou-Umsha Verlag, Frankfurt A.M. 1951.

2. Bussien, *Automobiltechnisches Handbuch*, Technischer Verlag H Cam. Ber-lin, 1953.

3. M. Bencini, *Dinamica del veicolo*, Tamburini, Milan, 1956.

4. C. Deutsch, *Dynamique des vehicules routiers*, Organisme National de Sécurité Routière.

5. W. Steeds, *Mechanics of Road Vehicles*, Iliffe & Sons, London, 1960.

6. G.H. Tidbury, *Advances in Automobile Engineering*, Pergamon Press, London, 1965.

7. F. Pernau, *Die entscheidenden Reifeneigenschaften*, Vortragstext, Eszter, 1967.

8. J.R. Ellis, *Vehicle Dynamics*, Business Books Ltd., London, 1969.

9. G. Pollone, *Il veicolo*, Levrotto & Bella, Turin, 1970.

10. H.C.A Van Eldik Thieme, H.B. Pacejka, *The Tire as a Vehicle Component*, Vehicle Res. Lab., Delft University of Technology, 1971.

11. M. Mitschke, *Dinamik der Kraftfahzeuge*, Springer, Berlin, 1972.

12. A. Morelli, *Costruzioni automobilistiche*, in *Enciclopedia dell'Ingegneria*, ISEDI, Milan, 1972.

13. A.J. Scibor Ryilski, *Road Vehicle Aerodynamics*, Pentech Press, Londra, 1975.

14. M.D. Artamonov, V.A. Ilarionov, M.M. Morin, *Motor Vehicles, Fundamentals and Design*, MIR, Moscow, 1976.

15. W.H. Hucho, *The Aerodynamic Drag of Cars. Current Understanding, Unresolved Problems, and Future Prospects*, in *Aerodynamic drag mechanism of bluff bodies and road vehicles*, Plenum Press, New York, 1978.

16. R. H. Macmillan, *Dynamics of Vehicle Collisions*, Inderscience Enterprises, Jersey, 1983.

17. E. Fiala, *Ingegneria automobilistica*, in Manuale di ingegneria meccanica, part 2, EST, Milano, 1985.

18. G.G. Lucas, *Road Vehicle Performance*, Gordon & Breach, London, 1986.

19. J.C. Dixon, *Tyres, Suspension and Handling*, Cambridge University Press, Cambridge, 1991.

20. T. D. Gillespie, *Fundamentals of Vehicle Dynamics*, Society of Automotive Engineers, Warrendale, 1992.

21. D. Bastow, G.P. Howard, *Car Suspension and Handling*, Pentech Press, London, and Society of Automotive Engineers, Warrendale, 1993.

22. W.F. Milliken, D.L. Milliken, *Race Car Vehicle Dynamics*, Society of Automotive Engineers, Warrendale, 1995.

23. J. Reimpell, H. Stoll, J.W. Betzler, *The Automotive Chassis: Engineering Principles*, Butterworth, Oxford, 2001.

24. W.F. Milliken, D.L. Milliken, *Chassis Design, Principles and Analysis*, SAE, Warrendale, 2002.

25. D. Karnopp, *Vehicle Stability*, Marcel Dekker, New York, 2004.

26. Rajesh Rqajamqni, *Vehicle Dynamics and Control*, Springer, New York, 2006.

INDEX

Printed in the United States
By Bookmasters